Biological Materials Science

Taking a unique materials science approach, this text introduces students to the basic concepts and applications of materials and biomedical engineering and prepares them for the challenges of the new interdisciplinary field of biomaterials science.

Split into three sections – Basic biology principles, Biological materials, and Bioinspired materials and biomimetics – the book presents biological materials along with the structural and functional classification of biopolymers, bioelastomers, foams, and ceramic composites. More traditional biomimetic designs such as VELCRO® are then discussed in conjunction with new developments that mimic the structure of biological materials at the molecular level, mixing nanoscale with biomolecular designs. Bioinspired design of materials and structures is also covered.

Focused presentations of biomaterials are presented throughout the text in succinct boxes, emphasizing biomedical applications, and the basic principles of biology are explained, so no prior knowledge is required. The topics are supported by approximately 500 illustrations, solved problems, and end-of-chapter exercises. PowerPoint slides and solutions for instructors are available online via www.cambridge.org/meyerschen.

Marc André Meyers, Distinguished Professor at the University of California, San Diego, is the author or co-author of three other books and approximately 400 papers. The recipient of important awards from Europe (Humboldt Senior Scientist Award, Heyn Medal from the DGM, and the J. S. Rinehart Award), China (Lee Hsun Lecture Award; Visiting Professor, Chinese Academy of Sciences), and the USA (Acta Materialia Materials and Society Award, TMS Educator Award, SMD/TMS Distinguished Scientist and Distinguished Service Awards, ASM Albert Sauveur Award, ASM Albert Easton White Award), he is a Fellow of TMS, APS, and ASM, and a member of the Brazilian Academy of Sciences. He is also the author of three fiction novels.

Po-Yu Chen, Assistant Professor of the Materials Science and Engineering Department at National Tsing Hua University, Taiwan, is a graduate of the University of California, San Diego. His current research is in the fields of biological (natural) materials, bioinspired/biomimetic materials, biomedical materials, and green and energy-related materials. He is the author or co-author of several highly cited review articles in biological and bioinspired materials. A member of the TMS Biomaterials Committee, he organized several bio-related symposiums and workshops at international conferences. He was the recipient of the *Materials Science and Engineering C* Young Researcher Award, the ASME Emerging Researchers in Biomedical Engineering Award in 2011, and the TMS Young Leaders Award, and he received the Distinguished Young Researcher Career Award from the Taiwan National Science Council.

"The union of the physical and biological sciences is in many respects one of the most exciting yet challenging aspects of scientific endeavor today. Nowhere is this more in evidence than in the area of biological materials science and engineering where many materials scientists struggle with the complex puzzle of biological form and function while biologists in turn have to deal with the invariably highly quantitative nature of the physical sciences and engineering. With this book, Meyers and Chen have delivered a true *tour de force* which takes the reader in clear and precise text from cells to virus-produced Li-ion batteries. This book is a must read for undergraduates, graduates and researchers alike in the rapidly expanding fields of biological, bioinspired and biomaterials science."

Robert Ritchie, *Lawrence Berkeley National Laboratory*

Biological Materials Science

Biological Materials, Bioinspired Materials, and Biomaterials

Marc André Meyers

University of California, San Diego

Po-Yu Chen

National Tsing Hua University, Taiwan

CAMBRIDGE
UNIVERSITY PRESS

University Printing House, Cambridge CB2 8BS, United Kingdom

One Liberty Plaza, 20th Floor, New York, NY 10006, USA

477 Williamstown Road, Port Melbourne, VIC 3207, Australia

314-321, 3rd Floor, Plot 3, Splendor Forum, Jasola District Centre, New Delhi - 110025, India

79 Anson Road, #06-04/06, Singapore 079906

Cambridge University Press is part of the University of Cambridge.

It furthers the University's mission by disseminating knowledge in the pursuit of
education, learning and research at the highest international levels of excellence.

www.cambridge.org
Information on this title: www.cambridge.org/9781107010451

First published 2014
Reprinted 2019

A catalogue record for this publication is available from the British Library

Library of Congress Cataloging in Publication data
Meyers, Marc A.
Biological materials science : biological materials, bioinspired materials, and biomaterials /
Marc André Meyers, University of California, San Diego, Po-Yu Chen, National Tsing
Hua University, Taiwan.
pages cm
ISBN 978-1-107-01045-1 (hardback)
1. Biomedical materials – Textbooks. 2. Biomedical engineering – Textbooks. I. Chen, Po-Yu.
II. Title.
R857.M3M49 2014
610.28–dc23

2014019407

ISBN 978-1-107-01045-1 Hardback

Additional resources for this publication at www.cambridge.org/meyerschen

The works of the Lord are great,
sought out of all them that have pleasure therein.
Psalms 111:2
Frontispiece of the new Cavendish Laboratory, University of Cambridge

Contents

Preface

The field of materials science and engineering (MSE) has undergone a tremendous development since it was defined for the first time in the 1950s. Materials science and engineering has supplanted traditional curricula centered on metallurgy, ceramics, and polymers. In the USA alone, there are over 50 MSE academic university departments. Materials science and engineering has initially merged metals, polymers, ceramics, and composites into a broad and unified treatment. Whereas the twentieth century was marked by revolutionary discoveries in physics and chemistry, the twenty-first century has been prognosticated to be dominated by biology. Indeed, medical and biological discoveries are bound to have a profound effect on our future. Consistent with the increasing demands of engineering students to acquire basic working tools in this domain, many engineering curricula are adding appropriate courses or modifying existing courses to address biological aspects. Within MSE, the nascent field of biological materials science encompasses three areas.

- Biological (or natural) materials: materials that comprise cells, extracellular materials, tissues, organs, and organisms.
- Biomaterials: synthetic materials used to correct, repair, or supplement natural functions in organisms.
- Biomimetics: this area encompasses the materials and structures inspired in biological systems and/or functions.

This book focuses on these three areas in a balanced manner. This is a necessity of space, and many curricula offer separate biomaterials courses. The book has 13 chapters, and the contents can be covered comfortably in one semester (one chapter per week).

This book was developed for courses aimed at seniors and first-year graduate students. The course has been taught at the University of California, San Diego, and at National Tsing Hua University. Solved examples in the text (approximately two per chapter) and end-of-chapter problems are an important part of the text, and serve as a learning tool and an opportunity to cement the knowledge gained by applying it to specific problems. We provide a solutions manual and PowerPoint presentations of figures and key concepts in each chapter, which are available online via www.cambridge.org/meyerschen.

We present the principles of biology and the connections between structures and properties in biological materials. The intended audience for this course are MSE and ME students with a sound MSE foundation but poor biology background. We use the materials science and engineering approach which is based on the correlation of structure with structural and functional properties. This approach is familiar to MSE and ME students.

Many courses in biomaterials devote the first half to explaining the principles of MSE and are designed for bioengineering and medical students. The opposite approach is implemented here. In Part I: Basic biology principles, we introduce the basic biology concepts that engineering students need to penetrate this area. Some of these concepts are rather basic for biology students, but provide important background material for engineering students.

In Part II: Biological materials, in a manner similar to classical MSE, which divides materials into metals, ceramics, polymers, and composites, we introduce biological materials in broad categories according to their structure and properties: biological ceramics (biominerals); biological polymers and their composites; biological elastomers; biological foams. This classification was introduced by Wegst and Ashby (1994), and is very useful for engineers, who can understand biological materials better through this familiar approach.

In Part III: Bioinspired materials and biomimetics, we present more traditional biomimetic designs such as VELCRO® and proceed towards new developments that mimic the structure of biological materials at the molecular level, mixing nano-scale with biomolecular designs. This is a unique aspect of this book, not treated heretofore in classrooms. Some of these bioinspired materials are already used in biomedical applications.

Boxes placed throughout the text discuss biomaterials, an important field of utilization of the concepts learned here.

Although this book has only two authors, it represents the efforts of our research groups at UCSD and National Tsing Hua University. In particular, our colleague J. McKittrick contributed greatly to this book via her collaboration over the past eight years. She is also co-author of four review articles whose material was used in different parts of the book. We may have inadvertently used some of the text generated by her, for which we apologize.

Former graduate students, G. Serra Guimarães, J. Kiang, A. Y. M. Lin, J. Li, Y. S. Lin, R. Menig, L. S. Morais, E. E. Novistkaya, D. Ren, and Y. Seki; postdoctoral fellow Dr. W. Yang; current students, I. H. Chen, D. Fernandes, M. I. Lopez, M. Porter, V. Sherman, and B. Wang helped by providing material for the book. B. Wang also undertook the arduous task of seeking figure permissions and assistance in proofreading, and W. Yang provided immense assistance throughout the entire project. The presence of our own research results is disproportionally high, but we tried to keep a balance throughout. This is a rapidly evolving field and we might have accidentally excluded important information. We thank our colleagues and graduate students that contributed to this book through research, discussion of the literature, and problem-solving. The field of biological materials is critically dependent on specimens, and we thank Jerry Jennings at Emerald Forest Gardens (toucan feathers and beaks), the San Diego Museum of Natural History (Brad Hollingsworth and Phillip Unitt), and Raul Aguiar, Rancho La Bellota (vulture wings). The esteemed friend and colleague of MAM, the foot and ankle specialist Dr. João Francisco Figueiró, shared with him an early biological experiment (a secret night visit to the cadaver ascribed to him, in our university days) and provided a number of radiographs for this book.

The generous input of and collaboration with colleagues globally has been very important in defining the coverage of this book. We particularly would like to thank Professors George Mayer (University of Washington), R. Ritchie (University of California at Berkeley), C. T. Lim (National University of Singapore), Carlos Elias (Military Institute of Engineering, Brazil), A. Miserez (Nanyang University, Singapore), and R. Roeder (Notre Dame University). Three towering figures inspired us to write this book: Y. C. Fung and R. J. Skalak, both pioneers in this field and both, coincidentally and fortunately for us, from UCSD; and M. F. Ashby, Cambridge University, who has preceded us in this endeavor and has entered this field with clarity and vision, implementing the "materials" approach that we follow.

This research was generously funded by the US National Science Foundation, Division of Materials Research (Grant 0510138 and Grant 1006931), The U C Labs grant No. 12-LR-239079, and the Taiwan National Science Council (NSC 100–2218-E-007–016-MY3 and NSC 101–2815-C-007–014-E).

Boxes

1 Evolution of materials science and engineering: from natural to bioinspired materials

1.1 Early developments

For one brief moment, as one of us (MAM) walked into one of the fabled rooms of the British Museum, he was handed a tool used by early hominids two million years ago. The stone had a barely recognizable sharp edge but possessed a roundish side that fit snugly into the hand. It could have been used to cut through meat, scrape a skin, or crack a skull (Fig. 1.1). It was a brief but emotional event until the zealous anthropologist removed it from the hand that eagerly clasped the artifact and imagined himself deep in the Olduvai Gorge, slicing through the hide of a gazelle that had been hunted down by the group. This connection is at the heart of this book.

The first materials were natural and biological: stone, bones, antler, wood, skins. Figure 1.2(a) shows an Ashby plot of strength vs. density, for early neolithic materials. These natural materials gradually gave way to synthetic ones as humans learned to produce ceramics, then glass and metals. Some of the early ceramics, glasses, and metals are also shown in the plot, and they provide added strength. These synthetic materials expanded the range of choices and significantly improved the performance of tools. The long evolution of materials, from the stone shown in Fig. 1.1 to the cornucopia of materials developed in the past century, is shown in Fig. 1.2(b). Contemporary materials are of great complexity and variety, and they represent the proud accomplishment of ten thousand years of creative effort and technological development.

Why, then, this resurgence of interest in natural (or biological) materials, if synthetic materials have, as clearly shown in Fig. 1.2(b), a much superior performance? We have used our ingenuity to the maximum, but one way to overcome this is to look to nature for new designs and concepts. The materials that nature has at its disposal are rather weak (as will be shown in Chapter 2), but they are combined in a very ingenious way to produce tough components and robust designs. The central idea in biomimetics is to produce materials using our advanced technology along with bioinspired designs that have evolved in nature for millions of years. It is difficult, almost impossible, to reproduce all the steps in biological materials, which involve cellular processes. The complexity of a single cell is dauntingly beyond our capability. We therefore study nature, its designs and solutions, and derive principles that we can apply to modern materials. This is one of the important purposes of this book.

Figure 1.1.

One of the earliest tools, a chopper ~2 million years old, from the Olduvai Gorge; British Museum. (Used with permission; © The Trustees of the British Museum.)

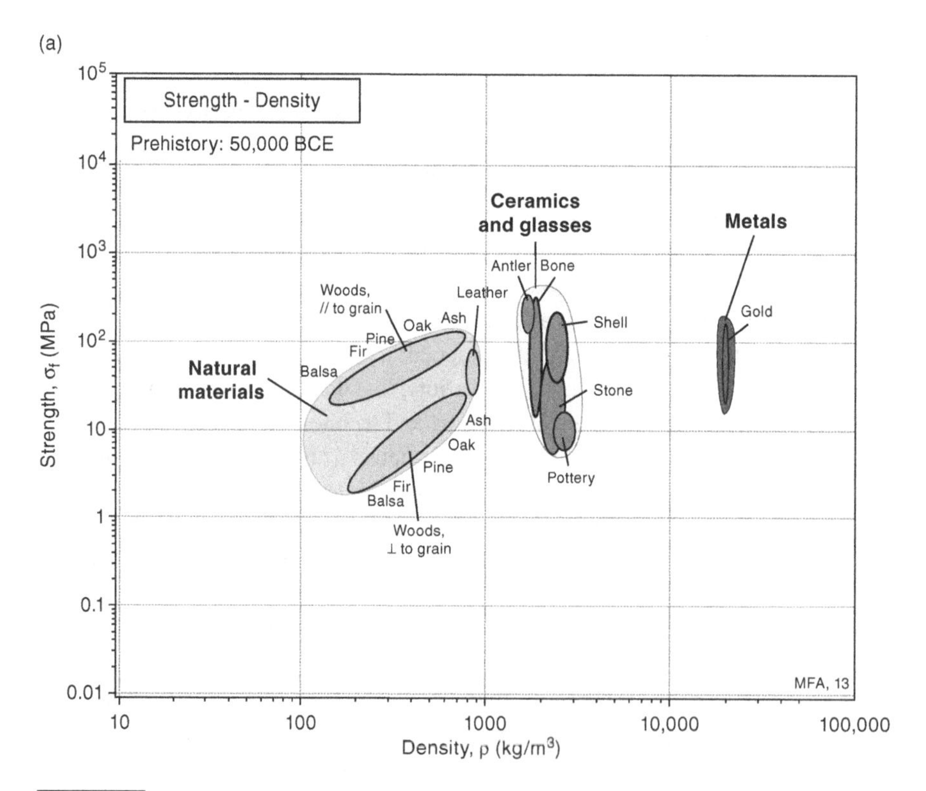

Figure 1.2.

Strength vs. density Ashby plots. (a) Prehistoric synthetic materials; (b) contemporary synthetic materials. (Figures courtesy of Professor Michael F. Ashby, Cambridge University.)

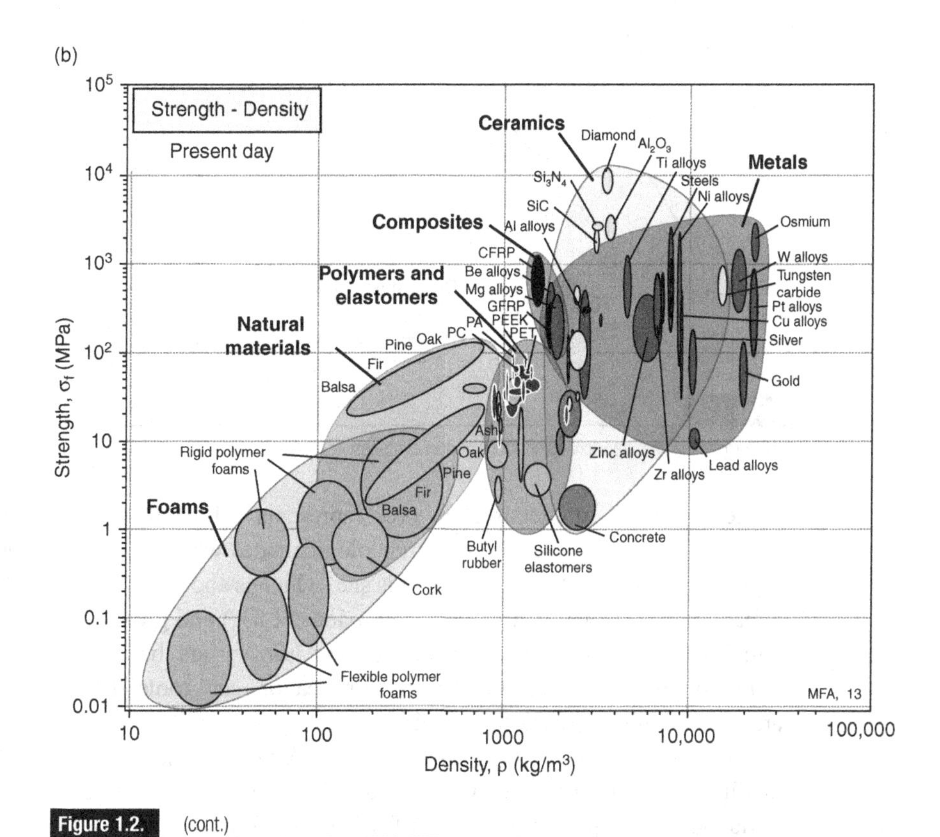

Figure 1.2. (cont.)

A brief historical overview shows how metallurgy gave rise to materials science and engineering, which expanded from primarily structural materials to functional and nanostructured materials starting in the 1970s, leading now to biological materials that serve as inspiration for complex hierarchical systems of the future.

1.2 Evolution of materials science and engineering

We can divide the evolution of materials science and engineering into three distinct phases.

1.2.1 Traditional metallurgy

Practiced over 5000 years and which dominated the field up to the first part of the twentieth century, traditional metallurgy can be represented by the metallurgical triangle (Fig. 1.3), which has extraction as the top vertex, with processing and properties as complementary components.

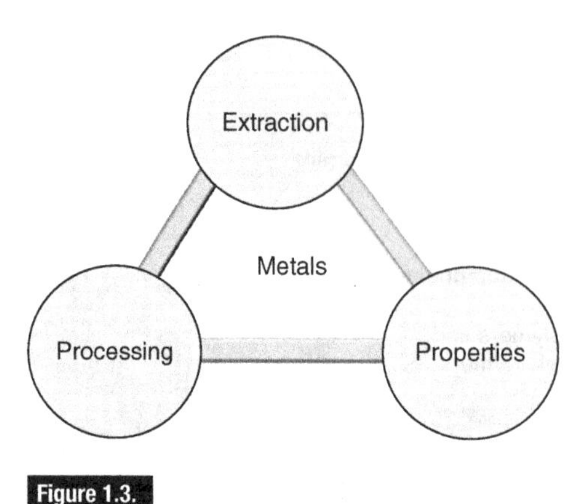

Figure 1.3.

Metallurgical triangle; traditional technology of past centuries.

The extraction of metals from ores represents a breakthrough in the civilizatory process. One of the oldest known archeological sites which shows evidence of mining is the "Lion Cave" in Swaziland. At this site, which is about 43 000 years old, paleolithic humans mined hematite, a reddish iron oxide (Fe_2O_3) and ground it to produce the red pigment ochre. Mines of a similar age in Hungary are believed to be sites where Neanderthals may have extracted flint for weapons and tools. The Egyptians also had large gold mining operations in Nubia.

Native copper, silver, and gold were certainly the first metals utilized. Their inherent ductility was a feature that was very attractive. The first vestiges of industrial-scale production of copper artifacts come from the Early Iron Age, 2700 BCE, from Jordan (Ben-Yosef *et al.*, 2010; Levy, Najjar, and Higham, 2010). The excavations reveal a layout that in many aspects predates modern industrial production by many centuries. From these humble beginnings, the synthesis and processing of materials often defined civilization, and the bronze, iron, and silicon eras are closely connected with the emergence of new materials. A team led by Professor Thomas Levy (UC San Diego) and Dr. Mohammad Najjar (Jordan's Friends of Archaeology) (Levy *et al.*, 2010), excavated an ancient copper-production center at Khirbat en-Nahas down to virgin soil, through more than 20 feet of industrial smelting debris, or slag. The factory had collapsed during an earthquake in about 2700 BCE. Buried in the rubble were hundreds of casting molds for copper axes, pins, chisels, bars, anvils, crucibles, along with metal objects and pieces of ancient metallurgical debris. Maps trace the copper production through about 70 rooms, alleyways, and courtyards. "This shows that the production of metal objects at Khirbat Hamra Ifdan was a highly specialized process performed by skilled crafts people," said Levy. The authors emphasize that the evidence of mass production found in the digs shows sophistication in mining, smelting, and fuel utilization, and demonstrates that Early Bronze Age leaders were able to plan, organize, and manage a large and technically skilled work force and train it to utilize complex technology. Analysis of the copper objects made at the ancient factory suggests that there was quality control at the factory.

A second dig discovered new artifacts, placing the bulk of industrial-scale production at Khirbat en-Nahas in the tenth century BCE, in line with the biblical narrative on the legendary rule of Kings David and Solomon. The research also documents a spike in metallurgical activity at the site during the ninth century BCE, which may also support the history of the Edomites as related by the Bible. Khirbat en-Nahas, which means "ruins of copper" in Arabic, is in the lowlands of a desolate, arid region south of the Dead Sea in what was once Edom and is today Jordan's Faynan district. The Hebrew Bible (or Old Testament) identifies the area with the Kingdom of Edom, foe of ancient Israel. Could these be King Solomon's fabled mines?

Box 1.1 Biomaterials

Biomaterials are as ancient as civilization, and there are reports of Egyptian mummies containing them. In modern times, the revolution brought on by Dr. J. Lister (in the 1860s), aseptic conditions of surgery, and the discovery of new materials propitiated this field, which is still expanding through innovation and the development of new procedures and devices.

Biomaterials may be classified in terms of the tissue response as follows.

- Biotolerant (e.g. stainless steel and polymethyl-methacrylate) materials release substances in non-toxic concentrations, which may lead to the formation of a fibrous connective tissue capsule.
- Bioinert (e.g. alumina and zirconia) materials exhibit minimal chemical interactions with adjacent tissue; a fibrous capsule may form around bioinert materials.
- Bioactive materials (e.g. tricalcium phosphate and Bioglass®) bond to bone tissue through bridges of calcium and phosphorus. However, the chemical bond between non-coated titanium implants and living tissue occurs through weak van der Waals and hydrogen bonds.

The structural classification of biomaterials follows the traditional lines of metals, polymers, ceramics, and composites. More complex arrays are usually called devices. Among metals, gold has been used as a biomaterial and is bioinert. Stainless steel (18wt.%Cr, 8wt.%Ni and 18–8 with Mo additions) fracture plates and screws led the way. Later, the composition (19wt.%Cr, 9wt.%Cr–Fe) became very successful. There are special stainless steel designations for bioimplant applications. For instance, the carbon level has to be very low to avoid embrittlement. This designation is called LC. For example, 304SSLC. In past years, titanium and titanium alloys, cobalt-based alloys (Vitallium), and an alloy exhibiting shape memory and super-elasticity effects, NITINOL (~50% Ti, ~50% Al), have found considerable applications. There is intense research activity in bioresorbable magnesium alloys. Implants made of these alloys dissolve at a prescribed rate so that second surgery for removal of the implant is not necessary. Magnesium is not toxic to the body.

Polymers found use as vascular implants, and a major breakthrough is the introduction of cloth prostheses made with Vinyon (a polyvinyl chloride and polyacrylonitide copolymer), Orlon, Dacron, and Teflon porous fabric that have enabled the formation of a neointima layer covering the inside wall of the implant, thus preventing blood coagulation. Polyethylene (both low density and high density) is used in many applications (e.g. lumens). It should be noted that the difference in density between LDPE and HDPE is minimal; however, the differences in mechanical response are

Box 1.1 (cont.)

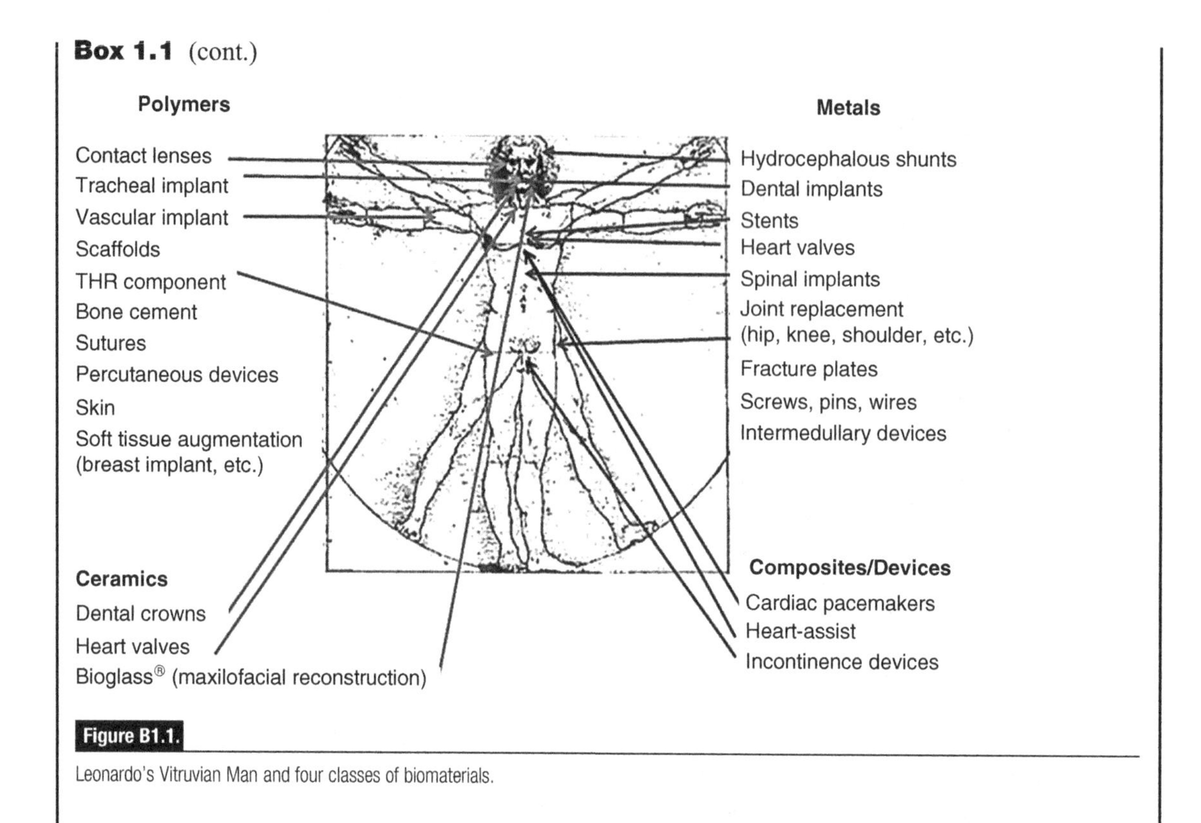

Figure B1.1.

Leonardo's Vitruvian Man and four classes of biomaterials.

striking, HDPE being a much more compact and organized structure with a higher yield stress and lower ductility. In the area of tissue engineering, biodegradable polymers have found great application and have led the way in the formation of scaffolds on which cells and tissue can grow, as they are reabsorbed by the body. Both biopolymers and synthetic polymers are used (Sonntag, Reinders, and Kretzer, 2012).

Ceramics have also found applications, primarily in dental reconstruction but also in scaffolds for bone (e.g. coral) and in total hip replacements because of the low wear rate.

Figure B1.1 shows selected applications of biomaterials in the human body in an illustrative manner.

1.2.2 The structure–properties–performance triangle

Created by Morris Cohen (Cohen, Kear, and Mehrabian, 1980) in the 1970s, the structure–properties–performance triangle emphasizes the connection between these three elements and presents MSE in a new light, with a novel approach unique to it. The unified approach to the study and utilization of metals, ceramics, polymers, and their composites as pioneered by M. Fine (see e.g. Fine and Marcus (1994)) comprises the second stage of evolution of MSE.

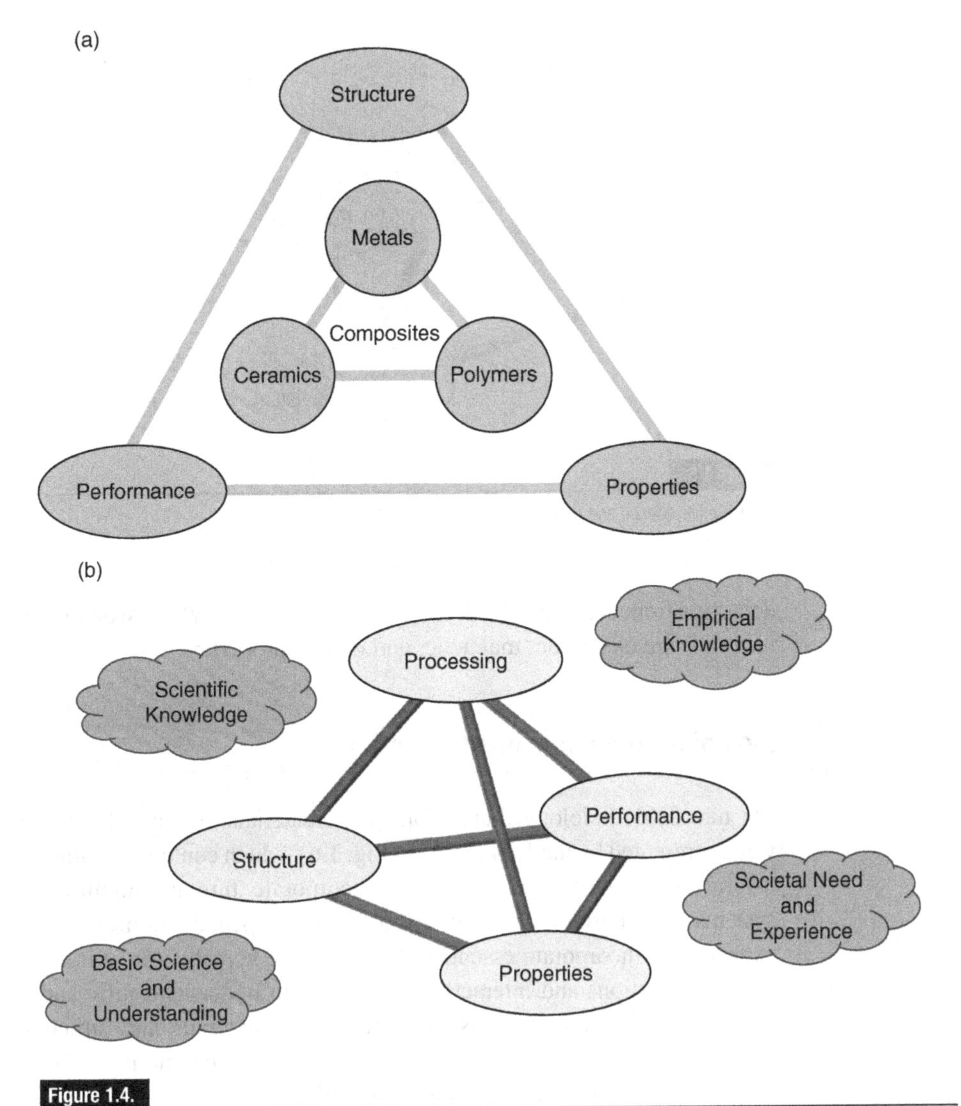

Figure 1.4.

The materials science and engineering revolution: a unified approach to metals, ceramics, and polymers. (a) The original Cohen structure–properties–performance triangle; (b) a modernized version.

The elements of the unified materials approach are shown in Fig. 1.4(a) in their original rendition (Cohen *et al.*, 1980). A more contemporary version of this schematic is shown in Fig. 1.4(b). This structure–property paradigm is still at the heart of MSE research.

1.2.3 Functional materials

In the 1990s, the tetrahedron proposed by G. Thomas, and which forms the cover of *Acta Materialia* (Fig. 1.5), emphasized the growing importance of functional materials, a

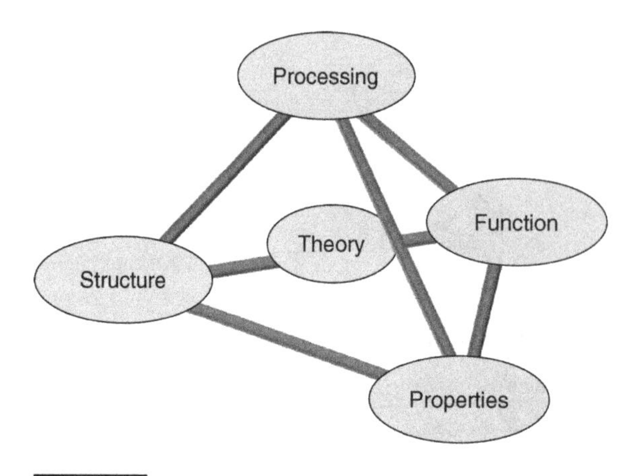

Figure 1.5.

The materials science and engineering (*Acta Materialia*) tetrahedron.

departure from the earlier focus in structural materials. What we denominate "functional" materials are electronic, magnetic, and optical properties.

1.3 Biological and bioinspired materials

The new field of biological and bioinspired materials, which is the theme of this book, is well represented by the *heptahedron* (Fig. 1.6), which contains features that are unique to natural materials and that we hope to incorporate, through biomimetics, into synthetic systems. It is based on the biological materials pentahedron created by Arzt (2006), expanded to incorporate essential elements. The heptahedron is indicative of the complex contributions and interactions necessary to understand fully and exploit (through biomimicking) biological systems. Biological materials and structures have unique characteristics that distinguish them from synthetic counterparts. Evolution, environmental constraints, and the limited availability of materials dictate the morphology and properties. The principal elements available are oxygen, nitrogen, hydrogen, calcium, phosphorous, and carbon. The most useful synthetic metals (iron, aluminum, copper) are virtually absent – only present in minute quantities and highly specialized applications. The processing of these elements requires high temperatures not available in natural organisms. The seven components are:

- *Self-assembly* – in contrast to many synthetic processes to produce materials, the structures are assembled from the bottom up, rather than from the top down. This is a necessity of the growth process, since there is no availability of an overriding scaffold. This characteristic is called "self-assembly."
- *Self-healing capability* – whereas synthetic materials undergo damage and failure in an irreversible manner, biological materials often have the capability, due to the

Figure 1.6.

Fundamental and unique components of biological materials. (Heptahedron inspired by Arzt (2006).)

vascularity and cells embedded in the structure, to reverse the effects of damage by healing.

- *Evolution and environmental constraints* – the limited availability of useful elements dictates the morphology and resultant properties. The structures are not necessarily optimized for all properties, but are the result of an evolutionary process leading to satisfactory and, importantly, robust solutions.
- *Hydration* – the properties are highly dependent on the level of water in the structure. Dried skin (leather) has mechanical properties radically different from live skin. There are some remarkable exceptions, such as enamel, but this rule applies to most biological materials and is of primary importance.
- *Mild synthesis conditions* – the majority of biological materials are fabricated at ambient temperature and pressure and in an aqueous environment, a significant difference from synthetic materials fabrication.
- *Functionality* – many components serve more than one purpose; for example, feathers provide flight capability, camouflage, and insulation; bones are a structural framework, promote the growth of red blood cells, and provide protection to the internal organs; the skin protects the organism and regulates the temperature. Thus, the structures are called "multifunctional."
- *Hierarchy* – there are different, organized scale levels (nano- to macro-scale) that confer distinct and translatable properties from one level to the next. We are starting

to develop a systematic and quantitative understanding of this hierarchy by distinguishing the characteristic levels, developing constitutive descriptions of each level, and linking them through appropriate and physically based equations, enabling a full predictive understanding.

The study of biological systems as structures dates back to the early parts of the twentieth century. The classic work by D'Arcy W. Thompson (Thompson, 1968), first published in 1917, can be considered the first major work in this field. He looked at biological systems as engineering structures, and obtained relationships that described their form. In the 1970s, Currey investigated a broad variety of mineralized biological materials and authored the classic book *Bones: Structure and Mechanics* (Currey, 2002). Another work of significance is Vincent's *Structural Biomaterials* (Vincent, 1991). The field of biology has, of course, existed and evolved during this period, but the engineering and materials approaches have often been shunned by biologists.

Materials science and engineering is a young and vibrant discipline that has, since its inception in the 1950s, expanded into three directions: metals, polymers, and ceramics (and their mixtures, composites). Biological materials are being added to its interests, starting in the 1990s, and are indeed its new future.

Many biological systems have mechanical properties that are far beyond those that can be achieved using the same synthetic materials (Vincent, 1991; Srinivasan, Haritos, and Hedberg, 1991). This is a surprising fact, if we consider that the basic polymers and minerals used in natural systems are quite weak. This limited strength is a result of the ambient temperature and the aqueous environment processing, as well as of the limited availability of elements (primarily C, N, Ca, H, O, Si, P). Biological organisms produce composites that are organized in terms of composition and structure, containing both inorganic and organic components in complex structures. They are hierarchically organized at the nano-, micro-, and meso-levels. The emerging field of biological materials introduces numerous new opportunities for materials scientists to do what they do best: solve complex multidisciplinary scientific problems. A new definition of biological materials science is emerging; as presented in Fig. 1.7; it is situated at the confluence of chemistry, physics, and biology.

Biological systems have many distinguishing features, such as being the result of evolution and being multifunctional; however, evolution is not a consideration in synthetic materials, and multifunctionality still needs further research. Some of the main areas of research and activity in this field are:

- Biological materials: these are the materials and systems encountered in nature.
- Bioinspired (or biomimicked) materials: approaches to synthesizing materials inspired by biological systems.
- Biomaterials: these are materials (e.g. implants) specifically designed for optimum compatibility with biological systems.
- Functional biomaterials and devices.

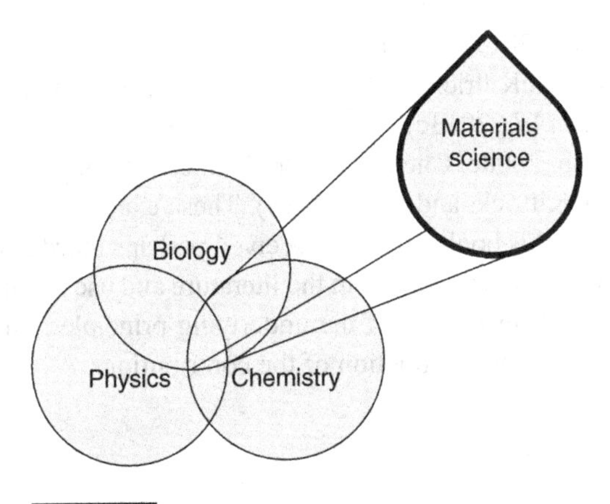

Figure 1.7.

Biological materials science at the intersection of physics, chemistry, and biology.

This book will focus primarily on the first and second areas, with short presentations (boxes) throughout the text focusing on specific biomaterials. Using a classical "materials" approach, we present the basic structural elements of biological materials (Chapters 3–5) and then correlate them to their mechanical properties. These elements are organized hierarchically into complex structures (Chapter 2). The different structures will be discussed in Chapters 6–11. In Chapters 12 and 13, we provide some inroads into biomimetics, since the goal of materials engineering is to utilize the knowledge base developed in materials science to create new materials with expanded properties and functions.

Although biology is a mature science, the study of *biological materials and systems* by materials scientists and engineers is recent. It is intended, ultimately:

(a) To provide the tools for the development of biologically inspired materials. This field, also called *biomimetics* (Sarikaya, 1994), is attracting increasing attention and is one of the new frontiers in materials research.

(b) To enhance our understanding of the interaction of synthetic materials and biological structures with the goal of enabling the introduction of new and complex systems in the human body, leading eventually to organ supplementation and substitution. These are the so-called *biomaterials*.

The extent and complexity of the subject are daunting and will require many decades of global research effort to be elucidated. Thus, we focus in this book on a number of systems that have attracted our interest. This is by no means an exhaustive list, and there are many systems that have been only superficially investigated. Our group has been investigating various shells, including abalone (Menig *et al.*, 2000; Lin and Meyers, 2005), conch (Menig *et al.*, 2001; Lin, Meyers, and Vecchio, 2006), and clams (Yang *et al.*, 2011a, 2011b); bird beaks (Seki, Schneider, and Meyers, 2005; Seki *et al.*, 2006); crab exoskeletons (Chen *et al.*, 2008a); bird feathers (Bodde, Meyers, and McKittrick,

2011); fish scales (Lin *et al.*, 2011; Meyers, 2012; Yang, Chao, and McKittrick, 2013a); antlers (Chen, Stokes, and McKittrick, 2009); armadillo, alligator, and turtle osteoderms (Chen *et al.*, 2011; Yang *et al.*, 2013c). This work has been reviewed in six extensive articles (Chen *et al.*, 2008b, 2008c; Chen, McKittrick, and Meyers, 2012; Meyers *et al.*, 2008b, 2011; Meyers, McKittrick, and Chen, 2013). These examples from our research have been used throughout this book and have received perhaps inordinate emphasis. In addition to those, we borrow extensively from the literature and use examples from what we judged were didactically illustrative of the underlying principles. Nevertheless, we can only cover in this book a minute fraction of the contributions.

Summary

- Early humans used exclusively natural materials, the majority of which were biological: wood, bone, hide, tendons and ligaments, horns, and antlers.
- The earliest vestiges of industrial-scale production of copper are from Jordan. This marked the beginning of the use of metals and alloys as a major class of materials.
- Biological materials have seven unique and defining features: (i) self-assembly; (ii) self-healing capability; (iii) evolution and environmental constraints; (iv) hydration; (v) mild synthesis conditions; (vi) functionality; (vii) hierarchy.
- Although our synthetic materials have properties that far exceed those of biological materials, the latter have ingenious designs that evolved through millions of years. Thus, we are developing a new class of bioinspired materials that apply the internal design principles of biological materials but use synthetic synthesis and processing methods.

Example 1.1 Describe the seven elements of the Arzt heptahedron for skin.

(a) Self-assembly: skin has a complex process of formation that starts at the molecular level. In the dermis, the collagen is produced by fibroblasts. Keratinocytes are cells in the epidermis that create keratin upon dying.
(b) Self-healing: if the skin is injured, cells converge on the site and reassemble the structure. The formation of scar tissue is essential to the survival of biological organisms.
(c) Evolution and environmental constraints: different species have evolved different skins that are adapted to the environment. Extremes are the rhinoceros skin, which resembles armor; the snake, which has a skin covered with flexible scales; the cat, which has a highly stretchable skin, making it especially valuable in the production of Brazilian Mardi Gras drums and "cuicas."
(d) Hydration: the mechanical response of skin is highly dependent on the degree of hydration. The stiffness increases with the removal of fluids. This is the reason why we apply moisturizers to the skin.

(e) Synthesis: effected at 1 atm and 310 K.
(f) Multifunctionality: skin is, in humans, the largest organ. It has several functions, the most important being retention of moisture inside of the body. Other functions are protection (it is, generally, much more tear resistant than the underlying tissues) and temperature regulation (embedded sweat glands). In some animals, it also has a camouflaging function.
(g) Hierarchy: skin is composed of three layers – the epidermis, the dermis, and the endodermis. The dermis is responsible for the mechanical properties. The collagen fibers are composed of fibrils, which comprise microfibrils. These in turn are formed by polypeptides. The polypeptides are formed by amino acids.

Example 1.2 In Fig. 1.2, there is a linear relationship between strength and density. Determine the equation.

Solution This is a log-log plot. We consider the exponents of 10 in a linear transformation; the vertical axis becomes −2, −1, 0, 1, 2, 3, 4, 5. The horizontal axis is now labeled 1, 2, 3, 4, 5. Taking the general form of a straight line expressed in an (x, y) coordinate system, we have

$$y = mx + b.$$

The slope is equal to ~2 and the intercept with the y-axis for $x = 0$ is equal to −4. The equation becomes

$$y = 2x - 4.$$

Expressing it again in decimal logarithm form yields

$$\log \sigma = 2 \log \rho - 4.$$

The slope m is obtained through two points.

Reorganizing the terms, we obtain

$$\sigma = 10^{-4} \rho^2.$$

Exercises

Exercise 1.1 Lucy is one of the earliest hominids (*Australopithecus afarensis*) who roamed Ethiopia between five and four million years ago. She was bipedal and her height was only 1 m. By using geometrical estimates, calculate her weight.

Johanna, Lucy's modern counterpart, is 1.70 m high and weighs 70 kg. From Fig. E1.1, estimate Lucy's femur bone diameter. How does it compare with the diameter of a modern human?

Exercise 1.2 The early chopper shown in Fig. 1.1 has a diameter of approximately 10 cm.

(a) Determine its weight, knowing that it is made of basalt (obtain the density from the web).

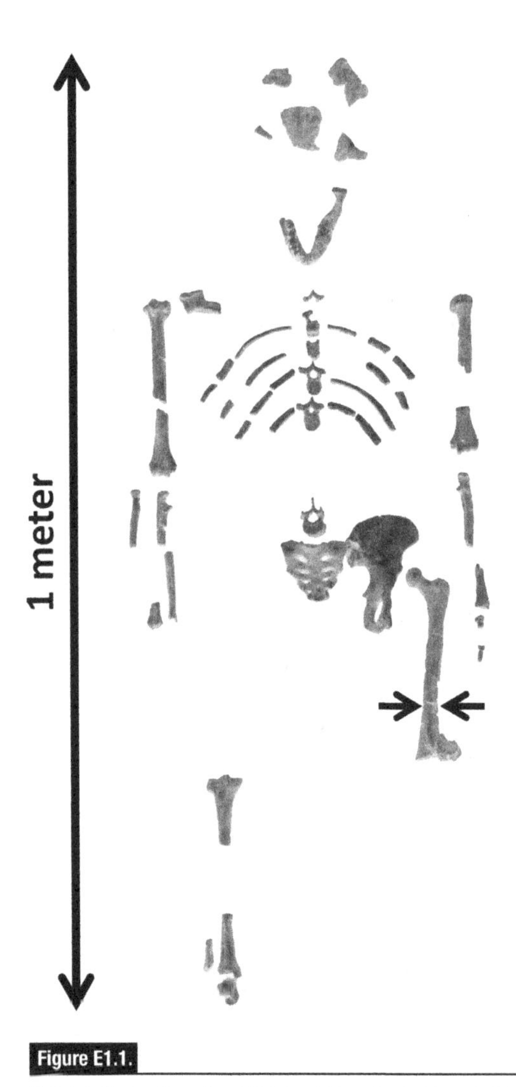

Figure E1.1.

Skeleton of Lucy. One of the earliest hominids, unearthed in East Africa (Ethiopia). Approximate age: 4–5 MY. (Adapted from http://commons.wikimedia.org/wiki/File:Lucy_blackbg.jpg. Image by 120. Licensed under CC-BY-SA-3.0, via Wikimedia Commons.)

(b) Estimate the velocity it can reach, thrust by the energetic arms of Lucy (see Fig. E1.1).
(c) What force can it generate if it is decelerated to zero over a distance equal to the skull thickness of Luciano, Lucy's suitor?
(d) Assuming that bone has a flexure strength equal to 120 MPa, establish whether the blow force is sufficient to crack Luciano's skull.

Exercise 1.3 Why are gold, copper, and silver three of the few metals that are found native (in metallic form)?

Exercise 1.4 The Fe–Ni alloy is also found in native form, and some important religious and archeological monuments use this alloy. What is the source of this material?

Exercise 1.5 What is the process by which metallurgy transformed ores into metals in prehistory? What was the source of energy?

Exercise 1.6 Discuss how each of the seven characteristics in the Arzt heptahedron of biological materials apply to bone.

Exercise 1.7 Of the natural (biological) materials listed in Fig. 1.2(a), how many are still in use nowadays?

Exercise 1.8 Compare the Ashby maps in Figs. 1.2(a) and (b). Only one natural (or biological) material is still at the top of its performance. Which one is it? Provide three examples of its current utilization. Can you think of ways in which these materials can be improved further?

Exercise 1.9 Compare the differences between biological (natural) materials and engineering materials.

Part I Basic biology principles

2 Self-assembly, hierarchy, and evolution

Introduction

A considerable number of books and review articles have been written on biological materials, and they constitute the foundation necessary to embrace this field. Some of the best known are given in Table 2.1. Table 2.2 lists some of the key review articles in the field. An important step was taken by D' Arcy Thompson with his monumental book, *On Growth and Form*, initially published in 1917 (Thompson, 1917). This book still constitutes exciting reading material and is a valiant attempt at representing biological shapes mathematically.

2.1 Hierarchical structures

It could be argued that all materials are hierarchically structured, since the changes in dimensional scale bring about different mechanisms of deformation and damage. However, in biological materials this hierarchical organization is inherent to the design. The design of the structure and of the materials are intimately connected in biological systems, whereas in synthetic materials there is often a disciplinary separation, based largely on tradition, between materials (materials engineers) and structures (mechanical engineers). We illustrate this by presenting four examples in Fig. 2.1 (avian feather rachis), Fig. 2.2 (abalone shell), Fig. 2.3 (crab exoskeleton), and Fig. 2.4 (bone).

Figure 2.1 shows the main shaft of the feather of a falcon. Feathers are keratinous. The detailed structure of keratin will be described in Chapter 3. The entire structure has a good stiffness/weight ratio. Weight minimization is of primary importance in flying birds, and the structure and architecture of feathers reflect this. The rachis has an external shell (called the cortex) and a core that is porous. If we image the internal foam at a higher magnification, we find that the cell walls are not solid membranes, but are composed of fibers themselves. These fibers act as struts and have a diameter of ~200 nm. This decreases the weight further, and is an eloquent example of a hierarchical structure.

Similarly, the abalone shell (Fig. 2.2) owes its extraordinary mechanical properties (much superior to monolithic $CaCO_3$) to a hierarchically organized structure, starting, at the nano-level, with an organic layer having a thickness of 20–30 nm, proceeding with single crystals of the aragonite polymorph of $CaCO_3$, consisting of "bricks" with

Table 2.1. Principal books on biological materials and biomimetism[a]

Author(s)	Year	Title
Thompson	1917	*On Growth and Form*
Fraser, MacRae, and Rogers	1972	*Keratins: Their Composition, Structure, and Biosynthesis*
Brown	1975	*Structural Materials in Animals*
Wainwright *et al.*	1976	*Mechanical Design in Organisms*
Vincent and Currey (eds.)	1980	*The Mechanical Properties of Biological Materials*
Currey	1984	*The Mechanical Adaptations of Bones*
Simkiss and Wilbur	1989	*Biomineralization: Cell Biology and Mineral Deposition*
Lowenstam and Weiner	1989	*On Biomineralization*
Fung	1990	*Biomechanics: Motion, Flow, Stress, and Growth*
Vincent	1990	*Structural Biomaterials* (revised edition)
Byrom	1991	*Biomaterials: Novel Materials from Biological Sources*
Fung	1993	*Biomechanics: Mechanical Properties of Living Tissues*
Neville	1993	*Biology of Fibrous Composites*
Fung	1997	*Biomechanics: Circulation* (2nd edition)
Feughelman	1997	*Mechanical Properties and Structure of α-Keratin Fibres*
Gibson and Ashby	1997	*Cellular Solids: Structure and Properties* (2nd edition)
Kaplan and McGrath	1997	*Protein-based Materials*
Elices	2000	*Structural Biological Materials: Design and Structure-Property Relationships*
Mann	2001	*Biomineralization: Principles and Concepts in Bioinorganic Materials Chemistry*
Currey	2002	*Bones: Structure and Mechanics*
Ratner *et al.*	2005	*Biomaterials Science: An Introduction to Materials in Medicine*
Forbes	2007	*The Gecko's Foot*
Gibson, Ashby, and Harley	2010	*Cellular Materials in Nature and Medicine*
Ennos	2012	*Solid Biomechanics*
Boal	2012	*Mechanics of the Cell* (2nd edition)

[a] Full details may be found in the References section.

Table 2.2. Principal review articles on biological materials and biomimetism

Author	Year	Title
Srinivasan *et al.*	1991	Biomimetics: advancing man-made materials through guidance from nature
Mann *et al.*	1993	Crystallization at inorganic-organic interfaces: biominerals and biomimetic synthesis
Kamat *et al.*	2000	Structural basis for the fracture toughness of the shell of the conch *Strombus gigas*
Mayer and Sarikaya	2002	Rigid biological composite materials: structural examples for biomimetic design
Whitesides	2002	Organic material science
Altman *et al.*	2003	Silk-based biomaterials
Wegst and Ashby	2004	The mechanical efficiency of natural materials
Mayer	2005	Rigid biological systems as models for synthetic composites
Sanchez, Arribart, and Giraud-Guille	2005	Biomimetism and bioinspiration as tools for the design of innovative materials and systems
Wilt	2005	Developmental biology meets materials science: morphogenesis of biomineralized structures
Mayer	2006	New classes of tough composite materials – lessons from natural rigid biological systems
Meyers *et al.*	2006	Structural biological composites: an overview
Lee and Lim	2007	Biomechanics approaches to studying human diseases
Chen *et al.*	2012	Biological materials: functional adaptations and bioinspired designs
Meyers *et al.*	2013	Structural biological materials: critical mechanics-materials connections

dimensions of 0.5×10 μm (microstructure), and finishing with layers of approximately 0.3 mm (mesostructure).

Crabs are arthropods whose carapace comprises a mineralized hard component, which exhibits brittle fracture, and a softer organic component. These two components are shown in Fig. 2.3. The brittle component is arranged in a helical pattern called a Bouligand structure (Bouligand, 1970, 1972). The Bouligand structure, also present in bones and other biological systems, is characterized by a characteristic sequence of layers in which each layer is offset at a specific angle to its neighbor. This stacking forms a helicoid, seen in the sketch, representing a total rotation of 180°. If the

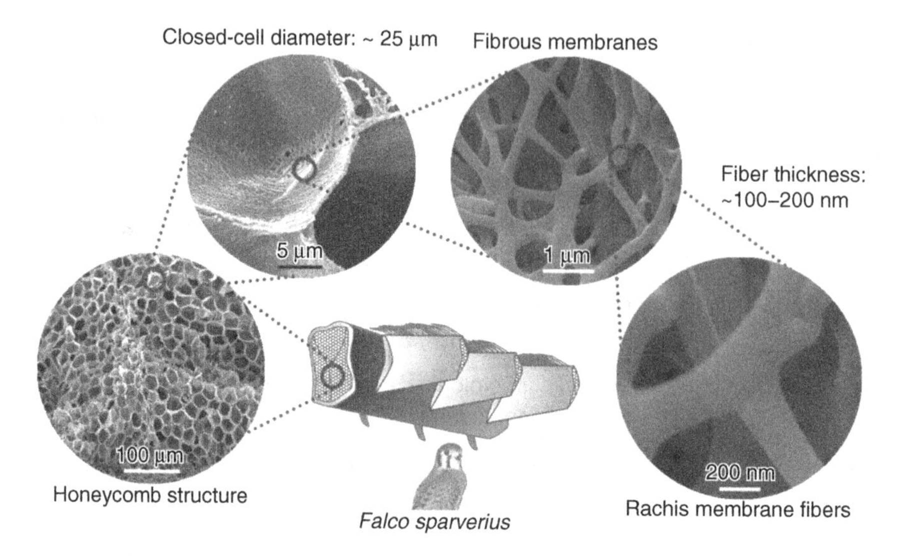

Figure 2.1.

Hierarchical structure of feather rachis. (Reprinted from Meyers *et al.* (2011), with permission from Elsevier.)

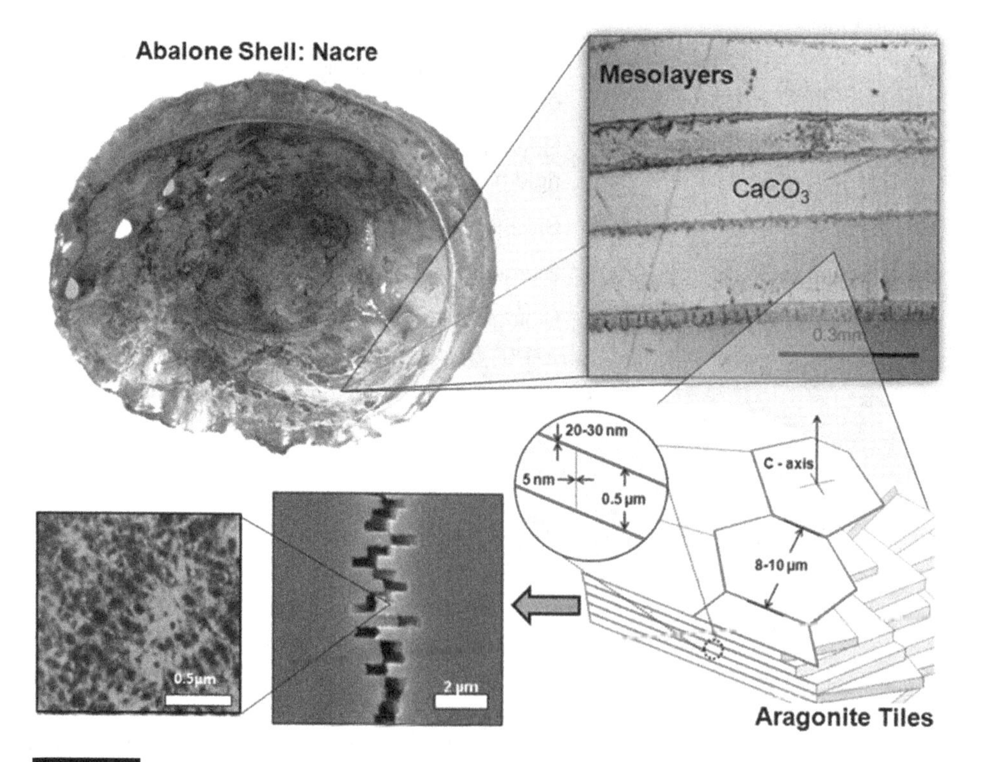

Figure 2.2.

Hierarchical structure of the abalone shell. (Reprinted from Meyers *et al.* (2008a), with permission from Elsevier.)

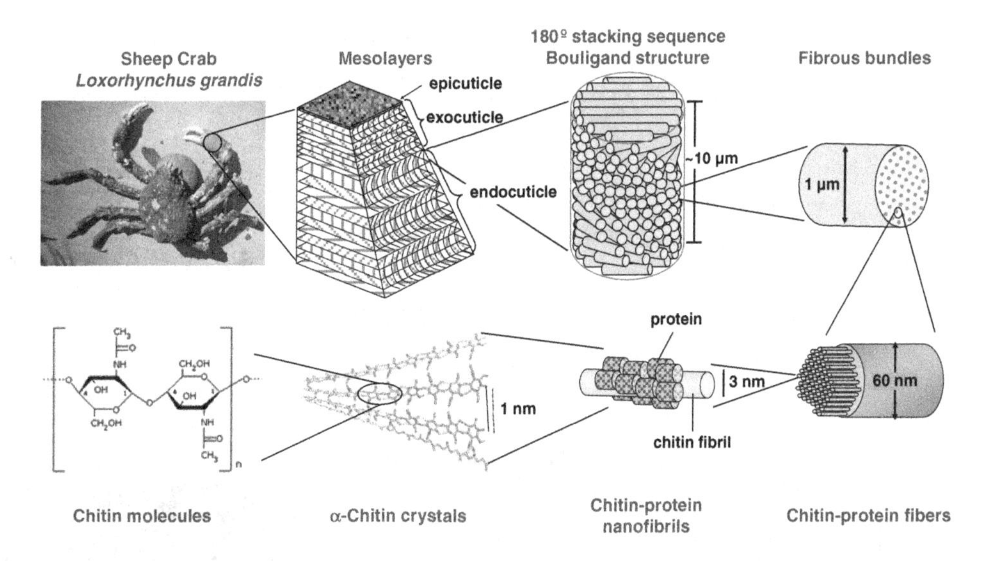

Figure 2.3.

Hierarchical structure of the crab exoskeleton. (Reprinted from Chen *et al.* (2008a), with permission from Elsevier.)

Bouligand structure is sectioned at an angle that is not normal to the layers, there appears, on the cut, a particular pattern of lines that can be confusing. This sequence of curved segments is also shown in Fig. 2.3. Each of the mineral "rods" (~1 μm diameter) contains chitin-protein fibrils with approximately 60 nm diameter. These in turn comprise 3 nm diameter segments. There are canals linking the inside to the outside of the shell; they are bound by tubules (0.5–1 μm) shown in the micrograph and in schematic fashion. The cross-section of hard mineralized component has darker spots as seen in the scanning electron microscope (SEM) micrograph.

Hierarchical structures can be defined as a group of molecular units/aggregates that are in contact with other phases, which in turn are similarly assembled at increasing length scales. Biological materials exhibit hierarchy at several to many length scales, depending on the complexity of the structure. Materials scientists can routinely probe down to the molecular level, which will be used in this book as the smallest scale. Other lengths include nano-, nano/micro-, micro-, micro/meso-, meso-, meso/macro-, and macro-scale.

In bone (Fig. 2.4), the building block of the organic component is the collagen, which is a triple helix with a diameter of approximately 1.5 nm. These tropocollagen molecules are intercalated with the mineral phase (hydroxyapatite, a calcium phosphate) forming fibrils that, in turn, curl into helicoids of alternating directions. These, the osteons, are the basic building blocks of bones. The volume fraction distribution between the organic and the mineral phase is approximately 60/40, making bone unquestionably a complex hierarchically structured biological composite. There is another level of complexity. The hydroxyapatite crystals are platelets that have a diameter of approximately 70–100 nm and a thickness of ~2–4 nm. They originally nucleate at the gaps between collagen fibrils. Not shown in Fig. 2.4 is the Haversian

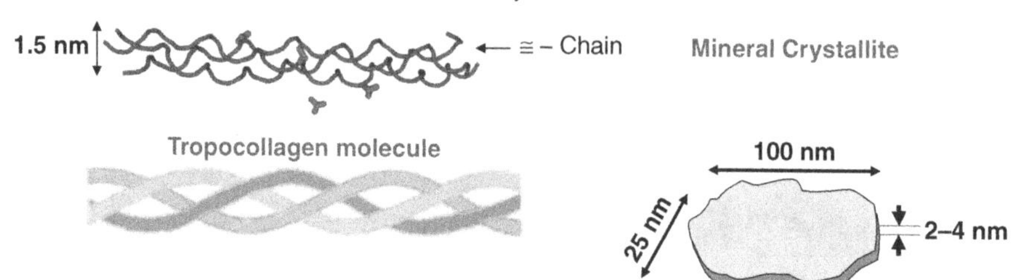

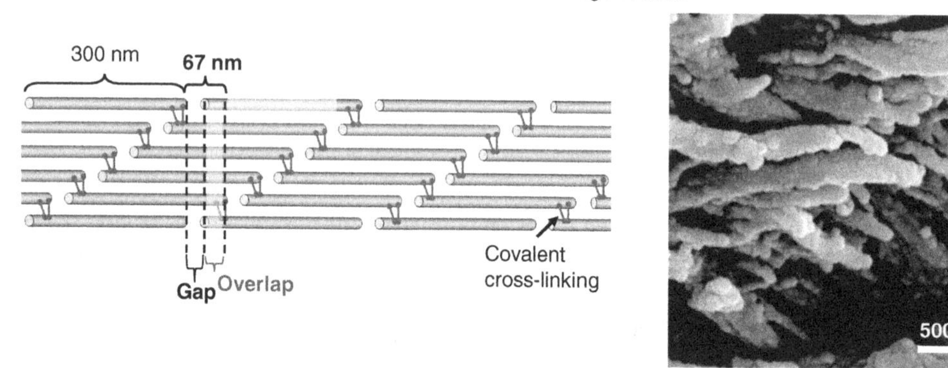

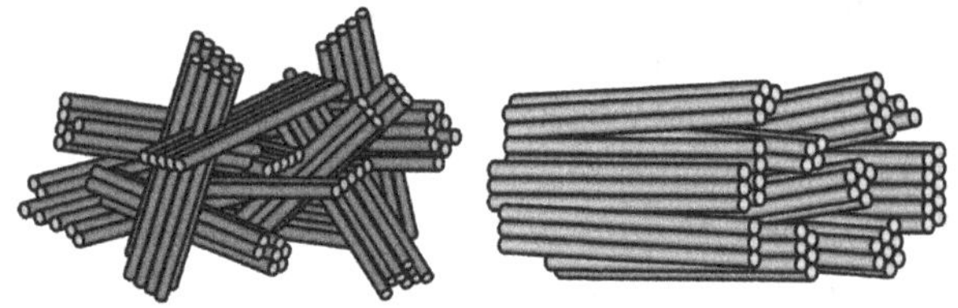

Figure 2.4.

Hierarchical structure of bone. Level I: tropocollagen (triple helix of α-collagen molecules) and hydroxyapatite form the basic constituents of bone. (From http://commons.wikimedia.org/wiki/File:1bkv_collagen.png Image by Nevit Dilmen. Licensed under CC-BY-SA-3.0, via Wikimedia Commons.) Level II: tropocollagen assembles to form collagen fibrils and combines with hydroxyapatite, which is dispersed between (in the gap regions) and around the collagen, forming mineralized collagen fibrils. Level III: the fibrils are orientated into several structures, depending on the location in the bone (parallel, circumferential, twisted). Level IV: cortical bone is lamellar – cylindrical- (cortical) and parallel-plate lamellae (cortical and cancellous) can be found depending on location. Level V: Bouligand structure of lamellae around osteon and struts in trabeculae; cancellous bone is flat lamellar bone. Level VI: light-microscope level showing osteons (organized cylindrical lamellae) with a central vascular channel and cancellous bone with porosity. Level VII: entire bone.

Level IV Lamella

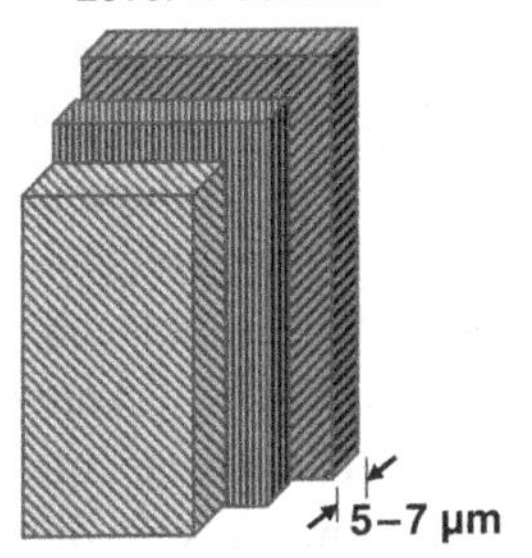

Level V Osteon & Trabecula

Osteon

Trabecula

100 μm

100–200 μm

Level VI Cortical & Cancellous bone

Cortical bone

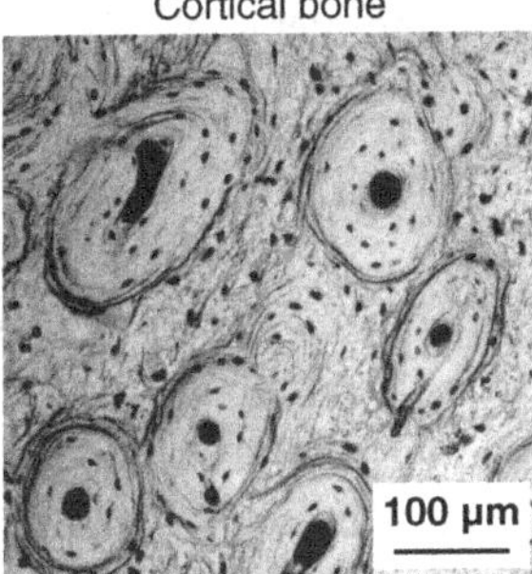

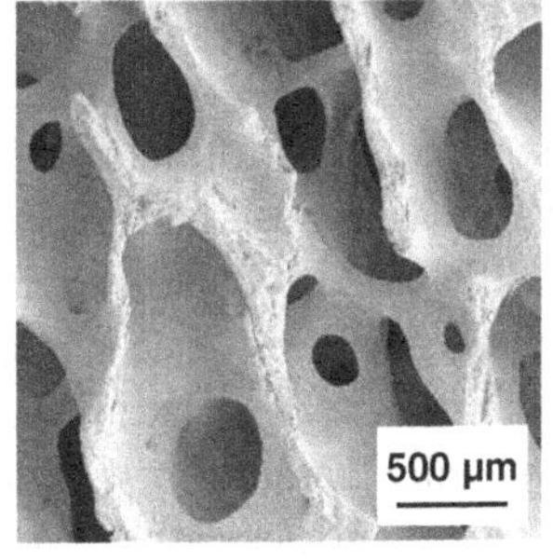

Level VII Whole Bone

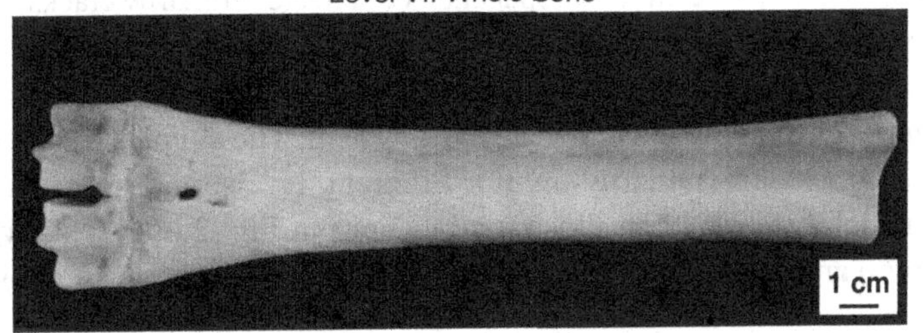

Figure 2.4. (cont.)

system that contains the vascularity which brings nutrients and enables forming, remodeling, and healing of the bone. Using bone as an example for a biological composite, the hierarchical structure is shown in Fig. 2.4, which is broken down into seven levels. Level I is the molecular arrangement of the collagen molecules – three α-helix chains twist together to form the tropocollagen molecule. Level II is the mineralized collagen fibrils (2–300 nm): the tropocollagen molecule and the mineral, hydroxyapatite. The tropocollagen molecules (300 nm length, 1.5 nm diameter) assemble to form collagen fibrils of ~100 nm in diameter. Hydroxyapatite (4 nm thickness, lateral dimension ~25–100 nm) is nucleated within and outside of the fibrils, which are held together by noncollagenous proteins (Fantner *et al.*, 2005). In Level III (10–50 μm), the fibrils further assemble into oriented arrays in plate-like structures (lamella). In Level IV, collagen fibers form lamellae 5–7 μm in thickness. In Level V, the lamellae assemble into concentric cylinders in the cortical bone, forming the basic unit, the osteon. This is also called a Bouligand structure of lamellae around the osteons. In this level, we also have the struts in trabecular bone; cancellous bone is flat lamellar bone. In Level VI we have the light-microscope level showing osteons (organized cylindrical lamellae) with a central vascular channel and cancellous bone with porosity; Level VII corresponds to the entire bone.

Osteons and parallel-plate lamellae (cortical and cancellous) can be found depending on the location. Embedded in the osteon are other microstructural features such as lacuna spaces (10–20 μm) and canaliculus channels (100 nm diameter). Bone cells (osteocytes) occupy the lacuna spaces that are connected by the canaliculi.

The stiffness, strength, and toughness of a structure depend on the level in the hierarchy and on the total number of levels in the hierarchy. There are two ways of looking at it.

(a) The first is to look at a biological structure and probe the mechanical properties, starting at the nano-scale and ascending in the scale of the test. In this manner we can test the different hierarchical levels. If we conduct such a thought experiment, we realize that the smallest scale is more perfect, i.e. the flaws are smaller. This will be explained in greater detail later in the book. Thus, the Young modulus and strength are higher. Usually, hardness tests done at the nano-scale (nanoindentation) provide higher values than those at the micro-scale (microhardness). As we move up, there are weak interfaces, sacrificial bonds, voids, cracks, etc., that contribute to a lowering of the Young modulus and strength (hardness). However, there are more effective mechanisms to retard the propagation of damage (primarily cracks). This results in an increase in toughness. This is represented schematically in Fig. 2.5.

(b) The second way of looking at it is to measure the macroproperties but change the number of hierarchical levels in the system. The simplest form of hierarchy is to consider self-similarity in the different levels, the so-called "Russian doll model," shown in an illustrative manner in Fig. 2.6(a). This was implemented by Ji and Gao (2010). The hierarchical levels (N) of bone are shown in Fig. 2.6(b)

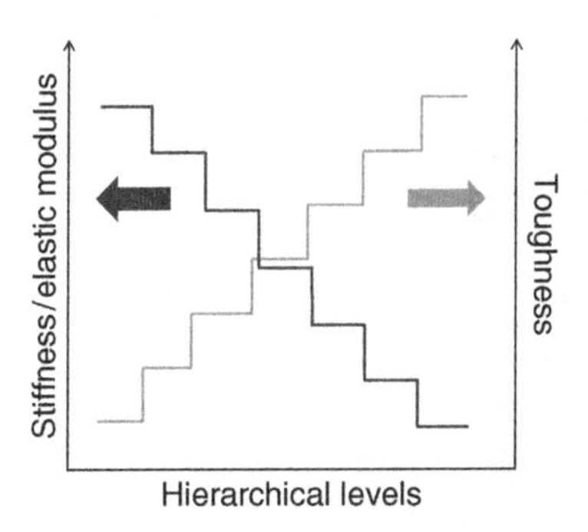

Figure 2.5.

Schematic representation of decrease in stiffness and increase in toughness as one marches up the hierarchical scale.

(Ji and Gao, 2010). There is a high aspect ratio (Φ) of the mineral phase in bone that is aligned in a soft matrix. In Figs. 2.6(c)–(e), the calculated stiffness, strength, and toughness are shown as functions of N. The influence of N is quite strong – the stiffness decreases as the strength and toughness increase. This gain in strength and toughness can be attributed to an increase in the number of crack-arresting mechanisms. The predicted values are based on the theory proposed by Ji and Gao (2010).

For the sake of didactics, we present in the following the equations used by Yao and Gao (2007) without deriving them. The derivation is quite involved, and it suffices here to know that they used a mathematical procedure to obtain quantitatively values of strength, the Young modulus, and fracture energy as a function of the hierarchical level. We start, at the smallest hierarchical level, with the Griffith equation, which expresses the theoretical strength in terms of the surface energy, γ, and of the maximum flaw size, or lateral dimension, h:

$$\sigma_{th} = \sqrt{\frac{E_0 \gamma}{h}}; \tag{2.1}$$

E_0 is the Young modulus of the mineral at the first hierarchical level: $E_0 = E_m$, where E_m is the mineral Young modulus, usually taken as 100 GPa. By setting the theoretical strength equal to $E/30$, and $\gamma = 1\ \text{J/m}^2$, we obtain a value for $h = 18$ nm.

Yao and Gao (2007) then expressed the mechanical properties at hierarchical level n as a function of the level $n - 1$. The total number of hierarchical levels of the biological system is given by N, whereas the level being calculated is denoted n; so, $1 \leq n \leq N$. By starting with $n = 1$, we can reproduce the plots in Fig. 2.6.

The Young modulus at level $n + 1$ is obtained as a function of level n, E_n:

$$E_{n+1} = \left[\frac{4(1 - \Phi^{1/N})}{G_p \Phi^{2/N} \rho^2} + \frac{1}{\Phi^{1/N} E_n}\right]^{-1}, \tag{2.2}$$

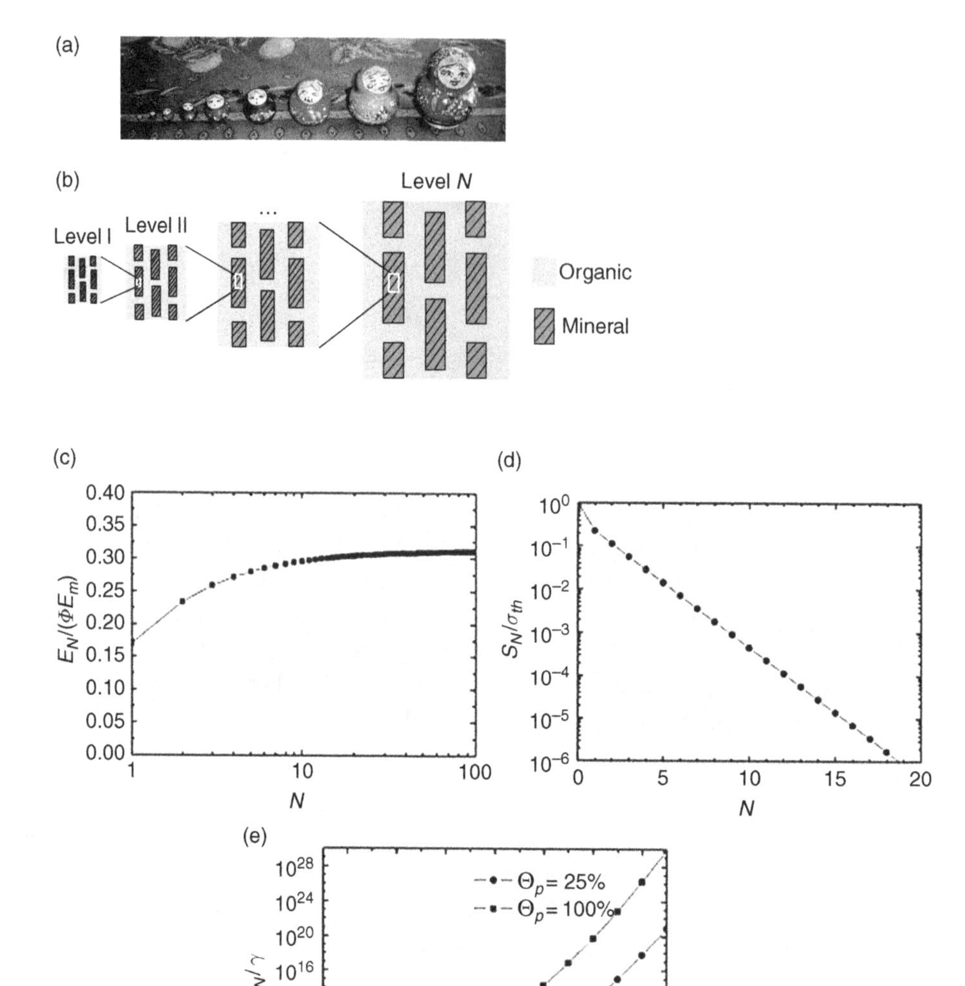

Figure 2.6.

Effect of varying the number of hierarchical levels (N) on properties. (a) Russian doll analog for first approximation to hierarchical structure in which levels are identical but with different scales; (b) N levels of hierarchy in bone. Calculated changes in (c) the Young modulus (normalized to the product of $\Phi = 0.45$ and $E_m = 100$ GPa), (d) strength (normalized to the theoretical value $E/30$), and (e) fracture energy (normalized to the surface energy $\gamma = 1$ J/m^2) with the number of hierarchical levels N of system. Θ_p is the effective strain measuring the deformation range of the soft component. Two values were used ($\Theta_p = 25\%$ and $\Theta_p = 100\%$) for the failure strain of protein. (Reprinted from Yao and Gao (2007), with permission from Elsevier.)

where Φ is the aspect ratio of the hard phase ($= \varphi \cdot \varphi \cdot \varphi \cdot \ldots = \varphi^n$); G_p is the shear modulus of the soft phase; the fraction of the soft phase is ρ.

The strength S_{n+1} is obtained from the strength S_n:

$$S_{n+1} = \Phi^{1/N} \frac{S_n}{2} = \Phi^{n/N} \frac{\sigma_{th}}{2^n}. \quad (2.3)$$

The simplified expression is possible because $S_0 = \sigma_0 = \sigma_{th}$, the theoretical strength of the material.

An effective fracture energy was defined as

$$\Gamma_{n+1} = (1 - \varphi_n) h_n S_n \Theta_n^p, \quad (2.4)$$

where $2h_n$ is the characteristic platelet width in level n; φ_n is the volume fraction of the hard phase at level n; Θ_n^p is the effective strain undergone by the soft phase at failure (25 and 100% in Fig. 2.6). Equation (2.4) requires the application of Eqn. (2.3) at each step.

Mechanical testing can probe only one or two levels (e.g. nanoindentation, Vickers hardness, nano-tensile testing, bulk testing). Bechtle, Ang, and Scheider (2010) defined hierarchical levels for nacre, enamel, sponge spicule, dentin, conch shell, cortical, and antler bone. They pointed out that there is a lack of correspondence between relating properties at one level to another. This remains a major challenge for the full understanding of the complex behavior of biological composites and for the design of future bioinspired materials. One important issue is the interfacial strength, since it determines the linkage efficiency between different layers and hierarchical levels.

2.2 Multifunctionality

Most biological materials are *multifunctional* (Srinivasan *et al.*, 1991), i.e. they accumulate functions such as:

(a) bone – structural support for body plus blood cell formation;
(b) chitin-based exoskeleton in arthropods – attachment for muscles, environmental protection, water barrier;
(c) sea spicules – light transmission plus structural;
(d) tree trunks and roots – structural support and anchoring plus nutrient transport;
(e) mammalian skin – temperature regulation plus environmental protection;
(f) insect antennas – mechanically strong and can self-repair; they also detect chemical and thermal information from the environment, and they can change their shape and orientation.

Another defining characteristic of biological systems, in contrast with current synthetic systems, is their *self-healing ability*. This is nearly universal in nature. Most structures can repair themselves after undergoing trauma or injury. Exceptions are teeth and cartilage that do not possess any significant vascularity. It is also true that brains cannot self-repair; however, other parts of the brain take up the lost functions.

2.3 Self-organization and self-assembly

In the traditional study of thermodynamics, we, materials scientists, restrict ourselves, for the most part, to isolated and closed systems. We know, from the second law of thermodynamics, that in an isolated system equilibrium is reached at maximum entropy. If t is defined as time, a system will approach equilibrium as

$$dS/dt > 0. \tag{2.5}$$

Thus, we learn that the entire universe marches inexorably toward entropy maximization. This translates into disorder because disordered states have a greater number of possible configurations, W (recall that $S = k \ln W$).

For closed systems, equilibrium is reached when the free energy is minimized. If pressure and temperature are constant, the Gibbs free energy is used; for systems at constant volume and temperature, the Helmholtz free energy criterion holds. However, we know that temporal evolution in nature starts in simpler and extends to more complex structures and leads to ever greater order and self-organization. This is clear from the world around us: subatomic particles aggregate to form atoms, atoms combine to form compounds, organic compounds form complex molecules, which eventually lead to life, life progresses from the simplest to the more complex forms, biological units combine in ever-increasing complex arrays, civilizations form and evolve. How can all this be reconciled with thermodynamics? The seminal work of I. Prigogine (Prigogine, 1962; Nicolis and Prigogine, 1989), the 1977 Nobel Laureate in Chemistry, is essential to our understanding. Prigogine developed the thermodynamics of irreversible systems and provided a link between thermodynamics and evolution. This work was actually preceded by Teilhard de Chardin, a French paleontologist and philosopher who advanced the law of complexity-conscience. In his classic work, *Le Phénomène Humain* (Teilhard de Chardin, 1970), he proposed boldly that complexity inexorably increases in the universe at the expense of a quantity which he named "physical energy." This increase in complexity results in an increase in conscience.

Most complex systems are indeed open and off equilibrium. Prigogine demonstrated that nonisolated (open) systems can evolve, by self-organization and self-assembly, toward greater order (Prigogine, 1962; Nicolis and Prigogine, 1989). Open systems exhibit fluxes of energy, matter, and exchanges in mechanical, electrical, and magnetic energy with the environment. The entropy variation can be expressed as having two components, one internal, i, and one due to the exchanges, e:

$$dS/dt = d_iS/dt + d_eS/dt. \tag{2.6}$$

If the system were isolated, we would have $d_eS/dt = 0$. For an open system, this term is not necessarily zero. There is no law dictating the sign of d_eS/dt, and it can be either positive or negative. The second law still applies to the internal components, *i*. Thus,

$$d_iS/dt > 0. \tag{2.7}$$

Thus it is possible to have a temporal decrease in S if the flux component contributes with a sufficiently negative d_eS/dt term: $dS/dt < 0$ if $d_eS/dt > -d_iS/dt$.

The physical interpretation of Eqn. (2.6) is that order can increase with time. This is a simple and incomplete explanation for the origin of complex phenomena. Irreversible thermodynamics of complex systems involves other concepts such as perturbation and symmetry breaking. Nicolis and Prigogine (1989) showed that nonlinearities combined with nonequilibrium constraints can generate multiple solutions through bifurcation and thus allow for more complex behavior in a system. The biosphere is an example of an open system, since it receives radiative energy from the sun and exchanges energy with the earth.

Self-organization is a strategy by which biological systems construct their structures. There are current efforts at this design strategy in order to manufacture synthetic structures. Among the first were Nuzzo and Allara (1983), who fabricated self-assembled monolayers (SAMs). This work, and other work by Whitesides (Whitesides 2002; Wu *et al.*, 2002) and Tamerler and Sarikaya (2007) (genetically engineered peptides for inorganics, or GEPIs) will be reviewed in Chapter 13. Biomimetics is laying the groundwork for biologically inspired self-assembly processes which have considerable technological potential.

As an example of a self-assembled structure, Fig. 2.7(a) shows a schematic rendition of a diatom. The end surface is shown in Fig. 2.7(b). This elaborate architecture is developed by self-assembly of the enzymes forming a scaffold upon which biomineralization takes place (Hildebrand, 2003, 2005). These orifices are each at the intersections of three lines: two spiral lines with opposite chirality and radiating lines. The rows of orifices radiating from the center exhibit "dislocations" because the number of spokes has to increase with the radius. These dislocations are marked with circles in Fig. 2.7(c). The orifices have a striking regularity and spacing, as shown in Fig. 2.8. Their diameter is approximately 2 μm, and they each have a ridge that serves, most probably, as reinforcement, such as in cast-iron components.

2.4 Adaptation

One of the unique characteristics of biological systems is the ability to adapt to their environment. During growth, biological systems actively respond to external stimuli and form architectures and microstructures with improved functionality. As mentioned earlier, they also have a self-healing capability, which can reduce the damage present in the system by various repairing mechanisms.

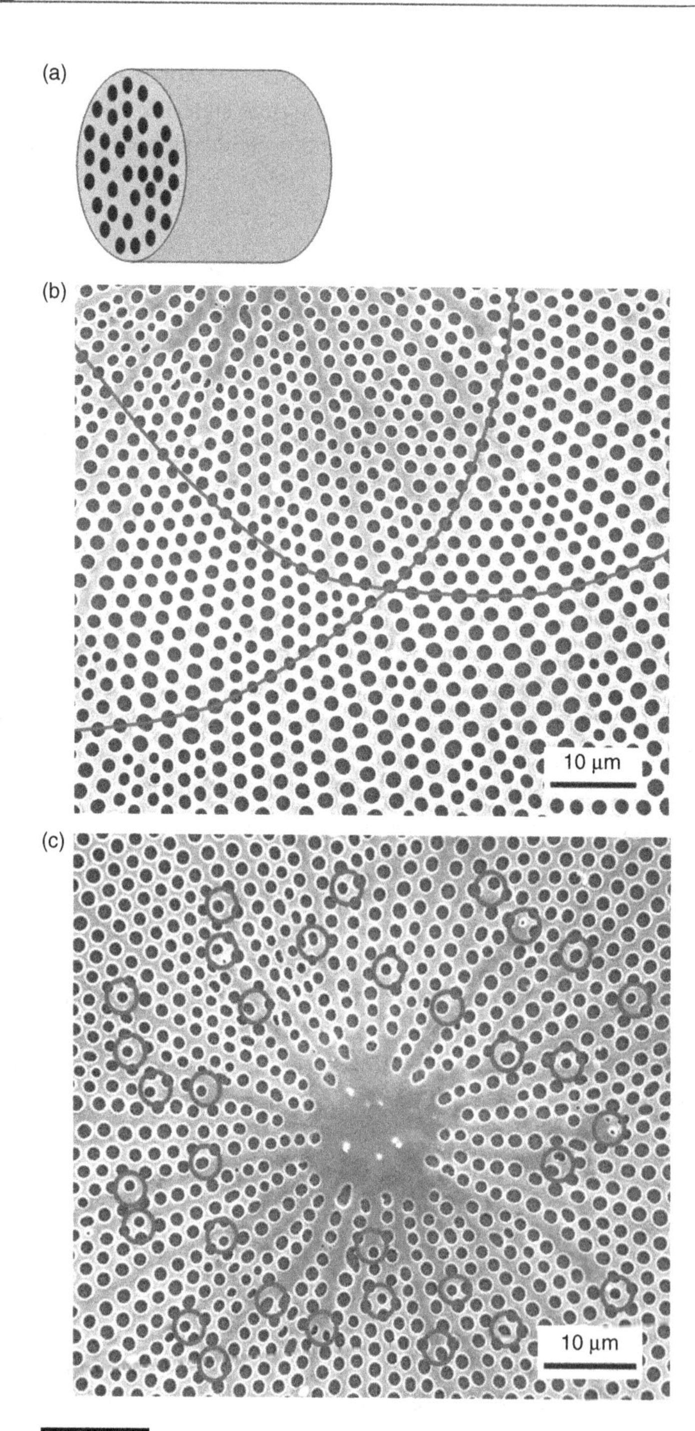

Figure 2.7.

(a) Cylindrical silica diatom. (b) End surface of cylindrical silica diatom showing self-organized pattern of orifices. Note the spiral patterns (both chiralities). (c) Center of cylinder with radiating spokes; extremities of "dislocations" are marked by circles. (Figure courtesy of E. York, Scripps Institution of Oceanography, UC San Diego.)

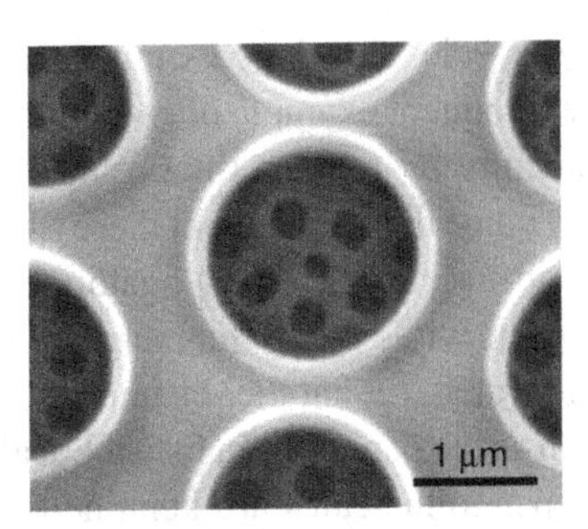

Figure 2.8.

Detailed view of orifice; back surface of diatom with hexagonal pattern can be seen in central orifice. (Figure courtesy of E. York, Scripps Institution of Oceanography, UC San Diego.)

The classic example of adaptation is bone, a living tissue which consistently undergoes modeling and remodeling. The famous Wolff's law, proposed in 1892, indicates that the external shape and internal architecture of bone are determined by the external stresses acting on it. The internal architecture of the trabecular bone undergoes adaptive changes along the principal stress trajectories, followed by secondary changes to the external cortical bone, typically becoming thicker and denser to resist the external loading. Such adaptation is also observed in the development of muscle. For example, the racquet-holding arm muscles of tennis players are usually much stronger than those of the other arm. Sprint athletes have quite different body figures compared with marathon runners.

Another example of adaptation to mechanical loading is the growth of trees. The tree branch has to adapt to increasing gravity that bends it downwards during the growth. On the lower side of the branch, the wood cells have to bear compressive loading, while the upper side of the branch is loaded under tension. The microstructure of the wood cells strongly depends on the loading conditions. The wood on the lower side of the branch (called compression or reaction wood) shows rounded cells in contrast to that on the upper side (named opposite wood), which has square-shaped cells. Such asymmetry in microstructure can also occur in the tree stem, for example, due to strong winds blowing predominantly from one direction.

2.5 Evolution and convergence

Evolution in general leads to the diversification of species, as they adapt, through selection, to different environments. However, there are outstanding exceptions showing that similar shapes and functions can form in different species. Well known to the general public is the Tasmanian tiger (*Thylacinus cynocephalus*), which was the largest carnivorous marsupial of modern times that became extinct in 1936 in Tasmania, an irreparable loss. Its skull had a striking resemblance to that of the gray wolf, a placental mammal.

Thus, one has to conclude that the two species developed similar features to achieve optimal predatory performance. Another eloquent example of evolutionary convergence is the ability to fly developed independently by reptiles (birds), mammals (bats), and insects. The wing evolved independently since the ancestors of these species do not have them. Less known is the protuberance in the nose of the alligators and hippopotami, which serve these two species well to breathe while totally submerged. Only the tip of the nose has to stay above the surface of the water. Yet another example is described by the scales of fish and of some mammals, such as the pangolin. In fish, the scales are a mixture of minerals and collagen, whereas the pangolin scales are keratinous. They serve the same purpose: flexible dermal armor.

The engineering characteristics of the features evolving in a convergent manner are important. Figure 2.9 shows the serrated teeth in three carnivores: a fish (great white shark), a reptile (Komodo dragon), and an extinct mammal (saber-toothed tiger). The

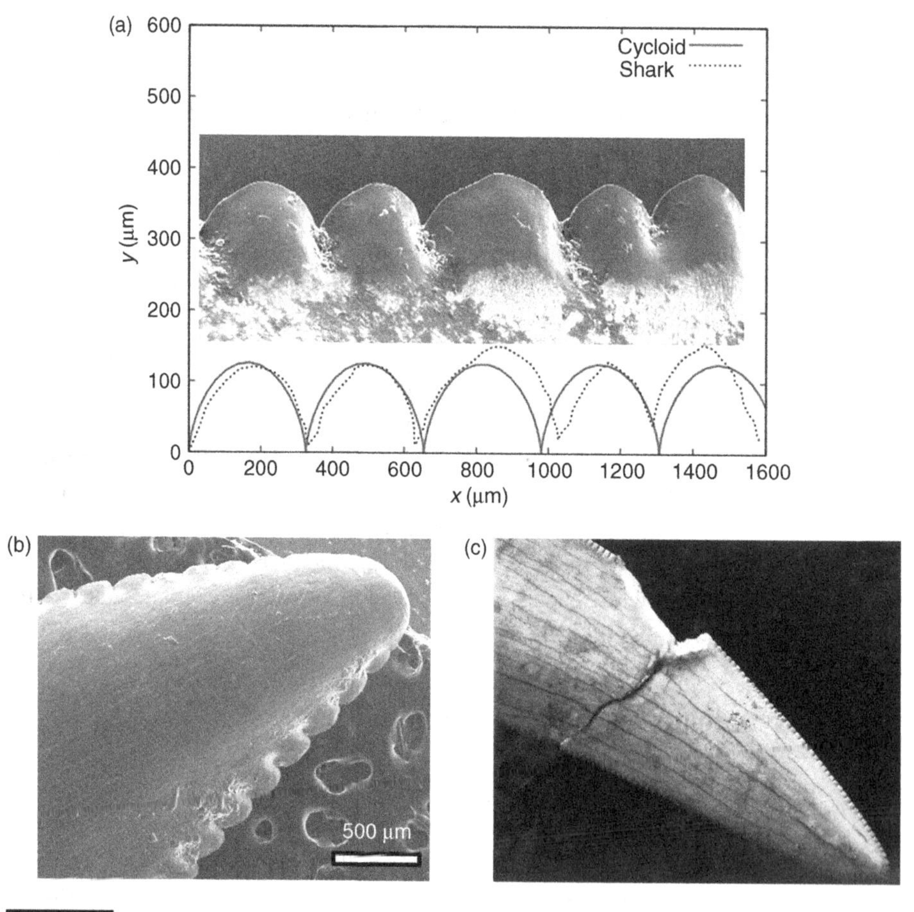

Figure 2.9.

Serrated teeth in (a) fish (great white shark); (b) reptile (Komodo dragon); (c) mammal (saber-toothed tiger, *Smilodon*).

function of the serrations, clearly known to us, is to ease the cutting process. This is done by trapping and stretching, until their failure, individual fibers (collagen and elastin) in the skin of the prey. It is interesting to observe that the dimensions of the serrations are approximately the same in these three species (0.2–0.3 mm) in spite of the considerable differences in size. This leads us to conclude that the dimensions of the individual biopolymer fibers determine the serrations. In the great white shark the serrations have a shape that closely resembles a cycloid. The parametric equations can easily be derived by an eager student. If we consider the radius of the circle to be r, we can follow the (x, y) coordinates of a point on the surface of the circle as it rolls on a flat surface (along Ox). This can be easily visualized as the trajectory of a mark made on a car tire. The parametric equations are given by

$$\begin{aligned} x &= r\theta - r\cos\theta, \\ y &= r - r\sin\theta. \end{aligned} \tag{2.8}$$

It can also be shown by the student that the trajectory over a full revolution of the circle is equal to $8r$. This is more tricky (but was solved in the 1600s). In Cartesian form, the equation becomes, after elimination of the angle θ,

$$x = r\cos^{-1}\left(1 + \frac{y}{r}\right) - \sqrt{y(2r - y)}. \tag{2.9}$$

Figure 2.9(a) shows the comparison of the tooth profile with the cycloid. The match is astonishingly close. We should note that we multiplied y by a factor $k = 1.2$ to obtain the correct match. Thus, we modified the original cycloid.

The last example of evolutionary convergence is the ability to adhere functionally to various surfaces in a dynamic manner. The gecko can attach to various surfaces and defy gravity by using a unique hierarchical structure in the foot pad to facilitate van der Waals' interactions (Autumn *et al.*, 2000, 2002). The gecko foot has V-shaped scansors, which consist of micro-sized bundles. These bundles, called setae, are made of keratinous material, ~2 μm in diameter, ~100 μm in length. The distal ends of the setae split up into hundreds of nanofibrils, termed spatulae. The tip of a gecko spatula has an asymmetric mushroom shape ~200 nm in diameter. However, the mechanism by which a gecko attaches itself is not unique and is widely found in nature. Beetles, flies, spiders, and other taxa have developed similar attachment devices (Arzt, Gorb, and Spolenek, 2003). Recent studies indicate that the dry adhesion through van der Waals' interactions between the nanofibrils terminations and the contact surface may also contribute to the adhesion of tree frog toe pad (Barnes, Perez-Goodwyn, and Gorb, 2005; Barnes, Oines, and Smith, 2006; Barnes, 2007) and abalone pedal foot (Lin *et al.*, 2009). Shown in Fig. 2.10 are SEM micrographs at similar magnifications comparing nanofibrils in (a) gecko, (b) tree frog, and (c) abalone foot. The nano/microstructural design in different animal species shows surprising similarity. This is indeed an example of how biological organisms utilize the same, often optimized, design strategies to fulfill their functional purpose, in this case adhesion. The

(a)

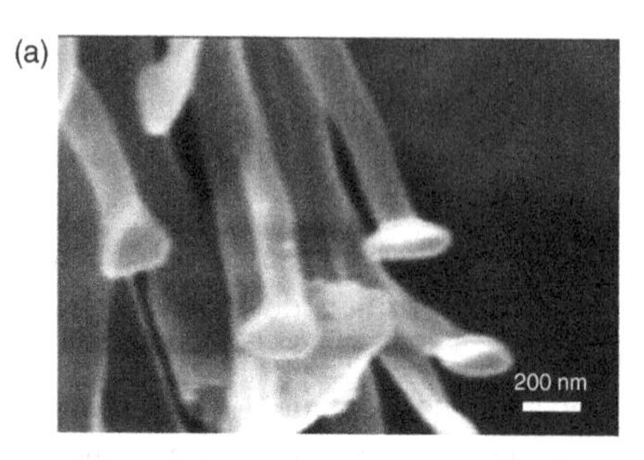

(b)

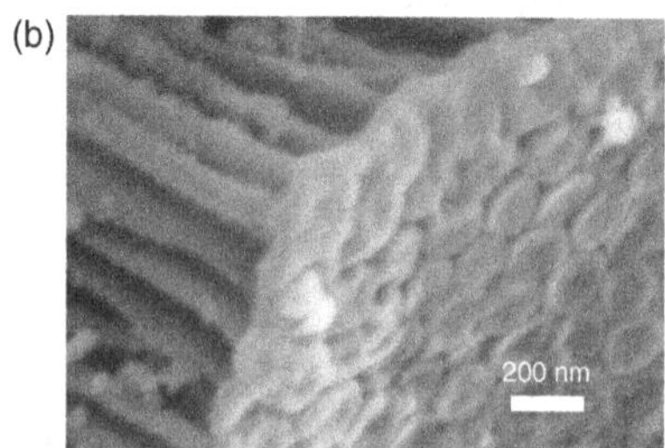

(c)

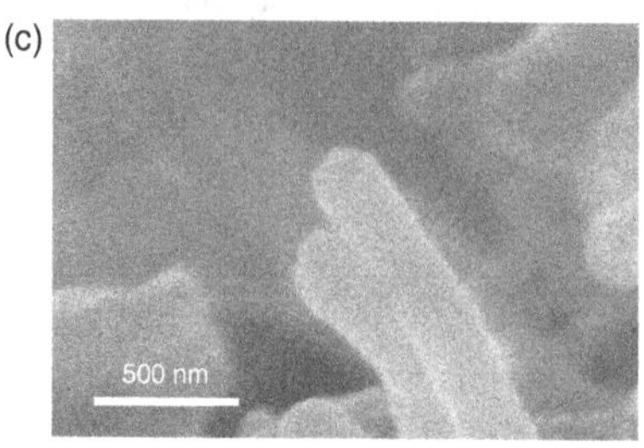

Figure 2.10.

Nanofibril structure found in (a) gecko foot (used with permission from Keller Autumn. Copyright © Kellar Autumn); (b) tree frog toe pads (reprinted from Meyers *et al.* (2011), with permission from Elsevier); (c) abalone pedal foot (Lin *et al.*, 2009). (Reprinted from Lin *et al.* (2009), with permission from Elsevier.)

hierarchical structure and adhesion mechanisms in nature will be discussed in detail in Chapter 11.

2.6 Ashby–Wegst performance plots

Mechanical property charts or maps (Ashby, 1989, 1992), more commonly known as Ashby maps, have become a convenient way of concentrating a large amount of information into one simple diagram. The first such map was proposed, for metals, by Weertman and Weertman (1970), and therefore the name Weertman–Ashby is sometimes used. They constitute a valuable design tool and have been extended to biological materials by Wegst and Ashby (2004). Two maps are presented in Fig. 2.11; they provide the Young moduli and strength as a function of density. There are several striking and defining features.

(a) The density of natural (biological) systems is low. It rarely exceeds 3 g/cm^3, whereas synthetic structural materials often have densities in the 4–10 g/cm^3 range.
(b) There is a broad range in Young moduli all the way from 0.001 to 100 GPa. This represents five orders of magnitude.
(c) The range of strengths is almost as broad as the Young moduli, varying over four orders of magnitude: 0.1 to 1000 MPa.

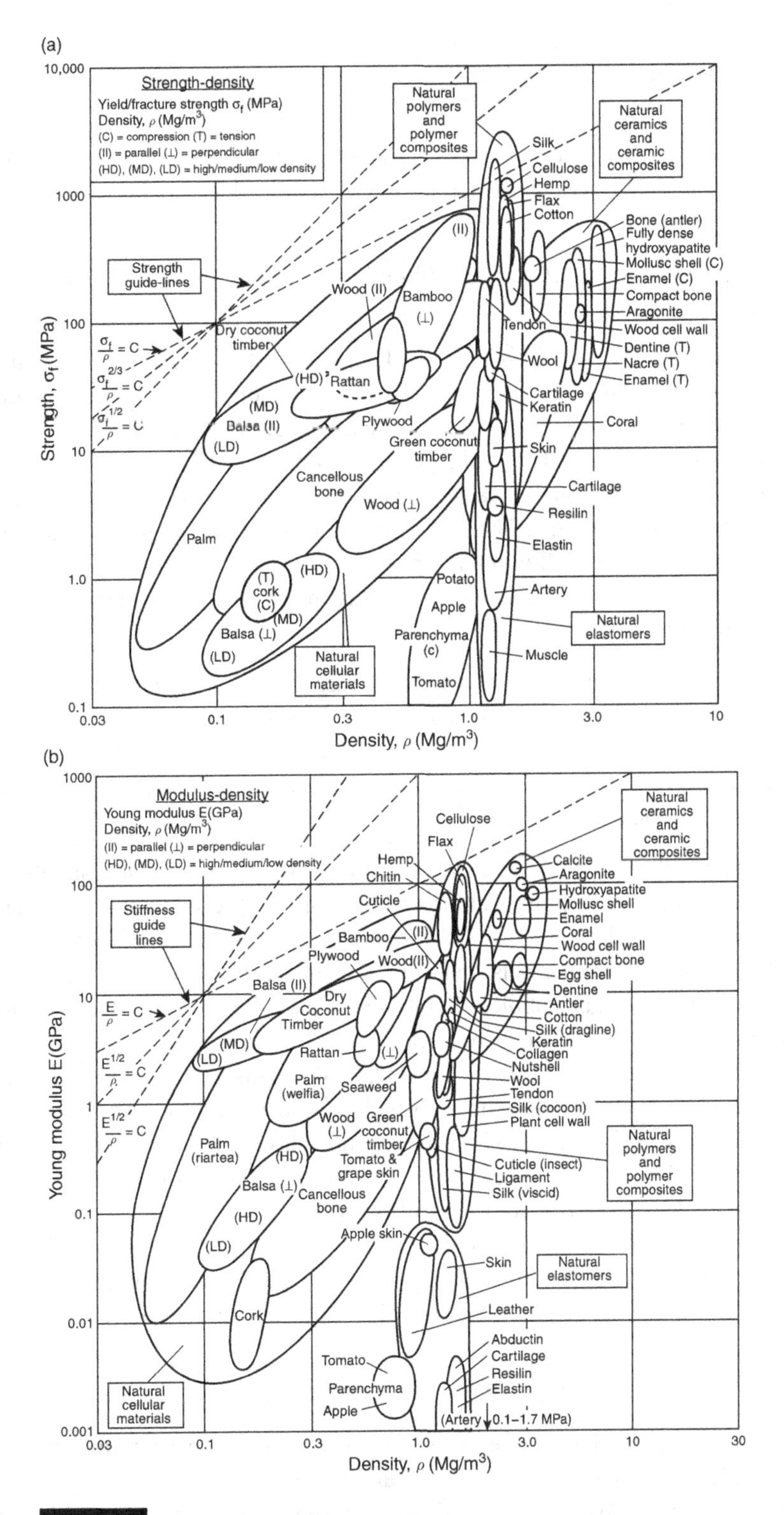

Figure 2.11.

Ashby plots for biological materials showing (a) elastic modulus and (b) strength as a function of density. (From Wegst and Ashby (2004), with kind permission from Professor Michael Ashby, Cambridge University.)

(d) There is an absence of metals, which require for the most part either high-temperature or high-electric-current processing. Nature does not have these variables at its disposal.

Wegst and Ashby (2004) classify biological (natural) materials into four groups, indicated by ellipses in Fig. 2.11.

Ceramics and ceramic composites These are biological materials in which the mineral component is prevalent, such as shells, teeth, bones, diatoms, and spicules of sponges.
Polymers and polymer composites Examples of these are the hooves of mammals, ligaments and tendons, silk, and arthropod exoskeletons.
Elastomers These are characteristically biological materials that can undergo large stretches (or strains). The skin, muscle, blood vessels, soft tissues in the body, and the individual cells fall into this category.
Cellular materials (or foams) Typical are the lightweight materials which are prevalent in feathers, beak interior, cancellous bone, and wood.

In this book we will follow this broad classification as it enables a logical and tractable description of very broad classes of biological materials.

Box 2.1 Bioresorbable metals

Biodegradable implants that have satisfactory mechanical properties present a significant benefit in certain applications, because the requirement for a second surgical procedure for removal is eliminated. The early use of magnesium alloys as a resorbable biomaterial dates from the 1930s (Verbrugge, 1934; McBride, 1938). One of the problems is that corrosion in the body was too rapid, leading to the accumulation of subcutaneous gas bubbles. Good biocompatibility was observed in clinical studies, although large amounts of hydrogen were accumulated as subcutaneous gas bubbles when the magnesium implants degraded too rapidly (Witte *et al.*, 2005). This led, in extreme cases, to gangrene.

The interest in magnesium alloys has been reactivated because new compositions have been developed which have much lower dissolution rates. These magnesium alloys have yield strengths in tension on the order of 200 MPa and a Young modulus of ~45 GPa. Apparently, the dissolution of magnesium is not harmful, since humans have approximately 20 g of magnesium in the body. The rapid dissolution of magnesium in the body can be controlled by a second phase in the structure, less amenable to dissolution. As the magnesium is dissolved, the scaffold left behind by the phase that dissolves retards the removal of magnesium; thus, the process can be controlled. The dissolution of Mg takes place through the following reaction:

$$Mg + 2H_2O \rightarrow Mg(OH)_2 + H_2.$$

The hydroxide is amorphous and is rather easily resorbed. The hydrogen can form bubbles if the rate of generation is higher than the rate at which the surrounding tissues can absorb it.

Degradable magnesium alloys have significant advantages over biodegradable polymers and ceramics as potential biomaterials because of their superior mechanical properties and excellent biocompatibility, especially in cardiovascular disease treatment. Conventional implant stents typically use corrosion resistant metals such as stainless steels, Ti-, Co-, and Cr-based alloys. The stents remain in the body permanently even though they are no longer needed. Biodegradable polymers (such as PLGA) and ceramic-based composites may not provide sufficient strength, ductility, and good biocompatibility for orthopedic applications. It is important for an orthopedic biodegradable implant to have a degradation rate that matches the healing or regeneration process of blood vessels in the biological system.

New processes are being developed through which the rate of resorption decreases: this includes alloying, surface treatments, and thermomechanical processing. Grain size reduction through ECAP (equal channel angular pressing) and hot rolling reduce the rate of dissolution of an AZ31 Mg alloy, as shown in Fig. B2.1 (Wang, Estrin, and Zuberova, 2008). The grain size after ECAP is ~2.5 μm, and the one after hot rolling (HR) is a little larger. The addition of rare earths to the alloys decreases the corrosion rate.

Commercial Mg alloys such as AZ31 are not desirable because the presence of Al has been linked to Alzheimer's disease. Mg-Ca and Mg-Ca-Zn alloys have great potential for the future. In these alloys, the Mg_2Ca phase forms a low-dissolution-rate component. The dissolution rate can be measured in vivo, but in-vitro experiments are valuable to produce preliminary data. Both SBF (synthetic/simulated body fluid, with an ion concentration close to that of human blood plasma) and HBSS (Hanks balanced salt solution, pH = 7.4) are used for this purpose. However, this is not quite the same environment as the alloy inside the body; it has been shown (Wang *et al.*, 2012) that compressive stresses affect the rate of degradation.

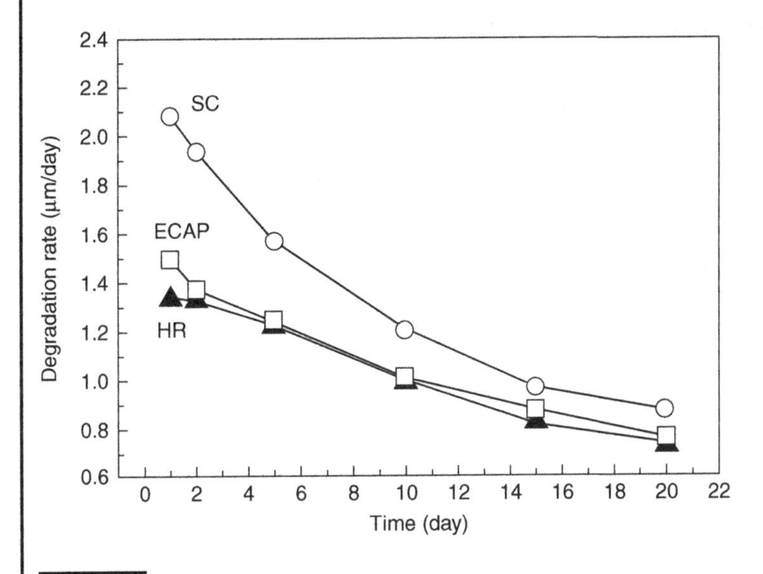

Figure B2.1.

Degradation rate of Mg alloy AZ31 (nominal composition: 3 wt.% Al and 1 wt.% Zn), in μm/day, and effect of grain size reduction by ECAP (equal channel angular pressing) and hot rolling (HR) in comparison with standard conditions (SC). (Reprinted from Wang *et al.* (2008), with permission from Elsevier.)

2.7 Viscoelasticity

Viscoelasticity is best understood if we consider the springs and shocks of automobiles. The shock absorbers dampen the rapid fluctuations of the vibrating spring. Otherwise, the vehicle would keep rattling without dampening after the wheel falls into a hole. In a similar manner, a purely elastic body will load and unload through the same force–displacement path with little energy loss in each cycle. On the other hand, a viscoelastic material, such as the one shown in Fig. 2.12, will load and unload through different paths. The area defined by the two paths is the energy absorbed in the process. If the paths are identical, this energy is zero and the response is elastic.

Polymers have considerable viscoelasticity, in contrast with metals and ceramics, where this effect is relatively unimportant at low homologous (T/T_m) temperatures. This is due to the sliding of chains, which are often connected by weaker bonds than the polymer backbone bonds, and by processes of chain reorganization under stress. This leads to a range of effects.

Creep Change in length with time as a function of temperature, after the material is stressed and then load (or stress) is held constant.

Stress relaxation Decrease in stress with time when a material is strained and then the straining is arrested (length held constant).

The simplest manner in which to represent viscoelastic materials is through springs and dashpots (shock absorbers). The spring is characterized by the following linear elastic equation:

$$\sigma = E\varepsilon. \tag{2.10}$$

The dashpot is, for an ideal (Newtonian) viscous material, given by

$$\sigma = \eta \frac{d\varepsilon}{dt}. \tag{2.11}$$

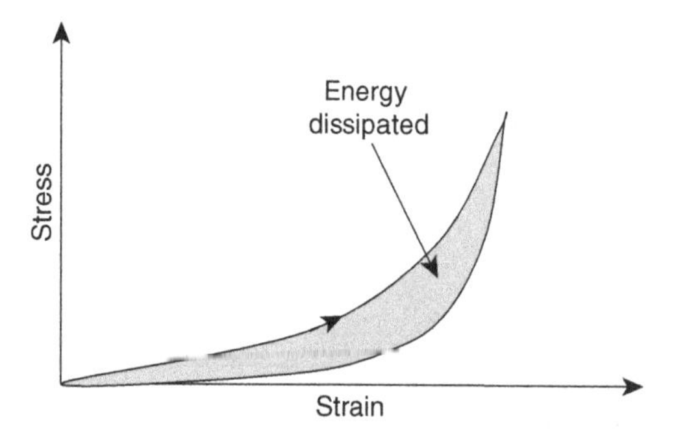

Figure 2.12.

Nonlinear stress–strain response of hypothetical biopolymer and viscoelastic response with energy dissipation.

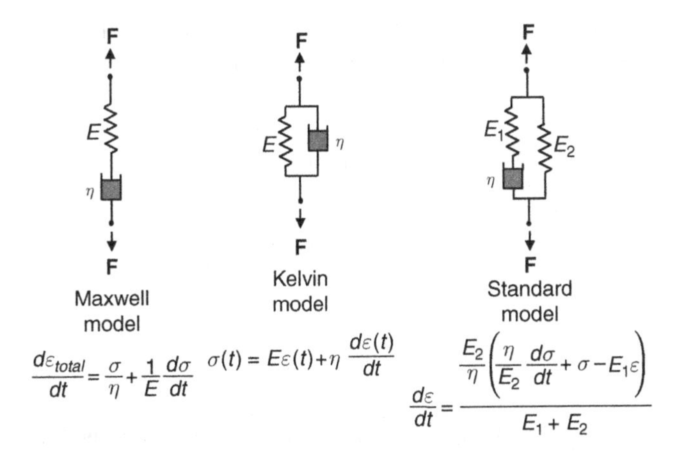

Figure 2.13.

Three common models for viscoelastic behavior; equations in terms of stress and strain are also given.

Figure 2.13 shows three configurations of springs and dashpots that are commonly used to describe viscoelastic behavior:

(a) a spring and a dashpot in series – the Maxwell model;
(b) a spring and a dashpot in parallel – the Kelvin model (also used in the automobile suspension);
(c) a Maxwell model in parallel with a spring – the so-called standard model.

With Eqns. (2.10) and (2.11) one can construct differential equations that represent the creep (constant stress) or stress relaxation (constant strain) according to the three models. Actual material behavior is often more complex, and additional springs and dashpots can be incorporated to represent the biopolymer response over a range of temperatures, stresses, and strain rates. For example, in the Maxwell–Weichert model, one additional spring–dashpot array is added in parallel to the standard model.

There is an apparatus in which viscoelastic materials (primarily polymers) are tested: the rheometer. Rheology is the study of the viscosity of fluids and "soft" solids. The origin of the word can be seen in the Greek proverb: *panta rei* (everything flows). One of the ways in which viscosity is measured is to subject the specimen to a sinusoidal variation in stress:

$$\sigma = \sigma_0 \sin \omega t; \tag{2.12}$$

$\omega = 2\pi f$, where f is the frequency of oscillation. The strain, if deformation is elastic, tracks the stress:

$$\varepsilon = \varepsilon_0 \sin \omega t. \tag{2.13}$$

If the material is totally viscous, the strain will be out of phase (by 90°) with the stress. In viscoelastic materials, the strain is out of phase by δ (this is called the phase lag):

$$\varepsilon = \varepsilon_0 \sin(\omega t + \delta). \tag{2.14}$$

We define two dynamic moduli,

$$E' = \left(\frac{\sigma_0}{\varepsilon_0}\right) \cos\delta, \tag{2.15}$$

$$E'' = \left(\frac{\sigma_0}{\varepsilon_0}\right) \sin\delta, \tag{2.16}$$

where E' is the storage modulus and E'' is the loss modulus. In a purely elastic material, $E'' = 0$; in a Newtonian viscous material, $E' = 0$. Viscoelastic materials have a mixed response, and the ratio E''/E' provides a good idea of the degree of viscoelasticity.

The articular cartilage is the classic example of viscoelasticity, although all biological materials containing biopolymers exhibit it. Both the collagen fibers and the interstitial fluid contribute to it. As the cartilage is loaded, the collagen fibers deform and the synovial fluid exudes into the joint. Figure 2.14 shows this phenomenon. This viscoelasticity is essential to the performance of the cartilage as a "shock absorber."

The results of cyclic nanoindentation tests conducted on porcine (pig) articular cartilage are shown in Fig. 2.15. The frequency range investigated represents the range encountered in actual life. Both moduli increase, as expected with frequency. This is a direct consequence of the strain-rate dependence of the modulus. The loss modulus is approximately one-third of the storage modulus, a significant deviation from elasticity.

Using a microelectromechanical systems (MEMS) device, Shen *et al.* (2011) were able to measure the viscoelastic properties of a single collagen fibril. They subjected the fibril to stress relaxation (by arresting the machine at 20% strain) and to creep, by keeping the load constant. The results, shown in Fig. 2.16, show significant stress relaxation and creep. The collagen fibril was kept wet. The best fit to the experimental results was obtained by adding a second dashpot–spring to the standard model. This is known as the Maxwell–Weichert model. Shen *et al.* identified two relaxation processes with characteristic times τ_1 and τ_2. The elastic modulus as a function of time was expressed as follows:

$$E(t) = E_0 + E_1 e^{-t/\tau_1} + E_2 e^{-t/\tau_2}. \tag{2.17}$$

The fibril range was 9.5–10.6 μm and the diameter range was 220–570 nm. At a strain of 20%, the elastic modulus $E_0 \sim 140$ MPa. The relaxation times were found to be $\tau_1 \sim 7$ s and $\tau_2 \sim 100$ s. The relaxation processes proposed are collagen and water

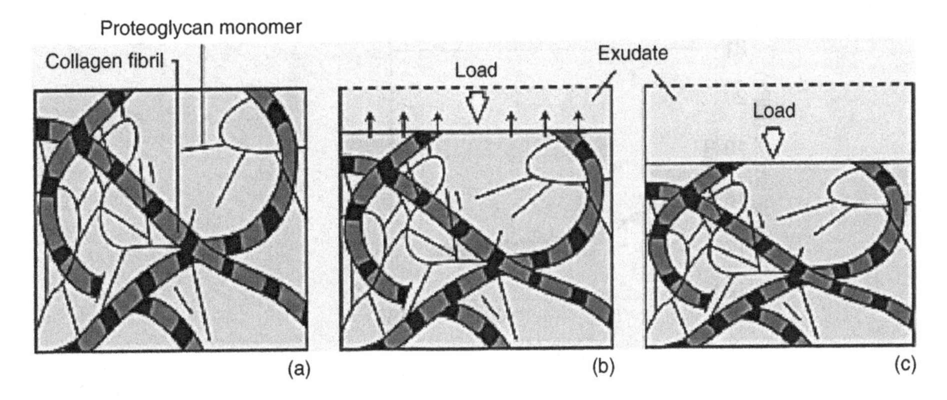

Figure 2.14.

Schematic sequence showing the compressive deformation of articular cartilage: (a) initial configuration; (b) loading with exudates flowing; (c) equilibrium. (From Pruitt and Chakravartula (2011). With permission from Cambridge University Press.)

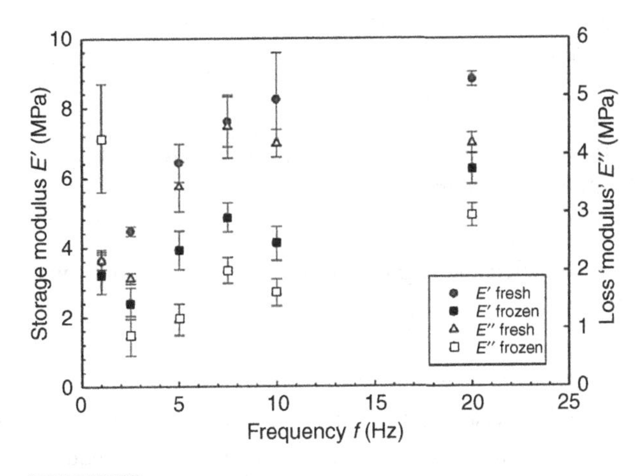

Figure 2.15.

Storage (E') and loss (E'') moduli for pig articular cartilage obtained through nanoindentation testing at frequencies varying from 1 to 20 Hz. (Reprinted from Franke *et al.* (2011), with permission from Elsevier.)

molecular rearrangements. When stress is applied, the hydrogen bonds enabled by water molecules slide. Collagen straightening can also take place. Other tests carried out on much larger specimens at the fiber level show different relaxation times: 2 s and 1500 s. This difference is attributed to the contribution of other proteins, such as proteoglycans, that connect fibrils to form fibers.

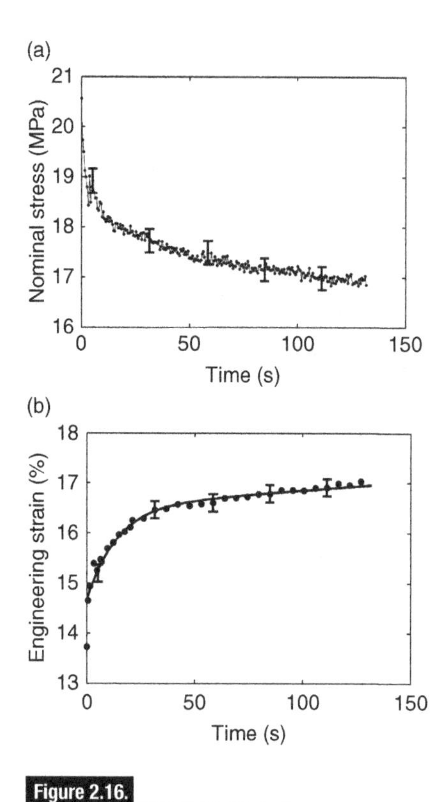

Figure 2.16.

(a) Stress relaxation and (b) strain change (at constant load) for single collagen fibril obtained using MEMS device. (Reprinted from Shen *et al.* (2011), with permission from Elsevier.)

In general, viscoelasticity is highest in cartilage and smooth muscle. It is lower in ligaments and tendons.

Example 2.1 A Maxwell element for a viscoelastic system simulating collagen fibril consists of a spring and a dashpot connected in series. If the cross-section is equal to 200 nm and the length is 10 μm, determine the stress as a function of time if the machine crosshead is stopped at a stress of 100 MPa. Given:

$$E = 0.1\,\text{GPa},$$
$$\eta = 10^8\,\text{Pa s}.$$

Using the Maxwell model,

$$\frac{d\varepsilon_{total}}{dt} = \frac{\sigma}{\eta} + \frac{1}{E}\frac{d\sigma}{dt},$$

$$\frac{\sigma}{\eta} + \frac{1}{E}\frac{d\sigma}{dt} = \frac{d\varepsilon_{total}}{dt} = 0,$$

$$\frac{d\sigma}{\sigma} = -\frac{E}{\eta}dt,$$

$$\int_{\sigma_0}^{\sigma_f} \frac{d\sigma}{\sigma} = -\frac{E}{\eta}\int dt,$$

$$\ln\frac{\sigma}{\sigma_0} = -\frac{E}{\eta}t,$$

$$\sigma = \sigma_0 e^{-(E/\eta)t},$$

$$\sigma = 100e^{-t}.$$

At $t = 10$ s, $\sigma = 100e^{-10} = 0.00454$ MPa; at $t = 10\,000$ s, $\sigma = 100e^{-10\,000} \approx 0$ MPa.

2.8 Weibull distribution of failure strengths

In the early twentieth century Weibull correlated the strength of rocks to statistical parameters. As early as the 1500s, Leonardo had already observed that long wires and ropes tended to be weaker than short ones. In the same vein, Weibull observed that large rocks were weaker than small rocks. This can be qualitatively explained by the following: a large body has a larger number of flaws than a small body. Additionally, a large body may contain, by virtue of its size, larger flaws. We know from fracture mechanics that the stress (σ) required to activate a large flaw ($2a$) is lower than for a small flaw. This is expressed by the fundamental fracture-mechanics equation:

$$K = Y\sigma\sqrt{\pi a},$$

where $2a$ is the flaw diameter, σ is the stress, K is the toughness, and Y is a parameter (~1.12). Let us consider the body shown in Fig. 2.17. It is composed of n elementary volumes V_0. From statistics, the probability of survival at a certain stress level σ, $P(V_0)$, is related to the probability of failure, $F(V_0)$, through

$$P(V_0) = 1 - F(V_0). \tag{2.18}$$

The probability of survival of the volume V composed of n elemental volumes is, at that stress level, given by

$$P(V) = P(V_0)P(V_0)P(V_0)P(V_0)\cdots P(V_0) = P(V_0)^n. \tag{2.19}$$

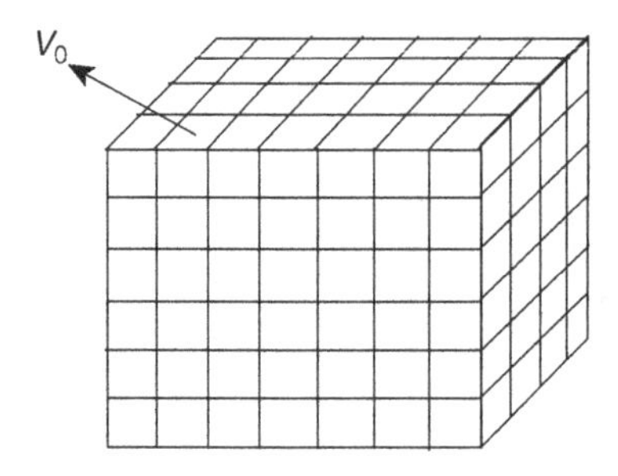

Figure 2.17.

Representation of a body as composed of n elementary volumes V_0.

The larger the volume, the lower is $P(V)$, since $P(V_0) < 1$. Weibull observed that $P(V_0)$ varies with the applied stress as

$$P(V_0) = \exp\left[-\left(\frac{\sigma - \sigma_u}{\sigma_0}\right)^m\right]. \tag{2.20}$$

Note that σ_0, m, and σ_u are material constants: σ_u is the stress below which the probability of failure is zero; σ_0 is a normalization parameter that renders the ratio unitless; and m is the slope of the $\ln[\ln[1/(P(V_0))]]$ vs. $\ln \sigma$ curve, which should be a straight line if Weibull statistics are obeyed.

It is often assumed that $\sigma_u = 0$. This reduces Eqn. (2.20) to

$$P(V_0) = \exp\left[-\left(\frac{\sigma}{\sigma_0}\right)^m\right]. \tag{2.21}$$

When $\sigma = 0$, $P(V_0) = 1$; when $\sigma \to \infty$, $P(V_0) = 0$.

The slope m is commonly known as the Weibull modulus. There is a direct relationship between σ_0 and $P(V_0)$. When $\sigma = \sigma_0$, $P(V_0) = 1/e$. This is approximately equal to 0.37.

If we now consider a volume $V = nV_0$, the Weibull equation takes the following form:

$$P(V) = \exp\left[-\frac{V}{V_0}\left(\frac{\sigma}{\sigma_0}\right)^m\right]. \tag{2.22}$$

The Weibull modulus m provides a good measure of the variability of the results. If the failure stress of all specimens is the same, m is infinite. In metals, m usually takes values between10 and 20. For fine (high-tech) ceramics, which are much more brittle, there is considerable variation between different specimens, and m ranges from 4 to 10. The low-end ceramics, such as brick, mortar, etc., show a greater level of variability, and in those cases $m < 3$. Biological specimens exhibit significant inhomogeneities. Their m values are characteristically in the range 4–10.

Example 2.2 The following experimental results were obtained in flexure testing of a bony fish scale. The strengths varied from 215 to 320 MPa (Table 2.3). We would like to fit the results into a Weibull curve and obtain the relevant parameters.

Before the analysis σ_i ($i = 1 \rightarrow N$, where N is the number of data points), we arrange the strengths from the lowest to the highest. We calculate two numbers from each data point, $\ln(\sigma_i)$ and $\ln\ln(1/(1 - F))$, where $F = i/(N + 1)$; the values after calculation are shown in Table 2.3.

After obtaining the two values of each data point, we make a "$\ln\ln(1/(1 - F_N)) - \ln(\sigma_i)$" plot, as shown in Figure 2.18(a).

From the "$\ln\ln(1/(1 - F_N)) - \ln(\sigma_i)$" plot in Fig. 2.18(a), we obtain the slope $m = 9.50$ and the $\ln(\sigma_i)$ value when $\ln\ln(1/(1 - F_N)) = 0$, which is $[\ln(\sigma_i)]_0 = 5.62$.

We express the Weibull equation as follows:

$$\left(\frac{\sigma}{\sigma_0}\right)_i = \sigma_i/\exp[\ln(\sigma_i)]_0.$$

Here, $\sigma/\sigma_0 = \sigma_i/\exp(5.62)$. Since the Weibull modulus (m) obtained from Fig. 2.18(a) is 9.50, the failure probabilities $F = 1 - P(V_0)$ and data are recalculated and listed in Table 2.4.

One plot can be made using strengths (σ_i) in the abscissa and F (in Table 2.4) as well as F_N (in Table 2.3) as ordinate as shown in Fig. 2.18(b). The Weibull fit in Fig. 2.18(b) was conducted with the F values shown in Table 2.4, and the points in Fig. 2.18(b) show the real failure probability of each strength data point for the scales. This plot shows that at the failure probability of 50% ($F = 50\%$), the strength is 265.4 MPa.

Summary

- In order to describe a biological material, it is necessary to present the various hierarchical levels of the structure, all the way from the nano-, micro-, and meso-, to the macro-level.
- The hierarchical structure of four biological materials is presented: feather rachis, nacreous part of shells, crab exoskeleton, and bone. Each one is uniquely different.
- The different levels contribute to the elastic modulus, hardness, strength, and toughness.
- Yao and Gao (2007, 2008) developed equations that describe the evolution of these properties with the number of hierarchical levels that applies to a composite

Table 2.3. Flexure strengths of a bony fish scale; values between 215 and 320 MPa

i	σ_i (MPa)	$\ln(\sigma_i)$	$F_N(V)$	$1/(1 - F_N)$	$\ln(1/(1 - F_N))$	$\ln \ln(1/(1 - F_N))$
1	215.56	5.37	0.077	1.083	0.080	−2.525
2	235.94	5.46	0.154	1.182	0.167	−1.789
3	238.76	5.48	0.231	1.300	0.262	−1.338
4	249.80	5.52	0.308	1.444	0.368	−1.000
5	253.67	5.54	0.385	1.625	0.486	−0.723
6	259.11	5.56	0.462	1.857	0.619	−0.480
7	266.92	5.59	0.538	2.167	0.773	−0.257
8	268.06	5.59	0.615	2.600	0.956	−0.046
9	276.12	5.62	0.692	3.250	1.179	0.164
10	284.44	5.65	0.769	4.333	1.466	0.383
11	290.05	5.67	0.846	6.500	1.872	0.627
12	319.66	5.77	0.923	13.000	2.565	0.942

Table 2.4. Failure probabilities (F) and survival probabilities (P) from data in Table 2.3 through application of the Weibull equation

i	σ_i (MPa)	σ/σ_0	$(\sigma/\sigma_0)^m$	P	F_N
1	215.56	0.781	−0.096	0.910	0.090
2	235.94	0.855	−0.226	0.800	0.200
3	238.76	0.865	−0.253	0.778	0.222
4	249.80	0.905	−0.389	0.680	0.320
5	253.67	0.919	−0.450	0.639	0.361
6	259.11	0.939	−0.551	0.578	0.422
7	266.92	0.968	−0.731	0.483	0.517
8	268.06	0.972	−0.761	0.468	0.532
9	276.12	1.001	−1.008	0.365	0.635
10	284.44	1.031	−1.337	0.262	0.738
11	290.05	1.051	−1.609	0.199	0.801
12	319.66	1.159	−4.051	0.017	0.983

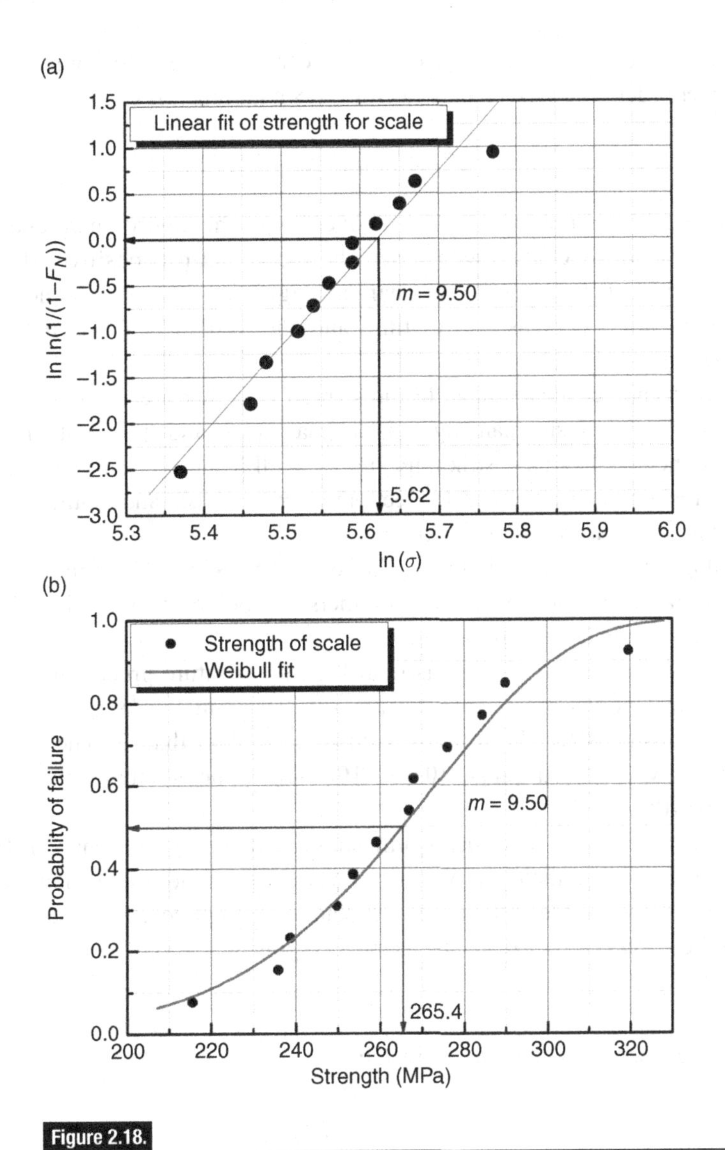

Figure 2.18.

(a) Linear fit of strength for the Weibull analysis. (b) Weibull plot showing the failure probabilities of the scale strengths. Experimental results, dots; model, continuous line.

composed of a hard phase embedded in a continuous soft phase. These phases represent the mineral and the biopolymer (e.g. collagen), respectively.

- Classical thermodynamics predicts that equilibrium is reached for a minimum of free energy in a closed system and for a maximum of entropy in an isolated system. Biological systems go from simpler to more complex, and this apparently opposes classical thermodynamics. However, there is a critical difference: most biological systems are open. These open systems can evolve, by self-organization, toward greater order. The total change in entropy S with time t is expressed as follows:

$$dS/dt = d_eS/dt + d_iS/dt,$$

where d_eS/dt is the exchange component with the environment and d_iS/dt is the internal entropy change with time. The second term is positive. However, if the first term is negative and has a larger magnitude than the second one, the total entropy dS/dt decreases with time and the system can march toward greater order.

- Convergence creates similar features in species that are not directly related. For instance, the Tasmanian tiger is a marsupial and the wolf is a mammal. Nevertheless, they have strikingly similar skull features because both are predators. Other examples are the serrated teeth in mammals, fish, and reptiles, which evolved as a necessity to cut through skin.
- Biological materials are divided in the Ashby–Wegst classification as: (i) bioceramics and ceramic composites; (ii) biopolymers and polymer composites; (iii) bioelastomers; (iv) cellular materials.
- Nature does not have at its disposal high-temperature processing methods. With some exceptions, no metals are present. Thus, in the Wegst–Ashby diagrams, their density is for the most part equal to or less than 3. Their Young moduli have a very broad range (0.001–100 GPa). The strengths vary from 0.1 to 1000 MPa.
- Viscoelasticity is a viscosity manifesting itself during deformation. Honey is the supreme viscous material. Viscoelasticity during load application leads to stress relaxation, when the strain (size of specimen) is kept constant, and creep, when the load (or stress) is constant. The elastic component is given by

$$\sigma = E\varepsilon,$$

and the viscous component is given by

$$\sigma = \eta \frac{d\varepsilon}{dt}.$$

The analog of springs and dashpots is used to represent this behavior. They can be in series (Maxwell), parallel (Kelvin), or a Maxwell model in parallel with a spring (standard model).

- Weibull distribution of failure stresses: the distribution of failure stresses can be obtained assuming that each elemental volume V_0 has the following probability of survival $P(V_0)$ at stress σ:

$$P(V_0) = \exp\left[-\left(\frac{\sigma - \sigma_u}{\sigma_0}\right)^m\right],$$

where m is the Weibull modulus, σ_u is the stress below which no failure occurs, and σ_0 is the characteristic stress corresponding to when $\sigma = \sigma_0$, $P(V_0) = 1/e$. For a volume $V = nV_0$,

$$P(V) = P(V_0)P(V_0)P(V_0)P(V_0)\cdots P(V_0) = P(V_0)^n.$$

Exercises

Exercise 2.1 Determine the density of bone if it is composed of 50 vol.% hydroxyapatite and 50 vol.% collagen. Density of HAP = 3.14 g/cm^3 and density of collagen = 1.03 g/cm^3.

Exercise 2.2 Using values from the Ashby plots (Fig. 2.11), determine the strength and Young modulus of bone and compare them with the values given in the plots. Use both Voigt and Reuss averaging techniques for properties, P:

$$P = P_1V_2 + P_2V_2,$$
$$\frac{1}{P} = \frac{V_1}{P_1} + \frac{V_2}{P_2}.$$

Exercise 2.3 Describe the different levels of the hierarchy of the lobster exoskeleton.

Exercise 2.4 Spider web is, perhaps, the strongest biological material. It is not given in the Ashby plot. Obtain its strength from the literature and put it on the plot. What is its elastic modulus?

Exercise 2.5 Explain how levels of hierarchy affect mechanical strength and toughness.

Exercise 2.6 How many times does a heart valve open and close in ten years? Make necessary assumptions about daily rhythm.

Exercise 2.7

(a) For the cycloid in Fig. 2.9(a), which has a period of 330 mm, determine the radius of the circle.
(b) If the highest point on each cusp is 130 mm, how would you correct the equation to reproduce the curves shown in the plot?
(c) Show that the trajectory from lowest to lowest point is equal to $8r$.

Exercise 2.8 Using the Kelvin model for bone, which is a mixture of collagen and hydroxyapatite, determine the strain as a function of time if it is subjected to a constant stress. This is called a creep test.

Given:

$$E_m = 100 \text{ GPa}; \eta = 10^9 \text{ Pa s}; \sigma = 100 \text{ MPa}.$$

Exercise 2.9 A Maxwell element for a viscoelastic system simulating collagen fibril consists of a spring and a dashpot connected in series. If the cross-section is equal to 200 nm and the length is 10 μm, determine the stress as a function of time if the machine crosshead is stopped at a stress of 100 MPa.

Given:

$$E = 1 \text{ GPa}; \ \eta = 10^9 \text{ Pa s}.$$

Exercise 2.10 Describe the hierarchical levels in the Achilles tendon.

Exercise 2.11 For the Maxwell model, obtain an expression for the stress as a function of time for a fixed length (stress relaxation).

Exercise 2.12 Obtain the strain as a function of time for a Maxwell model if the stress is kept constant (creep).

3 Basic building blocks: biopolymers

Introduction

Biological materials are more complex than synthetic materials. As seen in Chapter 2, they form hierarchical structures and are often multifunctional, i.e. one material has more than one function. We classify biological materials, from the mechanical property viewpoint, into "soft" and "hard." Hard materials provide the skeleton, teeth, and nails in vertebrates and the exoskeleton in arthropods. "Soft" biological materials build skin, muscle, internal organs, etc. "Hard" biological materials are strong in compression but brittle in tension. Conversely, "soft" biological materials are more amenable to be loaded in tension; the long fibers tend to buckle in compression. Table 3.1 provides the distribution (on a weight percentage) of different constituents of the body. We can see from this that we are about 60% water. This fact demonstrates, in eloquent fashion, that the mechanical properties of biological materials are strongly dependent on hydration.

Here are some examples of "hard" mineralized biological materials:

calcium phosphate (hydroxyapatite – $Ca_{10}(PO_4)_6(OH)_2$):	teeth, bone, antlers;
calcium carbonate ($CaCO_3$)(aragonite):	mollusc shells, some reptile eggs;
calcite:	bird eggs, crustaceans, molluscs;
amorphous silica ($SiO_2(H_2O)_n$):	spicules in sponges, diatoms;
iron oxide (magnetite – Fe_3O_4):	teeth in chitons (a weird-looking marine worm), bacteria.

Here are some examples of "hard" nonbiomineralized biological materials:

chitin:	arthropod and insect exoskeletons;
cellulose and lignin:	plant cell walls;
keratin:	bird beaks, horn, hair, nails.

Here are two examples of "soft" biological materials:

collagen:	organic component of bone and dentin, tendons, muscle, blood vessels;
elastin:	skin, lungs, artery walls.

Of the above, chitin and cellulose are polysaccharides. Lignin is something else. It is a cross-linked racemic macromolecule with molecular mass that ranges from 1000 to

Table 3.1. Occurrences of different biological materials in the human body

Biological material	Weight percentage in human body
Proteins	15–17
Lipids	10–15
Carbohydrates	~1
Minerals	~7
DNA, RNA	~2
Water	58–65

Table 3.2. Weight fractions of the six most common elements in the human body

Oxygen	43 kg (65 wt.%)
Carbon	16 kg (18 wt.%)
Hydrogen	7 kg (10 wt.%)
Nitrogen	1.8 kg (3 wt.%)
Calcium	1.0 kg (1.4 wt.%)
Phosphorus	0.78 kg (1.1 wt.%)

20 000 g/mol. It is hydrophobic and has an aromatic character. Collagen, keratin, and elastin are proteins (polypeptides).

If we look at the human body in terms of elements, some surprising facts evolve. For a person weighing 70 kg, the six most common elements account for a large fraction of the weight (69.5 kg!). See Table 3.2.

3.1 Water

Water is the most abundant substance on earth, covering 70% of the surface and is vital for all living organisms. Water plays an important and multifunctional role in our body. Water is the main constituent in the blood stream, transporting nutrients and oxygen into cells. It helps with digestion and metabolism, regulates body temperature, protects organs, lubricates joints, moisturizes lungs, assists with biomineralization, and acts as a solvent and buffer solution. Dehydration occurs when the water content in the body is too low; this often leads to severe symptoms, and may result in death.

Hydrogen bonds between water molecules give water unique physical and chemical properties, such as a higher melting point, boiling point, and heat of vaporization than most other solvents. Water is a polar solvent which can dissolve most biomolecules. Most chemical reactions in biological systems take place in the aqueous environment. Charged or polar molecules that dissolve easily in water are called "hydrophilic," and nonpolar molecules, such as lipids and waxes that cannot dissolve in water, are called "hydrophobic."

Water also plays an important role in the mechanical properties of structural biological materials. Water defines the structural characteristics and physical properties of proteins and organic substances, acting as a plasticizer in biological materials by enhancing the ductility and toughness. Without water, biological materials lose their original mechanical properties and become more brittle. In order to measure the realistic mechanical properties of biological materials, it is crucial to take the degree of hydration into account. Several examples will be given in the following chapters. It is easy to verify this. Untanned hides are stiff and hard; dehydrated muscles (jerky) are completely different from a juicy steak.

3.2 Nucleotides and nucleic acid

It is said that on 28 February 1953 Francis Crick interrupted the patrons having lunch at the Eagle pub in Cambridge, with a bombastic statement: "Watson and I have discovered the secret of life!" Indeed, this is close to the truth, but in the first part they forgot to give credit to some important contributors (Rosalind Franklin and Raymond Gosling) to the work. In honor of the discovery, a special ale, tasting slightly of vomit (a hydrous mixture containing live organic components with a peculiar smell), according to an expert, is still served at the pub.

Deoxyribonucleic acid (DNA) and ribonucleic acid (RNA) are the molecular repositories of genetic information. DNA contains genes required for the synthesis of functional biological molecules, such as proteins, RNA, and cellular components. The storage and transmission of genetic information are the functions of RNA. RNA can be classified into three types with different functions. Messenger RNAs (mRNA) carry genetic information to ribosomes in a process called transcription. Transfer RNAs (tRNA) translate the information in mRNA into a specific sequence of amino acids. Ribosomal RNAs (rRNA) are the major constituents of ribosomes, which synthesize proteins.

Nucleotides have three components: a nitrogen-containing (nitrogenous) base, a pentose, and a phosphate, as shown in Fig. 3.1(a). The nitrogenous bases are derivatives of pyrimidine and purine. Major purine and pyrimidine bases of nucleic acids are shown in Fig. 3.1(b). DNA and RNA both contain two purine bases, adenine (A) and guanine (G), and one pyrimidine base, cytosine (C). The other pyrimidine is different: it is thymine (T) in DNA and uracil (U) in RNA. Specific sequences of A, T, G, and C nucleotides in DNA provide genetic information.

The three-dimensional structure of DNA was discovered by Watson and Crick in 1953. It contains two DNA chains helically wound around the same axis to form a right-handed

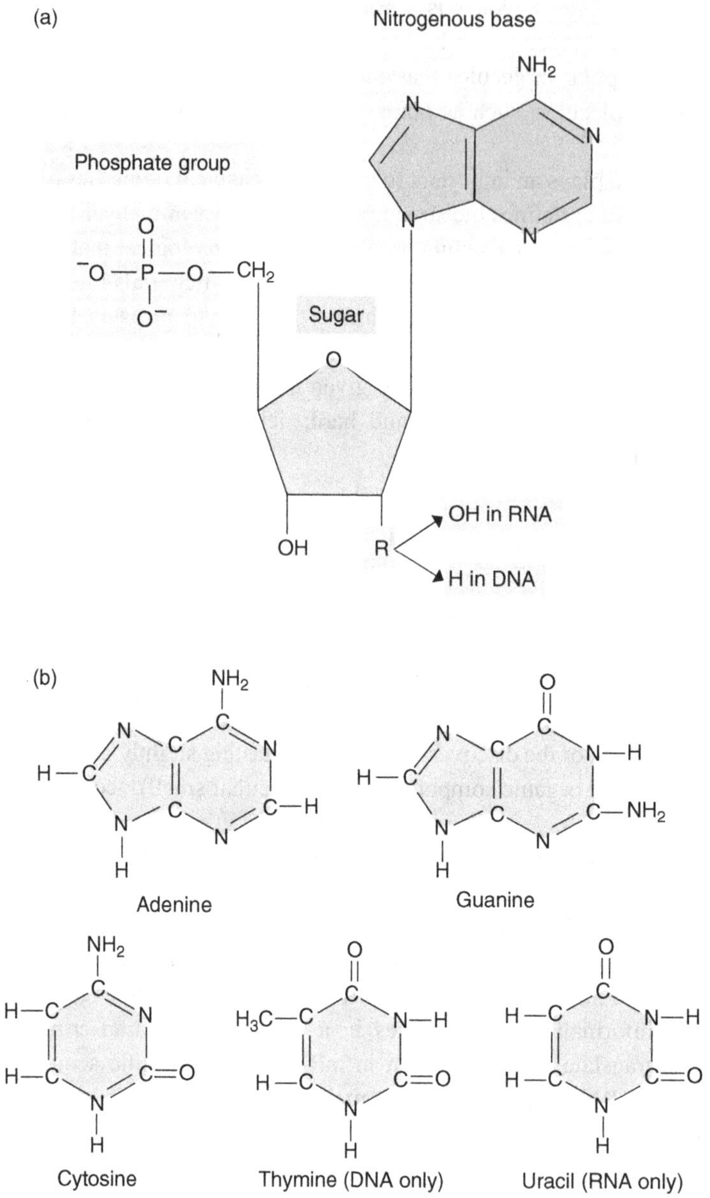

Figure 3.1.

(a) General structure of a nucleotide, which consists of a sugar (pentose), a phosphate group, and a nitrogenous group. (b) Molecular structure of nitrogenous bases: adenine (A), guanine (G), cytosine (C), thymine (T), and uracil (U).

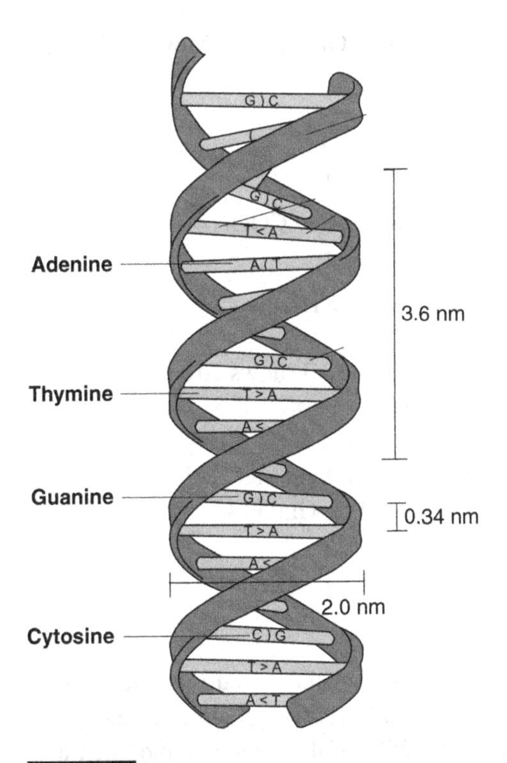

Figure 3.2.

The double helix structure of DNA. (Adapted from http://commons.wikimedia.org/wiki/File:DNA_structure_and_bases_PL.svg. Image by Messer Woland. Licensed under CC-BY-SA-3.0, via Wikimedia Commons.)

double helix, as shown in Fig. 3.2. Hydrogen bonds between base pairs permit complementary anti-parallel strands of nucleic acid. A bonds specifically to T and G bonds to C through hydrogen bonding. Three hydrogen bonds can form between G and C, symbolized G≡C; wheras only two can form between A and T, labeled A=T. The base pairs are vertically stacked inside the double helix, 0.34 nm apart, with 10.5 base pairs per turn. DNA is organized into a complex structure called a chromosome within cells. During cell division, these chromosomes are duplicated in the process of DNA replication, providing each cell with an identical set of chromosomes.

3.3 Amino acids, peptides, and proteins

3.3.1 Amino acids and peptides

In order to understand proteins fully, we have to start at the atomic/molecular level, as we did for polymers. Actually, proteins can be conceived of as polymers with a greater level of complexity. We start with amino acids, which are compounds containing both an amine ($-NH_2$) and a carboxyl ($-COOH$) group. Most of them have the following structure:

$$\begin{array}{c} H \\ | \\ RC - C - COOH \\ | \\ NH_2 \end{array}$$

where the R stands for a side-chain. Table 3.3 shows 20 common amino acids in proteins. They are classified into five groups:

Nonpolar: Gly, Ala, Val, Ile, Leu, Met
Uncharged polar: Ser, Thr, Cys, Asn, Gln, Pro
Aromatic: Phe, Tyr, Trp
Positively charged: Lys, Arg, His
Negatively charged: Asp, Glu

Post-translated amino acids are modifications occurring after protein translation. They play a role in many biological materials. Hydroxyproline is the key to providing stability to the collagen triple helix, and ~39% of the proline is modified to hydroxyproline in collagen. Allysine, derived from lysine, is also important in collagen cross-linking. DOPA (dihydroxylphenylalanine) is a post-translated amino acid that plays a role in mussel attachment to rocks.

The amino acids form linear chains similar to polymer chains; they are called polypeptide chains. These polypeptide chains acquire special configurations because of the formation of bonds (hydrogen, van der Waals, and covalent) between amino acids on the same or different chains. The two most common configurations are the α-helix and the β-sheet. Figure 3.3(a) shows how an α-helix is formed. The NH and CO groups form hydrogen bonds between them in a regular pattern, and this creates the particular conformation of the chain that is of helical shape. The backbone of the chain is shown as dark segment and comprises repeating segments of two carbon and one nitrogen atoms. In Fig. 3.3(b) several hydrogen bonds are shown, causing the polypeptide chain to fold. The radicals stick out. Figure 3.4 shows the hydrogen bond and the α-helix in greater detail. It is the hydrogen bonds that give the helical conformation to the chain, as well as the connection between the adjacent chains in the β-sheet (shown in Fig. 3.3(b)). This is shown in a clear fashion in Fig. 3.4(b). The hydrogen bonds are indicated by dotted lines. The period of the helix is 0.87 nm.

Another common conformation of polypeptide chains is the β-sheet. In this conformation, separate chains are bonded (Fig. 3.3(b)). Figure 3.4(a) shows two anti-parallel chains that are connected by a hydrogen bond (shown as dotted lines). We can see that the radicals (–R in Fig. 3.3(a) and boxes in Fig. 3.4(a)) of two adjacent chains stick out of the

Table 3.3. Twenty amino acids found in proteins

Name	Chemical formula
Alanine (Ala)	$^{+}NH_3-CH(CH_3)-C(=O)-O^-$
Leucine (Leu)	$^{+}NH_3-CH(CH_2-CH(CH_3)_2)-C(=O)-O^-$
Phenylalanine (Phe)	$^{+}NH_3-CH(CH_2-C_6H_5)-C(=O)-O^-$
Proline (Pro)	$NH-CH-C(=O)-O^-$ (ring: $NH-CH_2-CH_2-CH_2-CH$)
Serine (Ser)	$^{+}NH_3-CH(CH_2-OH)-C(=O)-O^-$
Cysteine (Cys)	$^{+}NH_3-CH(CH_2-SH)-C(=O)-O^-$
Glutamate (Glu)	$^{+}NH_3-CH(CH_2-CH_2-C(=O)O^-)-C(=O)-O^-$

Table 3.3. *(cont.)*

Name	Chemical formula
Lysine (Lys)	$^{+}NH_3-CH(CH_2-CH_2-CH_2-CH_2-NH_3^{+})-C(=O)-O^{-}$
Arginine (Arg)	$^{+}NH_3-CH(CH_2-CH_2-CH_2-NH-C(=NH)-NH_2)-C(=O)-O^{-}$
Asparagine (Asn)	$^{+}NH_3-CH(CH_2-C(=O)-NH_2)-C(=O)-O^{-}$
Aspartate (Asp)	$^{+}NH_3-CH(CH_2-C(=O)-OH)-C(=O)-O^{-}$
Glutamine (Gln)	$^{+}NH_3-CH(CH_2-CH_2-C(=O)-NH_2)-C(=O)-O^{-}$
Glycine (Gly)	$^{+}NH_3-CH_2-C(=O)-O^{-}$

Table 3.3. *(cont.)*

Name	Chemical formula
Histidine (His)	O ‖ $^{+}NH_3-CH-C-O^-$ CH_2 C N CH CH—NH
Isoleucine (Ile)	O ‖ $^{+}NH_3-CH-C-O^-$ $CH-CH_3$ CH_2 CH_3
Methionine (Met)	O ‖ $^{+}NH_3-CH-C-O^-$ CH_2 CH_2 S CH_3
Threonine (Thr)	O ‖ $^{+}NH_3-CH-C-O^-$ $CH-OH$ CH_3
Tryptophan (Trp)	O ‖ $^{+}NH_3-CH-C-O^-$ CH_2 CH C C CH CH C—NH CH CH
Tyrosine (Tyr)	O ‖ $^{+}NH_3-CH-C-O^-$ CH_2 C CH CH CH CH C OH

Table 3.3. (*cont.*)

Name	Chemical formula
Valine (Val)	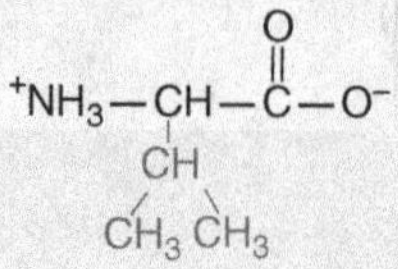 $^{+}NH_3-CH(CH(CH_3)_2)-C(=O)-O^-$

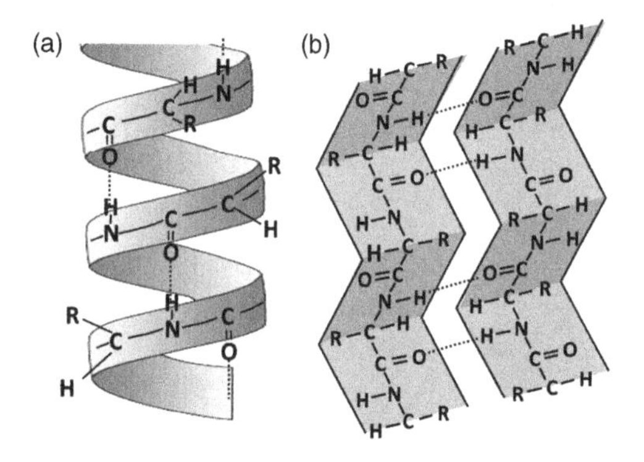

Figure 3.3.

(a) Structure of α-helix; dotted lines indicate hydrogen bonds; spacing of helix = 0.87 nm. (b) Structure of β-sheet with two anti-parallel polypeptide chains connected by hydrogen bonds (dotted lines).

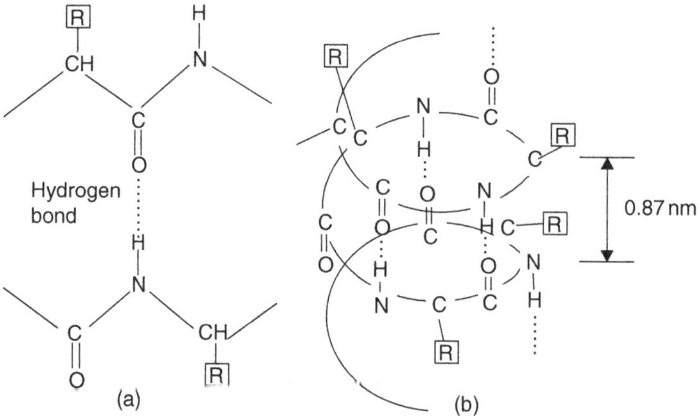

Hydrogen bond

(a)

0.87 nm

(b)

Figure 3.4.

(a) Hydrogen bond connecting a CO to a NH group in a polypeptide. (b) Successive hydrogen bonds on same polypeptide chain leading to formation of a helical arrangement.

Box 3.1 Joint replacement

Important joints in our body are of the ball-and-socket type: the hip and shoulder are prime examples. In standard engineering applications, one finds such ball-and-socket joints in a number of places: desk lamps, pen holders, leashes in surfboards, and automobiles. It basically consists of a spherical component and a cap into which the sphere fits snugly. The hip joint has a simple ball-and-socket geometry, the center of rotation being unchanged. Thus, early attempts were made, one of the first in 1908, to manufacture synthetic joints using ivory, by a certain Dr. Gluck. The successful hip joint replacement represents one of the great materials triumphs of the past half-century. The pioneer Sir John Charnley developed the surgical procedure and the early version of the modern total hip replacement during the 1950s.

In the USA alone, approximately 700 000 hip and 300 000 knee replacement operations are conducted every year, and this is a remarkable success story in terms of surgery and biomaterials development. The prediction is that by 2030 approximately 4 000 000 joint replacements (arthroplasties) will be conducted per year. It is claimed that the survival rate of these implants is 90% after ten years, although there is substantial evidence that the average life of implants is approximately ten years. The success is dependent on a number of factors, such as selection of patients, surgical procedures, and materials. In Brazil, there was a scandal in the early 2000s when it was discovered that femoral stems of lower-grade steel had been produced and implanted in numerous patients (obviously at the lower end of the socio-economical scale). There is a great variety of concepts and materials used.

In the total hip replacement (arthroplasty), THA in "acronimish," both the acetabular cup and femoral stem with ball are replaced. Figure B3.1(a) shows a schematic with the three principal components: the femoral stem, the ball, and the acetabular cup, a polymer insert that serves as a friction-bearing component. A typical surgery involves cutting the femoral head and reaming the acetabular socket, removing its cartilage. The femoral extremity is drilled and reamed to receive the stem. PMMA cement is prepared from a mixture consisting of polymer powder and monomer liquid. When it acquires the proper consistency, it is inserted in both the acetabular socket (for the cup) and the medular canal of the femur (for the stem). The two components are cemented and adjustments are made to ensure symmetric alignment and proper articulation. Figure B3.1(b) shows an X-ray in which the implant can be compared to the original joint. There are also other methods for the fixation of the femoral stem, one of them being a porous surface into which bone is expected to grow. This eliminates the use of the cement. Both procedures have problems.

Materials that have been used for the components include the following.

- *Acetabular cup* High-density polyethylene (HPPE), ultrahigh-density polyethylene (UHDPE), ultrahigh-molecular-weight polyethylene (UHMWPE). The *dernier cri* is cross-linked polyethylene (XPE). The importance of this component cannot be overestimated because friction generates wear with debris that creates an inflammatory response. It is for this reason that ceramic cups are so attractive as they decrease the wear rate even further. Figure B3.2 shows the wear rates of different cup–ball combinations. It is clear that the use of XPE has substantially reduced wear and that this is no longer a problem.

Box 3.1 (cont.)

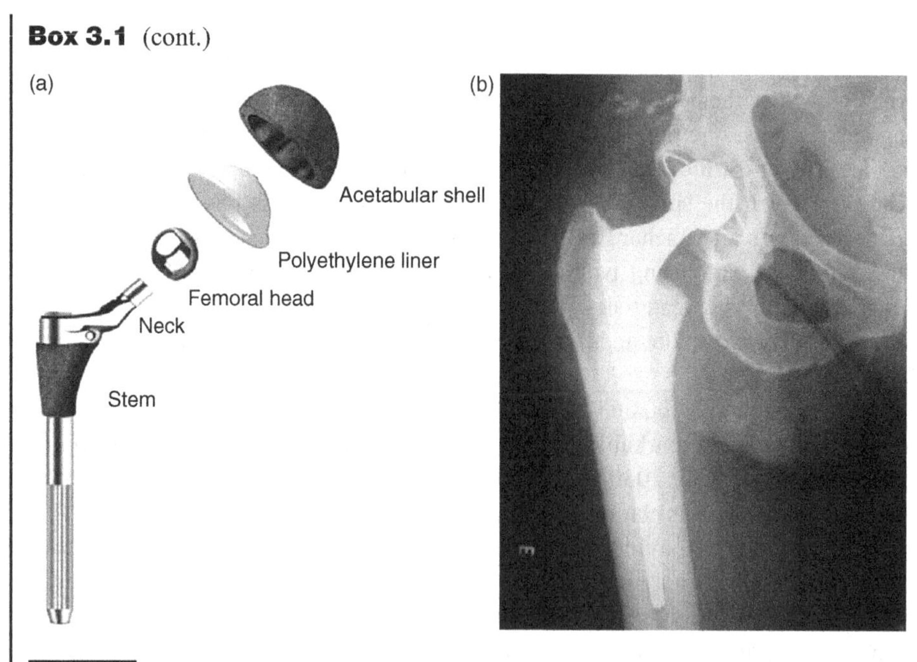

Figure B3.1.

(a) Schematic drawing of hip implant. (From Pruitt and Chakravartula (2011), Fig. 1.6, p. 15. Copyright © 2011 L. A. Pruitt and A. M. Chakravartula. Reprinted with the permission of Cambridge University Press.) (b) X-ray of hip implant.

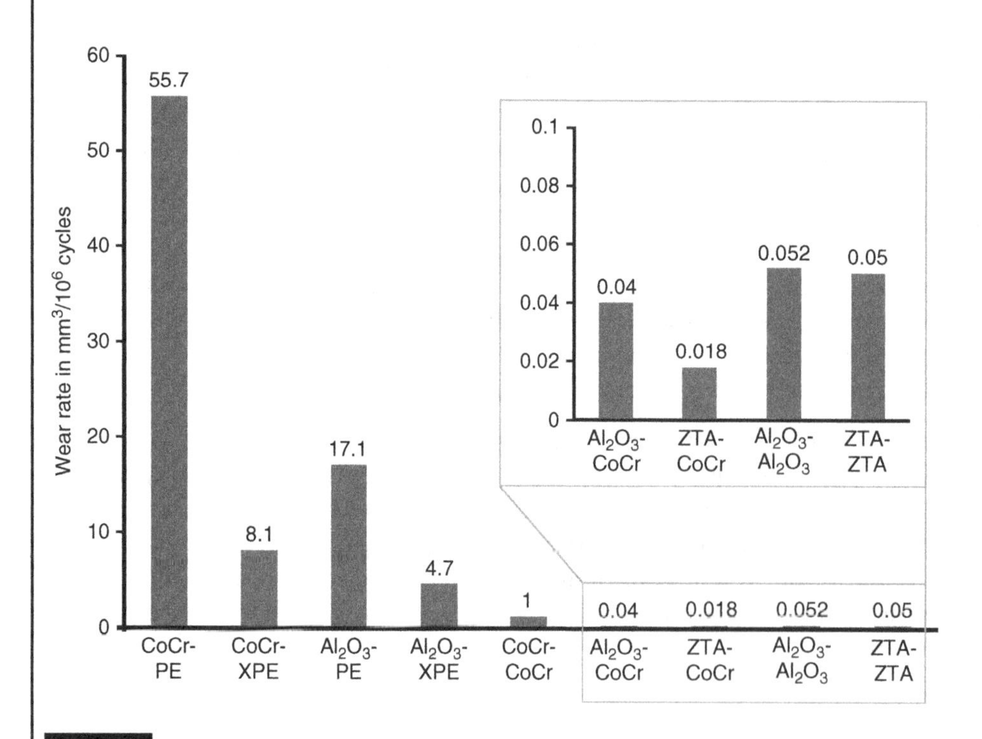

Figure B3.2.

Wear rate of various interfaces in total hip replacement. (Reprinted from Sonntag *et al.* (2012), with permission from Elsevier.)

- *Femoral head and stem* The early material of choice was 18Cr-8Ni-Mo stainless steel (316 SS). Other materials are Co-Cr alloys, Ti alloys, the ceramic ZrO_2 (partially stabilized zirconia), and composites. Femoral heads made of ceramics attached to a metallic stem are also used, since the wear rate is decreased.

A variety of surface treatment procedures have also been considered with diamond-like coatings, implantation, coating with porous and hydroxyapatite, etc. The main current problem with THA is at the implant–bone interface, since complex processes take place and contribute to the loosening of the acetabular and femoral components. This is an insidious problem, and progress in this respect is slow. The cementing of the components brings a series of problems which can be avoided by using cementless (biological) fixation. This is accomplished in the acetabular cup by backing the PE component by a metal foundation which is anchored in the bone. This metal may have a porous surface into which bone can grow. The coating can be applied by powder metallurgy techniques. It can also be coated with hydroxyapatite. However, other problems are created, and it seems that there is no clear advantage.

In the femoral head, stress shielding of the bone can lead to its remodeling and resorption. There are also other phenomena taking place, such as necrosis of the tissue due to the excessive temperature generated by the exothermic polymerization reaction.

The total knee arthroplasty (TKA) is more recent than the THA; one reason for this is the more complex nature of the articulation. The center of rotation is not fixed as in the case of a ball-and-socket articulation but moves forward as the knee is straightened. This is not easily accomplished, and simple hinged knees have been rather unsuccessful. The lateral mobility is also limited. One of the most successful designs is shown in Fig. B3.3(a).

Ultrahigh-molecular-weight polyethylene seems to be the best material for the bearing surface of the tibial component over which the top (femoral component) rides. The wear rate is decreased with cross-linking, similar to the case of the THA. The design shown in Fig. B3.3 has a metallic femoral component riding on a PE pad. It can be seen that the center of rotation changes when the two round terminations of the femoral component roll over the tibial component. The center of the femoral component has a hole that has the corresponding hill in the tibial component. This ensures alignment of the two parts, as well as some lateral and twisting motion. In order to prevent the sinking of the tibial component (plateau) into the trabecular bone, it is backed by a metal component with a protrusion for better fixation. The two components are most often cemented, but the usual problems occur. It is for this reason that porous surfaces have been developed for the tibial component, to allow bone to grow into the prosthesis. However, the immobilization time prior to walking is much higher in this case, bringing its own set of problems.

Other joints that are routinely replaced are the shoulder, finger (knuckles, or metacarpophalangeal joints), and elbow. The knuckle joints can be either a metallic or an elastomeric polymer. The temporomandibular joint (TMJ) can also be replaced, and a ball-and-socket system is used. Other joints that can be replaced are the ankle and wrist. However, they are complicated joints.

In summary, virtually all the joints in the body can be replaced, but the greatest success story is the total hip replacement.

Box 3.1 (cont.)

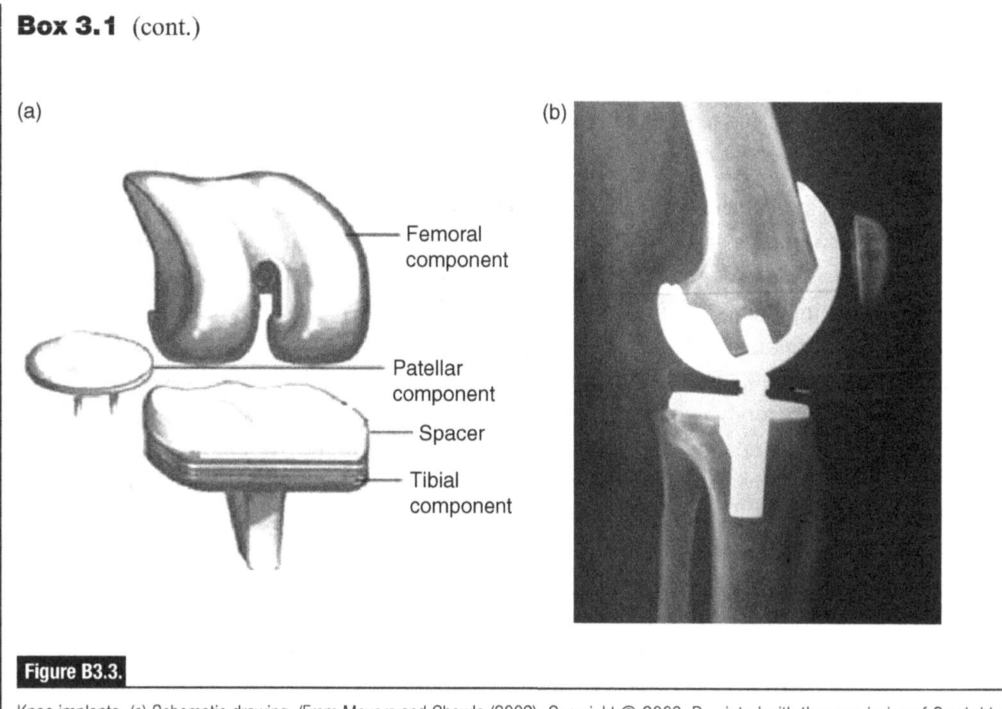

Figure B3.3.

Knee implants. (a) Schematic drawing. (From Meyers and Chawla (2009). Copyright © 2009. Reprinted with the permission of Cambridge University Press.) (b) Lateral view imaged by X-rays; note that polymer component is not visible. (From http://commons.wikimedia.org/wiki/File:PTG_P.jpeg#globalusage. Image by F. P. Jacquot. Licensed under the Creative Commons Attribution-ShareAlike 3.0 Unported License.)

sheet plane on opposite sides. Successive chains can bond in such a fashion, creating pleated sheets.

The three main fiber-forming polymers in nature are:

- polypeptide chains, which are the building blocks for collagen, elastin, silks, and keratins;
- polysaccharides, the building blocks for cellulose and hemicellulose;
- hybrid polypeptide–polysaccharide chains, the building blocks for chitin.

Polysaccharides will be discussed in Section 3.4.

3.3.2 Overview of protein structure

The structural organic constituents of biological tissues are either protein-forming polypeptides (primarily, in biological materials, collagen, keratin, elastin, resilin, fibroin, abductin) or polysaccharides (chitin, cellulose), as shown in Table 3.4. The proteins can

Table 3.4. Fibrous structural biopolymers found in living organisms with predominant amino acid residues
Gly = glycine; Pro = proline; Hyp = hydroxyproline; Cys = cysteine; Ser = serine; Ala = alanine; Asp = asparagine

	Dominant amino acids	Distribution
Proteins		
Collagen	35% Gly, 12% Pro, 10% Ala, 9% Hyp[a]	bone, teeth, vasculature, organs
Keratin	22% Cys, 13% Pro, 13% Ser[b]	hair, nails, claws, horn, hooves, feathers
Elastin	32% Gly, 21% Ala, 13% Pro[c]	skin, ligaments, vasculature, lungs
Resilin	37% Gly, 10% Asp, 11% Ala[d]	insect wing hinges
Abductin	68% Gly, 6% Ser[e]	hinged parts of bivalve molluscs
Fibroin	45% Gly, 30% Ala, 12% Ser[f]	silkworm silk
Chitin	$(C_8H_{13}O_5N)_n$	exoskeletons of crustaceans, insects, squid and octopus beaks
Cellulose	$(C_6H_{10}O_5)_n$	plants

[a] Rat-tail tendon (Brown, 1975).
[b] Sheep wool (Marshall and Gillespie, 1977).
[c] Franzblau (1971).
[d] Wing hinge, *Schistocera gregaria* (Kelly and Rice, 1967).
[e] Sea scallop (Kahler, Fisher, and Sass, 1976).
[f] *Bombyx mori* silk (Kelly and Rice, 1967).

be further divided into those that contain rigid, crystalline fibers (collagen, keratin) and those that contain amorphous, long-chain molecules (elastin, resilin, abductin, fibroin). The crystalline fibrils (and fibers) are molecular aggregates that align in a preferred direction. Groups of these fibers can curve or crimp, and in this manner the elastic modulus and strain to failure can be tailored. The stiffness of these crystalline biopolymers is generally orders of magnitude greater than the amorphous ones, but they have strains to failure orders of magnitude lower (Table 3.5) – the elastic energy is stored in the atomic bonds in the crystalline biopolymers, whereas it is stored as entropy changes in the elastomeric biopolymers (see Section 2.5 for mechanical properties discussion). Collagen and keratin

Table 3.5. Stiffness, tensile strength, and strain to failure for selected biological and synthetic materials[a]

	E (MPa)	σ_f (MPa)	ϵ_f
Biopolymers			
Abductin	1–4		
Cellulose	150 000	18	0.024
Collagen (along fibers)	1000	50–100	0.09
Elastin	0.3–1.5		>2
Fibroin	10 000	70	0.09
Keratin (wool)	5000	200	0.5
Kevlar 49 fiber	130 000	3600	0.03
PMMA	1600	15	0.013
Resilin	1.8–4	3	>3
Rubber	0.01–0.1	15	8.5
Silk			
• cocoon silk (*Bombyx mori*)	7	600	0.18
• spider silk (webframe and dragline, major ampullate, *Araneus diadematus*)	10 000	1100	0.27
• spider silk (viscid, catching spiral, *Araneus diadematus*)	3	500	
Tendon	2000–9000	50–100	0.09
Biominerals			
Aragonite ($CaCO_3$)	72 000		
Hydroxyapatite ($Ca_5(PO_4)_3(OH)$)	112 000	115	
Hydroxyapatite (from enamel)	83 000	60	
Magnetite (Fe_3O_4)	72 000		
Radular teeth (Fe_3O_4 from chiton)	90 000–125 000		
Silica ($SiO_2 \bullet nH_2O$)	600–2000		
Silicon carbide (SiC)	450 000	3500	
Zirconia (ZrO_2)	200 000	250	

Table 3.5. (*cont.*)

	E (MPa)	σ_f (MPa)	ϵ_f
Biocomposites			
Bone			
• Trabecular bone	800–14 000	1–100	0.03
• Cortical bone (longitudinal)	6000–20 000	30–150	
Nacre (abalone)	40 000–70 000	160	
Wood (oak)	10 000	100	

[a] Data from Meyers (2008b), p. 243.

have similar properties to cross-linked polymers, whereas elastin, resilin, and abductin have rubber-like properties.

3.3.3 Collagen

Collagen is a basic structural material for soft and hard bodies. It is present in different organs and tissues, and provides structural integrity. Fung (1993) compares it to steel in structures, not because of its strength, but because it is a basic structural component in our body. Steel is the principal load-carrying component in structures. In living organisms, collagen plays the same role: it is the main load-carrying component of blood vessels, tendons, bone, muscle, etc. In rats, 20% of the proteins are collagen. Humans are similar to rats in physiology and behavior, and the same proportion should apply.

Approximately 20 types of amino acid compositions of collagen have been identified, and their number is continuously increasing. In humans (and rats) the collagen is the same, called Type I collagen. Other collagens are named II, X, etc. The amino acid composition of different collagens differs slightly. This composition gives rise to the different types. Table 3.6 (Ehrlich and Worch, 2007) provides a few illustrative examples.

Figure 3.5 shows the structure of collagen. It is a triple helix, each strand being made up of sequences of amino acids. Each strand is itself a left-handed α-helix with approximately 0.87 nm per turn. The triple helix has a right-handed twist with a period of 8.6 nm. The dots shown in a strand in Fig. 3.5 represent glycine and different amino acids. Fiber-forming collagens organize themselves into fibrils and those, in turn, into fibers.

Figure 3.6 is a transmission electron microscopy (TEM) micrograph of tendon fibrils (Traub *et al.*, 1989). Each fibril has transverse striations, which are spaced approximately 67 nm apart. These striations are due to the staggering of the individual collagen molecules. The length of each collagen molecule is approximately 300 nm, which is about 4.4 times the distance of stagger, 67 nm. The gap between adjacent chains is about 35 nm ($67\times5-300 = 35$), and

Table 3.6. Amino acid composition (approximate percentages) from several collagens[a]

Collagen sources	Amino acid (%)																	
	Asp	Thr	Ser	Glu	Pro	HyPr[b]	Gly	Ala	Val	Met	Ile	Leu	Tyr	Phe	His	HyLys[c]	Lys	Arg
Sturgeon swim-bladder	6.9	3.8	5.8	11.4	12.8	11.8	27.7	11.6	2.3	1.4	1.7	2.6	0.5	2.5	0.8	1.9	3.5	10.0
Shark	6.4	2.4	3.3	11.0	13.3	8.8	25.4	11.4	2.7	1.8	2.7	2.6	7.2	2.1	1.7	0.9	3.7	8.6
Femur of ox	4.3	1.9	3.5	6.5	11.8	10.3	31.7	10.8	2.4	0.5	1.3	3.0	0.7	1.9	0.6	0.8	3.0	5.0
Porcine skin	5.1	1.7	3.5	11.4	13.3	12.6	22.1	8.7	2.1	0.5	1.1	2.9	0.5	1.9	0.7	–	3.5	8.4
Human tendon	3.9	1.5	3.0	9.5	10.3	7.5	26.4	9.0	2.1	0.5	0.9	2.1	0.3	1.2	1.4	1.5	3.5	16.0
Human bone	3.8	1.5	2.9	9.0	10.1	8.2	26.2	9.3	1.9	0.4	1.1	2.1	0.4	1.2	1.4	0.6	4.6	15.4

[a] Adapted from Ehrlich and Worch (2007).
[b] Hydroxyproline.
[c] Hydroxylysine.

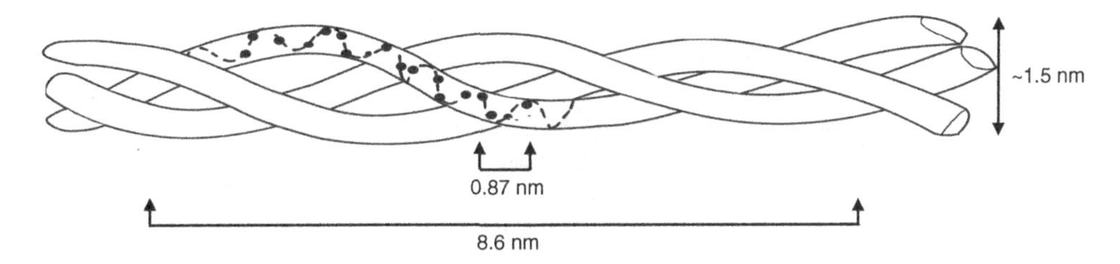

Figure 3.5.

Triple helix structure of collagen. (Adapted from Fung (1993); with kind permission from Springer Science+Business Media B.V.)

Figure 3.6.

Transmission electron micrograph of tendon fibrils. (Taken from Traub, Arad, and Weiner (1989), with kind permission from Stephen Weiner.)

the overlap is about 32 nm (300−67×4 = 32). Thus, there is a periodicity in the structure (35 nm + 32 nm), shown on the right-hand side in Fig. 3.7(b). Each repeating unit comprises five segments, and this periodicity produces a characteristic interference pattern that is observed as bands in electron microscopy, shown in Fig. 3.7(c). The collagen chains are directional, i.e. they have a "head" and a "tail," and the staggered arrangement allows for bonding between the "head" of one chain and the "tail" of the adjacent one, as shown in Fig. 3.7(b).

Figure 3.8(b) shows another clear example where the characteristic pattern of 67 nm is seen. In this case, the collagen was taken from a fish scale (*Arapaima*) and demineralized. The collagen is organized in fibrils with a diameter of ~100 nm. These strands can be individually pulled out. These fibrils are ubiquitous in biological organisms.

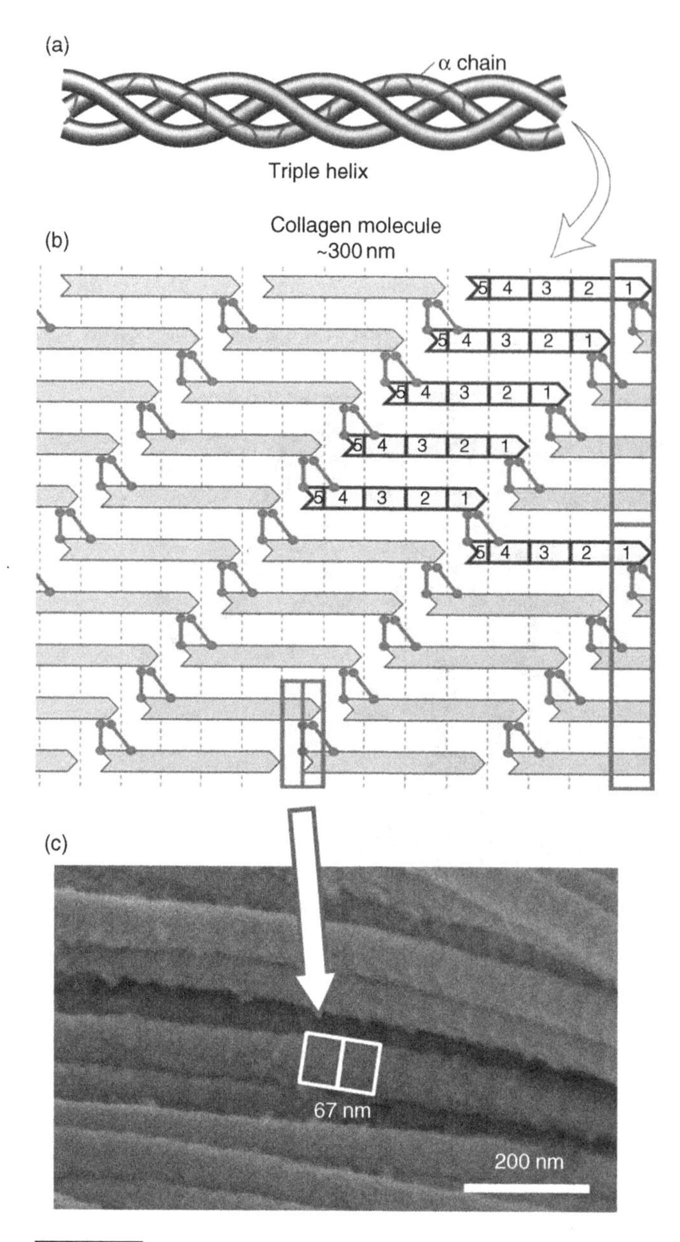

Figure 3.7.

Collagen composed triple helix, forming microfibrils which have 67 nm gaps. (a) Triple helix showing the amino acid chains in each strand. (b) General scheme with representation of 300 nm segments with gaps and overlaps, adding to 67 nm. (c) Collagen fibrils in chicken skin.

The pattern of 67 nm is known as the D-period. Groups of four to six triple helices are packed together forming microfibrils (shown in Fig. 3.8(a)). In modeling investigations, the amino acid chain is represented by the sequence glycene–proline–hydroxyproline. This is a simplified model for collagen.

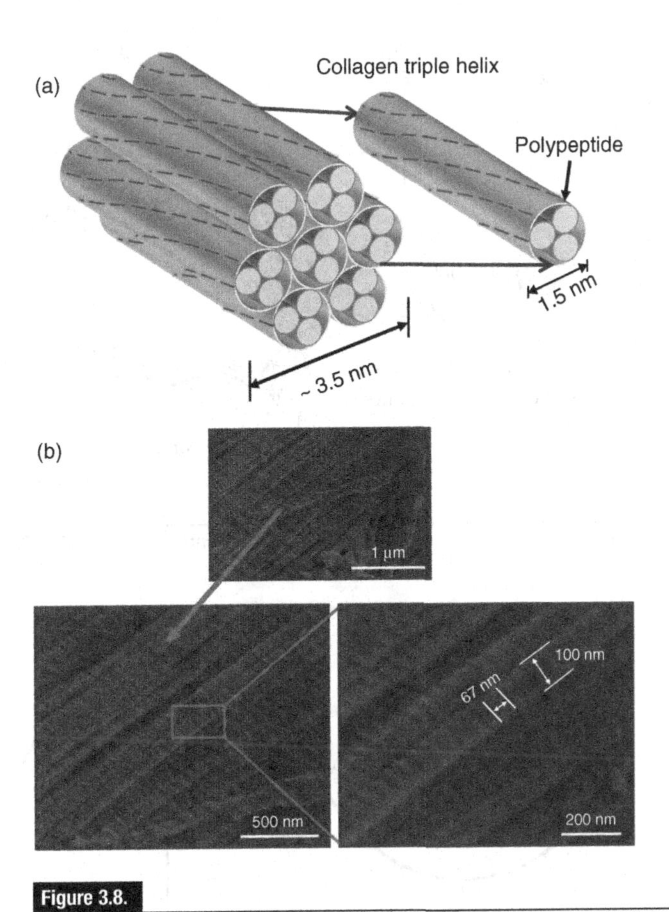

Figure 3.8.

(a) Collagen microfibrils; (b) collagen fibrils from demineralized fish (*Arapaima gigas*) scale seen in high-resolution SEM. (Figure courtesy Y. S. Lin.) Note: the sheath around the polypeptides is included to simplify the representation.

The collagen molecules, with a diameter of ~1.5 nm, form microfibrils with a diameter of ~4 nm, and these form fibrils with diameters of ~50 nm. There is some variation in different species, and this number can go from 20 to 100 nm. The collagen fibrils in an elephant may be larger than those in a mouse. The molecules attach to the adjacent molecules by intrafibrillar bonding. These fibrils form fibers. This is mediated by proteoglycans. Fibers are bundles of fibrils with diameters between 0.2 and 12 µm. In tendon, these fibers can be as long as the entire tendon. In tendons and ligaments, the collagen fibers form primarily one-dimensional networks. In skin, blood vessels, intestinal mucosa, and the female vaginal tract, the fibers organize themselves into more complex patterns, leading to two- and three-dimensional networks.

Figure 3.9(a) (Baer *et al.*, 1992) shows the hierarchical organization of a tendon, starting with tropocollagen (a form of collagen), and moving up, in length scale, to fascicles. There is a crimped or wavy structure shown in the fascicles that has an important bearing on the mechanical properties. Figure 3.9(b) shows an idealized

(a)

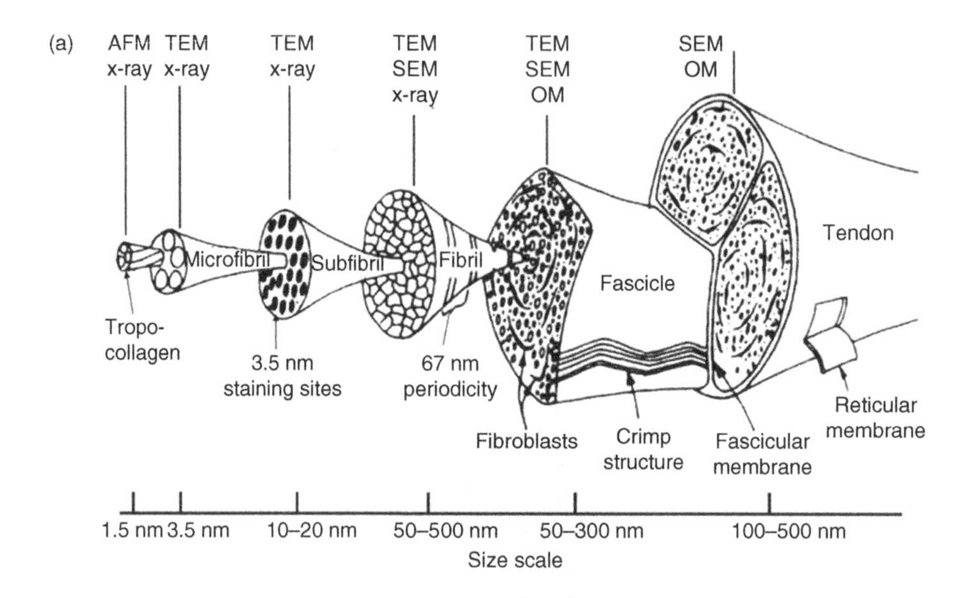

(b)

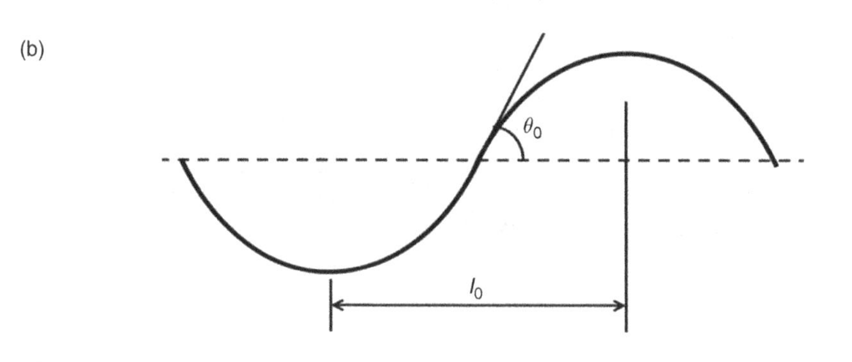

Figure 3.9.

(a) Hierarchical structure of tendon starting with collagen molecules and techniques used to identify it. (Taken from Baer, Hiltner, and Morgan (1992), with kind permission from Eric Baer.) (b) Idealized configuration of a wavy collagen fiber.

representation of a wavy fiber. Two parameters define it: the wavelength $2l_0$ and the angle θ_0. Typical values for the Achilles tendon of a mature human are $l_0 = 20–50$ μm and $\theta_0 = 6–8°$. These bent collagen fibers stretch out in tension. When the load is removed, the waviness returns. When the tendon is stretched beyond the straightening of the waviness, damage starts to occur. Figure 3.10(a) shows a schematic stress–strain curve for tendon. The tendon was stretched until rupture. There are essentially three stages:

Region I toe part, in which the slope rises rapidly. This is the physiological range in which the tendon operates under normal conditions.

Region II linear part, with a constant slope.

Region III slope decreases with strain and leads to failure.

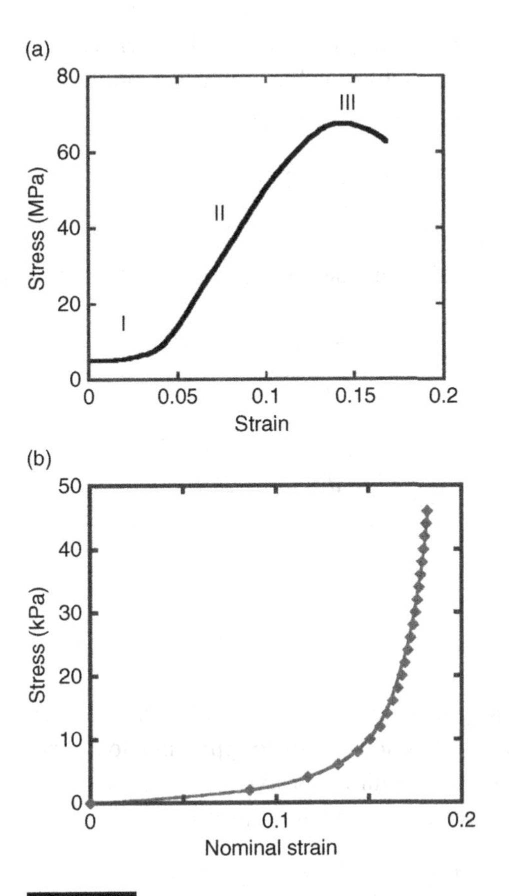

Figure 3.10.

(a) Stress–strain curve of collagen with three characteristic stages. I: toe; II: linear elastic; III: failure. (b) Stress–strain curve for ventricular papillary muscle, composed of titin and collagen. (Courtesy of A. Juskiel, Department of Bioengineering, UCSD.)

The elastic modulus of collagen is approximately 1–1.5 GPa, and the maximum strain is in the 10–20% range. The maximum strength is approximately 70–150 MPa. Cross-linking increases with age, and collagen becomes less flexible. Collagen often exists in combination with other proteins. An example is the ventricular papillary muscle; Fig. 3.10(b) shows the stress–strain curve. The passive tension in the cardiac muscle is governed by two load-bearing elements in the heart, collagen and titin. Titin has a spring-like region within the I-band of the sarcomere units. As the sarcomere units are extended, passive tension increases. At lower strains titin determines most of the passive tension in the heart. At higher strains, collagen begins to straighten and align with the axis of force, and hence it contributes increasingly to the passive tension. Figure 3.10(b) shows the characteristic toe region followed by a linear response.

It is possible to determine the maximum strain that the collagen fibers can experience without damage if their shape is as given in Fig. 3.9 with a ratio between amplitude and wavelength of r. We can assume a sine function of the form

$$y = k \sin 2\pi x/\lambda. \tag{3.1}$$

The maximum of y is reached when $x = \lambda/4$.

It is possible to find a relationship between θ_0, l_0, and the amplitude of the sine function:

$$\frac{dy}{dx} = \frac{2k\pi}{\lambda}\cos\frac{2\pi x}{\lambda}. \tag{3.2}$$

For $x = 0$,

$$\tan\theta_0 = \left.\frac{dy}{dx}\right|_{x=0} = \frac{2k\pi}{\lambda}. \tag{3.3}$$

Hence,

$$y_{\max} = \frac{l_0}{\pi}\tan\theta_0. \tag{3.4}$$

We can integrate over the length of the sine wave from 0 to 2π. However, this will lead to an elliptical integral of difficult solution. A simple approximation is to consider the shape of the wavy protein as an ellipse with major axis $2a$ and minor axis $2b$. The circumference is given by the following approximate expression:

$$L \approx \pi\left[\frac{3}{2}(a+b) - (ab)^{1/2}\right]. \tag{3.5}$$

In the sine function, we have two arms: one positive and one negative. Their sum corresponds, in an approximate manner, to the circumference of the ellipse.

The strain is equal to

$$\varepsilon = \frac{L-4a}{4a} = \frac{\pi\left[\frac{3}{2}(a+b) - (ab)^{1/2}\right] - 4a}{4a}. \tag{3.6}$$

Thus,

$$\varepsilon = \frac{\pi}{4}\left[\frac{3}{2}\left(1+\frac{b}{a}\right) - \left(\frac{b}{a}\right)^{1/2}\right] - 1. \tag{3.7}$$

The following ratio is defined: $b/a = 2r$. The corresponding strain is given by

$$\varepsilon = \frac{\pi}{4}\left[\frac{3}{2}(1+2r) - (2r)^{1/2}\right] - 1. \tag{3.8}$$

(a)

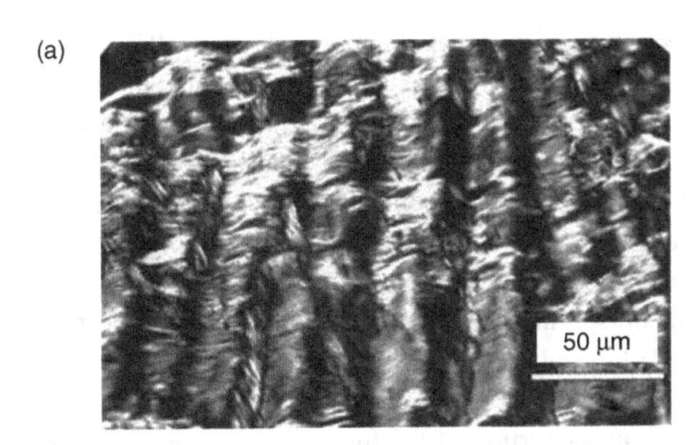

(b)

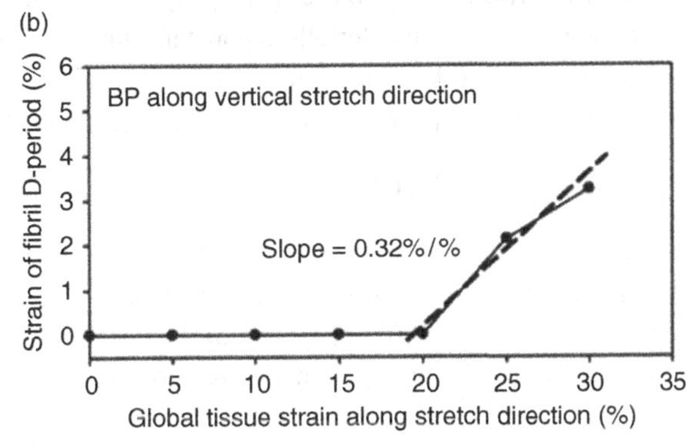

Figure 3.11.

(a) Crimped collagen structure in bovine pericardium; crimp period ~30 μm and amplitude ~15 μm. (b) Molecular strain (as determined from change in D-period) as a function of externally applied extension for bovine pericardium collagen. (Reprinted from Liao *et al.* (2005), with permission from Elsevier.)

Beyond this strain, the collagen will undergo bond stretching and eventually break.

The mechanical properties of collagen connective tissue are dictated, to a large extent, by the structure of the constituent collagen, which can form networks that are one, two, or three dimensional. In the case of planar soft tissues, the deformation is complex.

An example of this is the pericardium, which is a double-walled sac that contains the heart. It has a crimped collagen structure shown in Fig. 3.11(a). The crimp period of ~30 μm and the amplitude of ~15 μm were estimated by Liao *et al.* (2005). Synchrotron small-angle X-ray scattering (SAXS) was used to follow the molecular D-spacing (the 67 nm periodicity) of the collagen as a function of imposed stretching. The results are shown in Fig. 3.11(b). Using Eqn. (3.8) one can estimate the total nominal strain required to uncrimp completely and straighten the collagen structure shown in Fig. 3.11. The ratio

$$2r = \frac{b}{a} = \frac{15}{30} = 0.5. \quad (3.9)$$

The corresponding maximum nominal strain is given by $\epsilon = 0.22$. This is slightly higher than the strain at which molecular bond stretching is initiated in Fig. 3.11. Up to 0.2, the strains in the fibrils are negligible. Beyond this strain, the molecules stretch. However, not all external strain is accommodated by molecular bond stretching (the increase from 67 nm). Otherwise, the slope in the second stage would be equal to 1. We can speculate that part of the strain is accommodated by residual uncrimping of the structure (the last 0.02) and part by interfibril shear.

Another example illustrating the power of synchrotron X-ray scattering are the results (shown in Fig. 3.12) by Fratzl *et al.* (1998, 2004) on rat-tail tendon. The mechanical tensile force is also plotted. The toe and linear portions of the collagen response are clearly seen. In the bottom portion of the figure, the change in length in the molecular repeat length is shown. The D-repeat unit length is constant until a strain of 0.05. This indicates that the rat tendon is stiffer than the bovine pericardium. This is consistent with the parameters in the sine function reported by Fung (1990) for the rat-tail tendon:

$$l_0 = 100\ \mu\text{m}; \quad \theta_0 = 12°.$$

The corresponding parameters in Eqn. (3.7) are $a = 100\ \mu\text{m}$ and $b = 6.75\ \mu\text{m}$. The maximum nominal strain calculated from Eqn. (3.8) is 0.13. In Fig. 3.12, the molecular stretching of bonds starts for a strain of 0.05. The slope of the linear portion of the curve is equal to 0.4, indicating that we have concomitant molecular bond stretching, uncrimping of

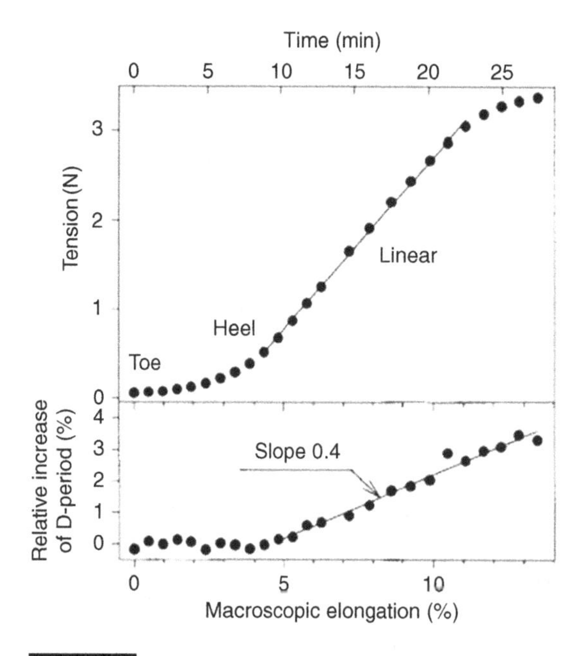

Figure 3.12.

Tension (top) and molecular strain (bottom) (as determined from change in D-period) as a function of externally applied extension for rat-tail collagen. (Reprinted from Fratzl *et al.* (1998), with permission from Elsevier.)

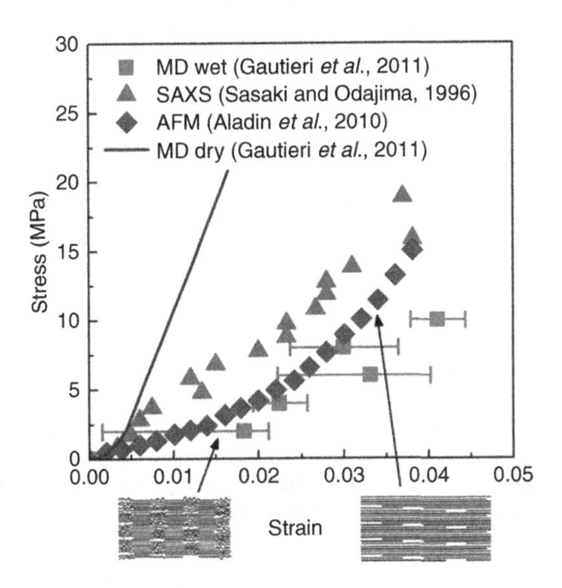

Figure 3.13.

J curve for hydrated and dry collagen fibrils obtained from molecular dynamics (MD) simulations, AFM, and SAXS studies. At low stress levels, significant stretching due to uncrimping and unfolding of molecules; at higher stress levels, the polymer backbone stretches. (Reprinted with permission from Gautieri *et al.* (2011). Copyright (2011) American Chemical Society. Depictions of collagen fibrils from Fratzl *et al.* (1998).)

the structure, and possibly shear between fibrils or fibers beyond this point. These results corroborate the bovine pericardium results in Fig. 3.12.

Collagen is the most important structural biological polymer, being the key component in many tissues (tendon, ligaments, skin, and bone), as well as in the extracellular matrix. The deformation process is intimately connected to the different hierarchical levels, starting with the polypeptides (0.5 nm diameter), to the tropocollagen molecules (1.5 nm diameter), to the fibrils (40–100 nm diameter), and fibers (1–10 μm diameter). Molecular dynamics computations (Gautieri *et al.*, 2011) of entire fibrils show the response, which is commonly known as a J curve; these computational predictions are well matched to atomic force microscopy (AFM) (Aladin *et al.*, 2010) and small-angle X-ray scattering (SAXS) (Sasaki and Odajima, 1996) and experiments (Fratzl, 2008), as seen in Fig. 3.13. The effect of hydration is also seen and is of great significance. The calculated density of collagen decreases from 1.34 to 1.19 g/cm^3 with hydration; this is accompanied by a decrease in the Young's modulus from 3.26 to 0.6 GPa. The strain at the stress level of 20 MPa is three times higher for the hydrated than for the dry collagen fibril.

Many tissues are also highly anisotropic. This is illustrated for vaginal tissue, shown in Fig. 3.14. Stress–stretch ratio curves for vaginal tissue are shown. They have clearly a J shape; differences between longitudinal and transverse stretches (anisotropy) are significant, with the maximum stretch being ~2.5 in the transverse direction and ~1.5 in the longitudinal direction. The viscoelasticity (difference in loading and unloading) is more marked in the transverse direction. This is the expansion direction in childbirth (Peña

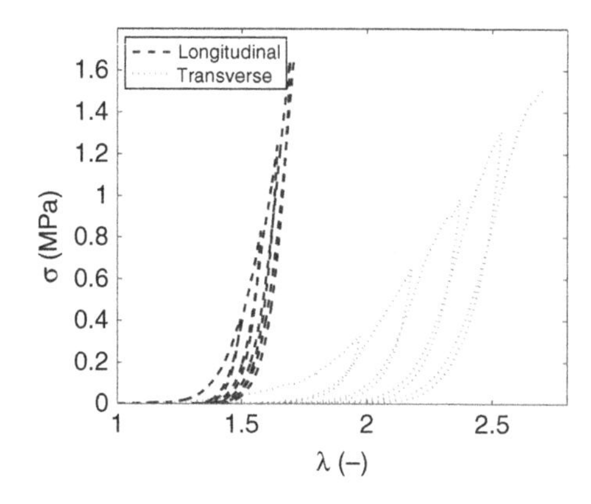

Figure 3.14.

Stress–stretch ratio curves for vaginal tissue; note J curves, differences between longitudinal and transverse stretches (anisotropy) and viscoelasticity (difference in loading and unloading), are more marked in the transverse direction. (Reprinted from Peña *et al.* (2011), with permission from Elsevier.)

et al., 2011). The anisotropy is generated by orientation-dependent crimping of the collagen. The stiff direction contains straighter collagen, whereas the more compliant orientation has more tortuous collagen fibers. This is also the case for skin, as will be shown in Chapter 9. The skin has stiff directions which are delineated by the so-called Langer lines.

Example 3.1

(a) Assume that the collagen fibers in rabbit skin have a curly structure with period ~50 μm and amplitude ~20 μm observed under SEM. Calculate the corresponding maximum normal strain after straightening the collagen fibers.

(b) The collagen fibers in rat tail have crimped period ~100 μm and amplitude ~25 μm. Estimate the maximum normal strain.

(c) Discuss how the ratio $2r$ affects the maximum normal strain of collagen fibers.

Solution

(a) The ratio $2r = b/a = 20/50 = 0.4$. The maximum normal strain is given by

$$\varepsilon = \frac{\pi}{4}\left[\frac{3}{2}(1+2r) - (2r)^{1/2}\right] - 1 = 0.15.$$

(b) The ratio $2r = b/a = 25/100 = 0.25$. The maximum normal strain is then

$$\varepsilon = \frac{\pi}{4}\left[\frac{3}{2}(1+2r) - (2r)^{1/2}\right] - 1 = 0.08.$$

(c) From the above examples, we conclude that the higher the ratio $2r$, the higher the maximum strain. The crimpy or wavy structure of collagen fibers can be stretched more compared to a straight one.

3.3.4 Keratin

Keratin is a structural protein that is found in most vertebrate animals. It is a fibrous protein that is produced in the integument (outer covering) of organisms and is typified by sulfur content. The integument provides the protective layer of animals and consists of two structural entities: the dermis and the epidermis. The dermis lies underneath the epidermis and is made mainly of collagen and elastin. The epidermis is the outer layer, produced by the dermis, and is made from epidermal cells. Keratin is produced by the keratinization process, in which epidermal cells die and build up at the outermost layer. It is usually classified as soft and hard originating from different mechanisms of biosynthesis. It is further classified into alpha and beta keratin, depending on its molecular structure. Alpha keratin, commonly known as mammalian keratin, is found in skin, wool, hoof, and horn; beta keratin, also known as avian and reptilian keratin, is found in claw, scale, feather, and beak (Mercer, 1961; Fraser *et al.*, 1972). The shells of turtles and porcupine quills are keratin, as is the flexible dermal armor of the pangolin. The shell of the toucan beak presented in Section 8.4 is made of beta keratin. The baleen in whales is also made of keratin. Whales fill their mouth with krill and expel the water using these long and flexible fibers. Whale baleen was prized in past centuries for the manufacture of corsets. Nowadays, less expensive polymers are used to reshape the body. It is also thought that the beaks of some dinosaurs were made of keratin.

The basic macromolecules that form keratin are polypeptide chains (Fig. 3.15(a)). These chains can either curl into helices (the α-conformation) or bond side by side into pleated sheets (the β-conformation). Mammals have about 30 alpha-keratin variants that are the primary constituents of hair, nails, hooves, horns, quills, and the epidermal layer of the skin. In reptiles and birds, the claws, scales, feathers, and beaks are beta keratin, which is tougher than the α-form, and is configured into a β-pleated sheet arrangement. The setae of the gecko foot, which provide the strong attachment of the feet to surfaces, are also composed of beta keratin.

Figure 3.15(b) shows the molecular structure of alpha keratin. Three distinct regions can be identified: the crystalline fibrils (helices), the terminal domains of the filaments, and the matrix. Isolated α-helix chains form a dimer (coiled coil) with sulfur cross-links, which then assemble to form protofilaments. The dimers have nonhelical N- and C-termini that are rich in cysteine residues and cross-link with the matrix. The protofilaments polymerize to form the basic structural unit, the intermediate filament (IF), with a diameter of ~7 nm and spaced ~10 nm apart. The IFs can be acidic (Type I) or basic (Type II). The IFs are embedded in an amorphous keratin matrix of two types of proteins: high sulfur, which has more cysteinyl residues, and high glycine–tyrosine proteins that have high contents of glycyl residues. The matrix has been modeled as an isotropic elastomer. A TEM micrograph of ram horn keratin is shown in Fig. 3.15(c) – the dark strand is the crystalline IF, which is surrounded by the lighter amorphous matrix.

The alignment of the IFs influences the mechanical properties. For example, the tensile strength of human hair (~200 MPa) is an order of magnitude greater than that of human

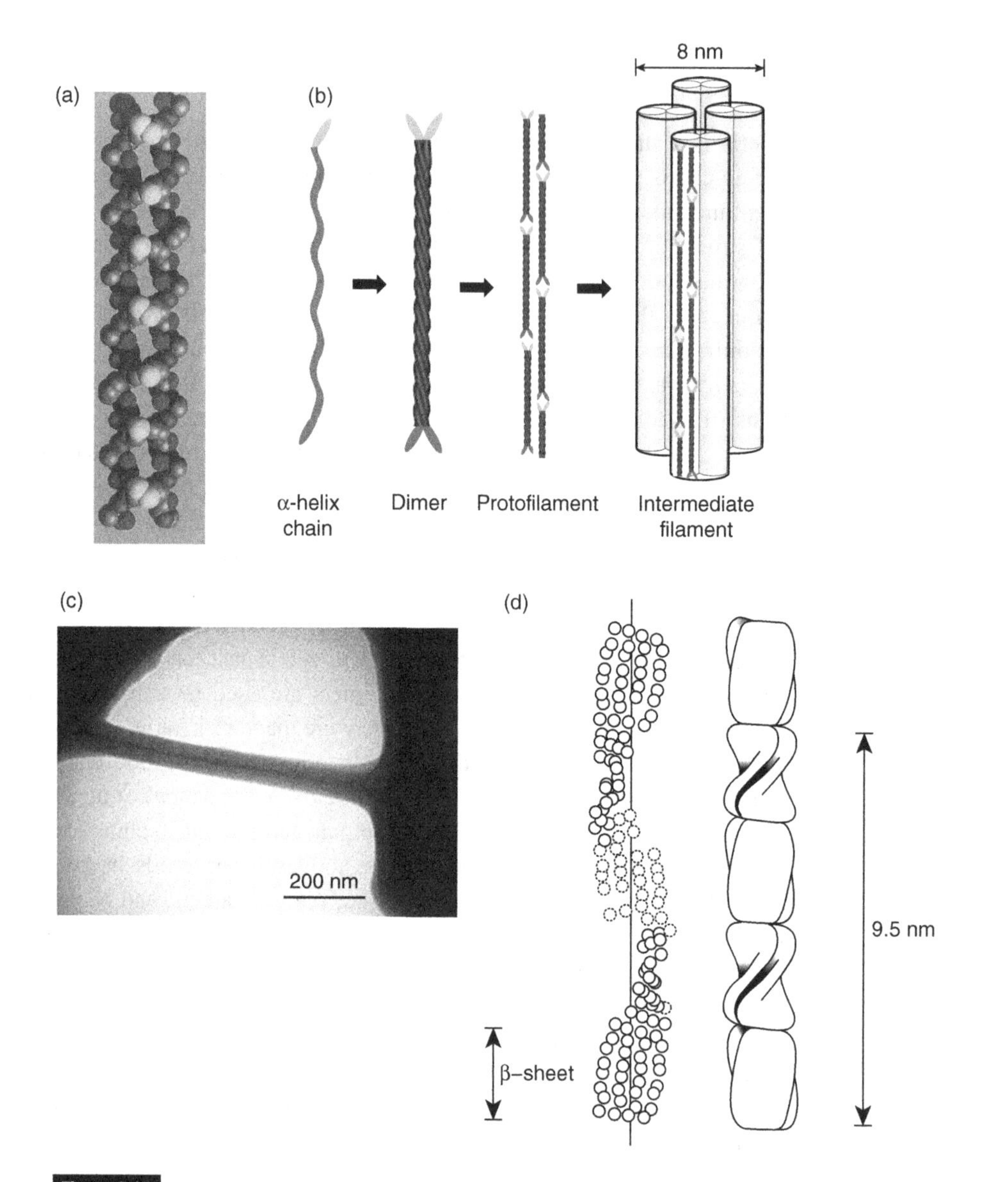

Figure 3.15.

(a) Molecular structure of alpha keratin: space-filling ball model. Gray = cysteine residue. (Reprinted from Parry and North (1998), with permission from Elsevier.) (b) Formation of alpha-keratin IF: two α-helices form a dimer, coiled coil; two dimers stagger and aggregate to form protofilaments; eight protofilaments associate into intermediate filaments. (c) TEM micrograph of alpha-keratin intermediate filament from a sheep horn. The strongly diffracting core of crystalline keratin is surrounded by an amorphous matrix. (Reprinted from McKittrick *et al.* (2010), with permission from Elsevier.) (d) Model for the arrangement of the β-sheet portions of the protein molecules in the filaments of avian keratin. (Reprinted from Fraser and Parry (1996), with permission from Elsevier.)

nail, due to the higher-order alignment of the keratin IFs in hair. The volume fractions of the matrix (amorphous) and crystalline fibers vary significantly in different materials.

The molecular structure of beta keratin comprises sheets with a pleated structure and the individual chains are in an anti-parallel chain configuration, as shown in Fig. 3.15(d). Positioned side by side, two or more protein strands (β-strands) link through hydrogen bonding. The linked β-strands form small rigid planar surfaces that are slightly bent with respect to each other, forming a pleated sheet arrangement. If the α-form is stretched, it will transform to the β-form, which is reversible up to about 30% strain.

A major difference between alpha keratin and beta keratin is the intermediate filament (IF). The IF of the alpha-keratin structure is based on an α-helix folding pattern. This is a coiled structure similar to collagen (three interwoven helices). These helices combine to form IFs, also called microfibrils, with a diameter of 8 nm. Figure 3.15(d) shows the microfibril of β-sheet. The folding pattern of beta keratin is β-sheet (Fraser and Parry, 1996), and the diameter of beta keratin is 4 nm. The IF of beta keratin has a smaller diameter, and the filament has a helical structure with a pitch of 9.5 nm and four turns per unit (Fraser and Parry, 1996). Interestingly, keratin undergoes a phase transformation (α→β-transition) under tensile load, which increases its elongation (Cao, 2002). This happens in wool and hagfish slime, as will be discussed in Chapter 9.

Keratin can be envisaged as a biological-fiber-reinforced composite consisting of a high-modulus fiber and a lower-modulus viscoelastic matrix. The matrix plays a role as a medium to transfer the applied load to the fiber, thus preventing crack propagation from local imperfections or points of rupture (Fraser and MacRae, 1980). Mineralization with calcium and other salts contributes to the hardness of keratin (Pautard, 1963). In mammalian keratin, the α-helix is aligned almost parallel to the filament; alpha keratin is mechanically linear elastic. While beta keratin is quite different from alpha keratin in molecular structure, the β-sheet also behaves linearly elastically and the mechanical behavior is similar. Alpha keratin changes the structure into the beta keratin during stretching (Cao, 2002). This change can be observed through X-ray diffraction. Generally, the stiffness of the β-sheet is higher than that of the α-helix. The mechanical behavior of both alpha keratin and beta keratin depends on moisture content. Increasing the hydration decreases the stiffness and the modulus (Fraser and Parry, 1996) because the matrix of keratin absorbs moisture. The mechanical properties of various keratinous materials will be presented in Sections 8.4 (hoof, horns, etc.), 9.7 (wool and hagfish slime), and 10.6 (feathers).

3.3.5 Elastin

Elastin is one of the most stable and insoluble proteins in the body. The structure of elastin is shown in Fig. 3.16 (Gray, Sandberg, and Forster, 1973). Each fibrous monomer is linked to many others and forms a three-dimensional network. There is a strong interaction between water and the protein, resulting in an inability to produce a good diffraction pattern (Gray *et al.*, 1973). Elastin can undergo over 200% strain

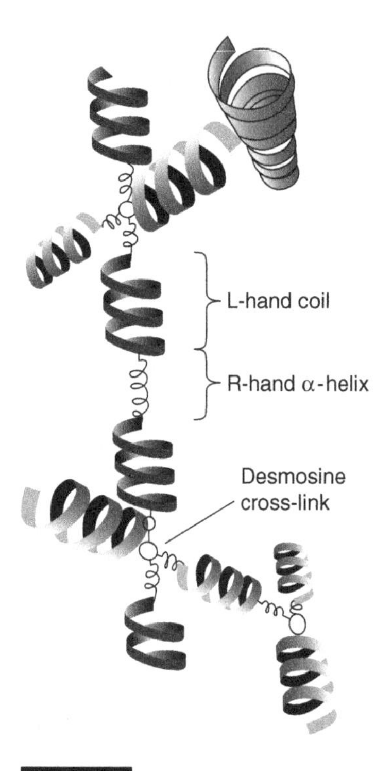

Figure 3.16.

Molecular model for elastin. (Reprinted by permission from Macmillan Publishers Ltd.: *Nature* (Gray *et al.*, 1973), copyright 1973.)

before the protein strands have completely straightened out (Aaron and Gosline, 1981). The elastin backbone is highly mobile. On stretching, the configurational entropy is reduced, which provides the restoring force upon unloading (Urry *et al.*, 1986). Elastin does not creep under an applied load. In contrast, the collagen molecule is much more rigid. Elastin is found in skin, walls of arteries and veins, and lung tissue. A prominent place is in the *ligamentum nuchae*, a long rope that runs along the top of the neck in horses and is constantly under tension. Other vertebrates have it too, but it is less pronounced. In this manner, similar to a cable in a crane, the horse can keep the head up without using muscles. The *ligamentum nuchae* plays a role similar to that in cables in a suspension bridge. It is a rather robust cylinder, ideally suited for mechanical property measurement. We will present the mechanical properties in Section 9.2.

3.3.6 Actin and myosin

These are the principal proteins of muscles, leukocytes, and endothelial cells. Muscles contract and stretch through the controlled gliding/grabbing of the myosin with respect to the actin fibers. Figure 3.17(a) shows an actin fiber. It is composed of

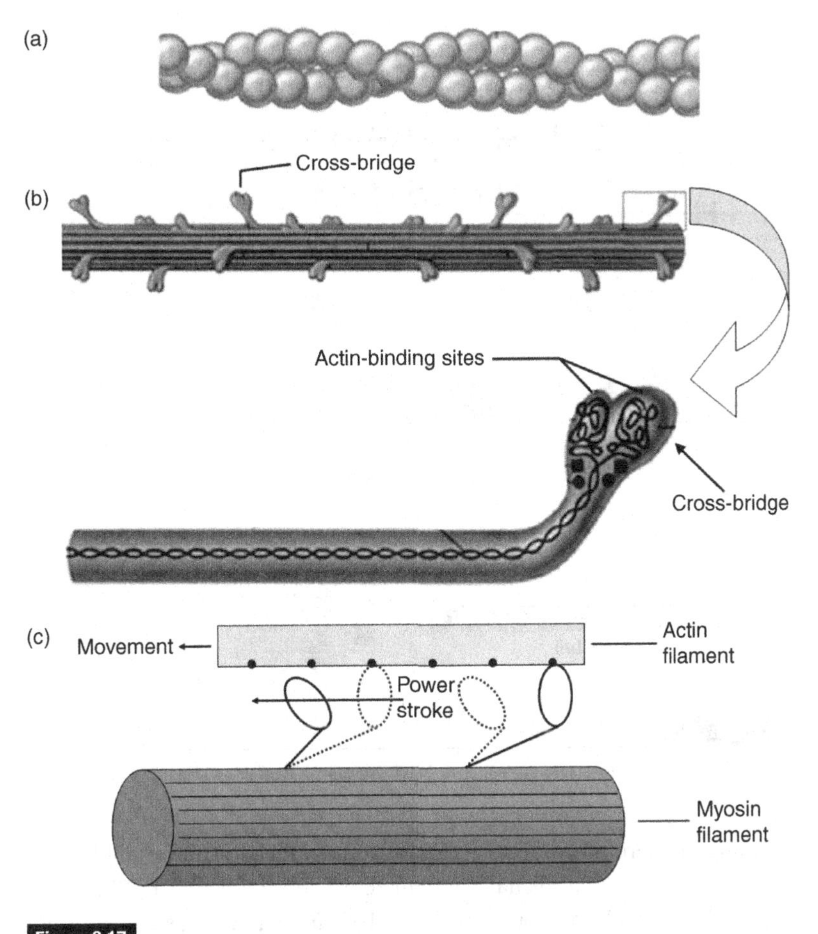

Figure 3.17.

Molecular structure of (a) actin and (b) myosin. (c) Action of cross-bridges when actin filament is moved to left with respect to myosin filament; note how the cross-bridges detach themselves then reattach themselves to actin.

two polypeptides in a helical arrangement. Figure 3.17(b) shows the myosin protein. It has little heart-shaped "grapplers" called cross-bridges. The tips of the cross-bridges bind and unbind to the actin filaments. Figure 3.17(c) shows the myosin and actin filaments, and the cross-bridges at different positions. The cross-bridges are hinged to the myosin and can attach themselves to different positions along the actin filaments, as the actin is displaced to the left. Thus, the muscles operate by a micro-telescoping action of these two proteins. They form sarcomere units which are actin–myosin arrays in the pattern shown in Fig. 3.18(a).

Figure 3.18 shows how the filaments organize themselves into myofibrils. Bundles of myofibrils form a muscle fiber. The Z line represents the periodicity in the myosin–actin units (called sarcomeres) and is approximately equal to 3 μm in the stretched configuration (Fig. 3.18(b)). It shortens when the muscle is contracted. This gives the muscle a

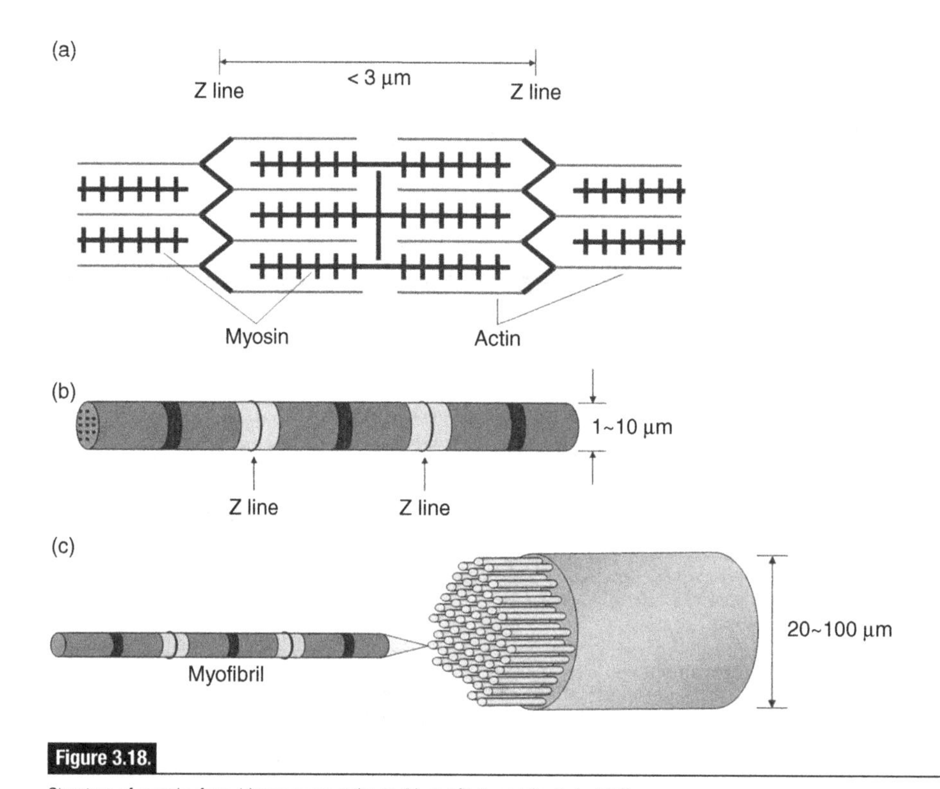

Figure 3.18.

Structure of muscle, from (a) sarcomere units, to (b) myofibril, and finally to (c) fibers.

striated pattern when observed at high magnification; it resembles a coral snake in the microscope. Myofibrils have a diameter of approximately 1–10 μm (Fig. 3.18(b)). They arrange themselves in bundles of 20–100 μm diameter, as shown in Fig. 3.18(c).

Example 3.2 A person is lifting a weight by contracting the biceps muscles. Assuming that each muscle fiber has the capacity to lift 300 μg, and that each muscle fiber has a diameter of 5 μm, what is the required cross-section of biceps muscle needed to lift a mass of 20 kg?

Solution The cross-section of each fiber is given by

$$A = \frac{\pi}{4} \times 5^2 = 19.625\ \mu\text{m}^2.$$

We can see from Fig. 3.19 that we need to apply a lever rule to calculate the force that the muscle has to exert. Distances given in Fig. 3.19 are typical. Students should check by measuring their arms. Equating the sum of the moments to zero,

$$\sum M_0 = 0,$$
$$F_1X_1 - F_2X_2 = 0.$$

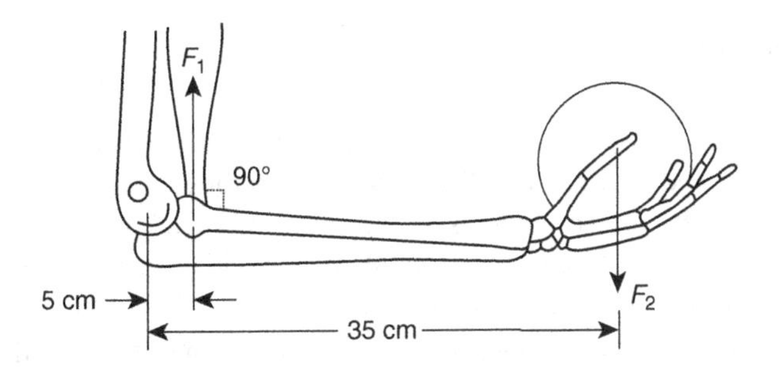

Figure 3.19.

Forearm and force F_2 exerted by weight, and reaction F_1 applied by biceps. (From Meyers and Chawla (2009). Copyright © Cambridge Univerity Press 2009. Reprinted with the permission of Cambridge University Press.)

Typical values are $X_1 = 5$ cm, $X_2 = 35$ cm. But

$$F_2 = 20 \times 9.8 = 196\,\mathrm{N}.$$

Thus

$$F_1 = \frac{196 \times 35}{5} = 1372\,\mathrm{N}.$$

The maximum force that each muscle fiber can lift is

$$F_f = 300 \times 10^{-6} \times 9.8 \times 10^{-3} = 2940 \times 10^{-9}\,\mathrm{N}.$$

The ratio F_1/F_f gives the number of fibers:

$$N = 4.66 \times 10^8.$$

The total area is equal to

$$A_t = N \times A = 91.425 \times 10^8\ \mu\mathrm{m}^2.$$

This may be converted into cm^2 to yield

$$A_t = 91.4\,\mathrm{cm}^2.$$

This is indeed a biceps with a diameter of

$$D = \left(\frac{4A_t}{\pi}\right)^{1/2} = 10.7\,\mathrm{cm}.$$

This corresponds to Arnold on steroids!

3.3.7 Resilin and abductin

Resilin has been the least studied, since only small portions are available from insects. It is extremely resilient and also has excellent fatigue properties. This basic resilience is the source of its name. Its elastic modulus in the range of stretch ratio (λ) from 1 to 2 is approximately 0.1–3 MPa (Elvin *et al.*, 2005). Weis-Fogh (1961a, b) first described the chemical and mechanical properties of resilin in the 1960s. He found that it could be stretched up to 300% and, when unloaded, regained its original shape. Additionally, it can maintain a load without relaxation. It has a low stiffness but a high-energy storage capability with a coefficient of resilience of up to 98%. Besides wing joints, it can be found in cuticular structures such as the abdominal wall of a termite queen, the thorax of flying insects, the material responsible for flea jumping, the element in the sound-producing tymbal mechanism in cicadas, and tendons in dragonflies. A locust can beat its wings over 20 million times during its lifetime; a flea can release stored energy in 1 ms; and a cicada can emit sound pulses over 400 million times during a lifetime. This remarkable protein is unique among the elastic proteins and demonstrates a wide functional diversity.

When swollen with water at pH 7, tendon resilin can be compressed to almost one-third and extended to three times its unstrained length before breaking (at 30 kg/cm^2 (2.94 kPa) unstrained, swollen area). The elastic modulus is 6–7 kg/cm^2 (0.588–0.686 kPa), as in most rubbers (Weis-Fogh,1961a).

Abductin is another low-stiffness, resilient elastic fiber that is found in the propulsion systems of swimming bivalve molluscs. A scallop propels its body through water by opening and closing the shell halves (valves) through constriction and release of the internal triangular hinge ligament (ITHL), in response to contractions of the abductor muscle in the animal. Kelly and Rice (1967) found that the ITHL has properties similar to elastin and resilin. The Young's modulus is ~4 MPa, compared with 0.6 MPa for elastin and resilin. Unusual for the other elastic proteins, there was ~3 wt.% mineral phase, presumably calcium carbonate, that provides the additional stiffness.

3.3.8 Other structural proteins

Proteins are the most abundant biological macromolecules, occurring in all cells. Thousands of proteins of different functionality can be found in a single cell. There is a great variety of proteins in biological systems that will not be covered here. Examples of relevance for us are lustrins (Shen *et al.*, 1997), identified with abalone shell organic layer, and silicatein (Cha *et al.*, 1999), found in sponge silica spicules.

Fibroin is found exclusively in spider and silkworm silk. The distinguishing amino acid sequence is (Gly-Ser-Gly-Ala-Gly-Ala)$_n$, which is in anti-parallel β-sheet arrangements. The Gly residue (50%) allows for tight packing; this makes silk a high-strength fiber. There are two types of spider silk – the radiating dragline web frame (major ampullate, MA) and the catching spiral (viscid) of the web. The silk fibrils consist of stiff β-sheet configured nanocrystalline regions surrounded by an amorphous matrix. The crystalline regions add to the strength, whereas the amorphous regions are responsible for the high toughness and extensibility. The mechanical properties of dragline silk and tangential web silk are quite different. The dragline silk has a high stiffness and failure strength, in contrast to the catching spiral with a much lower stiffness and strength. To capture prey in the web, a combination of silks is necessary – strong web frame coupled with a highly elastic, energy-absorbing spiral. The mechanical properties of spider silks are highly dependent on strain rate: the modulus, strength, and, surprisingly, the extensibility all increase with increasing strain rate – a phenomenon that does not occur in synthetic materials. Normally, with an increase in strain rate, a metal becomes stronger, but the strain to failure decreases. For spiders, this combination of properties results in a truly remarkable material. According to Gosline *et al.* (1999), the toughness of both MA and viscid spider silk is higher than that of Kevlar® fibers. This will be discussed in detail in Chapter 8, where we introduce the structure and present the mechanisms responsible for the incredible strength. Table 3.5 provides typical values of the mechanical properties.

3.4 Polysaccharides

Polysaccharides are long-chain molecules with the general formula $(C_6H_{10}O_5)_n$, where $40 \leq n \leq 3000$. The two structural polysaccharides cellulose and chitin are of interest as fibrous building blocks in plants (cellulose) and animals (chitin), respectively. Chitin is a polymer of N-acetylglucoseamine ($(C_8H_{13}O_5N)_n$) and is found in the exoskeletons of insects and crustaceans, the beaks of cephalopods (squid, octopus), and in the organic matrix in mollusc shells. The structure is shown in Fig. 3.20. It differs from cellulose by the substitution of a hydroxyl group by an acetyl amine group, resulting in a higher

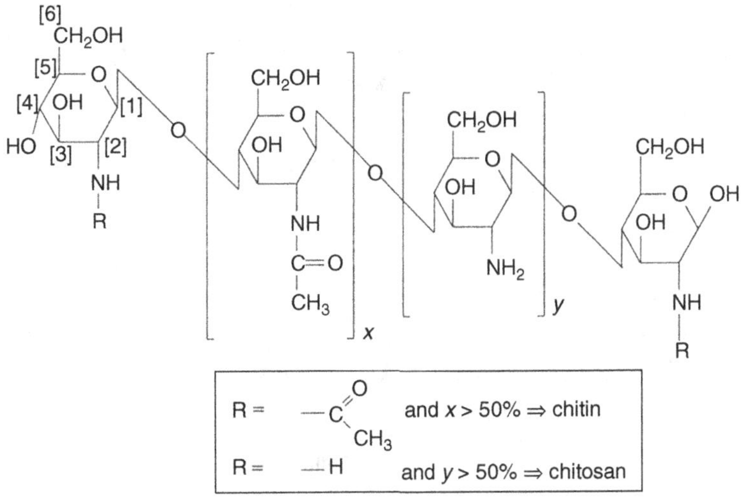

R = —C(=O)CH$_3$ and $x > 50\% \Rightarrow$ chitin

R = —H and $y > 50\% \Rightarrow$ chitosan

Figure 3.20.

Chemical structural representation of chitin and chitosan copolymer. (Reprinted from Kohr (2001), with permission from Elsevier.)

strength for chitin. It is considered to be functionally equivalent to the structural protein keratin. It is positively charged and bonds easily to other surfaces such as metal ions and proteins.

3.4.1 Chitin and chitosan

Chitin is the second most abundant natural polymer (after cellulose) on earth. It is a linear polysaccharide of β-(1–4)-2-acetamido-2-deoxy-D-glucose. The chemical structure of chitin is very similar to that of cellulose, with a hydroxyl group replaced by an acetamido group.

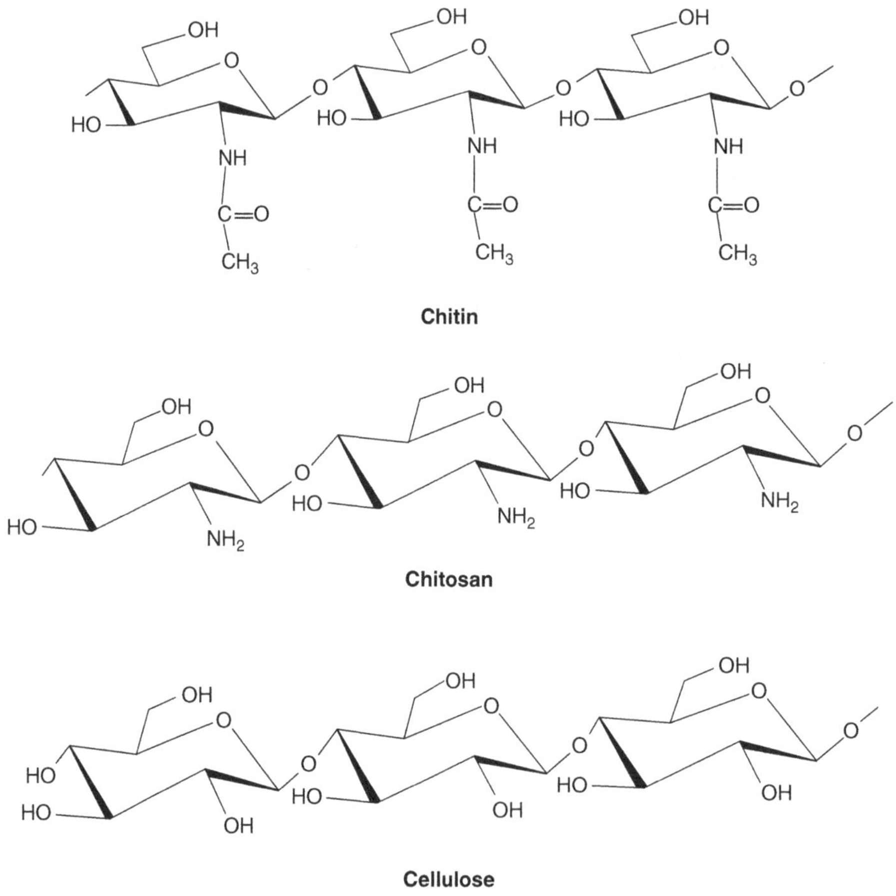

Chitin

Chitosan

Cellulose

Figure 3.21.

Chemical structures of chitin, chitosan, and cellulose. (Reprinted from Krajewska (2004), with permission from Elsevier.)

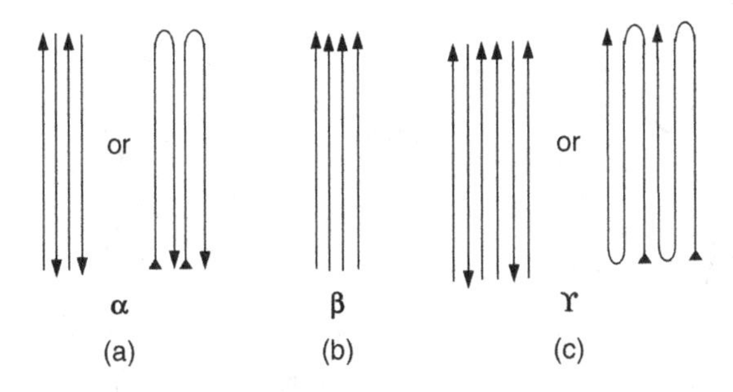

Figure 3.22.

Three polymorphic configurations of chitin: (a) α-chitin; (b) β-chitin; (c) γ-chitin.

Pure chitin with 100% acetylation does not exist in nature. Chitin tends to form a copolymer with its N-deacetylated derivative, chitosan. Chitosan is a polymer of β-(1–4)-2-amino-2-deoxy-D-glucose. The chemical structures of cellulose, chitin, and chitosan are shown in Fig. 3.21 (Kohr, 2001). When the fraction of acetamido groups is more than 50% (more commonly 70–90%), the copolymer is termed chitin. The copolymer consists of chitin and chitosan units randomly or block distributed throughout the polymer chain (Levi-Kalisman *et al.*, 2001).

Three polymorphic forms of chitins (α-, β-, and γ-chitins) have been differentiated due to their crystal structure, as shown in Fig. 3.22. α-chitin (the most abundant form found in nature) is arranged in an anti-parallel configuration, whereas β-chitin is organized in a parallel configuration; γ-chitin is a mixture of α- and β-chitin. The anti-parallel configuration gives α-chitin a highly ordered crystalline structure with strong hydrogen bonding between chitin chains (Fig. 3.23). The strong hydrogen bonding leads to the rigid and insoluble properties of α-chitin. Both α-chitin and β-chitin are crystalline. The lattice parameter along the b-axis of α-chitin (1.886 nm) is approximately twice that of β-chitin (0.926 nm); that along the c-axis is approximately the same. This can be seen in the electron diffraction patterns (bc-projection) shown in Fig. 3.24 (Rinaudo, 2006). The spacing between the diffraction spots of β-chitin is twice that of α-chitin (recall that $\lambda = 2d \sin \theta$, as in Bragg's law).

Chitin is widely distributed in fungal and yeast cell walls, mollusc shells, arthropod exoskeletons, and other invertebrates. It plays an important role as the structural component that provides support and protection to the organisms.

The cell walls of fungi are made mostly of chitin. Fungal chitin forms randomly oriented microfibrils typically 10–25 nm in diameter and 2–3 μm long. Chitin microfibrils are covalently linked to other polysaccharides, such as glucans, and form a chitin–glucan complex, which is the main structural component of fungal cell walls. The chitin content in fungi varies from 0.45% in yeast to 10–40% in some filamentous fungi species.

The presence of chitin has been reported in shells of the mollusc species. Despite the relatively small amount of chitin compared to the inorganic mineral (typically $CaCO_3$), chitin plays an important role, not only in the mechanical support, but also in the hierarchical control of the biomineralization processes. Studies on the organic matrix

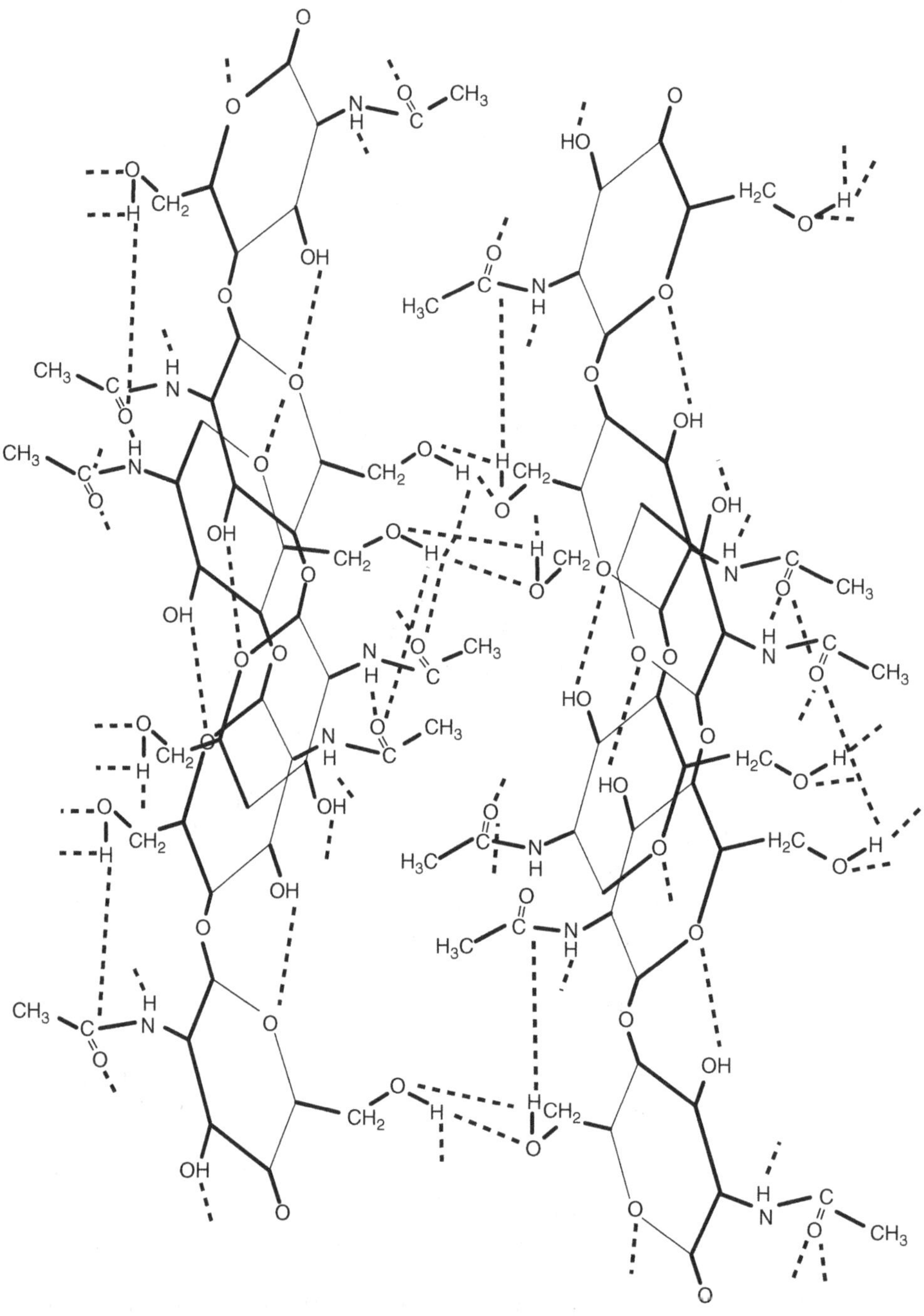

Figure 3.23.

The extensive hydrogen bonding between α-chitin chains. The inter- and intramolecular hydrogen bonds between two parallel cellulose chains are shown. (Reprinted from Kohr (2001), with permission from Elsevier.)

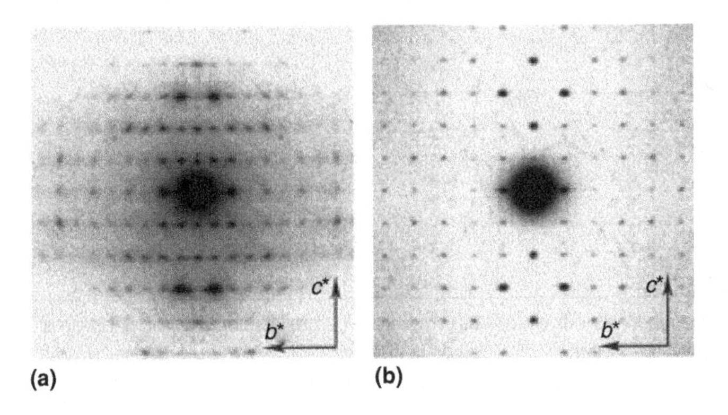

Figure 3.24.

Electron diffraction patterns of highly crystalline chitin: (a) α-chitin; (b) β-chitin. (From Rinaudo (2006), Fig. 3, p. 603; reprinted with permission from Elsevier.)

in the shell (Krajewska, 2004; Weiss and Schönitzer, 2006) revealed that the inter-lamellar sheets are composed of thin layers of β-chitin sandwiched between two thick layers of silk-like protein gel. The β-chitin is highly ordered at the molecular level and responsible for the overall shell formation. Weiss and Schönitzer (2006) studied the distribution of chitin in shells of mollusc using chitin-binding green fluorescent protein and confocal laser scanning microscopy. This will be discussed further for abalone in Section 6.2.2.1. The results showed that bivalve molluscs deposit and orient the chitin in a well-defined manner. Chitin distributes mainly in the hinge and edges of the shell and integrates the flexible region connecting two valves.

Chitin is also the main component in the exoskeletons of arthropods. The exoskeleton materials of arthropods are complex composites that are hierarchically structured and multifunctional, as shown in Fig. 2.3. The linear chitin chains align anti-parallel and form α-chitin crystals at the molecular level. Several of the α-chitin crystals which are wrapped by proteins form nanofibrils of about 2–5 nm in diameter and 300 nm in length. A bundle of chitin–protein nanofibrils then form chitin–protein fibers of about 50–100 nm diameter. These chitin–protein fibers align together to form planar layers which stack helicoidally. This structure is called a twisted plywood or Bouligand structure (Bouligand, 1970, 1972). In crustaceans, such as crabs and lobsters, there is a high degree of mineralization. The mineral is mostly calcium carbonate, which deposits onto the space of the chitin–protein network and gives rigidity to the exoskeleton. The multifunctionality and mechanical properties of arthropod exoskeletons as well as the Bouligand structure are discussed further in Section 8.3.

3.4.2 Cellulose

Cellulose is the most abundant natural polymer, and is the structural component of plant cell walls. It is a linear polysaccharide consisting of D-anhydroglucopyranose units (often

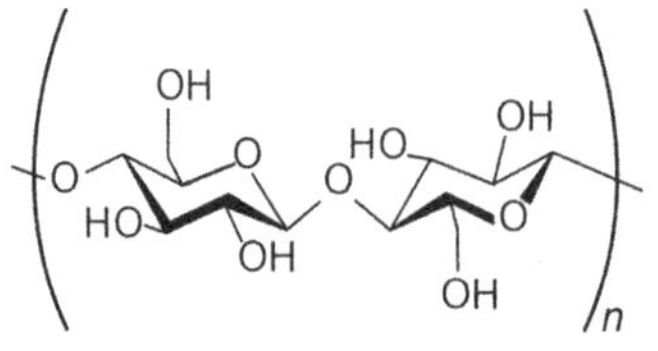

Figure 3.25.

The structure of cellulose: two units of a cellulose chain; the D-glucose residues are in β-(1–4) linkages. The rigid chain structures can rotate relative to one another.

abbreviated as anhydroglucose units or as glucose units for convenience) linked together by β-(1–4)-glycosidic bonds, as shown in Fig. 3.25 (Nelson and Cox, 2005).

Like amylase and amylopectin, the polysaccharides of starch, molecular cellulose is a homopolysaccharide consisting of 10 000 to 15 000 D-glucose units. The main difference between cellulose and other D-glucose-based polysaccharides is that the glucose residues in cellulose are in the β-configuration whereas those in amylase, amylopectin, and glycogen are in the α-configuration. This difference gives cellulose very unique physical and chemical properties. Most animals cannot digest cellulose, because they lack an enzyme to hydrolyze the β-(1–4) linkages. Some animals, particularly ruminants and termites, can digest cellulose with the help of a symbiotic microorganism which hydrolyzes the β-(1–4) linkages.

For cellulose, the most stable conformation is that each unit chair is turned 180° relative to its neighbors, yielding a straight, extended chain. When the cellulose is fully extended, all hydroxyl groups are capable of forming both intermolecular and intramolecular hydrogen bonds. The extensive hydrogen bonds produce stable supramolecular fibers with excellent mechanical properties. This property has made cellulose a useful material in civilization for millennia.

Cellulose forms a structure with regions of high order, i.e. crystalline regions, and regions of low order, i.e. amorphous regions. Naturally occurring cellulose (cellulose I) crystallizes in the monoclinic sphenodic structure. The unit cell of cellulose is shown in Fig. 33 of Bledzki and Gassan (1999); the lattice parameters are $a = 0.835$ nm, $b = 1.03$ nm, and $c = 0.79$ nm.

Pure cellulose is never found in nature. The cotton fiber is the purest natural source, containing more than 95% of cellulose and about 5% of other substances. More commonly, cellulose is associated with lignin and other substances (so-called hemicelluloses) in considerable quantities. The hemicelluloses are not forms of cellulose at all. They comprise a group of polysaccharides that remains associated with the cellulose after lignin has been removed. Depending on the species, wood contains on a dry basis about 40–55% cellulose, 15–35% lignin, and 25–40% hemicelluloses (Nevell and Zeronian, 1985). The plant cell wall is a composite of cellulose, lignin, and hemicelluloses, which provides strength, rigidity, and prevents the swelling of the cell. The plant cell wall is discussed in Section 10.3.

3.5 Lignin

Lignin is present in the cell walls of plants and contributes significantly to their compressive strength. Woods that are under compression have a significantly larger fraction of lignin. It is covalently bonded to hemicellulose.

3.6 Lipids

Lipids are naturally occurring molecules that perform important functions in organisms:

(a) they provide energy storage;
(b) they form the structural element of the cell membranes;
(c) they have important signaling functions.

The most important lipids are mono-, di-, and triglicerides, phospholipids, cholesterols, fat-soluble vitamins, sterols, and waxes. Figure 3.26 shows four important lipids. Lipids are characterized by hydrophobic and hydrophilic parts, and this has an important bearing on their function. The important phospholipid shown in Fig. 3.26 has a hydrophilic head on the right and a hydrophobic tail, consisting of the long molecules, on the left.

Phospholipids, in the presence of water, organize themselves into liposomes, micelles, or lipid bilayers. These configurations are formed to expose the hydrophilic heads to water and to shield the hydrophobic tails. Thus, these bilayers form the structural components of cells which contain the organelles, performing an infinity of functions. These vesicles have already been synthetically generated. This is shown in Fig. 3.27.

Triglycerides are an important component and are stored in the adipose tissue. Their breakdown (oxidation) generates significantly more energy (~9 kcal/g) than that of hydrocarbons and proteins (~4 kcal/g).

Lipids are also involved in the signaling among cells. G-protein coupled or nuclear receptors are responsible for these functions.

3.7 Formation of biopolymers

How are these biopolymers formed in organisms? This is a complex question with a complicated answer. Different proteins are produced by different methods, but the methodologies seem to have some parallel.

3.7.1 Collagen

We describe first the formation of collagen. The sequence is as follows:

(a) Transcription of mRNA in the cell nucleus (described in Chapter 4). There are 34 genes associated with collagen, each corresponding to a specific mRNA sequence. This leads to the formation of pre-pro-peptides.

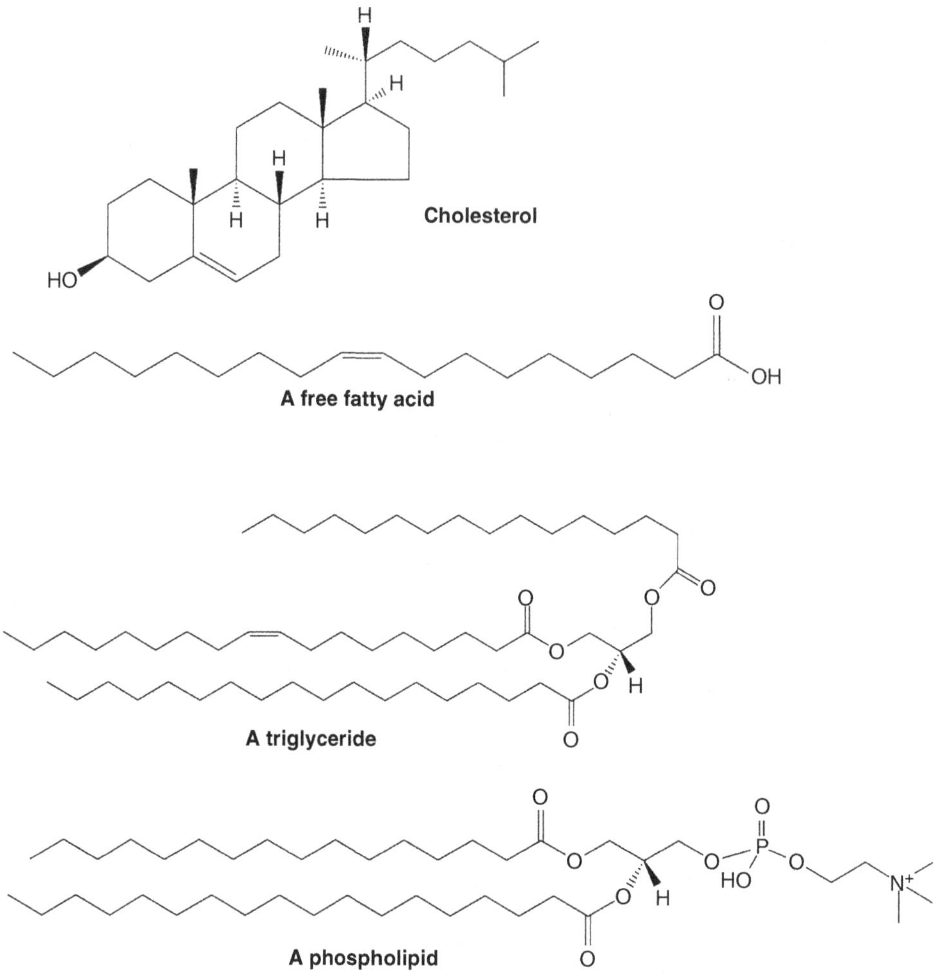

Cholesterol

A free fatty acid

A triglyceride

A phospholipid

Figure 3.26.

Common lipids: cholesterol; free fatty acid; triglycerides (note three tails); phospholipids (hydrophilic head and two hydrophobic tails). (From http://commons.wikimedia.org/wiki/File:Common_lipids_lmaps.png. Image by Lmaps at the English language Wikipedia. Licensed under CC-BY-SA-3.0, from Wikimedia Commons.)

(b) These pre-pro-peptides exit the nucleus into the cytoplasm, where procollagen is formed.
(c) Procollagen is encapsulated into a transport vesicle and taken to the Golgi apparatus.
(d) In the Golgi apparatus of the cell (see Chapter 4), oligosaccharides are added, and the chains are packed into a secretory vesicle and expelled from the cell.
(e) Outside the cell, enzymes act on procollagen, removing loose ends, leading to the formation of tropocollagen.

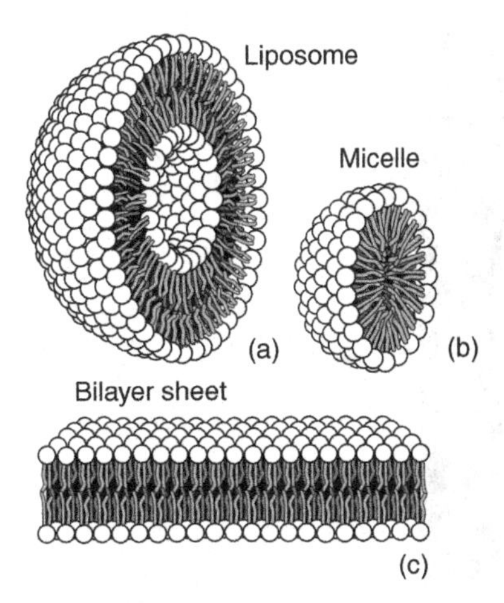

Figure 3.27.

Organization of phospholipids in an aqueous environment; the hydrophilic ends of lipids are represented by spheres and the two hydrophobic tails (fatty acyl chains) by wavy features: (a) liposome; (b) micelle; (c) lipid bilayer. (From http://commons.wikimedia.org/wiki/File:Liposome_cross_section.png. Image by Philcha, who released the image into the public domain.)

3.7.2 Keratin

The formation of keratin is somewhat different because it represents dead tissue. The outer layer of organisms consist of keratinocytes, cells that have undergone keratinization. These cells have a structural matrix of keratin and are almost waterproof, in contrast with cells inside the body, which exchange fluids with the environment. Figure 3.28 shows the keratin network in keratinocytes. Once these cells become cornified, the keratin forms in various ways, one of them being intermediate filaments.

3.7.3 Chitin

Chitin is synthesized from units of N-acetylglucosamine through a polymerization reaction catalyzed by an enzyme called chitin synthase. The degradation of chitin macromolecules is controlled by another enzyme named chitinase. Chitin synthases and chitinases play an important role in the chitin metabolism in arthropods.

Summary

- Biological materials can be considered as as either polymeric (soft) or polymer–mineral composites (hard).
- The hard component is primarily hydroxyapatite in vertebrates with an exoskeleton. In molluscs, arthropods, and reptile and bird eggs, it is calcium carbonate. In diatoms and sponge spicules, it is amorphous silicon.

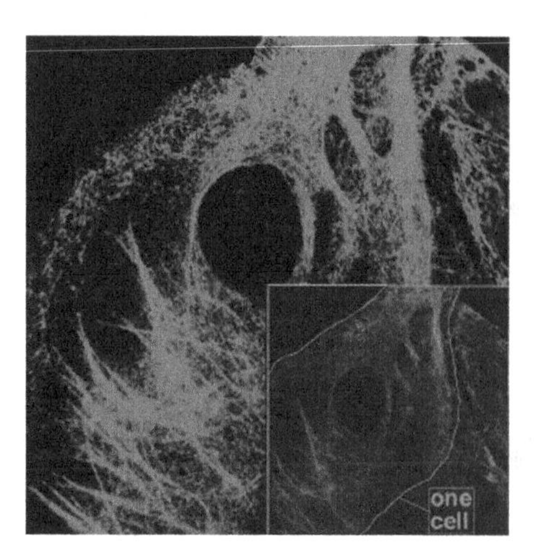

Figure 3.28.

Keratin fibers inside cell. (From http://en.wikipedia.org/wiki/File:KeratinF9.png. Image by Jamie Bush and John Schmidt. Licensed under the Creative Commons Attribution-Share Alike 3.0 Unported license.)

- The soft component comprises primarily collagen and elastin.
- There are also hard biopolymers, such as chitin (exoskeletons), cellulose (wood), and keratin (horn, hooves, nails, talons, beaks).
- The most common elements in the human body are, in order: oxygen (65 wt.%); carbon (18 wt.%); hydrogen (10 wt.%); nitrogen (3 wt.%); calcium (1.4 wt.%); phosphorus (1.1 wt.%).
- Water is of crucial importance in the mechanical properties of biological materials and especially of biopolymers. The tensile strength and elongation of collagen increase significantly with the degree of hydration.
- RNA and DNA are the molecular repositories of genetic information. DNA contains the nucleotides adenine (A), guanine (G), cytosine (C), and thymine (T). A bonds to T, C to G on separate chains of the double helix. In RNA, uranyl (U) replaces thymine (T).
- Amino acids form polypeptide chains, and these form proteins. They also form the polysaccharide chains, and these form chitin and cellulose. There are 20 amino acids, and they are the building blocks of bioploymers.
- In mammals, water is ~ 58–65 wt.% of the body; proteins are 15–17 wt.%; lipids (fat) 10–15 wt.%; minerals ~7 wt.%. DNA and RNA are only ~2 wt.%. There are thousands of different proteins.
- The most common proteins are collagen, keratin, elastin, resilin, fibroin, and abductin.
- The most common polysaccharides are chitin and cellulose.
- Collagen is the most important protein in the body; it has been called, due to its ubiquity, the "steel" of biological materials. It plays a key role in the mechanical strength of bone,

skin, ligaments, tendons, and arteries. It is characterized in high magnification by TEM, SEM, and AFM, by a 67 nm banding pattern similar to a coral snake.

- This banding pattern is the result of the structure, which is built up, hierarchically, from amino acid chains that form coils with a diameter of ~0.5 nm and are the polypeptide chains. Three polypeptides form a triple helix with ~1.5 nm diameter and a length of 300 nm. These, in turn, form fibrils with a diameter of 40–100 nm. These assemble into fibers with 1–10 μm diameter.
- The process of formation of collagen is as follows: the process starts in the cell nucleus, where pre-pro-peptides are formed, encapsulated in a vesicle, and transported to the Golgi apparatus. Procollagen is formed in the cytoplasm, encapsulated again, and expelled from the cell. Outside the cells, enzymes act on it and remove loose ends, turning it into tropocollagen.
- The process of formation of keratin is somewhat different because it represents dead tissue. An outer layer of keratinocytes, cells that have undergone keratinization, dies and forms keratin.
- In the same manner as polypeptides form proteins, polysaccharides form chitin, chitosan, and cellulose. Chitin exists in fungal and yeast cell walls, mollusc shells, and arthropods. Cellulose is the most abundant natural polymer.
- Lipids (fats) have three important functions: (i) they provide energy storage, 9 kcal/g, twice as much as hydrocarbons and proteins; (ii) they form the structural element of the cell membranes, the bilipid layer; (iii) they have signaling functions. The important phospholipids have hydrophobic tails and hydrophilic heads. The tails attract each other and form micelles, lipid bilayers, and liposomes.
- Muscles are composed primarily of the proteins actin and myosin. Muscles contract and stretch through the controlled gliding/grabbing of myosin with respect to the actin fibers. They assemble into the sarcomere units, which are actin–myosin arrays. These in turn form myofibrils. Bundles of myofibrils form muscle fibers, which in turn form fascicles.

Exercises

Exercise 3.1 Using the equation for the ellipsis, estimate the minor and major axes of a circular perforation having a diameter of 3 mm if the skin stretches to a maximum strain of 0.2 and 0.5 in the directions parallel to and perpendicular to the Langer lines, respectively. Make necessary assumptions.

Exercise 3.2 The stress–strain relationship for human skin is given in Fig. E3.2 for the directions parallel and perpendicular to the Langer lines. Using the following equation, calculate the parameters K_1, K_2, and ϵ_2 for the two cases:

$$\sigma = K_1\varepsilon^2 + H(K_2)(\varepsilon - \varepsilon_2),$$

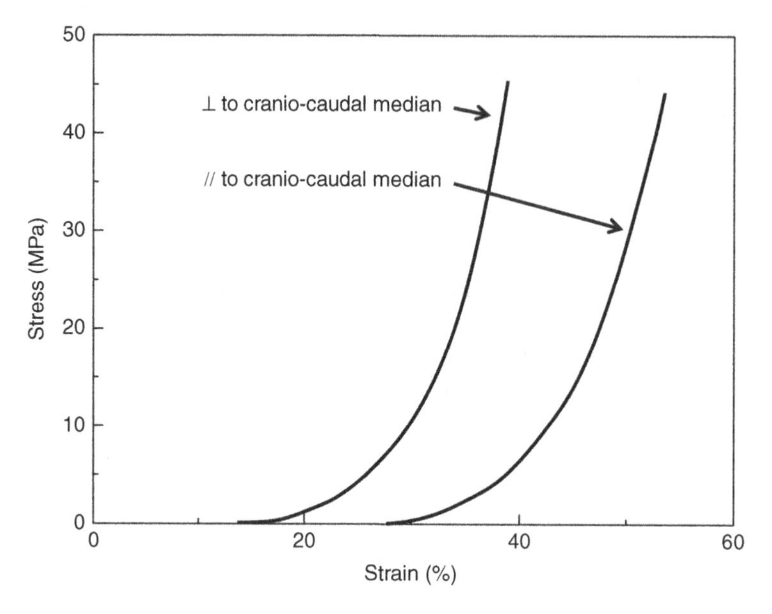

Figure E3.2.

Stress–strain curve for human skin.

where H is a Heaviside function. The equation has two terms: a quadratic one and a linear one (see more details in Chapter 9). Show the results graphically, comparing the experimental results and predictions.

Exercise 3.3 Fit the two curves found in Exercise 3.2 into the Fung equation (see Chapter 9, Eqn. (9.14)).

Exercise 3.4 Determine the minor axis of a crimped collagen structure if the wavelength is 100 μm and the maximum strain is 0.6.

Exercise 3.5 Assuming that the force required to break a hydrogen bond is equal to ~4 pN, calculate the stress required to open up the α-helix structure of the peptide chains in collagen. Hint: Do this by estimating the number of bonds per unit area in a plane.

Exercise 3.6 Read the paper Shergold *et al.* (2006).

(1) Derive Eqn. 10 from Eqn. 1.
(2) Can you express Eqn. 10 in terms of strain?
(3) Modify Eqn. 10 to incorporate the effect of strain rate. Use data from the paper.

Exercise 3.7 Collagen is one of the most important structural proteins in nature. What is the origin of the 67 nm periodicity? Explain the three stages of the J-curve behavior of collagen fiber under tensile loading.

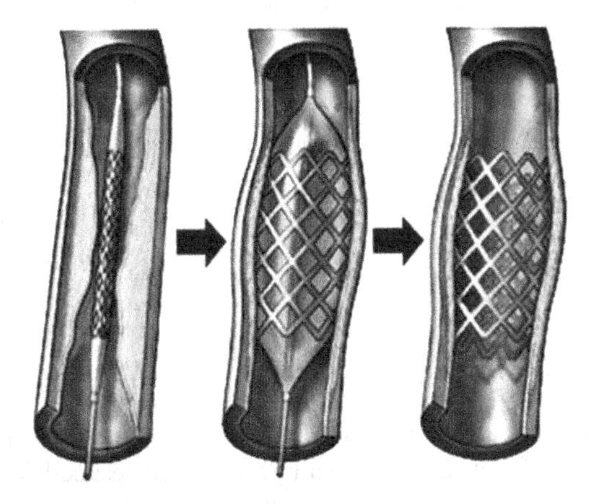

Figure E3.8.

Nylon micro-balloon used in the deployment of stent. (From Pruitt and Chakravartula (2011). Reprinted with the permission of Cambridge University Press.)

Exercise 3.8 A micro-balloon used for expanding a stent (see Fig. E.3.8) has a cylindrical shape inside the stent. It is made of nylon, with a diameter of 5 mm when completely expanded. The wall thickness is 30 μm. If the tensile strength of the nylon is 80 MPa, what is the maximum pressure that can be applied to the vessel?

Exercise 3.9 Collagen fibrils with a 200 nm diameter are subjected to tension. The fibrils are known to yield at a stress of 200 MPa (Shen *et al.*, 2008). What is the load that an Achilles tendon can take, if the diameter is 1.5 cm?

Exercise 3.10 Assuming that the collagen in the Achilles tendon is crimped and forms a zig-zag pattern with the tensile axis, what is the total strain that can be accommodated at complete stretching as a function of the initial zig-zag angle of the structure?

Exercise 3.11 A powerlifting competition took place in Socorro, New Mexico, that pitched a local strong man, *El Hulko*, a former Jamaican sprinter, *El Animal*, and a local professor, *El Loco*. Powerlifting is a brutal sport consisting of three simple elements: squat, bench, and dead lift.

(a) Present the biomechanical forces in the three events.
(b) Using measurements from extremely muscular individuals, obtain the theoretical maximum weights that can be lifted.
(c) Who got the third place (bronze medal) in the competition?

Exercise 3.12 A person is conducting squats. Calculate the force exerted by the quadriceps muscles to lift their weight. Make necessary assumptions using your own body. Each muscle fiber has the capacity to lift 300 μg and has a diameter of 5 μm. What is the maximum weight you can lift?

4 Cells

Introduction

The cell is one of the most complex machines known. Life is associated with cells, and there are researchers who feel that we will never be able to create a cell. The analogy with a Boeing 747 illustrates this. This airplane, one of the most complex machines ever built, has approximately six million parts (including rivets, screws, etc.), whereas the typical mammalian cell has 10 000 different proteins (~50 000 of each), for a total of 500 million parts (P. LeDuc, Carnegie-Mellon University, private communication). Added to these are the ATP molecules, of which there are approximately six billion per cell. Thus, the task of reproducing a cell, close to 100 Boeings, in a volume of approximately hundreds of cubic micrometers, is daunting.

How was the first cell formed? This is indeed the definition of life, because a cell can undergo mitosis, a process by which it generates another cell. There are different hypotheses, none of which have yet been substantiated.

- Astrobiology, a new field of study, explores the possibility of life in outer space; it is possible that life could have been transported to earth embedded in meteorites.
- The classic Miller–Urey experiments were conducted, in which Miller and Urey created a "primeval soup" and passed a high current through it, synthesizing more complex organic molecules. However, there is evidence that the early atmosphere was quite different. As an encouragement to graduate students, Miller was a rather testy one who asked Urey, then already a well-known professor, to set up such an apparatus. Reluctantly, Urey agreed and, to his surprise, the success came early and easily. This paper had far-reaching consequences. Alas, Miller never got a Nobel Prize for his experiment, but Urey did. Another warning for graduate students.
- The vent holes in the bottom of the ocean. It has been proposed these vent holes could have given rise to life.

Organisms are classified into prokaryotic and eukaryotic types. Prokaryotes are all unicellular and are divided into archaea and bacteria. Eukaryotes are the other organisms, including us, *homo sapiens*. Prokaryotic cells are simpler and smaller than eukaryotic cells, and they do not have a nucleus. The eukaryotic cells have diameters 10–100 μm, whereas the prokaryotic cells have diameters of 1–10 μm. Whereas bacteria are unicellular, humans contain 10^{13} cells. As a comparison, there are approximately 10^{10} humans on this planet. These cells control all the processes in our body.

Some cells can change shape or move, so that they can pursue foe and attack foreign invaders. Such is the case for macrophages, which chase hostile cells by moving through tissue.

4.1 Structure

Most biological cells are 1–100 μm in size, and they comprise many constituents. For instance, the red blood cell has a diameter of 8 μm. On the other side of the spectrum, neurons can have lengths of tens of centimeters, due to the long axon. Figure 4.1 shows (a) a red blood cell (erythrocyte), (b) a neuron, (c) osteoblasts, (d) keratinocytes, and

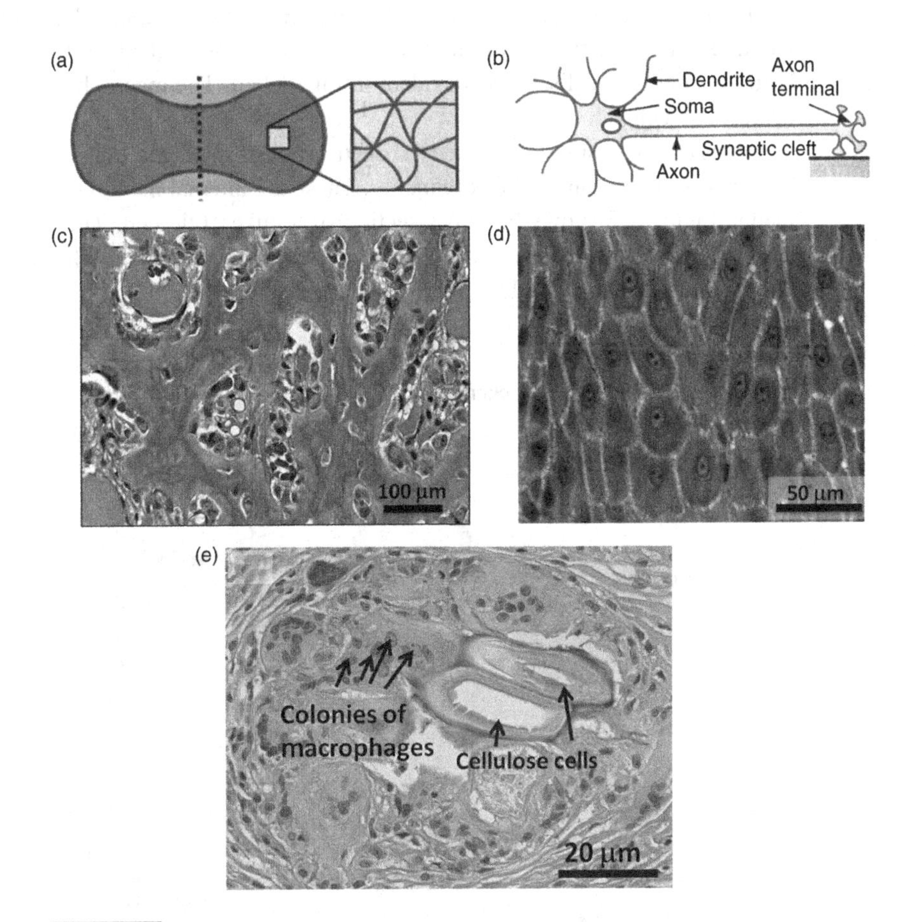

Figure 4.1.

Different cells. (a) Red blood cell (erythrocyte) with biconcave shape – note that it has no nucleus; spectrin network schematically shown. (b) Neuron (nerve cell) with long axon and dendrites. (From Boal (2012). Copyright © D. Boal 2012. Reprinted with the permission of Cambridge University Press.) (c) Osteoblasts inside bone; they remodel the bone by depositing hydroxyapatite. (d) Keratinocytes connected by a network of clamp-like features (white regions between cells). (e) Macrophages attacking foreign cells (in this case, cellulose cells from wood). (Parts (c), (d), and (e) used with kind permission of Dr. Sepi Mahooti.)

(e) macrophages surrounding cellulose cells. In Figs. 4.1(c), (d), and (e) the cell nuclei can be seen as dark circles in their center. The differences are clear.

The red blood cell has a characteristic biconcave shape. It does not possess a nucleus, which leaves it before it enters the circulatory system. A red blood cell has a mean life of 28 days in the human body. The neuron can have an axon of several centimeters, being part of the brain and nervous system. Osteoblasts remodel bone. They are responsible for constructing it, whereas osteoclasts dissolve it. Thus, bone is being constantly remodeled through the combined actions of osteoblasts and osteoclasts. Keratinocytes are located at the epidermis. Upon dying, they give rise to keratin. In the human skin, keratin forms microscopic follicles that continuously detach themselves from the body. We often complain that dogs and cats shed, but we do it as effectively. Figure 4.1(e) shows a fascinating event. In the center are macrophages surrounding cellulose cells, part of a wood chip that penetrated accidentally into a body. Plant cells tend to be considerably larger than animal cells. This is evident in Fig. 4.1(e). The macrophages attack foreign objects and cells.

Just as an organism is composed of organs, such as the heart, lungs, skin, etc., the cell contains "little" organs, called organelles. These are internal membrane-bound compartments. Figure 4.2 shows the cross-section of a eukaryotic cell. There is a distinction between animal and plant cells, the latter having the organelle chloroplast, key for photosynthesis, and vacuoles.

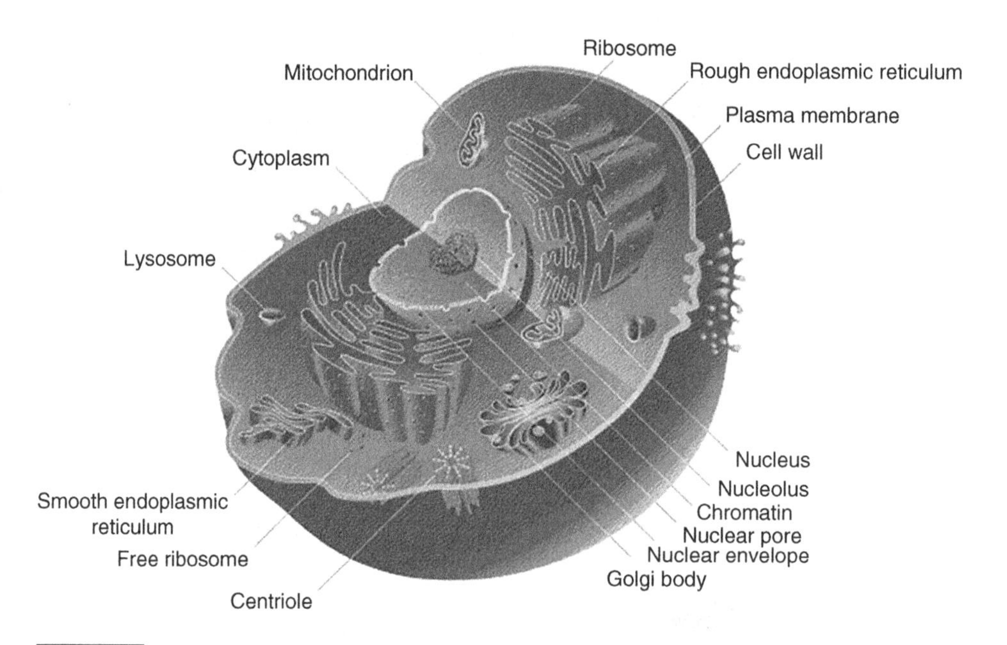

Figure 4.2.

Section of eukaryotic cell showing the principal components. (From http://commons.wikimedia.org/wiki/File:Eukaryotic_Cell_(animal).jpg. Image by Mediran. Licensed under CC-BY-SA-3.0, via Wikimedia Commons.)

The principal organelles shown in Fig. 4.2, together with their primary function, are as follows.

- Nucleus: contains most of the DNA and RNA.
- Golgi apparatus: a large organelle that processes proteins for inside and outside the cell.
- Cytoplasm: the fluid outside the nucleus and inside the cell.
- Ribosomes: RNA and protein complex required for protein production in cells.
- Mitochondria: the power generators of the cell that release the energy in the form of ATP.
- Vacuoles: store food and waste; they are particularly large in plants, where they are space-filling organelles.
- Lysosomes: digest organelles; they break down waste products by transforming them into simpler organic compounds.
- Endoplastic reticulum: transport network for certain molecules.
- Plasma membrane (cell wall): this is a lipid bilayer, as described in Chapter 3. However, it contains special "windows" that allow ion exchange.
- Cylia: hair-like features in the outside of the cell that are mechano-, chemo-, and thermosensitive. These tubules act like cellular antennae, sensing the environment.

Additionally, some cells have a flagellum, a tail-like fixture that enables it to swim. A well-known example is the spermatozoid. As an example of evolutionary convergence, tadpoles are similar to spermatozoids and move in water by an analogous mechanism.

In plant cells, the cell walls are more rigid and contain cellulose, lignin, and hemicelluloses. The plant cell walls have dimensions of ~0.2–10 μm. The lignin is hydrophobic and binds to the polysaccharides (cellulose and hemicelluloses), providing compressive strength to the cell walls.

The interior of the cell includes: a liquid phase (cytosol); a nucleus; the cytoskeleton, consisting of networks of microtubules, actin, and intermediate filaments; organelles of different sizes and shapes; other proteins (see, e.g., Bao and Suresh (2003)). The exterior of the cell is covered by a phospholipid bilayer membrane reinforced with protein molecules (see Fig. 4.3(a)).

Box 4.1 Cells and biomaterials

In 1665, Robert Hooke coined the term "cells" after observing them in an optical microscope. This was done in analogy to monks' cells. Robert Hooke is well known to materials scientists through another seminal contribution, namely Hooke's law of elasticity.

There are between 200 and 400 types of cells in the human body and they are prominently involved in the interaction with biomaterials. The average human comprises 75% water, and of the total 40 liters, 25 liters are inside cells. In many cases, the insertion of implants into the body requires cutting the skin and causing trauma to tissues and organs. The response of tissue to injury depends on the location and

Box 4.1 (cont.)

nature of the latter and on the biocompatibility of the implant. The term "biocompatibility" was proposed by Homsy *et al.* (1970) and has gained wide acceptance. The principal cells involved in the inflammatory and healing processes are:

- osteoblasts – immature bone-producing cells;
- chondroblasts – immature collagen (cartilage) producing cells;
- endothelial cells – lining of the cavities of heart and blood vessels;
- erythrocytes – red blood cells;
- granulocytes – blood cells containing specific granules;
- epithelial cells – surface of skin;
- macrophages ("big eaters," in Greek) – engulf and then digest cellular debris and pathogens;
- mesenchymal stem cells – can differentiate into a variety of cell types, including osteoblasts (bone cells), chondrocytes (cartilage cells), and adipocytes (fat cells);
- fibroblasts – produce extracellular matrix and collagen.

The process of healing has been divided into three or four stages.

Hemostasis Capillaries constrict to stop bleeding.

Inflammation Capillaries dilate. The walls enable leakage of plasma out of the capillaries, resulting in swelling. Simultaneously, the lymphatic system around the wound becomes plugged. This stage is characterized by redness, heat, and pain, in addition to the swelling. In some cases, when the tissue injury is accompanied by irritants such as debris or is penetrated by bacteria, the inflammation may prolong and lead to tissue destruction that is produced by collagenase, which is a protective enzyme that can digest collagen. Collagenase is released by granulocytes. Macrophages will also become present, eating bacteria.

Proliferative phase After two or three days, fibroblasts move to the area and start the regeneration process. The fibroblasts produce the collagen that will rebuild the tissue. Revascularization also occurs in this phase. The wound contracts to a fraction of its initial size. Epithelization takes place during this stage.

Maturation and remodeling Collagen aligns itself perpendicular to the injury, and the cells that are no longer needed in the healing process undergo apoptosis (death).

The body tends to reject any foreign object. We are all familiar with wood splinters and thorns embedded in the body that are gradually expelled. If the body cannot expel something it will tend to encapsulate it. Macrophages proliferate around the foreign object in an attempt to digest it. In the case of implants, one has to classify them into three groups: metals, ceramics, and polymers.

Metals tend to release ions into the body, and this is an irritant. Macrophages, also called foreign body giant cells, proliferate around the metallic implant. Of all implant materials, titanium is the most biocompatible because it forms a very stable layer of TiO_2, a ceramic, and this layer is inert. Other metals, such as Co–Cr, stainless steels, and other corrosion-resistant alloys also form a passivation layer on the surface, becoming quite inert. The capsule that forms around the implant is composed of connective tissue.

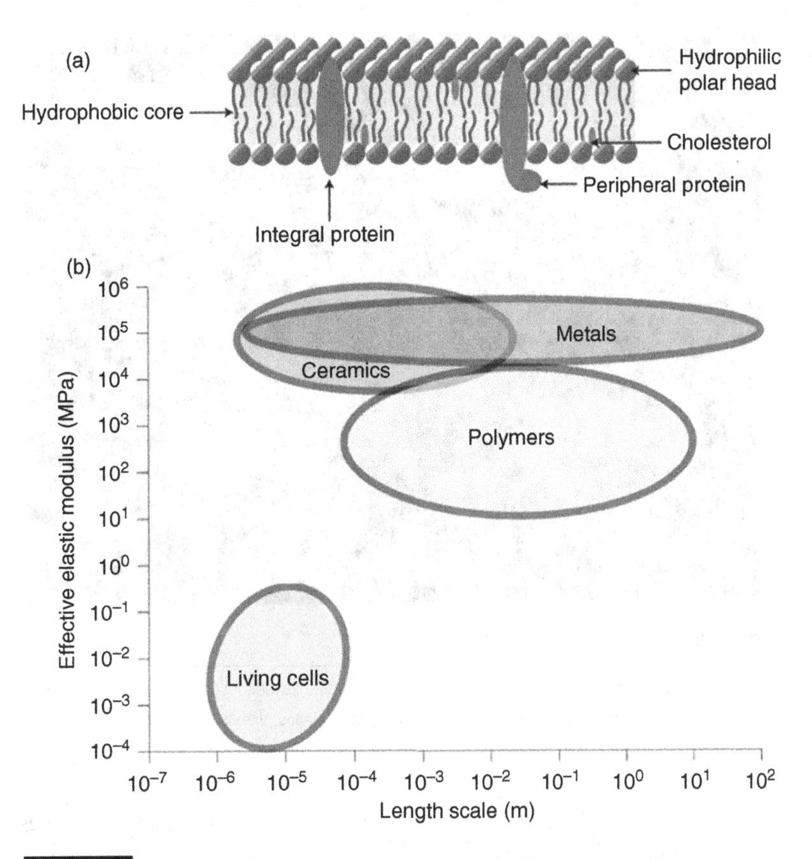

Figure 4.3.

(a) Cell wall structure consisting of lipid bilayer with hydrophobic tails toward the inside and transmembrane proteins embedded in the wall. (b) Elastic properties of cells compared to those of synthetic materials. (Reprinted by permission from Macmillan Publishers Ltd.: *Nature Materials* (Bao and Suresh, 2003), copyright 2003.).

4.1.1 Cytoskeleton

There is an interior structure of the cell that maintains its integrity. This scaffolding or "skeleton" is contained within the cytoplasm. By making changes in the cytoskeleton the cell can adapt, reorganize, and change shape. This structure can protect the cell, enable cellular motion, and play important roles in intracellular transport, cell signaling, gene expression, and cellular division. It also provides "tracks" with its protein filaments for the transport of organelles, molecules, vesicles, etc. Figure 4.4(a) shows a cell with the entire cytoskeleton. The three principal classes of protein fibers forming the cytoskeleton are:

- microfilament actin (5–9 nm diameter) – double helix (Fig. 4.4(b));
- microtubules (25 nm diameter) – long cylindrical protofilaments of tubulin subunits (Fig. 4.4(c));

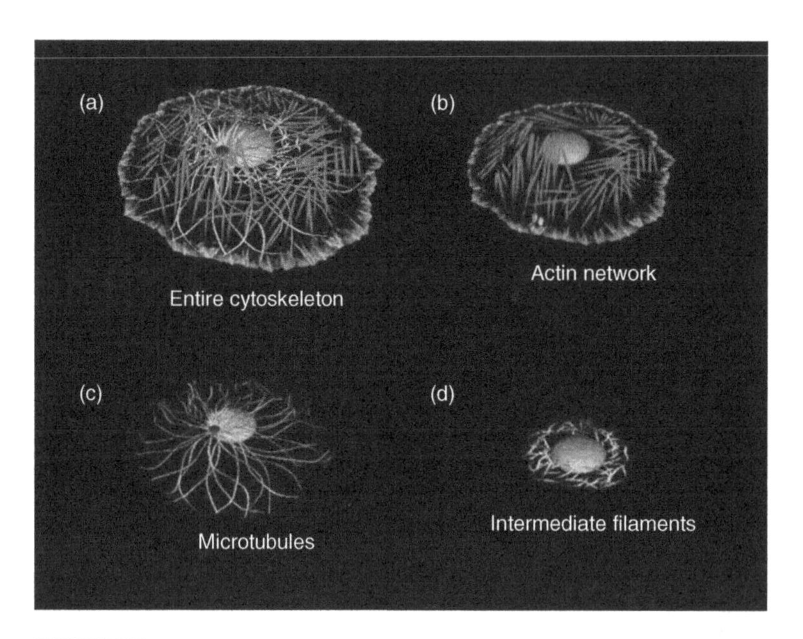

Figure 4.4.

Structural components of cell. (a) Entire cytoskeleton; (b) actin network; (c) microtubules; (d) intermediate filaments. (Figure courtesy of Vic Small, IMBA, Austrian Academy of Sciences, Vienna; with permission.)

- intermediate filaments (8–10 nm diameter) – two parallel helices/dimers forming tetramers (Fig. 4.4(d)).

A simple analog for the cystoskeleton is a circus tent in which some components are loaded in compression (the poles) and some in tension (the ropes). The microtubules have the largest diameter and have a greater ability to resist compressive forces (5–10 pN) exerted on the cell. The other two, actin and intermediate filaments, buckle under compression, as will be shown in Section 4.5, where we will introduce and derive the flexure formula and Euler's buckling formula, originally derived for large structures but applicable at the cell level. Therefore, they are thought to operate in tension. The microtubules act as circus tent poles, whereas the other two act as ropes. This type of structure is known as a "tensegrity" structure.

Figure 4.5(a) shows the cell centrosome, a central part which acts as the nucleation site for hundreds of microtubules that radiate toward the periphery. The microtubules are continuously either growing or shrinking in response to the cell's determination to change shape. This goes beyond the environmental stresses, but is also dependent on the nature of the cell.

The actin filaments are observed to be mostly present in the moving front of a cell. This can be seen in Fig. 4.5(b). These actin filaments do not move with the advancing front, but are stationary, in a "treadmill" action. The front of the filament grows by the addition of molecules, while the back dissolves and the filament is stationary with respect to the

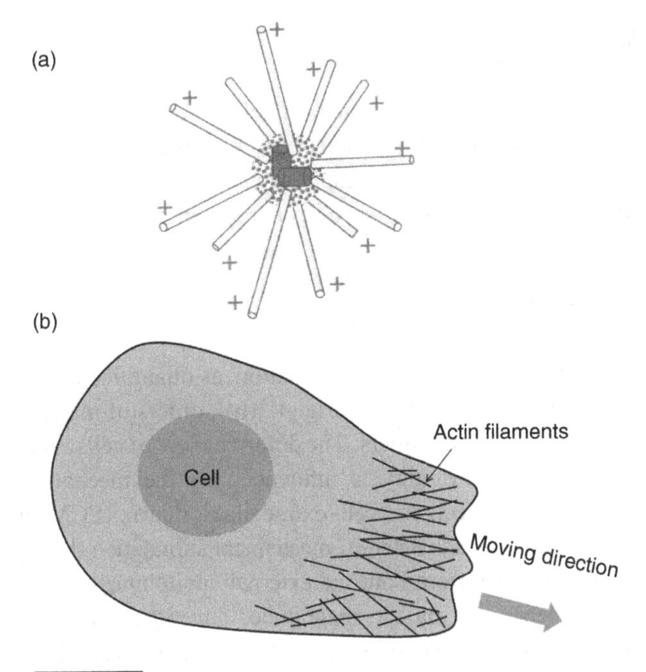

Figure 4.5.

Network of (a) microtubules radiating from the cell nucleus (from Boal (2012), copyright © D. Boal 2012, reprinted with the permission of Cambridge University Press) and (b) actin filaments concentrated in the forward region of a cell.

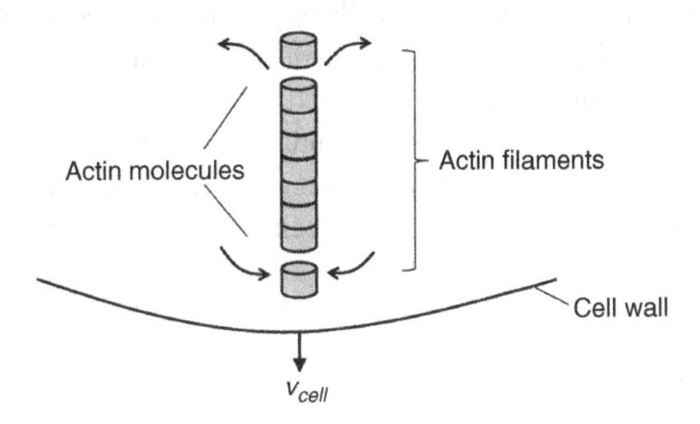

Figure 4.6.

Actin filament in the front part of a cell moving by actin molecules diffusing from the back to the front in such a manner that the molecule is stationary with respect to the substrate.

substrate. The velocity at which this diffusion occurs to and from the actin fiber is a function of the concentration gradient and the diffusion rate of the actin molecules. Figure 4.6 shows a schematic representation of the process. Actin molecules diffuse to the front of the filament, whereas at the back the inverse process takes place. In fibroblasts, the velocity is 0.05–0.1 μm/s.

4.1.2 Multifunctionality

Cells constitute the basic structural building blocks of living organisms and perform a variety of functions: self-replication, protection from the environment, acquisition of nutrients, movement, communication, catabolism of extrinsic molecules, degradation and renewal of aged intrinsic molecules, and energy generation.

4.2 Mechanical properties

The resistance of single cells to elastic deformation, as quantified by an effective elastic modulus, ranges from 10^2 to 10^5 Pa (see Fig. 4.3(b)), orders of magnitude smaller than those of metals, ceramics, and polymers. The deformability of cells is determined largely by the cytoskeleton, whose rigidity is influenced by the mechanical and chemical environments including cell–cell and cell–extracellular matrix (ECM) interactions.

Living cells are constantly subjected to mechanical stimulation throughout life. These stresses and strains can arise from both the external environmental and internal physiological conditions. Depending on the magnitude, direction, and distribution of these mechanical stimuli, cells can respond in a variety of ways. For example, within the body, fluid shear of endothelial cells activates hormone release and intracellular calcium signaling and stiffening the cells by inducing a rearrangement of the cytoskeleton. The mechanical compression of cells, such as chondrocytes, is known to modulate proteoglycan synthesis. Furthermore, the tensile stretching of cell substrate can alter both cell motility and orientation. As a result, to understand how cells mechanically respond to physical loads is an important first step in the further investigation of how the transmission and distribution of these mechanical signals are eventually converted to biological and chemical responses. Mechanical stimuli can also determine the differentiation of cells. For instance, a stem cell can become either an osteocyte (bone cell) or a chondrocyte (cartilage cell). This has been demonstrated by Gibson *et al.* (2010) and is being used to generate bone–cartilage junctions for joint repair. This is discussed in Chapter 12.

4.3 Mechanical testing

With the recent advances in molecular and cell biology, biophysics, nanotechnology, and materials science, several innovative and experimental techniques and equipments have been developed to probe the structural and mechanical properties of biostructures from the micro- down to the nano-scale. These techniques can not only perform direct mechanical probing and manipulation of single cells and biomolecules, but also allow such tests to be conducted under physiological conditions. Lim, Zhou, and Quek (2006b) summarized biomechanical testing and imaging instruments/techniques employed to study biological structures ranging from single biomolecules, to cells, to tissues, as shown in Fig. 4.7. Figure 4.7(a) shows some of the experimental techniques used for

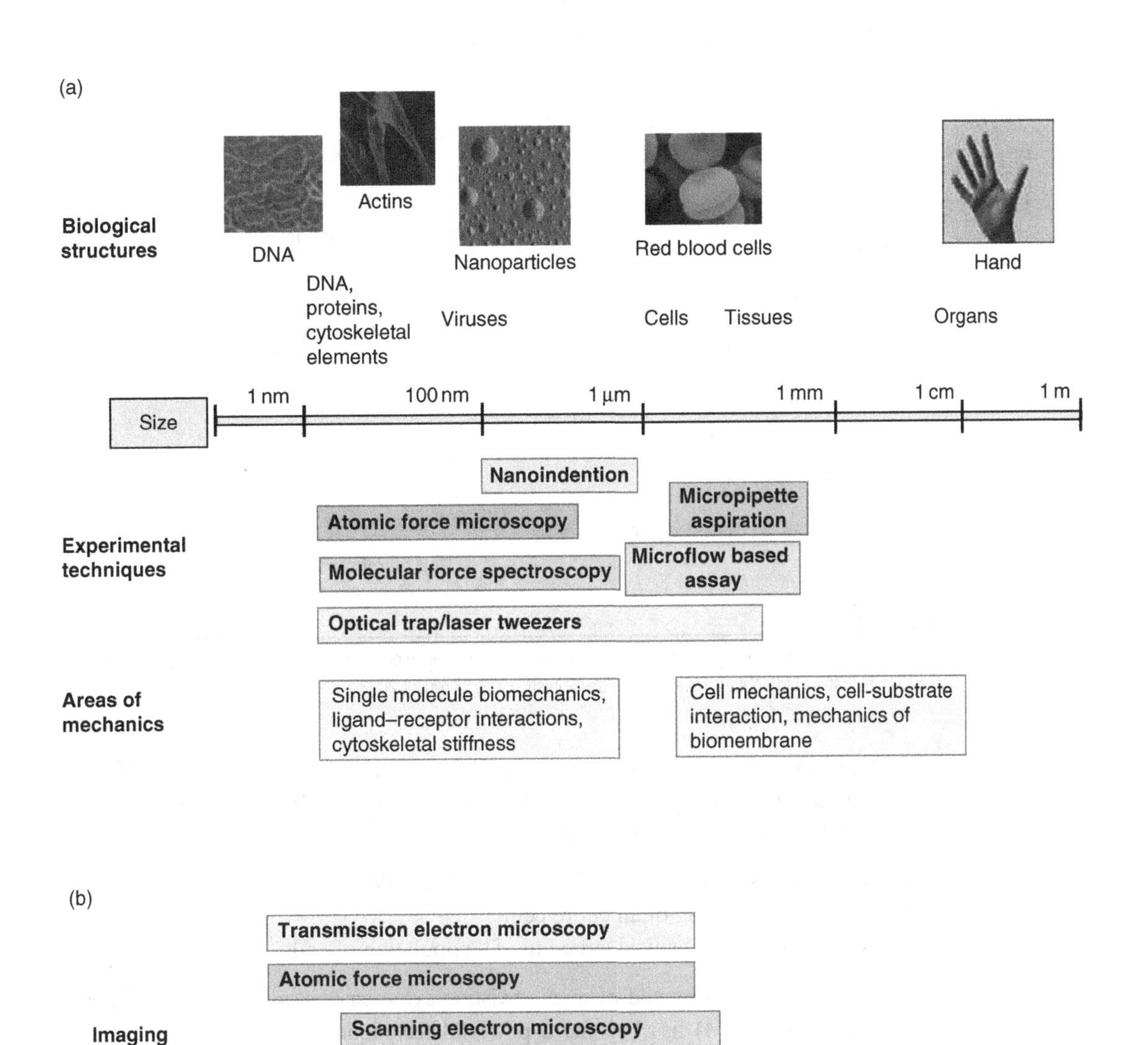

Figures 4.7

(a) Experimental techniques for conducting mechanical tests in single-cell and single-molecule biomechanics. (b) Imaging techniques that can be used to observe physical, biological, and biochemical changes occurring in biological structures during biomechanical tests of cells and biomolecules. (Reprinted from Lim *et al.* (2006a), with permission from Elsevier.)

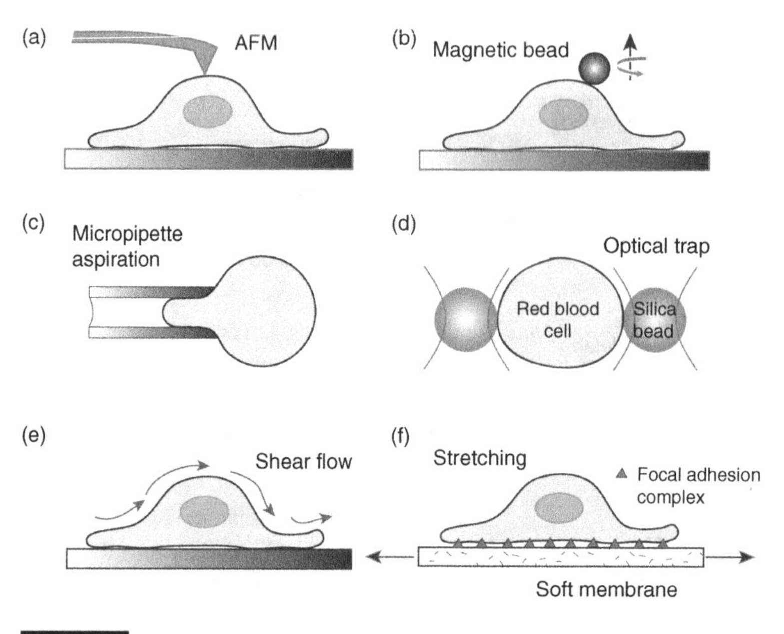

Figure 4.8.

Schematic representation of the three types of experimental technique used to probe living cells. (Reprinted by permission from Macmillan Publishers Ltd.: *Nature Materials* (Bao and Suresh, 2003), copyright 2003.)

conducting biomechanical tests in single cells and single molecules, and Fig. 4.7(b) shows the imaging techniques available for use in such tests. Bao and Suresh (2003) further classified the experimental techniques into three types:

type A local probes, in which a portion of the cell is deformed;
type B mechanical loading of an entire cell;
type C simultaneous mechanical stressing of a population of cells.

Atomic force microscopy and magnetic twisting cytometry belong to type A. Atomic force microscopy (AFM) has become a powerful technique for both imaging surface morphology and sensing force. A sharp tip mounted at the end of a flexible cantilever directly contacts the sample surface and generates a local deformation. The interaction between the tip and the sample surface induces a deflection of the cantilever, which can be calibrated to estimate the applied force. Figure 4.8(a) represents the deformation and unfolding of protein by AFM. Magnetic twisting cytometry (MTC) involves magnetic beads with functionalized surfaces. Magnetic beads are attached to a cell, and a magnetic field imposes a twisting moment on the beads, deforming a portion of the cell (Fig. 4.8(b)). The deformation generated by the magnetic beads is analyzed, and thus the mechanical properties of the cell can be obtained.

Micropipette aspiration (MA) and optical tweezers are common methods in type B processes. A suction pressure is applied through a micropipette to deform a single cell (Fig. 4.8(c)). The change in shape of the cell is recorded by video microscopy. By measuring the elongation into the pipette as a result of the suction pressure, the mechanical properties can be evaluated. Micropipette aspiration is widely used to study the mechanical response of blood cells. Optical tweezers, or laser traps, use a laser to control the particles in a medium. When a laser beam shines on a dielectric particle with a higher refractive index than that of the medium, the gradient force is higher than the scattering force. As a result, a net force pushes the particle toward the focal point of the laser. In order to deform a single cell, a laser trap is used with two microbeads attached to the opposite ends of a cell (Fig. 4.8(d)).

Shear-flow methods (Fig. 4.8(e)) and stretching devices (Fig. 4.8(f)) are type C methods used to study the mechanical response of an entire population of 10^2–10^4 cells. Shear-flow experiments are conducted with either a cone-and-plate viscometer or a parallel-plate flow chamber. The shear stress applied to cells for both cases can be quantified. Different stretching devices (uniaxial, biaxial, and pressure controlled) have been developed to deform cells. Cells are cultured on a thin polymer substrate, which is coated with extracellular matrix (ECM) molecules for cell adhesion. The substrate is then mechanically deformed, while maintaining the cell's viability in vitro. The effects of mechanical loading on cell morphology, phenotype, and injury can be examined.

Developments and advances in nanotechnology and biophysical techniques enable a better understanding of the pathophysiology and pathogenesis of human diseases that manifest structural and mechanical property changes. Recent works by Lim (2006) and Lee and Lim (2007) highlight some of the biomechanics research carried out on several types of diseases, such as malaria, sickle cell anemia, and cancer.

The human red blood cell, which has a biconcave shape with an average diameter of about 8 μm, is highly deformable. During the circulation in the narrow capillaries (about 3 μm in diameter), it undergoes severe deformation and transforms from a biconcave to a bullet shape. The cell fully recovers its original shape after it flows through these small capillaries. Blood cells can harden and lose their elasticity through diseases such as malaria and sickle cell anemia. The change in mechanical properties causes serious impairment of blood flow and results in severe anemia, coma, or even death. Lim, Suresh and co-workers (Dao, Lim, and Suresh, 2003; Lim *et al.*, 2004; Suresh *et al.*, 2005; Dao, Li, and Suresh, 2006) studied the mechanical response of living cells. Ingenious experimental techniques developed in collaboration with Lim established the load extension response of a single cell. From this, the researchers extracted the elastic response of the cell and how it is altered by disease. Figures 4.9(a) and (b) show the conceptual experiment (Dao *et al.*, 2006). Two glass beads (with diameters of approximately 5 μm) attach themselves to the ends of the cell by capillary forces. A laser beam is used to trap one of the beads. This is an optical (laser tweezer) trapping device. The glass slide under the cell is moved and the load is measured. The displacement of the cell is recorded. The actual photographs of the cells as they are being stretched are shown in Figs. 4.9(c) and (d). The information thus obtained is compared with computational simulations using both a finite element mesh and an approach based on actual spectrin fibrous molecules, developed by the MIT group (Fig. 4.9(e)).

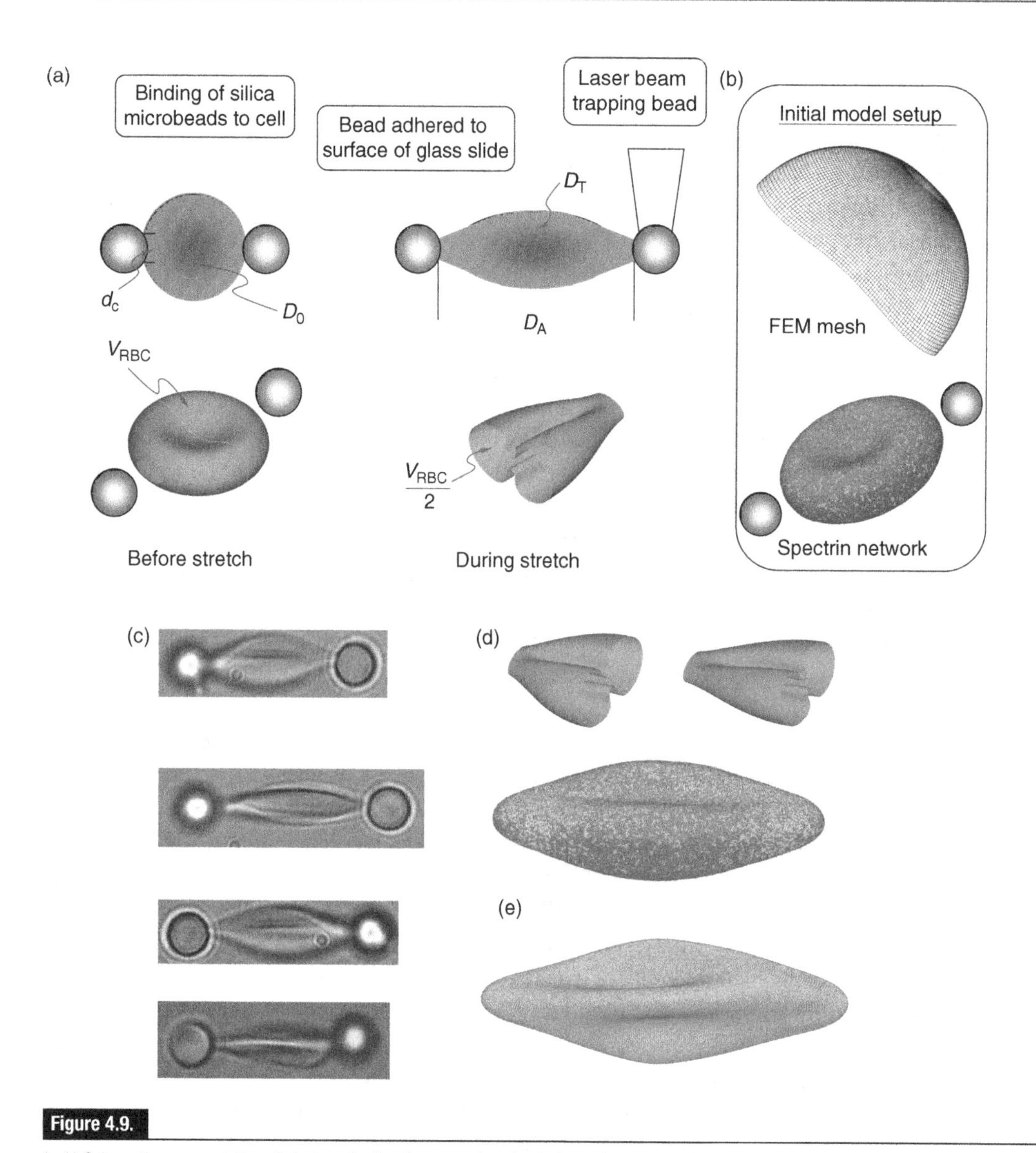

Figure 4.9.

(a, b) Schematic representation of single-cell extension apparatus; (c, d) photomicrographs of red blood cells in extended configuration; (e) spectrin fiber simulation of process; initial unstretched configuration shown. (Reprinted from Dao *et al.* (2006), with permission from Elsevier.)

Figure 4.10 shows the molecular-based and continuum models of red blood cell membrane (Dao *et al.*, 2006). Lim *et al.* (2006a) have published a comprehensive review on mechanical models for living cells. The spectrin network, forming the trellis of triangular units shown in Fig. 4.10(a), supports the lipid bilayer that constitutes the cell membrane. Figure 4.10(a) shows how transmembrane proteins can ensure transfer of matter into and out of the cell. The strength of the cell membrance is ensured by the

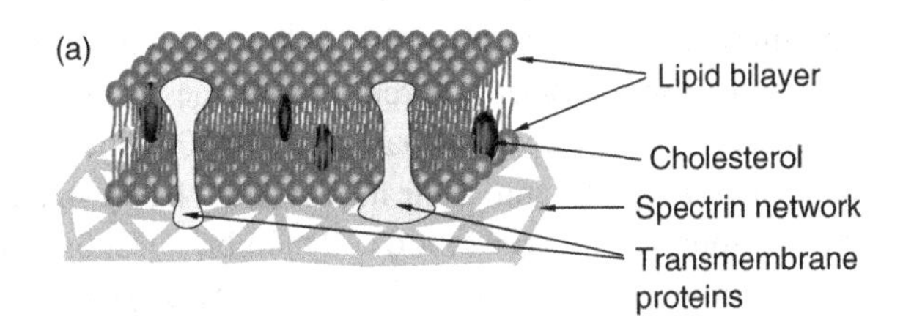

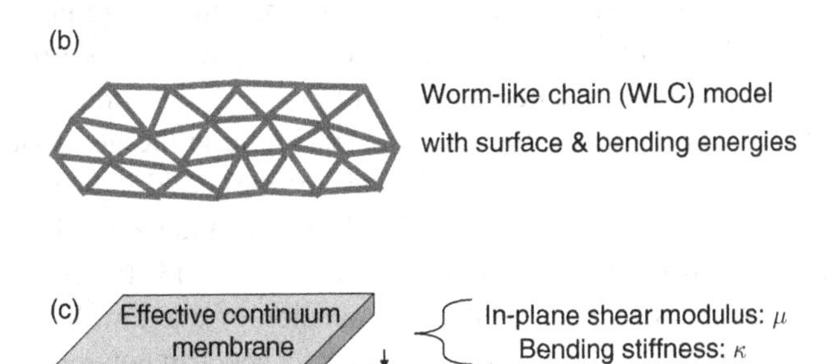

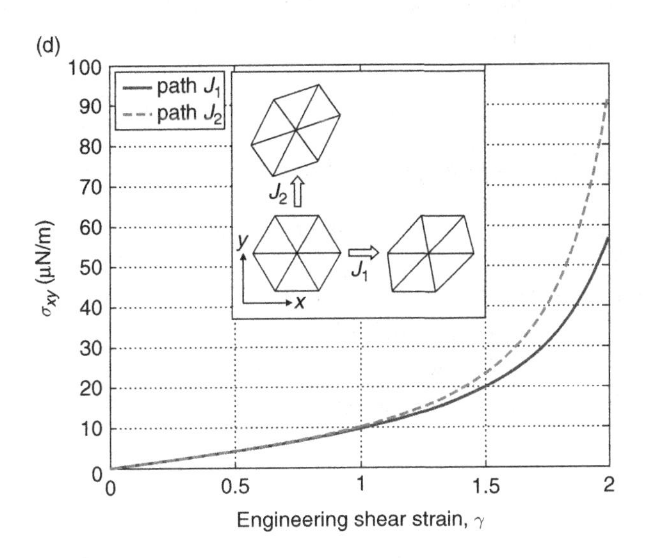

Figure 4.10.

Molecular-based and continuum models of red blood cell membrane. (a) Schematic drawing of the red blood cell membrane structure (not to scale). (b) Molecular-based model: worm-like chain (WLC) model. (c) Effective continuum membrane. (d) Large-strain response of "single-crystal" worm-like chain membrane (persistence length $p = 8.5$ nm; equilibrium length $L_0 = 87$ nm; maximum extension length (contour length) $L_{max} = 238$ nm) for two area-preserving shear paths J_1 and J_2, parameters defined in Section 9.1. (Reprinted from Dao *et al.* (2006), with permission from Elsevier.)

spectrin network. One of the models for this network is to consider each spectrin as a flexible "cable." The spectrin can be modeled by the worm-like chain (WLC) model (see Fig. 4.10(b)), which is presented in greater detail in Chapter 9. Indeed, the WLC model, as the name implies, assumes that the polymeric chain is a worm. The straightening of the worm requires a tensile force. The stiffness of the force–extension curve (its slope) increases with extension. Dao *et al.* (2006) incorporated bending stiffness and in-plane shear modulus into their model (Fig. 4.10(c)), and were able to predict the mechanical response of the red blood cell membrane, shown in Fig. 4.10(d). The response of the membrane has the characteristic J shape of many "soft" biological materials.

The detailed nature of the spectrin network is actually more complex that that shown in Fig. 4.10. It has been said that it is a "tensegrity" structure. Tensegrity structures are formed by a combination of members in tension (cables) and compression (pillars). A prime example of a tensegrity structure in architecture is the geodesic dome named after Buckminster Fuller, its creator. In the case of the red blood cell, it consists of a rigid central actin "mast" to which six spectrin fibers are attached at different locations. The structure is actually tridimensional because the central actin mast has a length of approximately 55 nm, whereas the lateral dimension of one unit hexagon is approximately 100 nm. Figure 4.11(a) shows a picture of a spectrin network; it is a top view, and the central masts are represented by open circles. The schematic arrangement is shown in Fig. 4.11(b); once the structure is stretched, the actin "mast" pivots (Zhu and Asaro, 2008) (Fig. 4.11(c)).

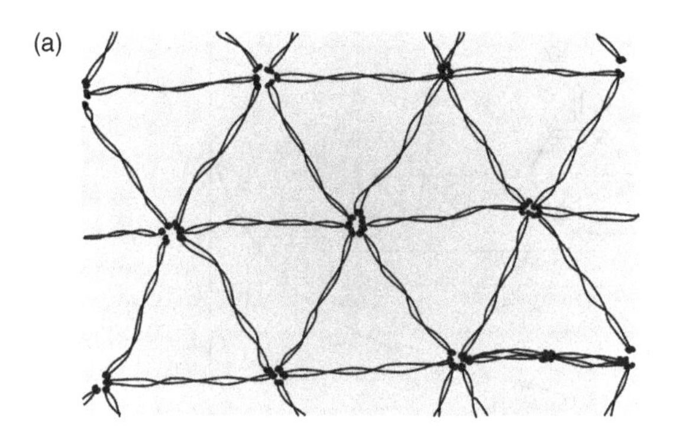

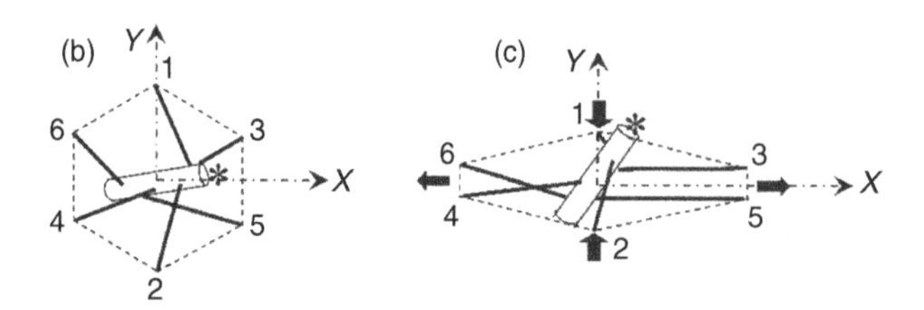

Figure 4.11.

(a) Spectrin network on the surface membrane of cell. (From Dr. W. Yang, with permission.) (b) Tridimensional view of spectrin network with central actin "mast" to which six spectrin filaments are attached. (c) Stretched configuration (in direction *OX*) showing rotation of central "mast." (Reprinted from Zhu and Asaro (2008), with permission from Elsevier.)

4.4 Cell motility, locomotion, and adhesion

Most cells have the ability to move. Some cells are specialized for locomotion, such as amebae and spermatozoids. In most cells, locomotion is usually repressed. However, this ability can be activated in wounds and oncogenesis. The mechanism of ameboid cell motility (crawling or gliding) involves the actin cytoskeleton. Actin filaments themselves are likely to be involved in the force-generating mechanisms. The major components of cell motility are shown in Fig. 4.12 (Mitchison and Cramer, 1996). Protrusion is the forward movement of the membrane at the front of the cell, as it separates itself from the substrate. Adhesion of the advancing portion of the cell is required for protrusion to be converted into movement along the substrate. Traction brings the cell nucleus forward. The last step in locomotion comprises two mechanistically distinct processes: de-adhesion and tail retraction.

The protrusion at the front of motile cells requires dense arrays of actin filaments. It seems that these filaments are organized with their barbed ends oriented preferentially in the direction of protrusion. The web of actin filaments is organized as an orthogonal cross-weave between two sets of filaments oriented at approximately 45° to the direction of protrusion.

Two mechanisms for generating protrusive force have been proposed.

(a) The action of motor proteins to drive protrusion: myosin I molecules contribute to this, just as in the case of muscles (Chapter 3); myosin I moves toward the actin filament barbed ends.
(b) Actin polymerization itself produces force (see Fig. 4.6). Polymerization of pure actin inside a lipid vesicle can deform the cell membrane; polymerization of other proteins also produces a membrane-deforming force.

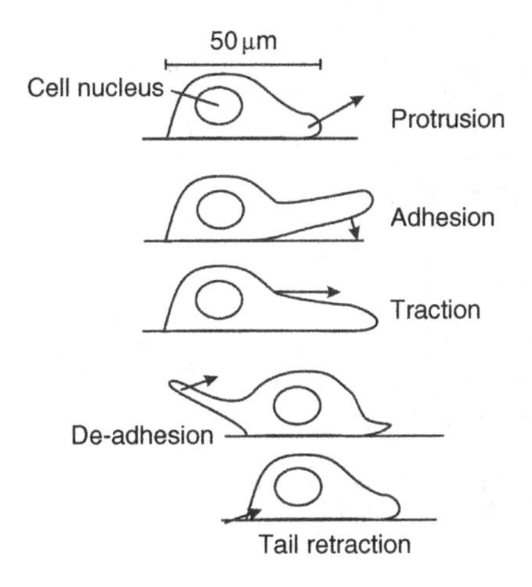

Figure 4.12.

Sequence motility showing elements involved in the locomotion of cells: protrusion, adhesion, traction, de-adhesion, and tail retraction. (Reprinted from Mitchison and Cramer (1996), with permission from Elsevier.)

Force production requires an energy source, and this derives from the chemical energy of nucleotide hydrolysis in the ATPase motor. In the polymer model, thermally driven motion creates the movement.

Cell motility is an important property, especially as it is involved in diseases such as cancer metastasis. The cancer cells separate themselves from the primary tumor, pass through the extracellular matrix in the body, and enter the circulatory system. Then, after being arrested, they penetrate through the blood vessel walls (extravasation) at a secondary location. This sequence is shown in Fig. 4.13. Indeed, the application of mechanics and materials science to the study of cancer cells will hopefully help in elucidating the basic mechanism. Suresh (2007) reviews the subject and presents a framework for this new field of investigation, which will incorporate cancer cell mechanics, motility, deformability, differentiation, and neoplastic transformation.

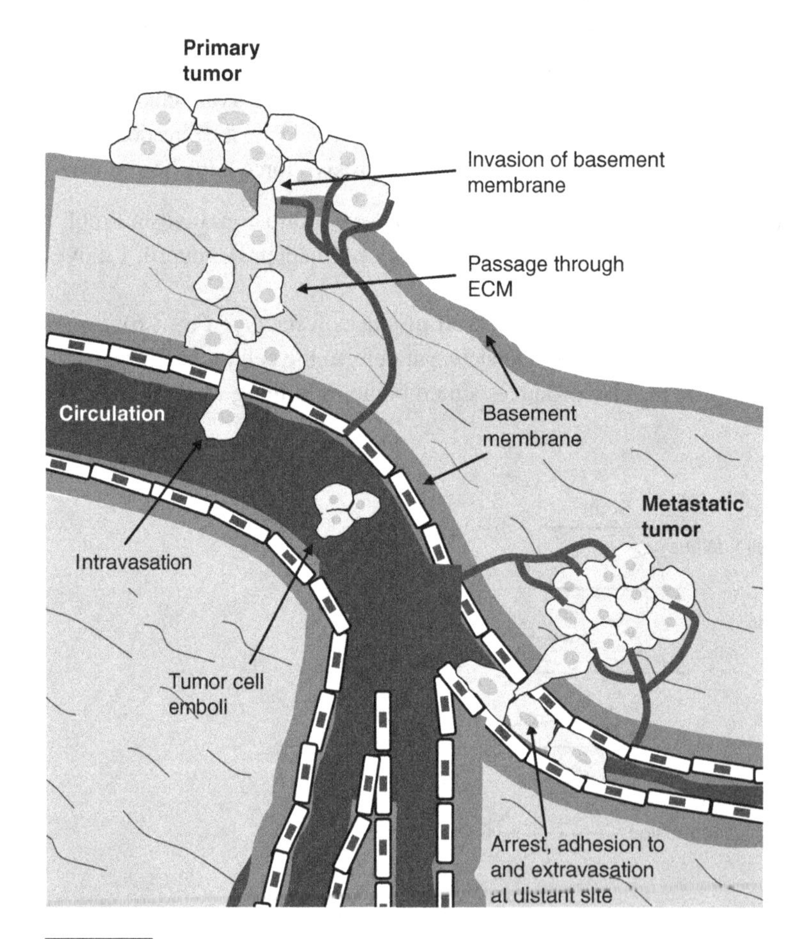

Figure 4.13.

Schematic diagram showing different stages of cancer metastasis: cancer cells penetrate through the ECM and into the blood stream, then extravasate into the secondary site, where metastasis takes place. (Reprinted from Lee and Lim (2007), with permission from Elsevier.)

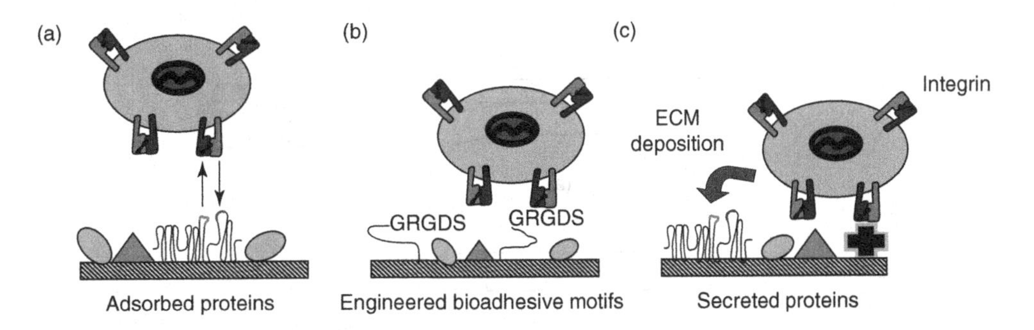

Figure 4.14.

Mechanisms controlling cell adhesion to biomaterials. (Reprinted from Garcia (2005), with permission from Elsevier.)

In the adhesion of cells to the extracellular matrix (ECM) or to other cells the protein integrin plays a key role. Integrin-mediated adhesion is a complex process. Integrins play an important role by providing anchorage that ensures cell survival and migration (motility). Integrins are important in how cells react to implants, in tissue engineering, and in how cells behave in cell arrays and biotechnology cell culture. Figure 4.14 (Garcia, 2005) shows schematics of cells attaching themselves to synthetic biomaterials whose surfaces were modified by different means:

(a) proteins adsorbed from solution, such as blood, plasma, or serum;
(b) ligands engineered at the surface, such as RGDs (arginine-glycine-aspartic acid);
(c) ligands deposited by cells, e.g. collagen deposition.

4.5 Flexure and compressive resistance of hollow and solid cylinders: application to microtubules

We will determine the deflection undergone by a beam when subjected to lateral and axial stresses. The first is the flexure formula and the second is the Euler equation. These are two famous equations from the mechanics of materials, and we will apply them, in this chapter, to microfilaments and microtubules in cells. However, these equations are general and will be utilized again in Chapter 10 (foams).

If you are an engineering student, you have probably derived these equations before. We can see the power of these equations by making a simple "desk" experiment: take a ruler (hopefully, you still own one, in this new age of computers!), and fix one end by applying a load (with your hand), leaving an overhang on the side. With your other hand, push the end of the ruler down. It will deflect. Now turn the ruler through 90°, with the sharp edge pointing down. Repeat the operation. Aha! The deflection is practically zero. Hence, the orientation of the cross-section with respect to loading is important. Next, set the ruler vertically on the table and push down on it, with the force parallel to the longitudinal axis. The ruler will first resist and then suddenly bow, with the deflection occurring in the direction perpendicular to the largest dimension in the cross-section.

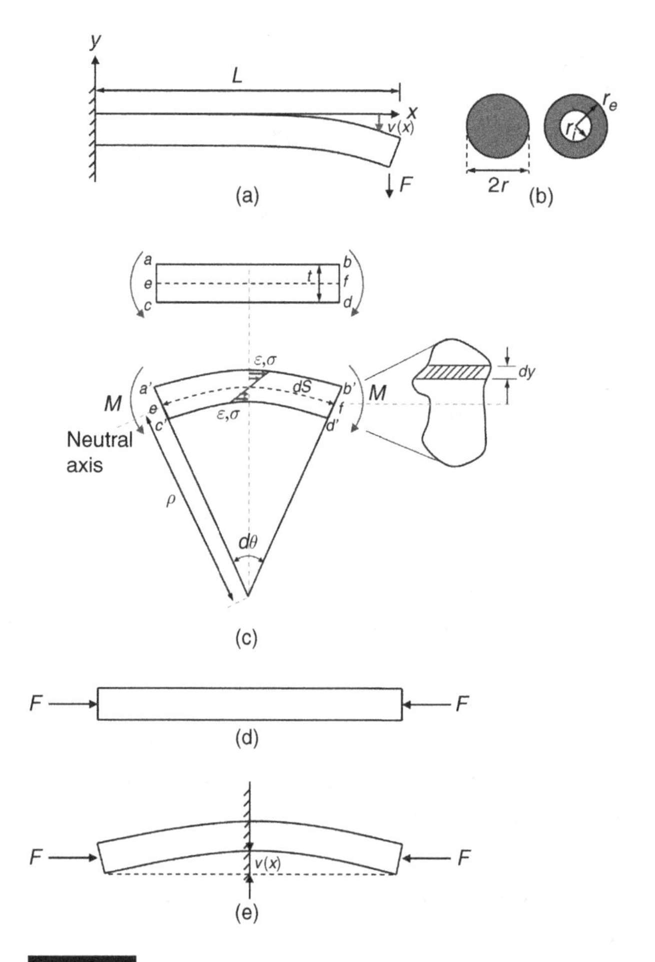

Figure 4.15.

(a) Deflection of a beam under a load perpendicular to the axis. (b) Two cross-sections: solid and hollow cylinders. (c) Calculation of internal stresses and strains for a beam subjected to bending moment *M*. (d) Beam with loading parallel to axis (column). (d) Deflection of column and splitting into two symmetrical parts; the Euler equation predicts the stress at which instability occurs.

These experiments illustrate the principles that we will present in this section. We understand these concepts intuitively, but the quantitative calculations are needed.

Figure 4.15 shows a beam subjected to a force F perpendicular to its axis. The deflection is $v(x)$. The force F creates a bending moment:

$$M = F(L - x). \tag{4.1}$$

Two cross-sections are shown in Fig. 4.15(b) representing solid and hollow beams. We know that the hollow beam will experience a smaller deflection if the cross-sectional area is the same as that of the solid beam. Let us prove it.

Figure 4.15(c) shows a section of the beam segment *abcd* subjected to a bending moment M, which we assume to have a general cross-section shown on the right-hand side of the figure. It is distorted to $a'b'c'd'$. This bending is opposed by internal stresses. We assume that plane sections remain plane. Thus, the strains vary linearly with distance from the neutral axis (the axis along which stresses are zero, passing through the centroid of area; thus, *ef* is constant). They are tensile on top and compressive at the bottom.

The applied moment M is balanced by the moment due to the resisting forces F_x:

$$\sum M = 0. \tag{4.2}$$

The forces can be expressed in terms of the strains, which vary linearly with distance from the neutral plane:

$$F_x = \sigma_x\, dA = -E\varepsilon_x\, dA \tag{4.3}$$

since $\varepsilon_x = -ky$ and $\sigma_x/\varepsilon_x = E$.

Taking the summation of the moments clockwise yields

$$M = -\int_A -Ekyy\, dA, \tag{4.4}$$

or

$$M = Ek\int_A y^2 dA. \tag{4.5}$$

We define now I_z, the second moment of inertia as follows:

$$I_z = \int_A y^2 dA. \tag{4.6}$$

Thus,

$$\sigma_x = -\frac{My}{I_z}.$$

The maximum tensile stress is reached when $y = y_{max}$. This value is made equal to $-c$. Thus,

$$\sigma_{max} = \frac{Mc}{I_z}. \tag{4.7}$$

Moment of inertia equations are derived in standard calculus classes. The values are given in Table 4.1 for different shapes.

We can now obtain the flexure formula. We consider the segment *ef*, which is constant, equal to dS. The angle $d\theta$ is shown at the bottom of Fig. 4.15(c). The curvature of the beam is given by

Table 4.1. Second moments of inertia for different cross-sections

Rectangle	$I = \frac{bh^3}{12}$
Triangle	$I = \frac{bh^3}{36}$
Circle (radius c)	$I = \frac{\pi c^4}{4}$
Hollow cylinder (radius c; wall thickness t)	$I = \frac{\pi c^3 t}{4}$
Ellipsis (major axis $2a$; minor axis $2b$)	$I = \frac{\pi b a^3}{4}$ or $I = \frac{\pi a b^3}{4}$

$$k = \frac{1}{\rho} = \frac{d\theta}{dS}. \tag{4.8}$$

Since $\rho\, d\theta = dS$, the strain can be expressed as

$$\varepsilon_x = \frac{du}{dS},$$

where du is the difference in length: $du = c'd' - ef$, i.e.

$$du = (\rho - y)d\theta - \rho d\theta = -y d\theta. \tag{4.9}$$

Hence,

$$\frac{du}{dS} = -y\frac{d\theta}{dS} = -\frac{y}{\rho}, \tag{4.10}$$

and the maximum compressive strain is given by

$$\varepsilon_x = -\frac{y}{\rho}. \tag{4.11}$$

The definition of the radius of curvature is given by

$$\frac{1}{\rho} = \frac{d^2v/dx^2}{\left[1 + \left(\frac{dv}{dx}\right)^2\right]^{3/2}} \cong \frac{d^2v}{dx^2}. \tag{4.12}$$

The approximation applies for small curvatures.

From Eqn. (4.12),

$$\frac{d^2v}{dx^2} = -\frac{\varepsilon_x}{y}. \tag{4.13}$$

From Eqn. (4.6),

$$\varepsilon_x = \frac{\sigma}{E} = -\frac{My}{IE}. \tag{4.14}$$

Thus,

$$\frac{d^2v}{dx^2} = \frac{M}{EI} = \frac{(L-x)F}{EI}. \tag{4.15}$$

Integrating yields

$$\frac{dv}{dx} = \frac{1}{EI}\int (L-x)F dx = \frac{F}{EI}\left(Lx - \frac{x^2}{2}\right) + C_1. \tag{4.16}$$

Apply the boundary conditions

$$\frac{dv}{dx} = 0 \text{ at } x = 0, \; C_1 = 0. \tag{4.17}$$

Integrating again yields

$$v = \frac{F}{EI}\left(\frac{Lx^2}{2} - \frac{x^3}{3}\right) + C_2. \tag{4.18}$$

Again, $v = 0$ at $x = 0$ and $C_2 = 0$, so

$$v = \frac{F}{EI}\left(\frac{Lx^2}{2} - \frac{x^3}{3}\right). \tag{4.19}$$

It is clear that the deflection is inversely proportional to I, the moment of inertia. By moving the mass as far as possible from the neutral axis, in the loading direction, we can minimize deflection. The moment of inertia is a very important geometric property of beams, columns, and shafts. Table 4.1 gives the values for different geometries. It can be seen that hollow tubes, sandwich structures, and foams effectively move the mass out, minimizing deflection.

We can now move on to the buckling equation. For this, we consider a compressive force F aligned with the beam. There are other configurations with different boundary conditions, but we will only derive the simplest case. Figure 4.15(d) shows this configuration. By dividing the beam into two symmetrical parts, as shown in Fig. 4.15(e), we have a geometry similar to the one in Fig. 4.15(c), except for the direction of the force. We apply the differential equation for the deflection again:

$$\frac{d^2v}{dx^2} = \frac{M}{EI}. \tag{4.20}$$

The moment is now due to the deflection of the beam, v, as shown in Fig. 4.15(e):

$$M = -Fv. \tag{4.21}$$

Thus,

$$\frac{d^2v}{dx^2} - \frac{Fv}{EI} = 0. \tag{4.22}$$

This is an ordinary differential equation of the form

$$\frac{d^2v}{dx^2} - \lambda^2 v = 0. \tag{4.23}$$

The following solution is given in standard texts:

$$v = A \sin \lambda x + B \cos \lambda x. \tag{4.24}$$

The boundary conditions are shown in Fig. 4.15(e), i.e. $v(0) = 0$ and $v(L) = 0$. Thus,

$$\sin \lambda L = 0, \ \lambda L = 0, \pi, 2\pi, \ldots,$$

$$\frac{F}{EI} = \lambda^2 = \frac{\pi^2}{L^2}, \frac{4\pi^2}{L^2}, \ldots \tag{4.25}$$

The Euler equation has the following form:

$$F = \frac{n^2\pi^2 EI}{L^2}. \tag{4.26}$$

The maximum value of the load F at buckling with a wavelength of $2L$ is reached for $n = 1$. There are other buckling modes, for $n = 2$, $n = 3$, etc.

Example 4.1 Determine the lateral deflections undergone by: (a) a 10 μm actin filament with a diameter of 12 nm; (b) a microtubule of the same length with an external diameter of 25 nm and a wall thickness of 1.5 nm. The force is 1 pN and is perpendicular to the long axis; $E = 1.5$ GPa.

Solution First, we calculate the moments of inertia:

$$I_{AF} = \frac{\pi c^4}{4} = 1015\,\text{nm}^4;$$

$$I_{MT} = \frac{\pi c^3 t}{4} = 2300\,\text{nm}^4.$$

The maximum deflections are given, for $x = L$ as follows:

$$v = \frac{F}{EI}\left(\frac{L^3}{2} - \frac{L^3}{3}\right) = \frac{FL^3}{6EI}v,$$

$$v_{AF} = \frac{10 \times 10^{-12} \times 10^3 \times 10^{-18}}{6 \times 1.5 \times 10^9 \times 1015 \times 10^{-9\times4}} = 1.07\,\mu\text{m},$$

$$v_{MT} = \frac{10 \times 10^{-12} \times 10^3 \times 10^{-18}}{6 \times 1.5 \times 10^9 \times 2300 \times 10^{-9\times4}} = 0.48\,\mu\text{m}.$$

Example 4.2 Calculate the buckling loads for the actin filament and microtubule from Example 4.1.

Solution Now the force is along the axis. We use the Euler equation for $n = 1$:

$$F = \frac{\pi^2 EI}{L^2} = \frac{1.5 \times 10^9}{10^2 \times 10^{-12}} \pi^2 I,$$
$$F_{AF} = 14.78 \times 10^{19} \times 1015 \times 10^{-9\times4} = 0.15\,\text{pN},$$
$$F_{MT} = 14.78 \times 10^{19} \times 2300 \times 10^{-9\times4} = 0.34\,\text{pN}.$$

The difference between the two cases is two fold, although the cross-sectional areas are almost the same:

$$A_{AF} = 113\,\text{nm}^2,$$
$$A_{MT} = 118\,\text{nm}^2.$$

4.6 From cells to organisms

A vast and complex ocean resides between an individual cell (itself complex beyond our apprehension) and organisms. It is being navigated by current researchers, and the challenges will consume this and future centuries. The hierarchy follows the sequence: cells, tissues, organs, organ systems, and organisms. We may extend this further, since societies form multi-organism systems.

There are four classes of tissues: connective, epithelial, muscle, and nervous. Organs comprise several types of tissue enclosed by epithelial tissue forming a "capsule" for the organ. Examples of organs are heart, liver, skin, brain, and intestines. They organize themselves into systems (or we classify them!), such as respiratory, digestive, and circulatory systems.

A vast number of cells perform a multitude of functions within organisms. We hereby give a few examples.

- Epithelial cells: form the boundaries of organs. They control the exchange of matter with the environment.
- Fibroblastic cells: produce the extracellular matrix and secrete collagen.
- Muscle cells: both skeletal and cardiac muscle are composed of these cells.
- Neurons: 10^{11} nerve cells comprise the human nervous system. Their length varies greatly and is dictated by the axon.
- Macrophages: fight infection.
- Red blood cells (erythrocytes): carry energy throughout the organism. The nucleus is extracted before they are launched into the circulatory system. The typical human has 10^{12} cells. They are the favorite drink of Count Dracula.
- White blood cells (leukocytes): fight infections and foreign objects in the body.

Summary

- The cell is probably the most complex machine known; it contains thousands of different proteins, for a total of approximately 500 million. Additionally, there are approximately six billion ATP (adinose triphosphate) molecules per cell. A typical human has 10^{13} cells! They control all the processes in our body.
- Most cells are 1–100 μm in size, although an ostrich egg has a diameter of ~200 mm.
- Cells have a cytoplasm, which includes the nucleus. The cytoplasm comprises cytosol, the gel-like substance enclosed within the cell membrane, and the organelles, including (in eukaryote cells) the nucleus. The cytoplasm contains a complex array of organelles or membrane-bound units that perform different functions: the Golgi apparatus, ribosomes, mitochondrium, lyposomes, cytosol, endoplastic reticulum, cell wall, cylia. Some cells have a flagellum, which enables them to swim.
- Red blood cells (erythrocytes) have a biconcave shape and a diameter of approximately 8 μm. Before entering the blood stream, the nucleus and organelles escape through the bilipid membrane.
- The membrane in animal cells is a bilipid layer, with the heads sticking out and the hydrophobic tails on the inside.
- The structural components of cells (cytoskeleton) consist mainly of three types of fibrils: actin fibers (5–9 nm diameter), microtubules (~25 nm), and intermediate filaments (~8–10 nm).
- Whereas the tubules can resist compressive forces on the order of ~5 pN in cells, the actin fibers and intermediate filaments are primarily tension members. One can visualize the structure of the cell as comprising members in tension and compression. This is the definition of a tensegrity structure, first implemented in architecture by Buckminster Fuller.
- Plants have cell walls that are much more rigid than animal membranes. These cell walls contain cellulose, which has a high tensile strength, and lignin, which gives it compressive strength.
- In red blood cells, the lipid bilayer is supported by a network of tetramers, proteins that form a triangular trellis and resist the tension due to the internal pressure.
- Three principal mechanical testing methods have been developed for cells: atomic force microscopy and magnetic twisting cytometry; micropipette aspiration; and shear-flow and stretching devices.
- Some cells can move: the basic mechanism involves protrusion (of the front), adhesion (of the front part), traction (advance of nucleus), de-adhesion (of back), and tail retraction. Each sequence advances the cell by a discrete amount.
- Cells organize themselves into tissues, which also contain the extracellular matrix produced by the cells. These in turn assemble into organs. The organs form organ systems and these form organisms.

- There are also simpler organisms, e.g. diatoms and bacteria, in which the above hierarchy is greatly simplified.
- At yet higher levels, we have families, organizations, and nations.

Exercises

Exercise 4.1 Assuming an average cell diameter of 15 μm, calculate the fraction of the human body that is extracellular material. Make necessary assumptions and justify them. Obtain the number of cells from the web.

Exercise 4.2 Using the spectrin network, calculate the changes in length undergone by the three sides of the trellis if the cell is stretched in uniaxial strain by 20%. Assume that each segment has an initial length of 100 nm.

Exercise 4.3 Calculate the pressure required to increase the volume of a red blood cell by 20%. Assume that all the strength is due to a spectrin network, that the actin "masts" stay perpendicular to the surface, and that the cell is spherical. Given: diameter of unit = 100 nm; response of spectrin is linear; for a stretch ratio of 3, a force of 20 pN has to be applied.

Exercise 4.4 Provide at least five methods to test the mechanical properties of cell(s) and explain the mechanisms.

(a) Determine the lateral deflections undergone by an 8 μm long actin filament with a diameter of 10 nm. The load is perpendicular to the axis of the filament and is equal to 7 pN.
(b) Find the deflection undergone by a microtubule having an external diameter of 25 nm if the cross-sectional area is the same as that of the actin filament.

Given: $E = 1.5$ GPa.

Exercise 4.5 Calculate the buckling loads for the two cases in Exercise 4.4 and explain the differences between them.

Exercise 4.6 What is the structural advantage for a microtubule to be hollow? Determine the flexural rigidity ratio, F/v, for the case of a microtubule with outer diameter 28 nm and inner diameter 12 nm, and compare it with that of a microfilament (solid cylinder) with the same cross-sectional area.

Exercise 4.7 Determine the pressure inside a cell at which a lipid bilayer will crack using the Griffith equation (explained in Chapter 6). The crack has a length of 20 nm and the thickness of the lipid bilayer is 4 nm. Given: $E = 1$ GPa, $\gamma_s = 5 \times 10^{-2}$ J/m^2.

Exercise 4.8 Determine the lateral deflections undergone by: (a) a 5 μm actin filament with a diameter of 10 nm; (b) a microtubule of the same length with an external diameter of 20 nm and a wall thickness of 1.5 nm. Given: the force is 0.2 pN and is perpendicular to the long axis; $E = 1.5$ GPa.

Exercise 4.9 Determine the lateral deflections undergone by an 8 nm diameter actin filament and by a microtubule with an external diameter of 15 nm and wall thickness of 1.1 nm, if the applied force is 2 pN. Given: $E = 1.3$ GPa.

Exercise 4.10 Calculate the buckling loads for the two cases in Exercise 4.9. The compressive loads are now aligned with the axis of the filament and microtubule.

5 Biomineralization

Introduction

Although there are over 80 minerals present in biological systems, the most important are hydroxyapatite (HAP) (mammals and fishes), calcium carbonates (shells, arthropods, corals), and silica (diatoms, sponges). These minerals seem to have evolved from approximately 560 million years ago.

Minerals are essential for providing compressive strength to biological systems, whereas biopolymers are primarily responsible for tensile strength. The combination of minerals and biopolymers leads to the formation of biological materials with mechanical properties tailored in terms of hardness, toughness, and anisotropy. The formation of minerals involves nucleation and growth, both mediated by biological components. The organic matrix mediates nucleation in many ways: by providing nucleation sites and by controlling the polymorphs. The growth is also mediated by organic compounds, and illustrative examples are given in Chapters 6 and 7. For instance, the rapid direction of growth for aragonite is the *c*-direction, and long needles are formed. In nacre, this growth is regulated by the periodic deposition of organic layers. In bone, the HAP crystals nucleate in the interstices of collagen fibrils and growth is also regulated: they reach sizes on the order of nanometers: 40–60 nm long, 20–30 nm wide, and platelets of a few nanometers thickness are formed in such a fashion.

5.1 Nucleation

Nucleation of minerals takes place, often, from a supersaturated solution. In nonideal solutions, the activities of the components are not equal to their mole fractions. Based on Raoult's law, an activity coefficient is defined for a component A^+ as follows:

$$f_A = \frac{a_A}{[A^+]}, \tag{5.1}$$

where a_A denotes the activity and $[A^+]$ is the concentration (mole fraction). The parameter that establishes the critical value for nucleation is the solubility product constant:

$$K_{SP} = a_A a_B = [A^+][B^-] f_A f_B. \tag{5.2}$$

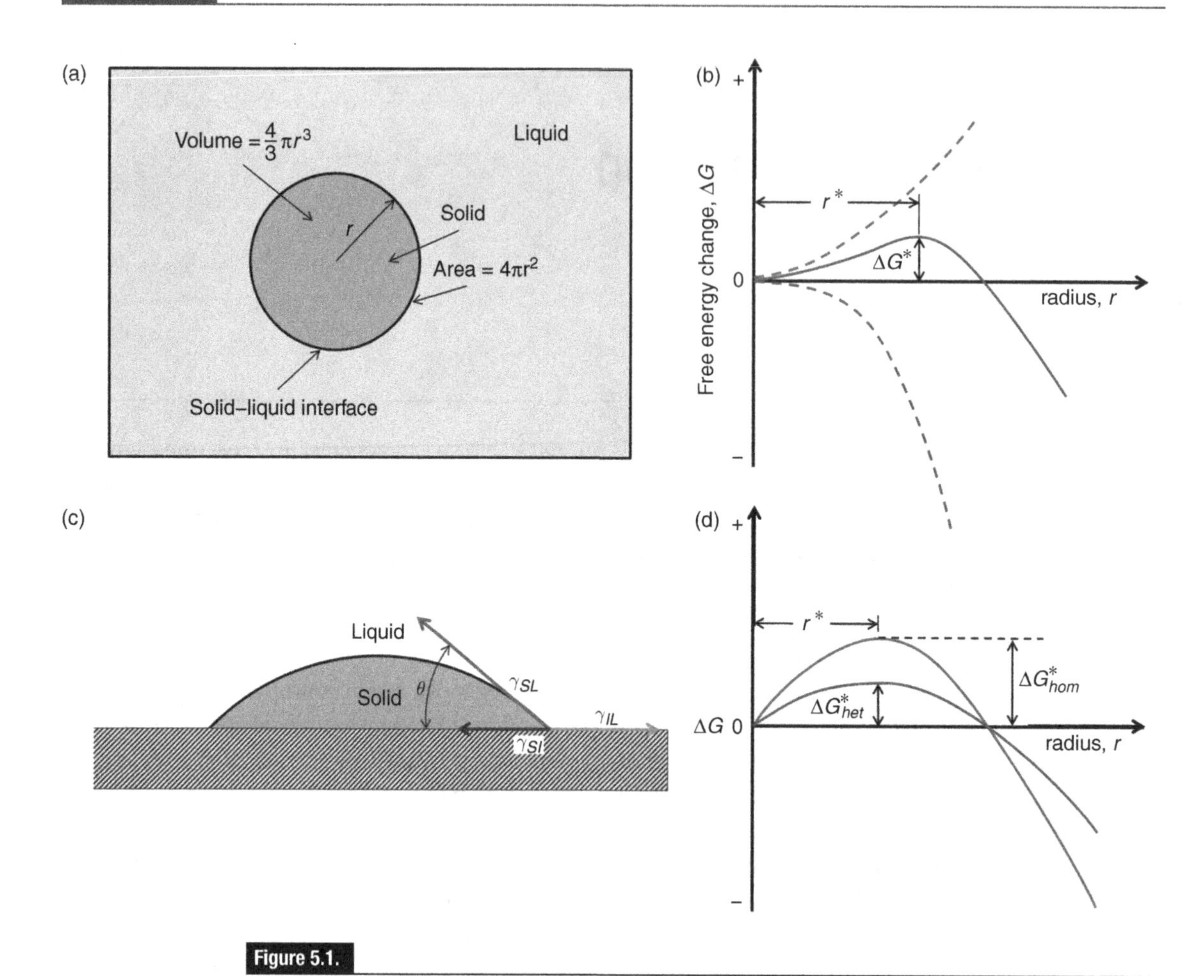

Figure 5.1.

(a, b) Homogeneous and (c, d) heterogeneous nucleation. (a) Homogeneous formation of a nucleus; (b) free energy vs. radius; (c) heterogeneous formation of a nucleus at a surface; (d) comparison of free energy vs. radius for homogeneous and heterogeneous nucleation.

The concentrations of the two species that form the mineral are $[A^+]$ and $[B^-]$, respectively. For calcium carbonate, we would have Ca^{2+} and CO_3^{2-}. In order to increase K_{SP} we do not necessarily have to increase the concentrations of the two components simultaneously; elevating one suffices. The two polymorphs of calcium carbonate have different values for K_{SP}: 4.7×10^{-9} kmole/m^3 for calcite and 6.9×10^{-9} kmole/m^3 for aragonite. However, in biological systems often the structure formed is aragonite. Two factors can contribute to this: (a) the presence of Mg ions and other molecules changes the equilibrium; (b) biological systems can drive up the Ca^{2+} ion concentration over the aragonite threshold.

The homogeneous nucleation is well described by the Becker–Doering theory, dating from *c.* 1925. It is a framework that helps us to understand the fundamentals only. Homogeneous nucleation is indeed extremely rare and only occurs under controlled laboratory conditions. The basic idea is that we have to overcome an activation barrier to initiate the process. The embryo that is formed is assumed to be spherical (radius r); it is shown in Fig. 5.1(a). It has a volume equal to $4/3\pi r^3$. The free energy of the product is below the reactants and there is a difference in free energy ΔG_V. On the other hand, and opposing nucleation, there is a surface forming, with an energy per unit area equal to γ. The surface area of the sphere is $4\pi r^2$. Thus, the total change in energy is given by

$$\Delta G = 4\pi r^2 \gamma - \frac{4}{3}\pi r^3 \Delta G_v. \tag{5.3}$$

Figure 5.1(b) shows the two curves and their sum. In order to find the critical value at which the embryo becomes a nucleus, we take the derivative with respect to r:

$$\frac{\partial \Delta G}{\partial r} = 0 = 8\pi r^* \gamma - 4\pi r^{*2} \Delta G_v. \tag{5.4}$$

Thus, the critical radius is given by

$$r^* = \frac{2\gamma}{\Delta G_v}, \tag{5.5}$$

and the critical energy (activation energy which has to be overcome) is given by

$$\Delta G^* = \frac{16\pi\gamma^3}{3(\Delta G_v)^2}. \tag{5.6}$$

The nucleation rate J is given by

$$J = n^* \beta; \tag{5.7}$$

n^* is the active population of nucleation sites, given by

$$n^* = N\exp\left(\frac{-\Delta G^*}{kT}\right), \tag{5.8}$$

where N is the potential number of nucleation sites, and β is the diffusion rate of the molecules to the embryo.

This critical energy is overcome by thermal activation. These values can be seen in Fig. 5.1(b). In real situations, the nucleation is heterogeneous, and this decreases significantly the critical energy that has to be overcome. The activation barrier is lowered, and the rate of nucleation therefore increases.

Another factor that plays a key role is the degree of saturation, since it provides a further decrease in the activation barrier. Figure 5.1(c) shows the nucleation at a surface. The nucleus is no longer spherical, but takes advantage of the low liquid–surface free energy of the nucleating surface. The decrease in the activation barrier is shown in Fig. 5.1(d).

In the presence of nucleation sites the activation energy is significantly reduced and the nucleation rate is increased accordingly. Figure 5.1(c) shows a solid surface onto which the structure nucleates. The shape and energetics of the nucleus are significantly altered. The decrease in the activation (or critical free energy difference) is shown in Fig. 5.1(d). The activation energy is reduced to

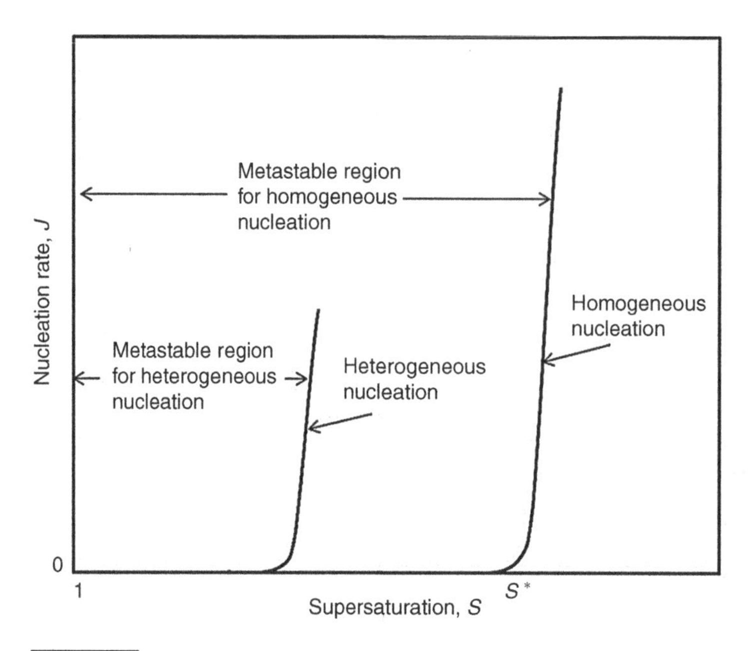

Figure 5.2.

Critical level of saturation for homogeneous and heterogeneous nucleation. (Reproduced based on Simkiss and Wilbur (1989).)

$$\Delta G^*_{het} = \Delta G^*_{hom}\left(\frac{1}{2} - \frac{3}{4}\cos\theta + \frac{1}{4}\cos^3\theta\right). \tag{5.9}$$

The angle of wetting, θ, is given in Fig. 5.1(c).

The degree of supersaturation required for homogeneous nucleation is also much larger than that for heterogeneous nucleation, as shown by Fig. 5.2. We can see that the rate of nucleation, which needs a critical level, increases with supersaturation for both cases. However, the critical level, S^*, is much higher for homogeneous nucleation.

5.2 Growth and morphology of crystals

The shape of crystals after nucleation is determined by two factors: the anisotropy of the surface energy and the growth kinetics.

(a) Anisotropy of surface energy: Wulff (1901) proposed a construction (a polar plot) of surface energy versus orientation. The normals to the vectors that define the smallest energy determine the shape of the crystal morphology. Figure 5.3(a) shows a schematic of a Wulff plot in which normals to the vectors are drawn passing through their extremities. We discuss these shapes for calcite and aragonite.

(b) Growth kinetics: the shape is determined by the attachment energies of the ions to the faces.

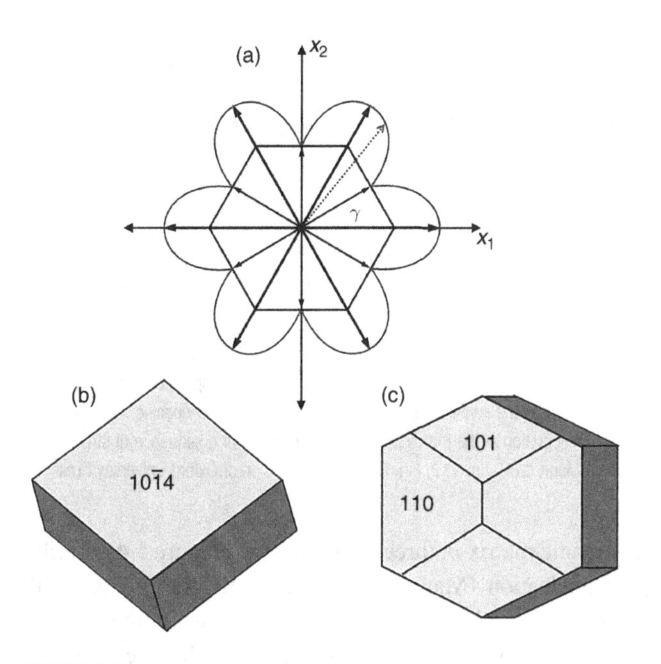

Figure 5.3.

(a) Two-dimensional Wulff plot showing equilibrium shape of crystal in plane. (b) Calculated shape of calcite crystal. (c) Calculated shape of aragonite crystal. (Adapted from de Leeuw and Parker (1998).)

Calcite The $\{10\bar{1}4\}$ surface is the most stable surface of calcite. Hydration of the surfaces has a stabilizing effect, and the growth and hydrated equilibrium morphology of the crystal obtained by molecular dynamics calculations (de Leeuw and Parker, 1998) agree with the experimentally found morphology. This leads to the geometry seen in Fig. 5.3(b), which is a parallelepiped.

Aragonite The same calculations predict a hexagonal basis for aragonite.

In contrast to the surface morphology, which is determined purely by the anisotropy of the surface energies, the growth morphology is obtained from the attachment energies. The growth morphology in aragonite shows the $\{101\}$ and $\{110\}$ faces in agreement with experiment. This is illustrated in Fig. 5.3(c).

The control of nucleation of inorganic materials in nature is achieved by the effect of activation energy dependency on organic substrate composition. Inorganic precipitation is controlled by the kinetic constraints of nucleation. Mann (2001) states that this activation energy may also depend on the two-dimensional structure of different crystal faces, indicating that there is a variation in complementarity of various crystal faces and the organic substrate. Weissbuch, Addadi, and Leiserowitz (1991) describe the auxiliary molecules which promote or inhibit crystal nucleation depending on their composition.

The morphology of the inorganic material created in nucleation is controlled through the interaction with the organic matrix. Activation energies can be influenced by the

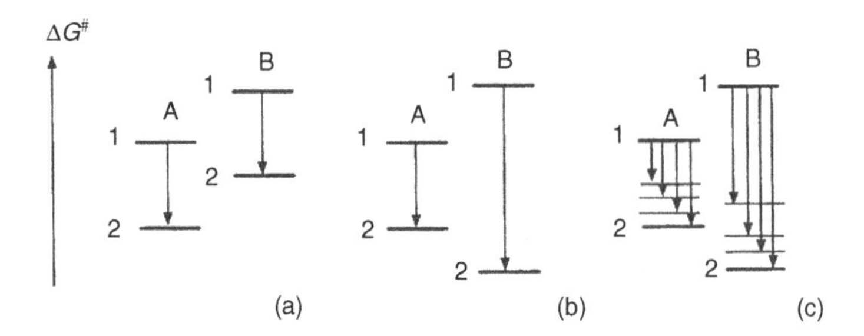

Figure 5.4.

Structural control by organic-matrix-mediated nucleation. (a) Promotion of nonspecific nucleation in which both polymorphs have the activation energies reduced by the same amount. (b) Promotion of structure-specific nucleation of polymorph B due to more favorable crystallographic recognition at the matrix surface. (c) Promotion of a sequence of structurally nonspecific to highly specific nucleation. (From Mann (2001) p. 110, Fig. 6.29. By permission of Oxford University Press, Inc.)

presence of an organic matrix in three possible ways. Figure 5.4 describes the possibilities of polymorphic nucleation (Mann, 2001). The activation energies of two nonspecific polymorphs A and B are shown in the presence (state 2) and absence (state 1) of the organic matrix. If A is more kinetically favored in the absence of the organic matrix, it is possible to examine the possibilities of organic effect on the activation free energy ($\Delta G^{\#}$) of various polymorphs with respect to each other. In the first case, both polymorphs are affected equally, thus A remains kinetically favorable. In the second case, the effect on the B polymorph is much larger than that for A and thus, when in the presence of the organic matrix, B is kinetically favorable. In the last case, we see a combination of the two preceding cases, in which the kinetic favorability of the two polymorphs is influenced by genetic, metabolic, and environmental processes.

The selection of the polymorph will also be determined by a transformation. This starts, in Fig. 5.4, with an amorphous mineral and continues through a series of intermediate structures that have the same composition but decreasing free energy (increasing thermodynamic stability (Mann *et al.*, 1993; Mann, 2001)). This cascade is shown in Fig. 5.5. The system will follow either the one-step route (A) or travel along a sequential transformation route (B) depending on the activation energies of nucleation, growth, and transformation. Addadi and co-workers (Addadi and Weiner, 1985; Addadi *et al.*, 1987; Addadi, Raz, and Weiner, 2003) proposed that the role of the solid-state amorphous precursor phase could be fundamental in the biomineralization process.

The composition of the complex structure in nacre and other biocomposites is mediated by the phase transformations which occur by surface dissolution of the precursor. The phase transformation is dictated by the solubility of the amorphous precursor into the crystalline intermediates, and the effect of these precursors on the free energies of activation of these interconversions. Thus an animal which is able to control its emission of molecular precursor (organic matrix, or soluble protein) will be able to control the growth and structure of its inorganic biocomposite. Addadi and Weiner (1985) and Addadi *et al.* (1987, 2003) demonstrated the stereo-selective

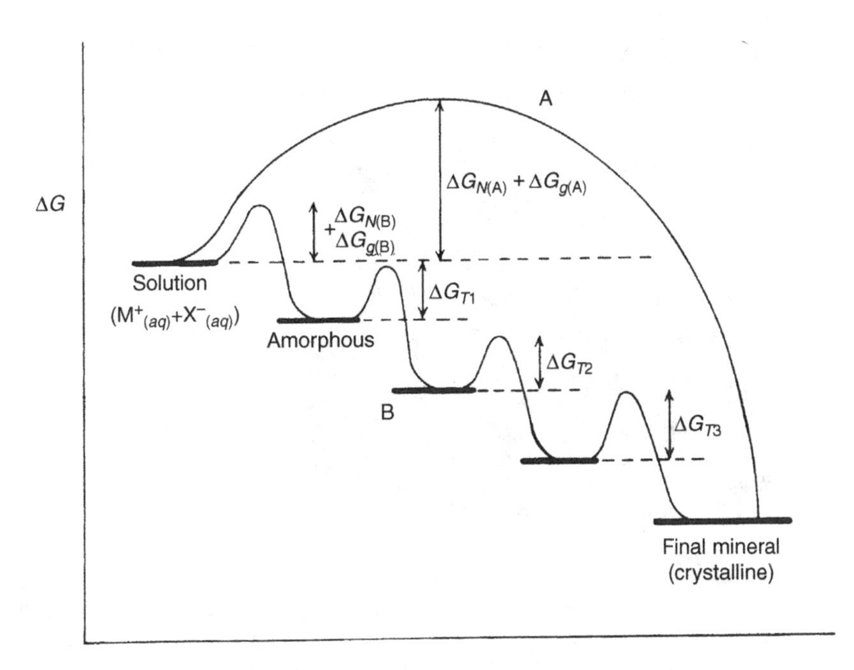

Figure 5.5.

Representation of activation energies of nucleation in the presence and absence of an organic matrix for two nonspecific polymorphs. (From Mann (2001) p. 60, Fig. 4.25. By permission of Oxford University Press, Inc.)

adsorption of proteins in the growth of calcite crystals resulting in a slowing down of growth in the *c*-direction and an alteration in the final shape of the crystal. This evidence of the influence of organics on inorganic crystal growth led them to examine the influence of proteins on the morphology of crystal growth.

Belcher and co-workers (Belcher, 1996; Belcher *et al.*, 1996) showed that a controlled phase transition between aragonite and calcite in nacre could be obtained in the laboratory with the use of soluble polyanionic proteins. They showed that biological phase transformation did not require the deposition of an intervening protein sheet, but simply the presence of soluble proteins. This was directly observed by Hansma *et al.* (2000) through atomic force microscopy. Mann *et al.* (1993) explained the role of soluble proteins as effective agents in the reduction of interfacial energies on the surface of the inorganic. An increase in hydrophobicity of the additive reduces its ability to control morphology and phase transition during crystallization. The effectiveness of the soluble proteins in the process of morphology control depends on their interaction with crystal surfaces in a way that is identical to that of an organic matrix (protein sheet). Thus, the effect of the protein sheet is to control crystal orientation with respect to the bonding energies of specific crystal phases.

Orme and co-workers (Teng *et al.*, 1998; Orme *et al.*, 2001) reported a dependency of calcite growth morphology on the selective binding of amino acids on the crystal step-

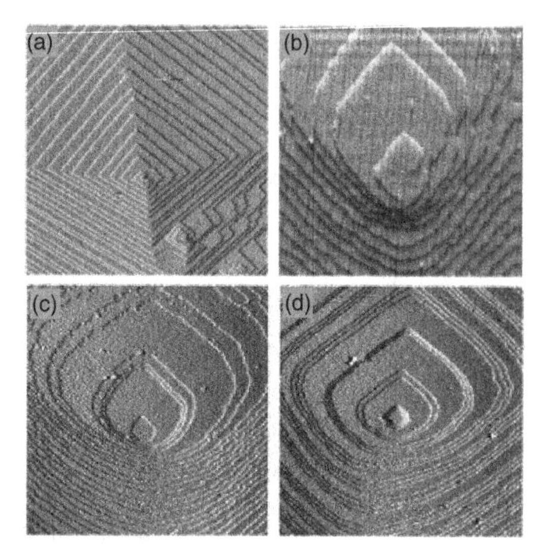

Figure 5.6.

Atomic force microscopy images showing (a) pure calcite growth hillock and (b–d) growth hillocks after the addition of supersaturated solutions of (b) glycine, an achiral amino acid, (c, d) aspartic acid enantiomers. Image sizes: (a) 3.5 × 3.5 μm; (b) 3 × 3 μm; (c), (d) 15 × 15 μm. (Reprinted by permission from Macmillan Publishers Ltd.: *Nature* (Orme *et al.*, 2001), copyright 2001.)

edges. Through in-situ atomic force microscopy (AFM) they were able to show that, in solution, amino acids bind to geometric and chemically favored step-edges, changing the free energy of the step-edge. Figure 5.6(a) shows the AFM image of pure calcite growth hillocks. When a supersaturated solution of glycine is introduced into the growth solution, it can be observed in Fig. 5.6(b) that the two acute steps become curved. By modifying the free energy of step-edges, preferential attachment of calcium ions onto specific locations can be controlled, resulting in macroscopic crystal shape manipulation. Similar results were obtained following the addition of aspartic acid enantiomers (Figs. 5.6(c) and (d)). The importance of this observation is the verification of such theories as those proposed by Mann *et al.* (1993), proving that the addition of various organic growth modifiers can change the rate and location at which calcite attaches to surfaces. In essence, this is an in-situ observation of nature's hand laying the bricks of self-organization and biomineralization.

5.3 Structures

One of the defining features of the rigid biological systems that comprise a significant fraction of the structural biological materials is the existence of two components: a mineral and an organic component. The intercalation of these components occurs at the nano-, micro-, and meso-scale, and often takes place at more than one dimensional scale. Table 5.1

Table 5.1. Principal components of common structural biological composites

Biological composites	Mineral			Organic				
	Calcium carbonate	Silica	Hydroxyapatite	Keratin	Collagen	Chitin	Cellulose	Other
Shells	×					×		
Horns				×				
Bones			×		×			
Teeth			×					×
Bird beaks				×				
Crustacean exoskeleton	×					×		×
Insect cuticle						×		×
Woods							×	
Spicules		×						×

exemplifies this for a number of systems. The mineral component provides the compressive strength, whereas the organic component contributes to the ductility. This combination of strength and ductility leads to high-energy absorption prior to failure. This is commonly referred to in materials science as "toughness."

The most common mineral components are calcium carbonate, calcium phosphate (hydroxyapatite), and amorphous silica, although over 20 minerals (with principal elements being Ca, Mg, Si, Fe, Mn, P, S, C, and the light elements H and O) have been identified. These minerals are embedded in complex assemblages of organic macromolecules which are in turn hierarchically organized. The most utilized are collagen, keratin, and chitin. Despite their main purpose, providing compressive mechanical strength, it should be noted that minerals also have other functions. For example, the magnetite in tuna, pigeons, and bacteria acts as a sensor of the earth's magnetic field. Biominerals are also used as gravity and inertia sensors. Some biominerals even have unique optical properties, e.g. single-crystal calcite is found in the compound eyes of trilobites. Silica in brittle-stars acts as a means to trap light, as will be discussed in Chapter 11.

Table 5.2 shows the minerals that have been identified in biological systems (Weiner and Addadi, 2002). The number of minerals is continuously increasing. The most common biominerals are calcium carbonate, silica, and calcium phosphate (hydroxyapatite).

Figure 5.7 shows the atomic arrangement of the calcium, phosphorus, and oxygen atoms in hydroxyapatite ($Ca_{10}(PO_4)_6(OH)_2$). The unit cell shown is quite complex and is the primitive hexagonal. Three primitive cells rotated by 120° form the hexagonal basis if they are brought together at their 120° angles ($3 \times 120° = 360°$). One can distinguish the PO_4 tetrahedra in the cell. Although one can see only five of them, there are actually six. The OH groups are at the four edges and their sum is $8 \times 1/4 = 2$. There should be ten

Table 5.2. Principal minerals found in biological systems[a]

Carbonates	calcite aragonite vaterite monohydrocalcite	amorphous calcium carbonate family hydrocerrusote protodolomite
Phosphates	hydroxyapatite (HAP) carbonated apatite (dahllite) francolite octacalcium phosphate whitlockite struvite	brushite amorphous calcium phosphate family vivianite amorphous pyrophosphate
Halides	fluorite hieratite	amorphous fluorite atacamite
Sulfates	gypsum celestite barite	jarosite calcium sulfate hemihydrates
Silicates	silica (opal)	
Oxides and hydroxides	magnetite goethite lepidocrocite ferrihydrite amorphous iron oxide	amorphous manganese oxide amorphous ilmenite todotokite birnessite
Sulfides	pyrite amorphous pyrrhotite hydrotroilite shalerite	galena greigite mackinawite wurtzite
Native element	sulfur	
"Organic minerals"	whewellite weddelite manganese oxalate calcium tartrate calcium malate earlandite guanine	uric acid paraffin hydrocarbon wax magnesium oxalate (glushinskite) copper oxalate (moolooite) ferric oxalate anhydrous sodium urate

[a] From Weiner and Addadi (2002).

calcium ions inside the primitive cell. The hydroxyapatite is also known simply as HAP, and the formula $Ca_{10}(PO_4)_6(OH)_2$ is often used to represent it.

Figure 5.8 shows two important crystal forms of calcium carbonate in biological systems, namely aragonite and calcite. Aragonite has an orthorhombic structure, as shown in Fig. 5.8

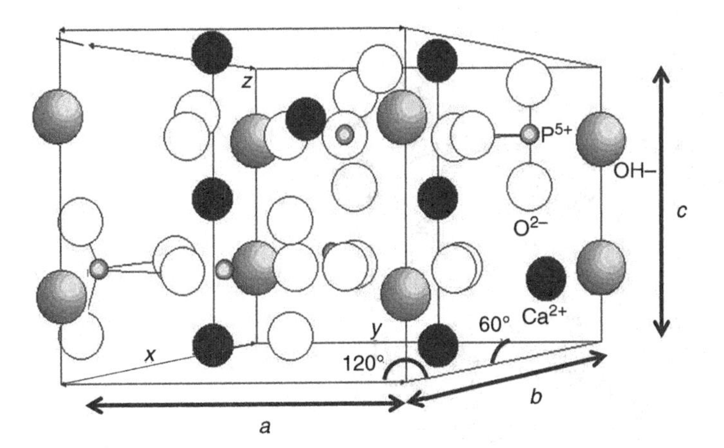

Figure 5.7.

Hexagonal (primitive) crystal structure of hydroxyapatite: $Ca_{10}(PO_4)_6(OH)_2$, where $a = b = 0.9432$ nm, $c = 0.6881$ nm. Symbols: Ca^{2+} = black; P^{5+} = small gray; O^{2-} = large white; OH^- = large gray.

(a). Figure 5.8(b) shows the crystal structure of calcite, which has a rhombohedral structure. It is important to recognize that the minerals do not occur in isolation in living organisms. They are invariably intimately connected with organic materials, forming complex hierarchically structured composites. The resulting composite has mechanical properties that far surpass those of the monolithic minerals. Although we think of bone as a biomineral, it is actually composed of 60% collagen and 30–40% hydroxyapatite (on a volume basis). If the mineral is dissolved away, the entire collagen framework is retained. In shells, the rapid growth direction for aragonite crystals is the *c*-direction.

Shown in Fig. 5.9 are the three principal structures (polymorphs) of calcium carbonate crystals formed shortly after nucleation. The shapes of these crystals are already indicative of the processes that take place in biological systems. The calcite crystals are cuboidal and reproduce in some fashion the structure. The three perpendicular growth directions are similar, leading to this shape. Indeed, calcite in shells forms equiaxed grains, as will be seen in Chapter 6. The aragonite crystals have a characteristic lenticular shape, and the rapid growth direction is the *c*-axis, forming a needle from which the other directions expand. This rapid growth direction is exploited by shells that are mostly aragonitic. It will be seen in Chapter 6 that there is constant mediation by the animal to control the growth along the *c*-direction. This is achieved by inserting periodic organic layers that, at the same time, enhance the toughness of the shell. The vaterite structure, on the other hand, forms "floral" shaped crystals that have irregular morphologies. It is the least common structure in biological systems, and is present in pearls and a small bone in the head of fish, the otolith. This bone is supposedly responsible for balance.

The structure of aragonite imaged along the *c*-axis through atomic resolution transmission electron microscopy is shown in Figure 5.10. This beautiful image, from

(a)

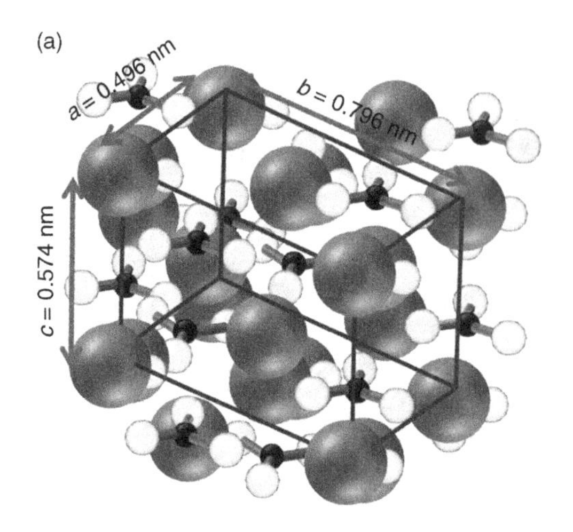

(b)

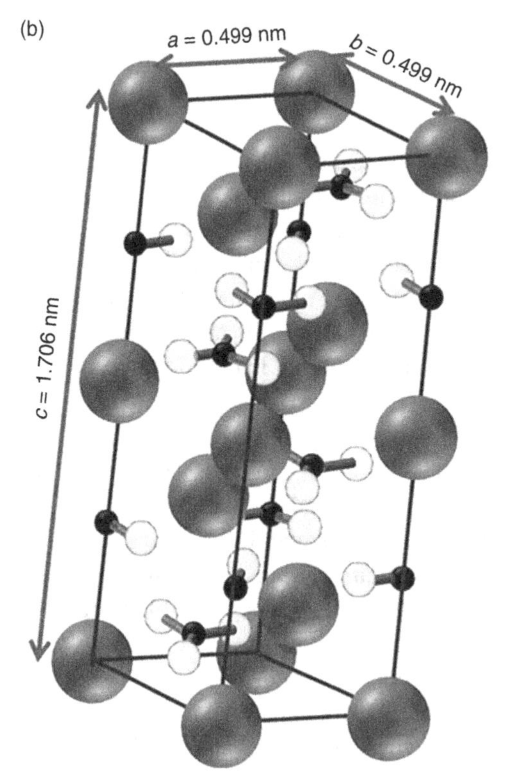

Figure 5.8.

Comparison of the two most common crystal structures of $CaCO_3$: (a) aragonite, orthorhombic, $a = 0.49614$ nm, $b = 0.79671$ nm, $c = 0.57404$ nm, and (b) calcite, rhombohedral, $a = b = 0.4989$ nm, $c = 1.7062$ nm. Symbols: Ca = large gray; C = small black; O = small gray). (Lattice parameters taken from Sass and Vidale (1957) and De Villiers (1971).)

Figure 5.9.

Three polymorphs of calcium carbonate: (a) cuboidal calcite (rhombohedra); (b) needle-like aragonite (orthorhombic); (c) floral vaterite (hexagonal). (Figure courtesy T. Kogure, with kind permission.)

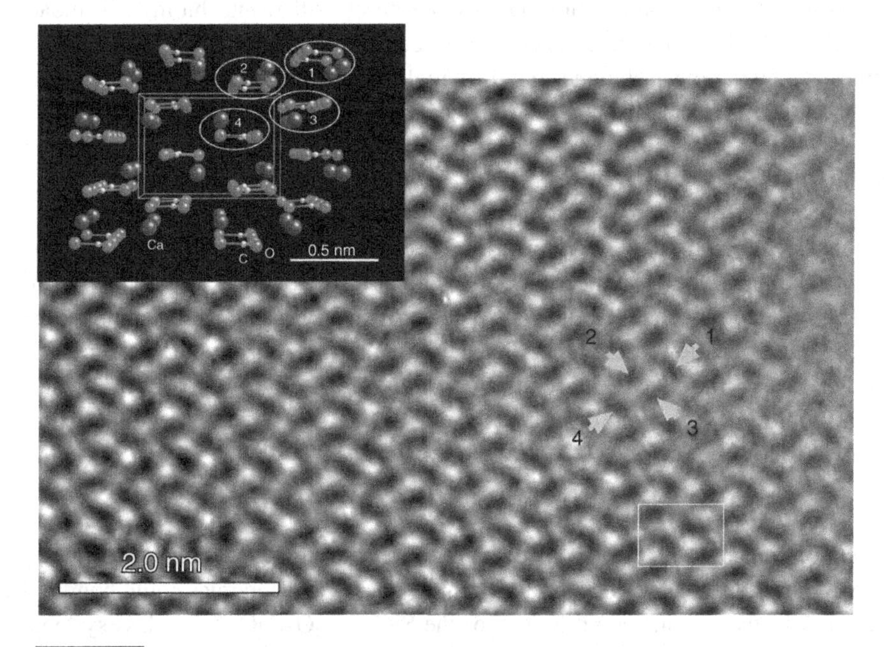

Figure 5.10.

Atomic resolution transmission electron micrograph of aragonite (orthorhombic); observation direction aligned with *c*-axis. Note the 1–2–3–4 parallelepiped defined in left-hand rendition and marked in the micrograph. Angles are 120° and 60°. (Figure courtesy T. Kogure, with kind permission.)

T. Kogure's lab in Japan, shows the basal plane of orthorhombic structure, with 90° angles, and a superposed pattern of molecular groups forming 60° and 120° angles. These groups form the crystalline shape shown in the inset.

Organic molecules in solution can influence the morphology and orientation of inorganic crystals if there is molecular complementarity at the crystal–additive interface. Phase transformations are believed to occur by surface dissolution of precursors which mediate the free energies of activation of interconversions. However, these principles are not yet well developed. Understanding the process by which living organisms control the growth and structure of inorganic materials could lead to significant advances in materials science, and open the door to novel synthesis techniques for nano-scale composites. Mann (2001) states that in order to address the question of nano-scale biologically induced phase transformations and crystallographic control, we must study the bonding and reactivity of extended organized structures under the mediation of organic chemistry. Here we can examine two important processes: nucleation and morphology.

Box 5.1 Ceramic biomaterials

Ceramics, glasses, and glass ceramics are inorganic/nonmetallic materials that have ionic or covalent bonding and have been widely applied in the biomedical field (Best *et al.*, 2008). Based on their interaction with the host tissues, bioceramics can be classified into bioinert or bioactive; bioactive ceramics can be further categorized as either resorbable or nonresorbable. These ceramic materials can be synthesized by various techniques into different forms, ranging from porous to dense structures in bulk, as granules, powders, fibers, and surface coatings.

Bioinert ceramics

Early applications of bioceramics emphasized the absence of toxic response and minimal tissue interaction, from which bioinert ceramics have been investigated and developed. High-density and high-purity alumina (Al_2O_3) has been used in load-bearing and wear-resistant hip prostheses and dental implants. It has good biocompatibility, excellent corrosion resistance, high strength, exceptional wear resistance, and a very low friction coefficient. Zirconia (ZrO_2) is another bioinert ceramic material commonly used as the articulating ball in hip prostheses. Other oxides, silica-, and carbon-based bioinert ceramics (carbon fiber, diamond-like carbon) have also been used as orthopedic and dental implants.

Bioglass®

The invention of Bioglass® can be dated back to 1967 during the Vietnam War when Professor Larry Hench and others discovered that various kinds of glasses and ceramics could bond to living bone. In the early 1970s, the specific compositions of the SiO_2-Na_2O-CaO-P_2O_5 glass systems with B_2O_3 or CaF_2 additions were reported by Hench *et al.* (1991). The distinguishing features of bioactive glasses are high Na_2O and CaO content (>40 wt.%), low SiO_2 content (<60 wt.%), and high CaO/P_2O_5 ratio. Many bioactive glasses are based on the 45S5 composition, representing 45 wt.% SiO_2, S being the network former and 5 coming from the 5:1 CaO/P_2O_5 ratio. Glasses with lower CaO/P_2O_5 ratios cannot bond well to bone. Hench further defined two classes of bioactive materials: class A materials possess both osteoconduction (bone growth along the bone/implant interface) and osteoproduction, whilst class B materials have osteoconduction only. A wide variety of bioactive glasses and ceramics has been developed and investigated since that time, and it is still a vigorous research area.

Glass ceramics

Following the work on Bioglass®, the apatite–wollastonite (A–W) glass ceramic was developed by Kokubo and co-workers (Kokubo, 1991), and this has been widely applied as a bone substitute. A–W glass ceramic is a composite of apatite particles reinforced by wollastonite ($CaO{\cdot}SiO_2$) with superior bending strength, elastic modulus, and fracture toughness compared with other bioactive glasses and glass ceramics. It is used as a load-bearing material under compressive loading, such as in vertebral prostheses and ilia crest replacement. A large number of studies on glass ceramics have been carried out which have led to various biomedical applications.

Calcium phosphates

The mineral component of bone is a calcium phosphate, a nonstoichiometric hydroxyapatite; calcium-phosphate-based synthetic ceramics have been widely used in the biomedical field. Hydroxyapatite (HAP, chemical formula $Ca_{10}(PO_4)_6(OH)_2$) has a theoretical Ca/P molar ratio of 1.667. The crystal structure of HAP can accommodate substitutions by cations (Na^+, Mg^{2+}, K^+, Mn^{2+}, Sr^{2+}) and anions (CO_3^{2-}, F^-, Cl^-), and ionic substitutions can affect the crystallinity, morphology, solubility, and stability of HAP. Tricalcium phosphate (TCP, $Ca_3(PO_4)_2$) is a biodegradable ceramic that can be replaced by bone during implantation. If the Ca/P ratio is lower than 1.67, TCP can be present after thermal processing. If the Ca/P ratio is higher than 1.67, CaO may be formed along with HAP. HAP with a Ca/P ratio less than 1 is not suitable for biological implantation. TCP with a Ca/P ratio of 1.5 is more rapidly resorbed than HAP. Biphasic calcium phosphates (BCPs) are mixtures of HAP and TCP that have been used as bone substitutes with controllable biodissolution rates. The higher the TCP content in BCP, the faster the dissolution rate.

Calcium-phosphate-based ceramics are limited to non-load-bearing biomedical applications due to their lower mechanical properties. Calcium phosphate coatings on metallic or other ceramic substrates and implants have received great interest. Many techniques have been utilized to deposit HAP, including plasma-spraying, sputtering, electrophoresis, and biomimetic routes. The most widely used commercial routes are those based on plasma-spraying, and many HAP-coated implants have been successfully developed. The properties of plasma-sprayed HAP coating depend on coating thickness, crystallinity, purity, porosity, and adhesion. Other approaches have been taken to synthesize composites of calcium phosphates and polymers or other matrices which are bioactive, biodegradable, and with enhanced mechanical performance. One recent advance is the development of functionally and compositionally graded, mineralized collagen–glycosaminoglycan (GAG) scaffolds for cartilage and ligament repair at the University of Cambridge, UK (Lynn *et al.*, 2005).

5.4 Origins and structures

There are three principal groups of biominerals: carbonates, phosphates, and silica. They seem to have appeared at about the same time, in the Precambrian, ~550 million years ago (MY). The tubular fossil *Cloudina* is often considered to be one of the first examples of a biomineralized skeleton. It consists of half-rings that were apparently constructed by the secreting gland of a worm. These half-rings were made of calcium carbonate. Figure 5.11 shows the evolution in the number of clades (a clade is a group consisting of a *species* and all its descendants) of the three groups. During the Cambrian (550–500 MY) there was a large increase in the number of species. For calcium carbonate, the earliest morphology of aragonite was the spherulitic/prismatic one that does not need to be mediated by proteins. This morphology is seen in abalone nacre after a mesolayer is formed and prior to the onset of the tiled configuration of nacre (Chapter 6). This morphology evolved during the Middle Cambrian to produce the following major classes of morphologies:

- aragonite fibers in a tangential arrangement (such as that observed in limpets);
- crossed-lamellar aragonite (such as that in conch);
- foliated calcite;
- nacre (abalone).

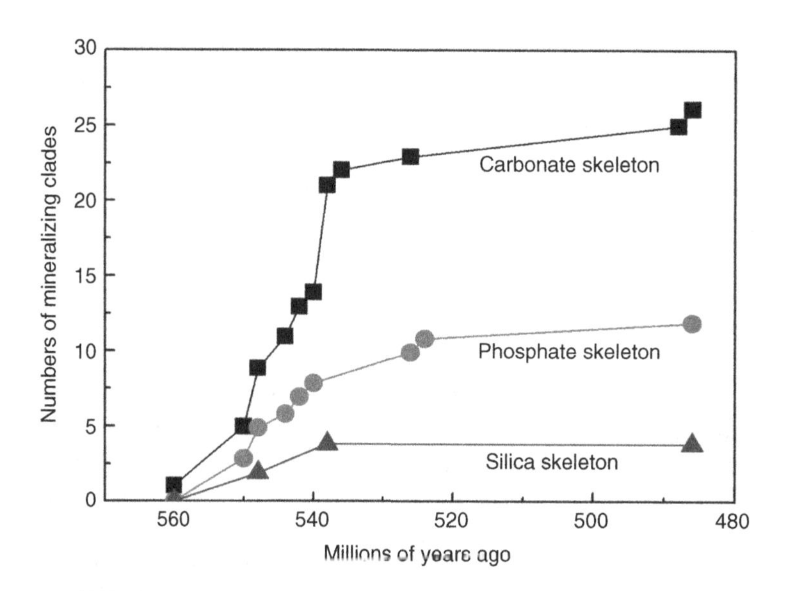

Figure 5.11.

Evolution in the number of clades (groups of taxa that appear to derive their biomineralization from a common ancestor) for the three principal biominerals: carbonates, phosphates, and silica. (From Runnegar and Bengtson (1992), with kind permission from Wiley.)

Table 5.3. Biominerals with corresponding living organisms[a]

	Chemical formula	Distribution
Barium sulfate	$BaSO_4$	algae (gravity sensor)
Calcium carbonate	$CaCO_3$	
Calcite (rhombohedral)		mollusc shells, bird eggs, sponge spicules, sea urchin spines
Aragonite (orthorhombic)		mollusc shells, corals, pearls
Vaterite (hexagonal)		fish otoliths (mineral in head), some freshwater pearls
Amorphous		arthropod exoskeletons, mollusc shells, plants
Calcium oxalate	CaC_2O_4	kidney stones, plants
Calcium sulfate	$CaSO_4$	jellyfish larvae (gravity sensor)
Dolomite	$CaMg(CO_3)_2$	sea urchin spicules and teeth
Ferrihydrate	$5Fe_2O_3{\cdot}9H_2O$	ferritin (animals), plants
Greigite	Fe_3S_4	*Crysomallon squamiferum* foot (gastropod living near hot vents in Indian Ocean)
Hydroxyapatite	$Ca_{10}(PO_4)_6(OH)_2$	bones, teeth, osteoderms
Iron (III) oxide/hydroxide	FeOOH	chitin and limpet teeth
Magnetite	Fe_3O_4	mollusc radula, bacteria
Pyrite	FeS_2	*Crysomallon squamiferum* foot (gastropod living near hot vents in Indian Ocean)
Silica (hydrated)	$SiO_2{\cdot}\,nH_2O$	diatom exoskeleton, sponge spicules

[a] Adapted from Mann (1988).

Some common biominerals (out of ~60 total) are shown in Table 5.3. Terrestrial animals have calcium carbonate or calcium phosphate minerals, whereas marine species can possess either calcium- or silicon-containing minerals. The calcium-containing minerals are crystalline, with some amorphous calcium carbonate found in the exoskeletons of crustaceans. Crystalline SiO_2 is not biogenetically synthesized; it occurs only in the amorphous, hydrated form ($SiO_2{\cdot}nH_2O$). Diatoms (phytoplankton) are the most ubiquitous animal found in the oceans and form beautiful and intricate $SiO_2{\cdot}nH_2O$ exoskeletons. Sea water contains ~3 ppm Si and ~400 ppm Ca, and it may be puzzling as to why calcium-containing minerals are not formed in diatoms. Using finite element modeling, Hamm *et al.* (2003) calculated the tensile fracture stress of a diatom frustule to be 540 MPa, much higher than that of aragonite ($CaCO_3$), at 100 MPa. Since the densities are similar, this implies that the thin, delicate cages of the diatoms are more structurally robust if fabricated from $SiO_2{\cdot}nH_2O$ than from $CaCO_3$. In addition, the key energetic issue is that silica autopolymerizes (i.e. there is no energy input), and therefore can be catalyzed with a minimum amount of organic materials. A supersaturated solution of $CaCO_3$ is required for nucleation, and this solution requires higher amounts of calcium than is the case for silicon in $SiO_2{\cdot}nH_2O$. Additionally, the cells have to exclude calcium from the cytoplasm because it is used as a second messenger in trace amounts, so there is likely to be high energy involved in transporting the calcium to the correct location. Silicic acid is generally nontoxic and can diffuse across membranes so less energy is required for transport (Hildebrand, 2008). Glass sponge spicules are also fabricated from $SiO_2{\cdot}nH_2O$.

Calcium carbonate ($CaCO_3$) is found as crystalline calcite (rhombohedral, $R\bar{3}c$), aragonite (orthorhombic, *Pmcn*), and vaterite ($a = 0.413$ nm, $c = 0.849$ nm; named after Hermann Vater, *c.*1850s) structures as well as in the amorphous form. Calcite is the more stable polymorph. Figure 5.8 compares the crystal structures of both polymorphs. Bird eggs, crustacean exoskeletons, some sponge spicules, and sea urchin spines are composed of calcite. Mollusc shells (nacre), cuttlebone, and corals are aragonitic; the former, or parts thereof, can also be calcitic. Magnesium additions into calcite form a modified dolomite, $Ca_{1-x}Mg_xCO_3$, which is stronger and harder than pure calcite and is precipitated in sea urchin teeth and spicules. Amorphous calcium carbonate is also found in some plants, sponges, sea urchins, molluscs, and sea worms (Addadi *et al.*, 2003).

In mammals, derivatives of hydroxyapatite ($Ca_{10}(PO_4)_6(OH)_2$) are the predominant minerals found in bone and teeth; it also forms the osteoderms of turtles, crocodiles, and armadillos. Hydroxyapatite has a hexagonal structure (*P*63/*m*) with lattice parameters $a = 0.9432$ nm and $c = 0.6881$ nm (Fig. 5.7). In bone, the minerals are nonstoichiometric, with 4–6% of the phosphate groups replaced by carbonate groups, making the structure more similar to the structure of dahlite. The bone crystals are in the form of platelets approximately 40–60 nm in length and 20–30 nm in width. The thickness of bone crystals measured from TEM (Jackson, Cartwright, and Lewis, 1978; Weiner and Price, 1986; Ziv and Weiner, 1994; Boskey, 2003) and small-angle X-ray scattering (SAXS) (Fratzl *et al.*, 1992; Watchtel and Weiner, 1994) varies from 1.5 nm for mineralized tendon to 4 nm for some mature bones. Recent atomic force microscopy (AFM) studies found that the bone crystals are longer than those observed by TEM, with widths and lengths ranging from 30 to 200 nm (Tong *et al.*, 2003; Hassenkarm *et al.*, 2004). This discrepancy may be due to breakage of the fragile crystallites during TEM sample preparation. Figures 5.12(a) and (b) show X-ray diffraction patterns from the minerals from cortical elk antler and bovine femur bone, respectively. The broadness of the peaks indicates that the crystals are

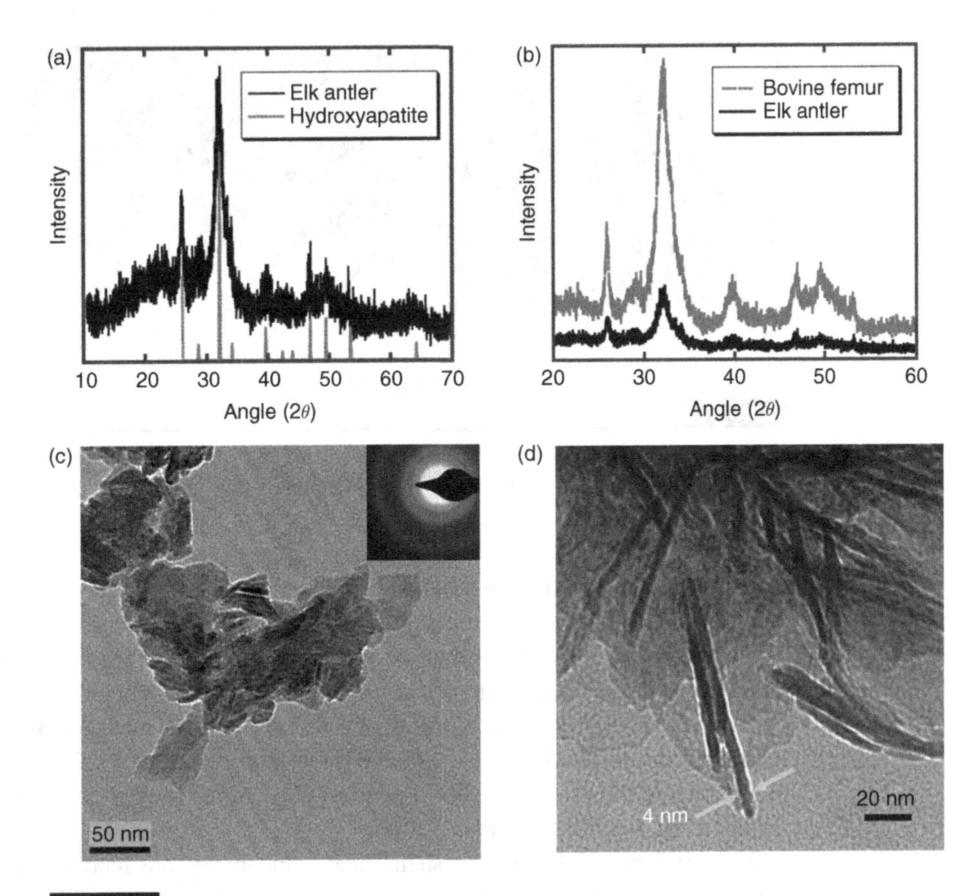

Figure 5.12.

X-ray diffraction patterns of elk antler compared to (a) pure hydroxyapatite and (b) bovine femur. (c) TEM micrograph of mineral crystallites derived from elk antler. (Inset: selected area electron diffraction pattern.) (d) High-magnification TEM micrograph showing the platelet-like crystallites with thickness ~4 nm. (Reprinted from Chen *et al.* (2009), with permission from Elsevier.)

nanocrystalline, which is confirmed by the TEM micrographs (Figs. 5.12(c) and (d)). The minerals have a plate-like morphology with a thickness of 4 nm and a lateral dimension of 40–150 nm. In teeth, a portion of the hydroxyapatite is converted to fluorapatite ($Ca_{10}(PO_4)_6F_2$), with fluorine replacing the hydroxyl groups, resulting in a harder mineral.

Magnetite (Fe_3O_4, $Fd3m$, $a = 0.8391$ nm) has two functional uses in marine species. It acts as a navigation sensor in the skulls of tuna and pigeons, for example, but it also has a high hardness and forms the teeth of some molluscs. The hardness of magnetite is found to be three times higher than that of either enamel or calcium carbonate, which is a useful property for the animals that use their teeth to scrape nutrients from rocks (Weaver *et al.*, 2010). Magnetite also exists in the chiton radula, described in Section 4.5.

A newly discovered gastropod, *Crysomallon squamiferum*, has been found near hot black-smoker chimney vents in the Indian Ocean. Tube worms, shrimp, and

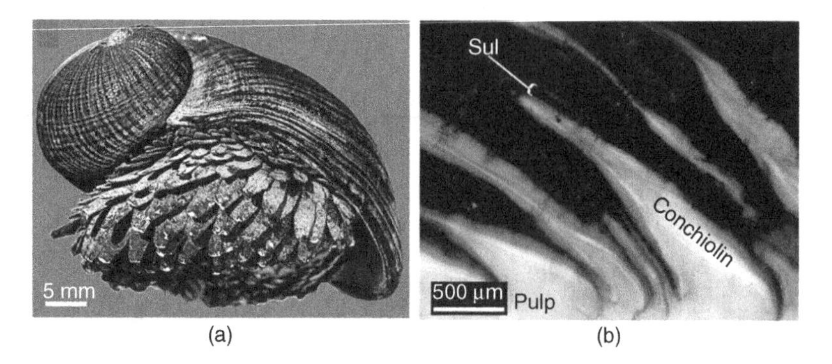

Figure 5.13.

The gastropod *Crysomallon squamiferum* (a) retracted in shell; (b) longitudinal sections of scales, viewed with light microscopy; "Sul" denotes sulfide layers. (From Warén *et al.* (2003). Reprinted with permission from AAAS.)

crabs also live there. Animals that live there must be able to survive in extreme environments – complete darkness, temperature variations of 2–400 °C, high pressures, and high concentrations of the sulfides that are emitted by the vents. The foot is covered with hundreds of scale-shaped sclerites consisting of proteins (conchiolin) and minerals – pyrite (FeS_2) and greigite (Fe_3S_4), both of which appear to be mediated by proteobacteria present on the surface (Warén *et al.*, 2003; Goffredi *et al.*, 2004). Figure 5.13 shows a photograph of the gastropod. The outermost layer of the shell is composed of greigite, similar to the sclerites (Yao *et al.*, 2010).

Biological materials such as shells, teeth, and bones are hierarchically structured composites of protein and mineral constituents with exceptional mechanical properties. Despite the complex hierarchical structures, the smallest building blocks of the mineral phase are usually on a nanometer length scale. As mentioned previously, bone minerals are platelets 40–100 nm in width and 2–4 nm in thickness. Hydroxyapatite minerals in enamel of tooth are rod- or needle-like crystals, 15–20 nm in thickness. The nano-sized minerals play an important role in enhancing mechanical performance by optimizing strength and maximizing fracture tolerance (Gao *et al.*, 2003). This so-called nano-size effect will be discussed in detail in Section 6.8.

Another distinct characteristic of biominerals is their high aspect ratio. Mineral crystals developed in biological systems often have a platelet-, rod- or needle-like shape with high aspect ratio (30–40 in bone minerals, ~10 in abalone nacre). Jäger and Fratzl (2000) developed a mechanical model to estimate the elastic modulus of biocomposites taking the aspect ratio of mineral phase into account. The elastic modulus E of the composite in the longitudinal direction can be expressed as follows:

$$\frac{1}{E} = \frac{4(1-\Phi)}{G_p \Phi^2 \rho^2} + \frac{1}{\Phi E_m}, \tag{5.10}$$

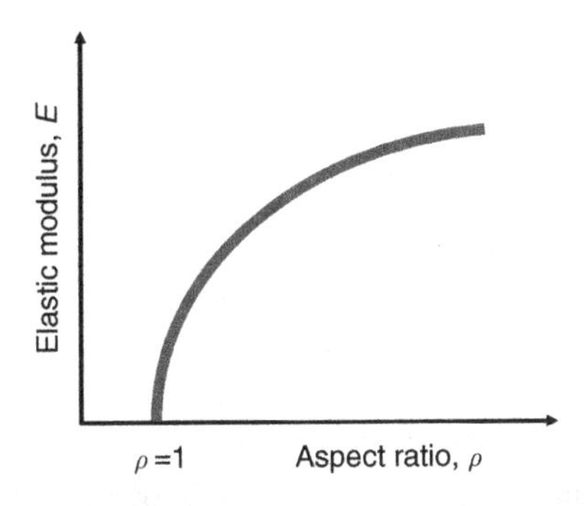

Figure 5.14.

Elastic modulus of mineralized biocomposite as a function of the aspect ratio of mineral crystals.

where E_m is the elastic modulus of the mineral, G_p is the shear modulus of the protein, Φ is the volume fraction of the mineral, and ρ is the aspect ratio of the mineral crystal. The detailed discussion and derivation are presented in Chapter 7. Equation (5.10) indicates that the elastic modulus of a biocomposite increases with increasing aspect ratio, and the dependence is shown in Fig. 5.14. The high aspect ratio of minerals compensates for the low modulus of the protein phase. The hard mineral constituents with high aspect ratio carry most of the load, and the protein constituents transfer stress between minerals via shear.

Example 5.1 Determine the densities of the three polymorphs of calcium carbonate: (a) aragonite; (b) calcite; (c) vaterite.
Atom masses: Ca 40.078 amu, C 12.01 amu, O 16.00 amu;
1 amu = $1/1000N_A$ kg $\approx 1.66054 \times 10^{-27}$ kg;
Z = lattice points per unit cell.

Solution (a) Aragonite (orthorhombic): $a = 0.496$ nm, $b = 0.796$ nm, $c = 0.579$ nm; $Z = 4$.
For each unit cell,

$$\rho = m/V = (m_{\text{Ca}} + m_{\text{C}} + 3m_{\text{O}}) \times Z/(a \times b \times c) = 2.9081\ \text{g/cm}^3.$$

(b) Calcite (trigonal-rhombohedral): $a = 0.4991$ nm, $b = 0.4991$ nm, $c = 1.7062$ nm; $Z = 6$; angle $\gamma = 120°$.
For each unit cell,

$$\rho = m/V = (m_{\text{Ca}} + m_{\text{C}} + 3m_{\text{O}}) \times Z/\left(a\frac{\sqrt{3}}{2} \times b \times c\right) = 2.711\ \text{g/cm}^3.$$

(c) Vaterite (hexagonal): $a = 0.7135$ nm, $c = 1.698$ nm; $Z = 12$.

For each unit cell,

$$\rho = m/V = (m_{\mathrm{Ca}} + m_{\mathrm{C}} + 3m_{\mathrm{O}}) \times Z/\left(a \times a\frac{\sqrt{3}}{2} \times c\right) = 1.6044\ \mathrm{g/cm^3}.$$

Example 5.2 Calculate the density of hydroxyapatite, $Ca_{10}(PO_4)_6(OH)_2$, knowing that the lattice parameter is based on a hexagonal rhombic prism with side a and height c, where $a = 0.9432$ nm, $c = 0.6881$ nm.

Solution Atom mass: Ca 40.078 amu, P 30.79376 amu, O 16.00 amu, H 1.00794 amu;
1 amu = $1/1000N_A$ kg $\approx 1.66054 \times 10^{-27}$ kg;
Z = lattice points per unit cell.
For hydroxyapatite, $a = 0.9432$ nm, $c = 0.6881$ nm, $Z = 2$. For each unit cell,

$$\rho = m/V = (10m_{\mathrm{Ca}} + 26m_{\mathrm{O}} + 6m_{\mathrm{P}} + 2m_{\mathrm{H}}) \times Z/\left(a \times a\frac{\sqrt{3}}{2} \times c\right) = 3.786\ \mathrm{g/cm^3}.$$

Example 5.3 A composite consists of 50 vol.% polymeric matrix (shear modulus 1 GPa) and 50 vol.% mineral reinforcements (elastic modulus 100 GPa). Calculate the elastic modulus of the composite if the mineral constituents are in the following geometries:

(a) sphere (aspect ratio = 1);
(b) platelet (aspect ratio = 10);
(c) needle (aspect ratio = 100).

Solution We use the following equation:

$$\frac{1}{E} = \frac{4(1-\Phi)}{G_p \Phi^2 \rho^2} + \frac{1}{\Phi E_m}.$$

(a) For spherical mineral particles,

$$\frac{1}{E} = \frac{4(1-0.5)}{1 \times 0.5^2 \times 1^2} + \frac{1}{0.5 \times 100} = 8.02;$$

E(sphere) = 0.125 GPa.

(b) For mineral platelets,

$$\frac{1}{E} = \frac{4(1-0.5)}{1 \times 0.5^2 \times 10^2} + \frac{1}{0.5 \times 100} = 0.1;$$

E(platelet) = 10 GPa.

(c) For needle-like minerals,

$$\frac{1}{E} = \frac{4(1-0.5)}{1 \times 0.5^2 \times 100^2} + \frac{1}{0.5 \times 100} = 0.0208;$$

E(needle) = 48 GPa.

The elastic modulus of composite increases significantly as the mineral aspect ratio increases. The elastic modulus saturates at the Voigt limit, ~50 GPa. Assuming a negligible elastic modulus for the polymeric matrix,

$$\frac{1}{E} \cong \frac{1}{0.5 \times 100}.$$

Summary

- More than 60 minerals occur in biological systems. Of these, calcium carbonate ($CaCO_3$), hydroxyapatite ($Ca_{10}(PO_4)_6(OH)_2$), and amorphous silica are the most important.
- Hydroxyapatite, also known as HAP, has a hexagonal structure that is quite complex. Each primitive cell has angles of 120°, and three of them join up to form the hexagon (120 + 120 + 120 = 360°).
- Calcium carbonate can exist in three allotropic forms: calcite (the most stable at ambient temperature and pressure), aragonite, and vaterite. Calcite and aragonite are common in biological systems, and vaterite is rarer. Aragonite is orthorhombic, calcite is rhombohedral, and vaterite has a hexagonal structure. These structures affect the shapes of the crystals that are formed. Crystals that are formed abiotically have characteristic shapes: aragonite forms needles with a rhombic base with angles of 60° and 120°; calcite forms cuboidal crystals; and vaterite forms floral arrangements.
- Calcium carbonate is the hard constitutent of many invertebrates. It is also the principal structural ingredient of coral.
- Vaterite appears in freshwater pearls and fish otoliths (a bone in the head).
- Silica is preponderant in diatoms, unicellular animals floating in the sea. It is what forms sponge spicules.
- Nucleation in biological systems is affected by the presence of organic substances. The theory of nucleation provides valuable guidance on the morphology of minerals in biological systems.

- Magnetite appears in two places: as a navigational sensor in the skulls of tuna and pigeons, and also in the chiton radula; it has a hardness three times that of enamel.
- Minerals in the deep vent holes in the bottom of the ocean have unique compositions. Iron sulfides and pyrite are present in the shell of the deep vent hole gastropod *Crysomallon squamiferum*.

Exercises

Exercise 5.1 Knowing the volume percentage of each constituent in bone (40 vol.% hydroxyapatite (HAP); 10 vol.% water; 40 vol.% collagen), calculate the weight percentages. Given: density of collagen = 1.03 g/cm^3; density of HAP = 3.116 g/cm^3.

Exercise 5.2 If the density of dry bone is 2.06 g/cm^3, and assuming that it is composed of 50 vol.% HAP and 50% collagen, determine the density of HAP.

Exercise 5.3 Using the Jäger and Fratzl equation (5.10), determine the Young modulus of bone for four different aspect ratios of the HAP platelets: 2, 20, 30, and 100. Assume that the volume percentage of HAP is 40%. Given:

E_{HAP} = 50 GPa;
E_{Coll} = 1 GPa;
G_{Coll} = 0.4 GPa.

Exercise 5.4 How do the values for the Young modulus of bones compare with predictions of the Voigt and Reuss averages?

Exercise 5.5 What are the distinguishing microstructural features in biominerals? Explain how these microstructural features affect the mechanical properties.

Exercise 5.6 A composite consists of 60 vol.% polymeric matrix (shear modulus 1 GPa) and 40 vol.% mineral reinforcements (elastic modulus 100 GPa). Calculate the elastic modulus of the composite if the mineral constituents are in the following geometries:

(a) sphere (aspect ratio = 1);
(b) platelet (aspect ratio = 10);
(c) needle (aspect ratio = 100).

Exercise 5.7 Explain how aragonite crystals tend to have a hexagonal base, whereas the structure is orthorhombic.

Exercise 5.8 Derive expressions for the volume of the orthorhombic, rhombohedral, and hexagonal unit cells exhibited by the three polymorphs of calcium carbonate (they are used in Example 5.1).

Exercise 5.9 What is the origin of the name aragonite (Google the question)?

Exercise 5.10 Calculate the critical radius for the nucleation of calcium carbonate crystals if their surface energy is ~1 J/m^2 (de Leeuw and Parker, 2008). The free energy difference is equal to 100 MJ/m^3.

Exercise 5.11

(a) Calculate the ratio between the heterogeneous and homogeneous critical energies for nucleation if the angles with the substrate are 10, 90, and 170°.

(b) Explain the differences and discuss the effect on the nucleation rate.

Part II Biological materials

Nature has evolved a palette of biological materials to address different structural requirements such as:

- hardness,
- toughness,
- stretchability,
- light weight.

The intricate and ingenious hierarchical structure is responsible for the outstanding performance. Toughness is conferred by the presence of controlled interfacial features, buckling resistance can be achieved by filling a slender column with a lightweight foam, and armor protection is accomplished by small dermal plates with unique attachment arrangements, resulting in controlled and prescribed flexibility. In Chapters 6–10 we present and interpret selected examples of biological materials. In addition to the structural requirements, there are also functional requirements such as adhesion and optical properties.

The number of elements and compounds that can be synthesized at ambient temperature and in aqueous environments is limited, and therefore the architecture of the structure is of utmost importance.

We introduce the different classes of biological materials in these chapters, following the Wegst–Ashby classification. These were defined in Chapter 2 (Fig. 2.11) as:

- biominerals (Chapters 6 and 7),
- biopolymers (Chapter 8),
- bioelastomers (Chapter 9),
- biocellular materials (foams) (Chapter 10).

For each class of biological material, we provide illustrative examples. The field of biominerals is so broad that we divide it into two chapters. Chapter 6 is solely devoted to silicate- and calcium-carbonate-based composites. Chapter 7 focuses on hydroxyapatite-based composites (mainly bone). We separate biopolymers from bioelastomers based on their response to stresses: biopolymers are more rigid, such as keratin and chitin. We also include tendons and ligaments (essentially collagen) in this chapter, as well as squid beaks.

Bioelastomers are able to withstand large deformations before failure. This distinction with biopolymers is somewhat arbitrary since the constituents are in some cases the same. For instance, collagen can form tendons and ligaments, which are relatively stiff, and skin, which is highly elastic. In Chapter 9 we present a number of constitutive models (equations) developed to describe the response of these highly stretchable bioelastomers, starting with models developed for rubber.

Many applications require a low weight/stiffness ratio and low-density materials such as foams find their use in a number of biological systems. The flexure, torsion, and buckling equations, expressed by well-established derivations from mechanics of materials (derived in Chapter 4 for cells), are universal in their applicability and are applied to biocellular materials in Chapter 10.

6 Silicate- and calcium-carbonate-based composites

Introduction

There are numerous silicate- and calcium-carbonate-based biominerals in biological systems. Their function is, for most of them, protection. They provide a hard shell beneath which the softer organism can survive and prosper.

Exceptions are the sponges. We will provide illustrative examples and describe the structure-mechanical property relations for a few representative examples. We first cover the most important silicates and then move to shells. The number of different species is staggering: over 100 000 living species bear a shell, ranging from a single valve to eight overlapping valves. There are roughly 1000 species of mussel bivalves.

For shells, we rely here primarily on the ones studied by the UC San Diego group: abalone, conch, giant clam, and ocean and river bivalve clams. They possess quite different and unique structures, and are representative of the large number of shell species.

We also describe other carbonate-based hard materials such as the sea urchin, the smashing arm of the mantid shrimp, and the ubiquitous egg shell.

6.1 Diatoms, sea sponges, and other silicate-based materials

6.1.1 Diatoms and radiolarians

Diatoms (the name comes from Greek: two halves) are unicellular algae that have a mineralized shell acting as protection. There are 10 000 species of diatoms, and their shells, called frustules, have a large number of shapes. Nevertheless, they have in common the pillbox construction with an overlap belt. Diatoms secrete a hydrated silica cage ($SiO_2 \cdot nH_2O$), which is not as stiff as calcite, thus it can undergo more deformation per unit load, making it more flexible. The Young modulus of the glass sponge spicule is ~40 GPa (Woesz *et al.*, 2006) and that of the diatom frustule is 22.4 GPa (Hamm *et al.*, 2003), whereas that for calcite is ~76 GPa. However, the tensile strength of silica is considerably higher than that of $CaCO_3$: 540 vs. ~100 MPa. Diatoms contain two valves with a regular set of perforations through which they filter nourishment from the ocean. Figure 2.8 provided close-up pictures of these perforations, which are circular. The diatoms capture 20% of atmospheric CO_2 – so they play a key role in the quality of the atmosphere.

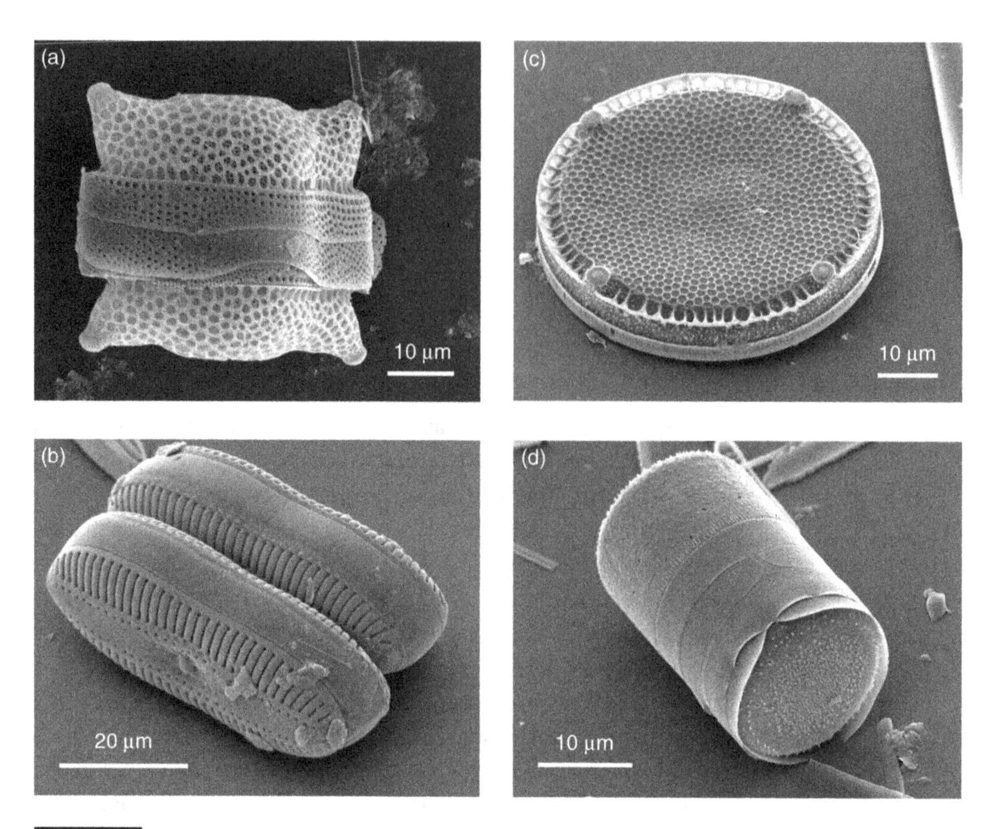

Figure 6.1.

Scanning electron micrographs of diatoms with different morphologies. (a) *Biddulphia reticulata.* The whole shell or frustule of a centric diatom showing valves and girdle bands. (b) *Diploneis sp.* This picture shows two whole pennate diatom frustules in which raphes or slits, valves, and girdle bands can be seen. (c) *Eupodiscus radiatus.* View of a single valve of a centric diatom. (d) *Melosira varians.* The frustule of a centric diatom, showing both valves and some girdle bands. (From http://commons.wikimedia.org/wiki/File:Diatoms.png#filelinks. Image by Mary Ann Tiffany. Licensed under the Creative Commons Attribution 2.5 Generic license.)

Figure 6.1 shows several types of diatoms. In diatoms, the silica is formed in the surface of the cell in a complex tridimensional network that is only partially understood. Each diatom species has a specific biosilica cell wall with regularly arranged slits or pores in the size range between 10 and 1000 nm. Biosilica morphogenesis takes place inside the diatom cell within a specialized membrane-bound compartment termed the silica deposition vesicle. It has been postulated that the silica deposition vesicle contains a matrix of organic macromolecules that not only regulate silica formation, but also act as templates to mediate the growth of the frustules and the creation of the holes and slits (nanopatterning). Using these biosilica-associated phosphoproteins, known as silaffins, Poulsen, Sumper, and Kröger (2003) were able to create a silica assembly with pores having ~100 nm diameters.

Progress toward the goal of synthetically creating frustules was reached when the genome of the marine diatom *Thalassiosira pseudonana* was established (Armbrust *et al.*, 2004) including novel genes for silicic acid transport and formation of silica-based cell walls. Based on this, Hildebrand and co-workers (Hildebrand, 2005, 2008; Hildebrand *et al.*, 2006) proposed that the first step is to identify cell wall synthesis genes involved in structure formation and stated that the completed genome sequence of *T. pseudonana* opens the door for genomic and proteomic approaches to accomplish this (Armbrust *et al.*, 2004). An approach that is also used in other organisms is to modify gene sequences or expression, introduce the modified genes into the diatoms, and to monitor the effect on structure. The ultimate goal of this approach is to produce genetically modified frustules that can be tailored to specific applications through biosilicification processes. Indeed, some progress has been made in this direction, illustrated in Fig. 6.2. Figure 6.2(a) represents the valve of a normal diatom; Fig. 6.2(b) shows the effect of treating the culture with 1,3-diaminopropane dihydrochloride (DAPDH). The silicification is altered, and the arrows show regions in which it has not occurred.

For example, the diatom, a single-celled marine organism that builds a hydrated, amorphous silica cage ($SiO_2{\cdot}nH_2O$) around itself may, at first glance, seem to involve two levels of hierarchy: the inorganic shell and the internal cell. However, at the microscopic scale, the cage structure can take on a surprising number of configurations. About

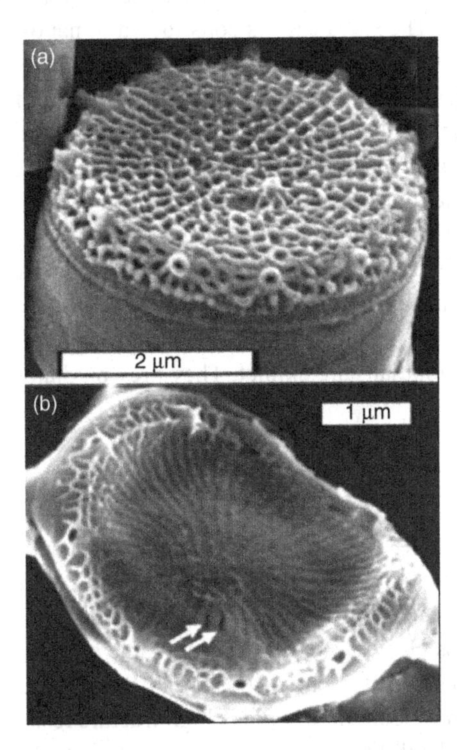

Figure 6.2.

Effect of the polyamine synthesis inhibitor DAPDH on valve formation in the *Thalassiosira pseudonana* diatom. (a) Valve from untreated culture; (b) valve from culture treated with 10 mm 1,3-diaminopropane dihydrochloride. Arrows denote areas where silicification has not occurred. (Reprinted with permission from Hildebrand (2008). Copyright 2008, American Chemical Society.)

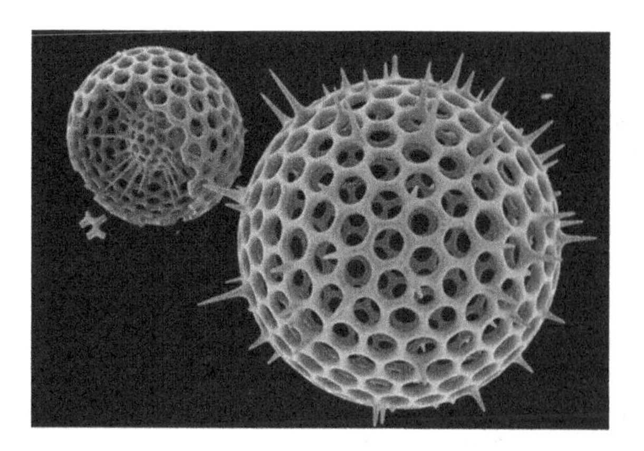

Figure 6.3.

SEM micrograph showing two radiolarians. The silica-made shells are perforated with holes and have little spikes (spicules). (Used with the kind permission of Michael Spaw.)

60 000 different diatoms have been identified, and marine scientists speculate this is only about 10% of the total number. The silica cage is constructed of nano-sized ribs, which are composed of 50 nm diameter particles; the cell itself is a complex arrangement of subcellular elements. Further probing will reveal ordered assemblies of proteins, lipids, and polysaccharides that form the subcellular constituents and nanoporous regions in the $SiO_2{\cdot}nH_2O$ cage (Hildebrand *et al.*, 2006). Thus, one must indicate the smallest length scale that will be used to define the number of hierarchical levels.

Radiolarians have some similarities to diatoms, since they also "float" in the ocean as zooplankton. However, they are ameboid protozoa. Their dimensions vary from 30 μm to 2 mm. Some are icosahedral shaped, as shown in Fig. 6.3. They often have spikes (for protection), in contrast to diatom frustules. Of the 10 000 diatom types, 90% are alive today, whereas 90% of the radiolarians are extinct. They prey on diatoms.

6.1.2 Sponge spicules

Sea sponges (*Porifera*) have fibrous skeletons that are classified depending on the chemical constituents of the skeleton. Skeletons made of calcium carbonate are in the class *Calcarea*, those with silica are in the class *Hexactinellidae* and those of protein fibers (spongin) are in the class *Demospongiae*. The inorganic sea sponge spicule is an excellent example of a well-designed ceramic rod. It can be up to 3 m in length and 0.8 cm in diameter (Woesz *et al.*, 2006) and displays high fracture resistance, and, in the case of *Hexactinellidae*, high flexibility.

Sea sponges often have long rods (spicules) that protrude outward. Their outstanding flexural toughness was first discovered by Levi *et al.* (1989), who were able to bend a 1 m rod, having a diameter similar to a pencil, into a full circle. This deformation was fully

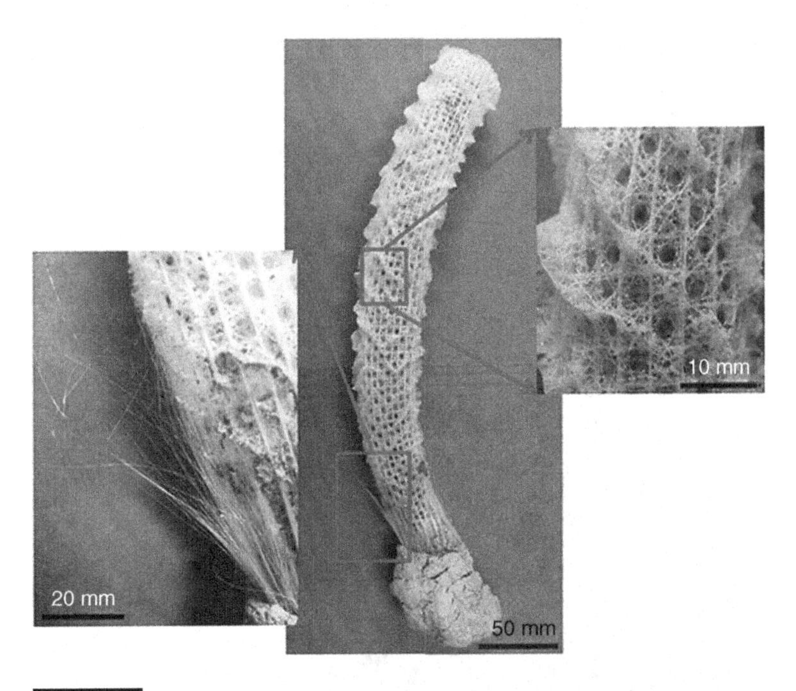

Figure 6.4.

Venus's flower basket (*Euplectella aspergillum*). Two shrimp live inside the cage and procreate; their offspring are able to escape. This is a symbiotic relationship whereby shrimp clean the cage and are fed. Left: spicules; right: detail of structure.

reversible. Additionally, these rods are multifunctional and carry light. The optical properties were studied by Aizenberg *et al.* (2005). They will be presented in Chapter 12.

The siliceous Venus's flower basket (*Euplectella aspergillum*) is shown in Fig. 6.4. It has an elaborate structure that appeals to mechanical engineers due to the regular arrays of longitudinal, radial, and helical fibers (both right and left hand) that provide strength with a minimum of silica. The basket holds as prisoners a pair of breeding shrimp that keep the basket clean while the latter protects them from predators. The offspring are small enough to escape; this is an example of a symbiotic relationship. The basket is held to the ocean floor by spicules radiating from the basket, which also have the unique concentric laminated structure. These fibers must flex with wave action but remain firmly embedded in the floor.

The structural hierarchy of the hexactinellid sponge spicule is a remarkable example of nature's ability to create sophisticated composites from relatively weak constituent materials. The spicules, which are found at the base of the silica basket (seen in Fig. 6.4), greatly resemble the fragile fibers used in modern fiber optics (Sundar *et al.*, 2003). They will be discussed in Chapter 12. As seen in Fig. 6.5(b), the microcomposite design of this natural rod creates remarkable toughness, especially in comparison to its industrial counterpart (Aizenberg *et al.*, 2005). Figure 6.5(a) shows the flexural stress as a function of strain. The flexure strength is between three to four times that of monolithic synthetic silica. Additionally, an important difference exists between the two: whereas the monolithic

(a)
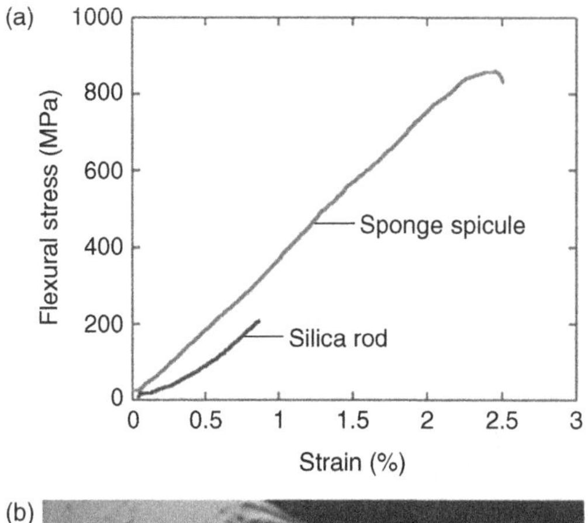

(b)
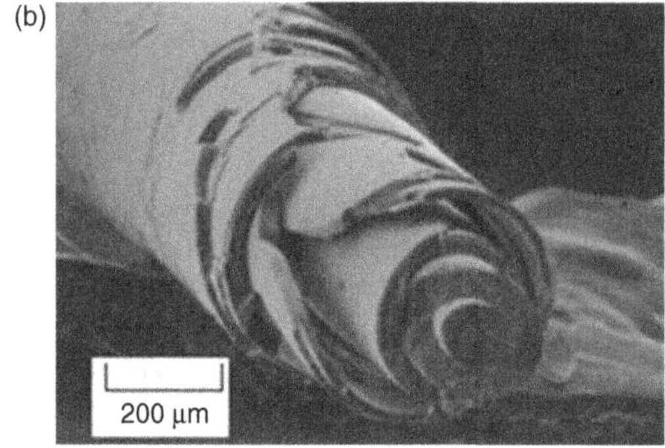

Figure 6.5.

(a) Flexural stress vs. strain for monolithic (synthetic) silica rod and sea spicule; (b) fractured spicule on sea sponge. (Figure courtesy G. Mayer, University of Washington.)

silica rod breaks in a single catastrophic event, the spicule breaks "gracefully" with progressive load drops. This is the direct result of the arrest of the fracture at the "onion" layers. These intersilica layers contain an organic component, which has been identified by Cha *et al.* (1999) as silicatein (meaning a silica-based protein). Figure 6.5(b) shows a fractured *Hexactinellida* spicule (much smaller than the one studied by Levi *et al.* (1989)), which reveals its structure. This spicule, which has been studied by Mayer and Sarikaya (2002), is a cylindrical amorphous silica rod and has an "onion skin" type structure which effectively arrests cracks and provides an increased flexural strength.

Each silica rod is composed of a central pure silica core of approximately 2 μm in diameter surrounded by concentric striated shells of decreasing thickness (see Fig. 6.6(a)) (Sundar *et al.*, 2003). The individual shells are separated by a thin organic layer (silicatein), which is marked by an arrow in Fig. 6.6(b). The mechanical toughness of the material is highly dependent on the striated layers, as they offer crack deflection and energy absorption at their interfaces (Levi *et al.*, 1989; Sarikaya *et al.*, 2001; Aizenberg *et al.*, 2005). The gradual reduction in the thickness of the layers as the radius is increased is clearly evident in Fig. 6.6(c).

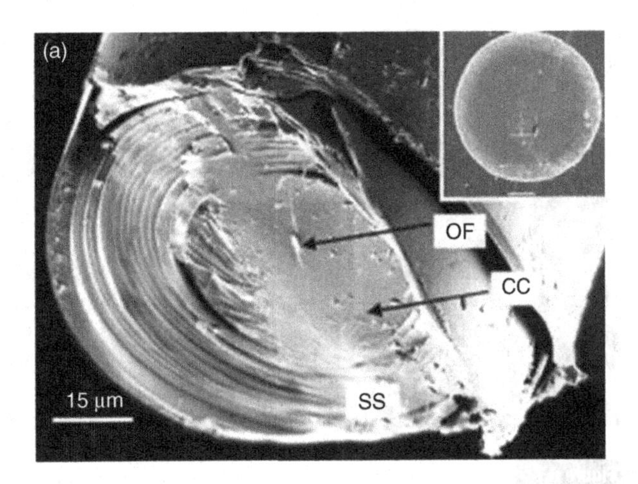

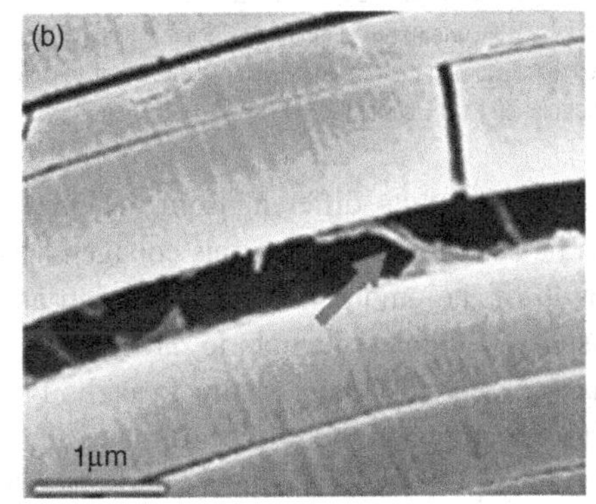

Figure 6.6.

Microstructure of a sponge spicule. (a) Fracture surface of a typical spicule in a strut, showing different levels; an organic polymer (silicatein) exists between the layers. (Reprinted by permission from Macmillan Publishers Ltd.: *Nature* (Sundar *et al.*, 2003), copyright 2003. (b) SEM of a fractured spicule, revealing an organic interlayer. (c) SEM of a cross-section through a typical spicule in a strut, showing its characteristic laminated architecture. (Taken from Aizenberg *et al.* (2005), with permission from Professor Aizenberg.)

Cha *et al.* (1999, 2000) demonstrated that silicatein can hydrolyze in spicules of the sponge *Tethya aurantia*, and condensed the precursor molecule tetraethoxysilane to form silica structures with controlled shapes at ambient conditions. This principle was used to generate bioinspired structures by using synthetic cysteine–lysine block copolypeptides that mimic the properties of silicatein (Cha *et al.*, 2000). The copolypeptides self-assemble into structured aggregates that can produce regular arrays of spheres and columns of amorphous silica.

Another fascinating spicule from a sea sponge is the *Hyalonema sieboldi*. This is also called the glass rope sponge, and it contains anchoring spicules that are remarkable for their size (up to 1 m), durability, flexibility, and optical properties. Figure 6.7(a) shows a basal spicule in *H. sieboldi*. It can be seen how it can be bent into a circle. Ehrlich and Worch (2007) describe their structure, with emphasis on the elaborate collagenous network,

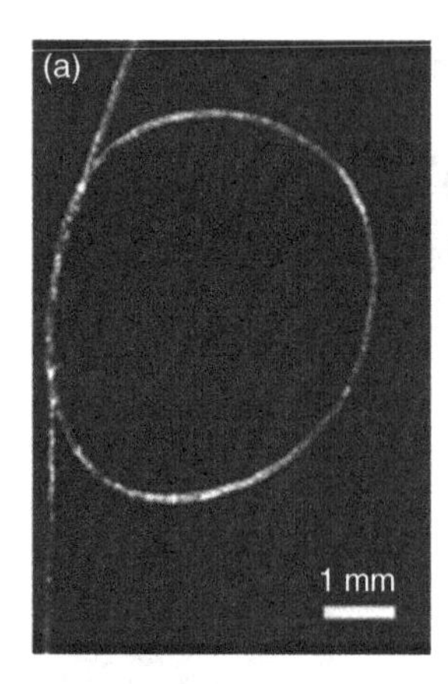

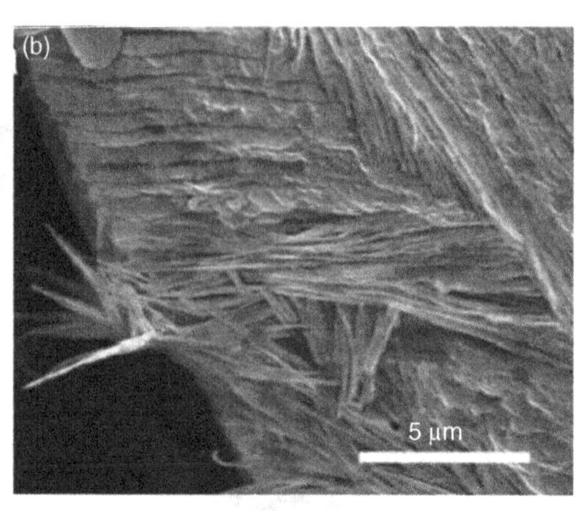

Figure 6.7.

(a) Unique flexibility of basal spicules of *H. sieboldi.* (Taken from Kulchin *et al.* (2009), with kind permission from Springer Science+Business Media B.V.) (b) SEM micrographs showing the twisted plywood orientation of collagen microfibrils. (Reprinted from Meyers *et al.* (2008b), with permission from Elsevier.)

which has a twisted plywood structure quite different from the one exhibited by the *Hexactinellida* sponge. This can be seen in Fig. 6.7(b); this collagen acts in a mediation role for the nucleation and growth of the amorphous silica. Ehrlich and Worch (2007) comment on the evolutionary aspects. Silicon is thought to have been a first stage in the inorganic to organic evolutionary process, leading finally to carbon-based organisms. Silicon is the most common of the elements on the surface of the earth after oxygen, and the ocean floors are covered with amorphous silica sediment, most of which results from living organisms. Ehrlich *et al.* (2007a,b) indicate that chitin is a component of the skeletal fibers of marine sponges, which have intricate elaborate structures.

In the calcareous sponge *Pericharax heteroraphis* the spicules behave as single crystals, verified by Laue diffraction that show a single orientation. High-resolution TEM and AFM studies show that the "single crystals" are actually composed of ~5 nm nanoclusters with organic matter between them, as shown in Fig. 6.8(a) (Sethman *et al.*, 2006). This is similar to the "single-crystal" tablets of aragonite in abalone nacre and in sea urchin spicules, which are also composed of nanocrystals. In Fig. 6.8(b), the conchoidal fracture surface characteristic of a single crystal is evident.

6.2 Mollusc shells

6.2.1 Classification and structures

Mollusc shells are primarily composed of calcium carbonate. In many species, they form structural arrangements that are "glued" together by biopolymer adhesives. Mollusc shells have evolved to incorporate various design strategies in the arrangement of calcium carbonate. Figure 6.9 shows a classification of the principal shell structures

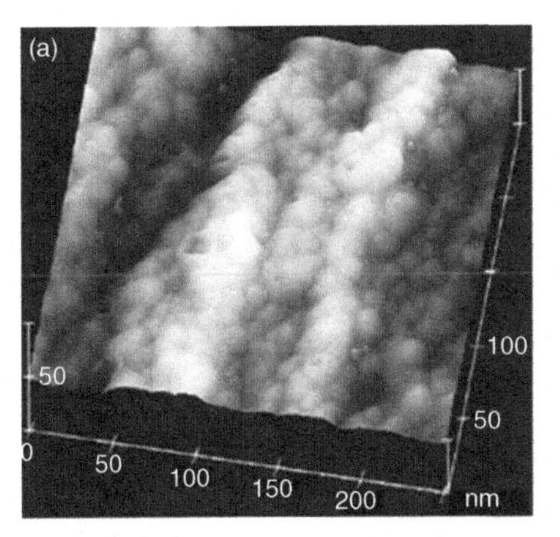

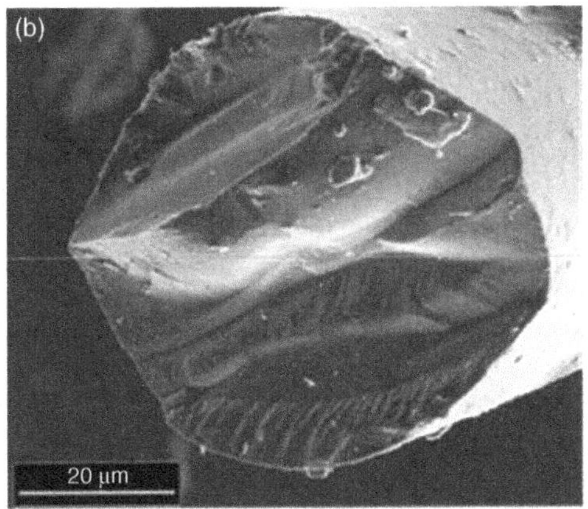

Figure 6.8.

(a) AFM height image of the conchoidal fracture surface of a spicule from the calcareous sponge *Pericharax heteroraphis*, revealing the nanocluster structure; (b) SEM micrograph of the fracture surface. (Reprinted from Sethman *et al.* (2006), with permission from Elsevier.)

according to Currey and Taylor (1974), who divided the microstructures of the shells into nacre (columnar and sheet), foliated (long thin crystals), prismatic, crossed-lamellar (plywood like), and complex crossed-lamellar. Kobayashi and Samata (2006) expanded this classification, identifying more than ten morphological types of bivalve shell structures. Among others, they described the structures as simple homogeneous, nacreous, foliated, composite prismatic, and crossed-lamellar. It should be noted that there is a significant variation in this classification. Often, researchers classify the structures into different names according to their own interpretations.

Shells have fascinated mankind since prehistory. They have been found in Neanderthal burial sites, evidencing the attraction they have exerted. Indeed, Aristotle and Pliny the Elder were among the first to write about shells; it seems that it was Aristotle who coined the name "mollusca," meaning soft bodied. Two aspects of seashells are of esthetic significance:

- the mother-of-pearl coloration, which results from the interference of visible light with the tiles that comprise nacre (~0.5 μm thickness, approximately equal to the wavelength);
- their multiple and intricate shapes.

D'Arcy Thompson showed that for many shells their shape is a derivative of the logarithmic spiral (Thompson, 1968). Figure 6.10(a) shows the top of an abalone shell. There are two spiral lines indicated: one follows the pattern of perforations which circulate the water and are closed as the shell grows; the second represents the markings of successive growth surfaces. The logarithmic spiral which exists in many shells is shown in Fig. 6.10(b). It can be understood as being formed by the aggregation of mineral, composed of two vectors: one in the radial direction ($d\vec{r}$) and one in the

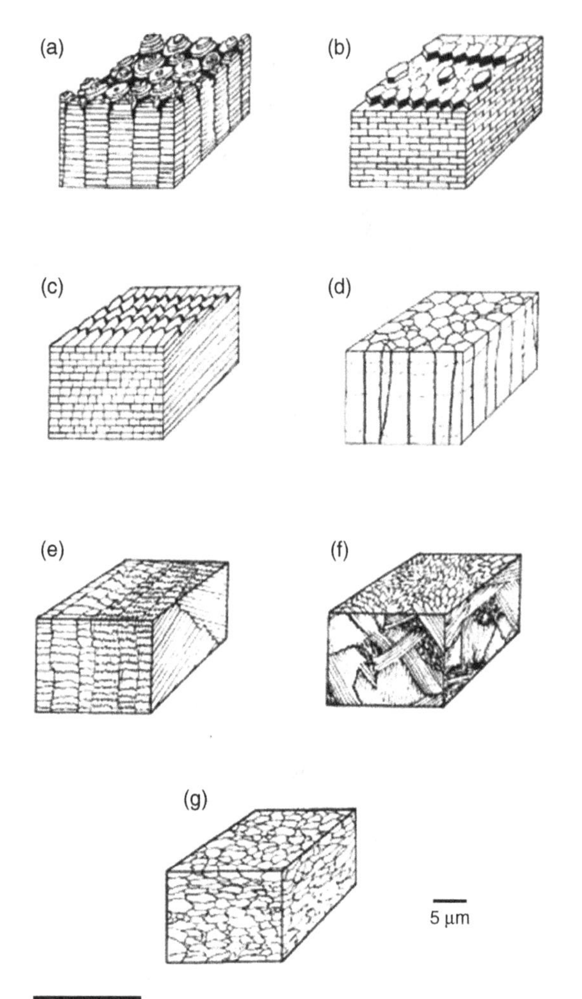

Figure 6.9.

Currey–Taylor classification of shell microstructures. Note the difference in size of the structural units between types. The blocks are oriented so that their vertical faces are in the thickness of the shell. The blocks would be loaded in the direction of their longer dimension. (a) Columnar nacre – aragonite. (b) Sheet nacre – aragonite. (c) Foliated calcite. (d) Prismatic calcite or aragonite. (e) Crossed-lamellar aragonite. Cross-foliated structure is similar but made of calcite. (f) Complex crossed-lamellar aragonite. (g) Homogeneous. (From Currey and Taylor (1974); used with permission from John & Wiley Sons, Inc.)

tangential direction ($d\bar{s}$). If the ratio of the magnitude of these two vectors is constant throughout the growth process, the angle α is unchanged:

$$\tan\alpha = \frac{dr}{ds}. \tag{6.1}$$

But we have

$$ds \cong r\,d\theta, \tag{6.2}$$

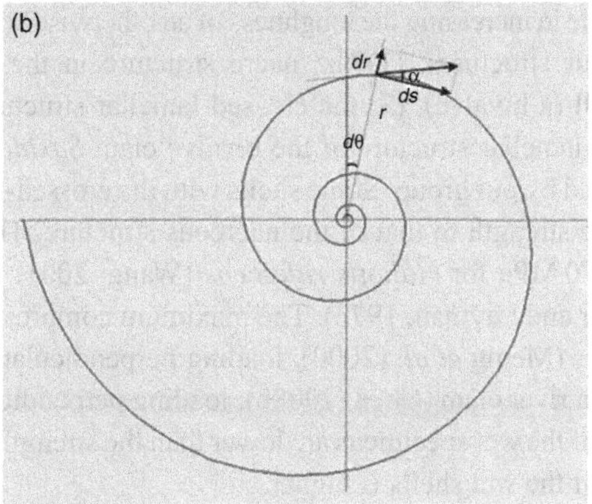

Figure 6.10.

(a) Abalone shell; the two lines indicate the logarithmic spiral fashion of growth. (b) The growth vector of a logarithmic spiral consists of two vectors: one in the tangential and the other in the radial direction. The ratio of the magnitude of these two vectors is constant throughout the growth process, and the angle α remains unchanged.

where r is the radial coordinate of a point along the curve and θ is its angular coordinate. Substituting Eqn. (6.2) into (6.1) we obtain

$$\frac{dr}{r} = d\theta \tan \alpha. \tag{6.3}$$

On integrating we get

$$\ln r = \theta \tan \alpha + C \tag{6.4}$$

or

$$r = e^{C} e^{\theta \tan \alpha}. \tag{6.5}$$

Equation (6.5) expresses the logarithmic spiral curve. Skalak, Farrow, and Hoger (1997) present a detailed analysis and apply the concept of a logarithmic spiral to other biological components, such as horns. In particular, the fascinating horns of antelopes are a beautiful illustration of this concept.

Other than their esthetic attributes, these shells provide the primary means of protection for the soft bodies of the animals they house. They are permanent encasements of body armor, which must be strong enough to withstand the impact and compression capabilities of the sea and the predators within it. Mollusc shells consist of one or more ceramic phases and a small fraction (0.1–5%) of proteins. These ceramic phases alone, i.e. calcium carbonate ($CaCO_3$), are not suitable as structural materials because of their inherent brittleness. However, when combined into these intricate natural structures a resulting biocomposite with outstanding mechanical properties is created. This is true for many biological materials (Mayer and Sarikaya, 2002). The micro- and macrostructure of these shells play a significant role in increasing the toughness of an otherwise brittle ceramic base.

We focus here on four structures: (1) the nacre structure in the abalone, (2) the Araguaia river clam shell (a bivalve), (3) the crossed-lamellar structure of conch, and (4) the complex crossed-lamellar structure of the bivalve clam *Saxidomus purpuratus*. These were all investigated by our group. Some shells with the crossed-lamellar structure can exhibit a comparable strength to that of the nacreous structure. The largest flexure strengths reported are 370 MPa for *Haliotis rufescens* (Wang, 2001) and 360 MPa for *Pinctada maxima* (Taylor and Layman, 1972). The maximum compressive strengths are 540 MPa for *H. rufescens* (Menig *et al.* (2000), loading perpendicular to lamellae) and 567 MPa for the Araguaia river clam (Chen (2008b), loading perpendicular to lamellae). In general, the strengths of the wet specimens are lower than the strengths of the dry ones, although the toughness of the wet shells is higher.

6.2.2 Nacreous shells

One might think that the study of nacre is of primarily scientific interest, in the sense of knowledge-driven research. Such is not the case. Pearls, in particular freshwater pearls, are an important product. The majority of Chinese freshwater pearls are raised in lakes and ponds in Zhejiang Province, which produced around 1500 tons of freshwater pearls in 2005, about 73% of the world freshwater pearl market. The production of pearls accounts for billions of dollars per year (Murr and Ramirez, 2012). Pearls owe their unique appearance to the thin layers (~0.4 μm) of aragonite.

The schematic of the longitudinal cross-section (Fig. 6.11(a)) of the abalone shell (*Haliotis*) shows two types of microstructure: an outer prismatic layer (calcite) and an inner nacreous layer (aragonite) as observed by Nakahara, Kakei, and Bevelander (1982). The two forms of $CaCO_3$ have the following structures: calcite (rhombohedral) and aragonite (orthorhombic). Calcite has a polycrystalline structure that is equiaxed. The polycrystalline nature of the layer with equiaxed grains ~50 μm is clearly observed under polarized light microscopy, as shown in Fig. 6.11(b).

On the other hand, nacre has a structure that is quite different. The structure of nacre (the inside portion of the shell, shown in Fig. 6.11(a)) within the shells of abalone

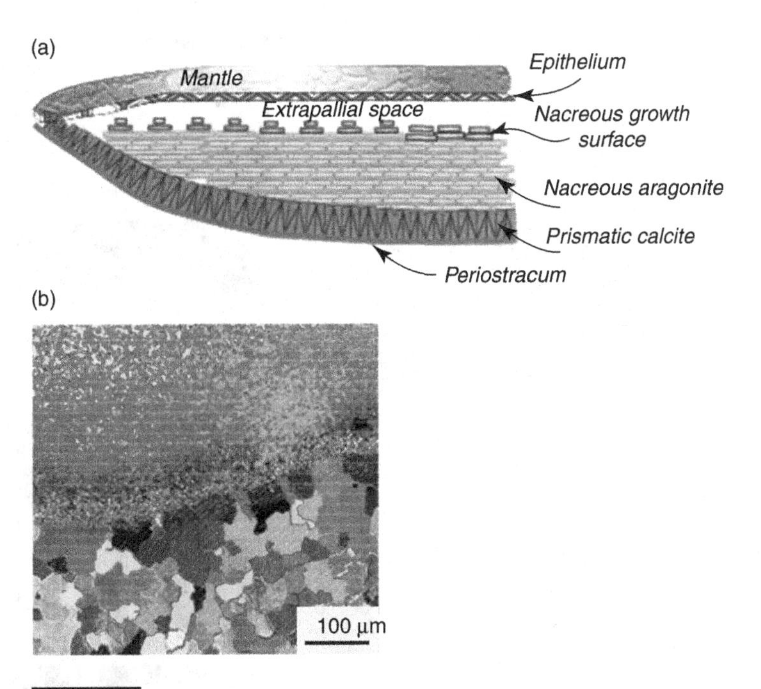

Figure 6.11.

(a) Structure of typical abalone shell. (Reprinted with permission from Zaremba *et al.* (1996). Copyright 1996, American Chemical Society.) (b) Calcitic layer in abalone shell; the polycrystalline nature of the layer with equiaxed grains having approximately 50 μm diameter is clearly seen under polarized light. (Used with permission from Schneider *et al.* (2012).)

consists of a tiled structure of crystalline aragonite that is often called "brick and mortar" because of its geometric similarity to masonry.

Figure 6.12(a) shows schematically the brick-and-mortar microstructure found in abalone nacre (Sarikaya, 1994), and Fig. 6.12(b) shows the layered structure in a TEM micrograph (Menig *et al.*, 2000). In Fig. 6.12(c) the "*c*-axis" orientation is shown. The staggering of tiles in adjacent layers is also clearly visible. The tiles have the *c*-direction of the orthorhombic cell perpendicular to the hexagonal side (Fig. 6.12(c)). Moreover, there is a very high degree of crystallographic texture characterized by a nearly perfect "*c*-axis" alignment normal to the plane of the tiles. The similarity of orientation between tiles on adjacent layers is demonstrated by the successive selected area diffraction patterns shown in Fig. 6.12(d) (Meyers *et al.*, 2008b). It will be shown in the following how the growth creates interconnected tiles having the same crystallographic orientation.

In the case of the abalone nacre, the mineral phase corresponds to approximately 95 wt.% of the total composite. The deposition of a protein layer of approximately 20–30 nm comprises intercalated aragonite platelets, which are remarkably consistent in dimension for each animal regardless of age (Lin and Meyers, 2005). However, there are differences when examining varying species of nacre-forming animals: the thickness of the tiles in the abalone shells is approximately 0.5 μm, whereas it is around 1.5 μm for a bivalve shell found in the Araguaia river (Brazil), thousands of miles from the ocean. In the case of abalone

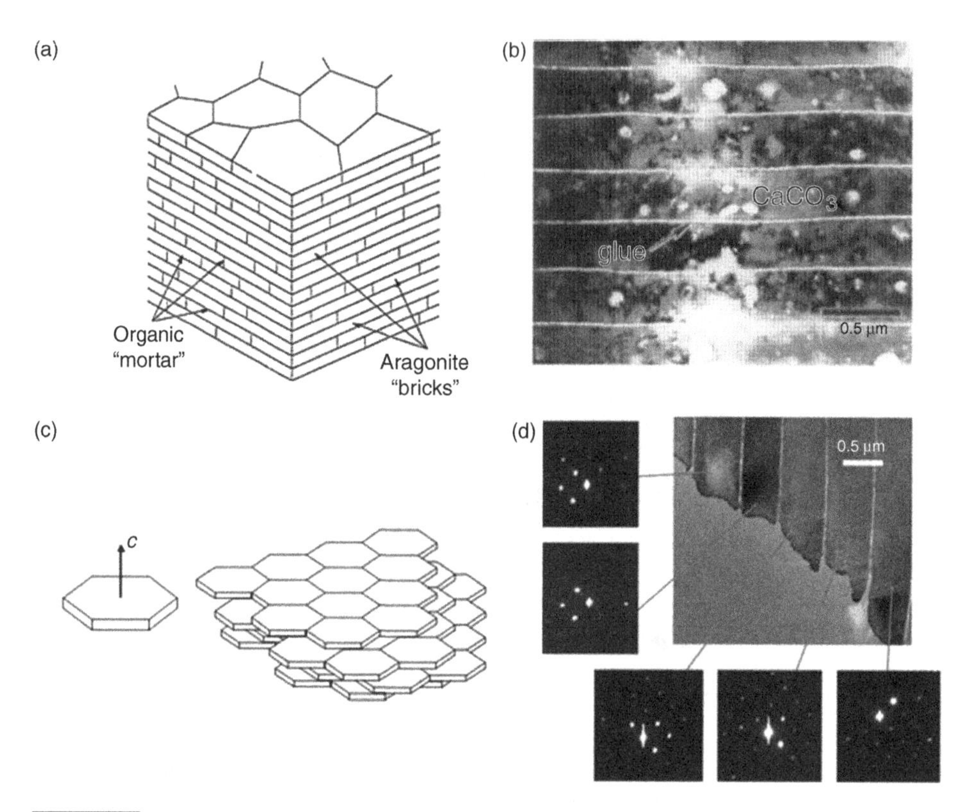

Figure 6.12.

"Brick-and-mortar" structure of nacre in the abalone shell. (a) Schematic drawing. (b) Transmission electron microscope (TEM) micrograph showing the aragonite layers ($CaCO_3$) of approximately 0.5 μm thickness and organic interlayer. (Reprinted from Menig *et al.* (2000), with permission from Elsevier.) (c) Overlap of tiles with *c*-direction marked (see Fig. 5. 3). (d) TEM with selected area diffraction patterns showing the same orientation for different layers ~0.4 μm thick, grown in the *c*-direction, exposing the (001) face. TEM selected area diffraction patterns of adjacent tiles show same crystallographic orientation. The organic layer between the tablets has a sandwich structure comprising a central core of chitin fibers and surface layers with pores 5–80 nm in diameter. (Adapted from Meyers *et al.* (2008b), with permission from Elsevier.)

nacre, the thickness of the tablets is on the order of the wavelength of light, and this creates the beautiful colored, iridescent surface of nacre. This is called mother of pearl. As will be seen later, the organic layer between the tablets has a sandwich structure comprising a central core of chitin fibers and surface layers with pores ~50 nm (5–80 nm) in diameter.

Box 6.1 Sutures, screws, and plates

The mechanical action of sewing (skin and other soft tissues), stapling, and attaching broken bones by screws, plates, and other devices is the most common utilization of biomaterials. Virtually every person is subjected to one of these procedures during his or her life. It is also the oldest – gold wire has been found attaching the teeth of Egyptian mummies. There is also ample evidence that closing of wounds by sutures was widely practiced 3500 years ago.

Sutures

One of the authors (MAM) vividly recalls hunting peccary in the hinterlands of Brazil. The guide, who lived in the bush with his wife, and numerous children and dogs, carried with him a wooden needle and some sort of thread. During the hunt, the sharp tusks of the peccary created huge gashes in one of the dogs. After the hunt was over, and two peccaries had been circled by dogs and duly shot, the guide patiently sewed the dog's wound as he lay stoically, only uttering an occasional whine. Thus, the procedure is simple.

There are essentially two types of sutures: biodegradable, which is absorbed by the organism, and nonabsorbable, which is permanent. The two types can also be classified into biological and synthetic, and as mono- or multifilament. The following are the most common types of suture.

Catgut Actually, this is made from sheep submucosa. It can be treated with chromic acid to increase the cross-linking and strength. The addition of chromic acid extends the life from 3–7 days to 20–40 days. The catgut suture consists of denatured (all cells removed)) collagen. Since the strength and ductility of collagen are so dependent on hydration, the needle and suture are kept in a physiological solution prior to use.

Synthetic absorbable suture The most common types are PGA (polyglycolic acid) and PGA/PLA (polylactic acid).

Nonabsorbable suture

- polyester (PET): heart valves, vascular prostheses;
- polyamide (Nylon 6, 6, 6): skin, microsurgery, tendon repair;
- stainless steel: abdominal and sternal closures, tendon repair.

In cases where the wound scar is not critical, staples are used. This is a more rapid procedure and produces more consistent healing. The principal material used is stainless steel, although titanium and PGA absorbable staples are also used. Figure B6.1 shows an array of staples closing the skin after a hernia operation.

Plates and screws

Although rudimentary and workshop-like in appearance, screws and plates are the workhorses of modern orthopedic surgery, and are routinely used to repair fractures, whereas in the past the procedure depended on the ability of the medical expert. The procedure of putting the bone parts in their correct position is called *reduction*, and consists of bringing the bones into alignment prior to immobilization. The principal advantage of surgery is that one is ensured that the bones are placed in the proper position. It is also necessary if the bone comminutes (breaks into many pieces; from *minutus*, meaning small in Latin) on fracture.

Figure B6.2(a) shows a broken metacarpal thumb bone as a result of a bicycle accident. The fracture is at ~45° to the longitudinal bone axis and was caused by compression, i.e. the hand hitting the pavement and absorbing the shock of the fall. The placement of two screws corrected the bone to its original configuration (Fig. B6.2(b)).

Box 6.1 (cont.)

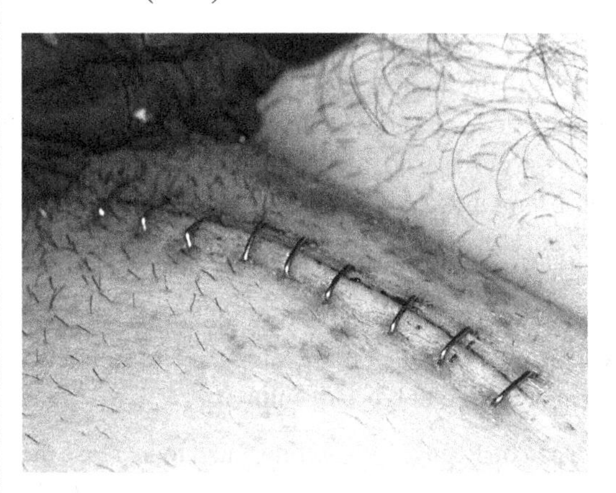

Figure B6.1.

Surgical staples in groin after hernia correction operation. (From http://commons.wikimedia.org/wiki/File:Surgical_staples3.jpg. Image by Garrondo. Licensed under CC-BY-SA-3.0, via Wikimedia Commons.)

(a)

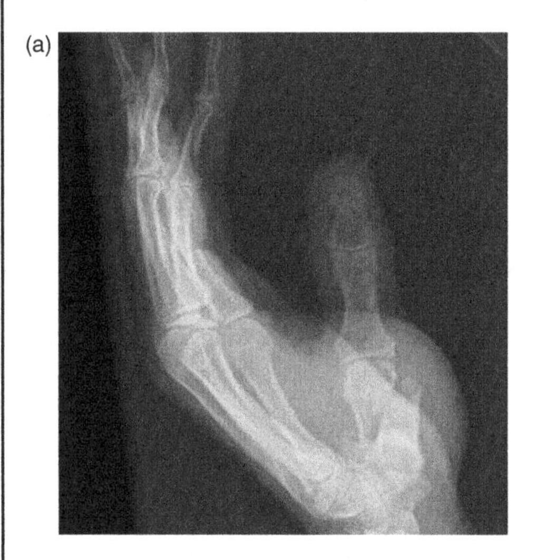

(b)

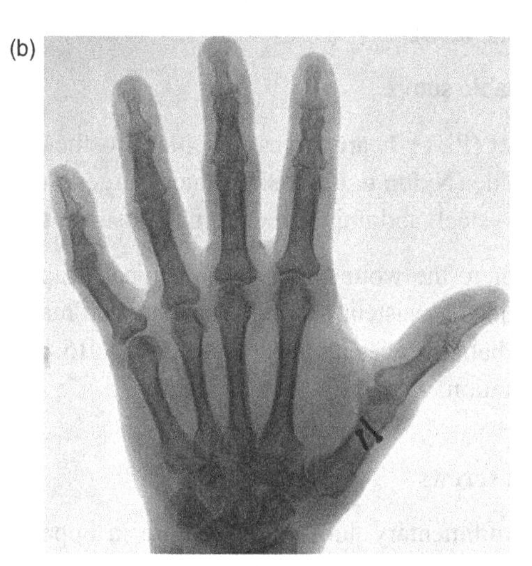

Figure B6.2.

(a) Broken thumb caused by bicycle accident; note the shear failure of the bone at ~45°. (b) Reconstituted thumb after repair surgery. (Figure courtesy of Professor Meyers.)

In long bones, fracture plates or intramedullary pins are inserted. The plates are used in connection with screws. Figure B6.3 shows the fracture in two arm bones: the ulna and the radius. The two bones were fixed using plates and three or four screws on each side (distal and proximal) of the fracture.

The principal material for screws, plates, and pins is stainless steel (316 SS, with 19% Cr and 9% Ni). Co-Cr alloys (Vitallium) and titanium alloys are also used. It is important not to have different metals in

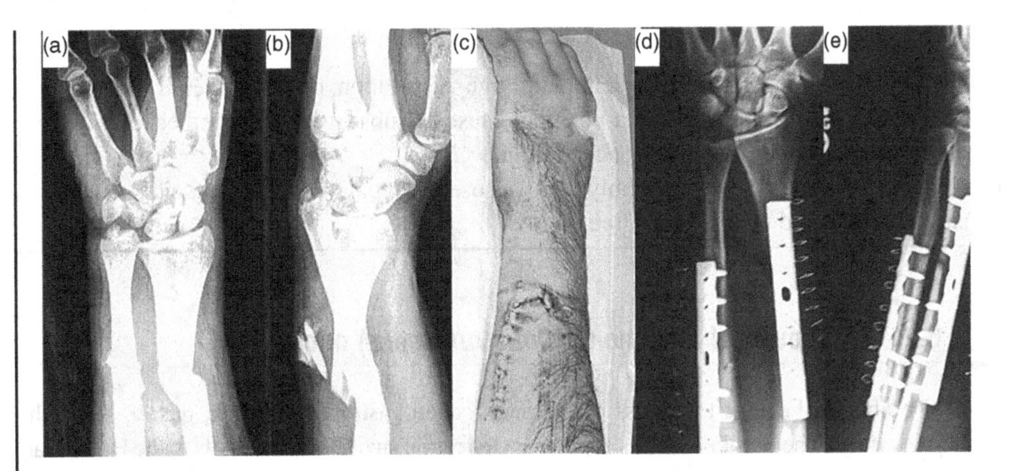

Figure B6.3.

Fracture plate and screws for bone repair. Fractured forearm – ulna (a) and radius (b). (c) Incision made in surgery. (d, e) Fracture plates and screws used for bone repair; note also array of staples. (From http://commons.wikimedia.org/wiki/File:Broken_fixed_arm.jpg by Sjbrown (public domain), from Wikipedia Commons.)

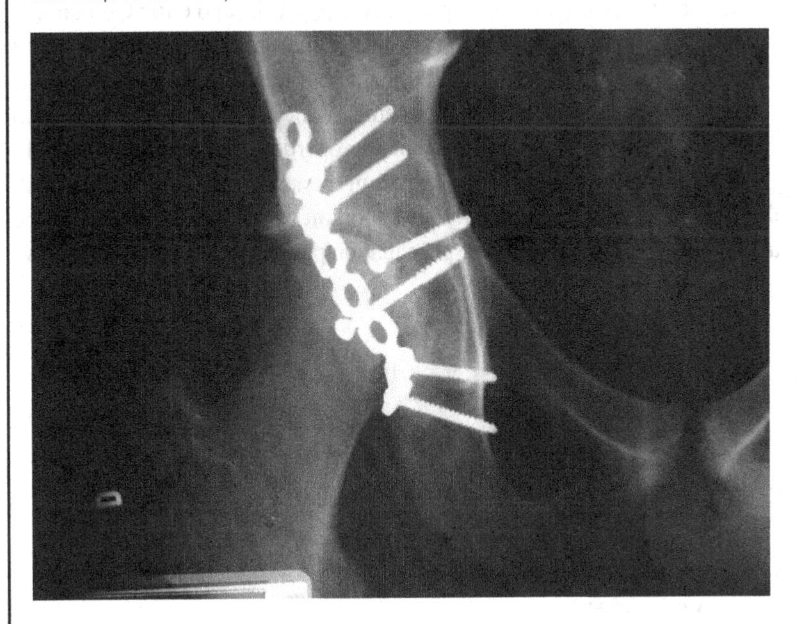

Figure B6.4.

Plate and screws repairing fractured pelvis; note proximal extremity of femur in background. (Figure courtesy of Dr. J. F. Figueiró.)

contact (such as a titanium plate and stainless steel screws) because this creates an electrochemical cell that accelerates corrosion.

A problem of considerable importance is stress shielding by the plates. The region that is subjected to less stress (we use the expression "stress shielding") by virtue of the great difference between the elastic

Box 6.1 (cont.)

moduli of bone and metal undergoes remodeling through dissolution of the mineral. The holes introduced into the bone by drilling are also a potential cause of subsequent fracture because of the decrease in load-bearing area and the stress concentration.

Figure B6.4 shows a plate and screw assembly that was used to repair a pelvis. Thus, it is not only long bones that can be repaired by this procedure.

6.2.2.1 Growth of abalone (*Haliotis rufescens*) nacre

The abalone belongs to a class of molluscs called gastropoda (Greek: *gaster*, stomach; *poda*, feet). The best-known gastropods are terrestrial snails. Abalone falls in the Haliotidae family in the genus *Haliotis*. Out of the 180 known species, *H. rufescens* (red abalone) is the most studied. The lustrous interior of the shell is called nacre (mother of pearl). Several hierarchical levels in the structure of abalone were shown in Fig. 2.2. The first level comprises mesolayers of ~300 μm in thickness, separated by ~20 μm organic, designated as the "green organic" or "brown organic" (Shepherd, Avalos-Borja, and Ortiz Quintanilla, 1995), which was identified simply as conchiolin. Typically, the age of the abalone is found by counting the "green" organic rings in the shell; the number deposited per year depends on the species. These layers were identified by Menig *et al.* (2000), but are not often mentioned in other reports dealing with the mechanical properties of abalone. It is thought that these thick organic layers form in abalone grown in the sea during periods in which there is little calcification. The mesolayers play an important role in the toughness of the abalone nacre, as can be seen from the fracture propagation path in Fig. 6.13, which shows

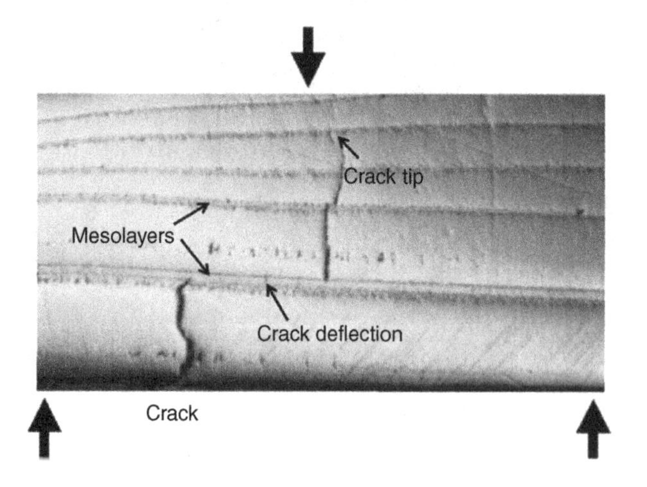

Figure 6.13.

Cross-section of abalone shell showing how a crack, starting at the bottom, is deflected by viscoplastic mesolayers between calcium carbonate lamellae. These are not the individual tiles, but rather layers around 300 μm (and not 400 nm!). (Adapted from Lin and Meyers (2005), with permission from Elsevier.)

a specimen subjected to flexure. The crack starts at the bottom and progresses relatively straight (at the meso-level); however, the mesolayers have a profound effect on its path and it is successively deflected.

The growth of the red abalone shell has been the subject of considerable study starting as early as the 1950s with Wada (1958, 1959), continuing with Watabe and Wilbur (1960), Bevelander and Nakahara (1969), and many others. The work by the UC Santa Barbara group (Fritz *et al.*, 1994; Belcher, 1996; Belcher *et al.*, 1996, 1997; Zaremba *et al.*, 1996; Shen *et al.*, 1997; Belcher and Gooch, 1998; Fritz and Morse, 1998; Su *et al.*, 2002) represents one of the most comprehensive efforts. Shell growth begins with the secretion of proteins that mediate the initial precipitation of spherulitic crystals (region E in Fig. 6.14), followed by a transition from spherulitic to tiled aragonite. There are at least seven proteins involved in the process. The periodic interruption of aragonitic tile growth in the form of mesolayers can possibly be attributed to sporadic interruptions in the animal's diet (Menig *et al.*, 2000; Lin and Meyers, 2005) and water temperature (Lopez *et al.*, 2011). In nacre there is a synergy of structural hierarchy pertaining to many different length scales. The following discussion begins with the formation of macro-scaled elements, followed by microstructural formation, and finally the growth of the nano-scale component of the shell. Figure 6.14 provides a macrostructural view of a cross-section of the inner nacreous layer. Organic bands approximately 20 μm thick can be seen separating larger, 300 μm thick, regions of nacre. These "mesolayers" mark interruptions in nacre growth, and thus are also called growth bands. The inorganic $CaCO_3$ undergoes morphological changes before and after the interrupting growth bands (Lin *et al.*, 2008). As seen in Fig. 6.14, five regions can be identified (direction of growth marked by arrow): tiled (A); block-like aragonite (B);

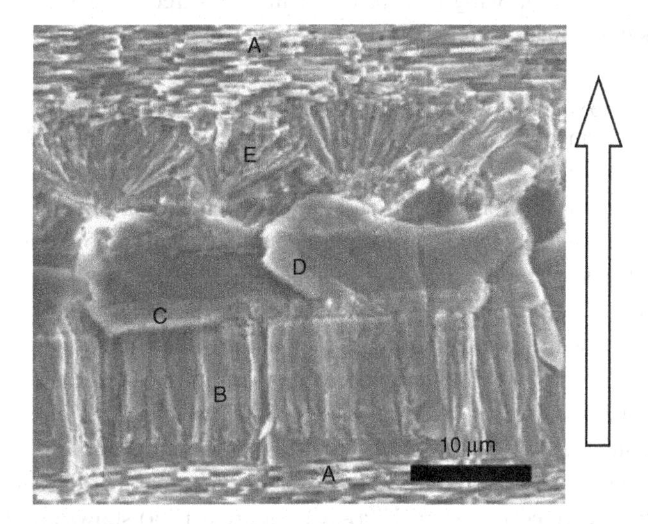

Figure 6.14.

SEM micrograph of fracture surface of *Haliotis rufescens* shell; the direction of growth is marked with an arrow. The mesolayer is composed of five regions A, B, C, D, and E. Growth bands (mesolayers) are darker lines and separate larger regions of nacre. (Adapted from Lin, Chen, and Meyers (2008), with permission from Elsevier.)

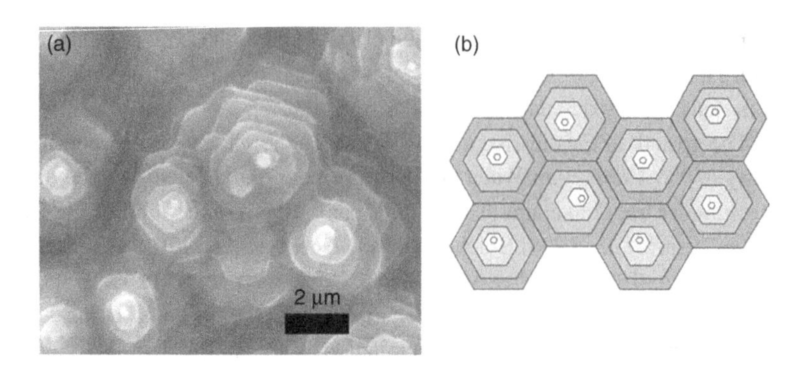

Figure 6.15.

Growth surface of nacre (the organic layer was removed by vigorous water jet on the surface). (a) SEM; (b) schematic drawing showing the same crystallographic orientation. (Adapted from Lin *et al.* (2008), with permission from Elsevier.)

organic/inorganic mix (C); organic (D); and spherulitic (E). The growth sequence is described in greater detail by Lin and Meyers (2005). In Fig. 6.14, the growth occurs from bottom to top. Prior to arrest of growth, the characteristic tiles are replaced by a block-like structure (B). This is followed by the massive deposition of the organic layer, which is initially intermediated with mineralized regions.

Figure 6.15(a) reveals the "Christmas tree" or "terraced growth" pattern described by Shen *et al.* (1997) and Fritz and Morse (1998). The surface of the sample was subjected to a water jet, which removes all traces of the organic layers and reveals the tiles in their entirety. A schematic drawing of adjacent "Christmas trees" is shown in Fig. 6.15(b). Each tile is smaller than the one below it. They have parallel sides, suggesting the same crystallographic orientation (parallel prism planes and basal plane).

The organic layers covering each tile layer may play an important role in providing the scaffolding for formation. First observed and described by Nakahara *et al.* (1982) (see also Nakahara (1991)), they exist and are in place before the growth of the aragonite tile is complete. Figure 6.16(a) represents the possible environment surrounding the aragonite tiles with the presence of the organic scaffolding. The calcium and carbonate ions can penetrate through the organic layer deposited by the epithelium. Electron microscopy of the resulting growth surfaces shows columns of sequential aragonite tiles (Fig. 6.16(b)). The presence of the organic layer, which was preserved, is seen in Fig. 6.16(b). It will be seen later that it contains holes that enable the ions to penetrate the space and ensure the lateral growth of tiles.

Atomic force microscopy (Schäffer *et al.*, 1997), transmission electron microscopy (Song, Soh, and Bai, 2003; Lin *et al.*, 2008), and scanning electron microscopy (Lin *et al.*, 2008) have been used to observe the existence of mineral bridges in abalone nacre; the results are presented in Figs. 6.17(a)–(d). Figure 6.17(a) shows an SEM image of a tile with the lighter regions marked by arrows identified by Song *et al.* (2003) as bridges. Figure 6.17(b) is a cross-sectional TEM micrograph showing the interconnecting mineral bridges. Figures 6.17(c) and (d) show cross-sectional regions of mineral bridges (marked by arrows) and nanoasperities, respectively.

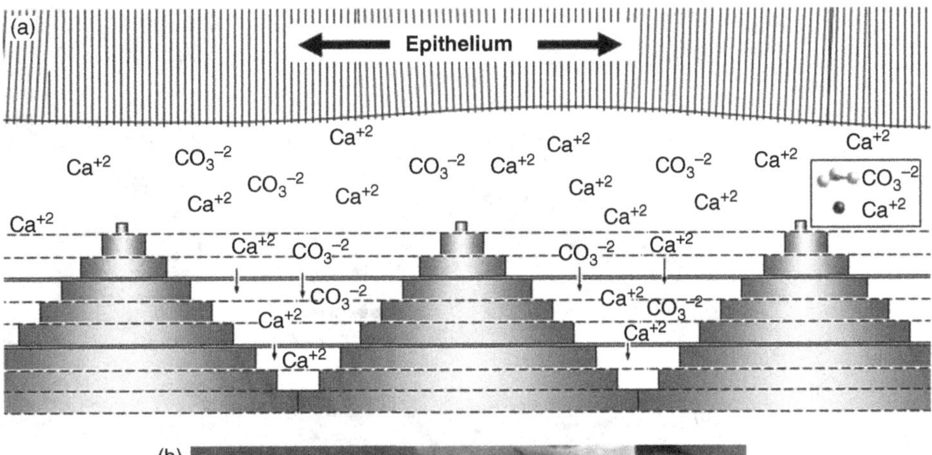

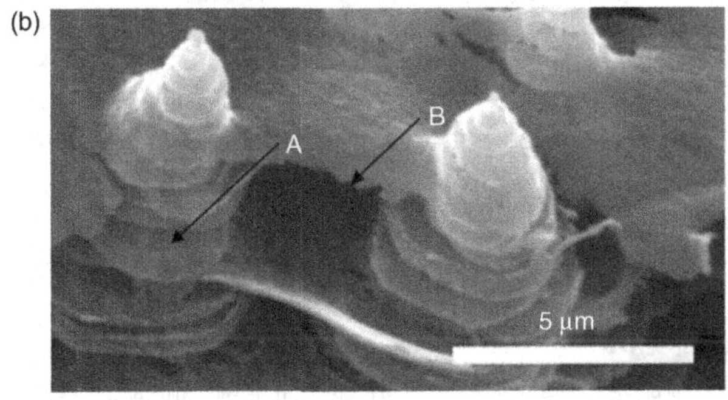

Figure 6.16.

Growth of nacreous tiles by terraced cone mechanism. (a) Schematic of growth mechanism showing intercalation of mineral and organic layers. (b) SEM of arrested growth showing partially grown tiles (arrow A) and organic layer (arrow B) (organic interlayer present). (Adapted from Lin *et al.* (2008), with permission from Elsevier.)

Figure 6.18(a) represents the sequence in which growth occurs through mineral bridges. (i) Organic scaffolding forms as interlamellar membranes between the layers of tiles arresting *c*-direction growth. (ii) A new tile begins growing through the porous membrane. (iii) The new tile grows in every direction, but faster along the *c*-axis. (iv) A new porous organic membrane is deposited, arresting the *c*-axis growth of the new tile while allowing continued *a*- and *b*-axis growth; also, mineral bridges begin to protrude through the second organic membrane while sub-membrane tiles continue to grow along the *a*- and *b*-axes, and eventually abut against each other; then a third layer of tiles begins to grow above the membrane. As shown, the bridges are believed to be the continuation of mineral growth along the *c*-axis from a previous layer of tiles. They protrude through the growth-arresting layers of proteins, creating a site on the covering organic layer where mineralization can continue. These mineral bridges are the seed upon which the next tile forms.

A detailed view of mineral bridges enabling growth through a permeable organic membrane is shown in Fig. 6.18(b). Holes in the organic nanolayer, which have been identified by Schäffer *et al.* (1997), are the channels through which growth continues.

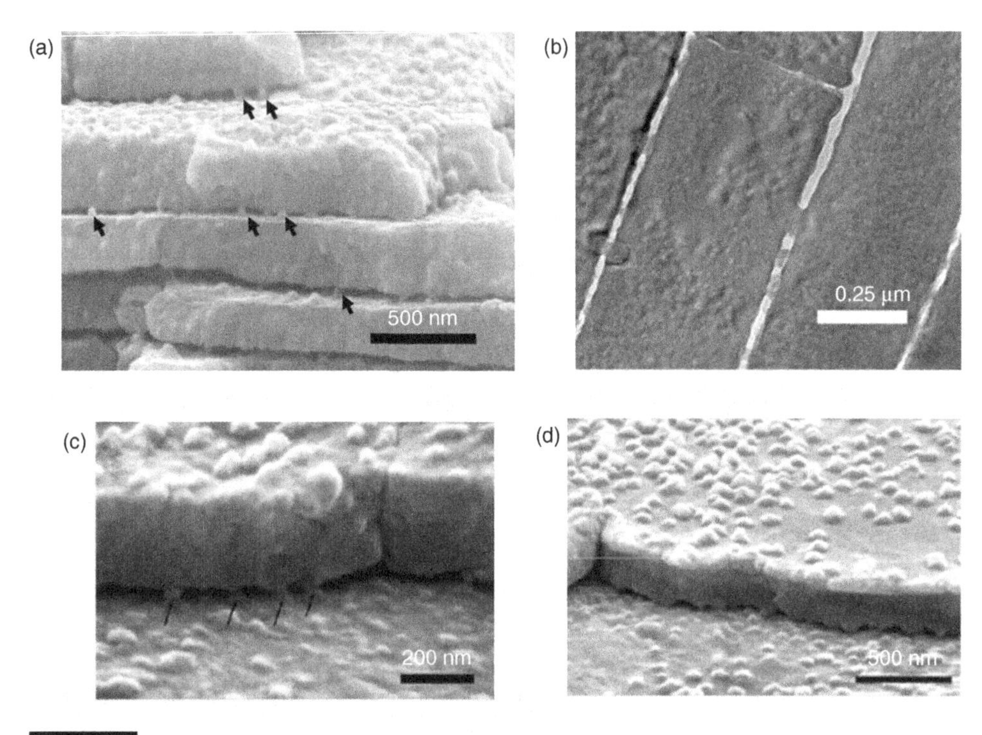

Figure 6.17.

Mineral bridges on tile surfaces. (a) SEM micrograph showing mineral bridges between tiles after deproteinization. (From Lin *et al.* (2008). (b) TEM micrograph of nacre cross-section showing mineral bridges. (c) Examples of mineral bridges (arrows). (d) Nanoasperities on the surface of an aragonite tablet. (Adapted from Lin *et al.* (2008), with permission from Elsevier.)

Mineral growth above the membrane is faster than growth in the membrane holes because of the increase in contact area with surrounding calcium and carbonate ions.

Since these holes are small (30–50 nm diameter), the flow of ions is more difficult, resulting in a reduction of growth velocity to $V_1 \ll V_2$ (Fig. 6.18(b)), where V_2 is the unimpeded growth velocity in the c-direction. The supply of Ca^{2+} and CO_3^{2-} ions to the growth front is enabled by their flow through the holes in the membranes. This explains why the tiles have a width to thickness ratio of approximately 20, whereas the growth velocity in the orthorhombic c-direction is much higher than in the a- and b-directions.

In gastropods, the nucleation of aragonite tiles occurs in the "Christmas tree" pattern previously described; bivalve mineralization, however, takes place with tablets offset with respect to layers above and below them.

Cartwright and Checa (2007) compared differences in microstructures between gastropods and bivalves and attributed them to variations in growth dynamics. In gastropods there are a large number of holes that enable the growth, and therefore a "Christmas tree" or terraced cone stacking of tiles is possible. In bivalves a smaller number of holes exist, most of which are filled with proteins and not mineral. There appears to be no direct evidence of mineral bridges. However, heteroepitaxy is required for the tiles to retain the

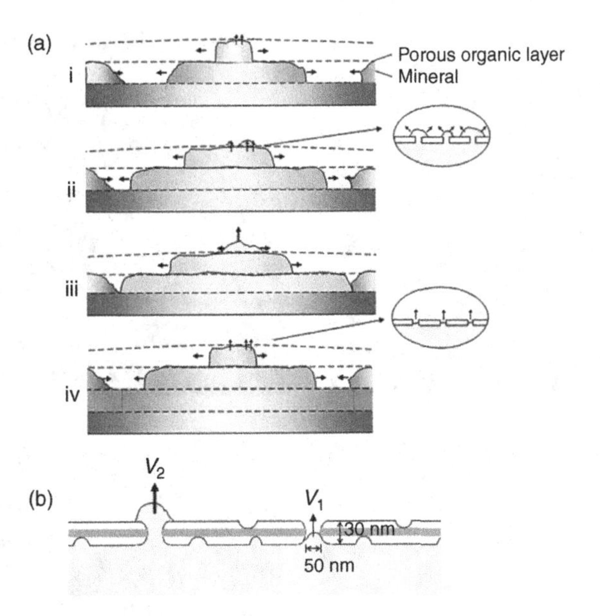

Figure 6.18.

(a) Growth sequence through mineral bridges. (b) Detailed view of mineral bridges forming through holes in organic membranes. (Adapted from Lin *et al.* (2008), with permission from Elsevier.)

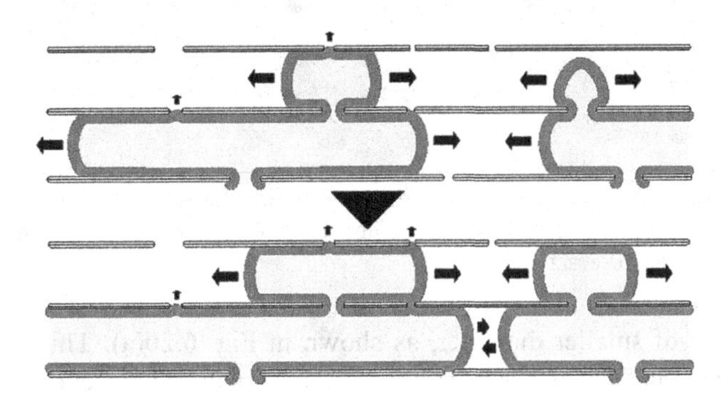

Figure 6.19.

Growth sequence in bivalve nacre. (Reprinted from Cartwright and Checa (2007), with permission from the authors and the *Journal of the Royal Society Interface.*)

same orientation. Cartwright and Checa (2007) suggest that there are more widely spaced bridges in bivalves, as shown in Fig. 6.19. There are two bridges per tile, causing the heteroepitaxial growth to dictate a random stacking of subsequent tiles.

The topology of the surface of the growing front reveals important aspects of growth. As shown in Fig. 6.20, the tablets grow with a terraced cone structure (Wise, 1970; Fritz *et al.*, 1994; Lin and Meyers, 2005; Lin *et al.*, 2008). Before a layer of aragonite tablets has grown to confluence, there are events that occur above this base layer,

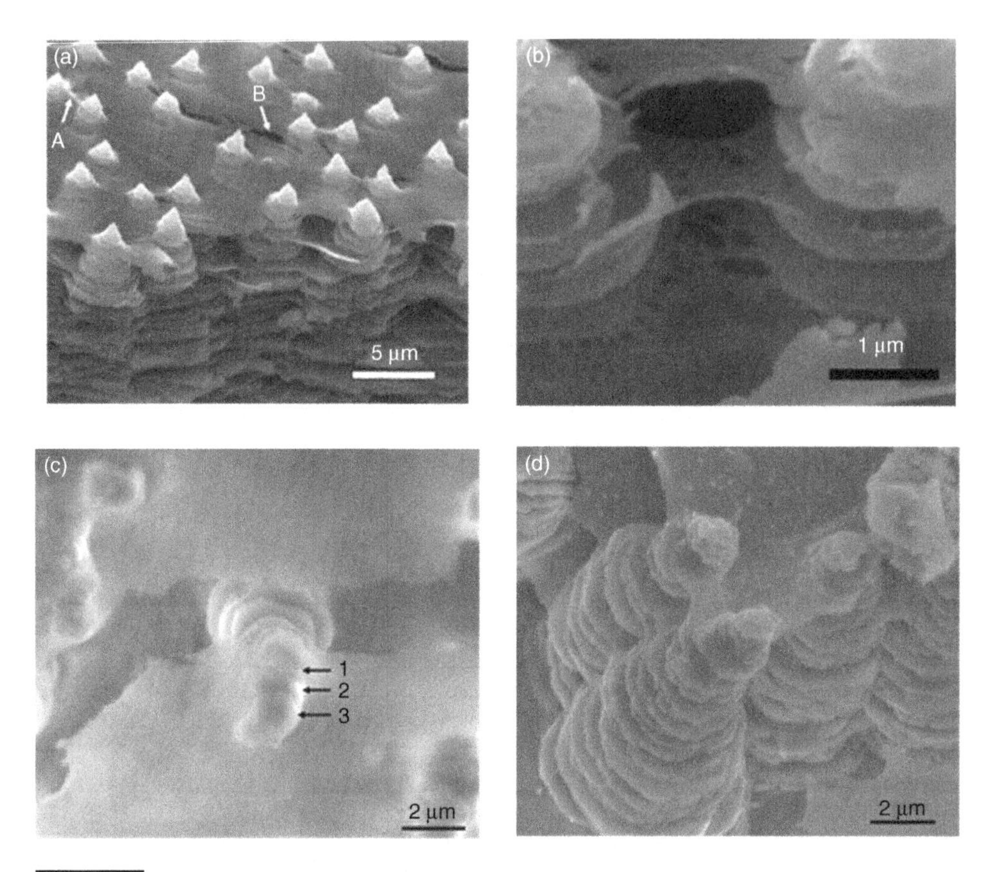

Figure 6.20.

Growth front of aragonite: (a) terraced cone structure and (b) close-up view, showing the porous organic layer. (Reprinted from Meyers *et al.* (2008a), with permission from Elsevier.) (c, d) Top views showing terraced cones. (Reprinted from Meyers *et al.* (2010), with permission from Elsevier.)

leaving stacks of smaller diameter, as shown in Fig. 6.20(a). This unique, ordered growth of the tablet stacks has been observed and analyzed by several groups (Erben, 1972; Fritz *et al.*, 1994; Lin *et al.*, 2008; Meyers *et al.*, 2008a). In Fig. 6.20(b), the organic sheet is observed to display a random network of pores. These pores expand when the organic layer is stretched at a velocity higher than the overall strain velocity; therefore the holes grow faster. Thus, nanometer-sized holes can grow easily to the sizes shown in Fig. 6.20(b). Figures 6.20(c) and (d) show the cracked top organic layer that reveals the stacks underneath. Arrows numbered 1, 2, and 3 indicate the top three layers in the stacks that are under the top organic layer. It was proposed by Checa, Cartwright, and Willinger (2009) that there is a top layer, or surface membrane, which protects the growing nacre surface from damage. This surface membrane contains vesicles that adhere to it on its mantle side, which secrete interlamellar membranes from the nacre side. Checa *et al.* (2009) observed that this top layer is thicker than the other ones and that it somehow forms the scaffold in which the tiles are mineralized.

The organic layer between the aragonite layers has a complex configuration, and was first discovered and analyzed by Grégoire (Grégoire, Duchateau, and Florkin, 1954; Grégoire, 1957, 1961; Bricteux-Grégoire, Florkin, and Grégoire, 1968) and subsequently studied notably by Wise (1970), Weiner's group from the Weizmann Institute in Israel (Weiner and Hood, 1975; Weiner, 1980, 1984; Weiner, Talmon, and Traub, 1983; Weiner, Traub, and Parker, 1984; Falini *et al.*, 1996; Addadi *et al.*, 2006; Nudelman *et al.*, 2006), and UC Santa Barbara (Schäffer *et al.*, 1997; Shen *et al.*, 1997; Fu *et al.*, 2005). A fibrous and porous two-dimensional network of insoluble proteins is composed of the polysaccharide β-chitin fibrils with ~8 nm diameter (Schäffer *et al.*, 1997; Bruet *et al.*, 2005). β-Chitin (described in Chapter 3) is a highly cross-linked biopolymer containing lustrin A as the major protein (Shen *et al.*, 1997). A silk hydrogel sandwiches the insoluble matrix and is composed of hydrophobic protein sheets in a β-configuration, similar to that of spider silk (Addadi *et al.*, 2006; Shen *et al.*, 1997). The silk-like proteins are amino acids rich in glycine and alanine (Grégoire *et al.*, 1954; Weiner *et al.*, 1983; Shen *et al.*, 1997; Marin and Luquet, 2005). At the surfaces of this two-dimensional chitin network, hydrophilic proteins, rich in aspartic acid, are in direct contact with the aragonite tablets (or tiles) (Weiner, 1980; Belcher, 1996; Belcher *et al.*, 1996b; Addadi *et al.*, 2006). Thus, they are the "glue" between the organic chitin-based layer and the aragonite tablets.

The organic layer consists of a core of randomly oriented chitin fibers sandwiched between acidic macromolecules. Figure 6.21 shows the configuration. The SEM micrograph in Fig. 6.21(a) shows a layer that was exposed after demineralization of the shell. The organic layer is not continuous, but contains holes, which can be seen even more clearly in Fig. 6.22(a). These holes have diameters of ~50 nm and have an important function in the structure and mechanical properties of aragonite. Figure 6.22(b) is a schematic showing the growth process.

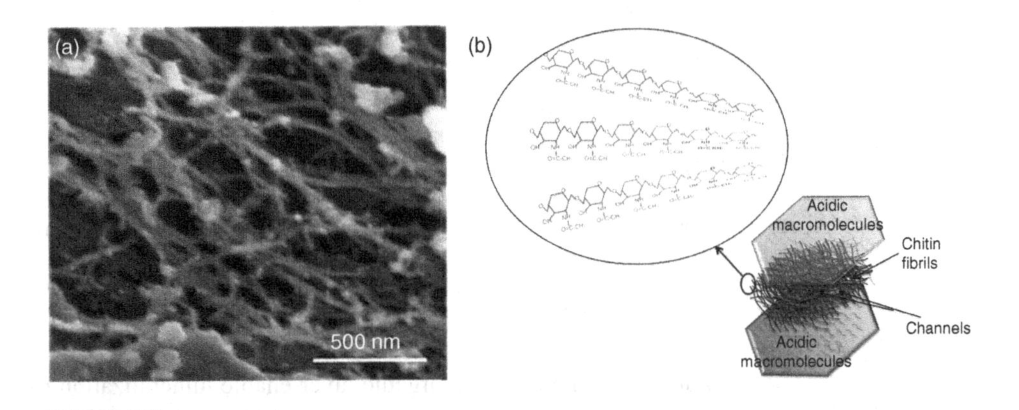

Figure 6.21.

(a) Randomly oriented chitin macromolecule fibrils. (b) Schematic representation of organic inter-tile layer consisting of central layer with randomly oriented chitin fibrils sandwiched between acidic proteins. (Reprinted from Meyers *et al.* (2010), with permission from Elsevier.)

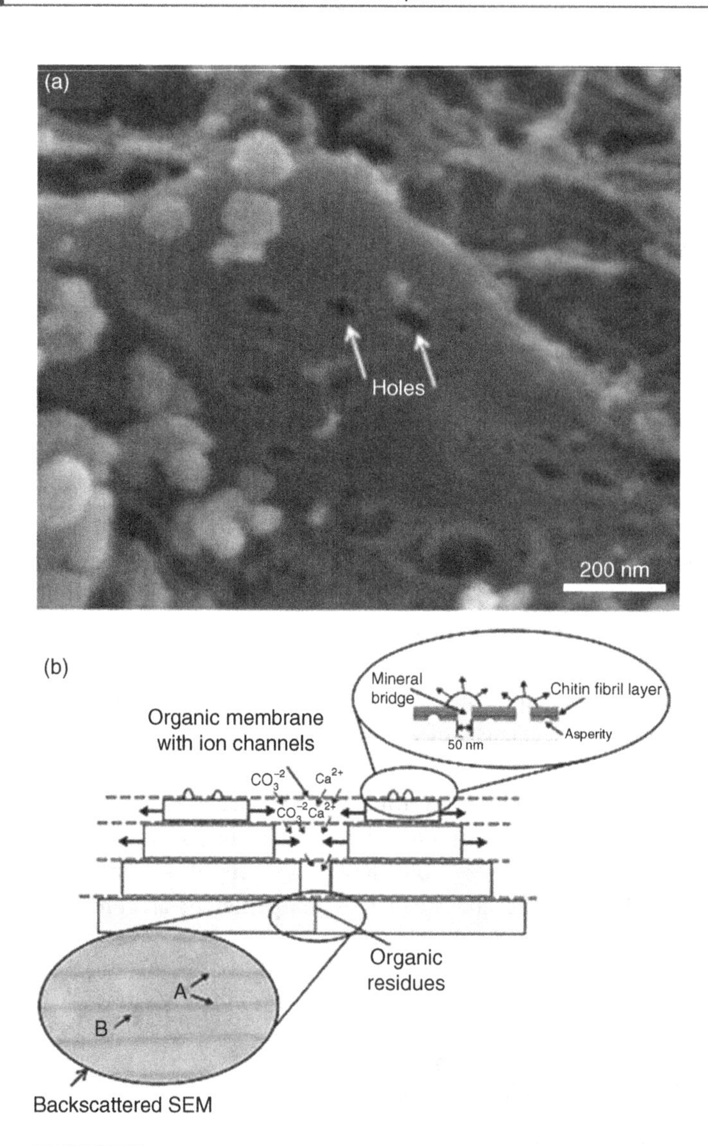

Figure 6.22.

(a) Thin inter-tile organic layer showing holes. (b) Proposed mechanism of growth of nacreous tiles by formation of mineral bridges; the organic layer is permeable to calcium and carbonate ions which nourish lateral growth as periodic secretion and deposition of the organic inter-tile membranes restricts their flux to the lateral growth surfaces. Arrows A designate organic interlayer imaged by SEM; arrow B designates lateral boundary of tile (schematic representation). (Reprinted from Meyers *et al.* (2010), with permission from Elsevier.)

In summary, the holes in the organic layer enable mineralization to continue from one layer to the next, as seen in Figs. 6.18 and 6.22(b). Hence, tiles on one stack have the same crystallographic orientation. The holes are filled with minerals and form the mineral bridges which connect the different layers. The process by which this sequence takes place is shown in a schematic fashion in Fig. 6.22(b). The SEM insert in this figure shows that the inter-tile layers (A, horizontal) are much clearer than the inter-tabular layers (B, vertical).

Figure 6.23(a) shows a more detailed view of the growth process. Two adjacent "Christmas trees" are seen. Their spacing, d, determines the tile size. Two growth velocities are indicated: V_{ab}, representing growth velocity in the basal plane (we assume that $V_a = V_b$), and V_c, the growth in the c-axis direction. Since the growth in the c-axis direction is mediated by organic layer deposition (with velocity V_1, as shown in Fig. 6.18), the real growth direction, V_c', is different from the apparent growth velocity, V_c (= V_2 in Fig. 6.18). It is possible to calculate the angle α of the Christmas tree if one has the growth velocities V_{ab} and V_c':

$$\tan \alpha = \frac{V_{ab}}{V_c'} \tag{6.6}$$

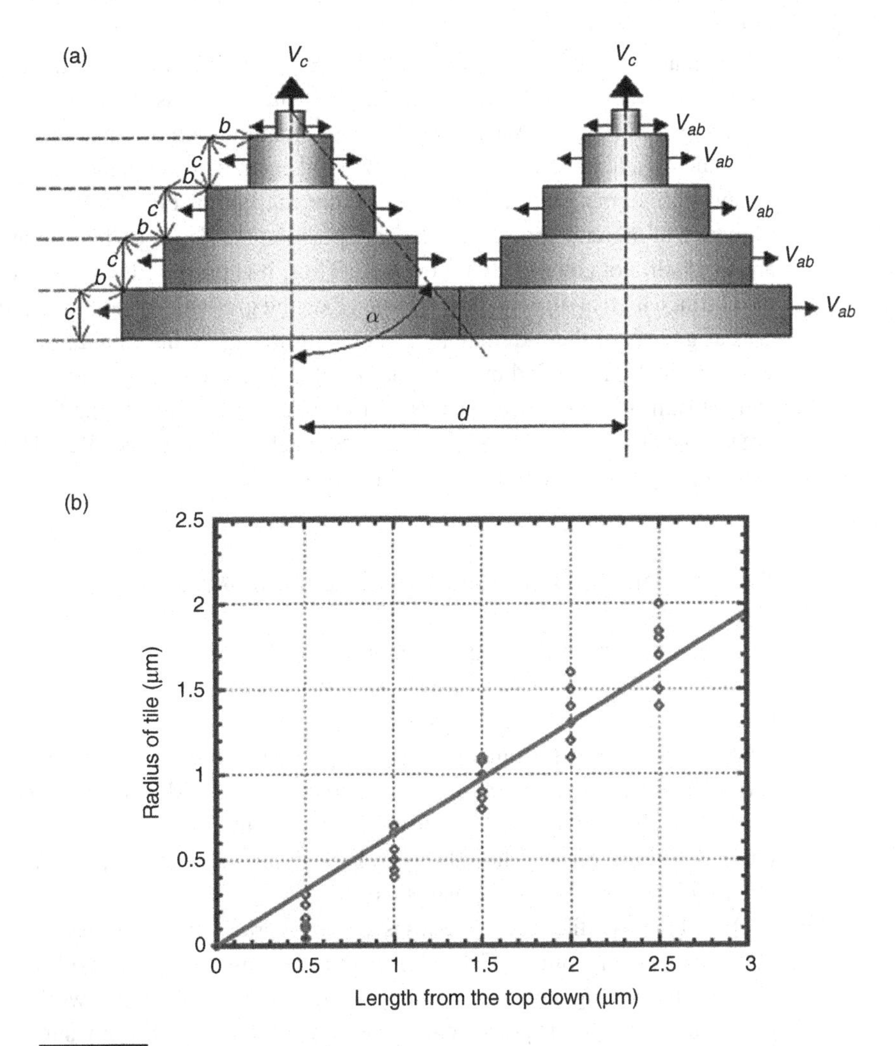

Figure 6.23.

(a) Calculation of semi-angle α of "Christmas tree" through velocities of growth in c- and a,b-directions; (b) measurements on growing tiles. (Adapted from Lin and Meyers (2005), with permission from Elsevier.)

The growth velocity V_c is, if the crystal is unimpeded, much higher than V_{ab}. Different estimates have been made, and these velocities depend on a number of factors.

Lin and Meyers (2005) estimated that $V_{ab} = 1.5 \times 10^{-11}$ m/s. They assumed that $V_c = 10V_{ab} = 1.5 \times 10^{-10}$ m/s. This enabled the estimation of the penetration time, t_p, the time taken to traverse the inter-tile organic layer. By adding this to the growth time, t_g, we obtain the velocity V_c'. The time required for growth in the c-direction to reach a value of b is equal to the time required for lateral growth to reach the distance $d/2$, half the tile spacing, so

$$V_c' = \frac{dx}{dt} = \frac{b}{t_p + t_g} = \frac{b}{t_p + 5V_{ab}d}. \tag{6.7}$$

Substitution of Eqn. (6.7) into Eqn. (6.6) leads to the determination of α, which has been found to vary between 17 and 34°.

The manner in which the chitin fibrils are generated from the internal portion of the mantle (epithelium) is shown in Fig. 6.24(b). This layer has channels in which the fibrils are assembled (Fig. 6.24(a)). They are subsequently (and periodically) squeezed out of the channels and penetrate into the extrapallial layer, being deposited on top of the stacks. A few fibrils embedded in the channels are marked in Fig. 6.24(a). The fundamental sequence of mineral and organic layer deposition that leads to the formation of the tiled arrays is still not completely understood. Thus, it is proposed that these channels create the chitin, which is subsequently deposited on the growing surface to retard the aragonite crystal growth in the c-direction. The cells in the epithelium contain microvilli, which were originally identified by Nakahara (1991). These microvilli are 100 nm in diameter and 400 nm in height, dimensions that correlate well with the pattern of holes in the organic layer and the thickness of the tiles, as shown in Fig. 6.24(c). Thus, we propose that a mechanism of templating is taking place.

6.2.2.2 Mechanical properties of abalone nacre

Abalone nacre has been a thoroughly studied system. The first studies from Currey (Currey and Taylor, 1974; Currey, 1976, 1977, 1980, 1984b; Currey and Kohn, 1976; Currey *et al.*, 2001) were followed by those from Jackson (Jackson, Vincent, and Turner, 1988, 1989), Heuer (Laraia and Heuer, 1989; Heuer *et al.*, 1992), Sarikaya (Sarikaya *et al.*, 1990; Sarikaya and Aksay, 1992; Sarikaya, 1994), Meyers (Menig *et al.*, 2000; Lin *et al.*, 2006; Meyers *et al.*, 2008a), and Evans (Evans, 2001a; Wang *et al.*, 2001). In this book, we cannot present the entirety of these contributions; thus, we will just emphasize the principal features and mechanisms.

One of the significant discoveries is that the work of fracture of nacre is approximately 3000× that of monolithic $CaCO_3$. It should be noted that this work of fracture is not identical to the toughness measured by Sarikaya *et al.* (1990). The work of fracture is the area under the load–displacement curve divided by twice the new surface area created:

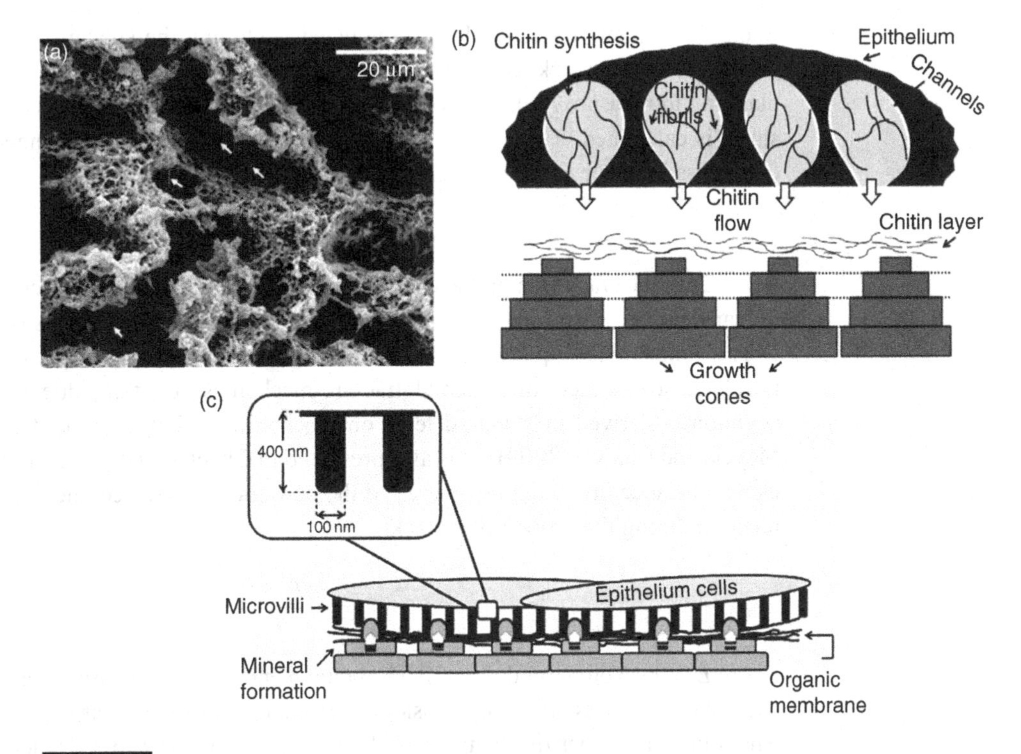

Figure 6.24.

(a) Array of channels and occasional chitin fibrils in epithelium (marked by arrows). (b) Hypothetical mechanism of generation of chitin fibrils and "squeezing" them onto the growth surface. (c) Model for regulating the thickness of tiles through microvilli in epithelial cells (which have a depth of ~400 nm); microvilli in epithelial layer of mantle facing growth surface of nacre; schematic rendition showing mechanism of deposition of inter-tile organic layer. (Reprinted from Lopez *et al.* (2011), with permission from Elsevier.)

$\left(\frac{\text{area under the load - displacement curve}}{2 \times (\text{new surface area created})}\right)$, and it is deeply affected by gradual, graceful fracture, whereas the fracture toughness does not incorporate this entire process. Thus, one should be careful when considering this parameter.

It is also known that water affects the Young modulus and tensile strength by reducing the shear modulus and shear strength of the organic matrix, which comprises less than 5 wt. % of the total composite. The toughness is enhanced by water, which plasticizes the organic matrix, resulting in greater crack blunting and deflection abilities. In contrast with more traditional brittle ceramics, such as Al_2O_3, or high toughness ceramics, such as ZrO_2, the crack propagation behavior in nacre reveals that there is a high degree of tortuosity.

Sarikaya *et al.* (1990) conducted mechanical tests on *Haliotis rufescens* (red abalone) with square cross-sections. They performed fracture strength σ_f (tension) and fracture toughness K_{Ic} tests on single straight notched samples in four-point and three-point bending modes, respectively, in the transverse direction, i.e. perpendicular to the shell plane. The fracture toughness is obtained by making a pre-crack in a sample under controlled conditions and geometry (ASTM E399 provides a detailed description on how fracture toughness tests should be conducted, including the specimen dimensions and geometry, initial crack size, and other parameters. It is a "must read" for anybody planning to measure K_{Ic}). The load at which the crack starts to grow is recorded and a stress is calculated. The essence of the test is that one obtains a plane-strain fracture toughness in Mode I loading (tension) that is defined as follows:

$$K_{Ic} = Y\sigma\sqrt{\pi a}, \tag{6.8}$$

where $2a$ is the crack size (in the case of an internal crack; for a surface crack, it is a), σ is the far-field stress, and Y is a geometry parameter that for central cracks is equal to 1.12.

This fundamental equation connects the toughness, K, to the strength and crack size. It is related to another fundamental fracture-mechanics equation, due to Griffith (this equation is derived in standard texts on mechanical behavior of materials (see, e.g., Meyers and Chawla (2009)). It is also presented in Chapter 2 (Eqn. (2.1)). It was derived using a balance involving the energy of the new surfaces created and the elastic energy released during the growth of a crack.

$$\sigma = \sqrt{\frac{E\gamma}{\pi a}}, \tag{6.9}$$

where E is the Young modulus and γ is the total energy per unit area of crack required to make it grow. It was originally considered as solely the surface energy γ_s; for materials in which there are other mechanisms of damage, it is equal to $\gamma_s + \gamma_p$, where $\gamma_p \gg \gamma_s$. By reorganizing Eqn. (6.9), we obtain

$$\sqrt{E\gamma} = \sigma\sqrt{\pi a}. \tag{6.10}$$

The similarity between Eqns. (6.8) and (6.10) is obvious. It is important to realize that, although both the stress and the crack size vary, the expression $\sigma\sqrt{\pi a}$ is a material parameter; this is clear from Eqn. (6.10). This parameter represents the resistance that a material has to the propagation of a crack.

Sarikaya *et al.* (1990) found a fracture strength of 185 ± 20 MPa and a fracture toughness of 8 ± 3 MPa m$^{1/2}$, and not the 3000-fold increase often quoted. This is an eight-fold increase in toughness over monolithic $CaCO_3$. The scatter is explained by the natural defects in the nacre and the somewhat curved shape of the layers. The K_{Ic} and σ_f values of synthetically produced monolithic $CaCO_3$ are 20–30 times less than the average value of nacre.

In the following we present the tensile and compressive strengths of abalone nacre with loading parallel and perpendicular to the planes of the tiles, and we use this information to infer the mechanisms of deformation and failure.

Compressive and tensile strengths in different directions

Menig *et al.* (2000) measured the compressive strength of red abalone and found considerable variation. Weibull statistics (Weibull, 1951) were successfully applied. This is within the range for synthetic ceramics. For greater detail on Weibull statistics, see Chapter 2

(Section 2.8). Presented in Fig. 6.25(a) are the results of tests on abalone nacre in quasistatic compression, with failure probabilities of 50% being reached at 235 MPa and 540 MPa with loading parallel and perpendicular to the layered structure, respectively.

Figures 6.25(b) and (c) shows the Weibull analysis of nacre in tension with loading parallel and perpendicular to layers. For loading parallel to the layers, the fracture strength at 50% is 65 MPa. It should be noted that in flexure tests higher values are obtained (around 170 MPa). However, when tension is applied perpendicular to the tiles, a surprisingly low value is obtained. The 50% fracture probability is only 5 MPa. The Weibull moduli in tension and compression are similar: 2 and 1.8–2.47, respectively. However, the difference in strength is dramatic and much higher than in conventional brittle materials. The ratio between compressive and tensile strength is on the order of 100, whereas for brittle materials it varies between 8 and 12. This difference is indeed

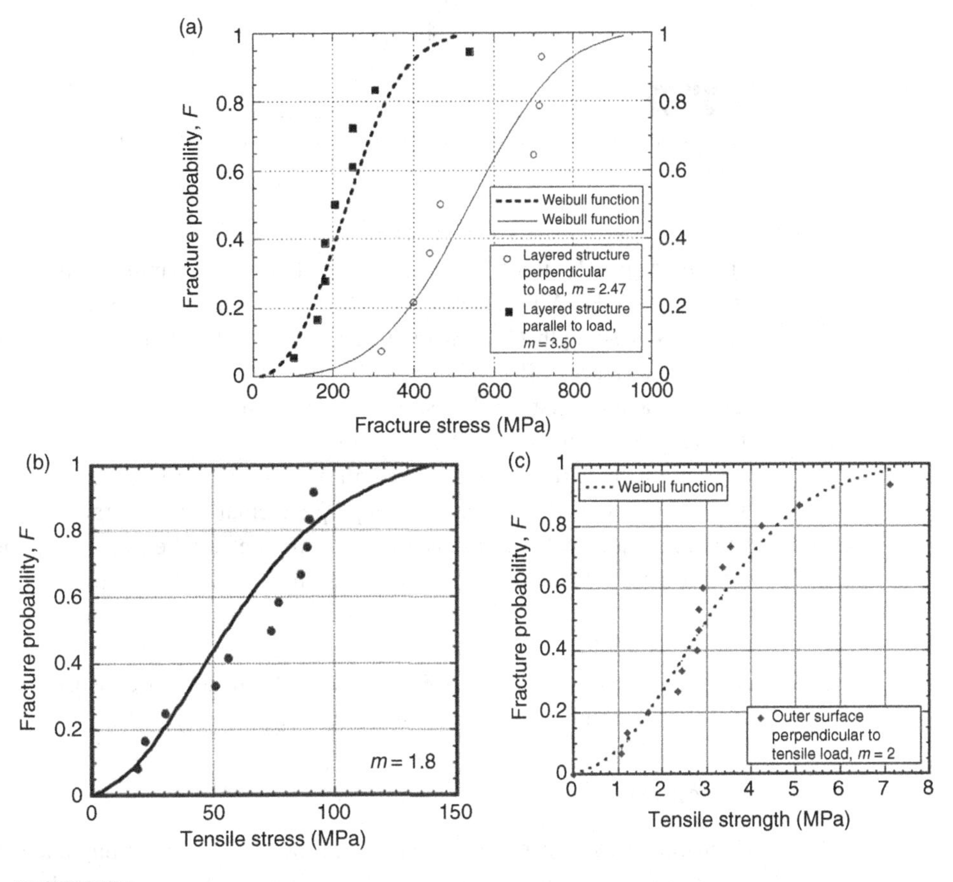

Figure 6.25.

Weibull distributions of strengths in abalone nacre. (a) Compressive strength in two orientations; (b) tensile strength parallel to the direction of tiles. (Reprinted from Menig *et al.* (2000), with permission from Elsevier.) (c) Tensile strength perpendicular to the tile plane. (Reprinted from Lin and Meyers (2009), with permission from Elsevier.)

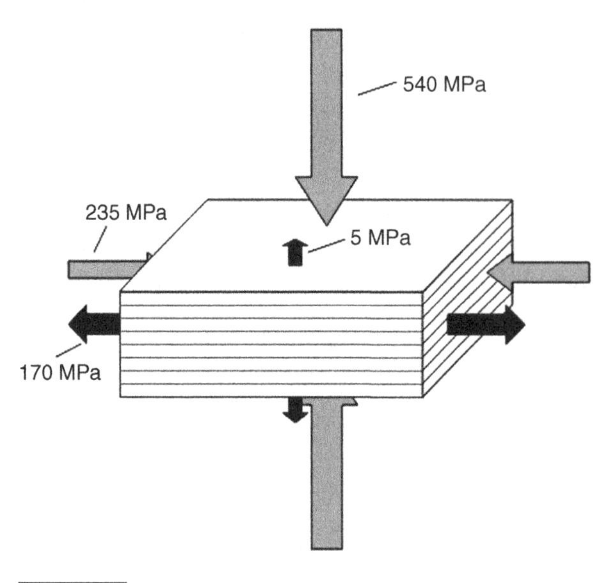

Figure 6.26.

Strength of nacre with respect to loading direction (tensile strength with loading direction parallel to tile planes measured by flexure).

striking, especially if we consider the tensile strength parallel to the layer plane, on the order of 140–170 MPa (Jackson *et al.*, 1988), which is approximately two-thirds the compressive strength. Other work, by Barthelat *et al.* (2006), found the tensile strength of nacre to be closer to 100 MPa, which is still just below half the compressive strength. It can be concluded that the shell sacrifices tensile strength in the perpendicular direction to the tiles to use it in the parallel direction.

Figure 6.26 summarizes the strength of nacre with respect to various loading directions. The unique strength anisotropy perpendicular to the layers (5 MPa vs. 540 MPa) is remarkable and will be discussed later. Another marked characteristic is the greater compressive strength when loading is applied perpendicular rather than parallel to the tiles. This is due to the phenomena of axial splitting and microbuckling (kinking) when loading is applied parallel to the tiles. The relatively small difference in tensile and compressive strengths (170 MPa vs. 235 MPa) in this direction of loading is directly related to the high toughness. Both of these aspects are discussed in the following.

Plastic microbuckling

Upon compression parallel to the plane of the tiles, an interesting phenomenon observed previously in synthetic composites was seen along the mesolayers (described in Section 6.2.2.1): plastic microbuckling. This mode of damage involves the formation of a region of sliding and a knee. Figure 6.27(a) shows a plastic microbuckling event. Plastic microbuckling, which is a mechanism to decrease the overall strain energy, was observed in

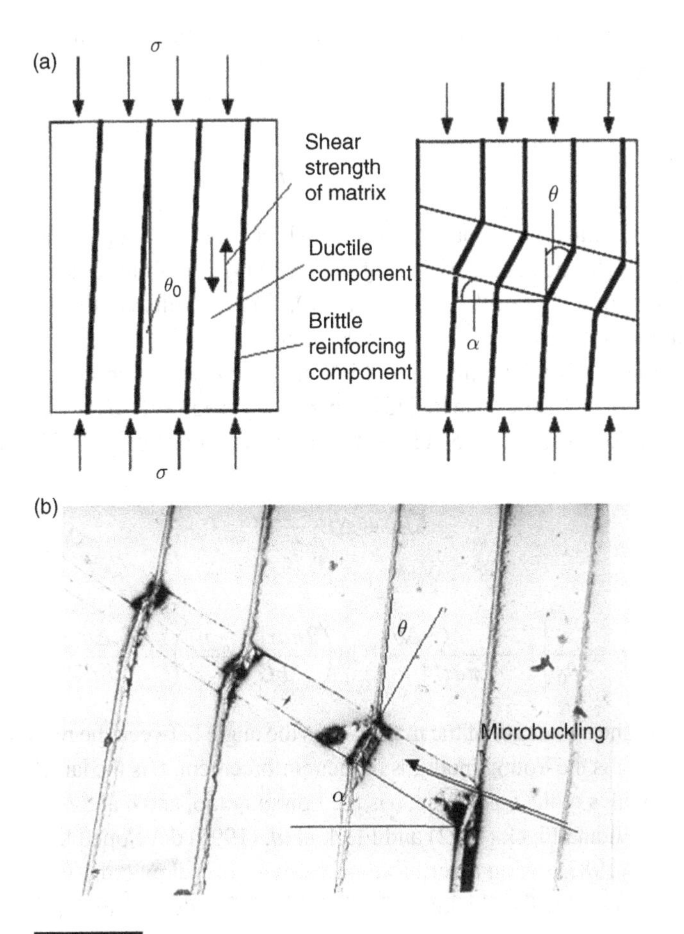

Figure 6.27.

Mechanism of damage accumulation in nacreous region of abalone through plastic microbuckling: (a) synthetic composite in which reinforcement is brittle and matrix is ductile; (b) nacre in which mesolayers are ductile and matrix is mineral which can undergo shear through sliding of tile layers.

a significant fraction of the specimens. It is a common occurrence in the compressive failure of fiber-reinforced composites when loading is parallel to the reinforcement. The coordinated sliding of layer segments of the same approximate length by a shear strain γ produces an overall rotation of the specimen in the region, with a decrease in length. Figure 6.27(b) shows a characteristic microbuckling region. The angle α was measured and found to be approximately 35°. The ideal angle to facilitate microbuckling, according to Argon (1972), is 45° (Fleck, Deng, and Budiansky, 1995).

The angle θ (Fig. 6.27(b)) varies between approximately 15° and 25° and is determined by the interlamellar sliding. These angles correspond to shear strains of 0.27 and 0.47. Hence, the rotation θ in kinking is limited by the maximum shear strain, equal to 0.45. If this kinking angle were to exceed this value, fracture would occur along the sliding interfaces. It is estimated the shear strain γ_0 undergone by the organic layers prior to failure is given by

$$\gamma_0 = \frac{\gamma}{f}, \tag{6.11}$$

where f is the fraction of organic layer, which has an approximate value of 0.05, providing $\gamma_0 \cong 9$. The results by Menig *et al.* (2000) are of the same order of magnitude as the ones reported by Sarikaya *et al.* (1990). These results are then applied to existing kinking theories (Argon, 1972; Budiansky, 1983).

The Argon (1972) formalism for kinking based on an energetic analysis can be applied. The plastic work done inside the band (W) is equated to the elastic energy stored at the extremities (ΔE_1) of the band and the energy outside the band (ΔE_2) that opposes its expansion:

$$\Delta E_1 + \Delta E_2 - W = 0. \tag{6.12}$$

This leads to

$$\sigma \cong \frac{\tau}{\theta_0}\left[1 + \frac{bG_c\Delta\theta}{2\pi a\tau(1-\upsilon)}\ln\left(\frac{2\pi a\tau(1-\upsilon)}{bG_c\Delta\theta}\right) + \frac{E_r\Delta\theta}{48\tau}\left(\frac{t_r}{b}\right)^2\right], \tag{6.13}$$

where τ is the shear strength of the matrix, θ_0 is the angle between the reinforcement and the loading axis, E_r is the Young modulus of the reinforcement, t_r is the lamella thickness, G_c is the shear modulus of the composite, υ is the Poisson ratio, and a and b are the kink nucleus dimensions. Jelf and Fleck (1992) and Fleck *et al.* (1995) developed this treatment further.

Budiansky (1983), using a perturbation analysis, developed the following expression for the ratio between the thicknesses of the kink bands and the spacing between the reinforcement units:

$$\frac{w}{d} = \frac{\pi}{4}\left(\frac{2\tau_y}{CE}\right)^{-1/3}, \tag{6.14}$$

where E is the Young modulus of the fibers and C is their volume fraction. It is interesting to note that Eqn. (6.14) predicts a decrease in w/d with increasing τ_y.

These formalisms for microbuckling were applied to our results and enable some conclusions to be drawn regarding the kink stress and spacing of the slip units. Figure 6.28(a) shows the predicted compressive kinking stress for abalone as a function of misalignment angle. It can be seen that the strength is highly sensitive to the angle θ_0. Figure 6.28(b), using the Budiansky equation adapted to the abalone geometry, shows the kink band thickness (w) as a function of strain rate. The results by Menig *et al.* (2000), carried out at different strain rates, confirmed the Budiansky prediction. Two parameters were used: the mesolayer and microlayer thicknesses. The experimental results shown in Fig. 6.28(b) fall in the middle, showing that both the mesolayers and platelets (microlayers) take part in kinking.

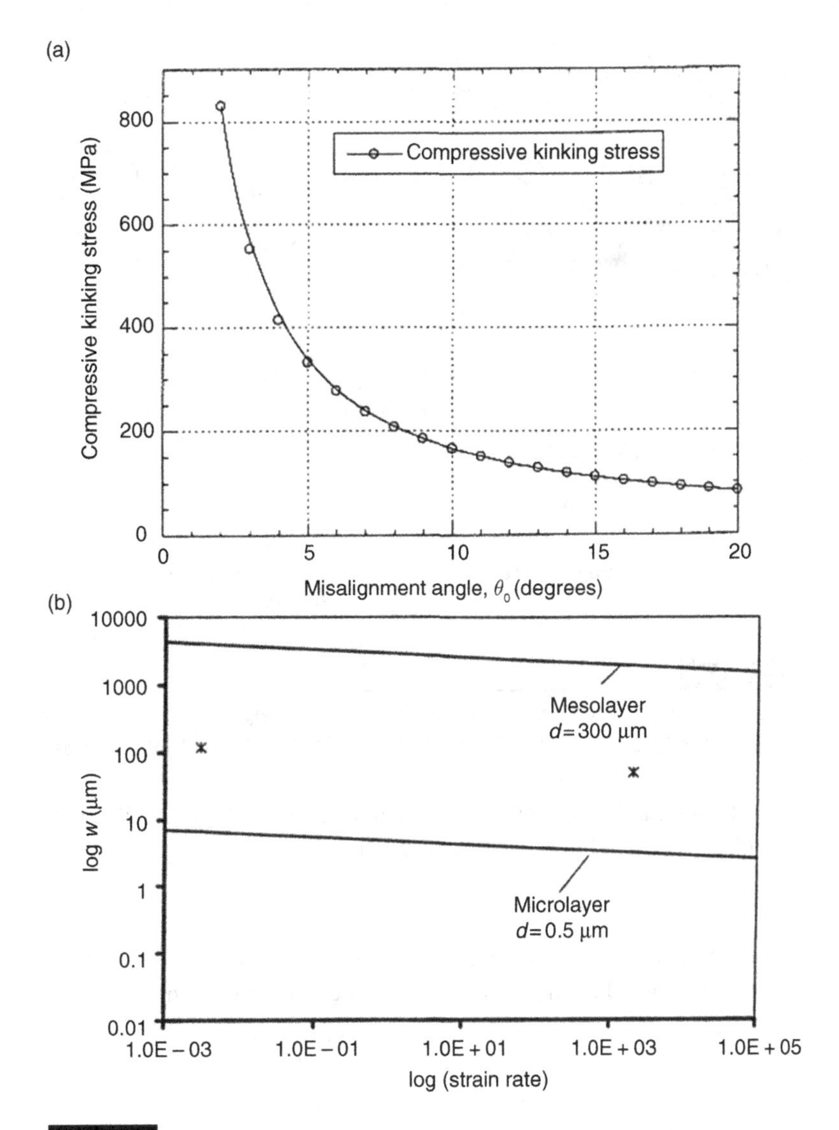

Figure 6.28.

Application to shell microbuckling. (Reprinted from Menig *et al.* (2000), with permission from Elsevier.) (a) Argon analysis for kink stress formation; (b) Budiansky formalism for the kink-band-thickness prediction.

Tensile strength parallel to tile direction

Figure 6.29 provides a schematic of nacre failure in tension testing with loading parallel to the tiles. The mechanism is tile pullout, as shown by the SEM in Fig. 6.29(a). Figure 6.29(b) shows schematically three tiles in a pullout mode. The force balance equation is as follows:

$$F_1 = F_2 + F_3. \tag{6.15}$$

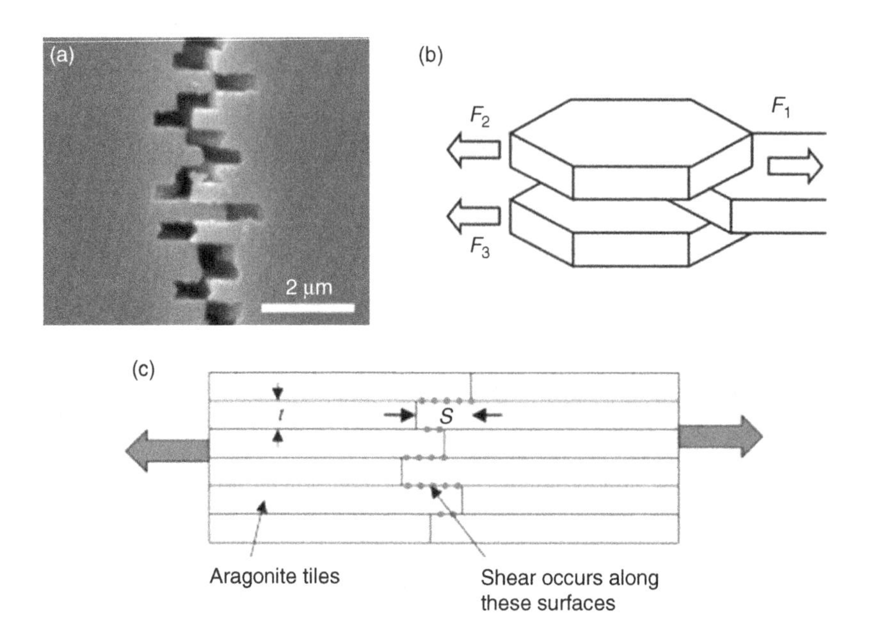

Figure 6.29.

Mechanisms of damage accumulation in nacreous region of abalone through tile pullout: (a) SEM showing pullout; (b) schematic of forces; (c) shearing surfaces. (Reprinted from Lin and Meyers (2009), with permission from Elsevier.)

The forces can be converted to normal stresses, σ_t. Figure 6.29(c) shows the resisting surfaces, subjected to a shear stress τ. Thus,

$$\sigma_t t = 2\tau S, \tag{6.16}$$

where S is the average resisting length and t is the tile thickness. Since S, measured from Fig. 6.29(a) is ~0.6 μm, and $t \sim 0.5$ μm, one has the following relationship:

$$\frac{\sigma_t}{\tau} = \frac{2S}{t} \sim 2.5. \tag{6.17}$$

This establishes a bound for the shear strength of the interface. It cannot exceed $\sigma_t/2.5$, otherwise the tiles will break in tension and the crack will propagate unimpeded through the nacre, and the toughening mechanism of tile pullout would no longer operate. The tensile strength shown in Fig. 6.25(b) is approximately 60 MPa. This gives a shear strength for the interface of ~24 MPa.

Mineral bridges: tensile strength perpendicular to the tile thickness

Meyers *et al.* (2008a) made observations indicating that the organic layer, while playing a pivotal role in the growth of the aragonite crystals in the *c*-direction (perpendicular to tile surface), may have a minor role in the mechanical strength. The tensile strength in the direction perpendicular to the layered structure can be explained by the combined presence of the mineral bridges and organic glue. These bridges, having a diameter of approximately 50 nm, have a tensile strength determined no longer by the critical crack size, but by the theoretical strength. Their number is such that the tensile strength of the tiles (parallel to the tile/shell surface plane) is optimized for a tile thickness of 0.5 μm, as shown by Lin and Meyers (2005). A higher number of bridges would result in tensile fracture of the tiles with loss of the crack deflection mechanism. This is a viable explanation for the small fraction of asperities that are bridges.

The tensile strength of the individual mineral bridges can be estimated by applying the fracture-mechanics equation to aragonite. Consistent with analyses by Gao *et al.* (2003), Ji and Gao (2004), and Ji, Gao, and Hsia (2004), the mineral bridges have sizes in the nanometer range. The maximum stress, σ_{fr}, as a function of flaw size, $2a$, can be estimated, to a first approximation, to be (see Eqn. (6.8))

$$\sigma_{fr} = \frac{K_{Ic}}{\sqrt{\pi a}}, \tag{6.18}$$

where K_{Ic} is the fracture toughness. Figure 6.30(a) provides a simple representation of such a crack. However, the strength is also limited by the theoretical tensile strength, which can be approximated as (Gao *et al.*, 2003)

$$\sigma_{th} = \frac{E}{30}. \tag{6.19}$$

We assume that $K_{Ic} = 1$ MPa m$^{1/2}$, $E = 100$ GPa, and that $2a = D$, where D is the specimen diameter. In Fig. 6.30(b), Eqns. (6.18) and (6.19) intersect for $a = 28$ nm ($D = 56$ nm). This is indeed surprising, and shows that specimens of this diameter and less can reach the theoretical strength. This is in agreement with the experimental results: the holes in the organic layer and asperities/bridge diameters are around 50 nm. Recent analyses (Song, Bai, and Bai, 2002; Song *et al.*, 2003; Gao *et al.*, 2003) also arrive at similar values.

It is possible to calculate the fraction of the tile surface consisting of mineral bridges, f. Knowing that the tensile strength is σ_t and assuming that the bridges fail at σ_{th}, we have

$$f = \frac{\sigma_t}{\sigma_{th}}. \tag{6.20}$$

The number of bridges per tile, n, can be calculated from

$$f = \frac{nA_B}{A_T}, \tag{6.21}$$

where A_B is the cross-sectional area of each bridge and A_T is the area of a tile. Thus,

$$n = \frac{\sigma_t A_T}{\sigma_{th} A_B}. \tag{6.22}$$

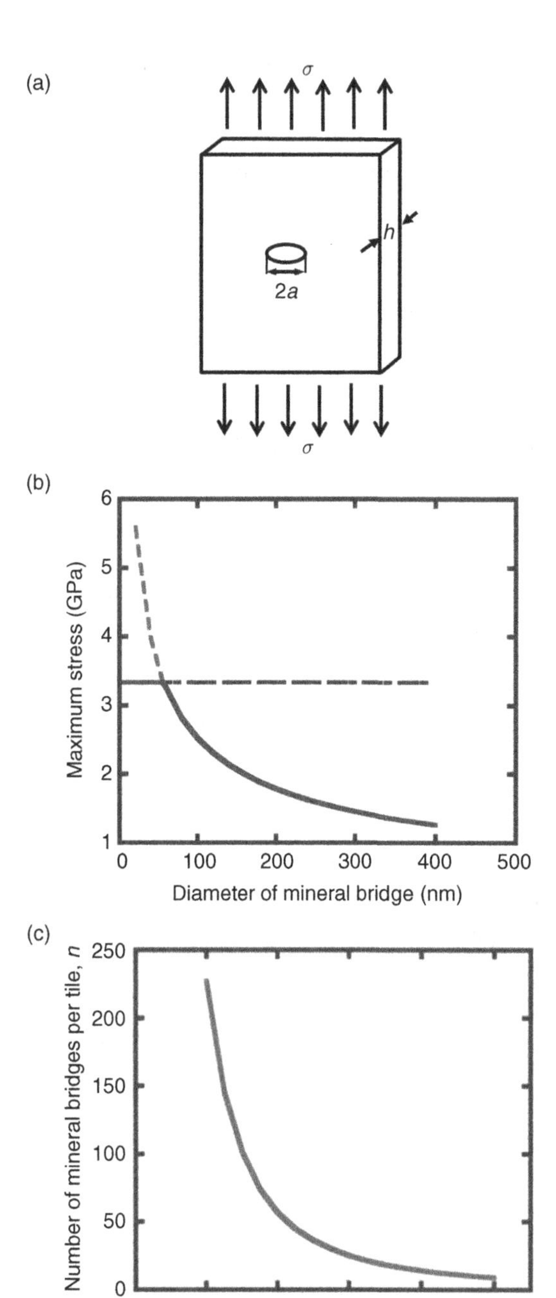

Figure 6.30.

(a) Schematic of mineral platelet with a surface crack (Griffith analysis). (b) Fracture stress as a function of crack length, $2a$. (c) Calculated number of mineral bridges per tile as a function of bridge diameter.

Assuming that the tiles have a diameter of 10 μm and that the bridges have a diameter of 50 nm (the approximate observed value), we obtain, for $\sigma_t = 3$ MPa and $\sigma_{th} = 3.3$ GPa, $n = 36$. Figure 6.30(c) shows the relationship between the mineral bridge diameter and the number of mineral bridges through Eqn. (6.22). The number of bridges calculated is surprisingly close to the measurements by Song *et al.* (2002, 2003): $35 \leq n \leq 45$. However, the interpretation of these is not clear. This result strongly suggests that mineral bridges can, by themselves, provide the bonding between adjacent tiles.

The number of asperities seen in Fig. 6.17 exceeds considerably the values for bridges calculated herein and measured by Song *et al.* (2002, 2003). The estimated density is 60/μm^2 (5000/tile). One conclusion that can be drawn from this is that a large number of asperities are indeed incomplete bridges and that these bridges are a small but important fraction of the protuberances.

Toughening mechanisms

The three models for the inter-tile region are shown in Fig. 6.31. Evans *et al.* (2001a) and Wang *et al.* (2001) proposed an alternative toughening mechanism: that nanoasperities on the aragonite tiles are responsible for the mechanical strength. These nanoasperities create frictional resistance to sliding, in a manner analogous to rough fibers in composite material. They developed a mechanism that predicts the tensile mechanical strength based on these irregularities. The asperities are represented in Fig. 6.31(a). Figure 6.31(b) shows the viscoelastic glue model, according to which the tensile strength is the result of stretching of molecular chains whose ends are attached to surfaces of adjacent tiles. Figure 6.31(c) shows the mineral bridge model, consistent with our observations. The sliding of adjacent tiles requires the breaking of bridges and the subsequent frictional resistance, in a mode akin to the Wang–Evans (Wang *et al.*, 2001; Evans *et al.*, 2001a) mechanism. It is possible

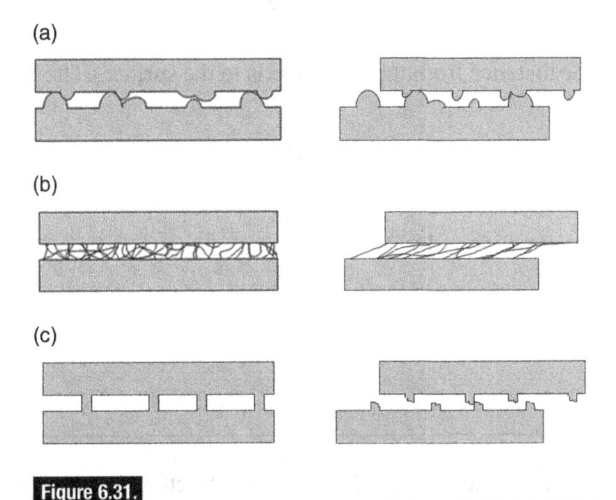

Figure 6.31.

Different models for sliding between tiles; inter-tile layer formed by (a) asperities; (b) organic layer acting as viscoelastic glue; (c) mineral bridges.

that all three mechanisms act in a synergetic fashion in which broken bridges act as asperities which are further reinforced by the viscoelastic organic glue (Su *et al.*, 2002).

Another significant mechanism of toughening is crack deflection at the meso-scale. The effect of the viscoelastic organic interruptions between mesolayers or even individual aragonite tiles is to provide a crack deflection layer such that it becomes more difficult for the cracks to propagate through the composite. This was shown in Fig. 6.13.

Example 6.1 Determine the load required to break a specimen of abalone in three-point bending if a crack with width $a = 1$ mm is introduced in the tension side. The specimen span between the two supports is 10 mm and the cross-section is square with dimensions of 2×2 mm. Given: $K_{Ic} = 10\ \mathrm{MPa\,m}^{1/2}$.

Solution We use the equation

$$K_{Ic} = Y\sigma\sqrt{\pi a}.$$

Thus, $\sigma = 16$ MPa, and the stress is related to the bending moment by (see Chapter 4)

$$\sigma = \frac{Mc}{I}.$$

The moment of inertia is given by

$$I = \frac{bh^3}{12} = \frac{16 \times 10^{-12}}{12}.$$

The bending moment is

$$M = \frac{\sigma I}{c} = 320\ \mathrm{Nm},$$

where $c = 1$ mm (the distance from the neutral axis to the surface).The load P is obtained from

$$M = \frac{PL}{4};$$

the load is therefore $P = 2.56 \times 10^5$ N.

6.2.3 Conch shell

Conch shells, with their spiral configuration, have a structure that is quite different from the abalone nacre. Figure 6.32(a) shows the overall picture of the well-known *Strombus gigas* (pink conch) shell. In contrast with the abalone shell, which is characterized by

(a)

(b)

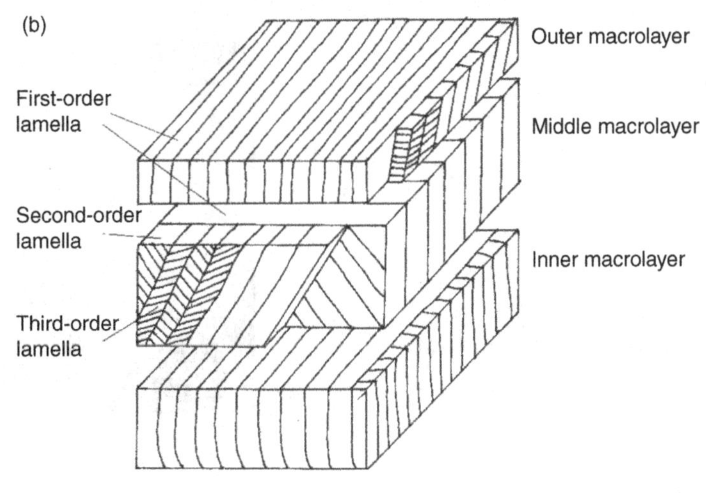

Figure 6.32.

Conch shell. (a) Overall view; (b) schematic drawing of the crossed-lamellar structure. Each macroscopic layer is composed of first-, second-, and third-order lamellae.

parallel layers of tiles, the structure of the conch consists of three macrolayers, which are themselves organized into first-order lamellae, which in turn comprise second-order lamellae. These are made up of tiles named, in Fig. 6.32(b), "third-order lamellae" in such a manner that successive layers are arranged in a tessellated ("tweed") pattern. Ballarini *et al.* (2005) described nano-scale components in this structure beyond the third-order lamellae. The three-tiered structure is shown in Fig. 6.32(b). This pattern, called *crossed lamellar*, is reminiscent of plywood or crossed-ply composites and has been studied extensively by Heuer and co-workers (Wu *et al.*, 1992). The crossed-lamellar microstructure consists of lath-like aragonite crystals (99.9% of weight) and a tenuous organic layer (0.1 wt.%). The "plywood" structure shown in Fig. 6.32(b) (Menig *et al.*, 2001) consists of three macroscopic layers: the inner (closest to the organism), middle, and outer layers, which are of relatively uniform thickness within the last whorl. This was further characterized by Hou *et al.* (2004). Kamat *et al.* (2000) showed that the

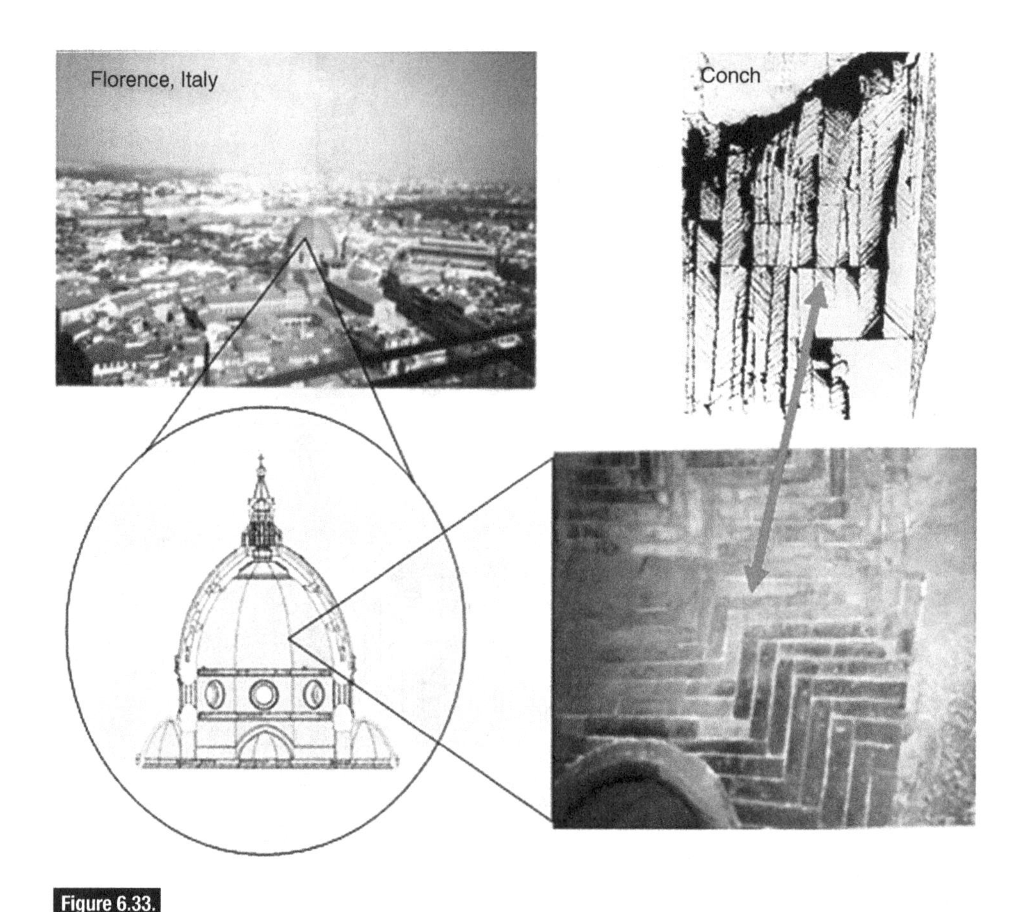

Figure 6.33.

Tessellated bricks on Brunelleschi's Duomo (Florence, Italy) and equivalent structure of conch shell.

result of this structure is a fracture toughness exceeding that of single crystals of the pure mineral by two to three orders of magnitude. An interesting analogy with a large dome structure is shown in Fig. 6.33. The Florence Duomo, built by the architect Brunelleschi, uses a tessellated array of long bricks which have a dimensional proportion similar to the tiles in conch. This arrangement provides the dome with structural integrity, which had not been possible before that time.

These three layers are arranged in a 0°/90°/0° direction. So-called first-order lamellae comprise each macroscopic layer and are oriented ±35–45° relative to each other. In each first-order lamella are long thin laths stacked in parallel, known as second-order lamellae. These second-order lamellae in turn consist of single-crystal third-order lamellae. Fine growth twins at an atomic scale layer each third-order lamella (Kuhn-Spearing *et al.*, 1996). The organic matrix with its 1 wt.% has only been observed by TEM as an electron dense layer that envelops each of the third-order lamellae (Weiner *et al.*, 1984). Kuhn-Spearing *et al.* (1996) measured flexural strength, crack-density evolution, and work of fracture for wet and dry specimens of the *Strombus gigas* conch shell. Four-point-bending tests in two

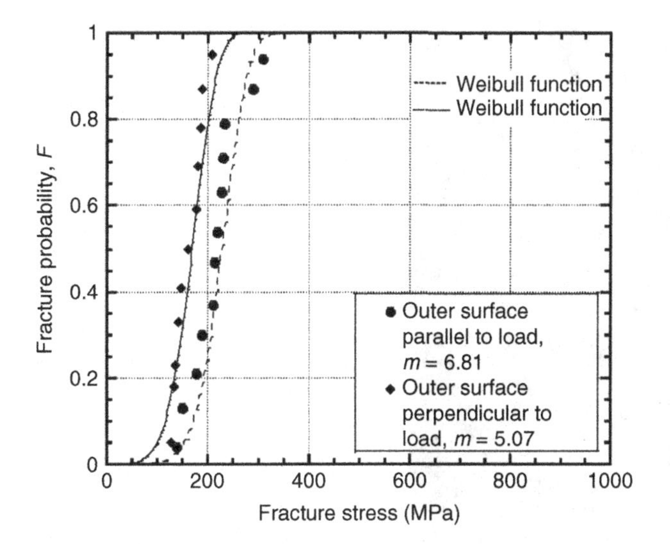

Figure 6.34.

Weibull analysis of conch shell in quasistatic compression. (Reprinted from Menig *et al.* (2001), with permission from Elsevier.)

different orientations were conducted: parallel and perpendicular to the shell axis. They report that the tensile strengths of crossed-lamellar shells are 50% lower than the strongest shell microstructure (nacre). Average apparent flexural strengths for dry and wet crossed-lamellar samples in the parallel orientation were found as 156 ± 22 MPa and 84 ± 49 MPa, respectively. The average apparent flexural strength of the perpendicular wet and dry samples was 107 ± 38 MPa. The results of Kuhn-Spearing *et al.* (1996) on the fracture of these shell structures suggest that the mechanical advantage is an increased fracture resistance, in addition to a previously observed increased hardness. The tests revealed a work of fracture of 4 ± 2 J/m^2 for dry and 13 ± 7 J/m^2 for wet samples tested in parallel orientation. These values are much higher than those reported for nacre (0.4 J/m^2 for dry and 1.8 J/m^2 for wet samples (Jackson *et al.*, 1988)). The increased fracture resistance for wet samples is correlated to a decreased interfacial strength that results in a more extensive cracking pattern. Menig *et al.* (2001) also performed a series of mechanical tests on the conch shell. Figure 6.34 presents the Weibull statistical analysis of conch in (a) quasistatic compression and (b) dynamic compression. In quasistatic compression the conch shell exhibited a failure probability of 50% ($F(V) = 0.5$) at 166 MPa and 218 MPa for the perpendicular and parallel direction of loading, respectively. In dynamic loading the 50% failure probabilities of the conch shell are found at 249 MPa and 361 MPa perpendicular and parallel to the layered structure, respectively.

The fraction of organic material in conch is lower than in abalone: ~1 wt.% vs. 5 wt.%. The strategy of toughening that has been identified in the conch shell is the delocalization of a crack by distribution of damage (Menig *et al.*, 2001). An example of how a crack is deflected by the alternative layers is shown in Fig. 6.35(a). The fracture surface viewed by SEM clearly shows the crossed-lamellar structure (Fig. 6.35(b)). The lines seen in the

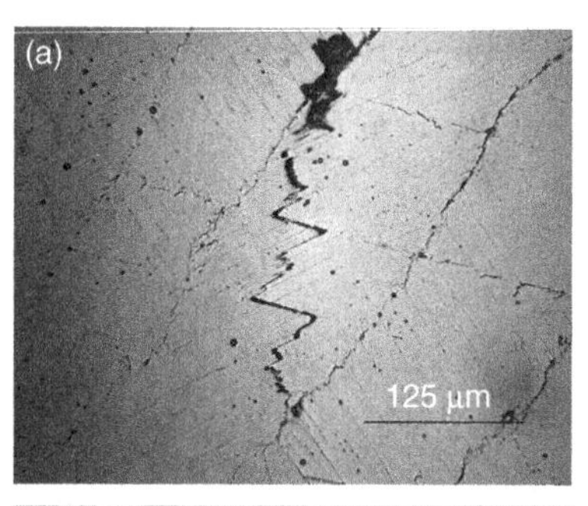

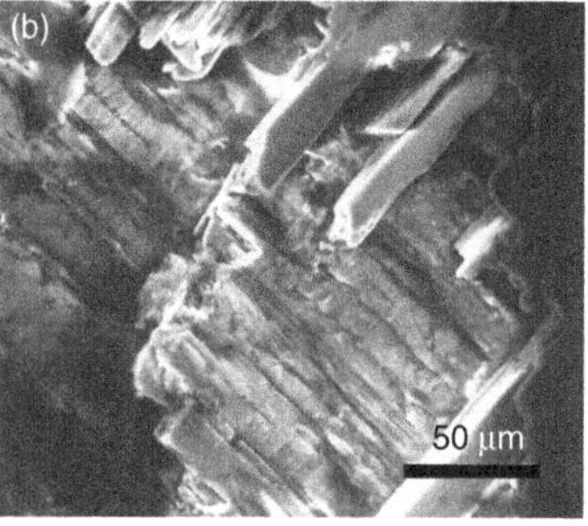

Figure 6.35.

Fracture patterns in conch shell: (a) crack delocalization shown in polished section; (b) scanning electron micrograph of fracture surface showing crossed-lamellar structure.

damaged surface of conch shown in Fig. 6.35(a) indicate sliding of the individual tiles. The absence of a clear crack leads to a significant increase in the fracture energy in comparison with monolithic calcium carbonate.

As with the abalone nacre, the structural hierarchies ranging from nano to macro are all responsible for the overall mechanical response of the shell. The crack deflection within the microstructure of tessellated tiles is only part of a larger crack delocalization mechanism. When the crack reaches the inner macrolayer it can again orient itself into an "axial splitting" configuration (seen in Fig. 6.36). The 0°/90°/0° architecture arrests the easy-to-form channel cracks and leads to additional channel cracking. Due to 45° second-order lamellar interfaces, the channel cracks (between first-order interfaces), which eventually penetrate the middle macroscopic layer, are deflected, and failure is noncatastrophic. It is this complex layered architecture that is responsible for improved toughness over that of nacreous structures (Weiner *et al.*, 1984).

Fractographic observations identify delocalized damage in the form of multiple channel cracks, crack bridging, crack branching, and delamination cracking. This increases the total crack area and frictional dissipation. The many interfaces between aragonite grains in the crossed-lamellar structure provide a multitude of places for energy dissipation.

Laraia and Heuer (1989) performed four-point-bending tests with *Strombus gigas* shells while the shell interior and exterior surfaces were the loading surfaces and were not

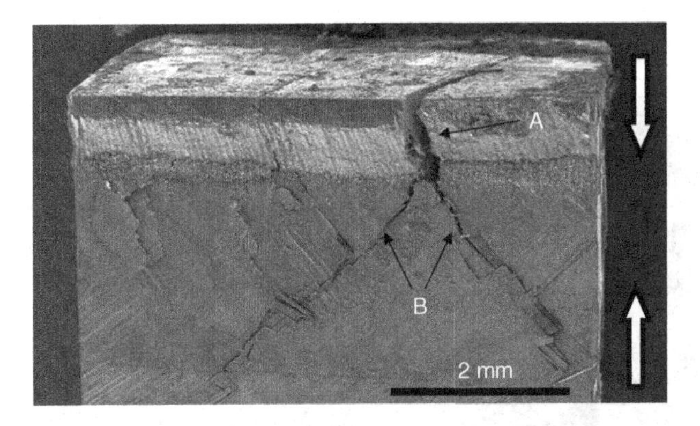

Figure 6.36.

Crack deflection by middle macrolayer in conch (SEM taken after testing); loading direction indicated. Multiple channel cracking and extensive microcracking in outer macrolayer of conch.

machined out. They found flexural strengths of about 100 MPa. With the exterior surface loaded in tension, the failure occurred catastrophically; however, when loaded with the interior surface in tension this kind of failure did not occur. This confirms the anisotropy of shells with crossed-lamellar microstructure, which leads to "graceful failure" in some orientations. Another indication for the anisotropy mechanical behavior of crossed-lamellar shells can be found in Currey and Kohn (1976). They found flexural strengths (in three-point bending) of shells of *Conus striatus* in the range of 70 to 200 MPa depending on the orientation.

Laraia and Heuer (1989) found that the resistance to crack propagation is due to several simultaneous toughening mechanisms. These are crack branching (i.e. the microstructure forces the cracks to follow a tortuous path), fiber pullout, microcracking (microcracks follow interlamellar boundaries), crack bridging, and microstructurally induced crack arrest. Kamat *et al.* (2004) found that the synergy between tunneling cracks and crack bridging was the source of an additional factor of 300 in fracture energy. They also carried out microindentation experiments. However, for applied loads from 0.1 to more than 10 kg the indentations failed to produce radial cracks when applied to polished interior shell surfaces. The damaged zones were elongated and the crack followed first-order lamellae.

The interaction of a crack with the different layers in conch is illustrated in Fig. 6.36. The crack, entering from the top, undergoes branching by bifurcation and delocalization as it enters the middle macrolayer. This mechanism does not operate at the micro-level, and this is evidence for the hierarchy of toughening.

Characteristic features of fractures at the micro-level parallel and perpendicular to the growth direction are shown in Fig. 6.37. It seems that the conch is designed for maximization of energy dissipation regardless of the cost in terms of crack formation. This allows the mollusc to survive the attack of a crab or another impact event. The mollusc can hide while repairing its damaged material. This design strategy is also desirable for

Figure 6.37.

Fracture surface of the *S. gigas* perpendicular growth direction.

armor; however, it is not necessarily appropriate for structural composites where lifetime and maintenance are also issues (Kamat *et al.*, 2000).

6.2.4 Giant clam

The giant clam (*Tridacna gigas*) can grow its shell to widths greater than 1 m and mass of over 340 kg (Rosewater, 1965). The large amount of shell material produced has made the giant clam of interest in both a contemporary, as well as an historical, context. Moir (1990) documented the use of this shell as the raw material for applications such as blades for wood-cutting tools in ancient and present-day Takuu Atoll dwellers of Papua New Guinea. The structure of the shell has a low level of organization in comparison to other shells, yet its sheer mass results in a strong overall system. The protective shell consists of two distinct regions: an outer white region and an inner translucent region.

The outer region acts as the animal's first line of defense against the harsh environment. This region appears to comprise approximately one-third of the shell thickness and is formed from dense structured layers of aragonite needles approximately 1–5 μm in length (Moir, 1990). Growth bands, which extend perpendicular to the direction of shell growth, are thought to contain a thin organic matrix, partially separating layers of the crossed-lamellar aragonite needles (Kobayashi, 1969). The structure of the outer region of the shell, presented in Fig. 6.38(a) (Lin *et al.*, 2006), somewhat resembles the microstructure of the middle macrolayer of conch shell, yet a considerable decrease in organization is observed. Growth bands form first-order lamellae, separating layers of second- and third-order lamellae perpendicular to the direction of growth. The second-order lamellae is composed of planes,

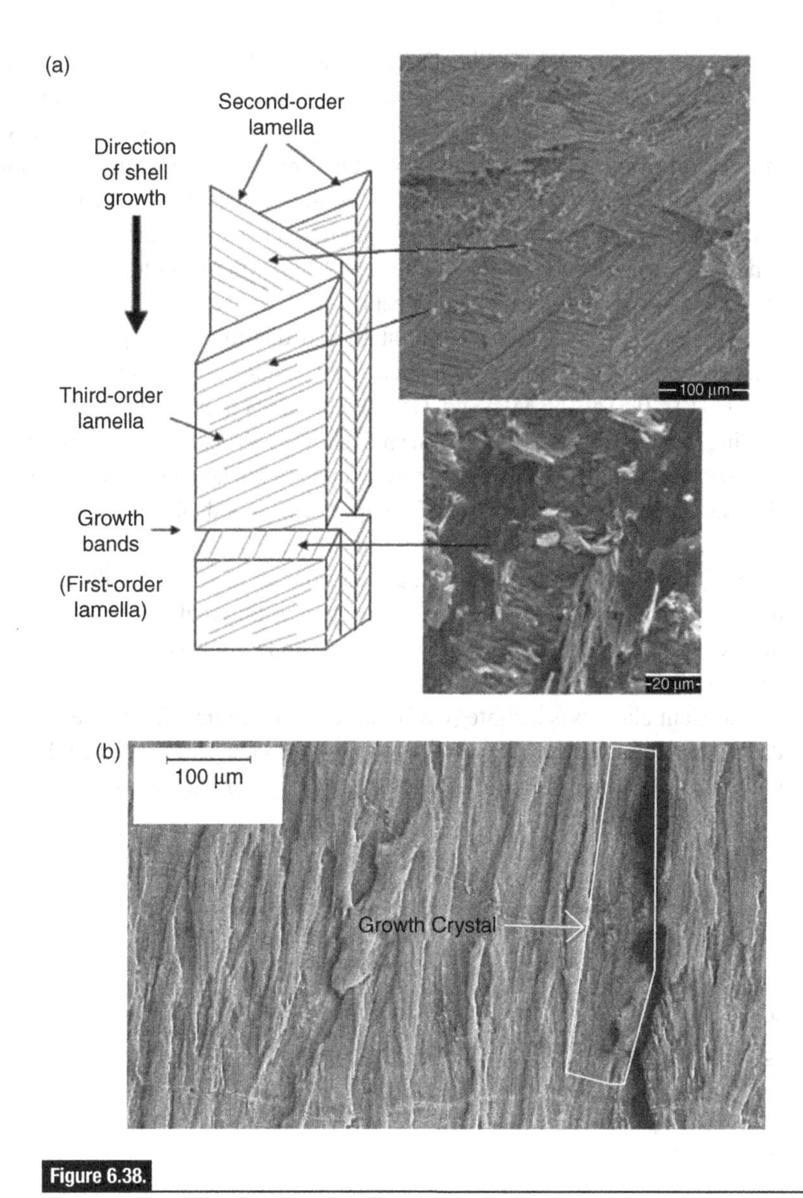

Figure 6.38.

(a) Schematic representation and SEM images of *T. gigas* shell (outer region). (b) Optical microscopy of polished cross-sectional specimen of *T. gigas* shell (inner region), with continuous single-crystal facilitating crack propagation.

parallel to the growth direction, which separate planes of needles (third-order lamellae) with alternating orientation. The directions of needles alternate between +60° and −60° to the direction of growth for each second-order lamella.

Within the inner region of the shell, the microlayered structure is also observed as continuous planes of growth bands. These layers separate approximately 3–7 μm of inorganic material and span normal to the direction of shell growth. Long single crystals of aragonite travel along the direction of growth and are not interrupted by growth bands. This inner region appears more transparent than the outer region and contains a high concentration of flaws traveling along the single columnar crystal interfaces. These flaws, in the form of microcracks, travel along the direction of growth, facilitating crack propagation along abutting interfaces of neighboring crystals. Figure 6.38(b) shows an optical micrograph of the microcracks along columnar crystal interfaces. The observed growth bands in the microstructure do not interrupt the growth of single crystals from one band to the next, and thus have a minimal effect on crack deflection.

Figure 6.39 presents the Weibull statistical distribution of giant clam in quasistatic compression. Whereas the conch shell from Section 6.2.3 had a failure probability of 50% ($F(V) = 0.5$) at 166 MPa and 218 MPa for the perpendicular and parallel direction of loading, respectively, the giant clam shell showed 50% failure probability at 87 MPa and 123 MPa for loading parallel and perpendicular to the layered structure, respectively. The abalone shell from Section 6.2.2.2 outperformed both the conch and the giant clam shells, with over twice the compressive strength in quasistatic loading. With failure probabilities of 50% being reached at 235 MPa and 540 MPa with loading parallel and perpendicular to layered structure, respectively, the abalone also exhibits the highest difference in strength between loading directions, consistent with the level of microstructure anisotropy.

The giant clam uses a strategy of rapid growth, unimpeded by periodic organic layer deposition, to create the *mosaic* sign of the shell. This occurs at a penalty of strength and toughness. The inner region fails at the crystal interfaces seen in Fig. 6.38 through a

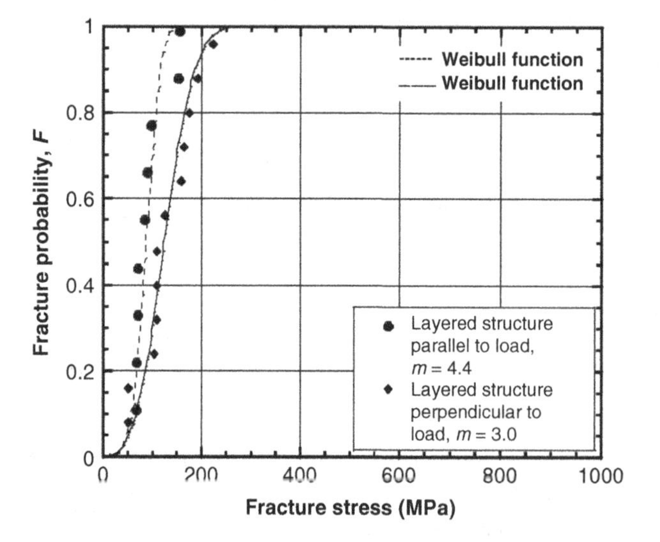

Figure 6.39.

Weibull analysis of *T. gigas* shells in quasistatic compressive loading. (Reprinted from Menig *et al.* (2001), with permission from Elsevier.)

mechanism of axial splitting. Initial microcracks within this region extend and coalesce under applied stress, resulting in the failure of the shell samples.

6.2.4.1 *Saxidomus purpuratus*

The strength and fracture behavior of *Saxidomus purpuratus* shells were investigated and correlated with the structure. The shells show a crossed-lamellar structure in the inner and middle layers and a fibrous/blocky and porous structure composed of nano-scaled particulates (~100 nm diameter) in the outer layer. Figure 6.40 shows the overall view of the *Saxidomus* structure with an external layer that is blocky/fibrous and an internal layer consisting of domains of approximately 50 μm. The crossed-lamellar structure of this shell is composed of domains of parallel lamellae with approximate thickness of 200–600 nm. These domains have approximate lateral dimensions of 10–70 μm with a minimum of two

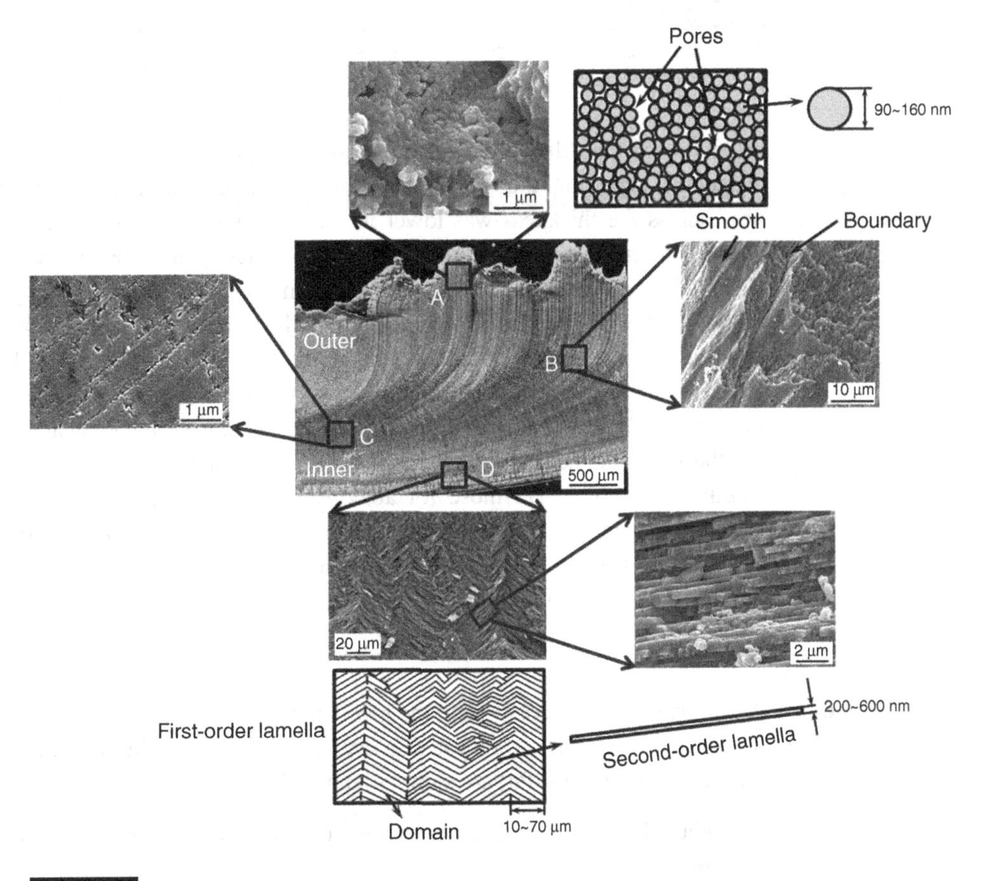

Figure 6.40.

Overall view (center) of section of *S. purpuratus* shell showing different morphologies in inner layer (bottom), middle layer (left), and outer regions (top and right). (Reprinted from Yang *et al.* (2011b), with permission from Elsevier.)

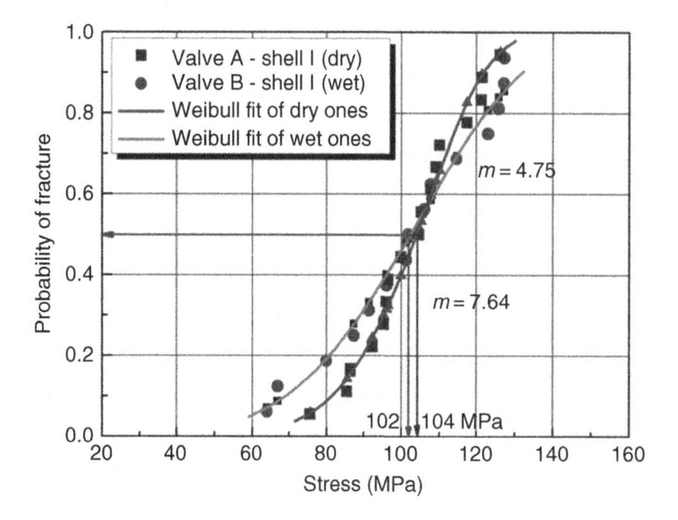

Figure 6.41.

Weibull plots of bending strengths from two valves of the same *S. purpuratus* shell. (Reprinted from Yang *et al.* (2011b), with permission from Elsevier.)

orientations of lamellae in the inner and middle layers. Neighboring domains are oriented at specific angles and thus the structure forms a crossed-lamellar pattern. The microhardness across the thickness was lower in the outer layer because of the porosity and the absence of lamellae. The tensile (from flexure tests) and compressive strengths were analyzed by means of Weibull statistics. The mean tensile (flexure) strength at probability of 50%, 80–105 MPa, is shown for two valves of the same shell in Fig. 6.41. The compressive strength (~50–150 MPa) is on the same order (Fig. 6.41). The compressive strengths were obtained along three loading orientations, and the results vary somewhat. Indeed, the *Saxidomus*, as well as most shells, has anisotropic mechanical properties that are the result of the aligned microstructure. The compressive and flexure strengths are significantly lower than those for abalone nacre, in spite of having the same crystal structure. The lower strength can be attributed to a smaller fraction of the organic interlayer. The fracture path in the specimens is dominated by the orientation of the domains and proceeds preferentially along lamella boundaries.

Example 6.2 A graduate student investigated the mechanical properties of clam shells under compression. The samples were tested in dry and hydrated conditions and the compressive failure strengths (in MPa) are summarized in Table 6.1.

(a) Calculate the average strength and corresponding standard deviation for each testing condition.

(b) Apply Weibull analysis and determine the Weibull modulus m, characteristic strength σ_0, and the strength at a probability of failure of 50%.

Table 6.1. Compressive strength of dry and wet *Saxidomus* shells

N	1	2	3	4	5	6	7	8	9	10	11	12	13	14	15	16	17	18	19	20
Dry	55	60	65	70	75	80	85	90	95	100	100	105	110	115	120	125	130	135	140	145
Wet	82	84	86	88	90	92	94	96	98	100	100	102	104	106	108	110	112	114	116	118

Solution (a) The average strength and standard deviation of dry and hydrated samples are:

$$\sigma_{avg}(\text{dry}) = 100 \pm 27.4\,\text{MPa},$$
$$\sigma_{avg}(\text{wet}) = 100 \pm 11.0\,\text{MPa}.$$

(b) The detailed derivation of Weibull analysis is presented in Chapter 2 (Section 2.8). The survival probability of a brittle material is given by

$$P(V_0) = \exp\left[-\left(\frac{\sigma}{\sigma_0}\right)^m\right],$$

where m is the Weibull modulus and σ_0 is the characteristic strength. The higher the value of m, the less is the material's variability in failure strength.
The failure probability can be written as

$$F(V_0) = 1 - P(V_0) = 1 - \exp\left[-\left(\frac{\sigma}{\sigma_0}\right)^m\right].$$

If N samples are tested, we rank their strengths in ascending order, and the probability of survival can be determined by

$$P_i(V_0) = (N + 1 - i)/(N + 1).$$

For example, if there are ten samples tested, the probability of survival of the first sample with the lowest strength is $P_1(V_0) = (10 + 1 - 1)/(10 + 1) = 10/11 = 91\%$.
The Weibull modulus can be obtained from the slope in the double logarithm for $1/P(V)$ (or $1/[1 - F(V)]$) and the logarithm for σ. The intercept at $\ln(\ln[1/(1 - F)]) = 0$ corresponds to the characteristic strength σ_0, which has a failure probability of 0.63 ($= 1 - 1/e$). The calculated $P(V)$, $F(V)$, $\ln \sigma$, and $\ln[\ln(1/P(V))]$ are summarized in Table 6.2. The results are plotted in Fig. 6.42. The Weibull modulus, characteristic strength, and strength at 50% failure probability for dry and hydrated samples are summarized in Table 6.3. Figure 6.43 shows the Weibull distribution with the preceding parameters superimposed on the data points:

$$F(V_0) = 0.5 = 1 - \exp\left[-\left(\frac{\sigma_{50\%}}{109.7}\right)^{3.73}\right] \Rightarrow \sigma_{50\%}(\text{dry}) = 100.6\,\text{MPa},$$

Table 6.2. Compressive strengths and calculation of failure and survival probabilities of dry and hydrated clam shells

N	σ_{dry} (MPa)	σ_{wet} (MPa)	*P*(*V*)	*F*(*V*)	ln σ_{dry}	ln σ_{wet}	ln[ln(1/*P*(*V*))]
1	55	82	0.95	0.05	4.01	4.41	−3.02
2	60	84	0.90	0.10	4.09	4.43	−2.30
3	65	86	0.86	0.14	4.17	4.45	−1.87
4	70	88	0.81	0.19	4.25	4.48	−1.55
5	75	90	0.76	0.24	4.32	4.50	−1.30
6	80	92	0.71	0.29	4.38	4.52	−1.09
7	85	94	0.67	0.33	4.44	4.54	−0.90
8	90	96	0.62	0.38	4.50	4.56	−0.73
9	95	98	0.57	0.43	4.55	4.58	−0.58
10	100	100	0.52	0.48	4.61	4.61	−0.44
11	100	100	0.48	0.52	4.61	4.61	−0.30
12	105	102	0.43	0.57	4.65	4.62	−0.17
13	110	104	0.38	0.62	4.70	4.64	−0.04
14	115	106	0.33	0.67	4.74	4.66	0.09
15	120	108	0.29	0.71	4.79	4.68	0.23
16	125	110	0.24	0.76	4.83	4.70	0.36
17	130	112	0.19	0.81	4.87	4.72	0.51
18	135	114	0.14	0.86	4.91	4.74	0.67
19	140	116	0.10	0.90	4.94	4.75	0.86
20	145	118	0.05	0.95	4.98	4.77	1.11

$$F(V_0) = 0.5 = 1 - \exp\left[-\left(\frac{\sigma_{50\%}}{104.9}\right)^{9.72}\right] => \sigma_{50\%}(\text{wet}) \;=\; 101.0\ \text{MPa}.$$

It can be seen that the characteristic and 50% failure strength for both conditions are not significantly different, yet the Weibull modulus for dry samples, which have a wider range of failure strength, is ~3.7, much smaller than that for hydrated samples ($m \sim 9.7$). The fracture of dry samples is more brittle, leading to its unpredictable mechanical behavior.

Table 6.3. Weibull parameters for the dry and wet shells

Specimen	m	σ_0 (MPa)	$\sigma_{50\%}$ (MPa)
Dry clam shell	3.73	110.7	100.6
Hydrated clam shell	9.72	104.9	101.0

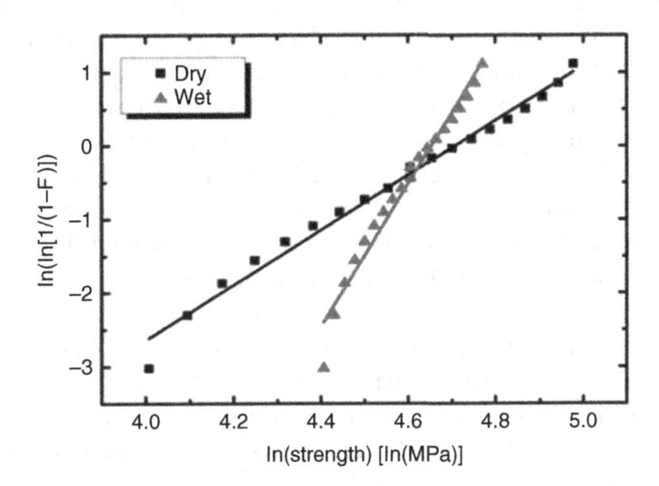

Figure 6.42.

Double logarithm of normalized failure probability vs. logarithm of strength for clam shells.

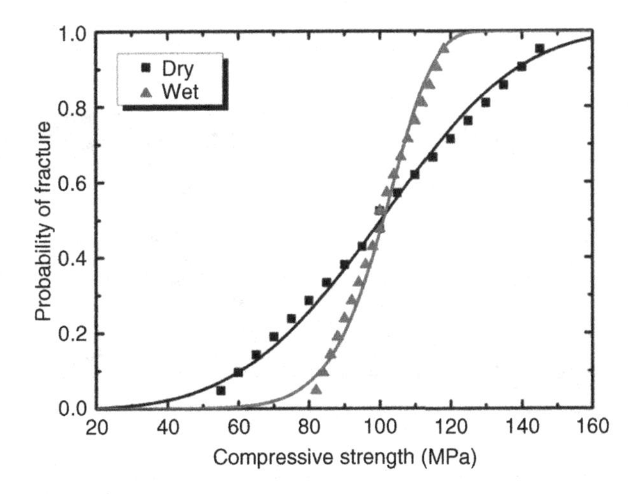

Figure 6.43.

Weibull plot of failure probability vs. strength.

6.2.4.2 Araguaia river clam

The Araguaia river clam is found to exist in the fresh water of the Amazon basin. In its natural environment it sits upright with its flat bottom base resting on the floor of a sandy river bed. Protruding upward, its shell makes a fin-like arc (Fig. 6.44(a)) cutting through the current of the moving river, allowing the capture of passing food. Although the environment of this freshwater bivalve differs greatly from that of the red abalone, their structures both consist of aragonite tiles. However, there are significant differences in this structure and, thus, differences in their mechanical response. The shell of the Araguaia river clam consists of parallel layers of calcium carbonate tiles, approximately 1.5 μm in thickness and 10 μm in length. This is three times thicker than abalone nacre, implying a higher inorganic to organic ratio. Furthermore, the uniformity of the tiles is far less apparent in the shell of the river clam than in the abalone. Although a uniaxial alignment is observed along the *c*-axis (the axis parallel to the direction of growth), the consistency of layer thickness is less pronounced than its saltwater counterpart. The wavy structure is observed in Fig. 6.44(c) and can be seen throughout. The greatest difference between the two structures, however, is at the meso-level. In contrast to the abalone shell, there were no observed mesolayers marking inorganic growth interruption. Figure 6.44(b) provides an optical view of the cross-section of the river clam shell. The missing growth bands and decreased organic composition lead to a more classically brittle ceramic. While mechanisms such as crack deflection and microbuckling were observed in the abalone nacre (Menig *et al.*, 2000), they were lacking in the river clam shell.

Three-point-bending and quasistatic compression tests were conducted in various orientations of shell microstructure. The compressive strength when loaded perpendicular to the layers is 40% higher than when it is loaded parallel to them. The 50% fracture

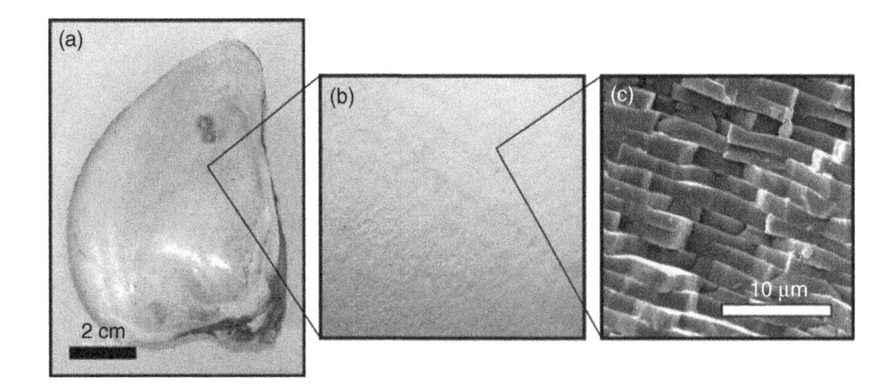

Figure 6.44.

Structural hierarchy of the Araguaia river clam. (a) Note the flat bottom, which ensures that the clam stays upright on the sandy river bed. (b) There are few or no observable mesolayers at the meso-scale. (c) Thick wavy tiles of 1.5–2 μm thickness and 10 μm length are observable at the micro-scale. (Reprinted from Chen *et al.* (2008b), with permission from Springer provided by Copyright Clearance Center.)

probability is found at 567 MPa for the perpendicular direction and at 347 MPa for the parallel. This is because, when loading is applied parallel to the tiles, they undergo splitting along the interfaces, resulting in lower strength.

The tensile strength, as obtained from flexure tests, is much lower (20–35 times) than the compressive strength: ~18 MPa. For the abalone, it was around 170 MPa in flexure tests. This ratio of compressive to flexural strength is far greater than that found in abalone nacre, and is characteristic of a brittle ceramic. This difference can be explained at two levels: (a) there are no organic mesolayers in the Araguaia clam; (b) the cracks propagate preferentially through the tiles.

6.3 Teeth of marine organisms: chiton radula and marine worm

It was discovered in 1962 by Lowenstam (1962) that the teeth in chitons (mollusc worms) contained iron oxide in the magnetite structure (Fe_3O_4). Chitons are marine molluscs whose dorsal shell is composed of not one, but eight separate plates. They belong to the class of *Polyplacophora* (from Greek: *poly* = many; *plako* = plates; *phorous* = bearing). These molluscs have a skirt around their periphery. They have also received the nickname of "coat-of-mail" shells.

The plates are composed of aragonitic calcium carbonate and are held together by a muscular girdle that surrounds the body (see Figs. 6.45(a) and (b)). Thus, the name chiton, also derived from Greek, meaning tunic. One unique aspect of the chiton that will be presented here is its radula, or raspy tongue. This radula is a conveyor-belt-like structure containing magnetite teeth. Figures 6.45 (c)–(f) depict the conveyor-belt appearance of the radula (Weaver *et al.*, 2010). Chitons derive their nourishment from

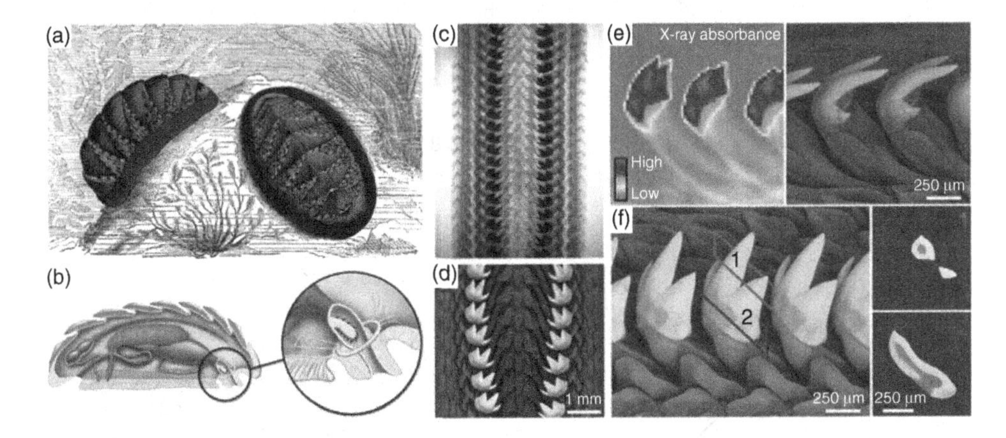

Figure 6.45.

Chiton, a mollusc containing eight plates. External (a) and internal (b) anatomy of chiton, with detail showing the radula, a rasping, toothed conveyor-belt-like structure used for feeding. Optical (c) and backscattered SEM (d) imaging and X-ray absorbance studies (e) reveal the nature of the electron density distribution of the tricuspid tooth caps. Cross-sectional studies through the teeth from *C. stelleri* (f) reveal a concentric biphasic structure. (Reprinted from Weaver *et al.* (2010), with permission from Elsevier.)

algae that form on rock, and therefore their teeth have to resist this action of grazing on the rock that can wear them. The teeth have three cusps and are attached to underlying cuticles that ensure proper alignment. This can be seen in Figs. 6.45(c) and (d). When the teeth are worn, new teeth form. Lowenstam and Weiner (1989) described the chiton teeth as containing magnetite. Its hardness is 9–12 GPa, the highest of any biomineral. In comparison, dental enamel has a hardness of 3.5–4.5 GPa. Ganoine, the hard layer present on fish scales, has a hardness of 4.5 GPa.

The structure of the teeth is similar to those of vertebrates, having an external layer that is harder and a more compliant core. The external layer is pure magnetite organized in parallel rods and the internal core is made of an enriched iron phosphate interspersed with chitin fibers; correspondingly, its hardness is lower, ~2 GPa.

This discovery was followed by another one, by Lichtenegger *et al.* (2002): the carnivorous marine worm *Glycera* has teeth that contain a copper mineral atacamite ($Cu_2(OH)_3Cl$). These minerals are contained in mineralized fibrils, as shown in the schematic picture of a tooth in Fig. 6.46(a). These mineralized fibrils are similar to the ones that form in dentin and bone. Again, we have a composite structure with hard fibers embedded in a softer protein matrix. The degree of mineralization varies along the tooth, and the hardness and the elastic modulus are directly related to the mineralization. Lichtenegger *et al.* (2002) used the Halpin–Tsai equation, well known in the composite field, to calculate the hardness and the Young modulus upper and lower bounds. The upper bound corresponds to loading parallel to the fiber direction; the lower bound corresponds to loading perpendicular to it. The calculated as well as experimental results

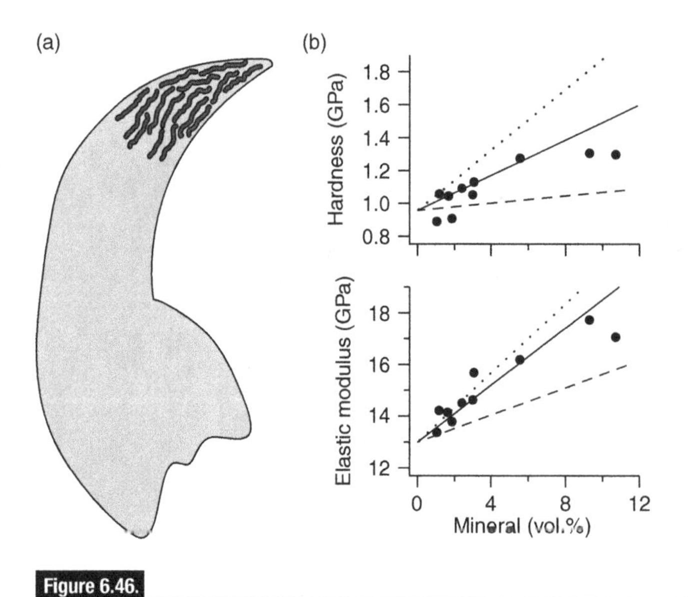

Figure 6.46.

(a) Schematic model of mineralized fibers in *Glycera* jaws. (b) Hardness and elastic modulus vs. mineral content. Dashed and dotted lines show the Halpin–Tsai boundaries. (Taken from Lichtenegger *et al.* (2002), with kind permission from Professor Galen D. Stucky, UC Santa Barbara.)

are shown in Fig. 6.46(b). The Halpin–Tsai upper and lower bounds are indicated by dotted and dashed lines in Fig. 6.46(b).

6.4 Sea urchin

Another interesting example is the calcite structure of the spines of the sea urchin (*Echinoidea*). Sea urchins are found in all marine environments. The spines can be up to 30 cm in length and 1 cm in diameter and either sharp or blunt (Magdans and Gies, 2004; Schultz, 2006). The spines are a highly Mg-substituted calcite, $Mg_{1-x}Ca_xCO_3$, $x = 0.02$–0.15 with crystallites 30–50 nm in diameter (Presser *et al.*, 2009). Substitution of Ca by Mg ions increases the strength of the crystallites; it tends to be at a higher concentration at the base than at the tip. The concentration of Mg is directly related to the water temperature – higher Mg concentrations are found at higher temperatures.

Figure 6.47 shows a sketch of the cross-section and an X-ray computer tomography micrograph along the length of a *Phyllacanthus imperialis* spine (Presser *et al.*, 2009). As shown, there is a gradient in porosity, with porosity increasing substantially from ~10% on the surface to ~60% in the medullary core. What can possibly be gained from such a configuration? Because the spines are used for protection, the compressive strength is more important than the tensile strength. The compressive force–displacement curve displays a graceful failure instead of a catastrophic failure typical of monolithic ceramics. Interestingly, the stress–strain curve resembles that of a classical cellular solid, as described by Gibson and Ashby (1997). The peak stress is related to the strength of the dense outer sheath, whereas the plateau region relates to the failure of the highly porous region, dependent on the density and other elastic properties of the solid material.

6.5 Shrimp hammer

In crustaceans (e.g. crabs, lobsters, shrimps), the exoskeleton is mineralized with $CaCO_3$ in the form of calcite and some amorphous $CaCO_3$, deposited within the chitin–protein matrix. The microstructure and mechanical properties of arthropod exoskeleton will be discussed in detail in Section 8.3, where we classify arthropod exoskeletons as polymer-based composites. There is a unique mantid shrimp which has a highly mineralized hammer that can crush the sturdiest hard-shelled preys.

The mantid shrimps are predatory on hard-shelled animals such as clams, abalones, and crabs, using their limbs as hammers. Although the smashing shrimp makes thousands of energetic strikes over months, the hammer is rarely damaged. The composition and structural features of the smashing limbs have been studied by Currey, Nash, and Bonfield (1982). Figure 6.48(a) shows a typical mantid shrimp and its smashing action. The smashing limb consists of the merus, the propodite, and the dactyl. The dactyl is the part used to strike the prey and is highly mineralized. Figure 6.48(b) is a schematic presentation showing the cross-section of propodite and dactyl. Both propodite and dactyl show three regions: soft tissue in the core, a layer of fibrous chitin cuticle in

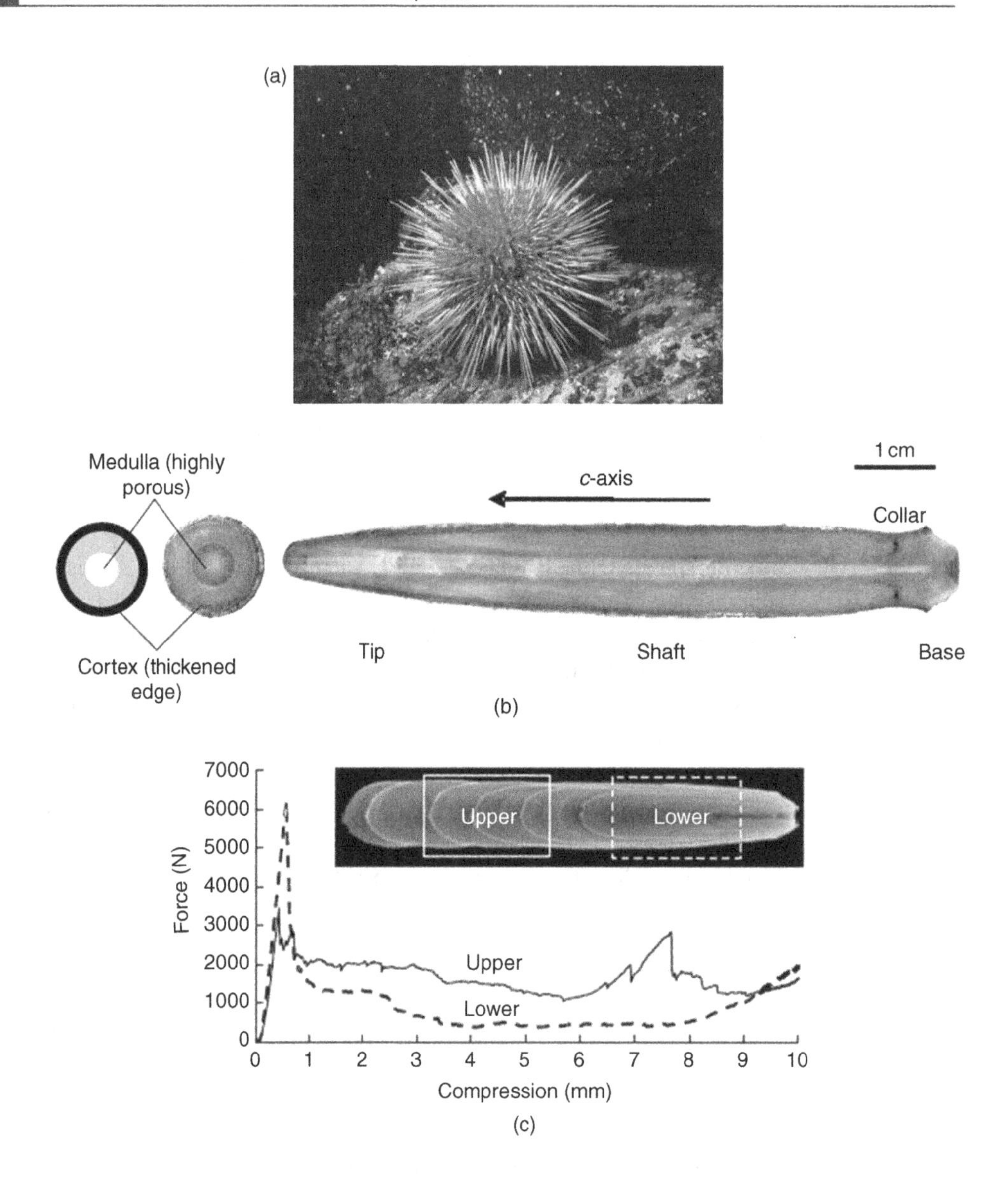

Figure 6.47.

(a) Sea urchin. (b) Cross-sectional and longitudinal views. (c) Compressive force–deflection curve, showing a peak load and then graceful failure during the plateau region. (Reprinted from Presser *et al.* (2009), with permission from Elsevier.)

between, and a layer of heavily calcified region on the outside. The dactyl, used to smash hard-shelled prey, has a thick, heavily calcified layer, while in the propodite this layer is much thinner. The microhardness measurements (Fig. 6.48(c)) indicate that the dactyl becomes much harder toward the outer surface. The increase in hardness is associated

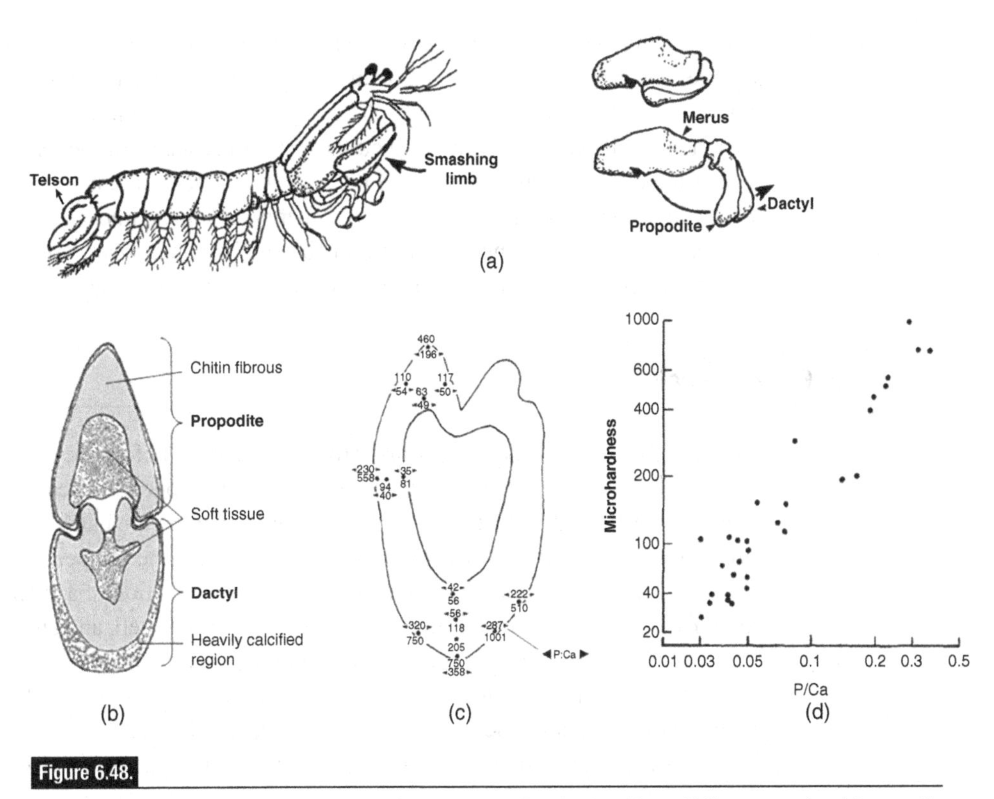

Figure 6.48.

(a) The smashing limb and typical smashing action of mantid shrimp, *Gonodactylus chiragra*. (b) The cross-section of the propodite and dactyl shows three regions: heavily calcified outer layer, fibrous region, and inner soft tissue. (c) Values for microhardness and values for P:Ca (multiplied by 1000). (d) Relationship between microhardness and the P:Ca ratio. (Reprinted from Currey *et al.* (1982), with kind permission Springer Science and Business Media.)

with the increased mineralization of the cuticle as well as the replacement of calcium carbonate by calcium phosphate. Figure 6.48(d) shows the relationship between the microhardness and the ratio of phosphorus to calcium. There is a strong and linear relationship between the hardness and the P:Ca ratio. The fibrous region between the hard outer layer and soft tissue not only absorbs the kinetic energy, but also prevents cracks from propagating through the cuticle. The outer layer has a sufficient thickness of heavily calcified cuticle with a significant amount of calcium carbonate replaced by calcium phosphate. The hammer of the mantid shrimp is well designed to break hard objects, and is an optimized biological ceramic composite.

The smashers have a highly developed club (dactyl heel) that can be propelled at accelerations up to $10g$, reaching speeds up to 23 m/s and generating forces of up to 1500 N. This is reported to be the fastest appendicular striker in the animal kingdom. The telson (tail fan) is another robust appendage that is battered during intraspecies fights. The telson is thumped by smasher clubs of an opponent until one or the other backs down.

The smasher limb is so robust that it can shatter glass aquaria, and may deliver up to 100 strikes per day. Currey *et al.* (1982) found there was a gradient in mineralization across the thickness of the cuticle, with a high concentration of calcium phosphates at the surface, resulting in a hard surface with a tough interior. Although the surface becomes pitted due to the hard strikes, it is replaced by molting every few months.

Taylor and Patek (2010) investigated the impact resistance of the telson through drop-weight impact tests with a small steel ball. The coefficient of restitution was found to be 0.56, with the ball losing most of its kinetic energy during the impact. They concluded that the telson is inelastic, acting like a punching bag and dissipating approximately 70% of the impact energy. A recent description of this material is provided by Weaver *et al.* (2012).

6.6 Egg shell

Oviparous animals produce eggs that protect the babies before they hatch. These eggs have mechanical properties that are adjusted to the environment encountered. They also have to be sufficiently weak to be penetrated or broken by the hatchlings. In the case of birds, the eggs are primarily calcite (96–98% by volume), the rest being hydrated organic material. Figure 6.49(a) shows a schematic of the cross-section of a chicken egg shell. It contains three layers: an inner membrane, the mineralized shell, and an outer epithelium. Figure 6.49(b) is an SEM micrograph showing the cross-section of an egg shell.

We are all familiar with an inner porous membrane from peeling boiled eggs. The structure of this membrane is shown in Fig. 6.49(c). It consists of a network of collagen fibers. The density can be calculated from Gibson and Ashby's (1997) equation (see Chapter 10):

$$\frac{\rho_c}{\rho_s} = C_1\left(\frac{t}{l}\right)^2, \tag{6.23}$$

where ρ_c is the density of the cellular structure and ρ_s is the density of the solid material, respectively. There are two characteristic dimensions: the cell size, l, and the strut thickness, t. C_1 is a proportionality constant. In Fig. 6.49(c), $t \sim 1$ μm and $l \sim 10$ μm, thus the relative density is a very low value. This inner membrane is quite elastic and has several functions, one being to ensure that shell fragments stay in place in the case of egg fracture. Thus, it serves as an inner scaffold. This can be readily seen by removing an egg from the refrigerator and hitting it on the counter until it fragments. The egg will maintain structural integrity. This additional mechanical property is achieved without significant weight penalty.

The calcite in eggs nucleates in the inside and forms radiating growth rods. This type of growth is also known as "spherulitic" and is also observed in polymers and in the aragonite growth in abalone after interruption and prior to the formation of regular tiles (Fig. 6.49(b)). This spherulitic growth is in the nacreous portion of the abalone shell, after growth interruption and the formation of a mesolayer (shown in Fig. 6.14). The crystals that grow toward the outside have a free path and therefore propagate until the external surface. These crystals grow in a spongy organic matrix. The external surface of the egg shell comprises an epithelium. The calcite crystals are shown in the SEM micrograph of Fig. 6.49(d).

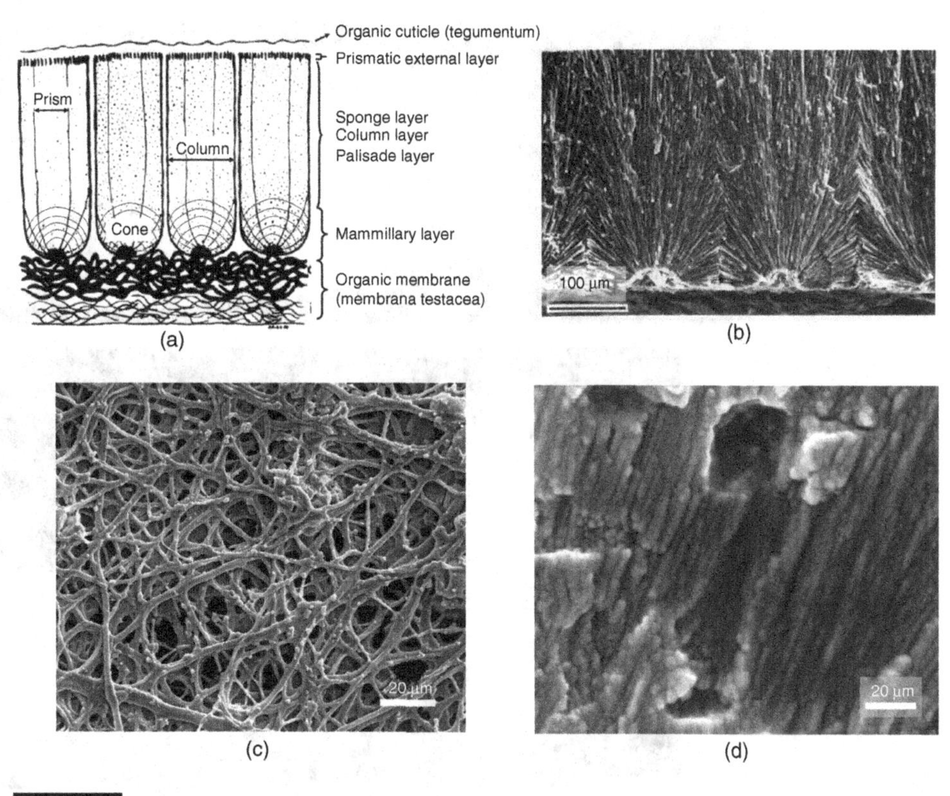

Figure 6.49.

(a) Cross-section of an egg shell showing inner membrane and growth of spherulites of calcite crystals. (Adapted from Wu *et al.* (1992), with kind permission from Elsevier.) (b) SEM micrograph showing cross-section of an egg shell. (Adapted from Silyn-Roberts and Sharpe (1988), with permission.) SEM micrographs of (c) structure of organic layer underneath shell; (d) calcium carbonate crystals growing from the internal surface toward the outside.

6.7 Fish otoliths

Otoliths are calcium carbonate biominerals present in the inner ear of vertebrates. They have a function in balance, movement, and sound detection. There are three pairs of otoliths: lapillus, sagitta, and asteriscus. It is very interesting that, in carp, two polymorphs exist: aragonite (in lapillus and sagitta) and vaterite (in asteriscus) (Ren *et al.*, 2013). They are shown in Fig. 6.50(a). Vaterite is a metastable phase of calcium carbonate. It is interesting that the two polymorphs can coexist in the same host. Similarly, calcite and aragonite coexist in many shells. A more detailed view of a salmon asteriscus is shown in Fig. 6.50(b).

6.8 Multi-scale effects

This chapter has shown that the strength of shells is due to a hierarchy of mechanisms. For nacre, the mesolayers, 0.3 mm apart, are separated by organic layers, as discussed in Section 6.2. The cracks are arrested and deflected at these interfaces. This approach provides

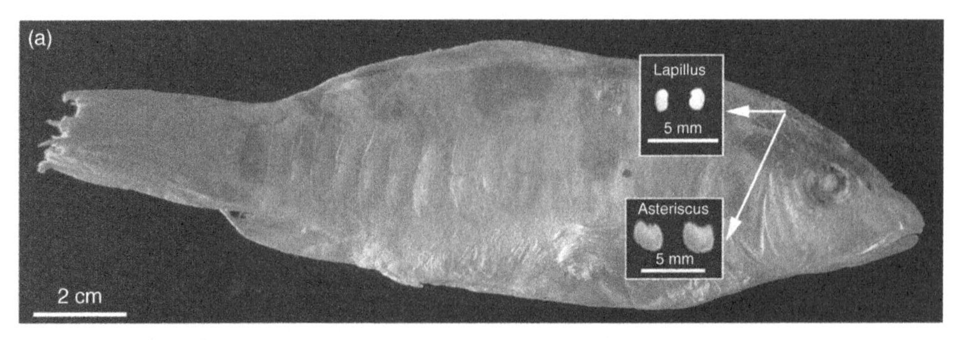

Figure 6.50.

(a) Carp otoliths lapillus and asteriscus. (Reprinted from Ren *et al.* (2013), with permission from Elsevier.) (b) Vaterite asteriscus from salmon. (Figure courtesy of Professor T. Kogure, with kind permission.)

a mechanics-based rationale for the toughening mechanics in these biological composites. At a lower scale, the micro-scale, tablets with ~0.5 μm thickenss provide barriers. And at the nano-scale, mineral bridges and the organic interlayers (~20–50 nm) contribute similar hierarchies which are seen in other silicate- and carbonate-based composites. The mechanics of crack propagation as well as toughening mechanisms involve, at all levels, interfaces that increase the energy required for damage evolution. Ballarini *et al.* (2005) argue that the Aveston–Cooper–Kelly (ACK) limit (Aveston, Cooper, and Kelly, 1971) estimates the length of the bridging zone and the amount of crack growth required to reach the ACK limit.

Summary

- Amorphous silica is present in diatoms, radiolarians, and sponge spicules.
- Diatoms float in the ocean, and they are responsible for 25% of CO_2 sequestration from the atmosphere. There are 100 000 species of diatoms, and, although they can be as long as 2 mm, most have sizes that are in the micrometer range.

- The genome of the marine diatom *Thalassiosira pseudonana* has been established, and it is possible to use genomic and proteomic approaches to accomplish genetic modification of the frustrule (external silica shell).
- Sponge spicules are glass fibers that have a structure akin to a two-dimensional onion. The rods are composed of concentric layers separated by a protein that was named "silicatein."
- The structures of mollusc shells, composed of calcium carbonate (both aragonite and calcite polymorphs) and a small fraction (~5 % or less) of organic constituents, have intricate microstructures that have been classified by Currey and Taylor (1974) as: nacre (columnar and sheet), foliated, prismatic, crossed-lamellar, and complex crossed-lamellar. Some shells have both calcitic and aragonitic parts, such as abalone.
- Many shells exhibit a logarithmic spiral configuration, which is the result of the growth pattern. This spiral can manifest itself primarily along a surface, such as in abalone, or in a conical pattern. The logarithmic spiral can be described by the equation

$$r = e^{C} e^{\theta \tan \alpha},$$

  where r is the radial coordinate of a point and θ is the angular coordinate.
- Nacre is present in a number of shells, including gastropods and bivalves. It is also present in oysters and pearls. In abalone, the external layer of the shell (periostracum) is calcite, which consists of equiaxed grains, and the inside is nacreous. Nacre consists of tiles that are approximately hexagonal in shape and construct a brick-and-mortar architecture. In abalone, the tiles have a thickness of approximately 400 nm and a diameter of ~10 μm. Between the tiles there is a thin organic layer (~20–50 nm) that acts as separation and a glue. This thickness is critical to the optical properties of nacre, and gives rise to the characteristic mother-of-pearl luster. This distance is approximately equal to the wavelength of light, and interference processes give rise to the unique coloration. There is a second hierarchical level to the structure. Seasonal fluctuations in water temperature and feeding lead to the formation of organic mesolayers that are spaced ~0.3 mm apart.
- Growth of nacre: the tiles have the orthorhombic c-axis perpendicular to the large dimension. This is also the direction of rapid growth in aragonite. The tiles form terraced cone arrangements at the growth front. The growth process is periodically retarded by the formation and deposition of an organic layer. This organic layer contains a network of chitin fibrils and has a pattern of holes, through which the mineral eventually penetrates, so that growth is resumed. The tiles in each terraced cone have the same crystallographic orientation, which can be established by transmission electron diffraction. From the angle of the terraced cone it is possible to calculate the relationship between the growth velocity in direction V_c' and the ones in directions a and b, V_{ab}:

$$\tan \alpha = \frac{V_{ab}}{V_c'}.$$

- The work of fracture of nacre is approximately 3000× that of monolithic $CaCO_3$. The work of fracture is related to the area under the stress–strain curve, and it is deeply affected by gradual, graceful fracture, whereas the fracture toughness does not incorporate this entire process. Thus, one should be careful when considering this number.

- The tensile fracture strength is 100–180 MPa and the fracture toughness is 8 ±3 MPa $m^{1/2}$. This is an eight-fold increase in toughness over monolithic $CaCO_3$.
- The strengths in the three directions are highly anisotropic: the tensile strength with loading direction perpendicular to the tiles is only 3–5 MPa, whereas the compressive strength is ~540 MPa. In the direction of the tile plane, the tensile strength is ~170 MPa and the compressive strength is ~235 MPa.
- When the specimens are compressed parallel to the tile planes, the phenomenon of plastic microbuckling takes place. This microbuckling can be interpreted using equations by Argon (for the stress at which buckling occurs) and Budiansky (for the width, *w*, of the buckling zone):

$$\sigma \cong \frac{\tau}{\theta_0}\left[1+\frac{bG_c\Delta\theta}{2\pi a\tau(1-\upsilon)}\ln\left(\frac{2\pi a\tau(1-\upsilon)}{bG_c\Delta\theta}\right)+\frac{E_r\Delta\theta}{48\tau}\left(\frac{t_r}{b}\right)^2\right],$$

$$\frac{w}{d}=\frac{\pi}{4}\left(\frac{2\tau_y}{CE}\right)^{-1/3}.$$

- The conch has a structure that is quite different and that is best described as crossed lamellar. It also has high mechanical properties.
- The bivalve clam *Saxidomus purpuratus* has the complex crossed-lamellar structure.
- Sea urchins have spines that consist of an Mg-substituted calcium carbonate. The replacement of calcium by magnesium confers greater strength. The core of the spines is porous, whereas the periphery is compact. Their compressive properties are of utmost importance, and the porous core might contribute to a decrease in tensile strength, a desirable feature, since pieces of the spine remain embedded in the predator, creating a painful feeling and inflammation.
- Egg shells are calcitic; the shape of the crystals is of divergent rods. This type of growth is also known as "spherulitic" and is also observed in polymers and in the aragonite growth after interruption and prior to the formation of regular tiles.

Exercises

Exercise 6.1 Find a shell on the web and apply the spiral logarithmic equation to it. Show the match between the shell and the equation, and provide an explanation.

Exercise 6.2 Assuming that the abalone creates one layer in 24 hours (based on Lin and Meyers (2005)) and that growth only occurs in six months per year, calculate the yearly increase in the diameter of the shell if self-similarity is maintained if the initial diameter is 10 mm and the initial thickness is 1 mm. Measure the ratio from the existing shell provided by Lin and Meyers (2005).

Exercise 6.3 The growth in the *c*-direction (V_c) is ten times the growth in the *a* and *b* directions for aragonite. However, this growth velocity is reduced to $0.01 V_c$ as the biomineralization is going through the holes in the organic layer. Assuming that this layer has a thickness of 50 nm, calculate the angle of the "Christmas tree" arrangement.

Exercise 6.4 Determine the maximum tensile stresses undergone by the *Saxidomus* shell specimens tested in three-point bending with a span of 20 mm. Data are given in Table E6.4. Find the avarage and the standard deviation.

Exercise 6.5 For the data in Table E6.4, make a Weibull plot and determine the modulus and the stress at a probability of failure of 0.5.

Exercise 6.6 (a) Determine the velocity that a shrimp hammer can reach if it accelerates at $10g$ (9.8 m/s^2) and the telson has a length of 3 cm and describes a circular motion with an angle of 60°. (b) Assuming that the hammer decelerates on a hard prey over a distance of 2 mm, what force can it exert? Estimate the mass of the "fist" from Fig. 6.48. Assume a density of 2 g/cm^3.

Exercise 6.7 Determine the density and elastic modulus of cancellous bone if the cell size is 500 μm and the strut width is 100 μm. Compact bone has a density of 2 g/cm^3 and

Table E6.4. Failure loads of *Saxidomus* shell specimens in flexure

Width (mm)	Thickness (mm)	Force (N)
3.56	1.18	20.63684
3.8	1.18	22.58107
3.74	1.18	16.94014
3.88	1.16	19.15261
4.02	1.2	21.56501
4	1.14	11.8906
3.54	1.24	18.58711
4.02	1.24	21.21456
3.94	1.24	11.27764
3.7	1.2	11.72666
4.02	1.24	22.33531
4.02	1.2	18.04434
4.08	1.28	15.68924
4.02	1.2	20.19495
3.84	1.22	9.77908

elastic modulus of 20 GPa. Use Eqn. (6.23) for the density and derive it appropriately. For E, use Eqn. (10.11).

Exercise 6.8 Explain with illustrations the growth mechanism of the brick-and-mortar structure in abalone nacre.

Exercise 6.9 What are the toughening mechanisms of abalone nacre at the meso-, micro-, and nano-length scales? Explain with illustrations.

Exercise 6.10 Sponge spicules are made mainly of amorphous silica yet have exceptional toughness compared with glass. Why? Explain with illustrations.

Exercise 6.11 Estimate the tensile strength of an abalone shell perpendicular to the tile layers. Given:

$E = 100$ GPa;
number of asperities/bridges per tile = 3500;
1% of asperities are bridges;
diameter of bridges = 50 nm.

Exercise 6.12 Calculate the tensile strength of abalone nacre parallel to the tile (or tablet) layers if the shear strength of the interfaces is 20 MPa, and the tiles have a thickness of 0.5 μm and a diameter of 10 μm.

Exercise 6.13 Express the Halpin–Tsai equation for composites and show how it can be applied to the carnivorous worm *Glycera*.

7 Calcium-phosphate-based composites

Introduction

Hydroxylapatite or hydroxyapatite (HAP) is a calcium-phosphate-based mineral of the apatite family. Its chemical formula is $Ca_{10}(PO_4)_6(OH)_2$. It can be found widely in nature and is the major component of bone, enamel, and dentin in teeth, antler, ganoid fish scales (in alligator gar and Senegal bichir), turtle shells, and armadillo and alligator osteoderms. It exists in minute quantities in the brain (brain sand), without significantly affecting its function. Thus, the expression "having sand in the head" is not without reason. The density of HAP is $3.15\,g/cm^3$. Nonstoichiometric minerals can exist with $Ca_{10}(PO_4)_6(OH, F, Cl, Br)_2$; if the OH group is replaced by F it is called fluoroapatite; if it is replaced by Cl, it is called chloroapatite. It can be occasionally used as a gem, and the cat's eye is a commonly known use.

In this chapter, we will concentrate on bone and teeth with emphasis on their structure and mechanical properties. They are HAP–collagen composites and their mechanical properties are the result of the complex interplay and hierarchy built by these structures. Selected calcium-phosphate-based bony tissues with unique functionalities, such as antler, turtle shells, alligator osteoderms, and fish scales, will also be described in the second part of this chapter.

Synthetic HAP is used in a significant number of biomedical applications. It can be produced by many techniques. One of the most interesting methods involves heating coral ($CaCO_3$) to high temperatures and, in the process, retaining its porous structure, which is favorable for bone growth.

7.1 Bone

Bone is a ceramic (calcium phosphate, or hydroxyapatite)–polymer (collagen) composite. It is the structural component of our body. The skeletal system is a composite material of a protein (mainly type-I collagen), a mineral phase (carbonated hydroxyapatite), and water, assembled into a complex, hierarchical structure. On a volumetric basis, bone consists of about 33–43 vol.% minerals, 32–44 vol.% organic, and 15–25 vol.% water. It also has other functions, but we will concentrate on the mechanical performance here. There are two principal types of bone: cortical (or compact) and cancellous (or trabecular, or porous). Cortical bone is found in long bones (femur, tibia, fibula, etc.). Cancellous bone is found in the core of bones and in flat bones.

The hierarchical structure of bone has been reviewed by several groups (Weiner, Traub, and Wagner, 1986; Weiner and Wagner, 1998; Rho, Kuhn-Spearing, and Zioupos, 1998; Currey, 2002; Fratzl and Weinkamer, 2007) and is shown schematically in Fig. 2.4. The skeletal system has multifunctionality. In general, it supports the body, protects the organs, produces blood cells, and stores mineral ions. There are some less familiar bony tissues in the animal kingdom that have unique functionalities, which have been recently reviewed by Currey (2010). For example, bird bones are designed for flying and have to be lightweight and flexible. Bones used as protective armor, such as the turtle shell (Krauss *et al.*, 2009; Rhee *et al.*, 2009), armadillo carapace (Chen *et al.*, 2011; Rhee, Horstemeyer, and Ramsay, 2011), and fish scales (Yang, 2013b, c), must be stiff and impact resistant. Mammalian ear bone (whale bulla) is extremely dense and highly mineralized in order to transmit acoustic signals (Currey, 1979, 1999). In this chapter, we focus on one of the most important functions of bone, which is the ability to resist fracture.

7.1.1 Structure

Figure 7.1(a) shows the structure of a long bone. The surface regions consist of cortical bone; the inside is porous and is called cancellous bone. Figure 7.1(b) is a photograph of a section of cancellous bone, showing a porous structure. The porosity of cancellous bone is typically in the range 75–95% and the apparent density (mass per unit volume of solid trabeculae) of cancellous bone ranges from 0.2 to 0.8 g/cm^3. The porosity reduces the strength of the bone, but also reduces its weight. Bones are structured in such a manner that strength is provided only where it is needed. This is presented in detail in Chapter 10. The mechanical strength is determined by the porosity and the manner in which this porosity is structured, since the struts can align such that the strength is optimized in the directions in which the stresses are highest. This remodeling was discovered by Wolff in the 1800s and is known as Wolff's law. The pores also perform other physiological functions and contain the marrow. The trabeculae morphology can vary from rod-like structures at lower apparent density (higher porosity) to plate-like structures at higher apparent density (lower porosity), and the trabeculae have a preferred orientation along the long axis of bone. Compact bone is denser (~2 g/cm^3) and often characterized by microstructural features called osteons. Osteons are composed of concentric lamellae surrounding a vascular channel. There are other prominent features in cortical bone, such as the vascular channels, lacunae spaces, and canaliculi. Lacunae spaces are where the osteocytes reside; they communicate with each other via small tubular channels called canaliculi. Figure 7.2 shows some important structural components in a typical cortical bone. Two types of vascular channels are present. Main vascular channels (or Haversian canals) run parallel to the long axis of bone, while the Volkmann's canals are transversely aligned to the osteons connecting the main vascular channels to each other. The porosity of compact bone is typically 5–10%.

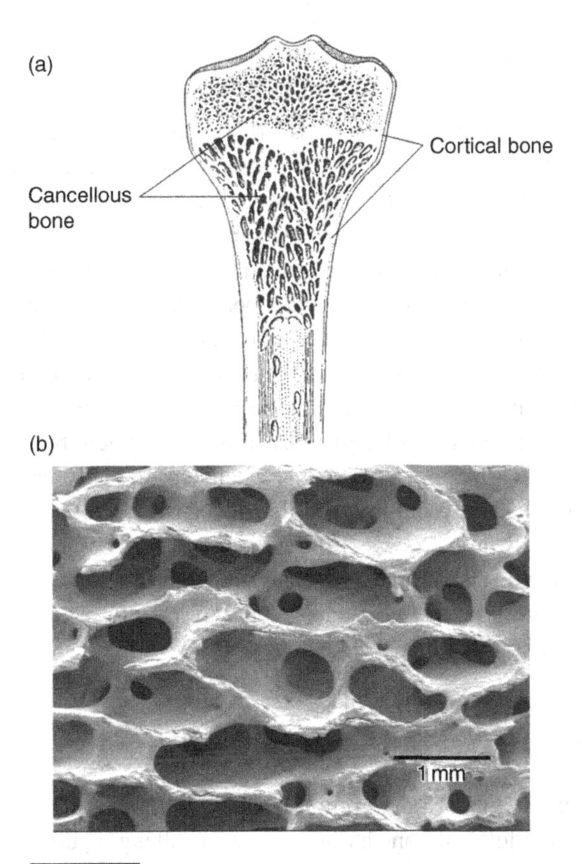

Figure 7.1.

(a) Longitudinal section of a femur. (From Mann (2001). By permission of Oxford University Press.) (b) Section of cancellous bone taken from a bovine femur.

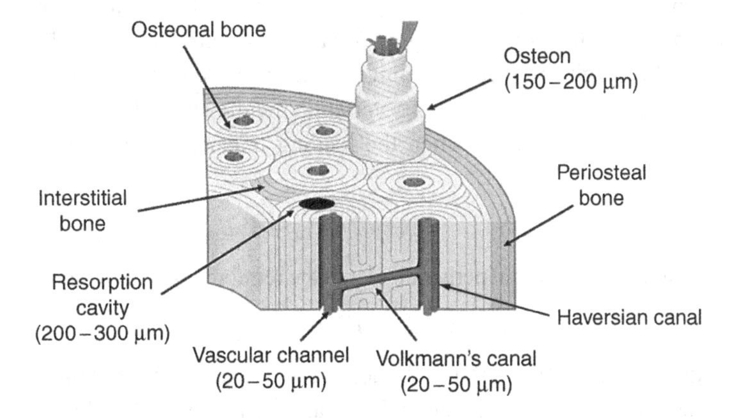

Figure 7.2.

Schematic of some important features in a typical compact bone: osteons; osteonal, interstitial, and periosteal bones; Haversian canal and Volkmann's canals; resorption cavity. (Reprinted from Novitskaya *et al.* (2011), with permission from Elsevier.)

7.1.2 Bone cells and remodeling

There are three types of bone cells: osteoblasts (bone-forming cells), osteoclasts (bone-destroying cells), and osteocytes (derived from osteoblasts). These bone cells form the basic multicellular units (BMUs). Bone is a very dynamic tissue, undergoing constant remodeling throughout the life span of vertebrate animals. Primary osteon is tissue initially forming on an existing bone surface during growth. Secondary osteon forms by a remodeling process when the BMUs resorb the existing bone and replace it with new bone. This remodeling follows a so-called ARF sequence (activation, resorption, formation) (Currey, 2002). Secondary osteon remodeling can arise in response to external mechanical load. Osteoclasts resorb surrounding bone tissue and create a cutting cone-shaped tunnel about 200 μm in diameter and 300 μm in length. The formation of new bone in the cavity is completed by osteoblasts (bone-forming cells), which deposit concentric lamellae on the internal surface. The resultant canal is called the Haversian canal, which contains blood vessels and nerves. The whole process (from initiation of the osteoclastic activity to the completion of the filling in) takes about 2–4 months for human bone (Currey, 2002). As the remodeling is a constant process, most adult bones are mainly composed of secondary osteons and interstitial bone (old, nonactive bone).

7.1.3 Elastic properties

The structure of bone is much more complex than most engineering composites, and modeling the elastic modulus of bone has long been a challenging task. There have been several attempts to model the elastic modulus of bone based on our knowledge of composites. Here, we introduce several selected mechanical models.

Bone is a composite of collagen, hydroxyapatite, and water. The elastic modulus of cortical bone varies from 7 to 24 GPa. This is much lower than that of single-crystal hydroxyapatite, which has a Young modulus of approximately 130 GPa and a strength of 100 MPa. Although collagen is not linearly elastic, we can define a tangent modulus; it is in the range of 1–1.5 GPa. The broad variation mentioned earlier is seen clearly. We have two limiting conditions: (a) when the loading is carried out along the reinforcement direction – the Voigt model:

$$E_b = V_{ha}E_{ha} + V_cE_c, \tag{7.1}$$

and (b) when it is carried out perpendicular to the reinforcement direction – the Reuss model:

$$1/E_b = V_{ha}/E_{ha} + V_c/E_c. \tag{7.2}$$

The indices b, ha, and c refer to bone, hydroxyapatite, and collagen, respectively. In bone, the orientation of the collagen fibers and mineral is difficult to establish. They are not all aligned, which adds to the complexity. Taking the given values and applying the Voigt and Reuss equations leads to, for 50 vol.% collagen,

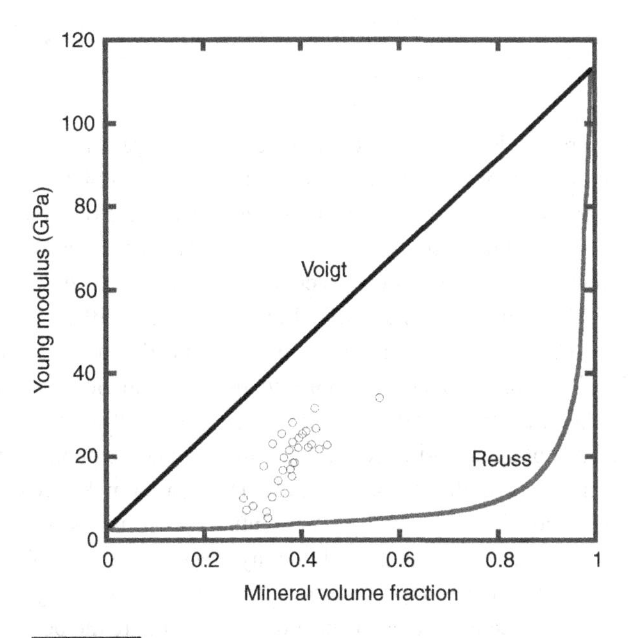

Figure 7.3.

Effect of mineral volume fraction on the Young moduli of bones from many animals. (Reproduced based on Currey (2002).)

$$E_b(\text{Voigt}) = 50\ \text{GPa},$$
$$E_b(\text{Reuss}) = 4\ \text{GPa}.$$

This is an unacceptably broad range, and more elaborate models, discussed in the following, are needed. Figure 7.2 shows the Young moduli for a number of bones with mineral fractions varying from 0.28 to 0.56. Experimental values (Currey, 2002) fall between the two limits set by the Reuss and Voigt averages. The hydroxyapatite content of bone varies from animal to animal, depending on function. Antlers have a low mineral content (~0.3), which leads to exceptional toughness and will be discussed in Section 7.2. For instance, an agile animal like a gazelle has bones that have to be highly elastic. Thus, the hydroxyapatite level is fairly low (around 50% by weight and quite a lot less by volume). Collagen provides the elasticity. On the other hand, a whale has bones with a much higher mineral content (~80% by weight). An aged professor is somewhere in between. Note that the density of hydroxyapatite is approximately twice that of collagen (~1 g/cm^3).

A more realistic model was proposed by Katz (1971). It takes into account the misorientation between the external loading axis and the collagen fibrils. Katz considered different orientations of collagen fibrils, each one with a fraction f_i and angle φ_i with the loading axis. The Young modulus is given by

$$E_b = \frac{E_c V_c (1 - \nu_c \nu_b)}{1 - \nu_c^2} + \sum E_{ha} V_{ha} f_i (\cos^4 \varphi_i - \nu_b \cos^2 \varphi_i \sin^2 \varphi_i), \qquad (7.3)$$

where υ_b and υ_c are the Poisson ratios for bone and collagen, respectively. The Katz equation is essentially a Voigt model that ascribes a contribution to E_b, decreasing rapidly with misorientation φ_i because of the fourth power dependence of the cosine of φ_i.

This was the basis of Jäger and Fratzl's (2000) model for the elastic modulus, which considers that minerals and collagen can overlap. The mineralized turkey leg tendon is a fascinating material; we have all encountered these long and stiff rods at Thanksgiving dinners, and a few of us have wondered why they are so different from chicken tendons. These mineralized tendons (connecting bone to muscles) are ideal specimens for the investigation of the strength of partially mineralized bone and for the establishment of their structures. The collagen fibrils are arranged in a parallel fashion, and the degree of mineralization increases with the age of the turkey. The minimum degree of mineralization of the turkey leg tendon is 0.15, which is substantially less than bone. The mineralized turkey tendons enable our understanding of how hydroxyapatite and collagen interact.

The physical model applied by Jäger and Fratzl (2000) is shown in Fig. 7.4. The collagen fibrils are arranged in concentric layers. Embedded into them are the bone platelets. These mineral crystals have nano-scale dimensions, the thickness being typically on the order of 1 nm, the length being approximately 60–100 nm. This agrees with the picture shown in Fig. 5.12(d). The partially mineralized bone can be modeled as a composite in which the reinforcement (hydroxyapatite crystals) is essentially rigid and the continuous collagen matrix carries the load. The degree of mineralization, Φ, is defined as:

$$\Phi = \frac{ld}{(l+a)(b+d)}, \qquad (7.4)$$

where a is the overlap between the crystals, b is the lateral distance between them, and l and d are the dimensions of the mineral platelets. All parameters are given in Fig. 7.4. A simpler equation was presented in Eqn. (5.10) of Chapter 5. We introduce a more complete analysis in the following.

The elastic modulus is considered as comprising contributions from four regions which are classified into two loading conditions: (a) the tensile regions A and B, and (b) the shear regions C and D, as shown in Fig. 7.5. The Young modulus, considering these four contributions, was found to be

$$E' = E/E_c = E_1 + E_2 + E_3 + E_4, \qquad (7.5)$$

Where E_c is the Young modulus of collagen and E_1 ,..., E_4 are the contributions from regions A to D, respectively. The four contributions are the four terms in the following equation:

$$\frac{E}{E_c} = \frac{d(l+a)}{ab} + \left(1 + \frac{l}{2a}\right) + \frac{\gamma(l-a)(l+a)}{4b^2} + \frac{\gamma a(l+a)}{2b(2b+a)}, \qquad (7.6)$$

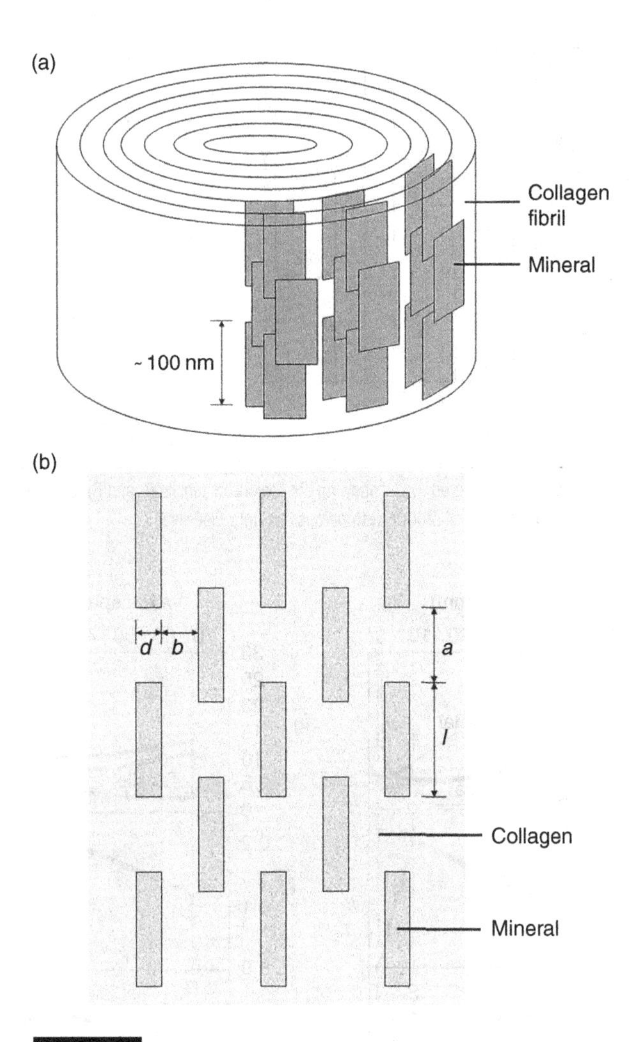

Figure 7.4.

(a) Schematic showing the three-dimensional arrangement of mineral platelets in a collagen matrix. (Reprinted from Jäger and Fratzl (2000), with permission from Elsevier.) (b) The staggered arrangement of mineral platelets. The dimensions of the mineral, *l* and *d*, and the distances between them, *a* and *b*, are indicated.

where is γ equal to $1/2(1+\upsilon)$ (recall that $G = E/2(1+\upsilon)$; so, $\gamma = G/E$). The predictions of the Jäger–Fratzl crystal model for two values of Φ (0.15, typical of mineralized turkey tendon, and 0.42, typical of bone) are shown in Figs. 7.6(b) and (a), respectively. A typical value of crystal thickness d = 3.5 nm was taken. The line marked S corresponds to the maximum in the elastic stress normalized to the maximum collagen stress, $(\sigma_{max}/(\sigma_{max})_c)$. The spacing between the platelets was varied. It can be seen that the optimal spacing corresponds to b = 4 nm for bone and b = 14 nm for mineralized turkey tendon. The predicted values of the normalized

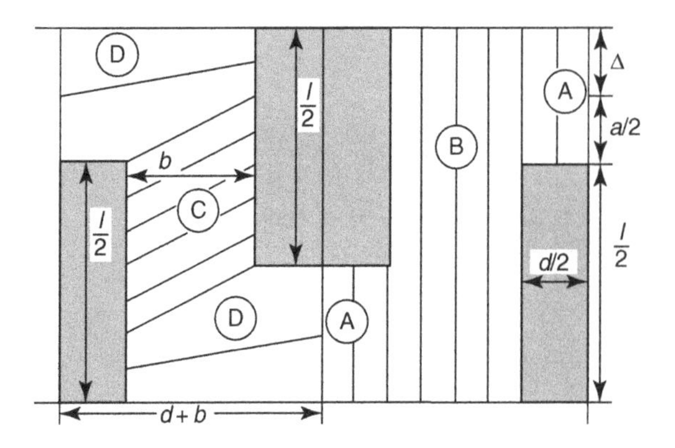

Figure 7.5.

Two adjacent elementary cells of the staggered model showing the regions of tensile (A and B) and shear (C and D) stresses, respectively. (Reprinted from Jäger and Fratzl (2000), with permission from Elsevier.)

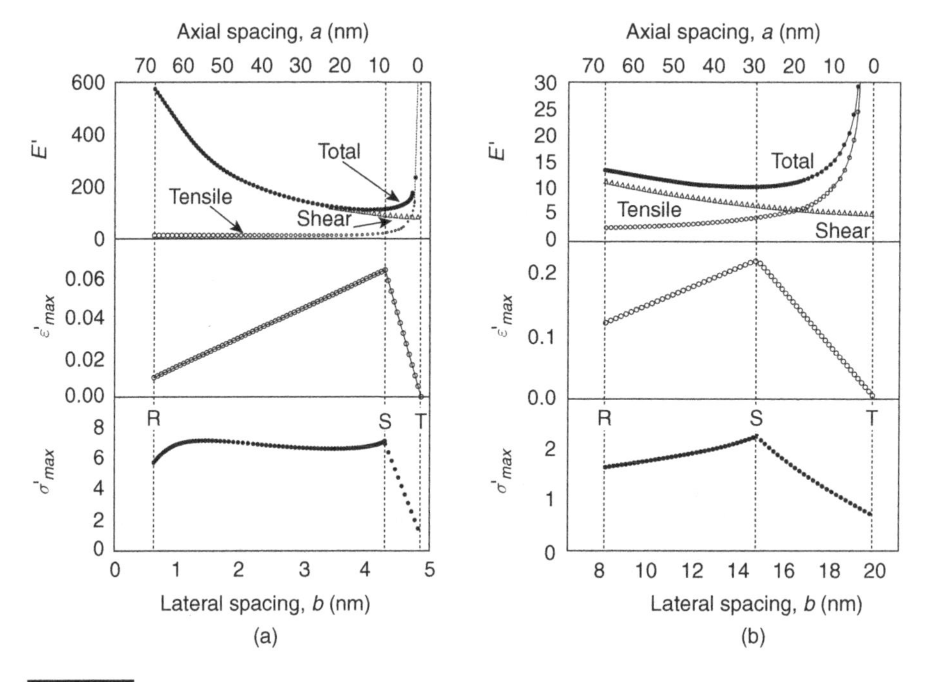

Figure 7.6.

Results for the normalized elastic modulus, E', the maximum strain, ε'_{max}, and the maximum normalized stress, σ'_{max}. (a) $\Phi = 0.42$; (b) $\Phi = 0.15$. (Reprinted from Jäger and Fratzl (2000), with permission from Elsevier.)

Young modulus E' (15 for $\Phi = 0.15$ and 150 for $\Phi = 0.42$) correspond to the experimentally obtained values; 8–15 and 200–400, respectively. Thus, we conclude that the stitching and spacing of mineral platelets is very important in determining the mechanical properties.

Example 7.1 Bone is a natural composite of mineral (mainly hydroxyapatite) and protein (mostly collagen) constituents.

(a) Given that the density of hydroxyapatite is 3.14 g/cm^3 and that of collagen is 1.35 g/cm^3, calculate the density of bone if the volume fraction of minerals is 0.45 and that of proteins is 0.55, neglecting the contribution from water.
(b) If the elastic modulus of hydroxyapatite is 100 GPa and that of collagen is 1 GPa, estimate the elastic modulus of bone. Using the Voigt and Reuss models, compare the calculated results with the values shown in the Wegst–Ashby plot in Fig. 2.11.

Solution (a)

$$\rho = \frac{M}{V} = \frac{\rho_{mineral} V_{mineral} + \rho_{protein} V_{protein}}{V_{mineral} + V_{protein}} = \frac{3.14 \times 0.45 + 1.35 \times 0.55}{0.45 + 0.55} \cong 2.16\ \mathrm{g/cm^3}.$$

(b) Voigt model:

$$E_{bone} = V^f_{mineral} E_{mineral} + V^f_{protein} E_{protein} = 0.45 \times 100 + 0.55 \times 1$$
$$\cong 45.6\ \mathrm{GPa}\ (\text{upper limit});$$

Reuss model:

$$1/E_{bone} = V^f_{mineral}/E_{mineral} + V^f_{protein}/E_{protein}\ ; \quad E_{bone} \cong 1.8\ \mathrm{GPa}\ (\text{lower limit}).$$

The elastic modulus of compact bone varies from 7 GPa to 30 GPa, within the upper limit (45.6 GPa) and lower limit (1.8 GPa) given by the Voigt and Reuss models, respectively.

Example 7.2 The mineral crystallites in bone have a platelet geometry of length 100 nm and thickness 3.5 nm observed from TEM micrographs. The mineral volume fraction is 0.4. Answer the following questions by applying the Jäger–Fraztl model.

(a) Calculate the overlap length a and lateral spacing b.
(b) Estimate the elastic modulus E' relative to collagen (E_c = 50 MPa). What are the contributions due to tensile and shear strains? Assume that the Poisson ratio is 0.3.
(c) Following (a) and (b), calculate the properties if the length is now reduced to 74 nm.
(d) If the length and thickness are kept the same (100 and 3.5 nm) and the mineral volume fraction
is increased to 0.5, calculate the properties.

Solution (a) The axial periodicity of the staggered collagen structure yields $(l + a)/2 = 67$ nm, so $a = 34$ nm. The mineral volume fraction Φ is given by

$$\Phi = \frac{ld}{(l+a)(b+d)} = \frac{100 \times 3.5}{134 \times (b+3.5)} = 0.4;$$

$b \approx 3$ nm.

(b)

$$E' = E/E_c = E_1 + E_2 + E_3 + E_4$$

$$= \frac{d(l+a)}{ab} + \left(1 + \frac{1}{2a}\right) + \frac{\gamma(l-a)(l+a)}{4b^2} + \frac{\gamma a(l+a)}{2b(2b+a)};$$

$$E_1 = \frac{3.5 \times 134}{34 \times 3} = 4.6;$$

$$E_2 = 1 + \frac{100}{2 \times 34} = 2.5;$$

$$\gamma = {}^1/_2(1+\nu) = 0.385;$$

$$E_3 = \frac{0.385 \times (100 - 34) \times 134}{4 \times 3^2} = 94.6;$$

$$E_4 = \frac{0.385 \times 34 \times 134}{2 \times 3 \times (2 \times 3 + 34)} = 7.3;$$

$$E' = E_1 + E_2 + E_3 + E_4 = 109.$$

The contribution from the tensile strain is $(E_1 + E_2)/E' = 7\%$, and the contribution from the shear strain is $(E_3 + E_4)/E' = 93\%$.

(c) $(74 + a)/2 = 67$nm; $a = 60$ nm. The mineral volume fraction Φ is given by

$$\Phi = \frac{ld}{(l+a)(b+d)} = \frac{60 \times 3.5}{134 \times (b+3.5)} = 0.4;$$

$b \approx 1.3$ nm;

$$E_1 = \frac{3.5 \times 134}{60 \times 1.3} = 6.0;$$

$$E_2 = 1 + \frac{74}{2 \times 60} = 1.6;$$

$$E_3 = \frac{0.385 \times (74 - 60) \times 134}{4 \times 1.3^2} = 106.8;$$

$$E_4 = \frac{0.385 \times 60 \times 134}{2 \times 1.3 \times (2 \times 1.3 + 60)} = 19.0;$$

$$E' = E_1 + E_2 + E_3 + E_4 = 133.4.$$

The contribution from the tensile strain is $(E_1 + E_2)/E' = 6\%$, and the contribution from the shear strain is $(E_3 + E_4)/E' = 94\%$.

(d) $a = 34\,\text{nm}$;

$$\Phi = \frac{ld}{(l+a)(b+d)} = \frac{100 \times 3.5}{134 \times (b+3.5)} = 0.5;$$

$b \approx 1.7\,\text{nm}$;

$$E_1 = \frac{3.5 \times 134}{34 \times 1.7} = 8.1;$$

$$E_2 = 1 + \frac{100}{2 \times 34} = 2.5;$$

$$E_3 = \frac{0.385 \times (100 - 34) \times 134}{4 \times 1.7^2} = 294.5;$$

$$E_4 = \frac{0.385 \times 34 \times 134}{2 \times 1.7 \times (2 \times 1.7 + 34)} = 13.8;$$

$$E' = E_1 + E_2 + E_3 + E_4 = 318.9.$$

The contribution from the tensile strain is $(E_1 + E_2)/E' = 3\%$, and the contribution from the shear strain is $(E_3 + E_4)/E' = 97\%$.

7.1.4 Strength

The prediction of the strength is much more difficult than that of the elastic modulus, because it involves a large number of mechanisms and hierarchies. Nevertheless, nothing escapes the attention of ambitious researchers, and valiant attempts have been made to develop models. One has to look at the two components and the hierarchy in which they arrange themselves (see Fig. 2.4) to understand the complexity of the task. The two major components, HAP and collagen, have radically different properties.

Lin, Chen, and Chang (2012) were able to fabricate HAP via a double sintering process with high density; the grain size was 193 nm and the relative density was 99.02%. They obtained excellent properties: a hardness of 4.86 GPa and a fracture toughness of 1.18 MPa $\text{m}^{1/2}$. However, the tensile strength is not expected to be high and is dependent on the size of flaws. Collagen, on the other hand, has a relatively high tensile strength, on the order of 50 MPa, at most. Chapter 9 will present the mechanical strength of collagen in detail. The tensile strength of bone "varies all over the map," as shown in Fig. 7.7(a), ranging from 20 MPa to 270 MPa. How can this be handled? We first look at the nano-scale. Why do the HAP crystals have dimensions of 4 nm × 40 nm?

Gao and Klein (1998) and Gao *et al.* (2003) proposed a conceptual framework that explains the lowest scale of the structural elements in hard biological materials (bone,

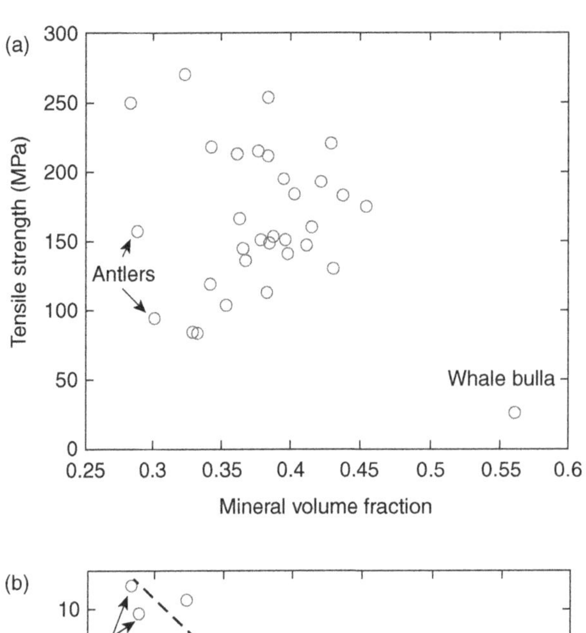

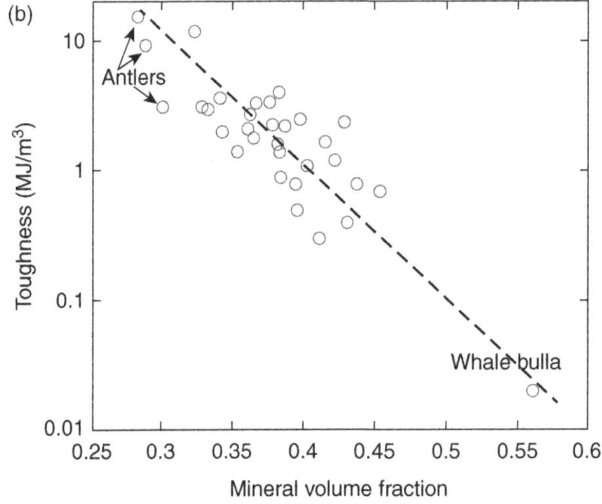

Figure 7.7.

Effect of mineral volume fraction on (a) tensile strength and (b) toughness (as measured by area under the stress–strain curve) of different bones. (Reproduced based on Currey (2002).)

teeth, and shells). They applied Griffith's criterion to the mineral components of hard biological materials (recall Eqns. (2.1) and (6.9)):

$$\sigma_f = \alpha\left(\frac{E\gamma}{h}\right)^{1/2}, \tag{7.7}$$

where γ is the surface energy and h is the thickness of the mineral. This expression was obtained by substituting a, the flaw size, by h, the dimension, since the flaw has to be smaller than the dimension. The schematic arrangement is shown in Fig. 6.30. They defined a parameter Ψ:

$$\Psi = \sqrt{\frac{\gamma}{Eh}}. \tag{7.8}$$

For the thumbnail crack shown in Fig. 6.30(a), the value of α is $\sqrt{\pi}$, and Eqn. (7.7) reduces to Eqn. (6.9). The theoretical stress, σ_{th}, has been defined, according to the Orowan criterion, for crystalline materials, as follows:

$$\sigma_{th} \cong \frac{E}{n}, \quad 2\pi < n < 30. \tag{7.9}$$

The analysis predicts a limiting value of h^* (or a) for which the strength is no longer size dependent by substituting Eqn. (7.9) and (7.7):

$$h^* \cong \alpha^2 \frac{\gamma E}{\sigma_{th}^2} = \frac{\alpha^2 n \gamma}{E}. \tag{7.10}$$

This critical value was calculated for bone, and a value of about 30 nm was obtained. This is the approximate size of the HAP tablets. Thus, having nano-scale HAP crystals optimizes the strength of these constituents. We have seen this before in Section 2.1.

Both the strength of bone and its toughness, roughly measured by the area under the stress–strain curve to failure, are dependent on the degree of mineralization, as evident in Fig. 7.7. The correlation between toughness and degree of mineralization, which was developed by Currey (2002) from a wide variety of animals, is much better than the one of tensile (three-point-bending) strength, because the latter is also affected by orientation and other factors. The point on the right-hand side denotes whale bulla. Antlers have a low degree of mineralization, and are on the left-hand side.

The longitudinal mechanical properties (strength and stiffness) are higher than the transverse ones. Cortical bone can be considered as orthotropic.

The tensile properties of fibrolamellar bone (compact bone without the Haversian canals) are different along the three directions, as seen in Fig. 7.8(a). The strength achieved in bone is therefore higher than in both hydroxyapatite (100 MPa) and collagen (~50 MPa), demonstrating the synergistic effect of a successful composite. The tensile strength along the longitudinal direction is three times as high as the one in the circumferential direction. This is the result of the alignment of the collagen fibrils, mineral component, and blood vessel cavities along the longitudinal axis. The strength (30 MPa) and fracture strain are lowest in the radial direction. The energy absorbed to fracture (area under the stress–strain curve) in the longitudinal direction is approximately 100 times the one in the radial direction. The bone is not, in real life, loaded radially in any significant way, and therefore strength is not needed in that direction.

Figure 7.8(b) provides the tensile and compressive stress–strain curves for cortical bone in the longitudinal and transverse directions (Lucas *et al.*, 1999). The anisotropy is

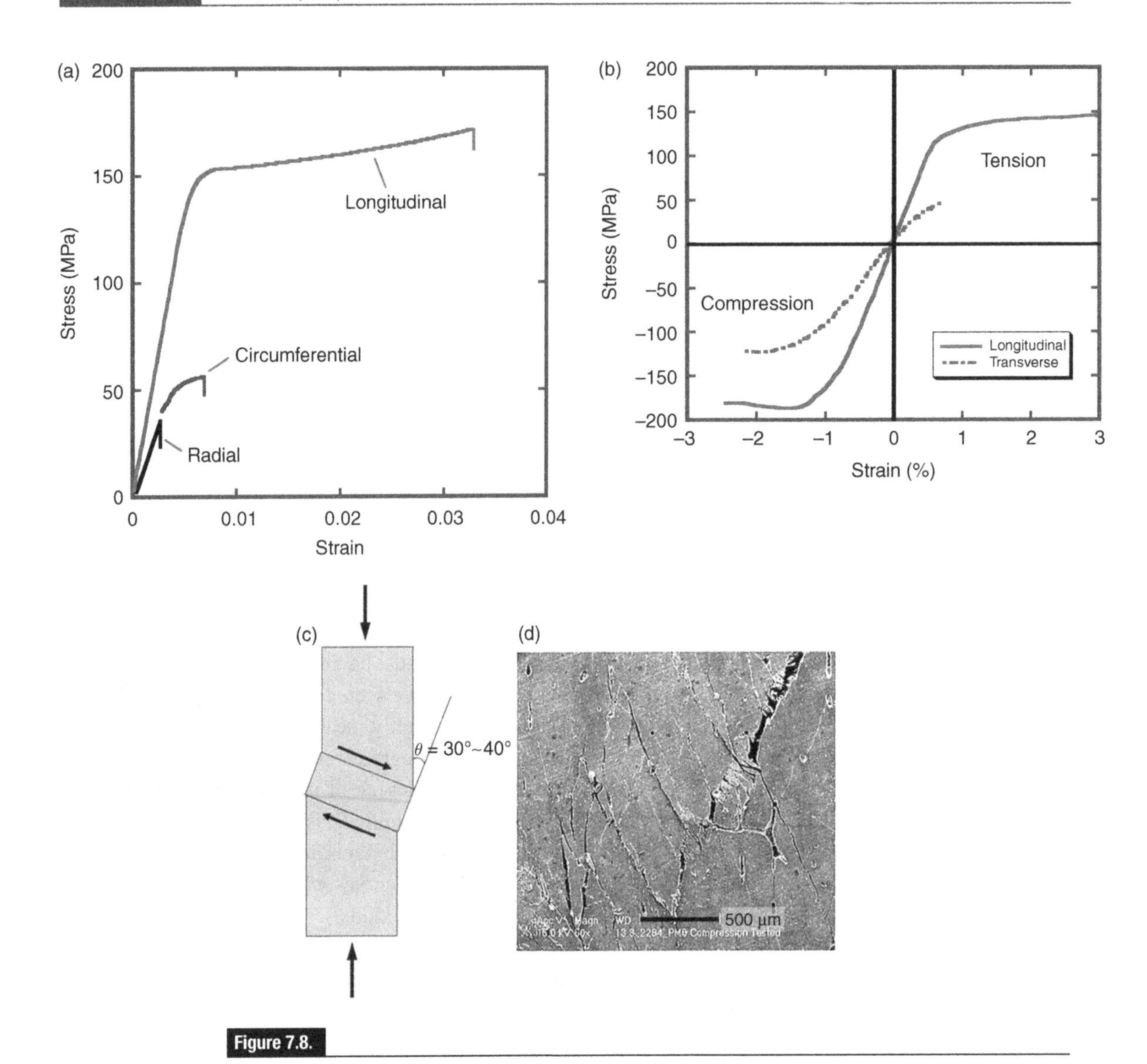

Figure 7.8.

Effect of orientation on tensile stress–strain curve of bovine fibrolamellar bone. (Reproduced based on Currey (2002).) (b) Tensile and compressive stress–strain curves for cortical bone in longitudinal and transverse directions. (Reprinted from Lucas, Cooke, and Friis (1999), with kind permission from Springer Science+Business Media B.V.) (c) Schematic showing plastic microbuckling. (d) Scanning electron micrograph of horse femur bone after compression test showing buckling. (Figure courtesy Professor K. S. Vecchio, UC San Diego, with kind permission.)

clearly visible. The bone is stronger in the longitudinal direction. It is also considerably stronger in compression than in tension. The compressive response of bone is characterized by a plateau with some softening after the elastic limit is reached (lower curves in Fig. 7.8(b)). This plateau is produced, at the structural level, by the formation of shear zones, which are the result of localized buckling of the fibrils. Plastic microbuckling is a well-known phenomenon when composites are loaded along the fiber axis and was first

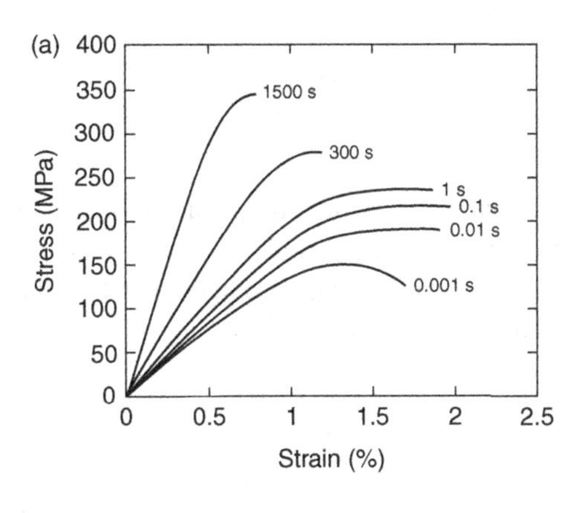

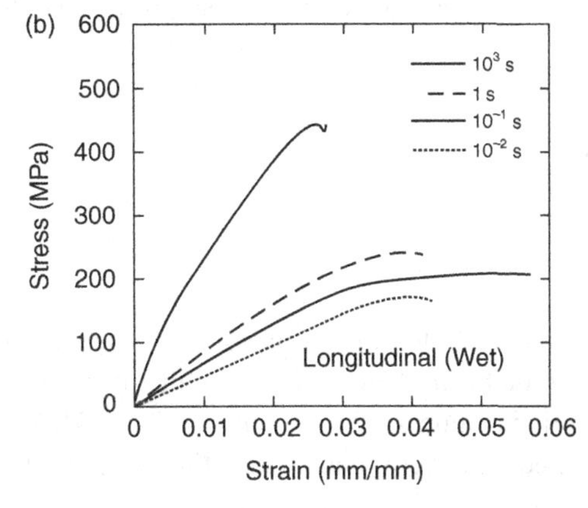

Figure 7.9.

Strain-rate dependence of tensile response of cortical bone. (a) Reproduced based on McElhaney (1966). (b) Adapted from Adharapurapu *et al.* (2006), with permission from Elsevier.

described by Evans and Charles (1976). This is shown in Fig. 7.8(c). The angle of these buckling regions with the compression axis varies between 30 and 40°. The equivalent process for bone is shown in Fig. 7.8(d). The buckling parameters, applied to abalone nacre as shown in Section 6.2.2.2, can also be applied to bone. The angles, as well as shear strain, can be obtained from the Argon and Budiansky–Wu equations (Eqns. (6.13) and (6.14)).

The mechanical response of bone is also quite strain-rate sensitive. As the velocity of loading increases, both the elastic modulus and the fracture stress increase. Hence, the stiffness increases with strain rate. The stress–strain curves for human bone at different strain rates are shown in Fig. 7.9(a) (McElhaney, 1966). Results by Adharapurapu, Jiang, and Vecchio (2006) also confirm the strain-rate dependency of cortical bone, as shown in Fig. 7.9(b). The Ramberg–Osgood equation is commonly used to describe this strain-rate dependence of the elastic modulus:

$$E = \frac{\sigma}{\varepsilon} = C\left(\frac{\dot{\varepsilon}}{\dot{\varepsilon}_0}\right)^d, \tag{7.11a}$$

where σ is the stress, ε is the strain, $\dot{\varepsilon}$ is the strain rate, C and d are experimental parameters, and $\dot{\varepsilon}_0$ is a normalization factor (added by us). This is actually a much simplified form of the original Ramberg–Osgood equation, which has the following strange form, difficult to explain to an intelligent student:

$$\varepsilon = \frac{\sigma}{C\dot{\varepsilon}^d} + a\sigma^N\dot{\varepsilon}^b. \tag{7.11b}$$

The parameters have the following values, for bone, according to Hight and Brandeau (1983):

$a = 2 \times 10^{-18}$;
$C = 24.5$ (units of GPa);
$d \sim 0.06$;
$N = 6.5$;
$b = -0.4$.

One can see that this equation is written in a somewhat cavalier fashion, because the terms that have exponents should be unitless. Thus, the strain rate should be divided by a reference value so that the ratio is unitless. This is done in Eqn. (7.11a). If one eliminates the second term by setting it equal to zero, one obtains Eqn. (7.11a). The following are also typical values (from other sources):

human cranium: $C = 15$ GPa, $d = 0.057$;
bovine cortical bone (longitudinal): $C = 12$ GPa, $d = 0.018$.

The strain-rate sensitivity of bone is primarily due to the collagen. Polymers have a high strain-rate sensitivity and thermal softening that are well represented by equations developed by Mooney and Rivlin (Mooney, 1940; Rivlin and Saunders, 1951), Treloar (1944), and Arruda and Boyce (1993). We will study them in Chapter 9. Arruda and Boyce used the following formulation to describe the strain-rate sensitivity of strength:

$$\tau_{AP} = \tau^*\left(\frac{\dot{\gamma}}{\dot{\gamma}_0}\right)^m, \tag{7.12}$$

where $\dot{\gamma}$ is the shear strain rate, m is the strain-rate sensitivity, τ_{AP} is the applied shear stress, and the other two terms are material parameters. This is a power law of the same nature as the Ramberg–Osgood equation. In this case, it is applied to the strength whereas the Ramberg–Osgood equation referes to the elastic modulus.

7.1.5 Fracture and fracture toughness of bone

7.1.5.1 Fracture of bone

The fracture of and fracture prevention in bone are of extreme importance. We know that bone strength decreases as porosity increases. This is one of the changes undergone by bone with aging. There are many fracture morphologies in bone, depending on the loading stresses, rate of loading, and condition of the bone. Figure 7.10 presents some of these modalities (Hall, 2003).

Greenstick fracture This occurs in young bone, which has a large volume fraction of collagen; it can break like a green twig. This zig-zag fracture indicates a high toughness.

Fissured fracture This corresponds to a longitudinal crack in the bone.

Comminuted fracture Many fragments are formed. This is typical of a fracture caused by impact at high velocities. Two factors play key roles. As the velocity of projectile is increased, its kinetic energy increases. This energy is transferred to the bone. The second factor is that at high velocities many cracks are produced simultaneously; they can grow independently until their surfaces intersect. This

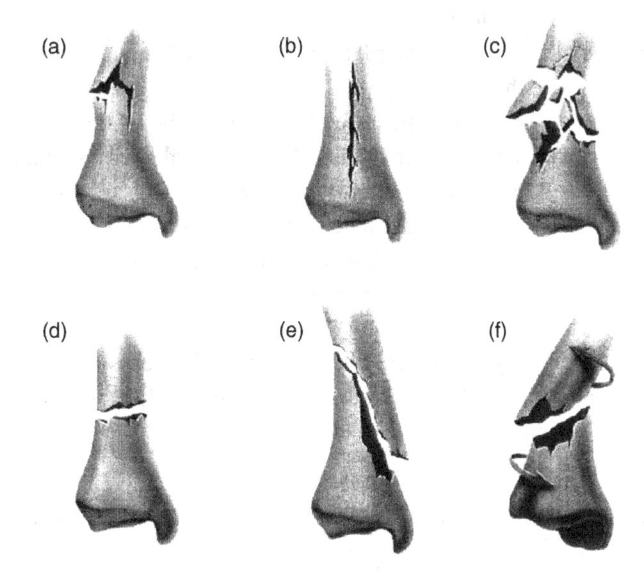

Figure 7.10.

Six modes of fracture in bone. (a) A greenstick fracture is incomplete, and the break occurs on the convex surface of the bend in the bone. (b) A fissured fracture involves an incomplete longitudinal break. (c) A comminuted fracture is complete and fragments the bone. (d) A transverse fracture is complete, and the break occurs at a right angle to the axis of the bone. (e) An oblique fracture occurs at an angle other than a right angle to the axis of the bone. (f) A helical (commonly but wrongly referred to as spiral) fracture is caused by twisting a bone excessively. (Adapted from Hall (2003).)

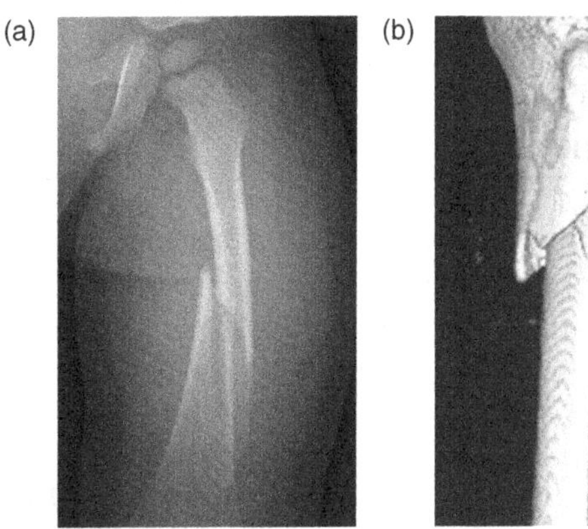
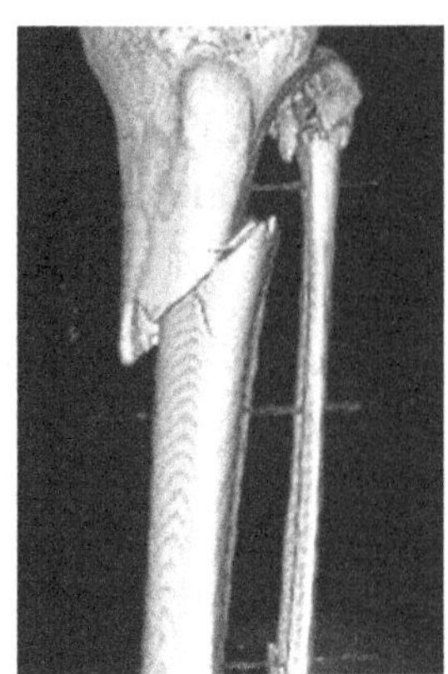

Figure 7.11.

(a) Helical fracture produced by torsion. (Image courtesy Teresa Chapman, MD, Seattle, WA.) (b) Shear fracture produced by compression in bone. (Figure courtesy Dr. J. F. Figueiró.)

is the reason why a glass, when thrown on the ground, shatters violently into many small fragments. An additional reason is that the bone becomes stiffer and more brittle as the strain rate is increased. This type of fracture is characteristic of bullet and shrapnel impact.

Transverse fracture This is a complete fracture approximately normal to the axis of the bone.

Oblique fracture Complete fracture oblique to the bone axis.

Helical fracture This fracture is caused by torsional stresses, and is known in the medical community as a spiral fracture. However, this name is not correct, as a helix describes the crack trajectory better than a spiral. Tensile stresses are highest along surfaces, making a 45° angle with the torsional stresses.

Figure 7.11 shows X-rays of helical and shear bone fractures. They are produced by different loading conditions: helical fractures are produced by torsion, whereas shear fractures result from compressive loading.

7.1.5.2 Single-value toughness and fracture toughness of bone

The fracture of and fracture mechanics in bone are of great interest and importance. Although there have been extensive studies, the fracture mechanics of bone is a complicated subject, and much remains to be understood due to its highly hierarchical structure. Most quantitative studies have focused primarily on "single-value" approaches, such as the work of fracture (W_f), the fracture toughness (K_c), the crack extension force (G_c), and the J integral (J_c).

The work of fracture (W_f) is obtained by dividing the area under the load–displacement curve measured during the test by twice the nominal fractured surface area. It is used to characterize the toughness of materials, but is flawed due to its dependence on both specimen size and geometry. Consequently, work of fracture results are not useful for comparing values determined in different studies, but they can demonstrate qualitative trends in materials behavior when the sample size and geometry are held constant.

The fracture toughness or stress-intensity factor (K_c) can be applied to linearly elastic materials, in which plastic deformations are limited to a small region near the crack tip. Facture toughness can be defined as one of three modes: mode I, opening mode; mode II, sliding mode; and mode III, tearing mode. It is a function of the applied stress, σ, and crack size, a, and the geometrical configuration of the crack:

$$K_{I,II,III} = Y\sigma\sqrt{\pi a}, \tag{7.13}$$

where Y is a parameter dependent on sample geometry and loading mode. The crack size a is defined for a crack starting at the surface. For an internal crack, the size is defined as $2a$.

In determining the fracture toughness, the minimum dimensions of the specimen are critical. For plane-strain conditions under which K_{Ic} tests are valid, the minimum thickness is given by

$$B \geq 2.5\left(\frac{K_{Ic}}{\sigma_y}\right)^2. \tag{7.14}$$

For bone, this value is on the order of a few millimeters and cannot be satisfied for smaller testing samples. In this case, the crack extension force (G_c) is more useful. The correct procedure for obtaining the fracture toughness of materials is described in great detail in ASTM E399.

Crack extension force (G_c), or the so-called critical strain-energy release rate, is defined as the change in potential energy per unit increase in crack area at fracture:

$$G_c = \frac{P^2}{2B}\frac{dC}{da}, \tag{7.15}$$

where P is the load, B is the specimen thickness, and dC/da is the change in sample compliance (C) with crack extension. For linear elastic materials, G and K are related by

$$G = \frac{K_I^2}{E'} + \frac{K_{II}^2}{E'} + \frac{K_{III}^2}{\mu}. \tag{7.16}$$

For mode I loading, Eqn. (7.16) simplifies to $G = K_I^2/E$ and $G_c = K_c^2/E$. The J integral (J_c) is the difference between the potential energies of identical bodies containing cracks of length a and $a + da$; in other words, it represents the change in potential energy for a crack extension da, i.e.

$$J_c = -\frac{1}{B}\frac{dU}{da}, \tag{7.17}$$

where U is the potential energy, a is the crack length, and B is the plate thickness. The J integral (J_c) can be related to the fracture toughness (K_c) as follows:

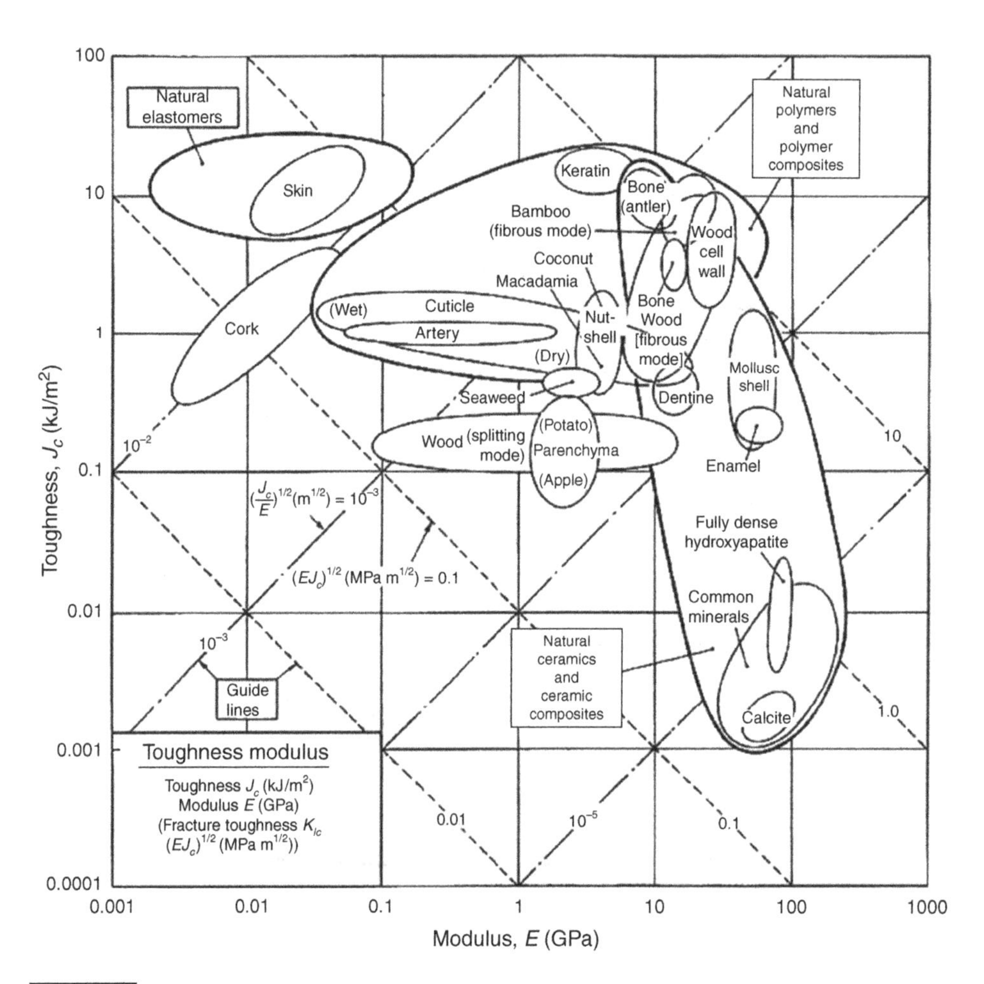

Figure 7.12.

Material property chart for natural materials, plotting toughness against Young modulus. Guidelines identify materials best able to resist fracture under various loading conditions. (From Wegst and Ashby (2004), reprinted by permission of Taylor & Francis Ltd.)

$$K_c = (E \cdot J_c)^{1/2}. \tag{7.18}$$

Figure 7.12 gives the J integral (J_c) for a number of biological materials as a function of elastic modulus (Wegst and Ashby, 2004). The plot shows lines of equal fracture toughness $(E\,J_c)^{1/2}$, drawn as diagonals. As we move toward the right, the toughness increases. Figure 7.12 provides a valuable insight into the toughness of biological materials. For pure HAP, $E \sim 100$ GPa, and the value of J extracted from Fig. 7.9 is ~0.01 kJ/m^2. This provides, as expected, $K_c \sim 1$ MPa m$^{1/2}$. As seen in Section 6.2, shells have toughnesses much superior to calcite, although the composition is similar. Similarly, bone has a toughness significantly higher than fully dense hydroxyapatite (equal, at most, to 1.18 MPa m$^{1/2}$, according to Lin *et al.* (2012)):

Table 7.1. Fracture toughness of bone

Type of bone	Direction[a]	K_{Ic} (MPa m$^{1/2}$)	Source
Bovine femur	longitudinal (slow)	3.21	Melvin and Evans (1973)
	longitudinal (fast)	5.05	
	transverse (slow)	5.6	
	transverse (fast)	7.7	
Bovine tibia	longitudinal (very slow)	3.2	Behiri and Bonfield (1980)
	longitudinal (slow)	2.8	
	longitudinal (fast)	6.3	

[a] Direction of crack propagation.

$$K_c \sim (20 \times 10^9 \times 2 \times 10^3)^{1/2} \sim 6\,\text{MPa}\,\text{m}^{1/2}. \tag{7.19}$$

Note that the Young modulus for bone was taken to be 20 GPa. The same plot shows that antler is slightly tougher than bone. Table 7.1 shows the toughness of bovine bones. These values are consistent with the range indicated in the Ashby plot of Fig. 7.12. Two aspects are important.

(a) The toughness is dependent on the crack propagation direction. The transverse toughness is higher than the longitudinal one.
(b) The toughness increases with strain rate. This is due to the combined effects of strain rate on the elastic modulus and strength of bone, discussed in Section 7.1.4.

Ritchie *et al.* (2006) obtained values for quasistatic fracture toughness of human cortical bone that they correlated to the principal contributing mechanisms that will be seen later in this section. The fracture toughness, K_{Ic}, varied in the following range:

$$1.7\,\text{MPa}\,\text{m}^{1/2} < K_{Ic} < 6.6\,\text{MPa}\,\text{m}^{1/2}.$$

Table 7.2 shows the contribution of these mechanisms. For bone, the most important mechanism is crack deflection, which contributes over 50% to the toughness. The second most important mechanism is the formation of bridges between uncracked ligaments, which contributes 1–1.5 MPa m$^{1/2}$.

Figure 7.13(a) is a schematic illustration showing the orientations used by standard fracture toughness measurements (ASTM E399). The first symbol designates the fracture plane orientation and the second the crack propagation direction. For example, L-C (longitudinal-circumferential) and L-R (longitudinal-radial) represent transverse cracking (across osteons) and C-L (circumferential-longitudinal) and R-L (radial-longitudinal) represent longitudinal cracking (splitting osteons) directions. Figure 7.13(b) shows experimental fracture toughness results in human cortical

Table 7.2. Contributions to fracture toughness of bone from different mechanisms[a]

Mechanism	Contribution to fracture toughness, K_{Ic} (MPa m$^{1/2}$)
Uncracked-ligament bridging	1–1.5
Crack deflection	3
Collagen-fibril bridging	0.1
Constrained microcracking	0.05
Total	**2–5**

[a] From Nalla, Kinney, and Ritchie (2003a,b); Nalla *et al.* (2004, 2005, 2006a); Launey, Buehler, and Ritchie (2010a).

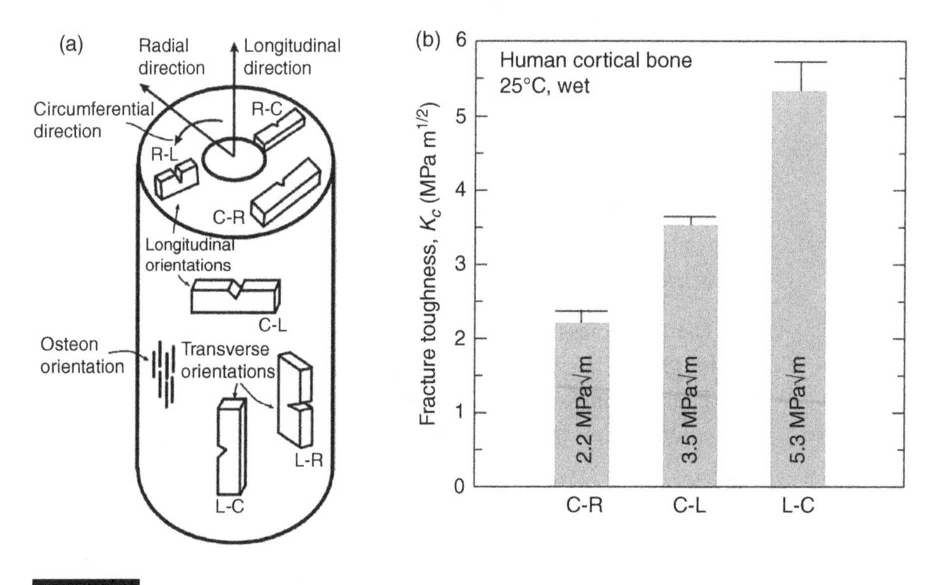

Figure 7.13.

(a) Orientation codes used by standard fracture toughness measurements. (Reprinted from Chen *et al.* (2012), with permission from Elsevier.) (b) Variation in fracture toughness with orientation in human humeral cortical bone. (Reprinted from Nalla *et al.* (2003a), with permission from Elsevier.)

bone with three different orientations (Nalla *et al.*, 2003a). The fracture toughness in the L-C orientation is significantly higher than that in the C-R and C-L orientations. Longitudinal toughness is lower than transverse toughness. This occurs because the crack propagates more readily along the fibrils. The corresponding toughening mechanisms will be discussed later.

It has been reported that fracture toughness also decreases with increasing mineral content (Wright, 1977; Currey, 1979; Currey, Brear, and Zioupos, 1996; Yeni, Brown, and Norman, 1998; Yeni and Norman, 2000) or porosity (Yeni *et al.*, 1997; Zioupos and

Currey, 1998; Evans *et al.*, 2001b). The cement line, the boundary between secondary osteons and the surrounding interstitial bone, plays a key role in the fracture of bone (Martin and Burr, 1982; Burr, Schaffler, and Frederickson, 1988; Yeni and Norman, 2000). Cracks have been observed to deflect and propagate along the cement lines, which are considered as a weak path for fracture. This weak path provided by the cement lines leads to the strong orientation dependency in bone fracture.

7.1.5.3 R-curve behavior and toughening mechanisms

The single-value approach, although useful, provides limited insights into the fracture behavior of bone because bone and other biological materials experience fracture-toughness enhancement during crack propagation. The resistance curve or "R curve" fracture-mechanics approach is more appropriate to explain the toughening mechanisms (Nalla *et al.*, 2005). The R curve graphically represents the resistance to crack propagation of the material as a function of crack growth, and it is widely used in fiber-reinforced composites. It can be experimentally determined by growing a crack of controlled size and measuring the stress necessary to propagate it. The rising R-curve behavior in bone has been studied in several groups (Vashishth, Behiri, and Bonfield, 1997; Vashishth, Tanner, and Bonfield, 2000, 2003; Malik *et al.*, 2003; Nalla *et al.*, 2004; Vasishth, 2004; Gao 2006; Yang, 2006a,b). R-curve behavior is the result of extrinsic toughening mechanisms (described in the following), and the crack resistance is determined in terms of the driving force required for crack extension (Δa).

Figure 7.14 shows the fracture toughness of horse leg bones (third metacarpal bone) (Malik *et al.*, 2003). As the crack grows, the toughness increases. The toughness starts as ~2 MPa m$^{1/2}$ and increases to a plateau of ~6 MPa m$^{1/2}$. Figure 7.15 shows the critical stress-intensity factor for human bone with different ages as a function of the crack size (Nalla *et al.*, 2006a). For young subjects (34–41 years), K increases considerably with

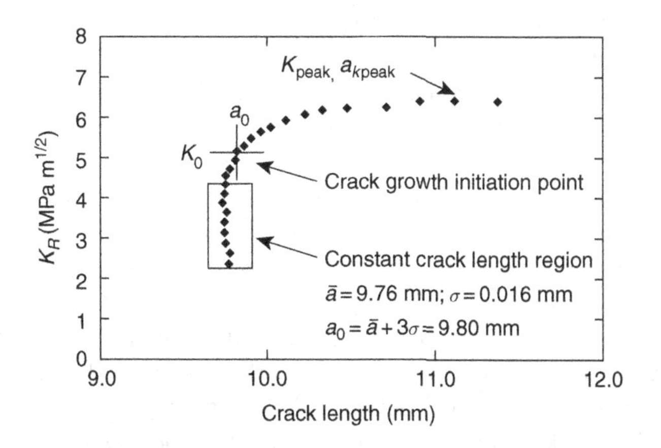

Figure 7.14.

Crack-resistance curve as a function of length for horse bone. (Reprinted from Malik *et al.* (2003), with permission from Elsevier.)

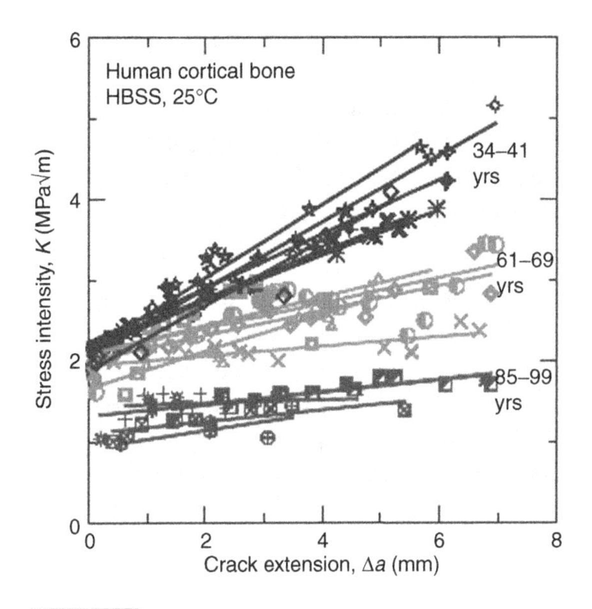

Figure 7.15.

Resistance curves for stable ex vivo crack extension in human cortical bone. Note the linearly rising R-curve behavior. (Reprinted from Nalla *et al.* (2006a), with permission from Elsevier.)

crack length a. This is clear R-curve behavior in bones, i.e. the toughness increases as the crack length is increased, and is analogous to the behavior observed in horse leg bone and shown in Fig. 7.14. Figure 7.15 also shows that the initial fracture toughness decreased with age. However, and more importantly, the increase in toughness with increasing crack length decreases with age, evidencing that the principal extrinsic toughening mechanisms cease to operate in old bone. Hence, a crack, once initiated, is more likely to stop in a young bone than in an old one.

Toughening mechanisms at varying hierarchical levels have been extensively studied and reviewed by Ritchie and co-workers (Nalla *et al.*, 2003a,b, 2004; Ritchie *et al.*, 2006; Yang *et al.*, 2006b; Launey *et al.*, 2010a,b), which provide valuable insights and understanding of the fracture and mechanical performance of bones. Crack toughening mechanisms in bone are shown in Fig. 7.16 (Launey *et al.*, 2010a), which can be classified into extrinsic (>1 μm) and intrinsic (<1 μm) toughening domains. In the extrinsic domain, Vashishth *et al.* (1997) established R-curve behavior in cortical bone in which the fracture toughness (K_{Ic}) increases with increasing crack length, resulting from microcracks that develop in the process zone wake around the crack tip. Collagen fiber bridging facilitates crack closure, thereby shielding the crack tip from additional stress. Uncracked ligaments bridge the main crack, helping to support the load and decrease the energy needed to propagate the main crack. Finally, crack deflection occurs where the Cook–Gordon crack deflection mechanism operates.

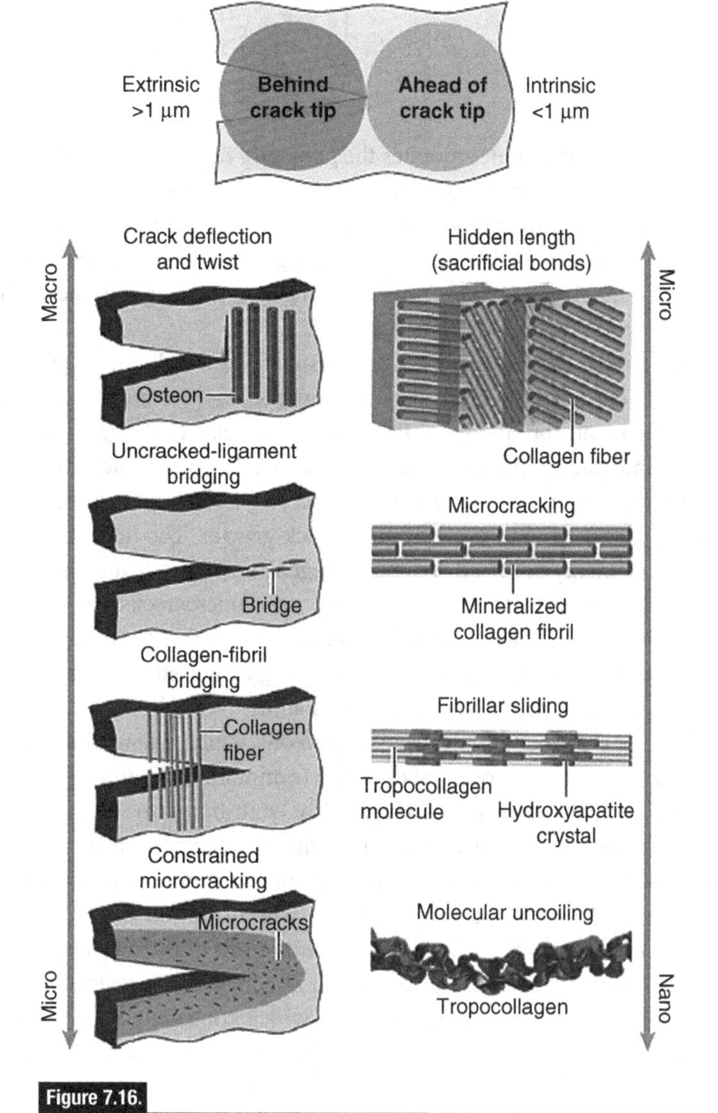

Figure 7.16.

Fracture-toughening mechanisms in bone are found at all hierarchical levels: (i) collagen molecular uncrimping/unkinking; (ii) collagen fibers sliding past each other; (iii) microcracking; (iv) separation of sacrificial bonds; (v) microcracking around crack tip; (vi) collagen fiber bridging; (vii) uncracked-ligament bridging; (viii) crack deflection. There are two prevailing domains: extrinsic (crack-tip shielding, length scale >1 μm) and intrinsic (plastic deformation, length scale <1 μm). (From Launey *et al.* (2010a), with the kind permission of Professor Ritchie.)

An analysis of crack bridging fibers has been performed by Li, Stang, and Krenchel (1993). The fiber bridging stress, σ_{fb}, is related to the normalized crack opening displacement, $\beta \equiv 2\delta/L_f$ (δ = crack opening displacement; L_f = fiber length):

$$\sigma_{fb}(\beta) = C\beta\left[2\left(\frac{C'\tau\alpha}{\beta}\right)^{1/2} - 1\right], \quad (7.20)$$

where C is a constant that incorporates the geometry of the crack and the volume fraction of the fibers, C' is a materials constant, α is the aspect ratio of the fibers (L_f/d_f), d_f = fiber diameter, and τ is the interfacial shear strength between the fibers and the protein matrix. Equation (7.20) indicates that the fiber bridging stress increases with an increase in the aspect ratio of the fibers. However, there is a limit to the length of the collagen fibrils, as minerals must be accommodated at the ends.

Extrinsic toughening mechanisms act to shield the crack from applied load (Ritchie 1988, 1999). Four types of extrinsic toughening mechanisms are present: crack deflection, uncracked-ligament bridging, collagen-fibril crack bridging, and microcracking (Fig. 7.16). Intrinsic mechanisms, on the other hand, typically act ahead of the crack tip and reduce stresses and strains through localized yielding and redistribution (Vashishth *et al.*, 1997, 2000), or may even promote crack growth. The intrinsic toughening arises, first, from the energy required for the collagen molecules to uncrimp and then, when extended, to slide past each other. Second, nano/microcracks can develop within the collagen fibrils. Third, it has been demonstrated that there are sacrificial bonds within or between the collagen molecules that dissipate energy when stretched and are reformed after the load is released (Thompson *et al.*, 2001; Fantner *et al.*, 2005).

In bone, crack deflection is caused by osteons which change the crack propagation with certain angle from the original direction (optimum condition = 90°). Crack bridging is provided by collagen fibers, which also contributes to crack-tip shielding. The uncracked-ligament bridging is another extrinsic toughening mechanism that provides crack-tip shielding. Microcracking is an intrinsic toughening mechanism that is initiated ahead of the crack and forms a process zone with dilatation which tends to "close" the crack. Nalla *et al.* (2006a) evaluated the contributions of the mechanisms to the fracture toughness of human cortical bone, as shown in Table 7.2. The most important toughening mechanism for bone is crack deflection, which contributes over 50% to the fracture toughness. The second most important mechanism is the uncracked-ligament bridging, which contributes 1–1.5 MPa $m^{1/2}$.

Figure 7.17(a) shows a crack, ahead of which there is a zone of damaged material consisting of cracks that are not connected. Bone owes part of its toughness to the formation of microcracks (daughter cracks) ahead of the main crack (mother crack). These microcracks form a process zone which decreases the stress concentration ahead of the crack tip. These microcracks tend to initiate in highly mineralized regions and do not grow to become macrocracks. Rather, they are arrested at internal obstacles, such as Haversian canals. As bridges form between these cracks, the crack front advances. The bridging by collagen fibers in the wake of the crack is another mechanism. This is shown in Fig. 7.17(b).

Although the R curves provide a means to characterize crack propagation, the underlying assumptions for such K_R calculations are based on linear elastic fracture mechanics

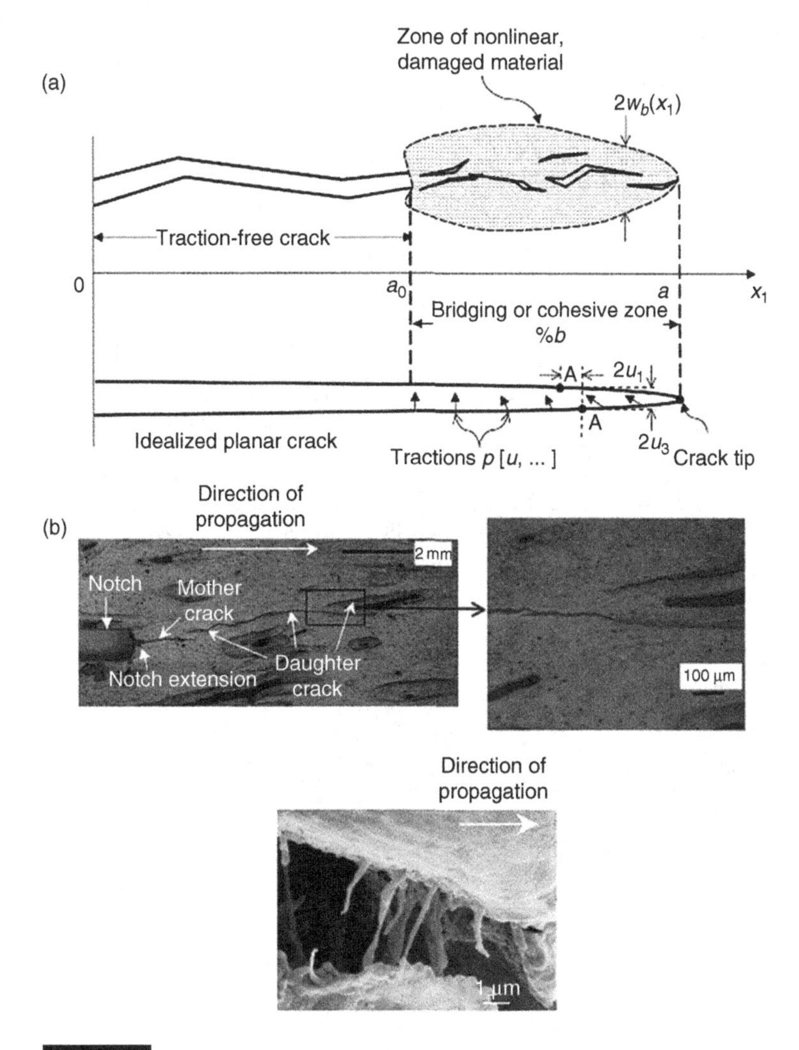

Figure 7.17.

(a) Schematic of discrete damage that has evolved into a single dominant crack tip damage zone. (b) Optical micrograph of a crack in human cortical bone. Note the formation of daughter cracks and corresponding uncracked ligaments and the bridging by collagen fibrils in the wake of a crack in human cortical bone. (Reprinted from Yang *et al.* (2006a), with permission from Elsevier.)

(LEFM), which cannot account for the energy associated with permanent deformation during bone fracture. The fracture toughness of materials with considerable permanent deformation, such as bone and antler, is usually underestimated. Koester, Ager, and Ritchie (2008) first applied a nonlinear fracture-mechanics approach based on the J integral to determine the R curve for human cortical bone. This approach accounts for the contribution from plasticity to the toughness, and provides a sound means to determine the R-curve fracture toughness in a material that undergoes multiple large-scale crack deflections. The effective stress intensity (K_{eff}) is calculated from the elastic modulus (E)

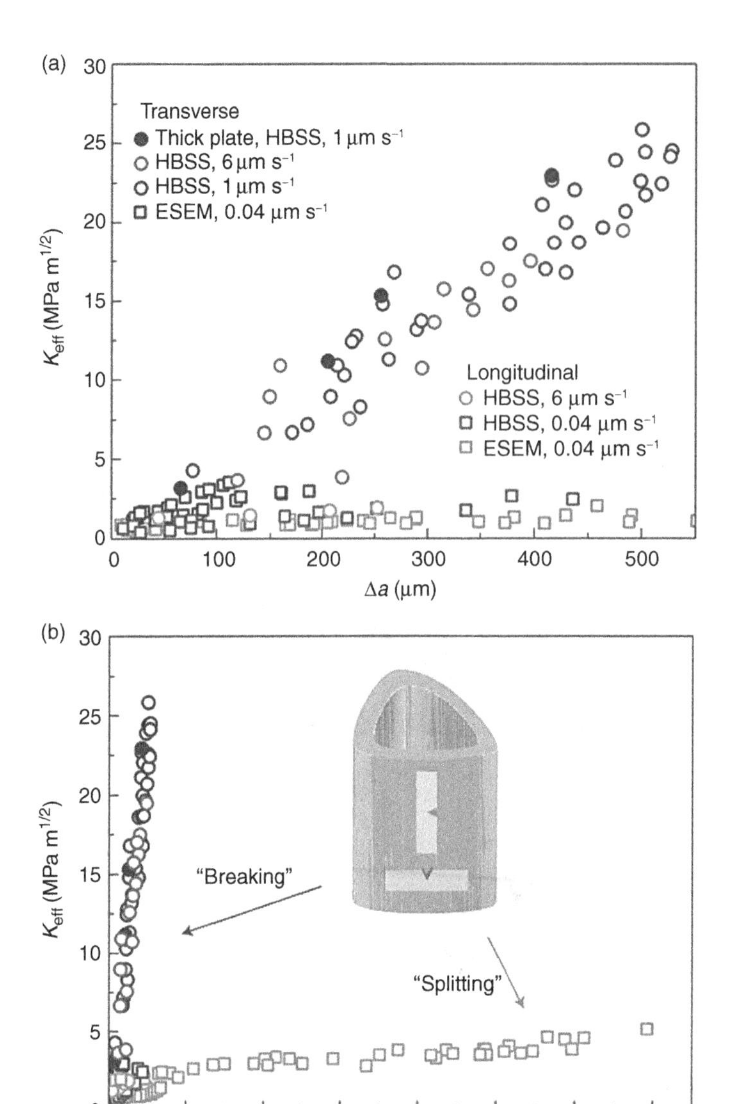

Figure 7.18.

(a) Fracture toughness resistance curve data for the transverse and longitudinal orientations in hydrated human cortical bone. (b) Nonlinear elastic fracture-mechanics measurements of the fracture toughness of bone show that resistance to crack propagation increases much more rapidly in the transverse (breaking) direction than in the longitudinal (splitting) direction. (Reprinted by permission from Macmillan Publishers Ltd.: *Nature Materials* (Koester *et al.*, 2008), copyright 2008.)

and the J integral (J_c) from Eqn. (7.18), defined earlier: $K_{eff} = (E \cdot J_c)^{1/2}$. They also used in-situ mechanical testing on small bending samples under environmental SEM to examine how physiologically pertinent short cracks (<600 μm) propagate in both the transverse (L-C) and longitudinal (C-L) orientations in human cortical bone. The R

Transverse orientation

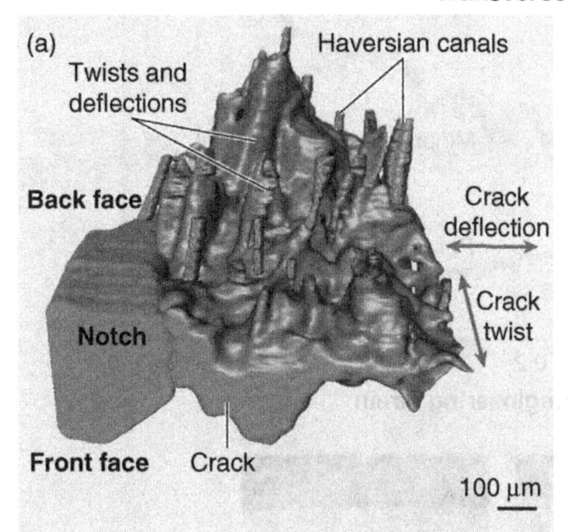

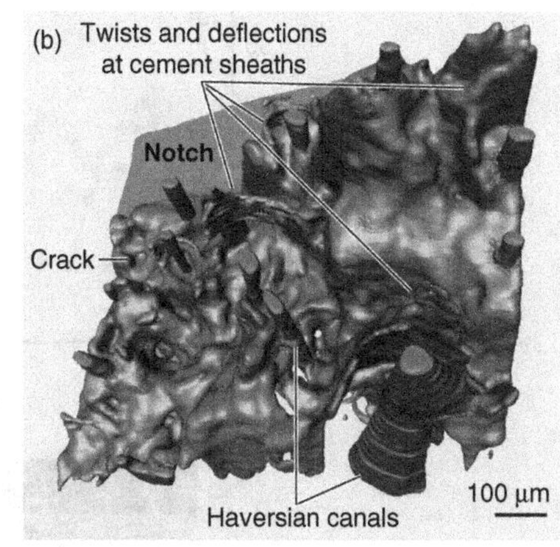

Longitudinal orientation

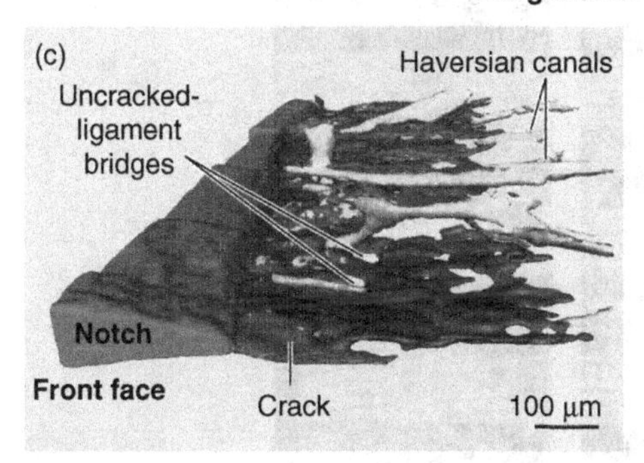

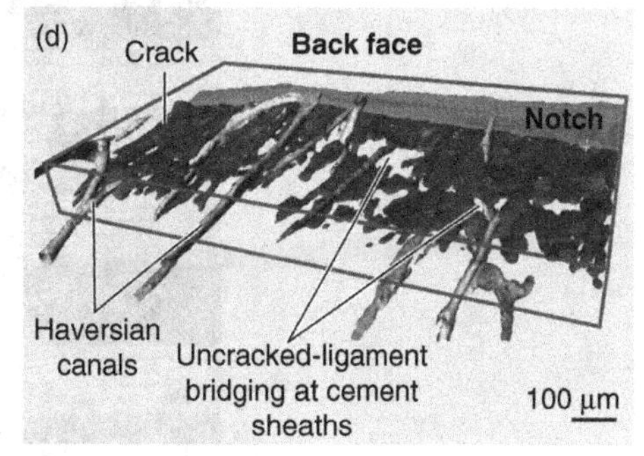

Figure 7.19.

Three-dimensional reconstructions in the transverse and longitudinal orientations from synchrotron X-ray computed tomography. (a), (c) Edge view of the notch and crack to show the shape of the crack paths. (b), (d) Oriented to highlight the dominant toughening mechanisms in the two orientations. (Adapted from Launey *et al.* (2010a), with the kind permission of Professor Ritchie.)

curves (Fig. 7.18) showed that the effective stress intensity in the transverse orientation was much higher than that in the longitudinal orientation, reaching a value of ~25 MPa m$^{1/2}$, much higher than the fracture toughness of bone previously reported in the literature. Synchrotron X-ray computed tomography images showed that the dominant toughening mechanisms were crack deflection (in-plane) and twisting (out-of-plane) in the transverse orientation, predominantly at the cement lines; see Fig. 7.19.

The toughening mechanisms of bone at the molecular and nanometer regimes were studied by Buehler and co-workers (Buehler and Wong, 2007; Buehler, 2008) using molecular dynamic simulation and theoretical analysis. See Fig. 3.13 for a typical J-shaped stress–strain curve. At low strains, molecular rearrangement occurs (toe region); in the heel region, hydrogen bonds break (which are reformed after the load is released); finally, the polymer backbone undergoes deformation with a substantial

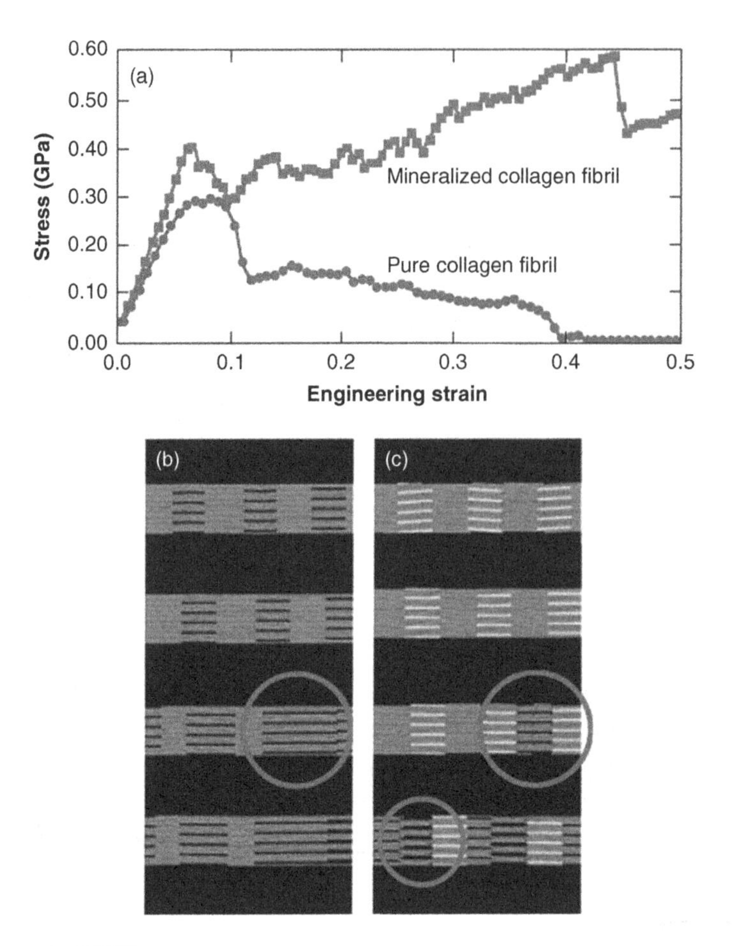

Figure 7.20.

(a) Computational result of the stress–strain response of a mineralized versus a nonmineralized collagen fibril, showing the significant effect that the presence of mineral crystals in collagen fibril has on the mechanical response. Snapshots of the deformation mechanisms of (b) collagen fibrils and (c) mineralized collagen fibrils are shown. Slip initiates at the interface between mineral particles and tropocollagen molecules. Repeated occurrence of slip (the circles denote local areas of repeated molecular slip) reduces the density, leading to the formation of nano-scale voids. (Adapted from Launey *et al.* (2010a), with the kind permission of Professor Ritchie.)

increase in stiffness. Figure 7.20(a) shows computationally generated stress–strain curves under tensile stress for pure (nonmineralized) and mineralized collagen fibrils. The mineralized collagen fibrils exhibit a higher strength and greater energy dissipation under deformation. Furthermore, the mineralized collagen fibrils show higher stiffness and significant softening at larger strains, with a sawtooth-shaped stress–strain curve due to repeated slip between the collagen and mineral. Deformation mechanisms of pure and mineralized collagen fibrils under increasing tensile stress are shown in Figs. 7.20(b) and (c), respectively. The mineralized collagen fibrils assemble into fibers or arrays, which are "glued" together by noncollagenous proteins, as shown in Fig. 7.21(a). At this

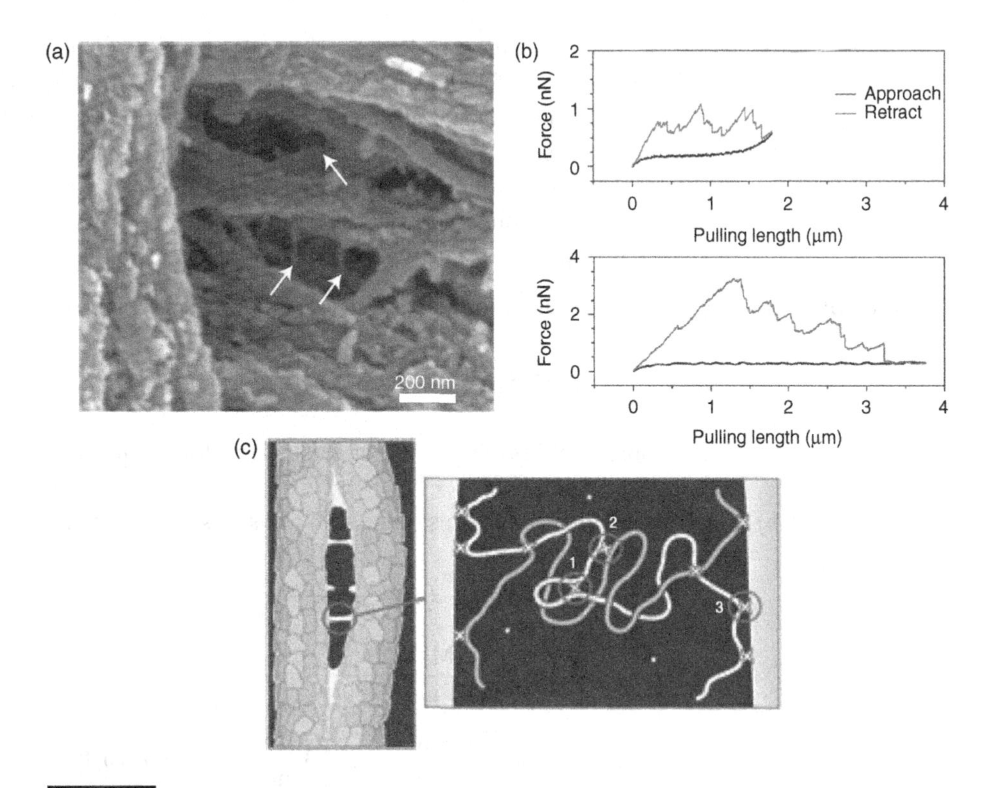

Figure 7.21.

(a) SEM micrograph showing individual collagen fibrils held together by glue filaments (arrows). (b) Representative force–extension pulling curves obtained from AFM measurement (upper curve: not all filaments were broken; lower curve: all filaments were broken), showing characteristic sawtooth shape, corresponding to successive fracture of polymer chains. (c) Possible deformation mechanism which involves breaking the sacrificial bonds in noncollagenous glue between mineralized collagen fibrils. (Reprinted by permission from Macmillan Publishers Ltd.: *Nature Materials* (Fantner *et al.*, 2005), copyright 2005.)

structural level, the toughness of bone has been attributed to the additional energy required to break sacrificial bonds in the "glue" (Hansma *et al.*, 2005; Fantner *et al.*, 2005). AFM pull-off force measurements (Fig. 7.21(b)) indicate that the bonds break at a fraction (0.1–0.5) of the force required to break the backbone of the macromolecules, and sawtooth force–displacement curves are observed. Figure 7.21(c) shows a possible toughening mechanism which involves breaking the sacrificial bonds in noncollagenous glue between mineralized collagen fibrils. At the largest length scales, the primary toughening mechanisms are constrained microcrack formation, collagen-fibril bridging, uncracked-ligament bridging, as well as crack deflection and twist as previously discussed in Fig. 7.16.

Example 7.3 (a) A fracture toughness measurement is performed on bone samples. If the fracture toughness (K_{Ic}) is approximately 5 MPa m$^{1/2}$ and the yield strength is 125 MPa, calculate the minimum sample thickness required for the plane-strain condition.

(b) Bone and many other biological materials have a significant amount of plastic deformation before fracture. Therefore, linear elastic fracture toughness measurements tend to underevaluate the toughness of bone. Apply nonlinear fracture mechanics (J integral) to estimate the effective fracture toughness. Given: specimen thickness $B = 4$ mm; width $W = 3$ mm; initial crack length $a = 1.5$ mm; elastic modulus of bone $E = 20$ GPa; stress-intensity factor $K = 5\ \mathrm{MPa\,m^{1/2}}$; Poisson ratio $\nu = 1/3$; area under force–displacement curve in plastic deformation region $A_{pl} = 40$ N mm.

Solution (a) Plane-strain condition for mode I fracture toughness measurement:

$$B \geq 2.5\left(\frac{K_{Ic}}{\sigma_y}\right)^2 = 2.5\left(\frac{5\ \mathrm{MPa\ m^{1/2}}}{125\ \mathrm{MPa}}\right)^2 = 4\,\mathrm{mm}.$$

The sample thickness should be at least 4 mm for a valid plane-strain fracture toughness measurement.

(b) J can be determined in terms of the sum of its elastic and plastic contributions:

$$J = J_{el} + J_{pl} = \frac{K^2}{E/(1-\nu^2)} + \frac{2A_{pl}}{B(W-a)};$$

$$J_{el} = \frac{(5\,\mathrm{MPa\,m^{1/2}})^2}{20\,\mathrm{GPa}/(1-1/9)} = 1.11 \times 10^3\ \mathrm{Pa\,m};$$

$$J_{pl} = \frac{2 \times 40\,\mathrm{N\,mm}}{4(3-1.5)\mathrm{mm}^2} = 13.33\frac{\mathrm{N}}{\mathrm{mm}} = 13.33 \times 10^3 \mathrm{Pa\,m};$$

$$J = J_{el} + J_{pl} = 14.44 \times 10^3 \mathrm{Pa\,m};$$

$$K_{eff} = (J \cdot E)^{1/2} = (14.44 \times 10^3 \times 20 \times 10^9)^{1/2}\,\mathrm{Pa\,m^{1/2}} \approx 17\,\mathrm{MPa\,m^{1/2}}.$$

The effective fracture toughness, which includes the contribution from plastic deformation, is over three times higher than the fracture toughness K_{Ic} determined using linear elastic fracture mechanics.

7.1.6 Fatigue

Bone fatigue is well documented. This phenomenon is commonly – and incorrectly – known in the medical community as "stress fracture." Repetitive loading above a threshold often generates this type of damage in athletes. It is attributed to the formation of microcracks which develop throughout the bone. These microcracks do not grow because of the internal barriers posed by the Haversian system. Figure 7.22 shows the stress–number of cycles (SN) curves for a number of bones (Currey, 2002). This is very similar to metal fatigue.

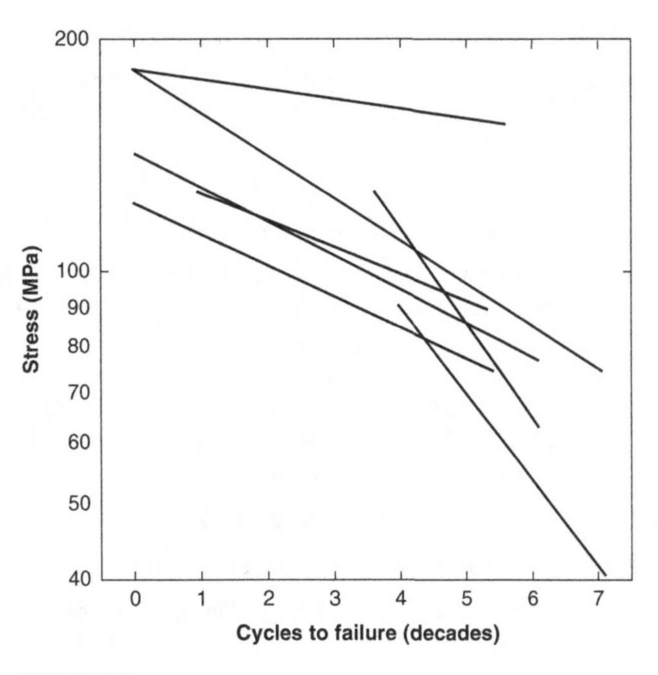

Figure 7.22.

Stress–number of cycles (SN) curves for a number of bones loaded under different conditions. (Adapted from Currey (2002).)

7.2 Antler

7.2.1 Structure and functionality

Antlers are bony protuberances that form on the heads of the Cervidae (deer) family; they have a chemical composition and microstructure similar to those of mammalian long bones. A comparison between elk antler and bovine (cow) femur bone is summarized in Table 7.3. Both antler and bone utilize the same basic building blocks, namely the type I collagen in the protein phase and the carbonated hydroxyapatite in the mineral phase, which is verified by amino acid analysis, X-ray diffraction, and TEM observation (Chen *et al.*, 2009). However, there are several distinct differences between the two. Antlers and skeletal bones have different functions. The primary functions of antlers are social display, defense against predators, and combat between male species (Henshaw, 1971; Lincoln, 1972, 1992; Clutton-Brock, 1982). Skeletal bones contain bone marrow, whereas antlers have no marrow. There exists a transition zone between cortical and cancellous bones in antlers, whereas there is no such transition zone in skeletal bone. The cancellous bone is well aligned and uniformly distributed through the entire antler. In bovine femur, the cancellous bone is mainly located in the femur head, and its density

Table 7.3. Comparisons between elk antler and bovine femur

Property	Antler bone	Bovine femur
Similarities		
Basic building blocks	protein: type I collagen mineral: carbonated hydroxyapatite	
Hierarchical structure	collagen/mineral < mineralized collagen fibrils < lamella < osteon/ trabecula < cortical/cancellous bone < antler/bone	
Differences		
Function	combat, defense, and social display	supports body, protects organs, and produces blood components, mineral storage
Bone marrow	no marrow	bone marrow
Transition zone	transition zone	no transition zone
Cancellous bone	uniformly distributed	localized in the head region
Mineral content (cortical bone)	57 wt.% 36 vol.%	67 wt.% 47 vol.%
Osteon type	primary	secondary
Cement line	no cement lines hypermineralized regions	cement lines

decreases progressively toward the central region of the femur, which correlates to the external loading conditions. Antler has lower mineral content and consequently lowest elastic modulus among mineralized tissues with a mineral content of ~50 wt.% (or ~30 vol.%), in contrast to the highly mineralized whale rostrum at 98 wt.% (56 vol.%) (Currey, 1984a). The mineral contents of mammalian, reptilian, and avian bones fall between these values.

Antlers consist mainly of primary osteons due to the limited time (1–2 months) available for bone remodeling (Skedros, Durand, and Bloebaum, 1995). Cross-sectional micrographs of cortical antler (Figs. 7.23(a) and (c)) are compared with those of bovine femur (Figs. 7.23(b) and (d)). In Fig. 7.23(a), osteons (Os) (100–225 μm diameter), Volkmann canals (Vo), vascular channels (Va) (15–25 μm diameter), and lacunae spaces (L) (~10 μm diameter) are observed, as indicated in the micrograph. In antler (Fig. 7.23(a)), the osteons are irregular in shape, and a high density of vascular channel, which may be related to the fast growth rate, can be observed. In bovine femur (Fig. 7.23(b)), the more uniform, circular shaped secondary

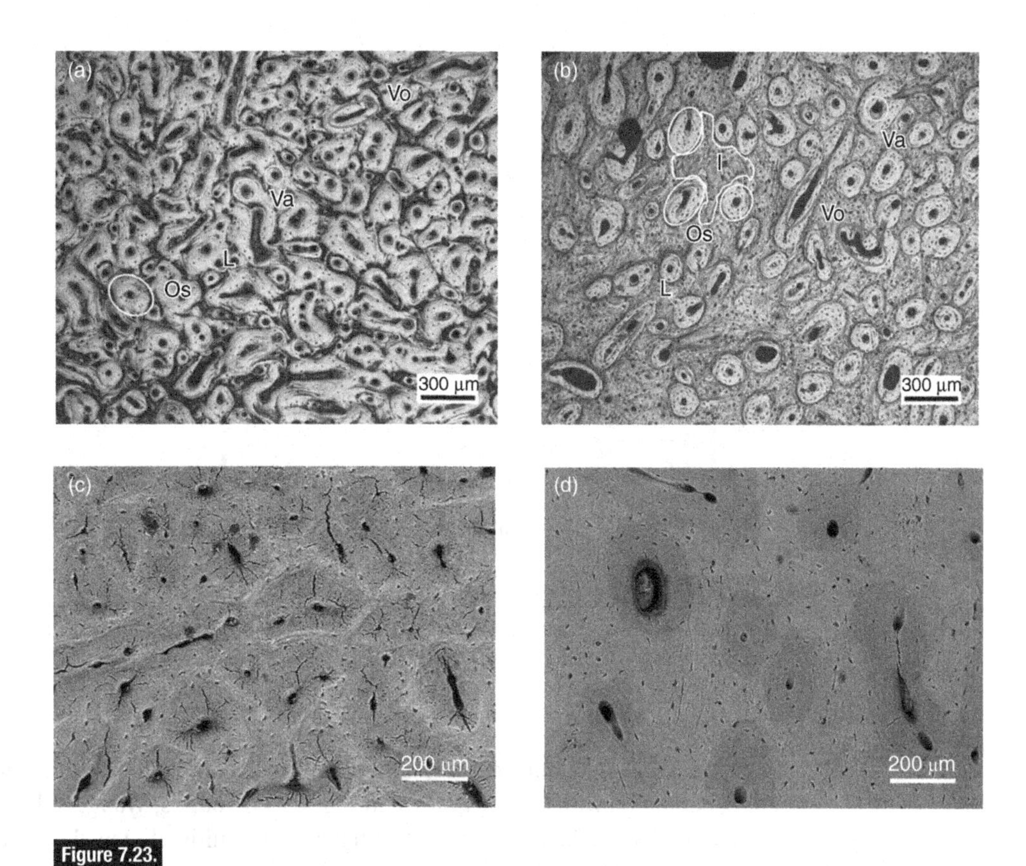

Figure 7.23.

Optical micrographs of cortical bone from the elk antler and bovine femur: cross-sectional area of (a) elk antler and (b) bovine femur (Os: osteons, Va: vascular canals, Vo: Volkmann canals, L: lacunae, I: interstitial bone), and backscattered electron (BSE) images showing cross-sectional microstructure of (c) elk antler and (d) bovine femur.

osteons are more sparsely distributed among interstitial bones. Figure 7.23(c) shows the backscattered electron (BSE) image of cortical antler. An irregular hypermineralized region (brighter region), which is 10–20 μm in width, surrounding the primary osteons is observed. The hypermineralized cement lines are considered to be the primary path for microcrack propagation and contribute to the fracture toughness in antler. A BSE scanning electron image of cortical bovine femur taken at the same magnification is shown in Fig. 7.23(d).

7.2.2 Quasistatic and dynamic mechanical behavior

The elastic modulus increases with mineral content in bony tissues. Antlers possess low mineral content and have lower elastic modulus compared with mammalian long bones, typically in the range of 7 GPa. In antler, the post-yield region of the

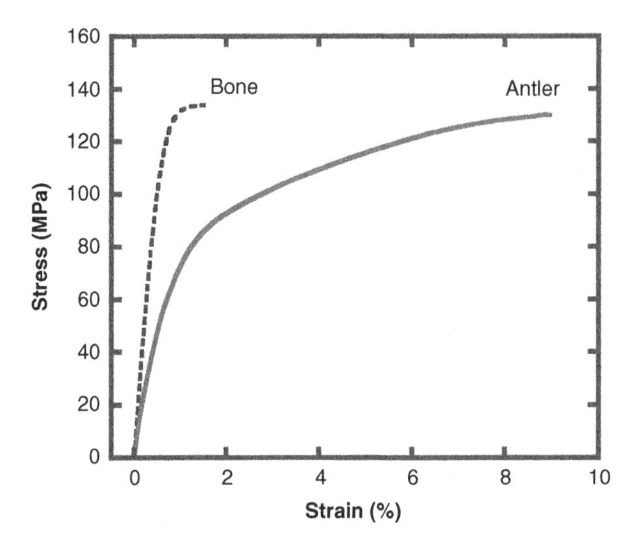

Figure 7.24.

Typical tensile stress–strain curves of bovine femur bone and red deer antler. (Reprinted from Zioupos *et al.* (1996), with permission from Elsevier.)

stress–strain curve is considerably longer and is accompanied by a gradual slope to failure, compared with bone, as shown in Fig. 7.24 (Zioupos, Wang, and Currey, 1996). Red deer antler and bovine femur bone have similar ultimate tensile strengths, which vary between 100 and 140 MPa; however the strain at failure (8–10%) and work of fracture ($6.2 \pm 0.6\,kJ/m^2$) for antler is four to five times greater than those for bovine femur (Currey, 1979; Currey and Brear, 1992). In bone, after the ultimate strength is reached, the stress drops rapidly until failure or increases slightly by about 10% to failure (Currey, 1989).

The main function of antler is combat between males, which reaches a high impact. The dynamic mechanical behavior of antler was investigated using the split Hopkinson pressure bar (SHPB) (Kulin *et al.*, 2010, 2011) with the strain rate on the order of $\sim 10^3\ s^{-1}$. The compressive mechanical behavior of antler as a function of strain rate varying from 10^{-3} to $10^3\ s^{-1}$ in either dry or rehydrated conditions is shown in Figs. 7.25(a) and (b), respectively. The compressive stress–strain curves clearly indicate the strain rate effect and viscoelastic behavior of antler. The strong strain-rate dependency of antler is in agreement with that of human cortical bone in tension previously reported by McElhaney (1966) and bovine femur bone in compression recently investigated by Adharapurapu *et al.* (2006). The strain-rate sensitivities of bone and antler are primarily due to the collagen, which has viscoelastic behavior. Antler has a higher volume fraction of protein than mammalian long bones and is expected to be more strain-rate sensitive.

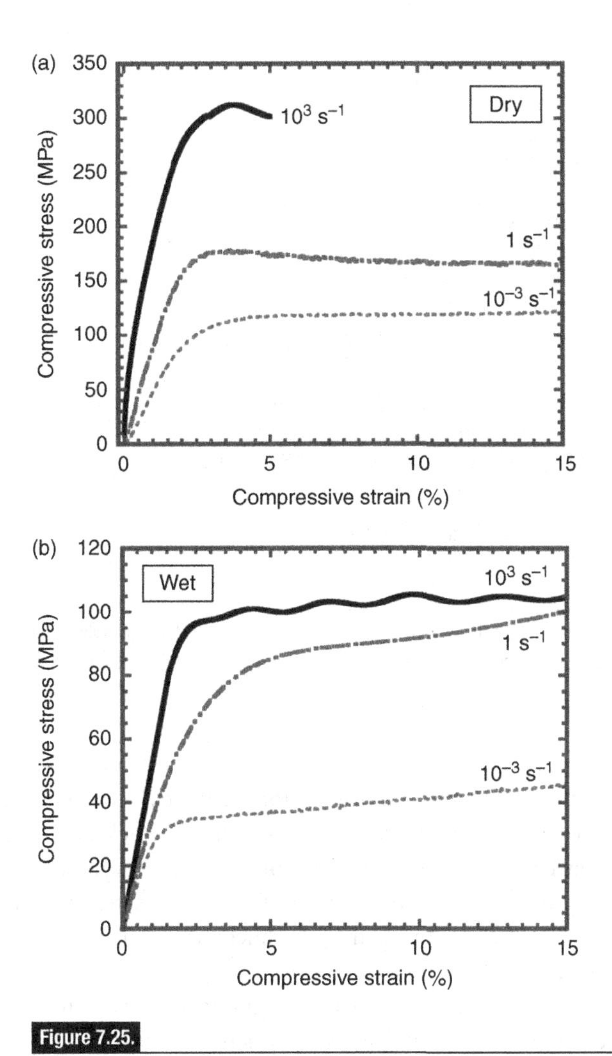

Figure 7.25.

Compressive stress–strain curves of antler as a function of strain rate varying from 10^{-3} to 10^3 s^{-1} in (a) dry and (b) wet conditions.

7.2.3 Exceptional fracture resistance

Antlers exhibit exceptional fracture resistance compared with other bones. They undergo large amounts of deformation during vigorous fighting yet are rarely found to be fractured or damaged. Launey *et al.* (2010b) investigated the fracture resistance of antler following the nonlinear fracture mechanics (in terms of the J integral) (Koester *et al.*, 2008) along with in-situ environmental SEM (ESEM) observations. Figure 7.26 shows the resistance curves for hydrated antler with short crack lengths ($\Delta a < 600$ μm) tested in transverse, *in-plane* longitudinal, and *anti-plane* longitudinal orientations. The results

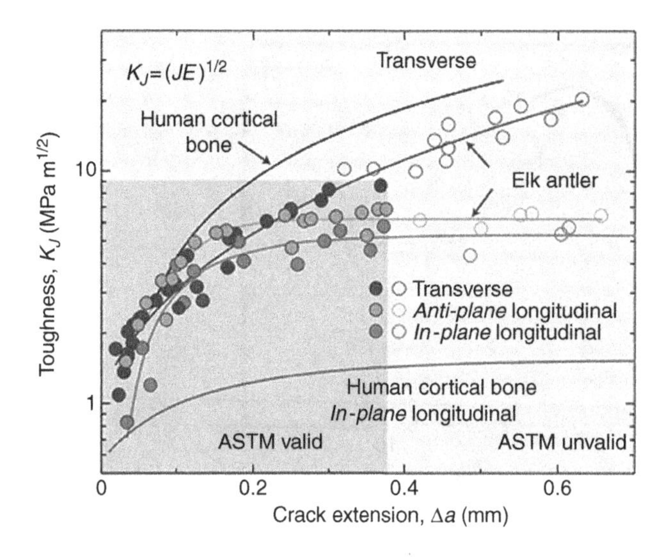

Figure 7.26.

Crack-resistance curves (R curves) showing resistance to fracture in terms of the stress intensity, K_J, as a function of crack extension, *a*, for hydrated antler and human compact bone in different orientations. (Reprinted from Launey *et al.* (2010b), with permission from Elsevier.)

are compared with those on human compact humerus bone (Koester *et al.*, 2008). The R curves for antler were terminated after about 600 μm of crack growth as none of the specimens broke in half. It is apparent that antler exhibits significant rising R-curve behavior, indicative of extensive toughening. The J values reach exceptionally high values of ~60 kJ/m^2, which is twice the toughness of human compact bone in the transverse orientation ($J \sim 30$ kJ/m^2).

The prominent toughening mechanisms in antler were identified by performing fracture toughness tests in situ in the environmental SEM (ESEM) (Launey *et al.*, 2010b). Sequential series of ESEM backscattered electron images of crack growth during in-situ R-curve testing in the in-plane longitudinal orientation are shown in Figs. 7.27(e)–(f). The crack trajectory is parallel to the long axis of the osteons, and the microcracks formed in front of the main crack (uncracked-ligament bridging) can be observed. The thin arrows indicate the uncracked-ligament bridges. Crack paths are consequently quite planar, with little evidence of deflection around the interface of osteons, resulting in much smoother fracture surfaces. In the anti-plane longitudinal orientation (Figs. 7.27(i)–(j)), the crack path is deflected around the hypermineralized regions surrounding the primary osteons in antler. The crack trajectory is the result of two competing factors: the external load forces the crack to propagate forward and the hypermineralized interface provides a preferentially weak path. The main crack in such orientations follows a much more tortuous route compared with that in the in-plane longitudinal orientation, which additionally contributes to the

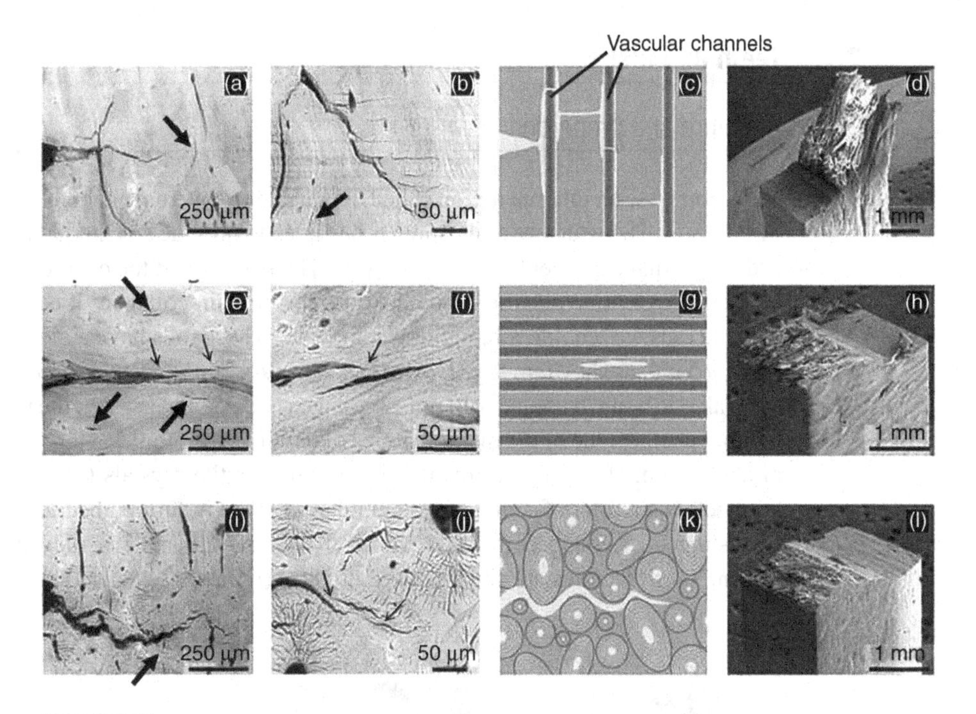

Figure 7.27.

Mechanisms for stable crack propagation and toughening in the transverse and longitudinal orientations of antler cortical bone. ESEM backscattered electron images of stable crack growth during in-situ R-curve testing in the ((a), (b)) transverse, ((e), (f)) in-plane longitudinal, and ((i), (j)) anti-plane longitudinal orientations. (d), (h), (l) SEM fractography images and (c), (g), (k) schematics of the crack trajectory for each orientation. The thin arrows indicate the uncracked-ligment bridges and the thick arrows designate microcracks. (Reprinted from Launey *et al.* (2010b), with permission from Elsevier.)

toughness. The toughness (in terms of both J and K_J) in the anti-plane longitudinal orientation is higher than that in the in-plane longitudinal orientation. In the transverse orientation, the prominent toughening mechanisms are crack deflection and twisting. As shown in the BSE images in Figs. 7.27(a) and (b), the crack deflects by as much as 90° at the interface between the osteons, resulting in a much more complicated crack path compared with those in the in-plane and anti-plane longitudinal orientations. The major crack travels through a long route in a crack extension Δa less than 500 µm.

The other characteristic of antler is its lower strength and much lower elastic modulus compared with human bone; this is associated with its extensive plasticity and intrinsically contributes to its toughness. The low yield strength in the longitudinal direction allows crack-tip plastic zones to form at lower stresses than in human bone (Zioupos, Currey, and Sedman, 1994; Koester *et al.*, 2008), which contributes to the large inelastic deformation and thereby to its exceptional toughness.

7.3 Teeth and tusks

7.3.1 Structure and properties

Teeth comprise an internal region called dentin and an external enamel layer, as shown in Fig. 7.28(a). The structure of the tooth is designed to provide an external layer that is hard and an internal core (dentin) that is tougher. The hardness of the enamel layer is due to a high degree of mineralization. Enamel does not contain collagen. It comprises hydroxyapatite rods woven into a fabric-like composite. These rods have a diameter of approximately 5 μm, as shown in Fig. 7.28(b) (Snead *et al.*, 2006). Dentin, on the other hand, is more akin to bone. It contains 30 vol.% collagen and 25 vol.% water, the remainder being hydroxyapatite. One of the major features of dentin is the tubules, which have a diameter of about 1 μm. They are surrounded by hydroxyapatite crystals (~0.5–1 μm diameter) arranged in a random fashion. These tubular units are in turn embedded in a composite consisting of a collagen matrix reinforced with HAP. This is called the intertubular region. These features are shown in Fig. 7.28(c) (Imbeni *et al.*, 2003).

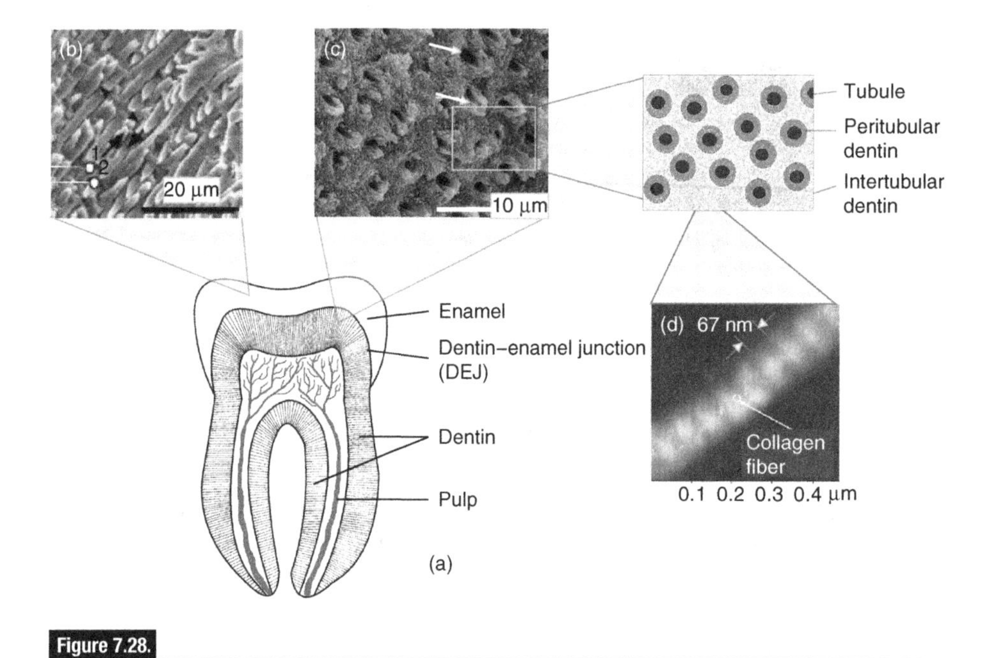

Figure 7.28.

Hierarchical structure of teeth. (a) Schematic showing enamel, dentin–enamel junction, dentin, and pulp. (b) Scanning electron micrograph of mouse tooth showing an etched image of mature enamel where the enamel rods weave past one another. (Reprinted from Snead *et al.* (2006), with permission from Elsevier.) (c) Scanning electron micrographs of dentin. (From Imbeni *et al.* (2003), with kind permission from John Wiley & Sons, Inc.) (d) AFM image of a collagen fiber. (Reprinted from Nalla *et al.* (2006a), with permission from Elsevier.)

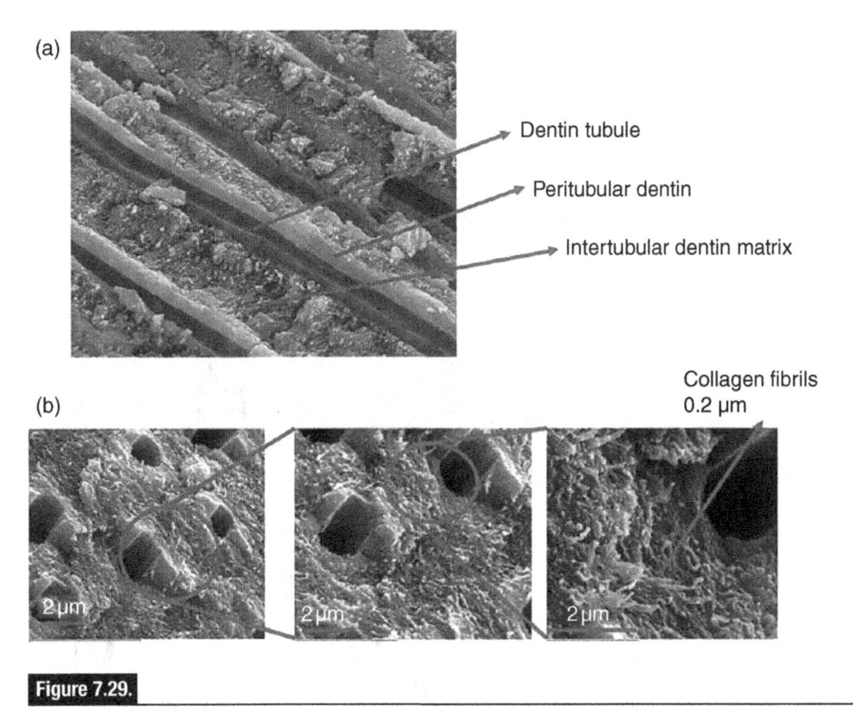

Figure 7.29.

(a) Longitudinally fractured tooth exhibiting tubules, peritubular dentin, and the intertubular dentin matrix; (b) transversely fractured tooth showing peritubular dentin (smoother region) and collagen fibers in intertubular matrix. (Figure courtesy Y.S. Lin.)

The longitudinal and transverse (to the tubules) fractures show the three regions very well: tubules, peritubular dentin, and intertubular dentin. The peritubular dentin forms a compact shell around the tubule, whereas the intertubular dentin contains a much greater amount of collagen. Indeed, a few collagen fibers can be seen in Fig. 7.29. The tubules are aligned as shown schematically in Fig. 7.30. This anisotropy of orientation of the tubules has an effect on the mechanical properties of dentin. Figure 7.30 shows that the compressive response differs for the transverse and longitudinal directions.

It is also interesting to note that the structure of teeth is common to most species. Fish, reptiles, and mammals share the same architecture of a harder external layer and a tougher core. The plot of Fig. 7.31 shows this effect clearly. The abscissa was normalized so that the through-thickness hardness of different species is plotted on the same scale. It can be seen that, in spite of the individual differences, the enamel or enameloid outside is shared by all teeth shown.

7.3.2 Fracture toughness and toughening mechanisms

Table 7.4 shows the hardness and fracture toughness of teeth (Nalla *et al.*, 2003b; Imbeni *et al.*, 2005), the latter obtained from cracks at the extremities of indentation

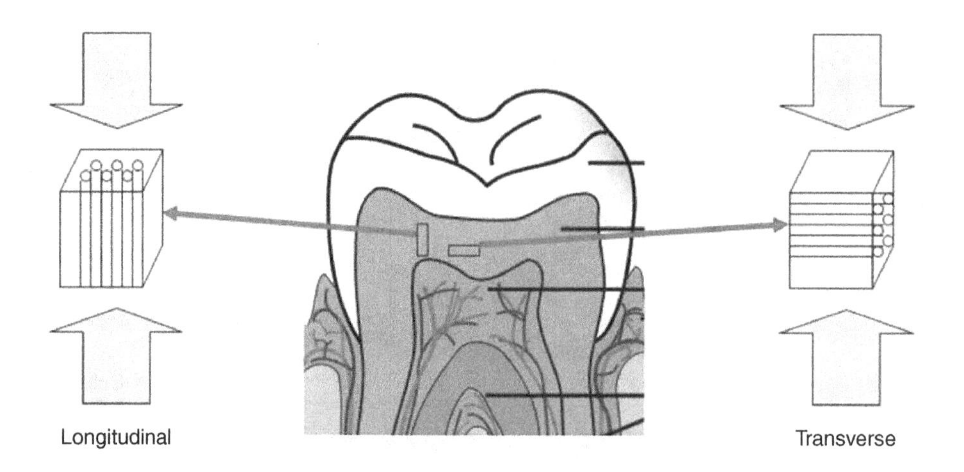

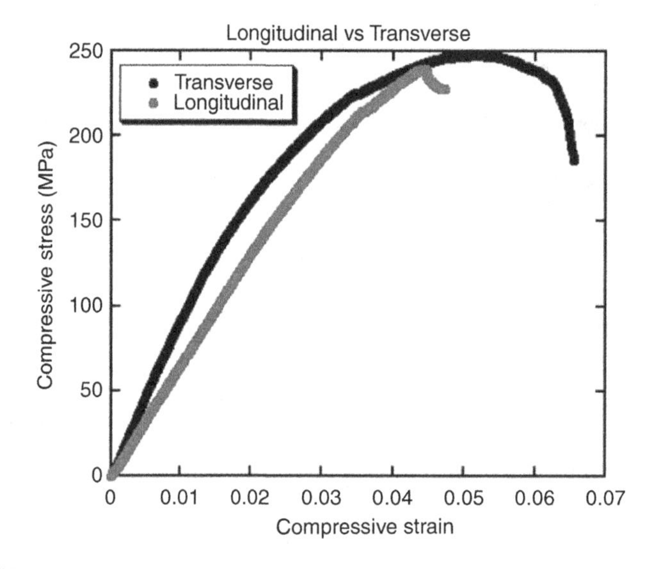

Figure 7.30.

Difference between compressive response in transverse and longitudinal directions: anisotropy induced by orientations of tubules. (Figure courtesy Y.S. Lin.)

through the Evans–Charles (Evans and Charles, 1976) technique. The toughness of enamel is lower than that of the dentin, whereas the converse is the case for the hardness. The high hardness of enamel is indeed significant: it is the hardest material in vertebrates, with the exception of iron oxide – in chitons. However, in contrast with bone, which has a vascular structure and can undergo self-repair and remodeling, both enamel and dentin are static, never repair, and cannot remodel. The changes in toughness and in hardness across the dentin–enamel junction (DEJ) are shown in Fig. 7.32.

Table 7.4. Mechanical properties of teeth[a]

	Enamel	Dentin
Fracture toughness, K_{Ic} (MPa m$^{1/2}$)	0.7–1.3	1–2
Hardness (GPa)	4	0.5
σ_{UTS} (MPa)		70–80
Young modulus, E (GPa)	60	perpendicular to tubules 5–6 parallel to tubules 13–17

[a] From Nalla *et al.* (2003b) and Imbeni *et al.* (2005).

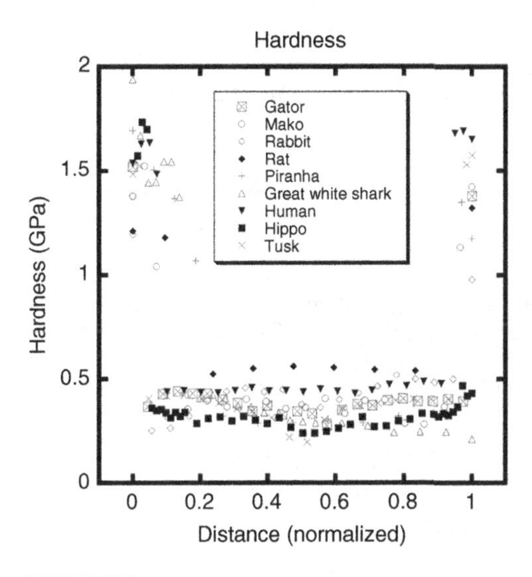

Figure 7.31.

Hardness across the cross-section of teeth for a number of species; note the harder surface (enamel or enameloid) and softer and tougher interior. (Figure courtesy Y.S. Lin.)

Figure 7.33 shows a crack initiated in enamel and with a trajectory perpendicular to the DEJ. Imbeni *et al.* (2005) observed that the interface never debonds. Rather, the crack penetrates it, travels a short distance (~10 μm) into the dentin, and subsequently stops. Figure 7.33 shows how uncracked ligaments arrest the crack.

Figure 7.34 is an in-situ synchrotron transmission diffraction measurement of deciduous bovine dentin carried out under an external load. The applied stress on the specimen produces an elastic strain in the HAP, which is determined by measuring the deformation of diffraction rings (shown in the insert). The slope is 24 GPa, much less than the modulus (120 GPa) for pure HAP; this is as expected because the porosity and collagen are also present in the tooth but are not load bearing. Thus, a considerable fraction of the deformation is taken up by the collagen.

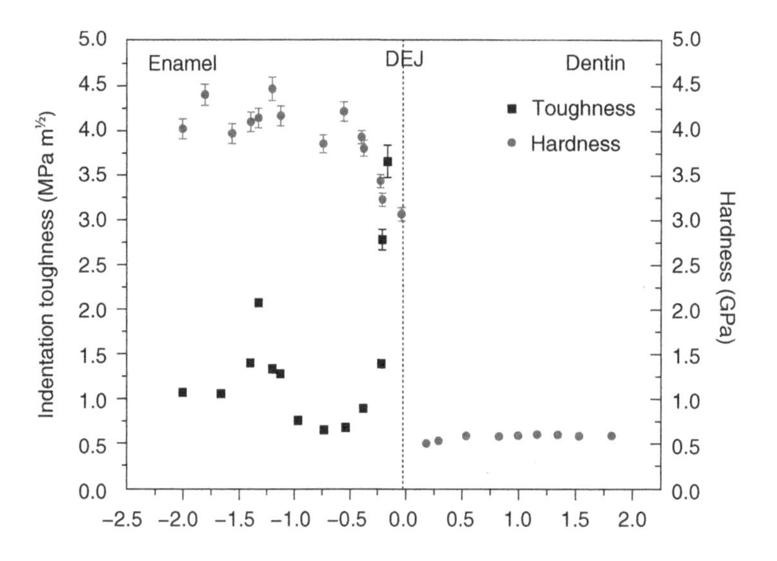

Figure 7.32.

Typical profiles of the Vickers hardness and indentation toughness across the dentin–enamel junction (DEJ) of human teeth. (Reprinted by permission from Macmillan Publishers Ltd.: *Nature Materials* (Imbeni *et al.*, 2005), copyright 2005.)

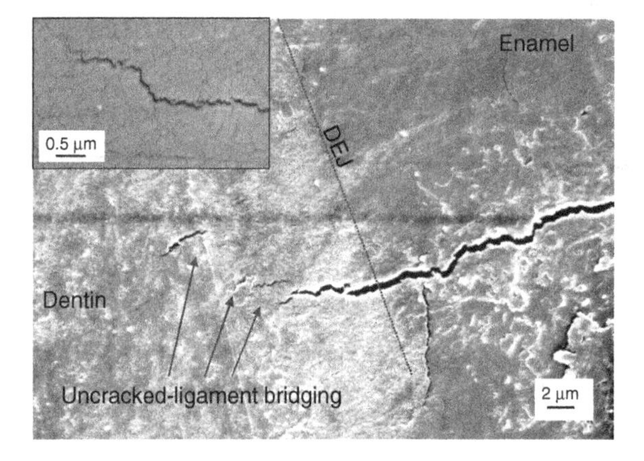

Figure 7.33.

Crack initiating in enamel arrested by the DEJ. (Reprinted by permission from Macmillan Publishers Ltd.: *Nature Materials* (Imbeni *et al.*, 2005), copyright 2005.)

Figure 7.35 shows in schematic fashion the four principal toughening mechanisms operating in dentin. The structure is fairly similar to bone, except that dentin is simpler. The mechanical properties in general and toughness in particular are quite anisotropic due to the presence of the tubules. The Young modulus parallel to the tubules is ~13 GPa, whereas it is ~5–6 GPa perpendicular to the tubules. Similarly, the fracture toughness is ~2.5 MPa $m^{1/2}$ parallel to the tubules and ~1.6 MPa $m^{1/2}$ perpendicular to the tubules. These mechanisms are akin to the ones in complex materials. Deflection (Fig. 7.35(a)) is

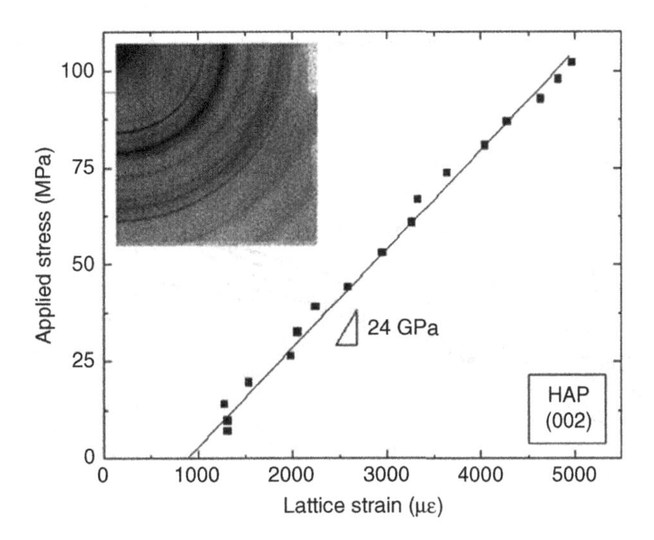

Figure 7.34.

Macroscopic applied compressive stress versus strain for deciduous bovine dentin. Synchrotron X-ray diffraction rings are shown in top left corner. (Used with the kind permission of Dr. A. Deymier.)

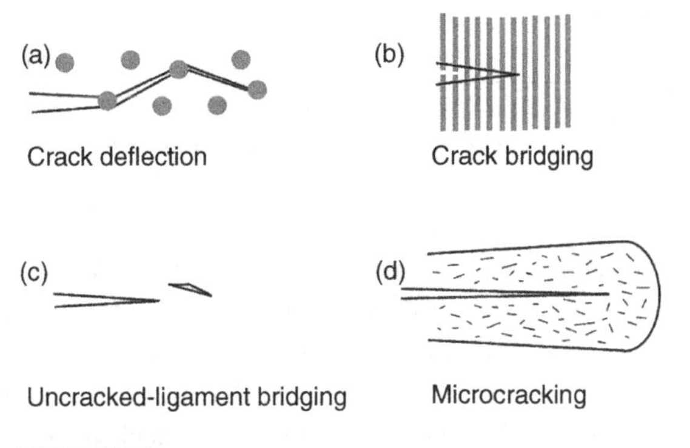

Figure 7.35.

Schematics of some possible toughening mechanisms in dentin: (a) crack deflection, (b) crack bridging (by collagen fibers), (c) uncracked-ligament bridging, and (d) microcracking. (Reprinted from Nalla *et al.* (2003b), with permission from Elsevier.)

caused by barriers which change the path of the crack. Deflection increases toughness by providing the initial barriers and by changing the crack angle from the applied stress from the optimum 90º condition. Crack bridging can be provided by the collagen fibers, and this allows crack-tip shielding (Fig. 7.35(b)). It can also result from the linkage of the principal cracks with microcracks ahead of the tip, producing uncracked ligaments (Fig. 7.35(c)). Microcracking is the damage that is initiated ahead of the crack and forms a process zone with dilatation which tends to "close" the crack (Fig. 7.35(d)). This mechanism has also been identified in bone. The contributions of the four different toughening mechanisms to K_{Ic} in dentin are listed in Table 7.4.

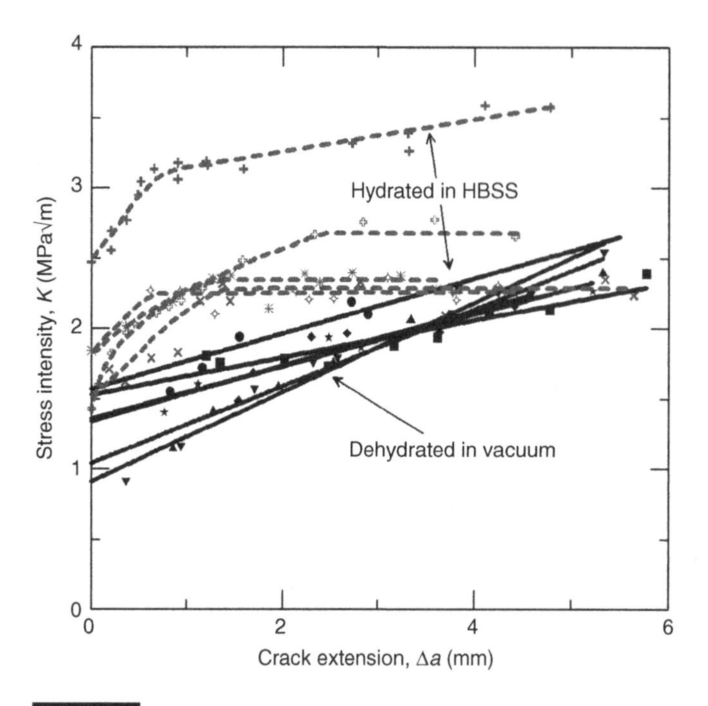

Figure 7.36.

K_R resistance curves for hydrated and dehydrated dentin. Note the significantly higher initial increase in toughness with crack extension for hydrated dentin. (Reprinted from Kruzic *et al.* (2003), with permission from Elsevier.)

The increase in K with crack extension, seen in Section 7.1.5 for bone and Section 7.2.3 for antler, is also observed in dentin. This is a direct consequence of the extrinsic toughening due to the formation of a process zone behind the crack front (called the "crack wake"). Figure 7.36 shows this response (Kruzic *et al.*, 2003). The different curves apply to specimens with different age, but the trend is clear. Another observation is that dentin hydrated in a standardized saline solution (HBSS) has a higher toughness. Nalla *et al.* (2006b) also discovered that dentin toughness increased in alcohol, auspicious news for hard drinkers.

The structure of dentin in elephant tusk is shown in Fig. 7.37. The tubules are seen as oval features. The insert in the figure shows the tubules and the collagen fibers radiating away from them. There is a difference between elephant tusk and human tooth dentin. In elephant tusk, the tubules are more elliptical and there is almost an absence of peritubular dentin. The presence of tubules confers a considerable degree of anisotropy to the structure. Figure 7.38 shows two mechanisms of toughening operating when the crack is running in the "anti-parallel" direction to the tubules. Uncracked ligaments are seen, marked by arrows in Fig. 7.38(a); microcracks in the vicinity of the cracks are seen in Fig. 7.38(b).

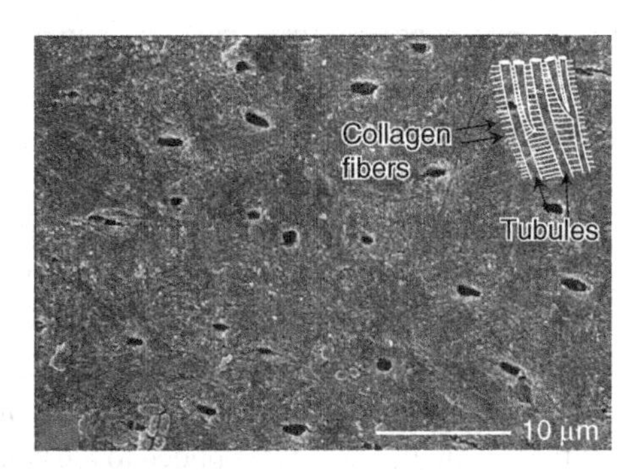

Figure 7.37.

Scanning electron micrographs of the typical microstructure of elephant tusk dentin. (Reprinted from Nalla *et al.* (2003b), with permission from Elsevier.)

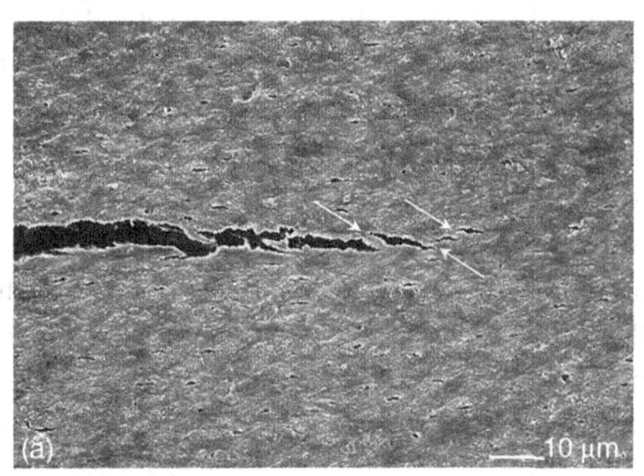

Figure 7.38.

Scanning electron micrographs of typical crack paths for the nominally "anti-plane parallel" orientation in the context of crack–microstructure interactions. The white arrows indicate (a) uncracked-ligament bridging, (b) microcracks in the vicinity of the crack. (Reprinted from Nalla *et al.* (2003b), with permission from Elsevier.)

Box 7.1 Dental materials and implantation

When P. Brånemark (Brånemark and Breine, 1964; Brånemark *et al.*, 1964; Brånemark, 1972a,b; Brånemark and Eriksson, 1972) was studying blood flow in bone marrow in the early 1960s, he never thought about the revolution that he was about to trigger. He used metallic hollow implants that traversed the tibia of rabbits and enabled observation, through a glass window, of the marrow. Upon removing these implants at the end of the experiment in order to reuse them, he observed that some of them were firmly attached to the bone and even sheared off. What would an average researcher have done? Use a few strong words and have the machine shop make new ones? Not Brånemark. He realized that some of the stainless steel implants had been replaced by titanium, and these were the ones that bonded to the bone. In a stroke of genius, he immediately abandoned his research, and dedicated the next 50 years to the study of osseointegration. The emergence of titanium dental implants was the result of this discovery. In the USA, close to one million teeth are implanted each year, and the majority of dentists are qualified for the new field of implantology. Even the UK, the Land of Ugly Teeth, is adhering to the practice.

The osseointegration of dental implants is critically dependent on their surface properties. Several investigations have analyzed the influence of implant surface properties for osseointegration (Brånemark and Breine, 1964; Brånemark *et al.*, 1964; Brånemark, 1972a,b; Brånemark and Eriksson, 1972). It has been shown that surface morphology, topography, roughness, chemical composition, surface energy, chemical potential, strain hardening, the presence of impurities, the thickness of the titanium oxide layer, and the presence of metal and nonmetal composites each have a significant influence on bone–tissue reactions. Osseointegration can occur only if the cells adhere to the biomaterial surface. At this phase, reorganization of the cytoskeleton and information exchange between cells and the extracellular matrix at the cell–biomaterial interface occur, generating gene activation and specific tissue remodeling. Both the morphology and roughness of the biomaterial's surface have an influence on cell proliferation and differentiation, extracellular matrix synthesis, local factor production, and even cell morphology.

Adhesion of osteoblasts onto implant surfaces is not sufficient to ensure osseointegration. The cells have to receive signals inducing them to proliferate.

Worldwide, over 250 million people lack teeth (this number could be as high as one billion); hence the potential of this technique is virtually limitless.

We describe in the following the procedure (in medical and dental terminology, it is called a "protocol") and some of the reasons why titanium alloys are so successful. The steps are as follows (see Fig. B7.1):

- An incision is made in the gum.
- The site is pre-drilled, after X-rays are taken that determine the direction of drilling.
- Pre-drilling is followed by drilling with the appropriate diameter (Fig. B7.1).
- The implant is inserted using a special torqueing instrument.
- The gum is reclosed for a period of three to six months in order to ensure maximum osseointegration prior to loading.

There are different diameters and shapes of the "studs." Some of them are shown in Fig. B7.2(a). Although most are cylindrical, there are also conical ones. The angles of the screws also vary. After osseointegration, the top portion is attached to the stud, as shown in Fig. B7.2(b). The prosthesis, a ceramic, is attached at the top as a final step.

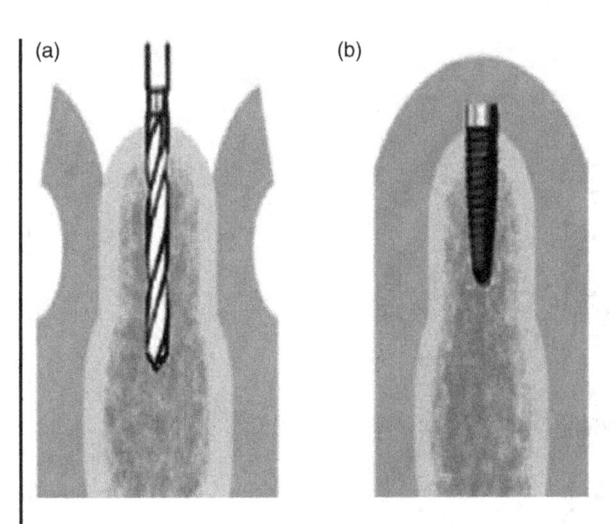

Figure B7.1.

(a) Pre-drilling the bone; (b) insertion of implant and reclosing of the gum tissue to allow osseointegration.

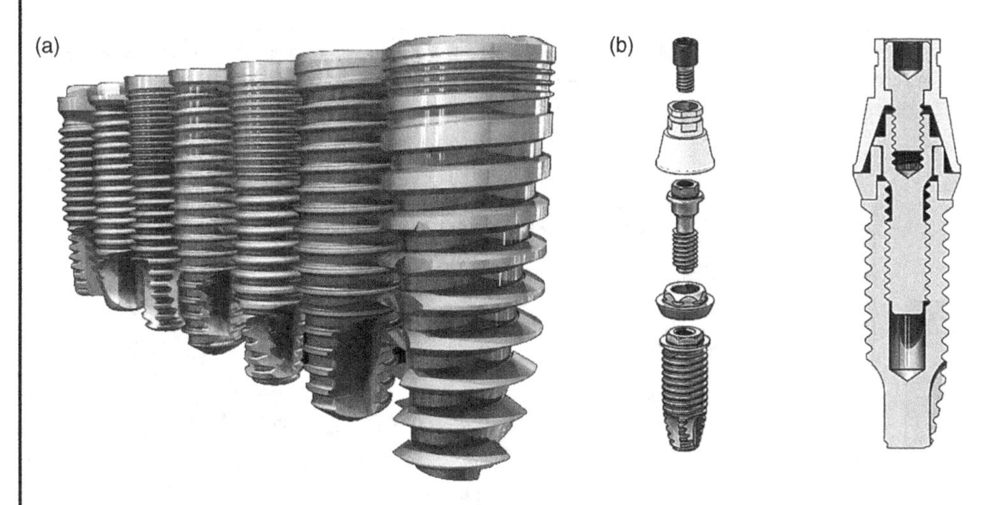

Figure B7.2.

(a) Different dental implant morphologies. (b) Implant assembly with screw at bottom and other components attached to it; a dental prosthesis is attached to top portion. (Figures courtesy Professor Elias, Instituto Militar de Engenharia, Brazil.)

The term osseointegration is a general one, indicating the attachment of the implant to the bone. The titanium alloy that is universally used for implants has a surface that is indeed an oxide. The exact nature of osseointegration is not well understood, but cells have to adhere to the implant surface. In-vivo experiments show that osteoblasts attach to the surface, as shown in Fig. B7.3. Bone cells are divided into osteoclasts and osteoblasts, the first creating empty volume and the second filling volume. The cytoskeleton of the cells can be seen, and fibers spreading out from them are attached. These cells are elongated

Box 7.1 (cont.)

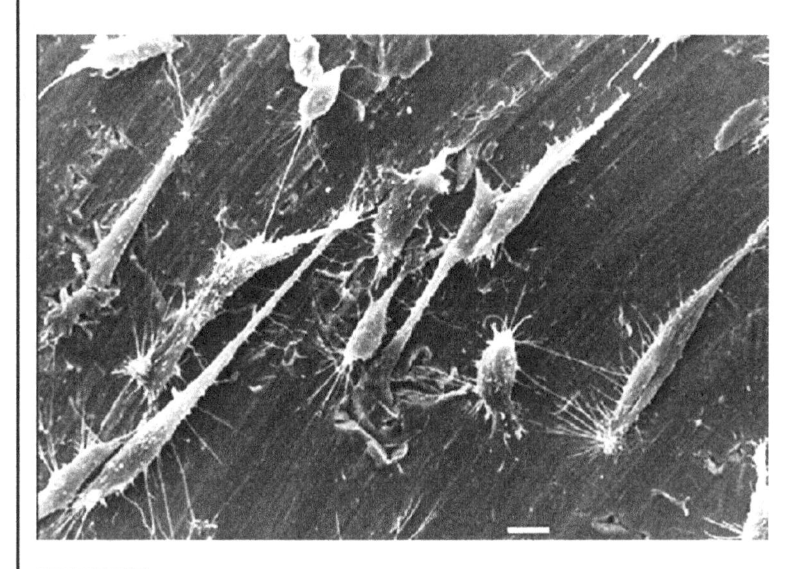

Figure B7.3.

Cells (osteoblasts) attaching to surface of dental implants; note cystoskeleton and alignment of cells corresponding to machining grooves. (Figure courtesy Professor Elias, Instituto Militar de Engenharia, Brazil. See also Menezes *et al.* (2003).)

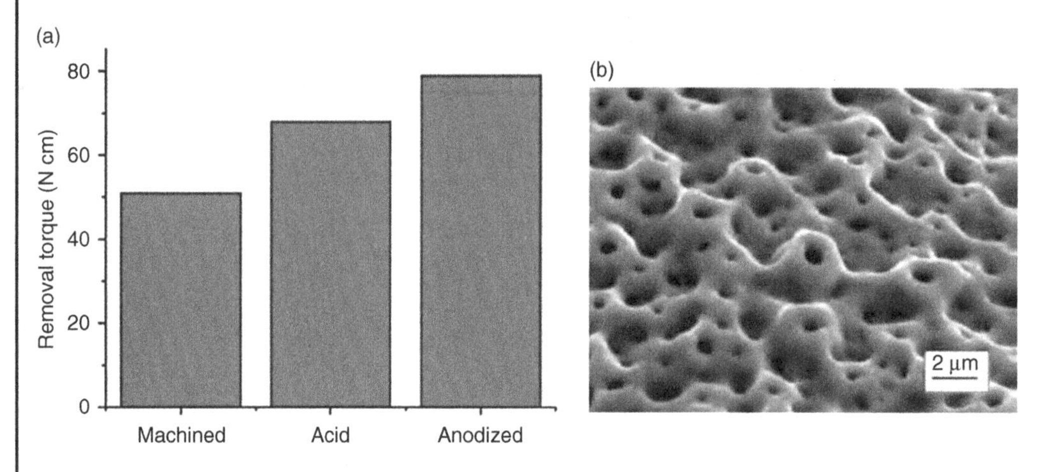

Figure B7.4.

(a) Removal torque of implants as a function of surface treatment. (b) Surface morphology for anodized titanium alloy implant; this surface contains titanium oxides with crystalline structure in the form of rutile and anatase. (From Elias (2011), with kind permission from Professor Elias, Instituto Militar de Engenharia, Brazil.)

along the machining groove orientation. It is clear that the roughness of the surface plays an important role in osseointegration. The anodization of the surfaces has been shown to improve the strength of the bonds. This is shown in Fig. B7.4(a), in which machined, acid etched, and anodized implants are compared. The removal torque is a good measure of osseointegration. This torque is measured and is ~50 N cm for the

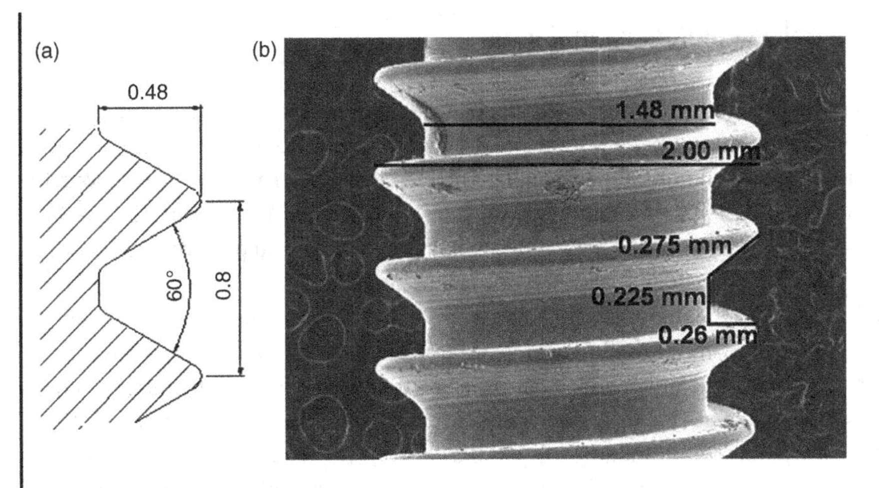

Figure B7.5.

(a) V threaded surface of implants. (From Elias (2011), with kind permission of Professor Elias.) (b) Buttress threaded surface of orthodontic mini-implants. (Reprinted from Morais *et al.* (2007), with permission from Elsevier.)

as-machined (i.e. no further surface treatment) implant, increasing to 80 N cm for the anodized condition. This is usually done in rabbits after the animal is sacrificed and (sometimes!) devoured. The anodized surface, seen in Fig. B7.4(b) shows an interesting pattern consisting of mini-volcanoes and holes that assist the attachment of cells (Elias, 2008; Elias *et al.*, 2013).

It is possible to correlate the removal torque (Morais *et al.*, 2007; Serra *et al.*, 2010), an external measurement dependent on the geometry of the implant, to the intrinsic shear strength of the interface. This is achieved by the application of simple mechanics equations.

The shear stress acting on the interface is equal to the force divided by its area, A. This force in turn is equal to the torque T divided by the moment arm (i.e. half the diameter of the implant, $D/2$). Thus,

$$\tau = \frac{2T}{nDA},$$

where n is the number of threads.

The resisting area per thread, A, can be obtained by summing the areas of the three sides of the screw, i.e. the flat bottom and the two angled sides:

$$A = A_1 + A_2 + A_3.$$

Figure B7.5 shows two types of screws. The conventional V thread (Fig. B7.5(a)) has an angle of 60° between the two sides. The buttress thread (Fig. 7.5(b)) is asymmetrical and has the leading edge making approximately 90° – with the longitudinal axis and the trailing surface making 45°. This geometry has the highest pullout resistance. The total area is computed by making approximations and assuming strips:

$$A = \pi(D_1L_1 + D_2L_2 + D_3L_3),$$

where D_1, D_2, and D_3 are average diameters, and L_1, L_2, and L_3 are the total lengths.

7.4 Other mineralized biological materials

In the preceding sections, we have introduced the structure and mechanical properties of typical bones and teeth. In nature, there are a wide variety of bones, teeth, and mineralized tissues that have unique functions. The mechanical properties and adaptations of some unique bony tissues were reviewed by Currey (2010) and Chen *et al.* (2012). For example, bird bones have to be lightweight with proper mechanical performance in order to fly efficiently. Manatees (*Trichechidae*, sometimes known as sea cows) have extremely dense bones without the cancellous portion that enable them to achieve neutral buoyancy and feed easily in shallow waters. Mammalian ear bones (ossicles) have quite different functions from other bones. In order to transmit sounds with minimal loss, ear bones have to be stiff. Fin whale bulla (ear bone) has very high mineral content (~86 wt.% ash) and Young modulus (~34 GPa) among mammalian bones, yet is very brittle with much less fracture resistance. There are some examples of animals that use mineralized tissues as protective armors. In the following sections, selected examples of mineralized biological materials will be reviewed.

7.4.1 Armadillo

The armadillo ("little armored one" in Spanish) is an indigenous mammalian from South and Central America and the southern part of North America. Armadillos are covered with bony armors, called osteoderms, which can protect them from predators. For example, the South American three-banded armadillo (*Troypeutes tricinctus*) can curl up into a ball that is completely covered by osteoderms. The nine-banded armadillo (*Dasypus novemcinctus*) in North America has nine accordion-like banded regions and can curl up into a ball with a certain amount of flexibility.

Figure 7.39(a) shows the hierarchical structure of the nine-banded armadillo osteoderm. The carapace is composed of osteoderms with two different morphologies – hexagonal and triangular, both covered by keratin. The hexagonal osteoderms appear on the pectoral and pelvic areas, whereas the triangular ones are along the torso. A photograph of the carapace, with some of the keratin removed to enable the underlying osteoderms to be seen, is shown in Fig. 7.39(b). The osteoderms are connected laterally by nonmineralized collagen fibers, called Sharpey's fibers. The osteoderms are attached to the skin by neurovascular foramen. Figure 7.40 is an optical micrograph showing the microstructure of the hexagonal osteoderms. The keratin component on the top epidermis layer is ~120 μm thick, followed by a thicker papillary layer (350 μm) composed of dense bone. Several small hollow spaces with layers of concentric lamellae with diameters 120–150 μm found at the bottom of the papillary layer correlate to osteons with vascular channels ranging from 10 to15 μm. The collagenous high-porosity region (reticular dermis layer) has larger cavities of diameters of 100–400 μm, which may be derived from the complex branched structure of the neurovascular channels. In the hypodermis layer, limited traces of blood vessels are observed;

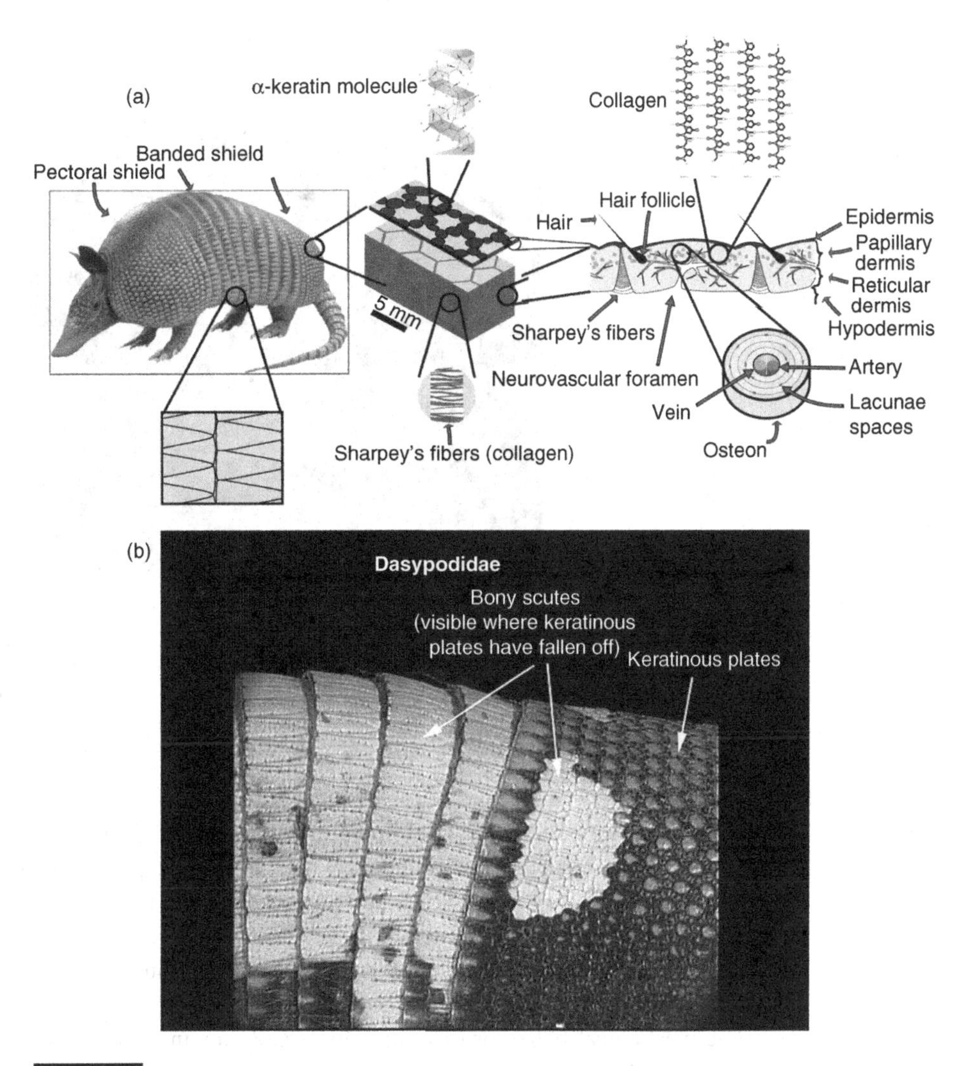

Figure 7.39.

(a) Hierarchical structure of the nine-banded armadillo. The carapace consists of hexagonal (pelvic and pectoral shield) or triangular (trunk) bony tiles (osteoderms). The osteoderms are covered by a water-repellant keratin epidermis, below which is the bone, consisting of both dense and porous (middle) regions. The tiles are held together by nonmineralized fibers (Sharpey's fibers) that provide flexibility to the carapace. (Reprinted from Chen *et al.* (2011), with permission from Elsevier.)
(b) Osteoderm with some of the keratin scutes removed to reveal underlying bone structure. (Used with kind permission of Professor Philip Myers.)

they are oriented parallel to the dermal layer. The osteoderm has a sandwich structure, with porous bone between layers of dense bone.

The Sharpey's fibers play an important role in connecting adjacent osteoderms and providing flexibility. The carapace can have a varying curvature: the stretched (Fig. 7.41 (a)) and curved (Fig. 7.41(b)) states are accompanied by extension and retraction of the

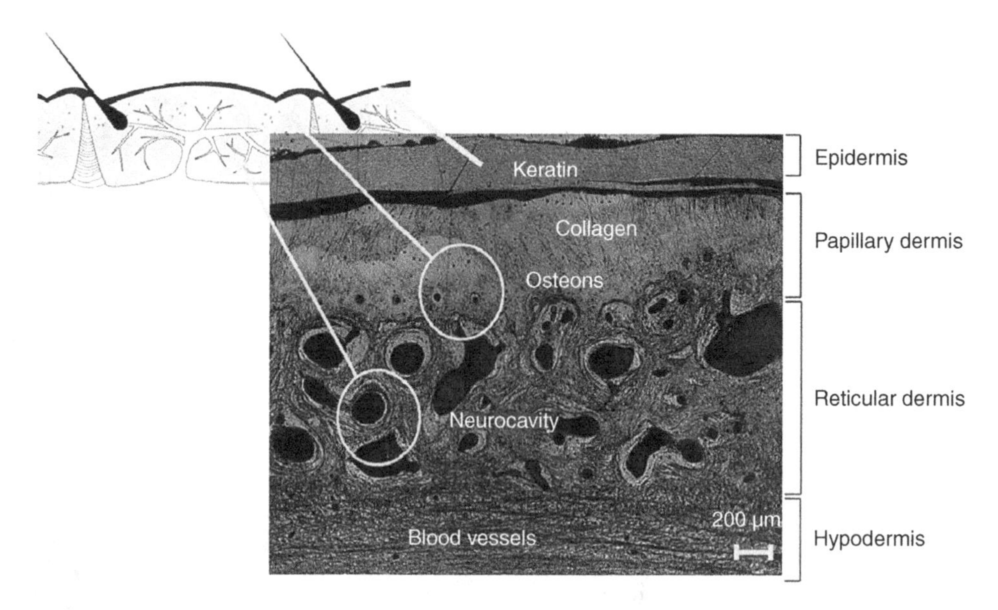

Figure 7.40.

Optical micrograph showing osteons and interior porosity at different depths in armadillo osteoderm. The epidermis consists of a keratin layer. The papillary dermis is dense bone that sandwiches the reticular dermis consisting of porous bone, arising from the complex branched vascular network (neurocavities) that stem from the vascular channels that attach the osteoderm to the dermal layer. The hypodermis is another layer of dense bone. (Reprinted from Chen *et al.* (2011), with permission from Elsevier.)

Sharpey's fibers, respectively. The inner dimensions (D_2) of the tiles are smaller than the outer ones (D_1). The curved configuration represents the typical armadillo carapace in nature. Figure 7.41(c) is an SEM micrograph showing the Sharpey's fibers. X-ray fluorescence mappings of calcium in Figs. 7.41(d) and (e) show that hexagonal and triangular osteoderms are calcium rich, whereas the junctions between the tiles (Sharpey's fibers) are nonmineralized.

The mechanical behavior of armadillo osteoderms in tension was investigated by Chen *et al.* (2011). Samples consisting of several osteoderms were tested. The tensile failure strength of the osteoderm is 16 MPa (hydrated) and 20 MPa (dry). In the hydrated condition, the failure was along the Sharpey's fibers, whereas the dry failure often occurred within the osteoderms. Under tensile stress, the mineralized tiles and the Sharpey's fibers can be considered to be under isostress conditions. The fibers undergo more strain than the bony tiles. One interesting aspect is that the shear strength of the (hydrated) tiles (the stress necessary to push out a tile from the carapace) is similar to the tensile strength of the hydrated samples (Figs. 7.42(a) and (b)). Figure 7.42(c) illustrates how the samples were loaded in tension and shear. The shear strength is typically half of the tensile strength for most materials. This is true for an isotropic material (properties the same in all directions). The maximum shear stress, τ_{max}, is equal to half the normal tensile or compressive stress in uniaxial loading. This unusual behavior is attributed to the stretching of the Sharpey's fibers.

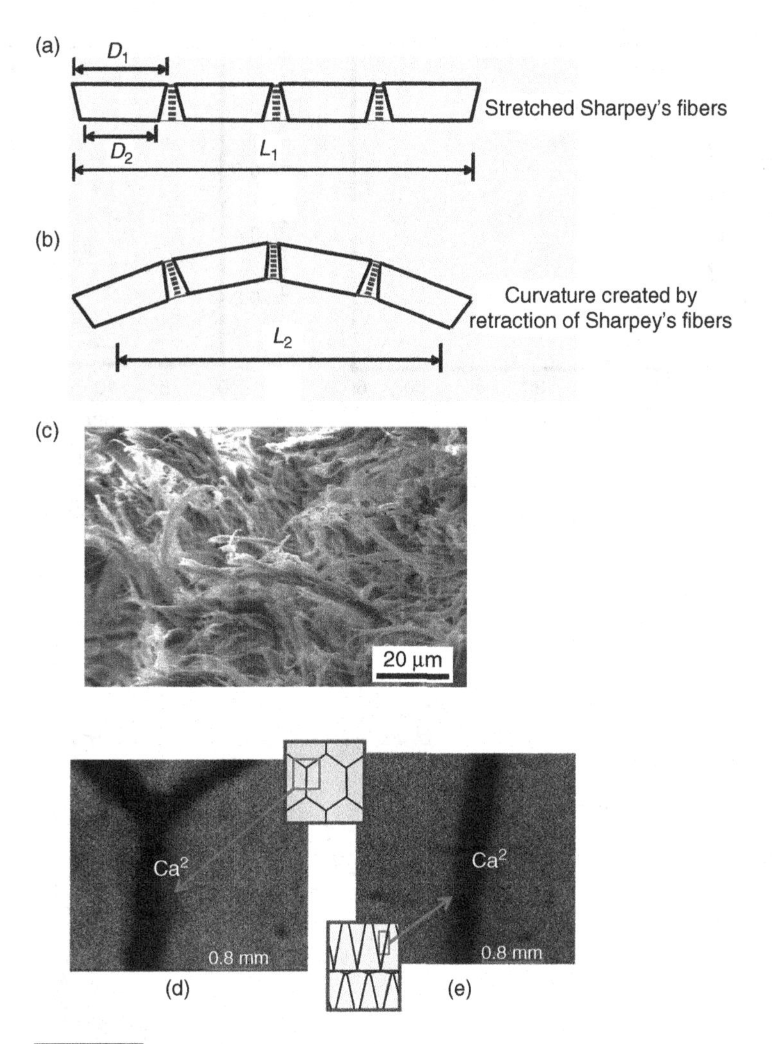

Figure 7.41.

Cross-sectional morphology of osteoderm scales of (a) the stretched and (b) the flexed carapace ($D_1 > D_2$). Sharpey's fibers create a variable curvature with retraction of the fibers. (c) SEM micrograph of fractured Sharpey's fibers from the armadillo carapace. (d) X-ray fluorescence images taken on the hexagonal tiles. The gray color corresponds to calcium. Calcium is not found between the tiles, indicating that the Sharpey's fibers are not mineralized, and (e) this is the same for triangular tiles. These images also show high calcium concentration in the tiles and little or no calcium between the tiles. (Reprinted from Chen *et al.* (2011), with permission from Elsevier.)

Rhee *et al.* (2011) examined the compressive properties and found the osteoderms behave as a cellular solid, having in the stress–strain curve an initial linear elastic region followed by a plateau region, attributed to the deformation of the interior porous bone, and a final upturn assigned to the crushing of the porous bone. The large plateau region, up to ~50% strain, demonstrates that the application of a compressive force will lead to extensive distortion of the osteoderm before failure.

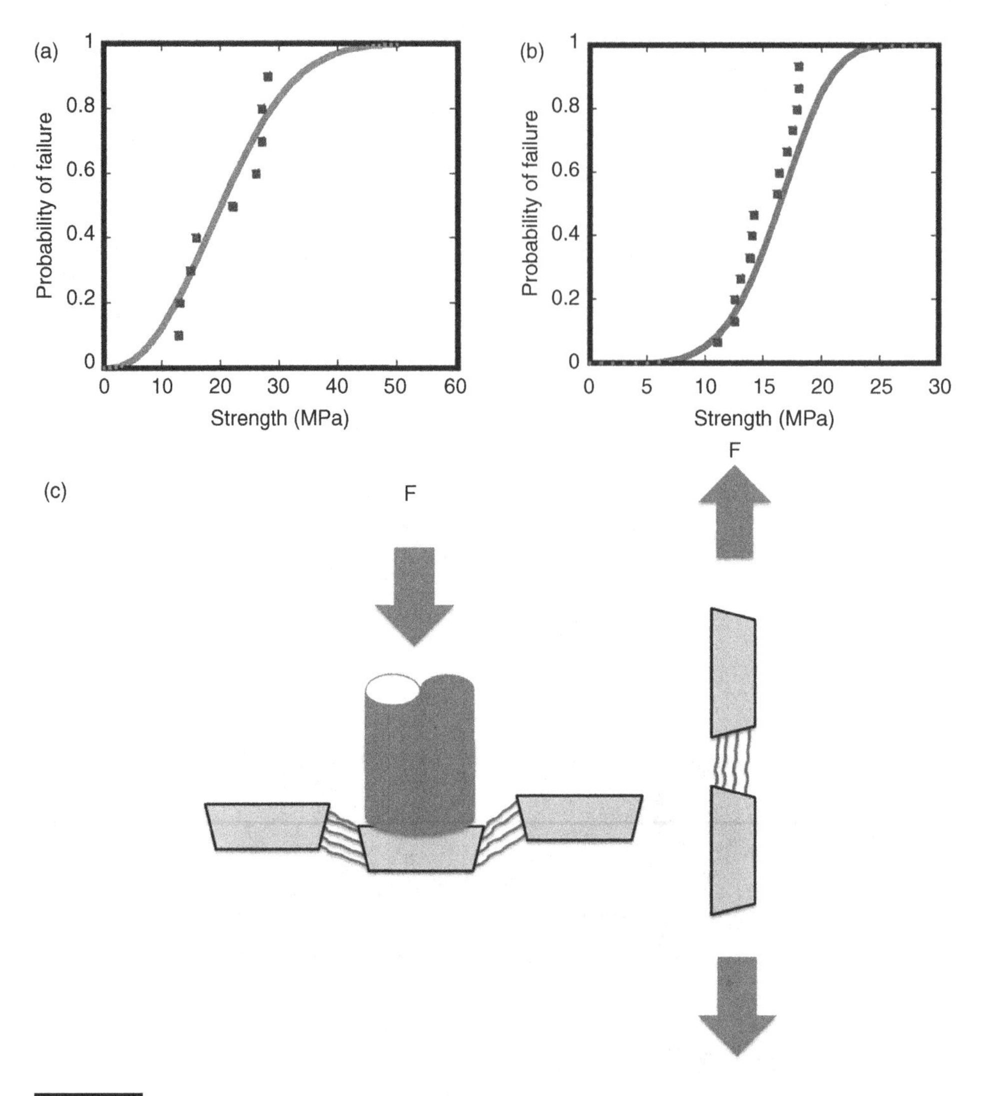

Figure 7.42.

Weibull probability distribution function as a function of failure strength for hydrated sample for (a) shear test results and (b) tensile test results. The mean stress is defined as a 50% probability of failure. (c) Configuration of the hexagonal tiles of the armadillo tested in shear (left) and tension (right). (Reprinted from Chen *et al.* (2011), with permission from Elsevier.)

7.4.2 Testudine

The order *testudine* includes turtles, tortoises, and terrapins, all of which are characterized by hard shells. The testudine shell consists of three parts: the carapace (the top shell), the plastron (the base of the shell), and the bridge (the lateral linkage between the carapace and the plastron), as shown in Fig. 7.43(a). The spine and rib bones are fused

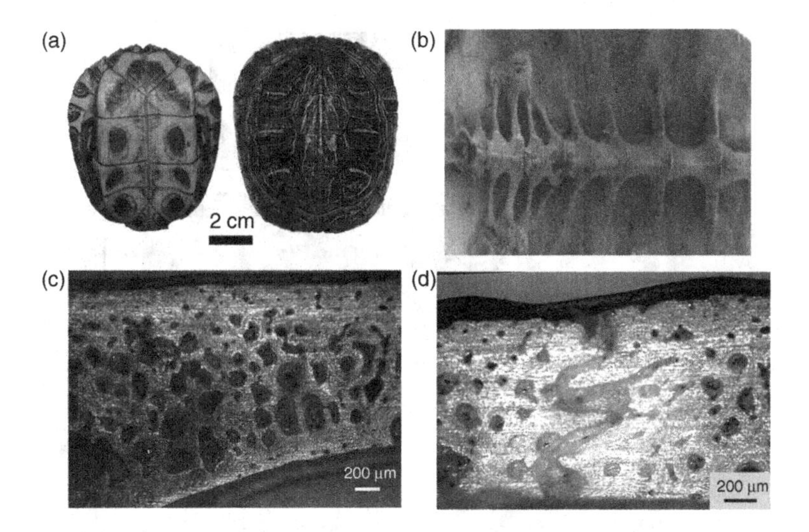

Figure 7.43.

Red-eared slider turtle carapace. (a) Overall picture of the shell and (b) micrograph showing that the skeleton (ribs) is fused to the carapace. (c) Cross-sectional micrographs showing trabecular (porous) bone surrounded by dense cortical bone. Pores in the trabecular bone, which comprise the core (baloney) of the sandwich. (d) "Zig-zag" suture (nonmineralized) between the osteoderms.

to the carapace, forming a rigid shell with limited flexibility (Fig. 7.43(b)). The carapace and plastron are composed of osteoderms, also called "scutes," which range from millimeter to centimeter diameters, depending on the species. There are about 38 scutes on the carapace and 12–14 on the plastron. The osteoderms are covered with keratin layers (for waterproofing) and have dense cortical bone on the top and bottom layers with porous trabecular bone sandwiched between, similar to the armadillo, as shown in Fig. 7.43(c). The osteoderms are connected by "zig-zag" shaped sutures, which serve to interlock the osteoderms together, as shown in Fig. 7.43(d). Leatherback turtles are known to dive to depths greater than 1000 m. The osteoderms in leatherback turtles have characteristic features (Fig. 7.44) (Yang, 2013b). The osteoderms are not covered with keratin but with skin. The porosity in leatherback turtle osteoderm is significantly high. The sutures are jagged and provide considerable flexibility, estimated to be ~15°, higher than other species. The flexibility of leatherback turtle carapace enables the contraction of the body at high hydrostatic pressures in deep sea conditions.

Figure 7.45(a) shows a schematic diagram of the sutures in the normal and displaced states (Krauss *et al.*, 2009). In Fig. 7.45(b), the three-point-bending load–displacement curves for the carapace with and without sutures show that samples with sutures result in a lower stiffness compared to those without sutures. The sutures give rise to effortless deformation under small loads, transferring to stiffer responses after locking under higher degrees of movement. Rhee *et al.* (2009) reported that the porous core of the turtle shell is made of closed-cell foam, causing the sandwich structure to undergo a nonlinear deformation, which leads to a higher specific energy absorption compared with the dense cortex alone. Achrai and Wagner

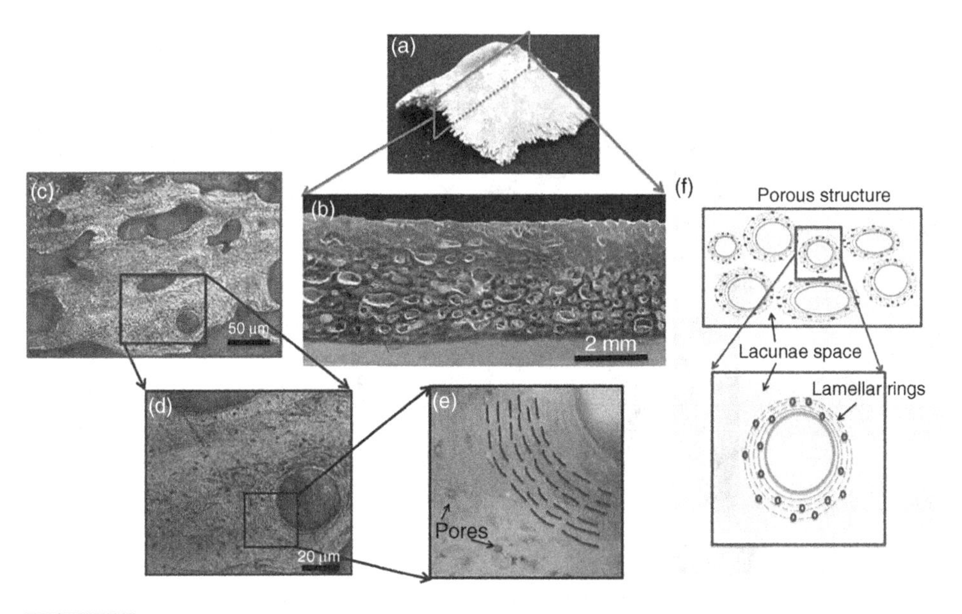

Figure 7.44.

Osteoderm in the leatherback turtle: (a) side view from the ridge osteoderm; (b) longitudinal section showing the porous structure; (c) porosity showing alignment of voids; (d) lamellar rings and lacunae around pores; (e) lamellar rings and small pores; (f) schematic drawing of the structure. (Figure courtesy W. Yang.)

(2013) revealed that the dorsal and ventral cortices of the sandwich structure exhibit different mechanical properties as a result of different fiber arrangements. Randomly oriented fibrillar network in the dorsal cortex can sustain sharp impact isotropically, while the plywood arrangement of fibers in the ventral cortex possesses anisotropic mechanical properties and is beneficial for structural support. The turtle shell appears to be a functionally graded material (FGM) in terms of composition, porosity, and mechanical properties.

7.4.3 Crocodilia

The order crocodilia contains notably crocodiles, alligators, and caimans. These ancient reptiles have long been considered as fierce carnivorous tetrapods with heavily armored skins. Although they seldom encounter predators, territorial fights among the same species can often be deadly because of their extremely high bite force, reaching ~10 kN. Thus, well-developed armor designs for excellent mechanical performance are demanded, along with some flexibility for speed and agility in order to capture prey. The dorsal sheath of crocodilians has been used as armored suits by ancient warriors since they were found to repel knives and arrows; they are even bulletproof under certain conditions.

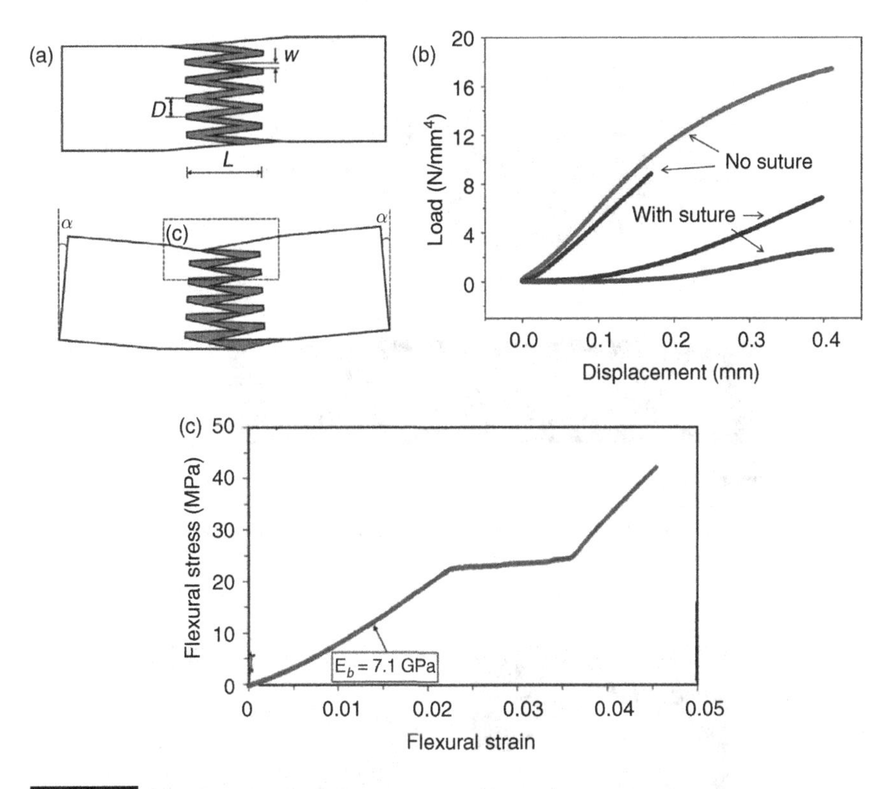

Figure 7.45.

(a) Mechanical deformation of red-eared slider turtle carapace. The sutures (nonmineralized) allow for some flexibility, but mostly the osterderms are interlocked with each other. (b) Force–displacement curves from two carapaces with and without sutures. (Taken from Krauss *et al.* (2009), with kind permission of John Wiley & Sons, Inc.) (c) Flexural stress–strain curve for a *Terrapene carolina* turtle carapace specimen. The plateau region corresponds to deformation of the interior trabecular bone, and the upturn results from crushing of the bone. (Reprinted from Rhee *et al.* (2009), with permission from Elsevier.)

An adult *Alligator mississippiensis* (American Alligator) armor has about 70 pieces of osteoderm (Fig. 7.46(a)), with different shapes according to the location (Figs. 7.46(b)–(e)). Each osteoderm contains a longitudinal (head-to-tail direction) keel in the middle. A cross-sectional image of the cervical osteoderm keel reveals the vascularity inside, with the sandwich structure of porous bone surrounded by compact bone (Fig. 7.47(a)). In the lateral edge of the osteoderm, where two neighboring pieces are connected, serrated suture joints can be observed (Fig. 7.47 (b)). The structural design and mechanical behavior of alligator osteoderms were investigated by Sun and Chen (2013) and Chen, Yang, and Meyers (2014). The microstructure of alligator osteoderm is nonuniform with varying mineral content and porosity, which results in graded mechanical properties across the osteoderm: a hard and stiff dorsal cortex gradually transforms to a more compliant ventral base. Figure 7.48 shows illustrations of three proposed deformation mechanisms in

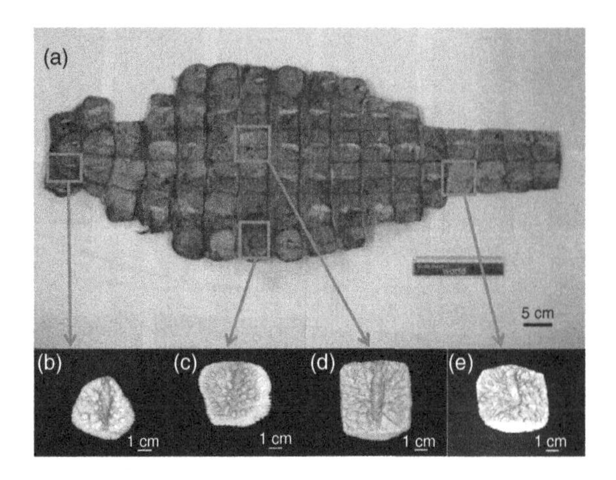

Figure 7.46.

(a) The overall view of the osteoderms of *A. mississippiensis*, with individual pieces from (b) cervical, (c,d) central, and (e) transverse terminal regions. (Figure courtesy C.-Y. Sun and P.-Y. Chen.)

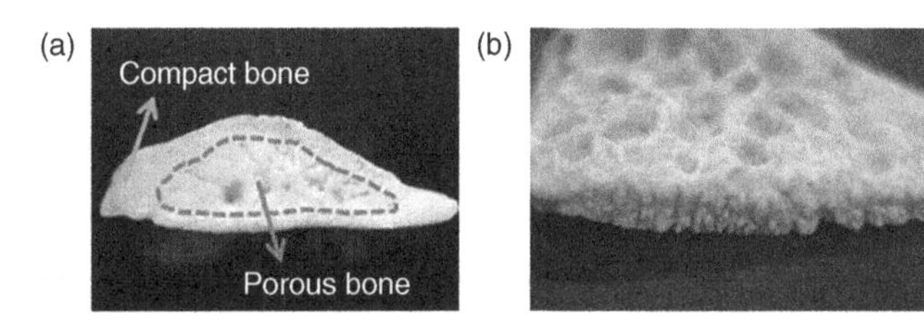

Figure 7.47.

(a) Cross-sectional view of the keel of a cervical osteoderm. The dashed line encircles the region with much higher porosity. (b) Serrated edge of an osteoderm. (Figure courtesy C.-Y. Sun and P.-Y. Chen.)

alligator dermal armors: (1) the flexibility provided by sutures and Sharpey's fibers can dissipate energy under small loads, similar to turtle shells; (2) deformations of the cellular foam interior absorb a certain amount of energy without the cortex cracking; (3) a combination of the hard dorsal region and the compliant ventral region with graded mechanical properties offers optimization in load redistribution and energy absorbance.

In summary, the mineralized armors share many similar structural features: (1) the main constituents of these mineralized tissues are bone, consisting of collagen fibers and hydroxyapatite minerals; (2) the bony plates are connected by soft tissues or joints; (3) they are covered by keratinous layers on the outer surface; (4) they are sandwich composites with a dense cortex and a porous core. A more detailed discussion on the mechanical design of natural flexible dermal armors can be found in the review by Yang *et al.* (2013b).

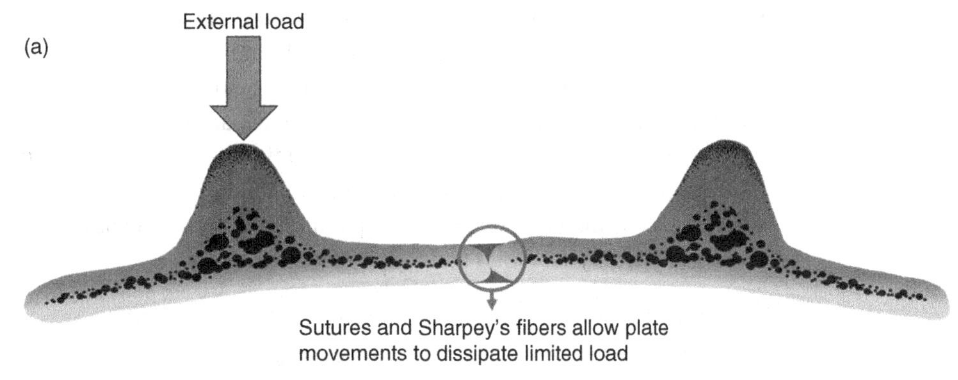

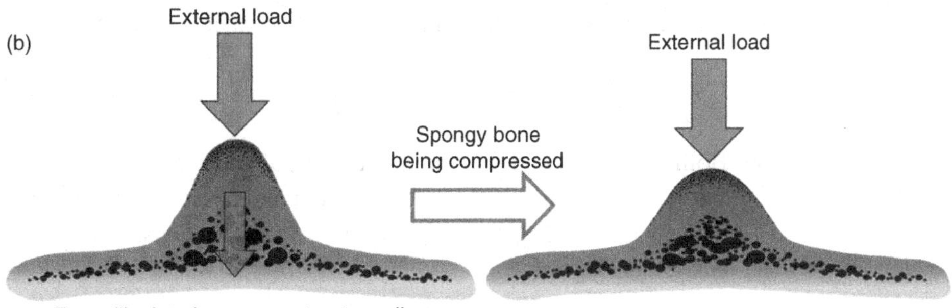

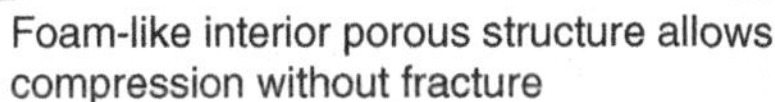

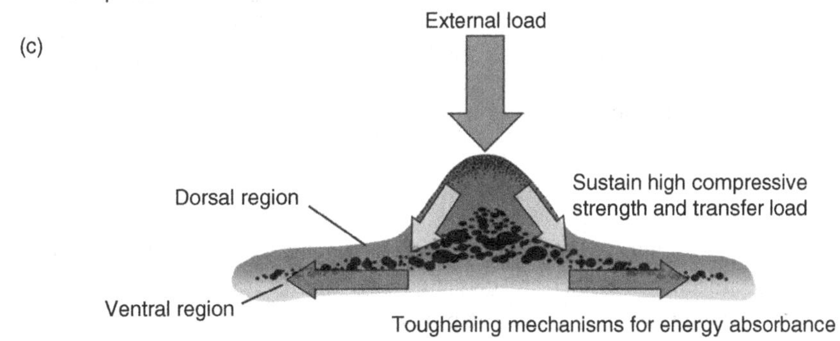

Figure 7.48.

Three deformation mechanisms of alligator dermal armor: (a) sutures and Sharpey's fibers provide flexibility; (b) sandwich structure absorbs energy; (c) graded mechanical properties from dorsal to ventral regions offer optimization in load redistribution and energy dissipation. (Figure courtesy C.-Y. Sun and P.-Y. Chen.)

Summary

- Hydroxyapatite, $Ca_{10}(PO_4)_6(OH)_2$, is the major constituent of bone. On a volume basis, the constituents are 33–48% HAP, 32–44% organic (primarily collagen), and 15–25% water. Since HAP has a density of 3.16 g/cm^3, the mass percentage of HAP is ~60%.

- The hierarchical structure of bone starts with HAP nanocrystals of ~100 × 4 nm platelets embedded between collagen fibrils, at the bottom. The mineralized collagen fibrils organize themselves into fibers and these organize into lamellae. At the meso-level, we have a system of channels (Haversian and Volkmann's canals), around which cortical bone grows into circular lamellae that have alternating fiber orientations (along a helicoid), generating a Bouligand pattern.
- The elastic modulus of bone, E_b, parallel and perpendicular to the growth can be represented by the Voigt and Reuss models, respectively:

$$E_b = V_{ha}E_{ha} + V_cE_c,$$
$$1/E_b = V_{ha}/E_{ha} + V_c/E_c.$$

- The Jäger and Fratzl model is more sophisticated and has the following form:

$$\frac{E}{E_c} = \frac{d(l+a)}{ab} + \left(1 + \frac{l}{2a}\right) + \frac{\gamma(l-a)(l+a)}{4b^2} + \frac{\gamma a(l+a)}{2b(2b+a)}.$$

- Both the elastic modulus and the strength of bone are highly strain-rate dependent by virtue of collagen, which is a biopolymer and has a high strain-rate sensitivity. These dependences are expressed by the following equations.
  Ramberg–Osgood:

$$E = \frac{\sigma}{\varepsilon} = C\left(\frac{\dot{\varepsilon}}{\dot{\varepsilon}_0}\right)^d.$$

  Arruda–Boyce:

$$\tau_{AP} = \tau^*\left(\frac{\dot{\gamma}}{\dot{\gamma}_0}\right)^m.$$

- The toughness of bone, which considerably exceeds that of monolithic HAP, is due to mechanisms that are classified into intrinsic and extrinsic. The fracture toughness is highly anisotropic. The principal toughening mechanisms are as given in Table 7.2. The mineral content in antler (36 vol.%) is lower than that in bovine femur (47 vol.%).
- Teeth comprise an internal region called dentin and an external enamel layer. The structure of the tooth is designed to provide an external layer that is hard and an internal core (dentin) that is tougher. The hardness of the enamel layer is due to a high degree of mineralization. Enamel does not contain collagen. It is composed of hydroxyapatite rods woven into a fabric-like composite. These rods have a diameter of approximately 5 μm.
- When the fracture toughness is dependent on crack size, linear elastic fracture mechanics cannot be applied and one has to use other testing methods, such as the R curve. Skeletal and antler bone exhibit R-curve behavior, with the resistance to crack propagation increasing with crack size. This is due to the operation of extrinsic toughening mechanisms.

- The increase in toughness with increasing crack length decreases with age, evidencing that the principal extrinsic toughening mechanisms cease to operate in old bone. Hence a crack, once initiated, is more likely to stop in a young bone than in an old one.
- There are six types of fracture that are identifiable though the morphology: *greenstick fracture*; *fissured fracture*; *comminuted fracture*; *transverse fracture*; *oblique fracture*; *helical fracture*.
- The osteoderm, as the name implies, is a "bony skin." Osteoderms are found on both mammals (cingulata, armadillos) and reptiles (testudines, crocodilia). The osteoderms are attached to the animals by blood vessels, or, in the case of testudines (turtles and tortoises), are fused to the skeleton.
- The osteoderms in armadillo, turtle, and alligator share similar structural features: (1) the main constituent of osteoderm is bone, consisting of collagen fibers and hydroxyapatite minerals; (2) the osteoderms are connected by soft tissues or joints; (3) they are covered by keratinous layers on the outer surface; (4) they are sandwich composites with a dense cortex and a porous core.
- The deformation mechanisms in alligator osteoderms are as follows: (1) the flexibility provided by sutures and Sharpey's fibers can dissipate energy under small loads, similar to turtle shells; (2) deformation of the cellular foam interior absorbs a certain amount of energy without cortex cracking; (3) combination of the hard dorsal region and the compliant ventral region with graded mechanical properties offers optimization in load redistribution and energy absorbance.

Exercises

Exercise 7.1 If the removal torque for the orthodontic mini-implant shown in Fig. E7.1 is 17 N cm, calculate the shear strength of the interface. Use the dimensions shown in Fig. B7.5.

Exercise 7.2 See Fig. E7.2.

(a) Determine the parameters in the Ramberg–Osgood equation for antler bone. How do these values compare with cortical bone?
(b) Express the strength vs. strain rate as a mathematical equation.
(c) If the mineral is insensitive to strain rate, why is such a dependence observed?
(d) What happens to the toughness as strain rate is increased?

Exercise 7.3 A fracture toughness specimen made from antler bone is tested in three-point bending. It has a thickness of 2 mm and the toughness is 4.5 MPa m$^{1/2}$. Is this a valid plane-strain fracture-toughness value? Given: $E_b = 20$ GPa; $r_y = 0.1$ mm.

Exercise 7.4 The J integral was determined for a bone specimen. The value is equal to 0.5 kN/m. Determine an effective toughness. Assume $E_b = 20$ GPa.

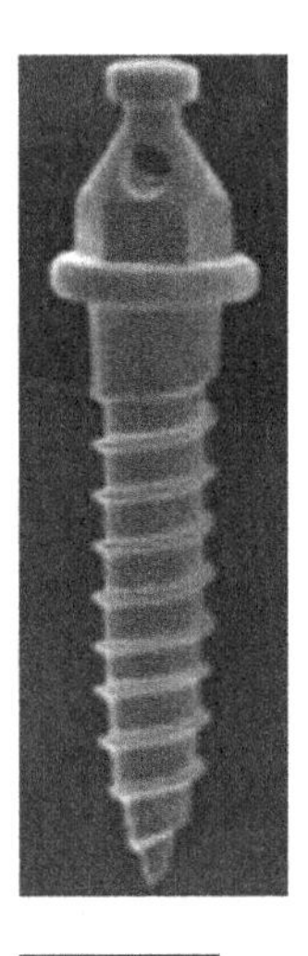

Figure E7.1.

Orthodontic mini-implant; dimensions given in Fig. B7.5.

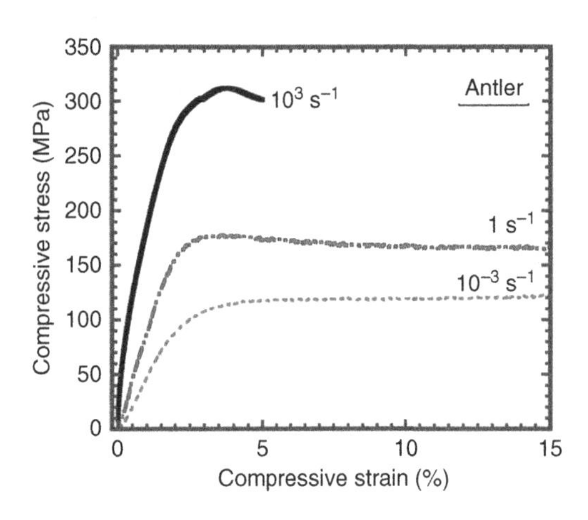

Figure E7.2.

Stress–strain curves for antler bone in compression at different strain rates. Strain rates are marked in the plot.

Exercise 7.5 The mineral content of antler bone (~35 vol.%) is lower than that of bone (~50 vol.%). What is the effect on the Young modulus? Calculate the Young modulus using the Voigt average. Given: E_{HAP} = 100 GPa; E_c –1 GPa.

Exercise 7.6 Calculate the elastic modulus for bone using the Katz equation. The fractions of HAP crystals are oriented to the tensile axis as in Table E7.6. Given: Poisson ratio for bone = 0.45; Poisson ratio for collagen = 1/3.

Table E7.6. Distribution of orientations of HAP crystals for bone

Fraction HAP crystal (%)	Orientation to tensile axis
30	15°
30	45°
30	90°
10	aligned

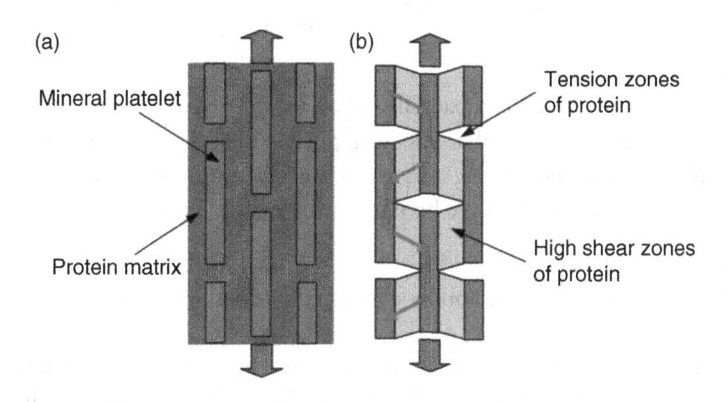

Figure E7.8.

Extension of bone and generation of breaks in collagen. (From Gao *et al.* (2003); copyright 2003, National Academy of Sciences, USA.)

Exercise 7.7 Bone characteristically consists of HAP platelets with a thickness of 1 nm and diameter of 50 nm. Assuming that the volume percentage is 50%, determine

(a) the Jäger–Fratzl degree of mineralization;
(b) the Young modulus.

Note: In Eqn. (7.6), $\gamma = [2(1 + \nu)]^{-1}$.

Exercise 7.8 Gao *et al.* (2003) applied the Jäger–Fratzl model (see Fig. E7.8) and obtained a maximum strength for bone using the Griffith equation. The mineral cracks as shown in Fig. E7.8.

(a) Calculate the maximum strength for the following two cases, using the Griffith equation. Given: (i) mineral platelets have thickness of 1 mm and diameter of 10 nm; (ii) mineral platelets have thickness of 1 nm and diameter of 50 nm.
Assume $\gamma_{surf} = 1$ J/m^2; $E_{HAP} = 100$ GPa.
(b) How do these values compare with the real strength of bone? Explain, using hierarchy of structure.

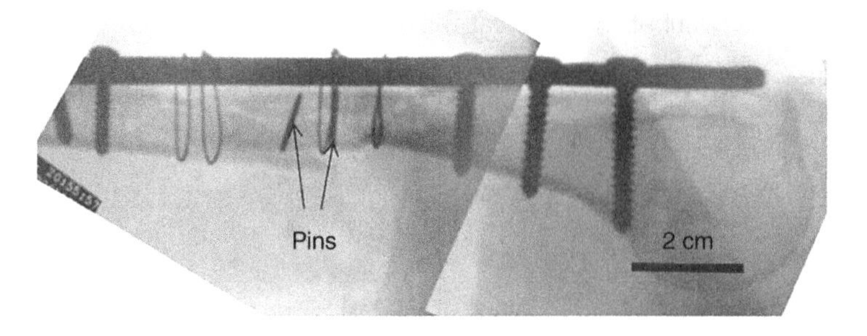

Figure E7.10.

Complex fracture in humerus repaired by stainless steel plate, screws, pins, and wires. (Courtesy Dr. J. F. Figueiró.)

Exercise 7.9 A skier takes a bad fall and his ski gets stuck in a tree while his body twists in the air. As a result, torsional forces are applied to his leg, which lead to the fracture of his tibia. Unfortunately, the bindings did not release due to improper maintenance. The ski is 2 m long, and the position of the boot is approximately in the center. Given:

tensile strength of bone = 80 MPa;
tibia dimensions (assume hollow cylinder): internal diameter, 24 mm; external diameter, 14 mm.

(a) Determine the type of fracture that is characteristic of this type of loading. Explain, using sketches. Hint: Take a piece of chalk and twist it.
(b) Calculate the torque applied to the bone at failure.
(c) What force has to be exerted by the tip of the ski to generate this torque?
(d) What is this type of fracture called in the medical profession? Comment on the appropriateness of the name.

Exercise 7.10 The stainless steel plate shown in Fig. E7.10 was inserted into a patient's humerus.

(a) Calculate the force required to bend it plastically (yield stress = 350 MPa).
(b) Calculate the force required to extract each of the screws. Make measurements on the X-ray photograph to calculate the necessary dimensions. Given: plate width = 1 cm.

Exercise 7.11 Calculate the force applied to the ball of the femur if a person jumps from a height of 2 m and decelerates the body by flexing the knees in such a manner that the speed of the body is reduced to zero after it has traveled 50 cm. Use sketches in your calculations.

Exercise 7.12 Stress shielding is a serious problem in some implants since bone remodels and the decrease of stress leads invariably to the weakening of the bone. Calculate the stresses in the femur head bone with and without an implant. Consider three cases:

titanium implant, E = 113 GPa;
stainless steel implant, E = 205 GPa;

carbon–polymer (polysulfone–PEEK) composite implant, $E = 30$ GPa.

Given: outer diameter of femur = 3 cm; inner diameter = 1.5 cm; $E_b = 20$ GPa.

Exercise 7.13 A bone has been drilled for insertion of screws. What is the effect on the tensile strength? Assume intact bone strength = 120 MPa.

Exercise 7.14 A carbon-fiber-reinforced polymer composite is being considered for the stem component in a total hip joint replacement. The goal is to create an implant with a Young modulus matched to the one of bone to minimize resorption due to stress shielding. The fibers are aligned with the stem axis. Calculate the volume fraction of carbon in the composite to achieve optimum match. Given:

$E_{\text{bone}} = 20$ GPa;
$E_{\text{carbon}} = 600$ GPa;
$E_{\text{polymer}} = 2$ GPa.

Exercise 7.15 What is Wolff's law? Briefly describe the process of bone remodeling using the examples of tennis players and archers.

Exercise 7.16 What are the three types of bone cells and their functions?

Exercise 7.17 What is the fracture resistance curve (R curve) behavior? List and explain at least three toughening mechanisms observed in bone at the micro- and meso-scales.

Exercise 7.18 Armadillos, crocodiles, and turtles all have armor. Explain the similarities and design strategies in natural armor materials.

Exercise 7.19. Estimate the life of a hip implant made of titanium if it contains initial flaws with length $2c = 200$ μm and a height $2a = 100$ μm. Assume that the forces applied on the artificial hip are as follows:

walking, $2W$;
running, $6W$,

where W is the weight of the person. The fatigue response of titanium is given in Fig. E7.19. The person is assumed to (a) walk three hours per day, (b) walk for three hours and jog for 20 minutes. Make all the necessary assumptions.

Exercise 7.20 A roof repairman fell and fractured the femur neck. After inspecting the fracture, the surgeon decided to install the contraption, shown in Fig. E7.20, called the Dynamic Hip Screw. Why didn't the surgeon opt for a total hip arthroplasty (even knowing that his fee would be higher for the latter)? Note: The individual experienced excruciating pain for several years and the plate was finally removed. He still carries it in his pocket as a reminder of the vagaries of life…

Exercise 7.21 An overactive graduate student decided to take sky diving lessons with results displayed in Fig. E7.21. The tibia fractured in two places. The fractures occurred

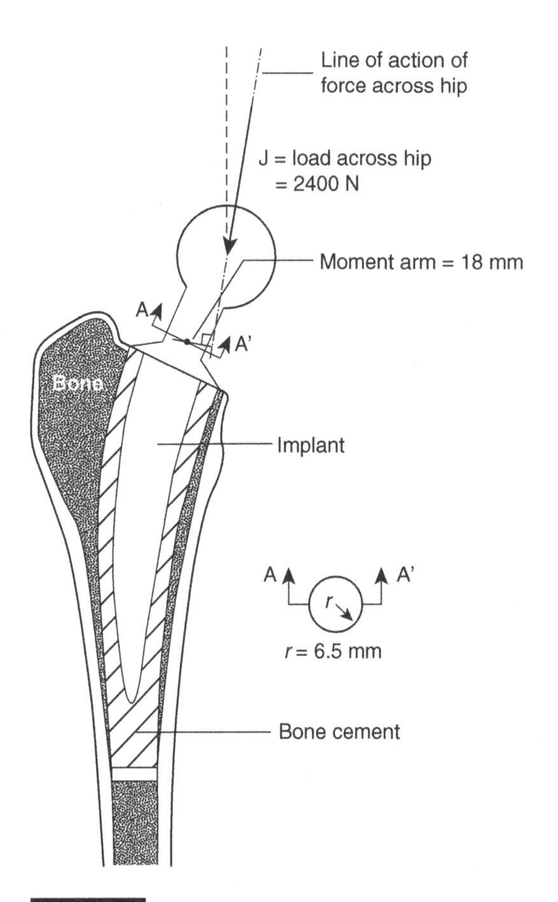

Figure E7.19.

Schematic representation of total hip replacement with dimensions. (From Meyers and Chawla (2009), Copyright © Cambridge University Press. Reprinted with permission of Cambridge University Press.)

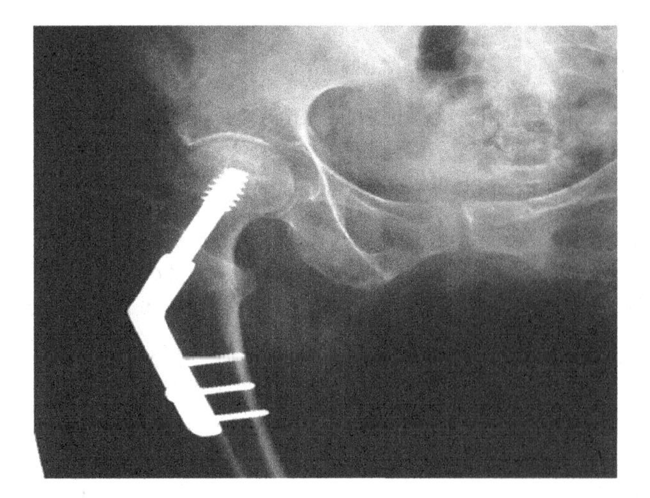

Figure E7.20.

Plate used to repair fractured femur neck (the narrow part below the ball). (Figure courtesy Dr. J. F. Figueiró.)

(a) 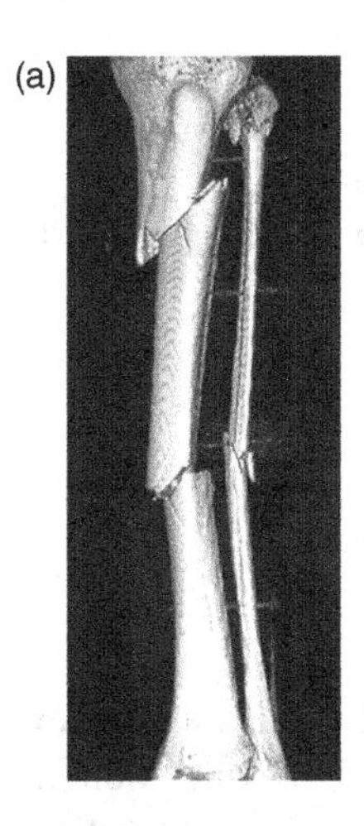 (b) 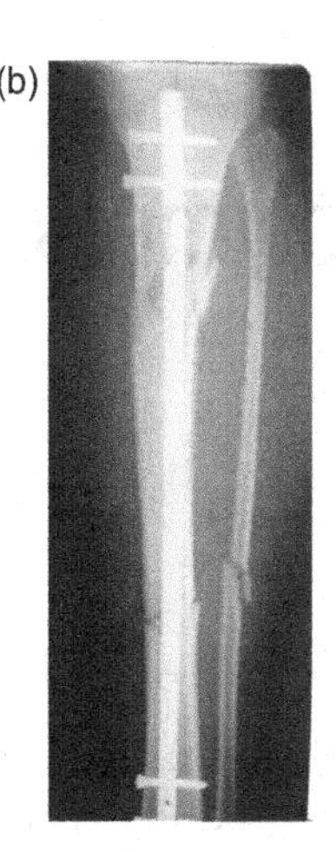

Figure E7.21.

(a) Fractured tibia and fibula; (b) intramedullary rod with fixation screws inserted to reduce tibia fracture. (Figure courtesy Dr. J. F. Figueiró.)

in both tibia and fibula at angles of ~45° with the bone axis. An intramedullary rod was inserted surgically to reduce the fracture.

(a) Why are the fractures at 45°?
(b) If the compressive strength of bone is ~150 MPa, what compressive load was applied to the bones? How many times body mass (70 kg) is this?
(c) What are the shear stresses on the intramedullary rod (at 45°) and in the screws? Which would fail first? Assume that the student weighs a hefty 90 kg and that the dynamic load equals three times her weight.

8 Biological polymers and polymer composites

Introduction

Chapters 6 and 7 covered the structure, at different hierarchical levels, of biological mineral-based composites (see the Ashby–Wegst classification in Fig. 2.11), and explained the mechanical properties in terms of it. There are also a number of biological composites that do not contain minerals, or in which minerals appear only in small proportions. This is what will be studied in this chapter. We differentiate Chapter 8 by covering the biological polymers that have extensive "stretchability" in Chapter 9.

We start with the very important components, tendons and ligaments that join bone and muscle, and bone and bone, respectively. They are primarily composed of collagen. An extraordinary biological material, spider silk, is introduced next. The strength level can reach and exceed 1 GPa. If we normalize this by dividing it by the density (~1 g/cm^3), we obtain a material that is considerably stronger than our strongest steels (~3 GPa, with density of 7.8 g/cm^3). Molecular dynamics computations explain how this high strength is accomplished with the weak hydrogen bond.

The arthropod exoskeleton is composed of chitin and, in the case of marine arthropods (crustaceans), chitin and minerals. The structure is presented in this chapter, for consistency. Next comes the very important keratin, which appears in nails, hair, wool, and bird feathers, beaks, and claws. We also present keratin-based materials such as hooves, nails, and the scales of the armored mammal, the pangolin.

Fish scales are layered composites of collagen, and their structures are presented. There are different classes of scales, each with its unique structure and mechanical performance. Their function is to provide protection while allowing the fish to retain flexibility. They can be considered as flexible dermal armor.

The beaks of squid have a graded hardness, which is produced by an ingenious process of tanning: no minerals are involved in their structure, and high values are obtained. The mandibles of invertebrates comprise another example of high hardness that is reached by peculiar ways. Finally, we discuss natural lignocellulosic fibers, which are finding a wealth of technological applications because of their high strength: sisal, bamboo, coconut fiber (known as coir), and piassava.

8.1 Tendons and ligaments

The function of a ligament is to connect bone to another bone. Tendon connects bone to muscle. The hierarchical structure of ligament is similar to that of tendon and is based on collagen fibers. The mechanical behavior of tendon and ligament are similar. Figure 8.1(a) shows the stress–strain curve of *ligamentum flavum* (the ligament between the lumbar vertebrae) from pig (Sikoryn and Hukins, 1990). The ligament has a high ductility and is able to transmit the stress until its ultimate level. Sikoryn and Hukins (1990) obtained a maximum stress of 2.6 MPa at a low strain rate and of 3.0 MPa at a high strain rate. The ligament strength seems to be independent of the strain rate. A small strain rate effect has been observed at the

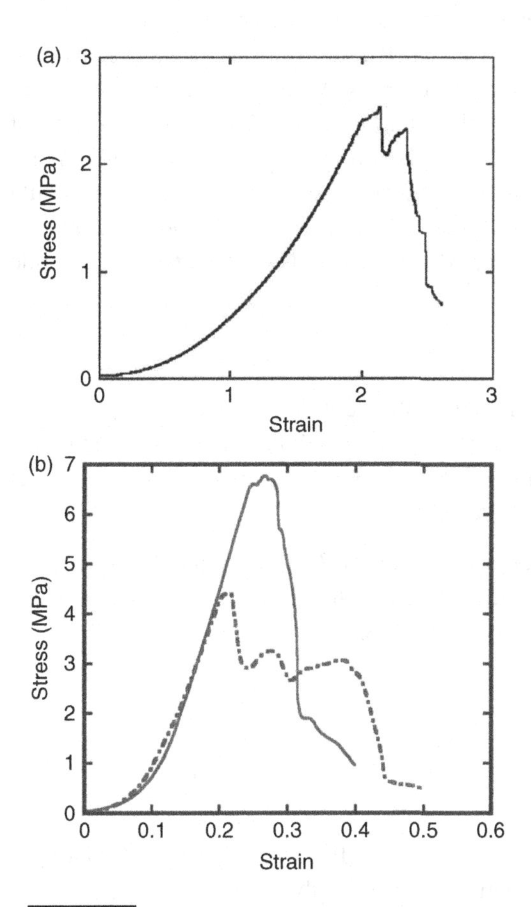

Figure 8.1.

Stress–strain curve of ligament from (a) pig (*ligamentum flavum*). (Taken from Sikoryn and Hukins (1990), with kind permission from John Wiley & Sons, Inc.) (b) Human (inferior glenohumeral ligament). Solid and dashed lines represent two different tests. (Taken from Bigliana *et al.* (1992), with kind permission from John Wiley & Sons, Inc.)

tendon of horse (Herrick, Kingsbury, and Lou, 1978). Nachemson and Evans (1968) reported that the maximum stress for human *ligamentum flavum* had a value 4.4 ± 3.6 MPa. Although the maximum stress for *ligamentum flavum* in pigs and humans is on the same order, the strain at maximum stress is not. The maximum stress and strain decrease significantly with age for humans. For young subjects (13 years), the maximum stress was 10 MPa, decreasing to 2 MPa for older subjects (79 years). The strain at maximum stress is about 2 for pig *ligamentum flavum*, as shown in Fig. 8.1 (a). For human *ligamentum flavum*, the strain values are lower (0.5 ± 0.2). It should be mentioned that Chazal *et al.* (1985) reported a much higher maximum stress of 15 ± 5 MPa and much lower strain at maximum stress of 0.21 ± 0.04. The *ligamentum flavum* is located between laminae of adjacent vertebrae. These values are very (and suspiciously) low, and are indeed not representative of many tendons and ligaments. Rogers *et al.* (1990) report ultimate tensile stresses in ovine anterior cruciate ligament ranging from 60 ± 3 to 123 ± 15 MPa, ultimate specific extension from 37 ± 7 to 93 ± 20%. There are also reports on tendons (Wren *et al.*, 2001) that have strengths in the range of 71 MPa (at a velocity of 1 mm/s) to 86 MPa (at a velocity of 10 mm/s).

Figure 8.1(b) shows the stress–strain curves for the inferior glenohumeral ligament (from shoulders) from human (Bigliana *et al.*, 1992). Two different failure mechanisms were observed: the abrupt failure which occurs at the bone–ligament interface, and the step-like failure which occurs within ligaments (Bulter *et al.*, 1984; Pollock *et al.*, 1990; Bigliana *et al.*, 1992).

Example 8.1 Determine the safety factor built into the Achilles tendon of a person weighing 80 kg, assuming a cross-sectional area of 1.5 cm^2, if the person can jump up to a height of 1 m, then land with a deceleration time of 0.3 s. Assume that the tensile strength of the tendon is 60 MPa. Dimensions are given in Fig. 8.2.

Solution We first calculate the relationship between T, the tension in the Achilles tendon, and F, the force exerted on the ground. We assume that the person is standing on the ball of the foot.

Setting the sum of moments equal to zero,

$$\sum M_B = 0,$$

$$BC \times F - AB \times T = 0,$$

$$T = \frac{BC \times F}{AB}.$$

The forces and distances are defined in Fig. 8.2. We now calculate F for the static and dynamic cases. For the static case, we simply have

$$F_s = 80 \times 9.8 = 784\,\text{N}.$$

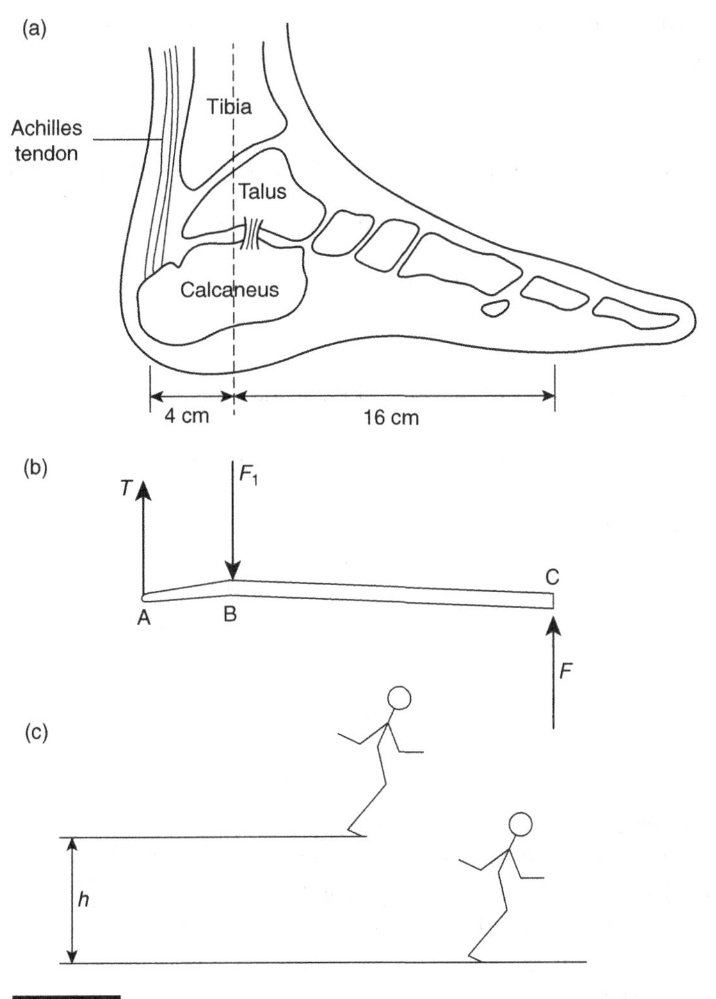

Figure 8.2.

(a) Structure of foot with Achilles tendon shown; (b) force T on tendon; (c) jump from a height h. (From Meyers and Chawla (2009), Copyright © Cambridge University Press. Reprinted with permission of Cambridge University Press.)

For the dynamic case, we have to consider the kinetic energy gained by the person when jumping down from a height of 1 m. The potential energy is converted into kinetic energy:

$$mgh = \frac{1}{2}mv^2.$$

The velocity is given by

$$v = (2gh)^{1/2} = 4.43 \text{ m/s}.$$

In order to find the dynamic force, F_d, we set the impulse equal to the change in momentum:

$$mv - m \times 0 = F_d \times t.$$

The deceleration time is given: $t = 0.3$ s. Thus, $F_d = 707$ N. The total force is given by

$$F = F_s + F_d = 1492\,\text{N}.$$

From Fig. 8.2, we obtain values of AB and BC, so

$$T = 5968\,\text{N}.$$

Assuming a round section, the area of the tendon

$$A = \frac{\pi}{4}(1.5 \times 10^{-2})^2 = 1.7 \times 10^{-4}\,\text{m}^2.$$

Thus, the stress

$$\sigma = \frac{T}{A} = 35.1\,\text{MPa}.$$

The safety factor

$$SF = \frac{60}{35.5} = 1.7.$$

This is indeed a small number, and a weakened Achilles tendon could easily rupture. Indeed, this happened to one of the co-authors (MAM) while playing soccer (his last game). The tendon was operated upon and reconnected through stitches. The foot was immobilized in the stretched position for four months, enabling the tendon to repair itself. Interestingly, the operated tendon now has a cross-section twice as large as the other one. Hence, nature somehow remembers the trauma and overcorrects for it. The same thing happens in bones. The healed portion becomes stronger than the original bone.

8.2 Spider and other silks

Silk is composed of two proteins: fibroin (tough strands) and sericin (gummy glue). Figures 8.3(a) and (b) show silk with and without the sericin layer, respectively. The mechanical properties (strength and maximum elongation) can vary widely, depending of the application intended by the animal. For instance, spiders produce two types of silk. The first is the dragline, used in the radial components of the web. This is the structural component, and has high tensile strength (0.6–1.1 GPa) and a strain at failure of about 6%. The tangential components, called spiral, are intended to capture prey, and are "soft" and "sticky." Their functions are:

(a)

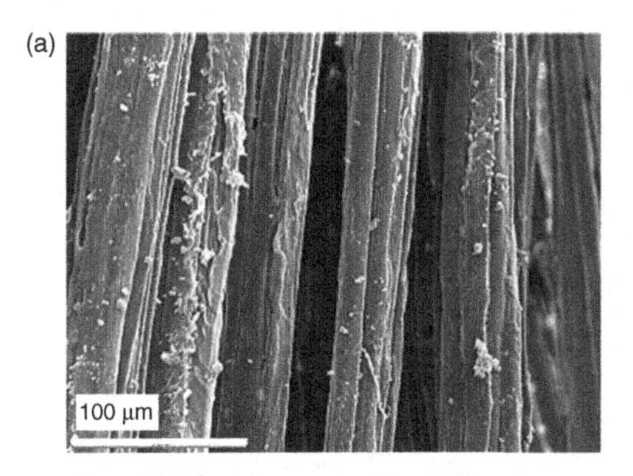

(b)

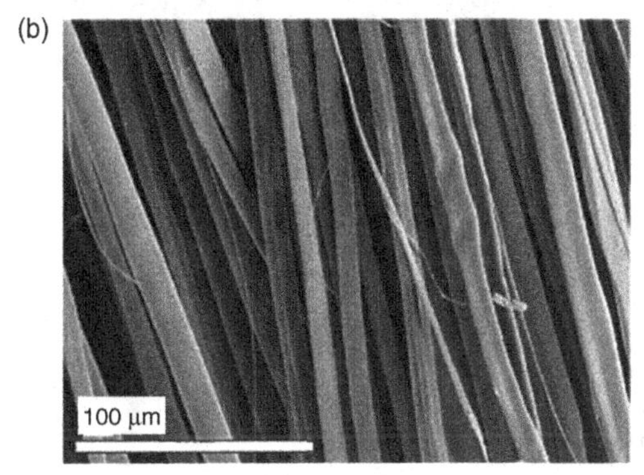

Figure 8.3.

SEM micrograph of *B. mori* silk fibers (a) with and (b) without the gum-like sericin proteins. (Reprinted from Altman *et al.* (2003), with permission from Elsevier.)

(a) the support, or safety dragline, silks, which create the spokes and frame of a web as well as the lines from which a spider hangs, and
(b) the viscid silks, which are used as the prey-catching spirals of a web.

The silks produced by orb-web-spinning spiders exhibit mechanical properties that are superior to almost all natural and man-made materials. Although orb-webs evolved over 180 million years ago (Selden, 1989; Shear *et al.*, 1989), the investigation of the physical structure and properties of spider silk began only in the late 1970s, starting with Denny (1976). Since then, the mechanical properties of various spider silks have been well described by Gosline, DeMont, and Denny (1986), Kaplan *et al.* (1991, 1994), Vollrath (2000), and many others. The properties of *Araneus diadematus* and *Bombyx mori* silk in comparison to various other materials are listed in Table 8.1 (Cunniff *et al.*, 1944; Pins *et al.*, 1977; Gosline *et al.*, 1999; Perez-Rigueiro *et al.*, 2000). The removal of sericin increases the strength of the silk.

Spiders are capable of producing a variety of silks, each with distinct functions and mechanical characteristics. Support and dragline silk is spun from the major ampullate (MA) gland of the spider and is often referred to as the MA silk. Because of its function,

Table 8.1. Tensile mechanical properties of spider silks and other materials

Material	Stiffness (GPa)	Strength (GPa)	Extensibility	Toughness (MJ m^{-3})	Hysteresis (%)	Source
Natural fibers						
Araneus MA silk	10	1.1	0.27	160	65	Gosline *et al.* (1999)
Araneus viscid silk	0.003	0.5	2.7	150	65	Gosline *et al.* (1999)
Nephila clavipes silk	11–13	0.88–0.97	0.17–0.18			Cunniff *et al.* (1944)
Bombyx mori cocoon silk	7	0.6	0.18	70		Gosline *et al.* (1999)
B. mori silk (w/ sericin)	5–12	0.5	0.19			Perez-Rigueiro *et al.* (2000)
B. mori silk (w/o sericin)	15–17	0.61–0.69	0.4–0.16			Perez-Rigueiro *et al.* (2000)
B. mori silk	10	0.74	0.2			Cunniff *et al.* (1944)
Rat-tail collagen X-linked	0.4–0.8	0.047–0.072	0.12–0.16			Pins *et al.* (1977)
Tendon collagen	1.5	0.15	0.12	7.5	7	
Bone	20	0.16	0.03	4		
Wool, 100%RH	0.5	0.2	0.5	60		Gosline *et al.* (1999)
Elastin	0.001	0.002	1.5	2	10	
Resilin	0.002	0.003	1.9	4	6	
Synthetic materials						
Synthetic rubber	0.001	0.05	8.5	100		
Nylon fiber	5	0.95	0.18	80		
Kevlar 49 fiber	130	3.6	0.027	50		Gosline *et al.* (1999)
Carbon fiber	300	4	0.013	25		
High-tensile steel	200	1.5	0.008	6		

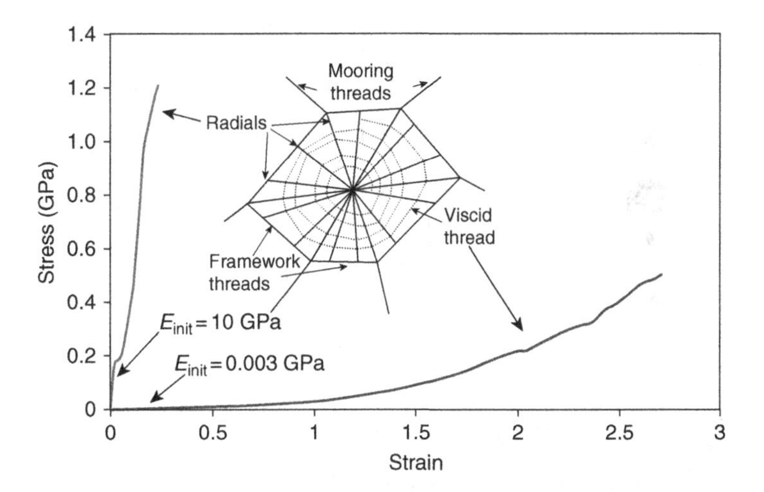

Figure 8.4.

Stress–strain curves for MA gland silk and viscid silks from the spider *Araneus diadematus*. (Reproduced based on Gosline *et al.* (1999).)

MA silk is much stiffer and stronger then the viscid silk fibers which compose the spirals of the web. It can have an elastic modulus on the order of 10 GPa, and a maximum strength of over 1 GPa (Kaplan *et al.*, 1991; Gosline *et al.*, 1999; Vollrath, 2000). On the other hand, viscid silk exhibits remarkable extensibility, withstanding up to 500% strain before failure in some species (Vollrath, 2000). Its elastic modulus is over three orders of magnitude lower then MA silk yet both have surprisingly similar and incredible toughness. The stress–strain curves of both silks are presented in Fig. 8.4 (Gosline *et al.*, 1999); a dramatic difference in properties can be seen.

It seems intuitive that the function of the web is to absorb the kinetic energy of a flying insect then relax without an elastic recoil that could reject the prey from its entanglement (Gosline, Denny, and DeMont, 1984). The microstructure of the MA silk provides a support framework which is able to produce the required viscoelastic response. It can be thought of as a semi-crystalline material, with an amorphous region composed of disordered protein chains connected to protein crystals (Wainwright *et al.*, 1976; Kaplan *et al.*, 1994; Gosline *et al.*, 1999). These crystals form from layers of anti-parallel amino acid sequences, which are known as β-pleated sheets. Figure 8.5 provides an illustration of the microstructure consisting of amorphous proteins cross-linked through crystalline blocks. The silk is spun from multiple fibers of this semi-crystalline material.

The combination of a hard crystalline region and an elastic amorphous region results in a viscoelastic material with a notable strain-rate sensitivity. Increased strain rates result in increased yield stress, tensile strength, and breaking strain (Elices *et al.*, 2005). High strain rates reflect the natural condition under which the material is loaded, i.e. the impact of a fast-moving insect. When the material is loaded past its yield stress, a region of strain hardening is observed (Garrido *et al.*, 2002). The plot in Fig. 8.6 shows this strain-hardening region under which most of the energy

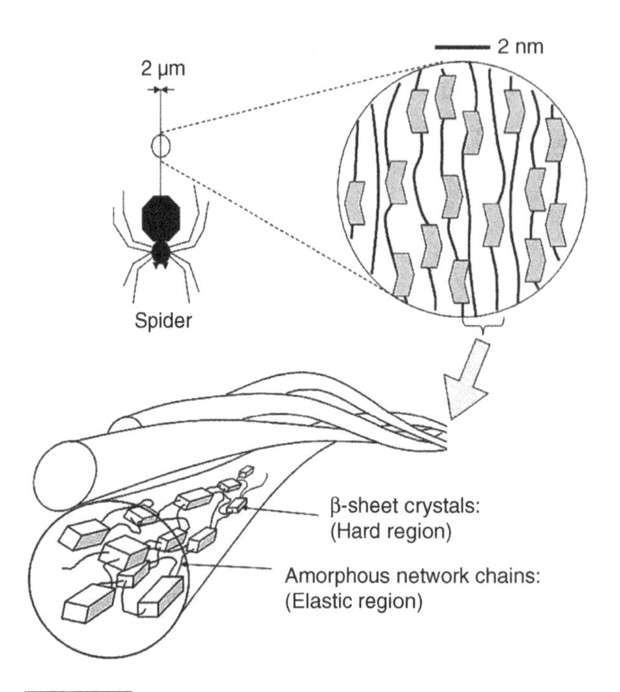

Figure 8.5.

Schematic representation of the hierarchical microstructure of spider silk.

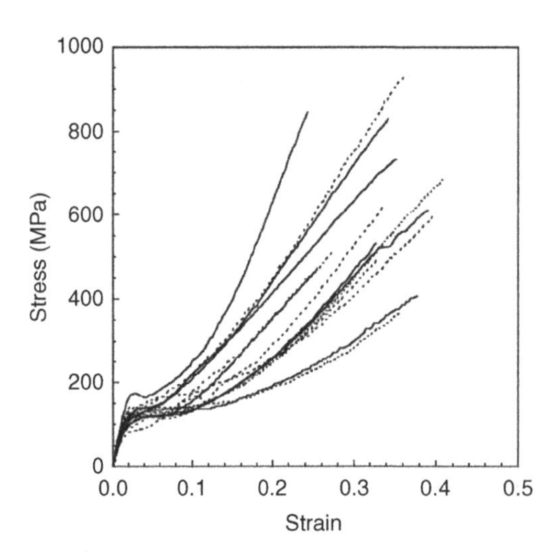

Figure 8.6.

Strain hardening in MA dragline spider silk. (Reprinted from Garrido *et al.* (2002), with permission from Elsevier.)

absorption occurs. This may be due to the rearrangement of the β-pleated sheet blocks in the amorphous network. The different curves represent different specimens; the figure also illustrates the variability between different specimens, a characteristic of biological materials.

8.2.1 Adhesive in spider web

Orb-weaving spiders have evolved to produce adhesive droplets on their webs to capture prey. The mechanical design of the adhesive coating on the webs has been revealed by Sahni, Blackledge, and Dhinojwala (2010) and Lee (2010). The adhesive droplets consist mainly of glycoprotein, along with other viscous small molecules and salts. Figure 8.7(a) shows the components of the capture thread of the spider *Larinioides cornutus*: (1) viscous coat of salts, (2) glycoprotein granule, and (3) axial silk fiber. Silk threads were immobilized on a glass surface, and a conical glass probe 10 μm in diameter was brought into contact with single glue droplets (Fig. 8.7(b)) then retracted at constant speeds (Fig. 8.7(c)). The critical pull-off forces at varying stretching rates were measured. The force–extension responses showed a high pull-off rate dependency (Fig. 8.7(d)). Critical pull-off forces increased from 60 μN at a rate of 1 μm/s to ~400 μN at 100 μm/s. Glycoprotein behaves as a viscoelastic solid, a property that is important in enhancing the adhesion of these almost invisible capture-silk threads. At high extension rates, corresponding to the impact of fast-flying insects, the adhesive forces of the glue droplets are dramatically enhanced due to the high-strain-rate effect, providing maximum adhesion to capture prey effectively. At low extension rates, similar to the movements of trapped insects, the glue droplets behave like an elastic rubber band, keeping the prey from escaping. The multifunctional design observed in the spider web adhesive droplets is providing inspiration of novel synthetic adhesives.

8.2.2 Molecular dynamics predictions

The response of silk and spider thread is fascinating. As one of the toughest materials known, silk also has high tensile strength and extensibility. It is composed of β-sheets (10–15 vol.%) of nanocrystals (which consist of highly conserved poly-(Gly-Ala) and poly-Ala domains) embedded in a disordered matrix (Keten *et al.*, 2010). Figure 8.8 shows the J-shape stress–strain curve and molecular configurations for the crystalline domains in silkworm (*Bombyx mori*) silk (Keten *et al.*, 2010). Similar to collagen, the low-stress region corresponds to uncoiling and straightening of the protein strands. This region is followed by entropic unfolding of the amorphous strands and then stiffening from the β-sheets. Despite the high strength, the major molecular interactions in the β-sheets are weak hydrogen bonds. Molecular dynamics (MD) simulations, shown in Fig. 8.9, illustrate an energy dissipative stick-slip shearing of the hydrogen bonds during failure of the

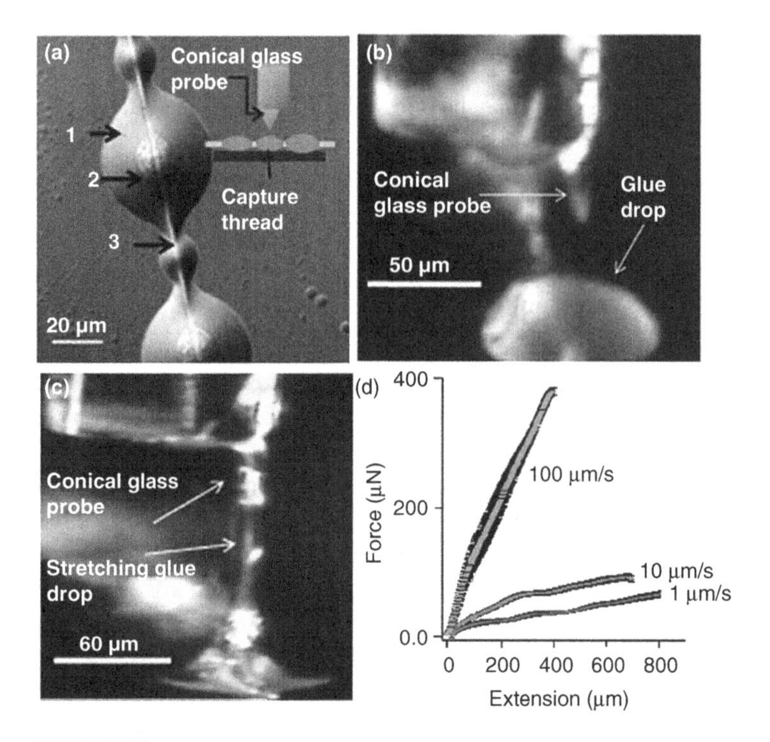

Figure 8.7.

(a) The components of the capture thread of the spider *L. cornutus*: (1) viscous coat; (2) glycoprotein granule; (3) axial thread. Inset shows a schematic of the single-drop-pulling experiments, wherein single glue drops of a capture thread were stretched using a conical glass probe while force responses were recorded. (b) Conical glass probe approaching a single glue drop. (c) Stretching of a single glue drop using the conical glass probe. (d) Force responses when single glue drops were stretched at different rates until separation from the glass probe. (Reprinted by permission from Macmillan Publishers Ltd.: *Nature Communications* (Sahni *et al.*, 2010), copyright 2010.)

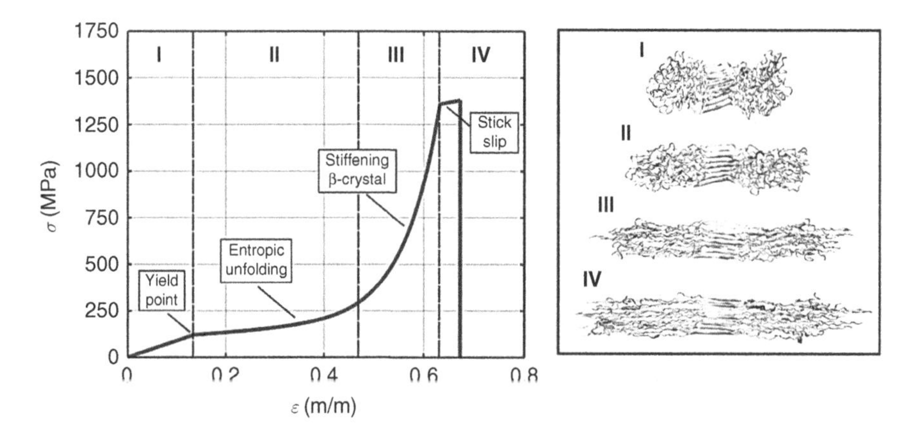

Figure 8.8.

Stretching of dragline spider silk and molecular schematic of the protein, fibroin. At low stress levels (stage II), entropic effects dominate (straightening of amorphous strands), and at higher levels the crystalline parts sustain the load (stages III and IV). (Reprinted by permission from Macmillan Publishers Ltd.: *Nature Materials* (Keten *et al.*, 2010), copyright 2010.)

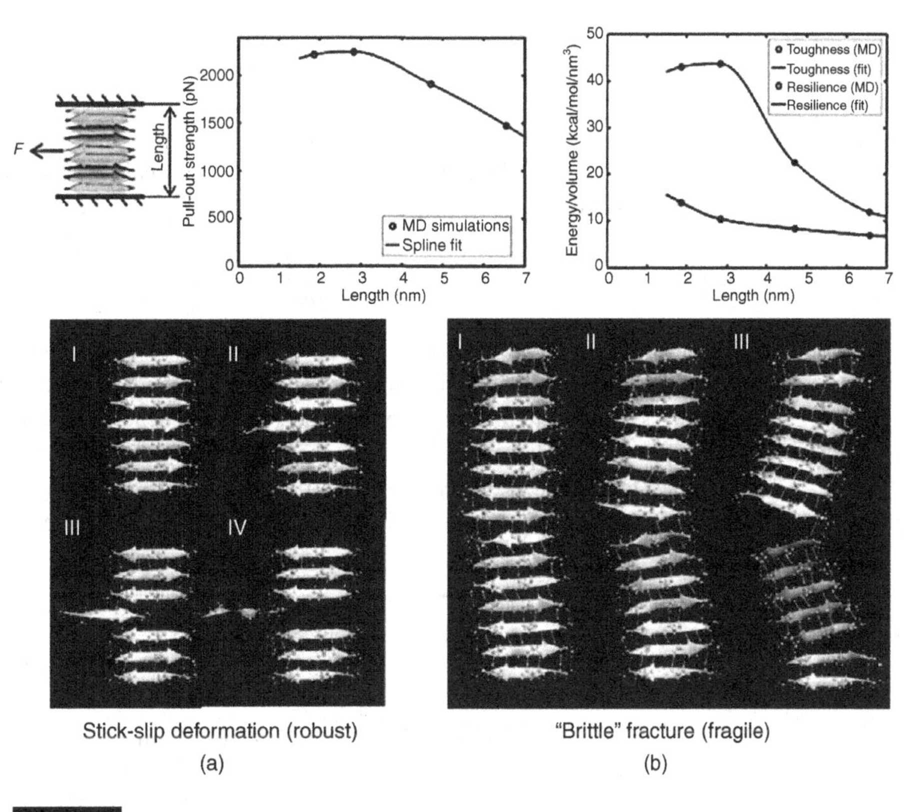

Figure 8.9.

β-sheet crystals being sheared; in short crystal, layers can be pulled out (a), whereas in longer crystal bending occurs (b). (Reprinted by permission from Macmillan Publishers Ltd.: *Nature Materials* (Keten *et al.*, 2010), copyright 2010.)

β-sheets. For a stack having a height $L \leq 3$ nm (left-hand side of Fig. 8.9), the shear stresses are more significant than the flexure stresses and the hydrogen bonds contribute to the high strength obtained (1.5 GPa). However, if the stack of β-sheets is too high (right-hand side of Fig. 8.9), it undergoes bending with tensile separation between adjacent sheets. The nano-scale dimension of the β-sheets allows for a ductile instead of brittle failure, resulting in the high toughness values of silk. Thus, size effects significantly affect the mechanical response, which changes the deformation characteristics of the weak hydrogen bonds. Keten *et al.* (2010) also matched their MD simulations to a simple mechanics of beams analysis and introduced a nondimensional term $s(L)$ that denotes the contribution of shear. This term is related to the ratio between the bending rigidity D_B and the shear rigidity D_T:

$$s(L) = \frac{3D_B}{L^2 D_T}, \tag{8.1}$$

where L is the length of the nanocrystal. The rigidities are, for a rectangular cross-section (b and h are lateral dimensions of the β-sheet stack in Fig. 8.9(b)):

$$D_B = EI = \frac{Ebh^3}{12},$$
$$D_T = GA_S = Gbh, \tag{8.2}$$

where I is the moment of inertia, E is the Young modulus, G is the shear modulus, and A_S is the cross-sectional area. Substituting Eqns. (8.2) into (8.1), we obtain

$$s(L) = \frac{Eh^2}{4GL^2} \sim \frac{h^2}{2L^2}. \tag{8.3}$$

When the beam is short, $s(L)$ is large and shear dominates. This is a well-known fact in the mechanics of beams. For this situation, all the hydrogen bonds in a sheet undergo the same shear stress. As the length L increases, $s(L)$ decreases and bending becomes dominant. The hydrogen bonds undergo tension, with the ones farther from the neutral axis having the highest tensile forces. Keten *et al.* (2010) argue that this will make the crystal fracture. Thus, the short nanocrystals can resist applied loads that create stresses higher than 1 GPa.

Example 8.2 (a) Using the methodology of Keten *et al.* (2010), calculate the bending and shear rigidities for silk crystals. Given: $G = 4.6$ GPa; $E = 22.6$ GPa; $b = 5$ nm; $h = 7$ nm;

$$D_B = \frac{Ebh^3}{12} = 3.23 \times 10^{-24}\ \text{Nm}^2;$$
$$D_T = Gbh = 1.61 \times 10^{-7}\ \text{N}.$$

(b) Find the critical length L below which the molecules are pulled out by shear without flexural fracture of the stack ($s > 1$);

$$s(L) = \frac{3D_B}{L^2 D_T} = \frac{3 \times 3.23 \times 10^{-24}}{1.61 \times 10^{-7} L^2} = 1$$

So, $L = 7.76$ nm.

8.3 Arthropod exoskeletons

Arthropods are the largest phylum of animals, including the trilobites, chelicerates (spiders, mites, and scorpions), mariapods (millipedes and centipedes), hexapods (insects), and crustaceans (crabs, shrimps, lobsters, and others). All arthropods are covered by a rigid exoskeleton, which is periodically shed as the animal grows. The arthropod exoskeleton is multifunctional: it not only supports the body and resists mechanical loads, but also provides environmental protection and resistance to desiccation (Neville *et al.*, 1975; Vincent, 1991, 2002; Vincent and Wegst, 2004).

The exoskeleton (cuticle) of arthropods is a laminated biological composite composed of chitin fibers embedded in a protein matrix. In the crustacean species, the exoskeleton shows a high degree of mineralization, typically calcium carbonate in calcite form with a small amount of amorphous $CaCO_3$ deposited within the chitin–protein matrix. However, the mineral phase is often absent in the exoskeletons of insects.

8.3.1 Crustaceans

The arthropod exoskeleton is a multilayered structure that can be observed under an optical microscope, sometimes even with the naked eye. The outermost layer is the epicuticle, a thin waxy layer which is the main waterproofing barrier. Beneath the epicuticle is the procuticle, the main structural part, which is primarily designed to resist mechanical loads. The procuticle is further divided into two parts: an exocuticle and an endocuticle. The mesostructure of the lobster exoskeleton is shown in Fig. 8.10(a). The exocuticle (outer layer) and endocuticle (inner layer) are similar in structure and composition. The difference between exocuticle and endocuticle is that the exocuticle is stacked more densely, the endocuticle being sparsely stacked. The spacing between layers varies from species to species. Generally, the layer spacing in the endocuticle is about three times thicker than that in the exocuticle. The hardness and stiffness decrease with distance from the external surface (Fig. 8.10(b)). This is characteristic of many biological structures: a hard surface is backed by a more compliant but tougher foundation. This is due to the dense twisted plywood structure in exocuticle compared to the coarse twisted plywood structure in endocuticle.

Raabe and co-workers extensively studied the structure and mechanical properties of the exoskeleton of American lobster, *Homarus americanus* (Raabe *et al.*, 2005a, b, c, 2006; Sachs, Fabritius, and Raabe, 2006a, b; Romano, Fabritius, and Raabe, 2007). They observed the unique honeycomb-type arrangement of the chitin–protein fibers surrounding pore canals, as shown in Fig. 8.11. The honeycomb structure of lobster is quite stable after chemical and heat treatments, and could be made suitable for medical applications, for example tissue engineering (Romano *et al.*, 2007). The through-thickness mechanical properties of the American lobster exoskeleton were studied using both micro- and nanoindentation techniques (Raabe *et al.*, 2005c; Sachs *et al.*, 2006a). The flat fracture surface of *Homarus americanus* in Fig. 8.11 shows clear evidence of the brittle structure. The hierarchical structure can be seen in the honeycomb-type arrangement of the chitin–protein fibers.

The most characteristic feature of arthropod exoskeletons is their well-defined hierarchical organization, which reveals different structural levels, as shown in Fig. 2.3 for crab. At the molecular level is the polysaccharide chitin. Several chitin molecules arrange in an anti-parallel fashion forming α-chitin crystals. The next structure level consists of 18–25 such molecules, wrapped by proteins, forming nanofibrils of about 2–5 nm in diameter and about 300 nm in length. These nanofibrils further assemble into bundles of

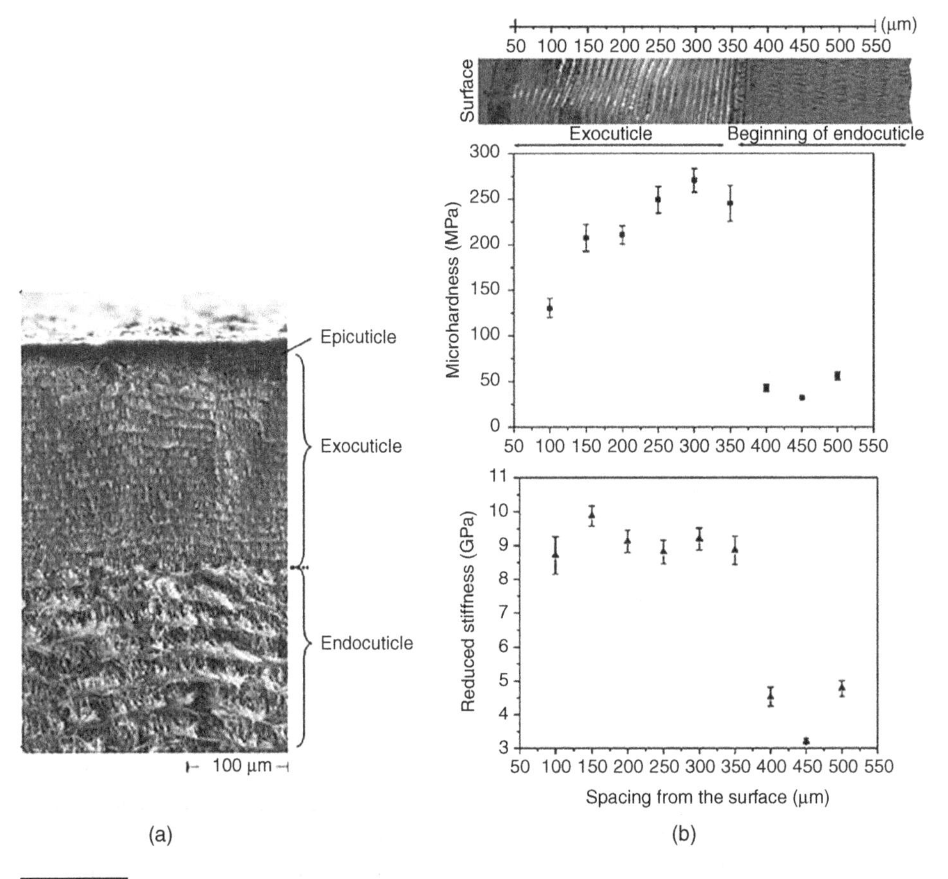

Figure 8.10.

(a) A through-thickness SEM micrograph shows that the exocuticle has higher stacking density than the endocuticle. (b) The hardness and the reduced stiffness through the thickness of the lobster exoskeleton. (Reprinted from Raabe *et al.* (2005c), with permission from Elsevier.)

fibers, about 50–300 nm in diameter. The fibers then arrange parallel to each other and form horizontal planes.

These planes are stacked in a helicoid fashion, creating a twisted plywood or Bouligand structure (Bouligand, 1970, 1972; Giraud-Guille, 1984, 1990, 1998). A stack of layers that have completed a 180° rotation is referred to a Bouligand or twisted plywood layer, which further forms the exocuticle and endocuticle. The Bouligand (helical stacking) arrangement provides structural strength that is in-plane isotropic (*xy* plane) in spite of the anisotropic nature of the individual fiber bundles. In crustaceans, the minerals are mostly in the form of crystalline $CaCO_3$, deposited within the chitin–protein matrix (Lowenstam, 1981; Lowenstam and Weiner, 1989; Giraud-Guille and Bouligand, 1995; Mann, 2001). The highly mineralized Bouligand arrangement provides strength in the in-plane (or in-surface) direction, and can be considered as the hard or brittle component.

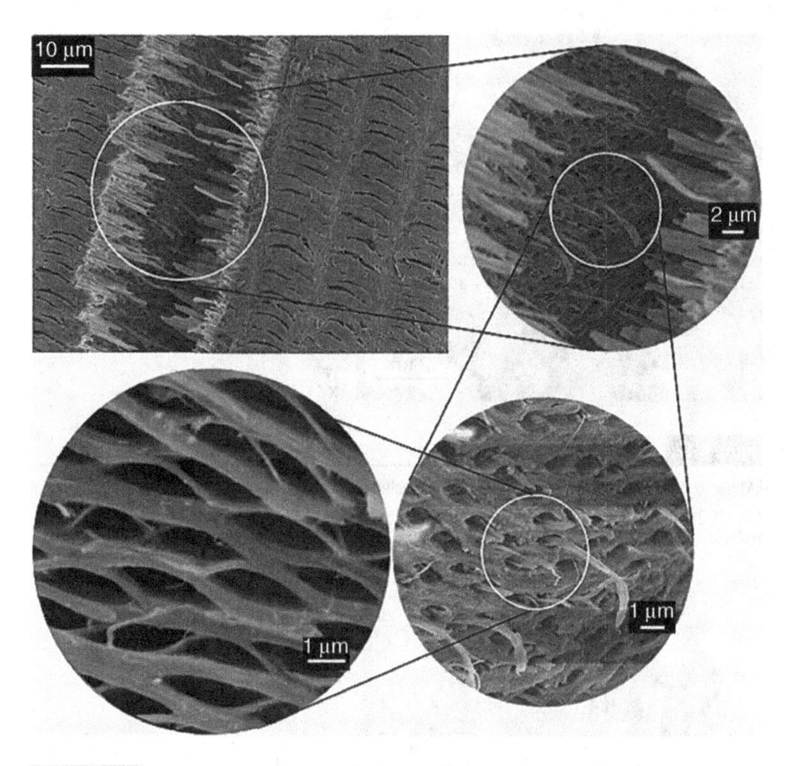

Figure 8.11.

Scanning electron micrographs taken from fractured specimens of lobster *Homarus americanus* showing the hierarchical structure. The honeycomb-type arrangement of the chitin–protein fibers is visible at higher magnifications. (Reprinted from Raabe *et al.* (2005b), with permission from Elsevier.)

In the vertical direction (*z*-direction), the crab exoskeleton in Fig. 8.12, shows well-developed, high-density pore canals (spacing between canals is ~2 μm) containing tubules penetrating through the exoskeleton. Greater detail is shown in Fig. 8.13(a). These tubules are hollow and have a flattened configuration (~2–3 μm wide) that twists in a helical fashion. They play an important role, not only in the transport of ions during the mineralization of the new exoskeleton after the animals molt, but also in the enhancement of mechanical properties in the *z*-direction. These tubules are organic materials and can be considered as soft or ductile components which stitch the fibrous layers together and provide toughness to the structure. A region where separation was initiated by tensile tractions is shown in Fig. 8.13(a). The tubules are stretched and fail in a ductile mode. The neck cross-section is reduced to a small fraction of the original thickness. Figure 8.13(b) shows the top view of the fracture surface (*xy* plane); the tubules are the protruding lighter segments. Many of them are curled, which is evidence of their softness and ductility. It is thought that this ductile component helps to "stitch" together the brittle bundles arranged in the Bouligand pattern and to provide toughness to the structure. It also undoubtedly plays a role in keeping the exoskeleton

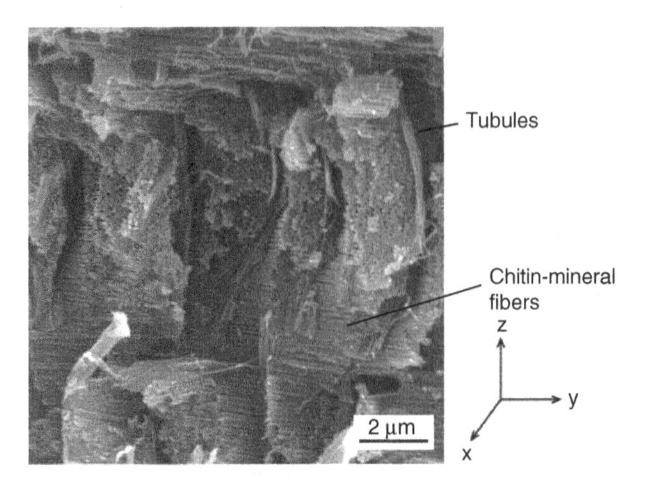

Figure 8.12.

SEM micrograph of fracture surface showing the twisted plywood (Bouligand) structure of crab exoskeleton. (Reprinted from Chen *et al.* (2008b), with permission from Elsevier.)

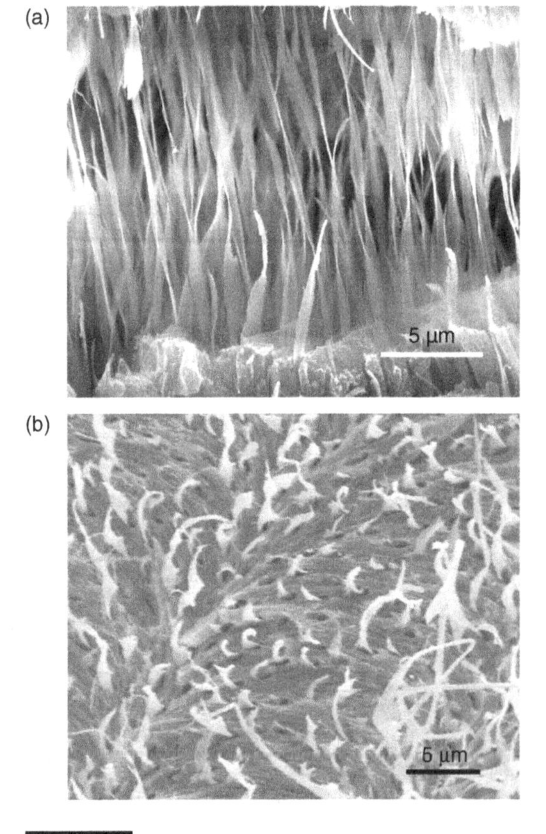

Figure 8.13.

SEM micrographs of ductile fracture surface showing the tubules in the *z*-direction in crab exoskeleton: (a) side view showing the necked tubules; (b) top view showing fractured tubules in tensile extension. (Reprinted from Chen *et al.* (2008a), with permission from Elsevier.)

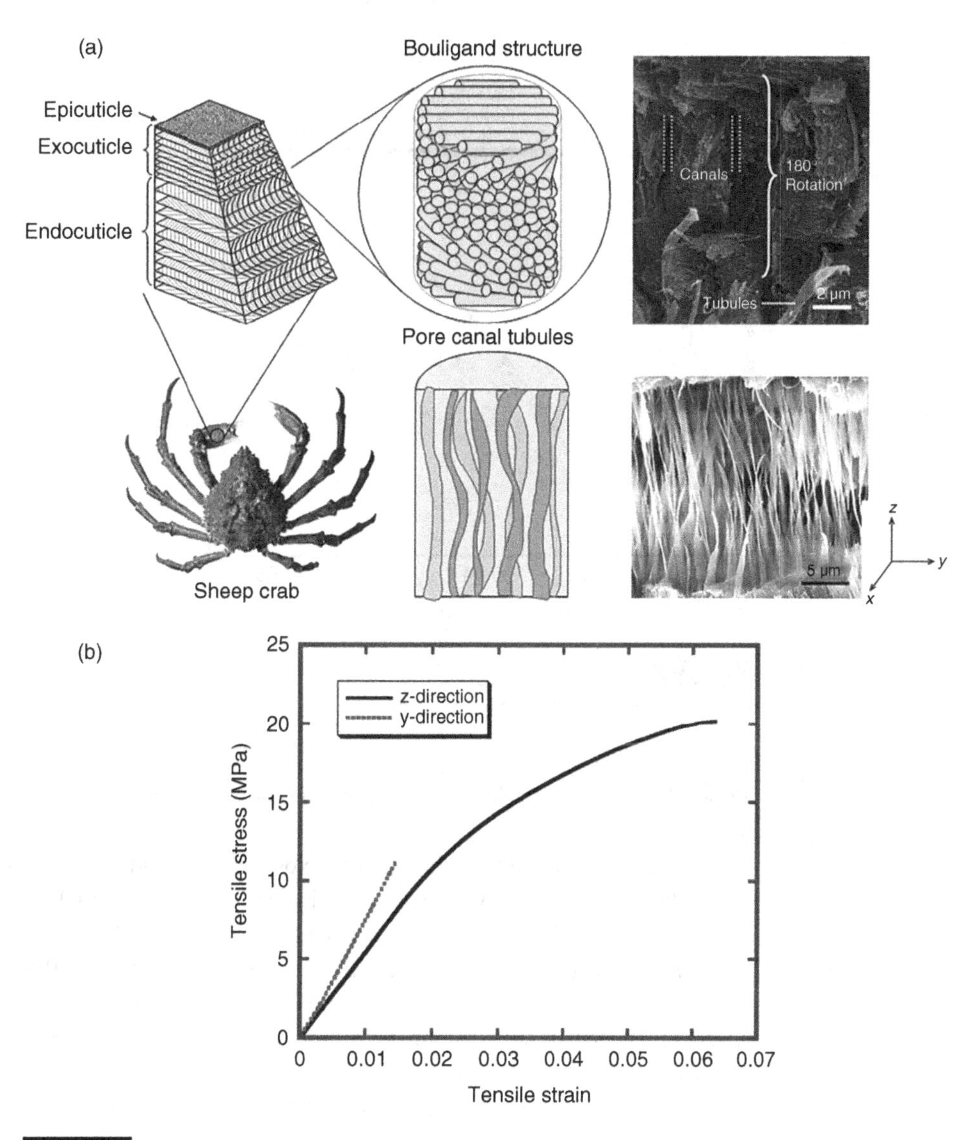

Figure 8.14.

(a) Mechanical design of the crab exoskeleton: mineralized chitin–protein fibers arranged in the Bouligand structure provide in-plane (*xy*) stiffness, and ductile pore canal tubules in the *z*-direction stitch the layers together and enhance toughness. (b) Tensile stress–strain responses parallel and perpendicular to the exoskeleton surface. (Reprinted from Chen *et al.* (2008a), with permission from Elsevier.)

in place even when it is fractured, allowing for self-healing. The mechanical design of crab exoskeleton as an anisotropic, tough natural composite is shown in Fig. 8.14(a); this recapitulates the information given in Fig. 2.3, with emphasis on the meso- and microstructures. The anisotropy of the tensile response is seen in Fig. 8.14(b); the

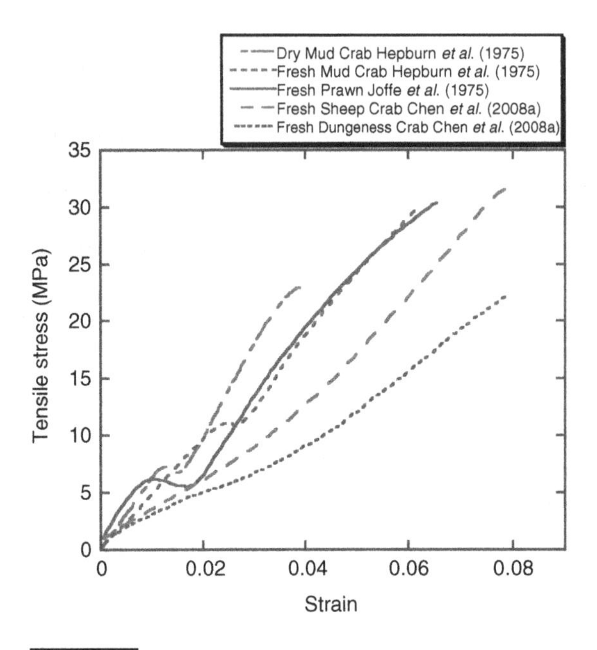

Figure 8.15.

Tensile stress–strain curves for crustacean exoskeletons.

action of the chitin tubules in enhancing the toughness perpendicular to the surface is evident.

The mechanical properties of crustacean exoskeletons (mud crab, *Scylla serrata*, and the prawn, *Penaeus mondon*) were first investigated by Hepburn and Joffe (Hepburn *et al.*, 1975; Joffe *et al.*, 1975), followed by Raabe and co-workers (American lobster, *Homarus americanus*) (Raabe *et al.*, 2005a, b, c, 2006; Sachs *et al.*, 2006a, b; Romano *et al.*, 2007) and Chen *et al.* (2008a) (sheep crab, *Loxorhynchus grandis*, and Dungeness crab, *Cancer magister*). The tensile stress–strain curves for various exoskeletons are shown in Fig. 8.15. The results from Hepburn *et al.* (1975) show a unique discontinuity (load drop) in the low-strain region. They suggested that this discontinuity is associated with the brittle failure of the mineral phase. When exoskeleton specimens are stretched, brittle failure of the mineral phase occurs at a low strain, leaving the chitin and protein phases to bear the load. Table 8.2 shows the mechanical properties of crustacean exoskeletons (Hepburn *et al.*, 1975; Joffe *et al.*, 1975; Chen *et al.*, 2008a). The dried exoskeleton material is rigid and brittle compared to that in the hydrated state.

Melnick, Chen, and Mecholsky (1996) studied the hardness and toughness of exoskeleton of the stone crab, *Menippe mercenaria*, which exhibits a dark color (ranging from amber to black) on tips of chelae and walking legs. The dark material was much harder and tougher than the light-colored material from the same crab chelae (Table 8.3). Scanning electron micrographs (Fig. 8.16) showed that the dark exoskeleton material has

Table 8.2. Mechanical properties of crustacean exoskeletons

		UTS (MPa)	Young modulus (MPa)	Fracture strain (%)	Source
Sheep crab (*Loxorhynchus grandis*)	wet	29.8 ± 7.2	467 ± 92	6.9 ± 1.8	Chen *et al.* (2008a)
	dry	12.5 ± 2.3	735 ± 65	1.7 ± 0.3	
Mud crab (*Scylla serrata*)	wet	30.1 ± 5.0	481 ± 75	6.2	Hepburn *et al.* (1975)
	dry	23.0 ± 3.8	640 ± 89	3.9	
Prawn (*Penaeus mondon*)	wet	28.0 ± 3.8	549 ± 48	6.9	Joffe *et al.* (1975)
	dry	29.5 ± 4.1	682 ± 110	4.9	

Table 8.3. Mechanical properties of stone crab, *Menippe mercenaria*[a]

	Hardness (GPa)	Fracture strength, σ_f (MPa)	Fracture toughness, K_{Ic}(MPa m$^{1/2}$)
Black	1.33	108.9	2.3
Yellow	0.48	32.4	1.0

[a] From Melnick *et al.* (1996) Copyright © 1996 Materials Research Society. (Reprinted with the permission of Cambridge University Press.)

a lower level of porosity, and this may relate to the tanning effect. It can also be more highly mineralized.

A hierarchical multi-scale mechanical modeling was used to elucidate the structural and mechanical design of the lobster exoskeleton (Nikolov *et al.*, 2010). Figure 8.17 is an overview showing key structural features at varying length scales and the methods used to model the mechanical anisotropy. The twisted plywood structure provides isotropic in-plane strength, prevents microcrack propagation, and plays a role in energy dissipation during impact loadings. The nano-sized $CaCO_3$ minerals and mineral–protein interfacial interactions stiffen the exoskeleton further. The multifunctional, hierarchically structured arthropod exoskeleton will hopefully provide inspiration for the novel design of high-performance composites.

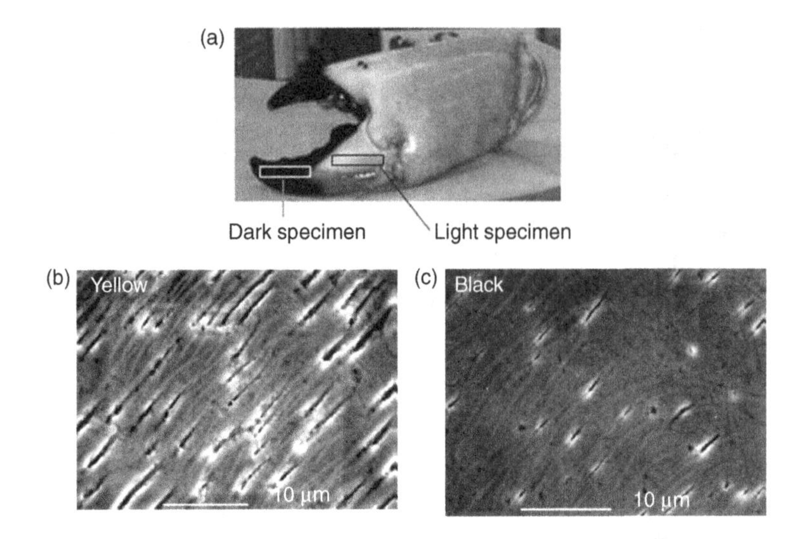

Figure 8.16.

(a) Stone crab, *Menippe mercenaria*, chelae showing dark- and light-colored regions. (b) SEM photograph showing high level of porosity in yellow exoskeleton material. (c) SEM photograph showing lower level of porosity in black exoskeleton material. (From Melnick *et al.* (1996) Copyright © 1996 Materials Research Society. Reprinted with the permission of Cambridge University Press.)

8.3.2 Hexapods

Insects are hexapods (i.e. they have six feet), and are the most diverse group of animals on earth, including over a million species and representing more than half of all known living organisms. Insect exoskeletons demonstrate great diversity and functional adaptation. Figure 8.18 is a section cut from the hind leg of a large adult grasshopper (locust). The layers can be clearly seen as a succession of darker and lighter regions, which constitute the endothelium. Daily layers are created (different day and night layers are formed) by the epidermal cells, on the bottom. The top layer is the epicuticle and is, in terrestrial insects, waterproofed by lipids. Its thickness is ~1–2 μm and is harder than the endocuticle. This layer is created before molting, whereas the endocuticle is created after molting. Figure 8.19 is a schematic representation with the principal components labeled. In locusts, the deposition that occurs during the day consists of a layer of parallel fibers. Nightly deposition is in a helical sequencing of fibers, creating a Bouligand (1972) pattern. Figure 8.19 shows a 180° rotation in the night layer. The cuticle is multifunctional: it supports the insect, and gives it its shape, means of locomotion, waterproofing, and a range of localized mechanical specializations such as high compliance, adhesion, wear resistance, and diffusion control. It can also serve as a temporary food store and is a major barrier to parasitism and disease, e.g. fungal invasions (Vincent and Wegst, 2004). Arthropods have pore canals that join the epidermal cells to the epicuticle. These hollow tubes transport waxes to the

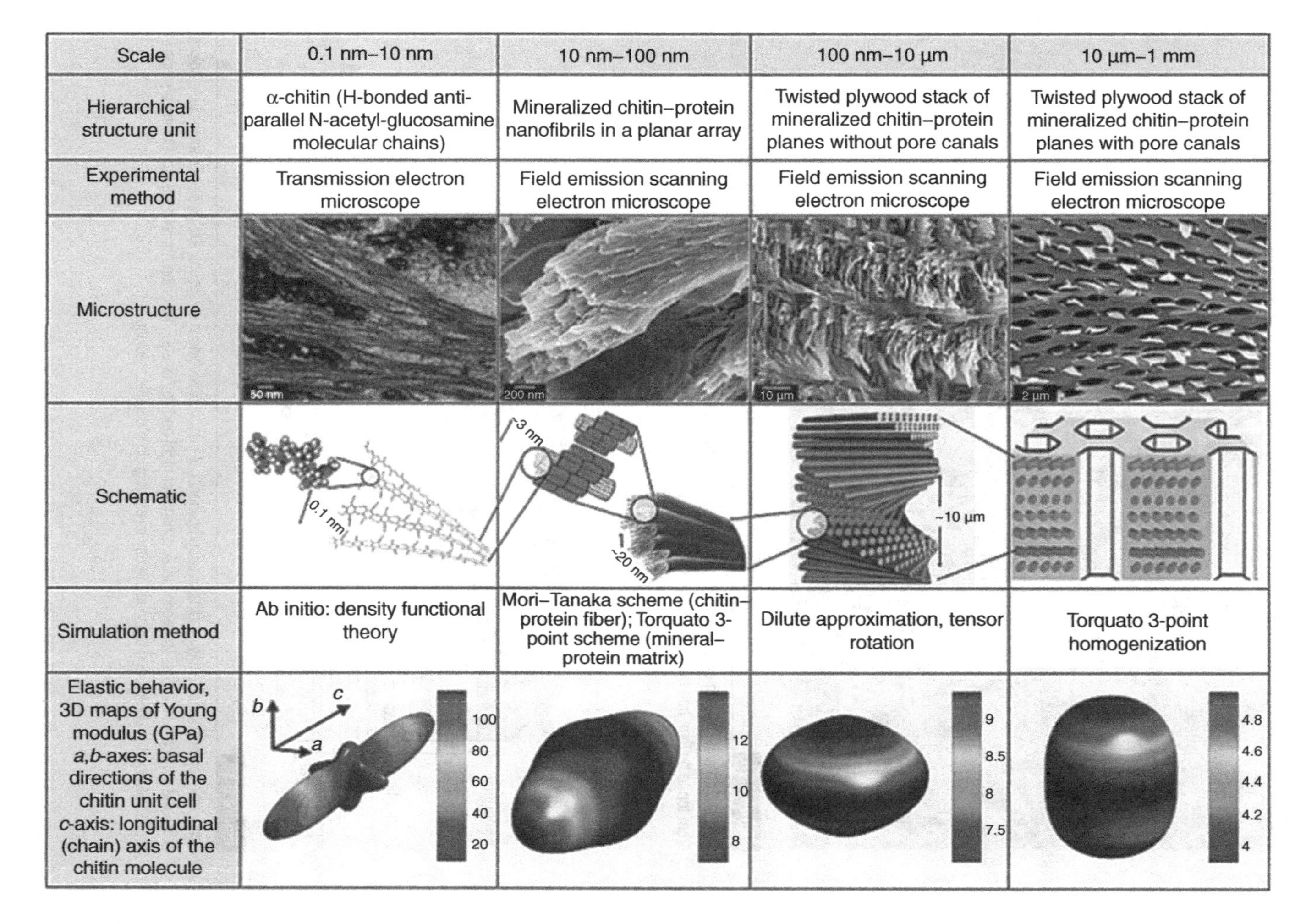

Figure 8.17.

Overview of key structural features at different hierarchical length scales and the corresponding mechanical modeling methods. (Reprinted from Nikolov *et al.* (2010), with permission from Elsevier.)

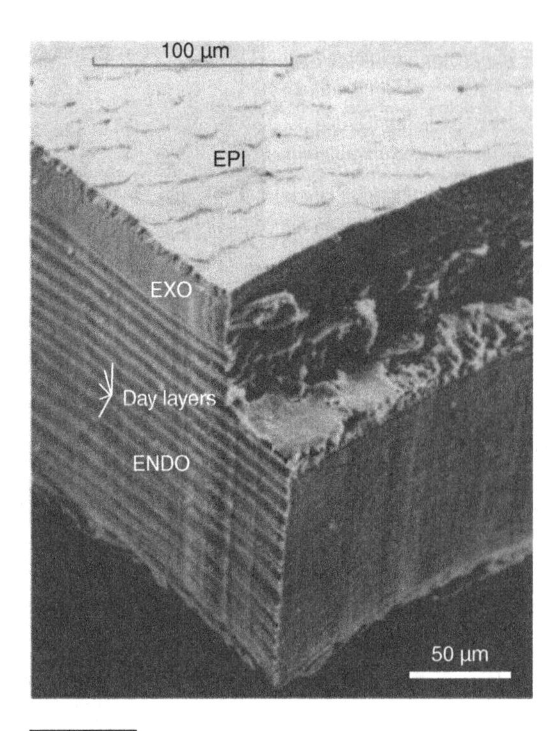

Figure 8.18.

SEM micrograph showing exoskeleton of adult grasshopper. Epicuticle (EPI), exocutile (EXO), endocuticle (ENDO), and day layers (from top to bottom) can be observed. (Reprinted from Hughes (1987), with permission from Elsevier.)

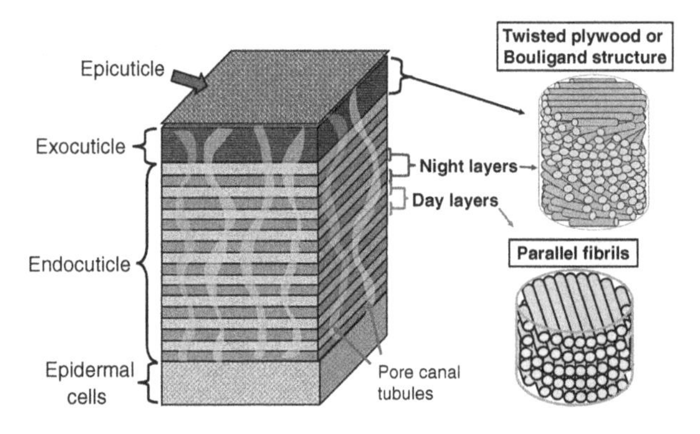

Figure 8.19.

Schematic representation showing structural components of insect exoskeleton: epicuticle, exocuticle, endocuticle, epidermal cells, and pore canal tubules. The endocuticle consists of night and day layers. The chitin fibrils in the night layers arrange into a twisted plywood structure, and those in the day layers align in a parallel stacking.

Figure 8.20.

Scanning electron micrograph showing successive layers with orientation close to (but not equal to) 90° in the wing case (elytron) of a scarab beetle. (Used with permission of M. Jackson.)

epicuticle, ensuring that it is waterproof. Several of these pores are shown in Fig. 8.19. The pores canals are "squeezed" by the chitin fibers and take on the flattened configuration, which twists according to the helical shape of the layers. This is similar to the structure displayed by the crab in Figs. 8.12 and 8.13(a).

The succession of layers in the wing case (elytron) of a scarab beetle is shown in Fig. 8.20. It should be noted that they are not quite orthogonal, providing a "twist." The thickness of the exo- and endocuticles in insects can vary considerably: as thin as 1 μm in the hindgut over gills of larvae and as thick as 200 μm in the elytra of large Coleptera (Neville, 1993). Figure 8.21(a) shows the microstructure of the lateral section of elytra in three types of beetles, revealing a well-organized cellular structure (Dai and Yang, 2010). The center portion has pores around which the chitin fibers wrap. This is especially clear in Fig. 8.21(a). This structural arrangement increases the stiffness/weight.

The mechanical properties of insect exoskeletons show a great range of variation. This range is needed for the different functions and is highly dependent on the degree of hydration. A fascinating aspect is the hardness of some regions, which can reach extraordinary values, which is associated with an increase in concentration of the metals Zn, Mn,

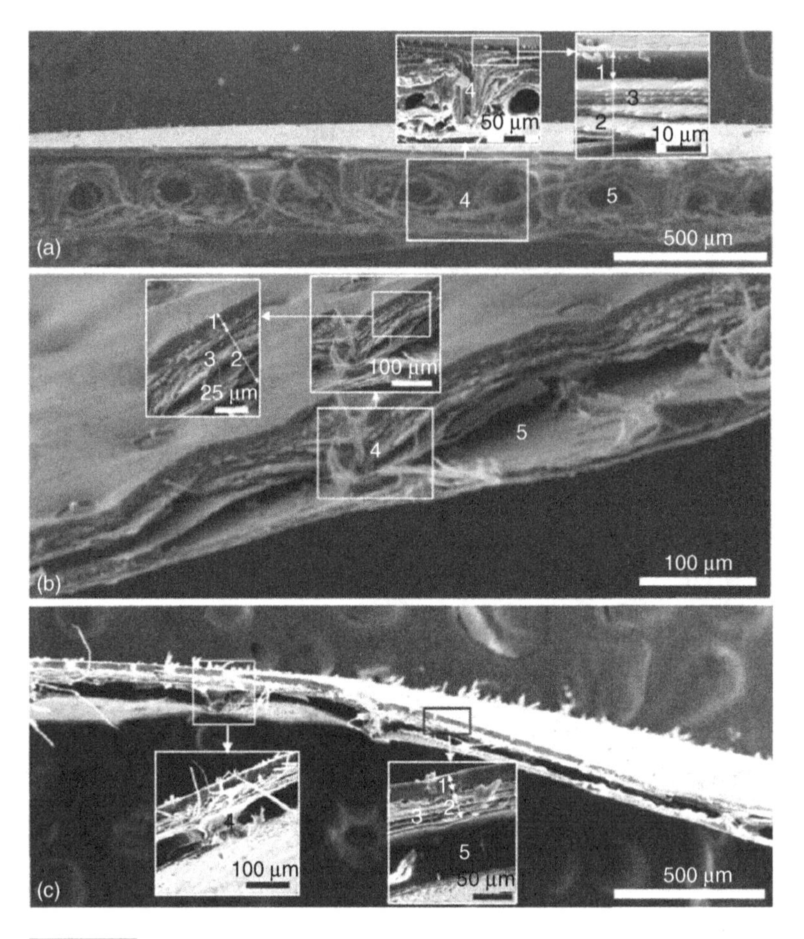

Figure 8.21.

Scanning electron micrographs showing microstructure of lateral section in three beetle's elytra. (a) *Caligula japonicas*; (b) *Allomyrrhina dichotoma*; (c) *Protaetia brevitarsis*. (1: epicuticle; 2: exocuticle; 3: fiber layers; 4: bridge pier connecting the exocuticle and endodermis; 5: cavity.) (Reprinted from Dai and Yang (2010), with permission from Elsevier.)

and Fe. The microhardness of the mandible of ground beetles (*Scarites subterraneous*) is comparable with enamel, ~5 GPa (Dickinson, 2008). It is not known what form the metal has in the mandible. It is probable that it is present as a mineral. The Young modulus varies from 1 kPa to 20 GPa, for adult insect wing, tibia, and elytron. This is understandable, since the cuticle is a composite of chitin nanofibers that are among the stiffest of the natural fibers (E = 150 GPa). "Compliant" cuticles contain 40–75% water, whereas stiff cuticles contain only 12% water. The tensile strength of femoral cuticle in locust is 60–200 MPa and the failure strain is as high as 5% (Vincent and Wegst, 2004).

One of the important components in the efficient strengthening of the organic matrix by the chitin nanofibrils is the bonding between the two. If the fibers were only weakly

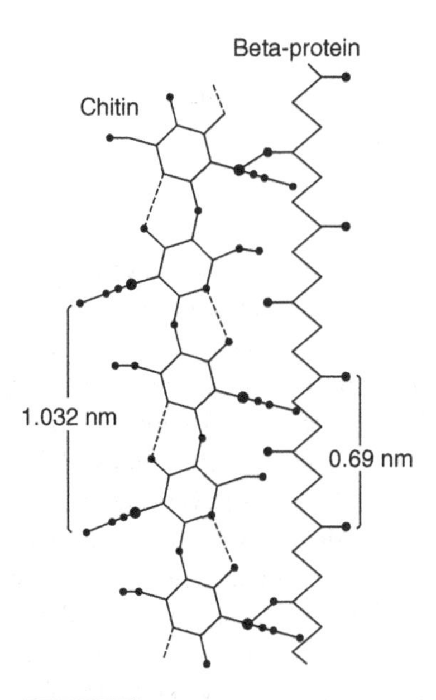

Figure 8.22.

Schematic showing how chitin and protein can bond in cuticle. Every fourth amino acid residue of the chitin molecule (4 × 1.032 = 4.128 nm) nearly coincides with every sixth amino acid residue of the protein 0.69 × 6 = 4.14 nm). (Reprinted from Vincent and Wegst (2004), with permission from Elsevier.)

bonded this strengthening effect would not occur, such as in polymer synthetic composites. Figure 8.22 shows a chitin chain, aligned with the β-pleated protein. There is an almost exact coincidence between nearly every fourth amino acid residue in the chitin chain with every sixth amino acid residue in the protein. This can be verified by multiplying 1.032 nm by 4 (4.128 nm) and 0.69 nm by 6 (4.14 nm). These sites are proposed to provide hydrogen bonds. Vincent and Wegst (2004) estimated the shear strength of the chitin–protein interface by assuming that each bond has a strength of about 30 pN. Considering nanofibrils with a diameter of 3 nm and length of 300 nm containing 19 chains, they obtained a shear strength of approximately about 30 MPa, or about half of that measured for carbon fibers in a resin matrix. This is indeed significant, and explains the excellent stiffness and strength of the cuticle.

Box 8.1 Biomedical adhesives and sealants

An adhesive is a material capable of holding or bonding substances together. Adhesives can come from either natural or synthetic sources and may comprise two or more constituents that harden or cure by chemical reactions or other mechanisms. Adhesives include glue, paste, cement, fixative, and bonding agent. A sealant is a viscous material that has little or no flow characteristics. It is used to bond substances, fill gaps, and prevent the penetration of air, water, and dust. Other required properties of sealants include insolubility, corrosion resistance, stability, and mechanical strength. Adhesives and

Box 8.1 (cont.)

sealants have been applied in the biomedical field, ranging from soft-tissue adhesives, for example wound closure, to hard-tissue adhesives, such as bone and tooth cements.

Soft-tissue adhesives

Most soft-tissue adhesives are intended to be temporary and are removed or degraded after wound healing is accomplished. Cyanoacrylates are fast-acting adhesives that cure rapidly in the presence of moisture. Cyanoacrylates are more easily recognized by the trade names like "Super Glue" or "Krazy Glue," and have been widely used after their invention by Harry Coover in 1951. Initially, methyl cyanoacrylate was used but was replaced by n-butyl (veterinary and skin glues) or 2-octyl cyanoacrylates (medical grade glue), which are more acceptable in biomedical applications. However, the reports of enhanced cancer risk and toxicity have limited the applications of cyanoacrylates.

Protein glues include gelatin-resorcinol-formaldehyde (GRF) glue and fibrin sealants. GRF glue was developed in the 1960s as a less toxic adhesive than methyl cyanoacrylate and was used in soft-tissue adhesion. However, toxicity and technical problems have limited its application. Fibrin sealant is made up of fibrinogen and thrombin to create a fibrin clot. It is hemostatic, biodegradable, adheres to connective tissues, and promotes wound healing. Despite the adhesive strength of fibrin sealant not being as high as cyanoacrylate, it is sufficient for most clinical applications. Fibrin sealants have been applied in cardiovascular, plastic, neurological, and ophthalmic surgery, and wound-dressing scaffold.

Hydrogels, PEG for example, have been developed to be a new class of synthetic sealants. They are used in lung, blood vessels, and dura mater. Another approach in developing novel biomedical adhesives and sealants includes the synthesis of mussel-inspired underwater adhesives (e.g. DOPA protein) and gecko-inspired nano-patterned surfaces (see Chapter 12).

Hard-tissue adhesives

Attachment of prostheses to calcified tissues (bone, cartilage, tooth) can be achieved by cements through mechanical interlocking mechanisms. Room-temperature polymerizing (cold-curing) methyl methacrylate systems (PMMA) are extensively used for orthopedic implant fixation. Dental cements are fast-setting pastes formed by mixing solid powder and liquid together. They are either resin cements or acid–base cements. Ideal dental cements require them to be nonirritant to pulp and gums, strong bonding with enamel and dentin, resistant to dissolution, and to have good thermal and chemical stability. Common dental cements include zinc phosphate ($Zn_3(PO_4)_2$), zinc polyacrylate, glass ionomer cement (GIC), resin-based sealants, and composite resin filling materials.

8.4 Keratin-based materials

Keratin consists of dead cells that are produced by biological systems. The basic structure is provided in Chapter 3 (Section 3.3.4). The most common examples are hair, feathers, porcupine quills, nails, claws, hooves, horns, beaks, and pangolin scales. We will briefly describe the latter four.

8.4.1 Hoof

The complex design of equine (horse) hoof wall consists of two structural elements: tubules and intertubular material (Bertram and Gosline, 1986; Kasapi and Gosline, 1996, 1997, 1999). Figure 8.23(a) shows the schematic of the equine hoof wall. Hollow tubules occupy half of the wall and are parallel to the surfaces. The intermediate filament, a fibrous structure of keratin, is embedded in the keratin matrix and is composed of α-helical protein bundles with a diameter of 8 nm. In the innermost wall, the intermediate filament is mainly oriented along the tubule axis and placed horizontally in the intertubular material, shown in Fig. 8.23(b). In the mid-wall region, the intermediate filament in the intertubular material is arranged in a helical fashion with angles from 0° to 33°. This complex design of the hoof wall provides a wide range of the mechanical properties in 0% to 100% of humidity condition. As the water content increases, the Young modulus decreases. This effect can be dramatic. Figure 8.24 shows the initial Young modulus plotted as a function of water content. The mechanical role of the tubules and orientation of the intermediate filaments are to control the crack propagation process and enhance fracture toughness of the wall. Figure 8.25 shows that the J integral of the hoof wall is not very sensitive to the rate of load application. The average J integral is 12 ± 3 kJ/m^2; the hoof wall prevents brittle failure in the entire strain-rate range investigated: 1.6×10^{-3} to 70 s^{-1}. This covers most situations encountered by ungulates.

Hooves are the most similar biological material to horns and have been the subject of the largest body of scientific literature. Hooves contain tubules ~220 × 140 μm in the

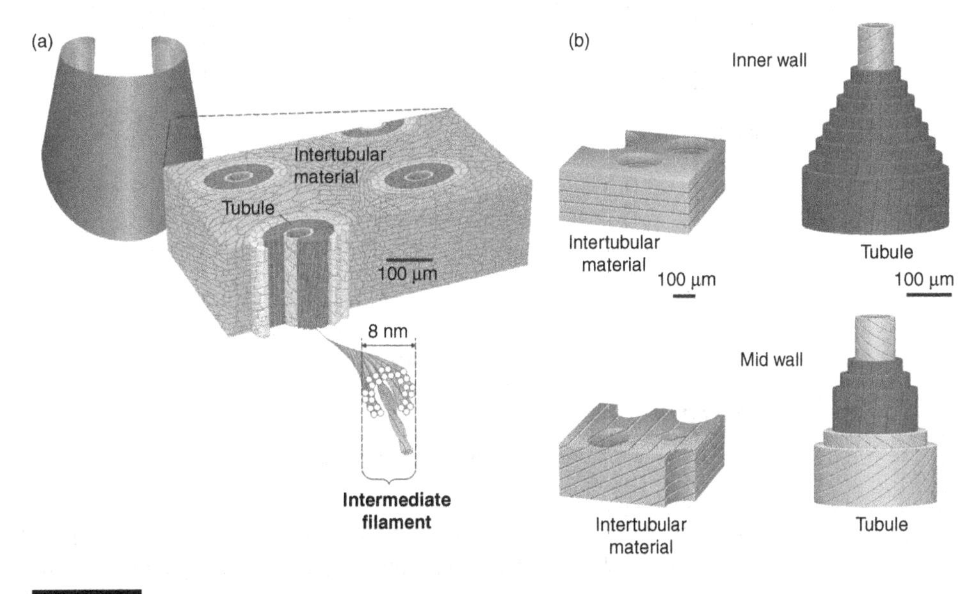

Figure 8.23.

(a) Schematic of hoof wall; (b) orientation of intermediate filament in inner wall and mid wall. (Adapted from Kasapi and Gosline (1999), with permission of *The Journal of Experimental Biology*.)

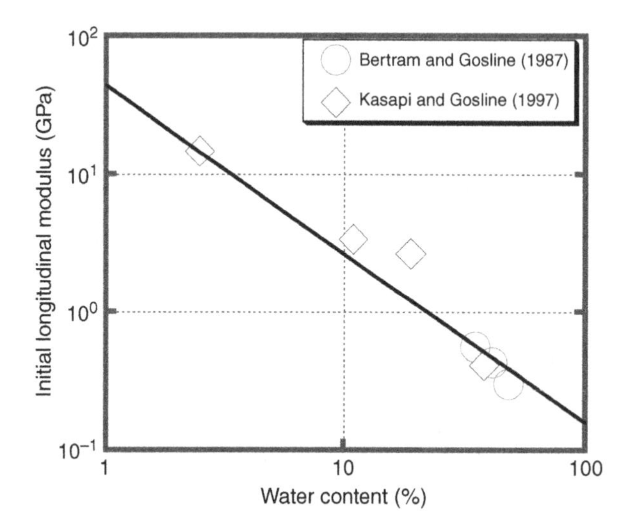

Figure 8.24.

Initial Young modulus as a function of water content. (Reproduced based on Bertram and Gosline (1987) and Kasapi and Gosline (1997).)

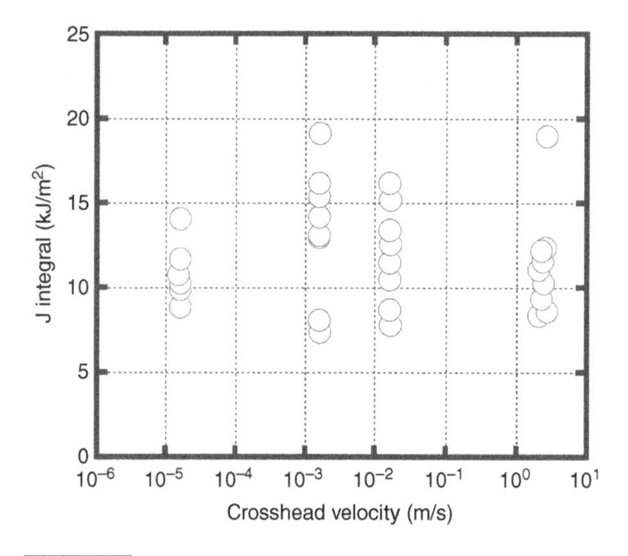

Figure 8.25.

J integral of hoof wall at different strain rates. (Reproduced based on Kasapi and Gosline (1996).)

major and minor axes, respectively, with a medullary cavity of ~50 μm. These tubules are oriented in the longitudinal direction (parallel to the leg). The keratin forms in circular lamellae (~5–15 μm thick) surrounding the tubules, as shown in Fig. 8.23(a) (Kasapi and Gosline, 1999). It was concluded that the tubules serve only a mechanical function – to increase crack deflection, thereby increasing the toughness, making the equine hoof a highly fracture resistant biological material. Hooves must support large compressive and impact loads and provide some shock absorption from the impact. The most thorough studies have been from Gosline and co-workers (Bertram, 1986, 1987; Kasapi and Gosline, 1997, 1998, 1999). The hoof keratin is described as a nano-scale composite comprising intermediate filaments (IFs) as fiber-oriented reinforcement of a hydrated keratin matrix, as shown in Fig. 8.23. Bertram and Gosline (1987) measured the effect of hydration on tensile and fracture properties. They found the stiffness to decrease in the hydrated condition, ranging from 14.6 GPa (ambient) to 0.4 GPa (100% relative humidity, RH). Water penetrates the intertubular matrix as well as the amorphous polymer surrounding the keratin fibers, acting as a plasticizer, thereby reducing the density and stiffness of the material (Vincent, 1990). This might have saved Europe from the onslaught of Mongolian hordes in the thirteenth century. The horses moved well over dry or frozen terrain, and the soft and water-soaked European soil was a deterrent for them, since the hooves were not protected by horseshoes.

The J-integral toughness was found to be maximum at 75% RH (22.8 kJ/m^2). Kasapi and Gosline (1997, 1999) tested stiffness, tensile strength, and work of fracture, in fully hydrated conditions, to correlate IF volume fraction and alignment with mechanical properties. They found that the stiffness increased toward the outer hoof wall, ranging from 0.30 GPa at the inner region to 0.56 GPa on the outer surface of the hoof wall, despite the porosity increase in that direction. The increase in elastic modulus was attributed to an increase in the volume fraction of keratin fibers. Further studies revealed that the stiffness reinforcement was due to IF volume fraction rather than IF orientation. In the tubular material, the IFs are aligned in the tubule direction. However, they are aligned more perpendicular to the tubule direction in the intertubular matrix. These different orientations help resist crack propagation through crack redirection, suggesting that the hoof wall structure evolved to maximize the fracture toughness (Kasapi, 1999).

Bovine hooves are similar to equine hooves in both structure and properties (Franck *et al.*, 2006; Clark and Petrie, 2007), as shown in Table 8.4. Baillie and Fitford (1996) and Baillie *et al.* (2000) described the bovine hoof structure as comprising tubules embedded in intertubular material. Franck *et al.* (2006) determined the tensile, compressive, and bending strengths and the stiffness. These values are similar to those for equine hooves, considering the slightly different moisture content. Clark and Petrie (2007) found the fracture toughness for bovine hooves (J integral = 8.5 kJ/m^2) to be lower than for equine hooves (J integral = 12.0 kJ/m^2) (Bertram and Gosline, 1986). This might be related to the fact that equines are designed for high speed, whereas bovines, having powerful horns, have different defense mechanisms.

The structural differences found between bovine and equine hooves appear mainly to affect the toughness. The bovine tubule wall is thinner and the keratin cells in the intertubular material are oriented more parallel to the tubules than in the equine hoof. Accordingly, the intertubular IFs are more aligned in the direction of the tubules compared to those of equine hooves. Finally, in the bovine hoof the interaction between tubular and intertubular material appears to be stronger than that in the equine hoof,

Table 8.4. Comparison of mechanical properties of keratinized structures

	Elastic modulus (GPa)	Tensile strength (MPa)	Bending strength (MPa)	Toughness (MJ/m^3)	Work of fracture (kJ/m^2)	Moisture content (wt.% water)	Reference
Oryx horn (*Oryx gazella*)	4.3		212		19	n/a	Kitchener (2000)
	6.1	137				0	Kitchener (1987)
	4.3	122				20	Kitchener (1987)
	1.8	56				40	Kitchener (1987)
Waterbuck horn (*Kobus ellipsiprymnus*)	3.3		245		20	n/a	Kitchener (2000)
Sheep horn (*Ovis canadensis*)	4.1		228		22	n/a	Kitchener (2000)
	9					0	Warburton (1948)
	1.5					20	Warburton (1948)
	2.20		127.1	56		10.6	Tombolato *et al.* (2010)
	0.81		39.1	12		34.5	Tombolato *et al.* (2010)
Bovine hoof	0.4	16.2	14.3			29.9	Franck *et al.* (2006)
Equine hoof	2.6	38.9	19.4			18.2	Bertram and Gosline (1987)
	0.2					41% RH[a]	Wagner, Wood, and Hogan (2001)
	0.3–0.6	6.5–9.5				100% RH	Kasapi and Gosline (1997)

[a]RH = relative humidity.

indicating a stronger interface. These differences account for the higher fracture toughness of the equine hoof compared with that of the bovine hoof.

8.4.2 Horn

Horns appear on animals from the bovidae family, which includes cattle, sheep, goats, antelope, oryx, and waterbuck, and are tough, resilient, and highly impact resistant. In the case of male bighorn sheep, the horns must be strong and durable as they are subjected to extreme loading impacts during the life of the animal and, unlike antlers, will not grow back if broken. Horns are not living tissue – there are no nerves and they do not bleed when fractured. On the living animal, horns encase a short bony core (*os cornu*) composed of cancellous bone covered with skin, which projects from the back of the skull. The horn is not integrated to the skull and can pull away if the hide is removed. The skin covering the bony core is a germinative epithelium that generates new cells to grow the horn. There are quite a variety of horn shapes and sizes, from the stumpy horns on domestic cattle to the extravagant forms seen on the greater kudu (*Tragelaphus strepsiceros*), blackbuck (*Antilope cervicapra*), and the Nubian ibex (*Capra nubiana*). Unlike other structural biological materials (e.g. bone, tusk, teeth, antlers, mollusc shells), horn does not have a mineralized component and is mainly composed of alpha keratin. The shape of these horns, often following a logarithmic spiral/helicoidal pattern, was described in detail by D'Arcy Thompson (1917, Chapter 7) and modeled by Skalak *et al.* (1997). The shapes are the result of differences in the growth velocity along the circumference of the base of the horn, of which the germinative epithelium generates the keratinocytes.

The structure of rhinoceros horn is similar to that of hoof; it has a laminated structure of tubules, as shown in Fig. 8.26 (Ryder, 1962). Horn tubules are placed more closely to each other than hoof tubules. The mechanical properties of the oryx horn have been studied and composite theory has been applied to predict the stiffness (Kitchener and Vincent, 1987). Viscoelastic behavior of the gemsbok horn has been investigated at

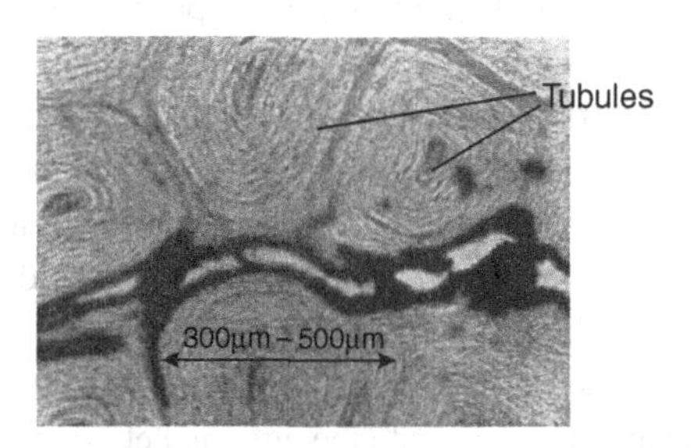

Figure 8.26.

Transverse section of rhinoceros horn showing six tubules (magnification ×220). (Reprinted by permission from Macmillan Publishers Ltd.: *Nature* (Ryder, 1962), copyright 1962.)

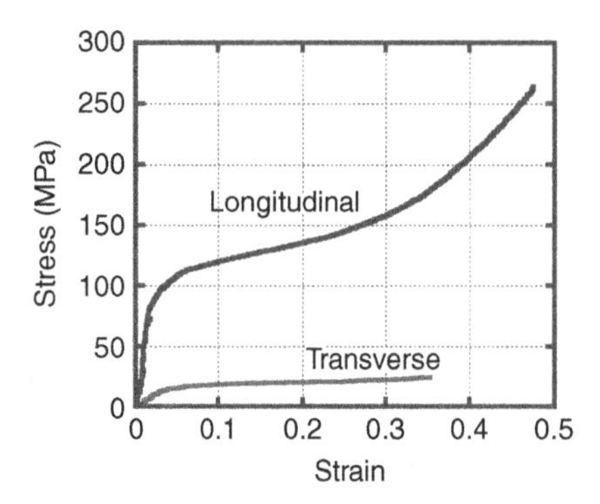

Figure 8.27.

Stress–strain curves of horn. (Reproduced based on Druhala and Feughelman (1974).)

different water contents (Kitchener, 1987a). Figure 8.27 shows stress–strain curves of horn along transverse and longitudinal directions (Druhala and Feughelman, 1974). There is a great degree of anisotropy with the strength in the longitudinal direction (parallel to the tubules) being approximately ten times that in the transverse direction (perpendicular to tubules).

Figure 8.28 shows the hierarchical structure of horn from a desert bighorn sheep. At the lowest level, two polypeptide chains (types I and II) (Fraser and MacRae, 1980), which belong to the family of related proteins, form two-strand coiled-coil molecules approximately 45 nm in length and 1 nm in diameter. These molecules are helically wound and assemble into microfibrils (called intermediate filaments, IFs) forming "superhelical" ropes of 7 nm in diameter (Feughelman, 1997). The α-helices are mainly parallel to the long axis of the ropes. These IFs are embedded in a viscoelastic protein matrix. This matrix is composed of two types of proteins: high-sulfur protein, which has more cysteinyl residues, and high-glycine-tyrosine proteins, whcih have high contents of glycyl residues (Fraser and MacRae, 1980). These lamellae are flat sheets that are held together by other proteinaceous substances. Long tubules extend the length of the horn interspersed between the lamellae. The resulting structure is a three-dimensional, laminated composite that consists of fibrous keratin and has a porosity gradient across the thickness of the horn. Optical micrographs of transverse and longitudinal sections of the horn are shown in Fig. 8.29. The transverse section in Fig. 8.29(a) shows a lamellar structure with elliptically shaped porosity interspersed between the lamellae. The lamellae are ~2–5 μm thick, with the pore sizes ranging from 60–200 μm along the long axis of the pores. This porosity results from the presence of tubules that extend along the length of the horn, as shown in the longitudinal section in Fig. 8.29(b).

Kitchener and co-workers (Kitchener 1987a, b, 1988, 1991, 2000; Kitchener and Vincent, 1987) were the first to provide insights into the fighting behavior of various species in the bovidae family. Mechanical property measurements (strength, stiffness, work of fracture – see Table 8.4) reveal that horns are capable of high-energy absorption before breaking and that hydration is important for decreasing the notch sensitivity. The maximum

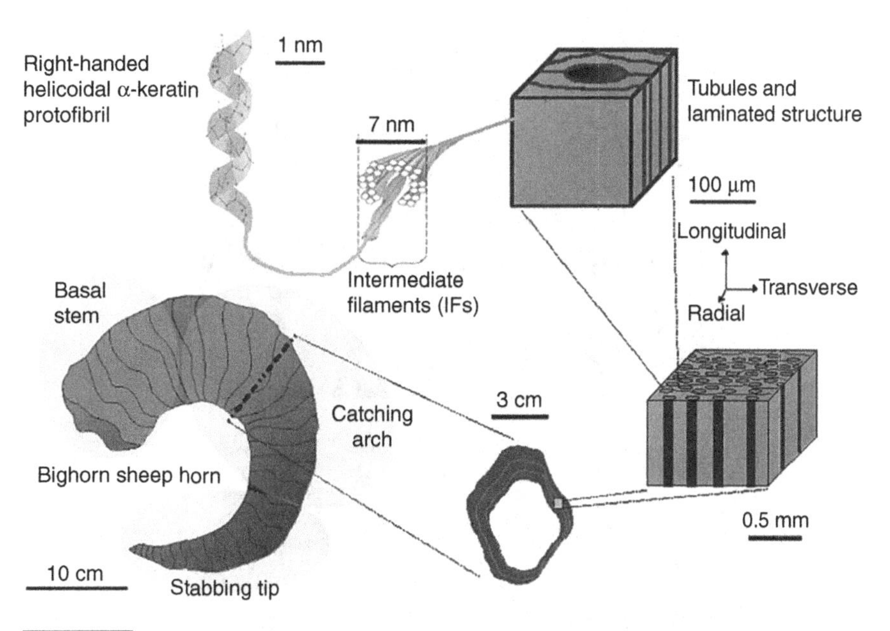

Figure 8.28.

Hierarchical structure of bighorn sheep horn (*Ovis canadensi*): the horns grow in spiral fashion, with ridges on the surface that correspond to seasonal growth spurts. Horns are composed of elliptical tubules, embedded in a dense laminar structure. Each lamina has oriented keratin filaments interspersed in a protein-based matrix. These filaments are two-strand coiled-coil rope polypeptide chains (intermediate filament type I and II), helically wound to form "superhelical" ropes 7 nm in diameter. (Reprinted from Tombolato *et al.* (2010), with permission from Elsevier.)

impact force of a bighorn sheep can be calculated from the mass, velocity, and deceleration: 3400 N (Kitchener, 1988). This generates compressive and tensile stresses in the horn of ~4 and 1.4 MPa, respectively. He further calculated the critical crack length for crack propagation to be ~60% of the transverse dimension of the horn, indicating the superior flaw sensitivity of the material. The specific work of fracture was found to vary along the length of the horn: for fresh waterbuck (*Kobus ellipsiprymnus*) horns they were found to range from 10 to 80 kJ/m^2, and for mouflon (*Ovis musimon)* horn they ranged from 12 to 60 kJ/m^2 (increasing from the base to the outer tip). This work of fracture is greater than most other biological and synthetic materials (antler, 6.6 kJ/m^2; bone, 1.6 kJ/m^2; glass, 5 J/m^2; mild steel > 26 kJ/m^2) (Kitchener, 1987a). This was attributed to crack arrest and deflection mechanisms such as delamination and keratin fiber pullout. Kitchener and Vincent (1987) examined the effect of hydration on the elastic modulus of horns from the oryx (*Oryx gazella*). They considered the structure of the horn as a chopped fiber composite, where the crystalline alpha-keratin fibers (40 nm long) were embedded in an amorphous keratinous matrix. Applying the Voigt model (Eqn. (7.1)) and using a chopped fiber composite analysis with a volume fraction of fibers of 0.61, they predicted a value of the elastic modulus close to the experimental value, indicating that a fibrous composite model of horn keratin is a reasonable assumption. The elastic and shear modulus decreased significantly with an

(a)

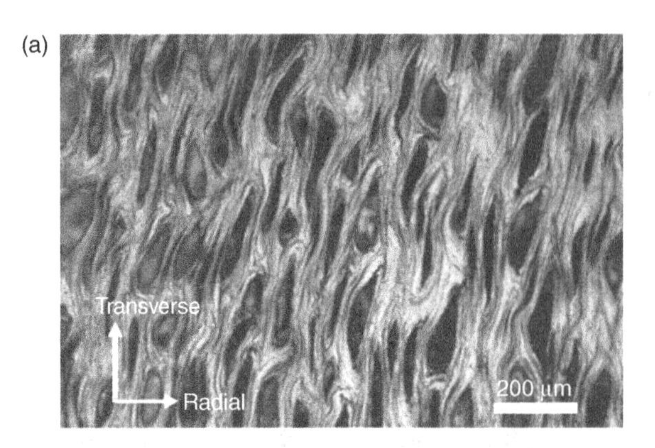

(b)

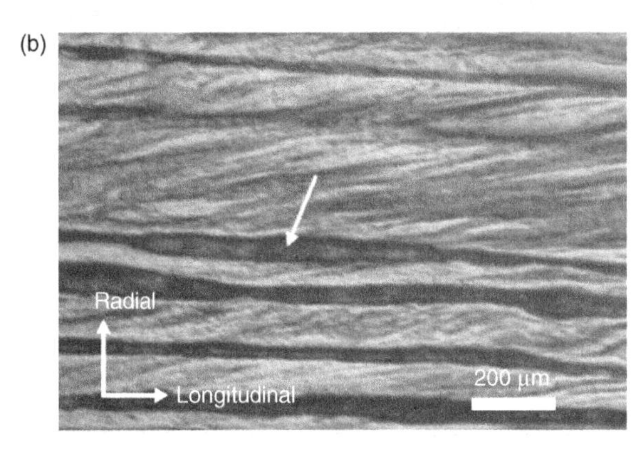

Figure 8.29.

Optical micrographs of the ambient dried horn: (a) cross-section showing the dark elliptical-shaped tubules and (b) longitudinal section showing the outline of the parallel tubules (arrow points to a tubule). (Reprinted from Tombolato *et al.* (2010), with permission from Elsevier.)

increase in the moisture content (Kitchener, 1987b; Kitchener and Vincent, 1987). The keratin fibers were not affected, rather the matrix swelled with the water, which decreased the elastic and shear moduli. Warburton (1948) had earlier determined on sheep horn that moisture severely decreased the elastic modulus, reducing it by as much as 75% with 20 wt.% water (compared with dry horn).

The rhinoceros horn is another example of biological structural material composed of alpha keratin. Ryder (1962) observed that the tubules, 300–500 μm in diameter with a medullary cavity of ~20 μm × 60 μm in the major and minor axis, respectively, in the horn were slightly coarser than that of equine hoof (20 μm × 40 μm), and there was little intertubular material. Hieronymus, Witmer, and Ridgely (2006) found that, unlike the horns of other ungulates, the rhinoceros horn does not have a bony core. Thus, it can be sawed off to protect the animal against poaching. The rhinoceros horns consist of tubules embedded in the amorphous keratin matrix, as shown in Fig. 8.30.

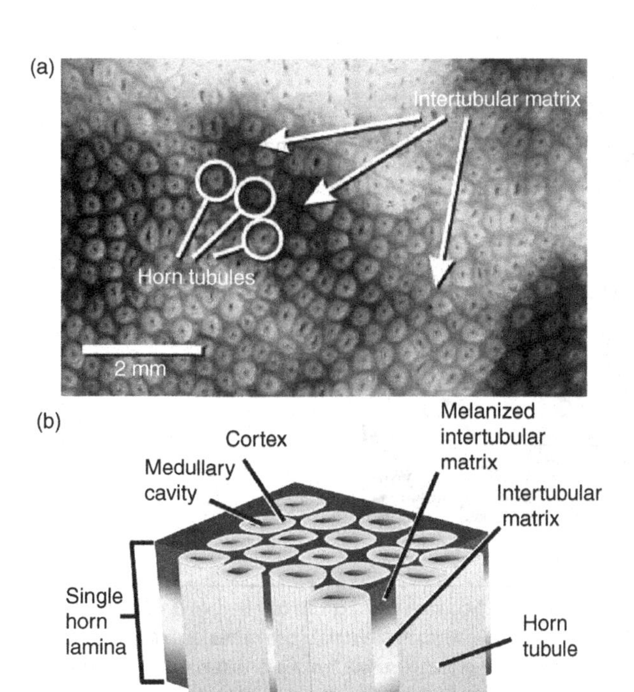

Figure 8.30.

Optical micrograph of the white rhinoceros horn, showing tubules and the intertubular matrix, along with illustration indicating microstructural features. (Reprinted from Hieronymus *et al.* (2006), with permission from Elsevier.)

Tombolato *et al.* (2010) studied the microstructure, elastic properties, and deformation mechanisms of desert bighorn sheep, *Ovis canadensis*. Compression and bending tests were performed in both hydrated and ambient dried conditions. The elastic modulus and yield strength are anisotropic and correlated with the orientation of tubules. Three-point-bending tests showed that the elastic modulus and strength are higher in the longitudinal orientation (tubules parallel to the long dimension of the specimen) than those in the transverse orientation (tubules perpendicular to the long dimension of the specimen). An optical micrograph taken from the central region of the hydrated samples tested in three-point bending in the longitudinal direction is shown in Fig. 8.31(a). Toughening mechanisms, such as delamination and crack bridging, are observed. In Fig. 8.31(b), the fracture surface from a longitudinally oriented sample shows numerous tubules surrounded by concentric lamellae. The compressive deformation mechanisms of horn in three orientations (longitudinal, transverse, and radial) are summarized in Fig. 8.32. Lamellar microbuckling is observed in longitudinal and transverse orientations, while collapse of tubules is the main deformation mechanism in the radial orientation. Trim *et al.* (2011) investigated the mechanical behavior of bighorn sheep horn under tension

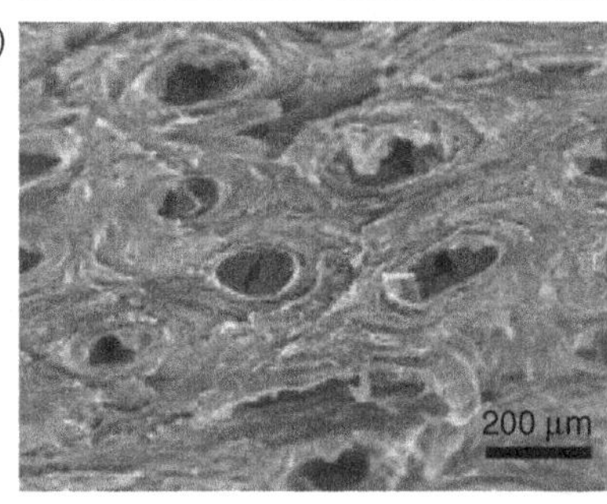

Figure 8.31.

Micrographs of a ram horn. (a) Optical micrograph taken from the central region of the hydrated samples tested in three-point bending in the longitudinal direction. Toughening mechanisms of delamination and crack bridging are observed. (b) SEM micrographs of the three-point-bending test fracture surface from a longitudinally oriented sample in ambient dried conditions. The microstructure is characterized by numerous tubules, with concentric rings surrounding them. (Reprinted from Tombolato *et al.* (2010), with permission from Elsevier.)

and compression in hydrated and dry conditions. They found that tensile failure occurred by matrix separation followed by fiber pullout (Fig. 8.33(a)). The horn keratin failed in a brittle manner in the dry condition, while wet horn keratin was much more ductile. Compressive failure occurred by microbuckling followed by delamination (Fig. 8.33(b)), in agreement with Tombolato *et al.* (2010).

Lee *et al.* (2011) investigated the dynamic mechanical behavior of a wide range of biological materials (abalone nacre, elk antler, armadillo carapace, bovine femur, and steer and ram horns) and compared them with synthetic composites using a drop-weight impact testing system. The impact strengths of horns are the highest among biological materials, as shown in Fig. 8.34, confirming the exceptional energy-absorbing capability of horn.

8.4.3 Beak

Bird beaks serve a variety of purposes: eating and probing for food, fighting, courtship, grooming, killing prey, and as a heat exchanger. There is a wide variety in the morphology, color, and size but all have mandibles (bone) that project from the head which are covered by a beta-keratin layer. Birds usually have either short, thick beaks or long, thin beaks. Exceptions are toucans and hornbillls, which have both long and thick beaks. The Toco Toucan (*Ramphastos toco*) has the largest beak among the species. The outside

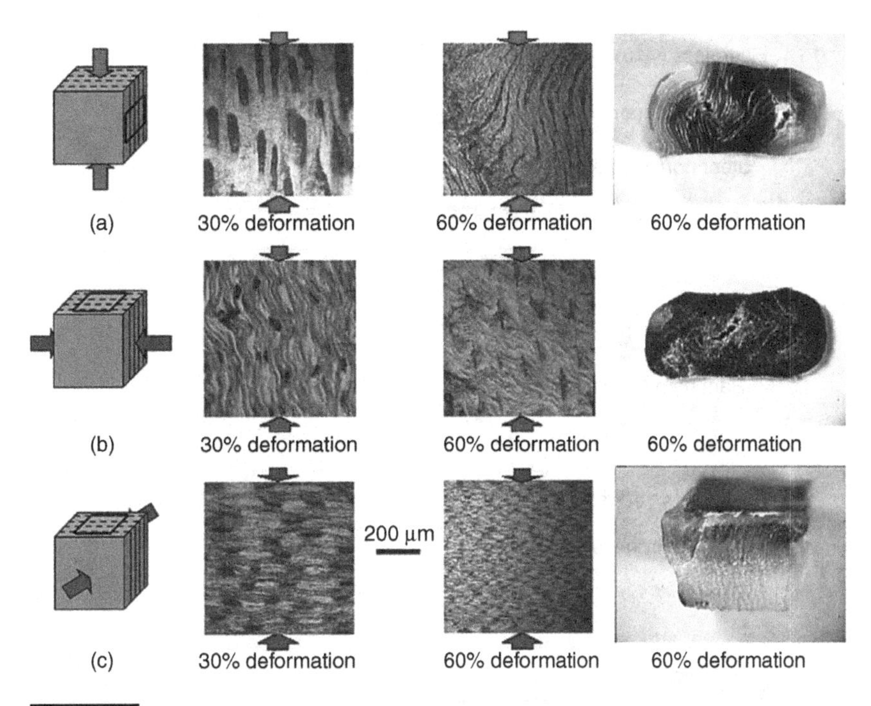

Figure 8.32.

Optical micrographs of the deformation of ambient dried horn (*Ovis canadensi*) at 30% and 60% strain for (a) longitudinal, (b) transverse, and (c) radial directions. Pictures on the right show the accumulated damage in the samples. (Reprinted from Tombolato *et al.* (2010), with permission from Elsevier.)

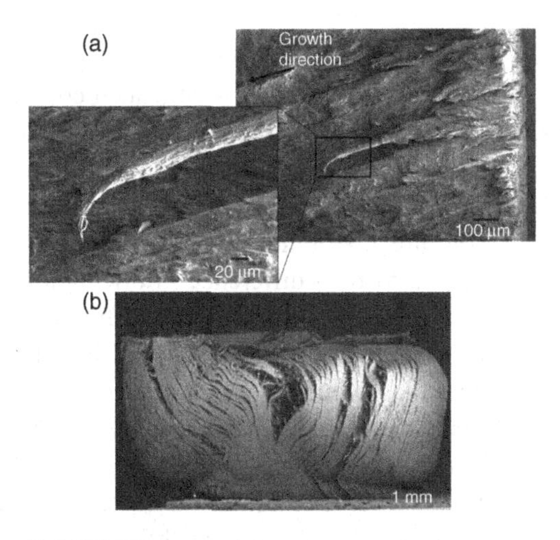

Figure 8.33.

(a) SEM micrographs of fracture surface of horn keratin tested in tension showing fiber pullout; (b) microbuckling and delamination after compressive deformation. (Reprinted from Trim *et al.* (2011), with permission from Elsevier.)

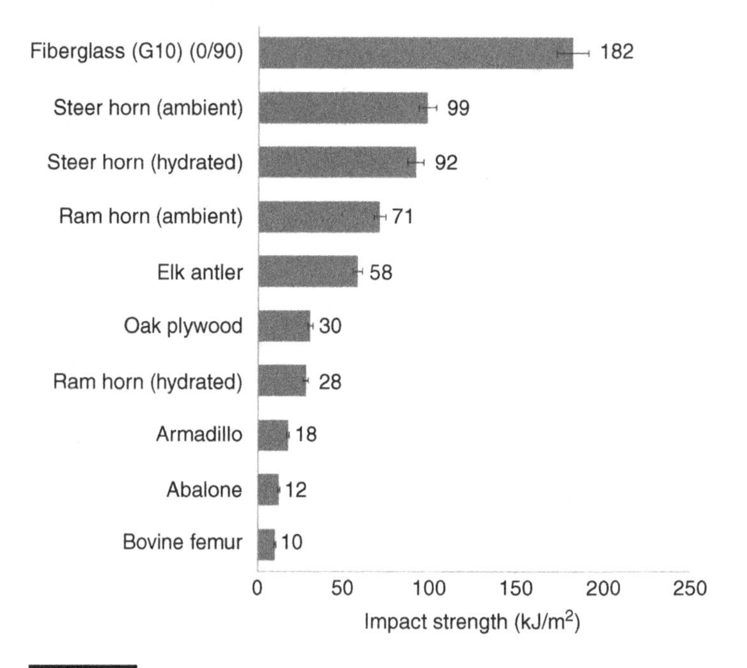

Figure 8.34.

Comparisons of impact strength among structural biological materials and synthetic composites from drop-weight tests. (Reprinted from Lee *et al.* (2011), with permission from Elsevier.)

shell of the beak consists of beta keratin, and the inside is filled with a cellular bone. This internal foam has a closed-cell structure constructed from bony struts with thin membranes.

Figures 8.35(a)–(c) show photographs and schematics of the toucan beak (Seki *et al.*, 2006; Seki, Bodde, and Meyers, 2010). The keratin shell consists of polygonal tiles 30–60 μm in diameter and 2–10 μm thick (Fig. 8.35(c)). TEM images of the longitudinal and transverse sections are shown in Fig. 8.35(d). The wavy keratin tile boundaries, traced by black lines for greater clarity, are shown in the longitudinally sectioned beak keratin. The IFs are distributed in the amorphous keratin matrix, indicated by arrows. There appears to be a difference in orientation of the IF from layer to layer, similar to a 0°/90° laminated composite. The stiffness of the beak keratin was found to be mechanically isotropic in the transverse and longitudinal directions (Seki *et al.*, 2005). The surface tiles exhibit a layered structure and the tiles are connected by organic glue. The intermediate filaments, embedded fibers in the keratin matrix appear to be aligned along the cell boundaries. These tiles undergo a peculiar behavior known in metallurgy as a ductile-to-brittle transition. As the strain rate is increased, the yield strength increases significantly. In this region, the fracture transitions from inter-tile (tile pullout) to trans-tile (tile fracture). This is due to the existence of two competing failure processes with different strain-rate sensitivities.

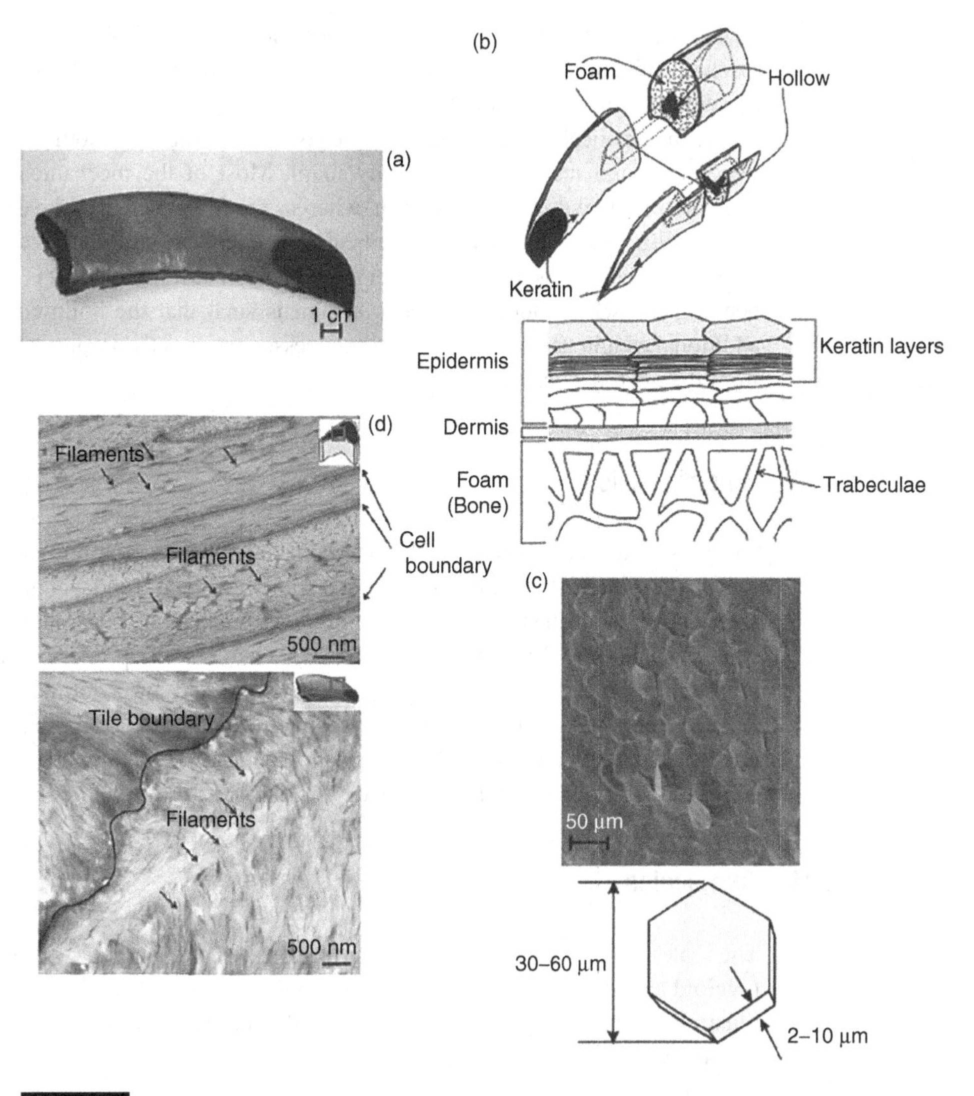

Figure 8.35.

(a) Toco Toucan beak. (b) Schematic representation of the cross-section of the beak – the keratin layer is 500 μm thick. (c) SEM image of the keratin tiles on the surface of the beak. (d) TEM micrograph of the transverse cross-section (top) and longitudinal surface (bottom) showing the keratin intermediate filaments. (Figure courtesy Y. Seki.)

Seki *et al.* (2005) found the toucan beak to have a bending strength (Brazier moment) that is considerably higher than if all the mass were concentrated in the shell as a solid hollow cylinder by applying the analysis developed by Karam and Gibson (1995a, b). Seki *et al.* (2005, 2010) showed that the internal cellular core serves to increase the buckling resistance of the beak and demonstrated a synergism between the two components that provides the stability in the bending configuration. Thus, there is clearly an advantage in having internal foam to support the shell. This is the same conclusion that is reached with quill and feather studies regarding the role of the internal foam. This will be presented in greater detail in Chapter 10.

The mechanical behavior of bird beaks is governed by both the ductile keratin integument and the semi-brittle bony foam. Most of the mechanical loading on the beak is carried by the exterior keratin, whereas the foam increases the energy absorption and stabilizes the deformation of the beak to prevent catastrophic failure. In the case of the toucan, the beak is mainly used for apprehension of food so that it is designed to resist bending moments. Indeed, the beak design is such that the hollow core provides an additional weight gain, since the bending stresses are directly proportional to the distance from the neutral axis.

8.4.4 Pangolin scales

An unusual armor is found on the pangolin, a small insectivore that lives in the rain forests of Asia and Africa. It ranges from 0.4 to 1 m in length and weighs up to 18 kg. The exterior of the animal is covered with nonmineralized keratin scales, shown in Fig. 8.36(a), which weigh up to 20% of the total animal. When curled up, these scales extend from the body producing a barrier of razor-sharp edges and are a formidable defense. The tensile response of the keratin scales is shown in Fig. 8.36(b). It has been reported that a pride of lions toyed with a balled-up pangolin for several hours before giving up. The edges of the keratin scales are also sharp and can cut a potential predator.

8.5 Fish scales

The scales of fish are classified into four groups: placoid, ganoid, cosmoid, and elasmoid (cycloid and ctenoid). These are shown in Fig. 8.37 (Yang *et al.*, 2013c) together with illustrations indicating their arrangement and overlap. Placoid scales (Figs. 8.37(a) and (b)) are typical of sharks or rays. They have a surface structure that generates small-scale vorticity in water, thereby decreasing drag. The ganoid scales (Figs. 8.37(c) and (d)) are composed of a thin surface layer of ganoine, akin in hardness to tooth enamel, riding on a softer but tougher bony scale. These are the hardest scales in fish, and are characteristic of the alligator gar (Figs. 8.37(c) and (d)) and Senegal bichir. The elasmoid class consists of two kinds of scales, cycloid and ctenoid. They have similar shapes, but there are significant differences. For example, the outer surface of the cycloid is smooth, whereas the ctenoid has a comb-like outer surface. Some fish can have both cycloid and ctenoid scales as armor.

A number of studies on fish scales have been conducted including those on the structural arrangement, collagen formation, and orientation (Olson and Watabe, 1980; Zylberberg and Nicolas, 1982; Zylberberg, Bereiter-Hahn, and Sire, 1988), and the mechanical properties (Bruet *et al.*, 2008; Lin *et al.*, 2011). The fish scales provide protection from the

(a)

Exposed
Embedded
1 cm

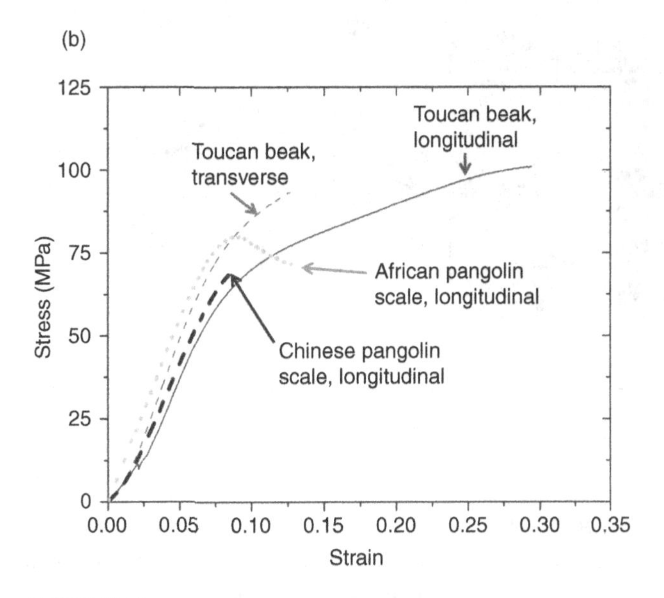

Figure 8.36.

(a) Pangolin and its keratinous scales; (b) tensile response of pangolin scales compared to keratin from the toucan beak. (Figure courtesy W. Yang and B. Wang, UCSD.)

environment and predators. Most fish scales have similar material components to other hard tissues such as bones and teeth. They are mainly composed of type I collagen fibers and calcium-deficient hydroxyapatite, which is also found in bones and teeth. The collagen fibers are usually densely packed lamellae with different orientations from layer to layer,

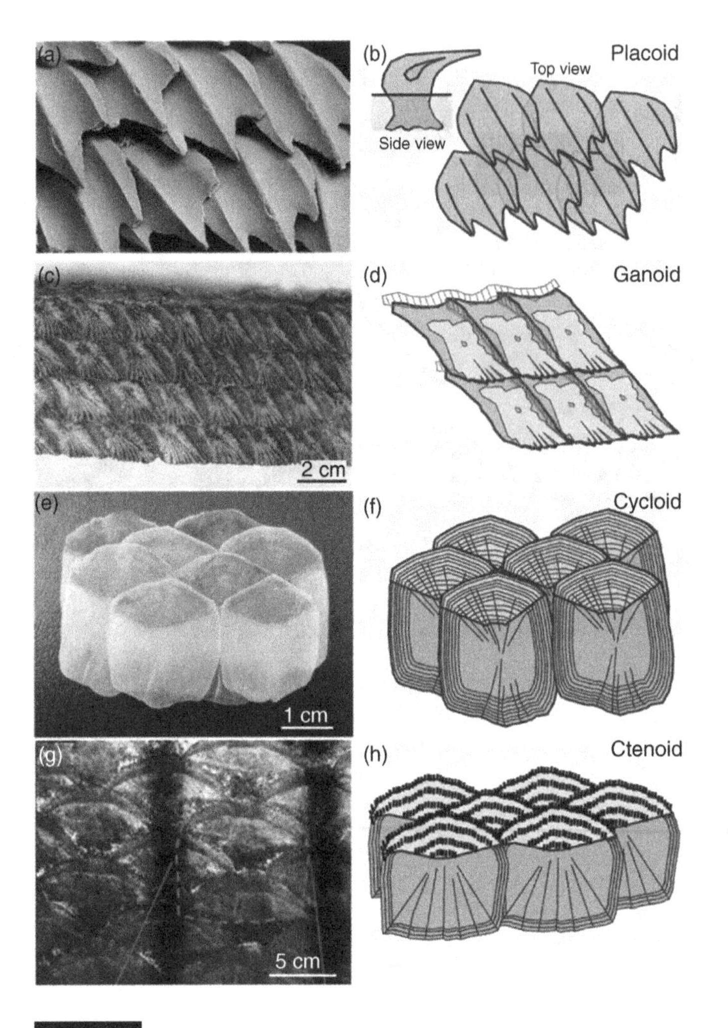

Figure 8.37.

Different types of fish scales and their overlaps: (a), (b) placoid; (c), (d) ganoid; (e), (f) cycloid; (g), (h) ctenoid. (Cosmoid scales not shown.) (Figure courtesy W. Yang.)

forming a plywood structure (Zylberberg *et al.*, 1992; Weiner and Wagner, 1998). Such is the case for the *Pagrus major* (sea bream) scales reported by Ikoma *et al.* (2003). Figure 8.38 shows a TEM micrograph for *P. major* in which several layers are imaged. The different alignment of the sequential layers is clearly seen. Both orthogonal and double twisted plywood patterns are reported. This plywood structure has been proposed to form a Bouligand arrangement. The concept of the Bouligand arrangement was discussed in Section 8.3.1 and will be discussed further at the end of this section.

A fascinating fish is *Polypterus senegalus*, a smallish fish, ~100 mm long, living in estuaries and muddy river bottoms in Africa. It is an example of a living fossil, such as the coelacanth, thought to be extinct since the late Cretaceous but rediscovered in 1938.

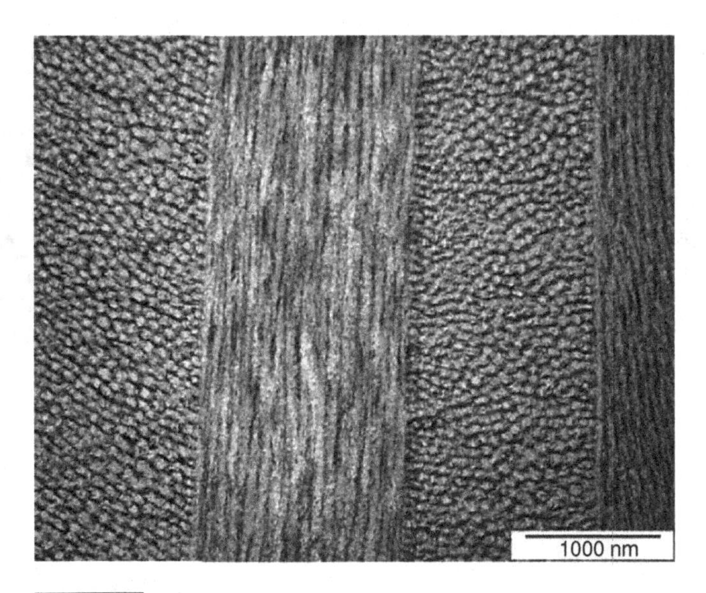

Figure 8.38.

Transmission electron micrograph of the cross-section of scale of *Pagrus major*. (Reprinted from Ikoma *et al.* (2003), with permission from Elsevier.)

Bruet *et al.* (2008) have recently demonstrated that it has a multilayered dermal armor. Following this, Song, Ortiz, and Boyce (2011) analyzed the threat-protection mechanisms offered by the dermal layer against biting attacks from the same species. Nanoindentation measurements on the scales show that distinct reinforcing layers with different moduli and hardness offer a graded protection mechanism. Figure 8.39(a) shows four different organic–inorganic nanocomposite material layers (from outer to inner surface): ganoine, an enamel-like material (thickness 10 μm), dentin (thickness 50 μm), isopedine (thickness 40 μm), and a bone basal plate (thickness 300 μm). The nanoindentation hardness (Fig. 8.39(c)) is seen to vary systematically with the composition of the layer, being highest for ganoine (4.4 GPa) and lowest for bone (<1 GPa). The more collagenous dentin layer has a lower mineral content than ganoine, but is more mineralized than the bone basal plate with a nanoindentation hardness of 1.2 GPa. The third layer, isopedine, consists of superposed orthogonal collagenous layers forming a laminate structure. The bone is under this layer.

On the other side of the size spectrum is the *Arapaimas gigas*, a freshwater fish that lives primarily in Amazon basin lakes and can weigh as much as 200 kg. Figure 8.40 shows the hierarchical structure of the scales, which are quite large and can be up to 10 cm in length. The collagen fibrils form fibers, which, in turn, form bundles with diameters on the order of 1–5 μm. The latter are aligned and organized into lamellae with an average thickness of 50 μm. The collagen fiber alignment has been described by Onozato and Watabe (1979) and Zylberberg *et al.* (1992), among others. The fibril orientation varies from layer to layer. A rough

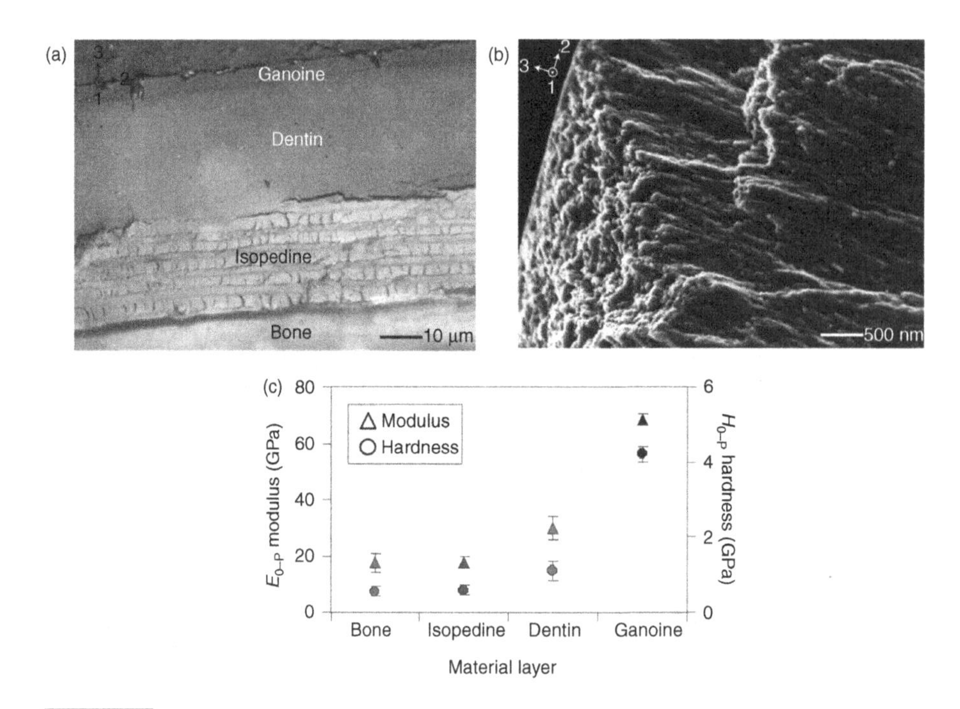

Figure 8.39.

(a) Cross-section showing laminate composite structure of living fossil *Polypterus senegalus* with inside layer comprising bone and outside layer of ganoine, one of the hardest biominerals. (b) Microstructure of ganoine, similar to enamel, containing less than 5% collagen. (c) Variation in nanoindentation hardness from the inside (bone) to the outside (ganoine). (Reprinted by permission from Macmillan Publishers Ltd.: *Nature Materials* (Bruet *et al.*, 2008), copyright 2008.)

external layer, ~600 μm thick, is highly mineralized with hydroxyapatite. As shown later, this structure can undergo significant nonelastic deformation prior to failure, providing considerable toughness.

A cross-section of a scale is shown in Fig. 8.41(a), confirming the uniformity of the layer thickness. The cracks present in some of the layers are due to the drying process. When the collagen fibrils are parallel to the plane of observations the cracks can no longer be seen. Figure 8.41(b) shows the microindentation hardness across the cross-section. The external layer (550 MPa) is harder than the proximal layer (200 MPa), consistent with its higher degree of mineralization.

SEM observation of the scales reveals layers which consist of collagen fibrils with diameters of ~100 nm and the characteristic banding pattern with periodicity of 67 nm is seen in Fig. 8.42. These collagen fibrils were revealed by demineralizing the scales. Adjacent lamellae have the fibrils rotated at angles that vary between 40° and 80°. This is revealed in the SEM micrograph of Fig. 8.43(a). It can be understood that the 75° rotation does not bring the structure into a crossed-lamellar configuration into which layers have the same orientation. This is the essence of the Bouligand arrangement, which produces properties in the surface that are close to isotropy. Thus, the scale is a

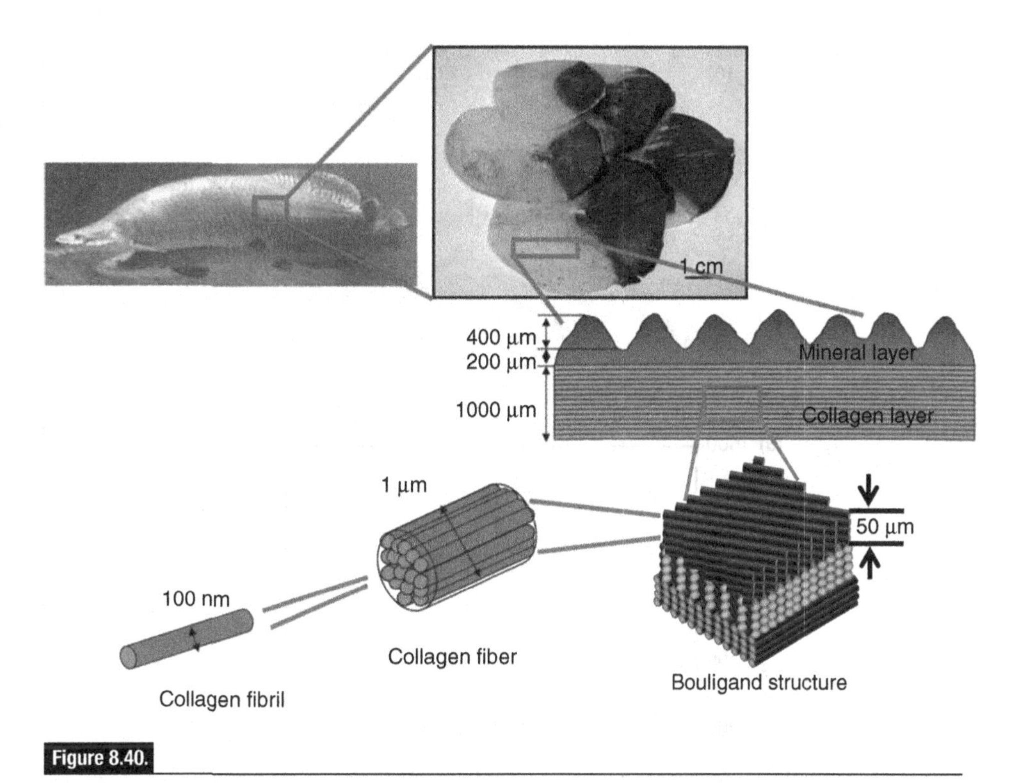

Figure 8.40.

Hierarchical structure of *Arapaimas gigas* scales. (Reprinted from Lin *et al.* (2011), with permission from Elsevier.)

collagen composite laminate. This structure is analogous to that observed in other fish scales, apart from the dimensions. In *Poecilia reticulata* (guppy), the lamellae are ~1 μm thick; in *Carassius auratus* (goldfish), they are around 5 μm (Zylberberg and Nicolas, 1982). The angles between layers have also been found to vary from 36° for some teleosts (Zylberberg and Nicolas, 1982; Zylberberg *et al.*, 1988, 1992), to 90° for *Pagrus major* (sea bream) (Ikoma *et al.*, 2003), to less than 90° for *Hemichromis bimaculatus* (jewelfish).

Arapaimas gigas scales play an important role in protecting this large Amazon basin fish against predators such as the piranha. Piranha teeth form triangular arrays that create a guillotine-like cutting action that is highly effective in slicing through fish tissue. Piranhas are primarily piscivores, as opposed to the carnivorous nature portrayed by the popular media. It was demonstrated that the puncturing ability of the piranha teeth could not penetrate the *A. gigas* scales (Meyers *et al.*, 2012). The ability of *A. gigas* to resist the attack of piranha is clearly the result of the hierarchical structure displayed in Fig. 8.40.

Figure 8.43(b) shows the typical tensile stress–strain curves in the hydrated condition. The Young modulus is ~120 MPa. The strength of the hydrated samples ranges from 15 to 30 MPa. The failure strain is 30–40%, indicating that the scales are quite flexible in water, a requirement for swimming. The tensile strength obtained here is in agreement with the test results of Torres *et al.* (2008), which yielded a value of 22 MPa. Their curves

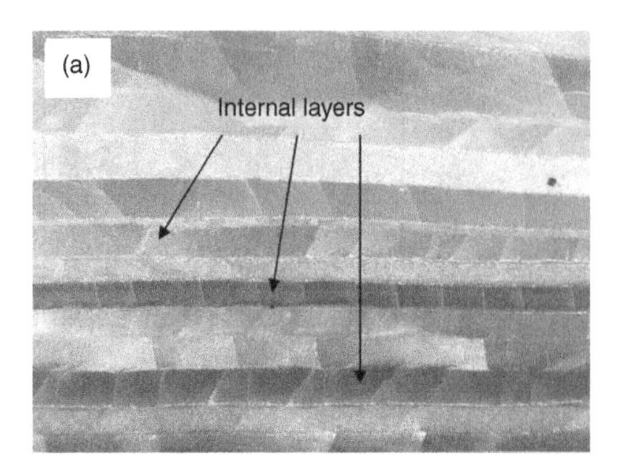

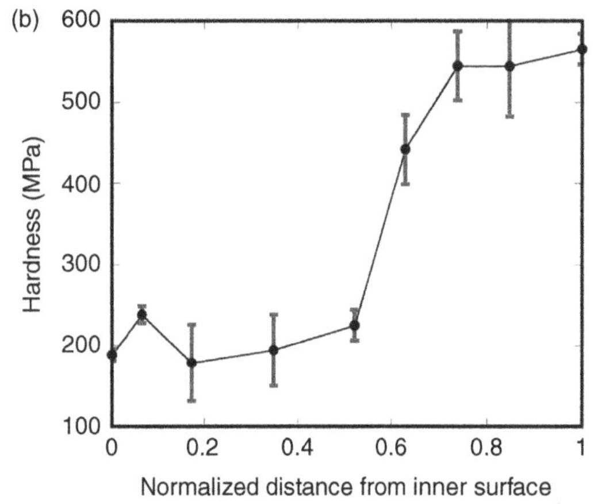

Figure 8.41.

Arapaimas gigas scale: (a) SEM image of the cross-section showing the laminated structure in the internal layers; (b) microindentation through the cross-section from inner surface to outer surface showing that hardness is higher in the external layer. (Reprinted from Lin *et al.* (2011), with permission from Elsevier.)

also exhibited stress drops, evidence, as in our experiments, of "graceful" failure. Several local discontinuities correspond to either the breakage of different layers or the sliding between layers. They are marked by arrows in Fig. 8.43(b). Ikoma *et al.* (2003) reported that the tensile strength of *P. major* scales was 93 MPa, which is higher than our results. The strength of the *A. gigas* scale in the dry condition is about one-half of that of the *P. major*. It is inferred from the low ductility (~5%) that the *P. major* scales were tested in the dry condition, which would increase the tensile strength. The tensile curve also exhibited a pseudo-plastic behavior, i.e. the curve shows nonlinearity prior to failure.

The fracture surface (Fig. 8.43(a)) of a demineralized scale clearly shows the different collagen-fibril orientations in the lamellae. The angle between two adjacent lamellae seems close to 90°, thus forming a plywood structure. However, the twisted plywood structure (e.g. angle of 60°) is a definite possibility as well. Figure 8.43(a)

Figure 8.42.

Collagen fibers in *Arapaimas* scales after demineralization process; note 67 nm periodicity in structure. (Reprinted from Lin *et al.* (2011), with permission from Elsevier.)

shows pulled out or distorted collagen fibrils damaged during the tensile testing. Tensile fracture not only breaks the collagen fibers, but also tears the collagen fibers away from each other. The fracture mechanism seems to be a combination of the collagen fracture and the pulling out of the collagen fibers in a single layer. Tensile testing of the scales carried out in dry and wet conditions shows that the strength and stiffness are hydration dependent. As is the case in most biological materials, the elastic modulus of the scale is strain-rate dependent. The strain-rate dependence on the Young modulus can be expressed by the Ramberg–Osgood equation (Eqns. (7.11a) and (7.11b)):

$$E = C\left(\frac{\dot{\varepsilon}}{\dot{\varepsilon}_0}\right)^d, \tag{8.4}$$

where d is equal to 0.26, which is approximately ten times higher than that of bone (Hight and Brandeau, 1983). Again, we warn the reader that the strain rate should be divided by a reference value to result in a unitless number that can then be raised to any power: $\dot{\varepsilon}/\dot{\varepsilon}_0$. The high value of d is attributed to the higher fraction of collagen in the scales and to the high degree of hydration (30% H_2O).

8.6 Squid beak

Squids are molluscs in the class cephalopods. They have a head, mantle, eight arms, and two tentacles. They have some uniquely interesting aspects due in part to the fact that they do not contain minerals. Their mantle, beak, and suckers will be presented here.

The mantle of the squid is a hollow cylinder. Biopolymers are strong in tension but cannot resist compression without buckling. The mantle is rigid because of the internal

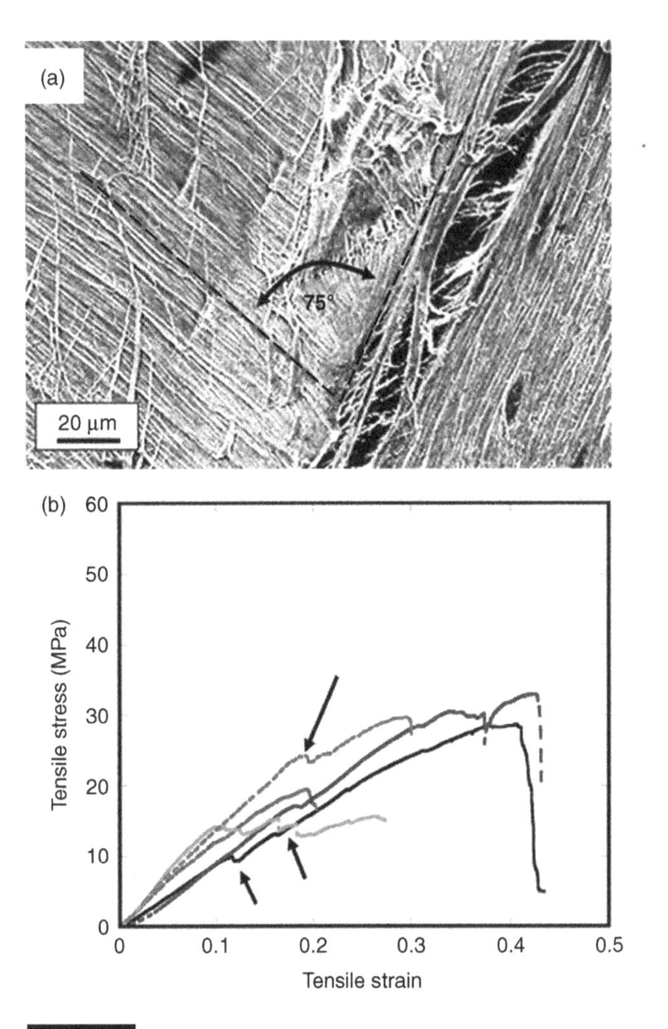

Figure 8.43.

(a) SEM image of scale fracture surface showing different orientations of the collagen fibers (75°). (b) Tensile stress–strain curve for hydrated scale of *Arapaimas gigas* at 10^{-4} s^{-1}; arrows indicate partial failure of collagen layers. (Reprinted from Meyers *et al.* (2012), with permission from John Wiley & Sons Inc.)

water pressure. The principal locomotion method is through an ingenious water-jet mechanism. The water stored inside the mantle is expelled through an orifice by the contraction of the muscles that form a special pattern of orientations (Fig. 8.44). There are inner and outer layers of collagen fibers that are called tunics. These collagen fibers are oriented at ±27° to the longitudinal axis of the mantle. The collagen layers sandwich the central region, which is the muscle. This muscle, upon contraction, creates the water jet. The muscle fibers are oriented in two directions: radially, connecting the outer and inner collagen tunic, and circumferentially. The propulsion takes place by the contraction of the circumferential muscle fibers and consequent reduction of the cavity volume. The

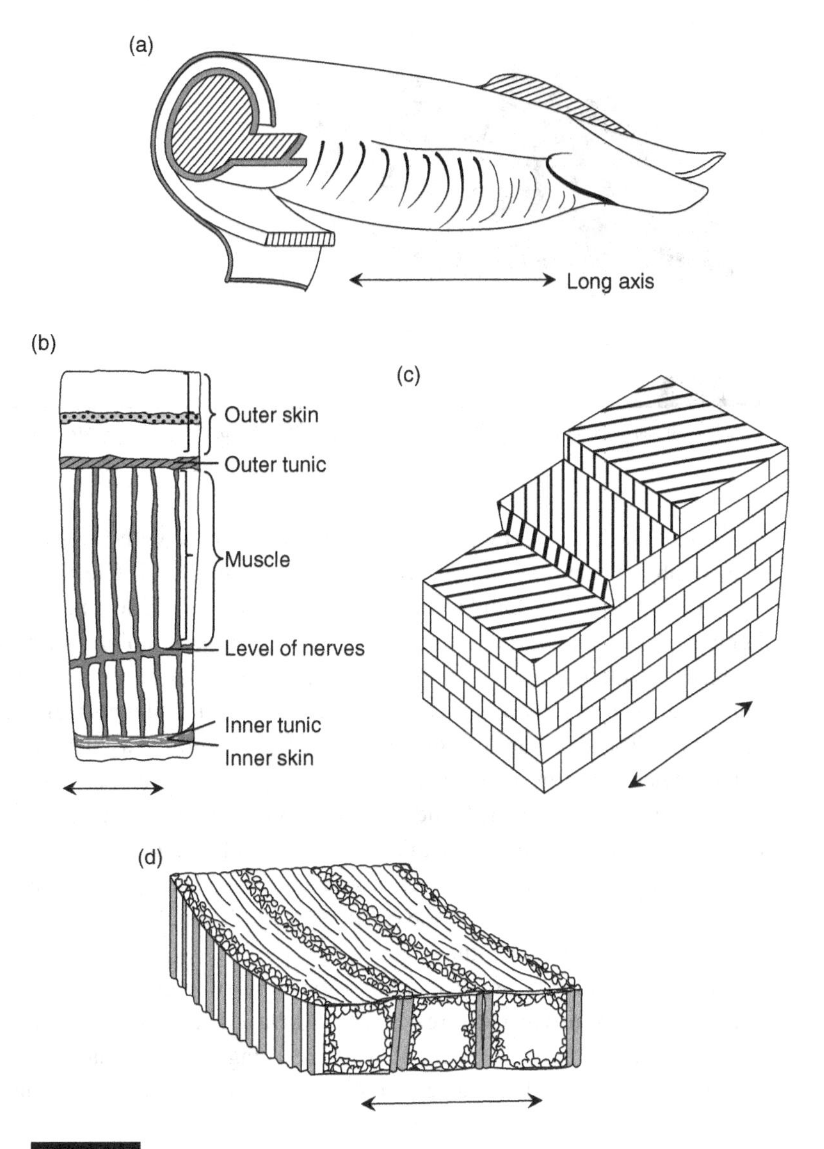

Figure 8.44.

Structure of squid (*Lolliguncula brevis*). (a) Cross- and longitudinal sections of body; (b) cross-section showing muscle sandwiched between inner and outer tunic; (c) outer tunic consisting of a pattern of collagen molecules forming a cross pattern; (d) muscles forming a pattern that is symmetrical with respect to the longitudinal axis. (Reproduced based on Johnson, Soden, and Trueman (1972).)

length of the mantle is unchanged, and therefore the wall thickness increases correspondingly. The jet propulsion in the squid can create significant velocities. For a 350 g *Loligo* squid (Johnson *et al.*, 1972), this velocity is on the order of 2 m/s.

The squid beak is another component with unique characteristics. The necessity of the beak to tear flesh from its prey in the absence of a biomineral poses a significant

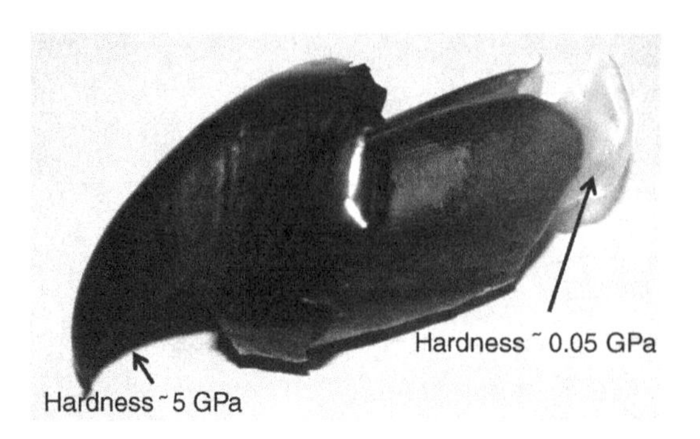

Figure 8.45.

Squid beak (black) embedded in body (lighter); the nanoindentation hardness of two regions is marked.

challenge. The beak somewhat resembles a parrot's beak, with two sharp parts. The beak of a Humboldt squid (*Dosidicus gigas*) is shown in Fig. 8.45. The beak is embedded into the softer tissues of the squid head. The beak is much darker, and the coloration gradually lightens as one moves away from its tip. This change in color (the dark part is called "tanned") is directly related to the presence of proteins in the structure. Whereas the lighter part is primarily a network of chitin in water, the tanned region contains hydrophobic post-translated amino acids that have been identified as "DOPA" (Miserez *et al.*, 2008). When the entire beak and surrounding region are dried, the Young modulus is the same: 5–10 GPa. However, in the hydrated condition, the Young modulus of the tanned region corresponding to the beak tip is ~5 GPa, while it drops to 0.05 GPa in the farthest regions. This indicates that there is a gradient in mechanical stiffness provided by the gradual change in hydration of the chitin network. It should be noted that there are other proteins mixed with the chitin.

The squid tentacles are yet another example of ingenious design to accomplish necessary performance (Miserez *et al.*, 2009a). They have "suckers" which attach themselves to the prey by suction, while resisting any shear action that would allow the prey to slide off. The suckers contain sharp edges along the periphery which can penetrate into the flesh of the prey. The suckers also have a gradient in mechanical properties. In this case, however, the gradient is accomplished by the controlled presence of aligned ducts. The section of the sucker in Fig. 8.46 shows the elongated protein-based fibers with pores between them. The greater the porosity, the lower the elastic modulus.

8.7 Invertebrate jaws and mandibles

Insect mandibles ("jaws") are a pair of attachments near the mouth that are multifunctional: they gather, manipulate, and process food, they catch and attack prey, build nests, and, in some ants, provide propulsion. The mandibles have evolved from legs and move with a horizontal motion unlike the jaws of vertebrates. Figure 8.47 is an SEM

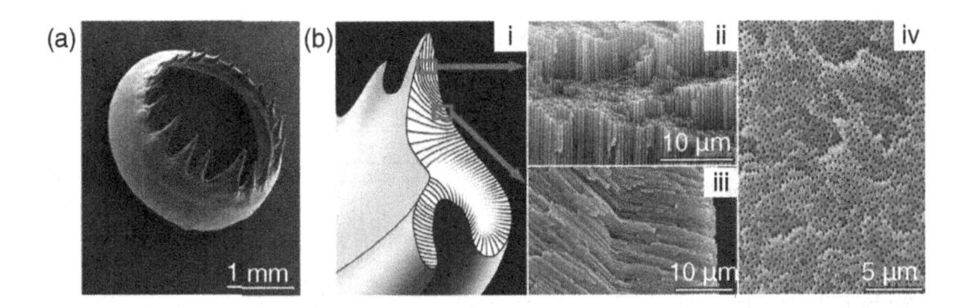

Figure 8.46.

(a) Squid sucker; (b) sketch showing array of protein-based fibers with voids. (Reprinted from Miserez *et al.* (2009a), with permission from John Wiley & Sons Inc.)

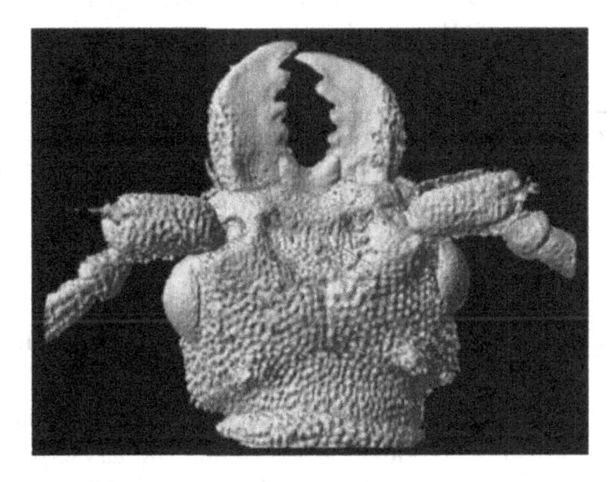

Figure 8.47.

Head and mandibles of a bleach beetle (*Priacma serrata*). (Reprinted from Hörnschemeyer, Beutel, and Pasop (2002), with permission from John Wiley & Sons Inc.)

micrograph of the mandibles of a bleach beetle (*Priacma serrata*) that are as long as the head, showing four teeth on each side. Mandibles, like the rest of the cuticle, are composed of waxes, polysaccharides, and proteins. Chitin provides strength and sclerotin (a protein) gives the hardness. They have a large variety of morphologies – some insects have no teeth on the mandible (nectar feeding), some have serrated-like teeth (grasshoppers), and some have needle-like teeth in piercing or sucking insects such as tree bugs. They are the hardest part of the insect integumentary system and are found to be twice as hard as the rest of the exoskeleton (Hillerton, Reynolds, and Vincent, 1982). This is attributed to the presence of metals such as zinc, manganese, iron, and, in some cases, calcium (Hillerton *et al.*, 1982; Hillerton and Vincent, 1982; Quicke *et al.*, 1998; Schofield, Nesson, and Richardson, 2002; Cribb *et al.*, 2008, 2010). Zinc, in particular, is connected with a 20% increase in hardness (Cribb *et al.*, 2008). The "tools" of an insect (mandible, ovipositor, claws) undergo sclerotization (tanning), which involves extensive

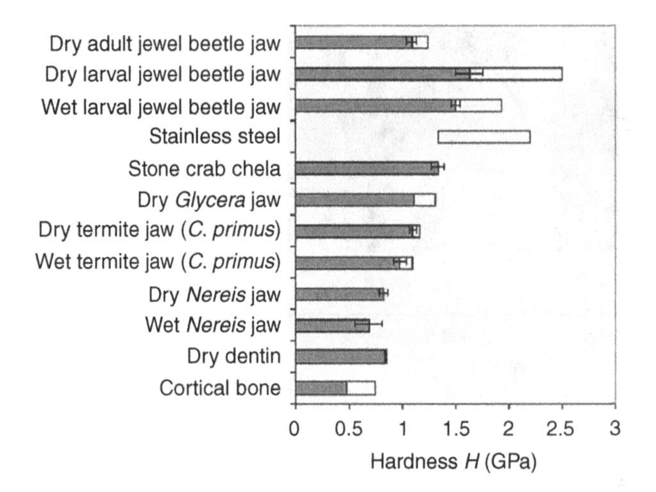

Figure 8.48.

Comparison of the hardness of insect mandibles with other materials. Light boxes represent maximum values, whereas the bars indicate standard errors. Dark boxes represent the mean values. (Reprinted from Cribb *et al.* (2010), with permission from Elsevier.)

cross-linking of the proteins, which thereby darkens and hardens the material. It is during the sclerotization process that the transition elements are incorporated.

One interesting study has been carried out on the comparison of adult and larval jewel beetle (*Pseudotaenia frenchi*) mandibles. The dry larval mandible (Cribb *et al.*, 2010) was found to have a higher hardness than the adult one, despite not having any metals incorporated into the structure, as shown in Fig. 8.48. The larvae must bore through wood to emerge and thus develop a hard mandible that is crucial to their survival. This work provides evidence that a purely organic material has a hardness that is higher than cortical bone and even approaches values for stainless steel.

Mandibles can snap at incredible speed and with surprisingly high forces. The speed of the trap-jaw ant (*Odontomachus bauri*) has been measured at up to 64 m/s with an acceleration of 100 000*g* and a force that exceeds the body mass (12–14 mg) (Patek *et al.*, 2006). Within the animal kingdom, this is the fastest measured predatory strike. Additionally, by striking their mandibles on a surface, these ants can jump vertically ~8 cm and horizontally ~40 cm to evade predators. The mechanical properties of the mandible of a ground beetle (*Scarites subterraneus*) were tested using nanoindentation (Dickinson, 2008) and compared with those of the abdomen, as shown in Fig. 8.49. The hardness (~1 GPa) and reduced elastic modulus (~30 GPa) of the mandible were much higher than those at the abdomen (hardness ~0.35 GPa; reduced modulus ~9 GPa). This may be attributed to an increased level of heavy metals (such as Zn and Mn) and halogens incorporating into the nano-scale structure of the exoskeleton during maturation and likely correlate to the different functions of fighting and protection of the two areas.

The proboscis is an organ that can be inserted into tissue (plant or animal) to suck out nutrients. It is a modification in which the mandible and maxilla (upper jaw) are

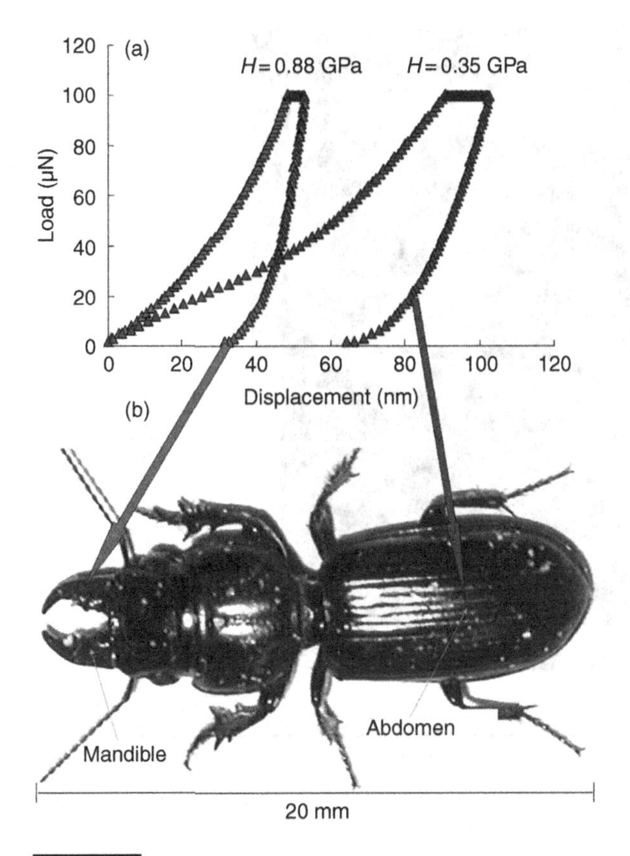

Figure 8.49.

(a) Representative load–displacement curves and hardness values (H) generated from nanoindentation of mandibular and abdominal exoskeleton of *Scarites subterraneus* beetle. (b) *S. subterraneus*, showing the two regions where nanoindentation tests were performed. (Figure courtesy Dr. M. Dickinson.)

combined to form a needle-like appendage. Insects of the order Hemiptera (e.g. cicadas, aphids) and Diptera (mosquitoes) employ the proboscis to pierce plants to suck out the sap – and the female mosquito needs blood to lay eggs. Butterfly proboscises are long and tubular, having a sharp tip, and are coiled under the face when not in use.

Figure 8.50 shows the proboscis of the mosquito (*Culex pipiens*). The proboscis is composed of outer sheath, which is used to detect the surrounding environment, such as temperature and chemical balance, while there are two tubes which enter its unsuspecting prey. One of them is terminated with an inner stylet that is used to pierce the skin and draw blood while the other injects an anticoagulant to keep the blood flowing. Figure 8.50 shows that there are serrations on the edge of the stylet, possibly designed to reduce nerve stimulation during a bite by increasing the efficiency of the cutting edge (Ikeshoji, 1993). This is in congruence with Oka *et al.* (2002), who concluded that the initial bite of a mosquito is painless because of the highly serrated proboscis. They used

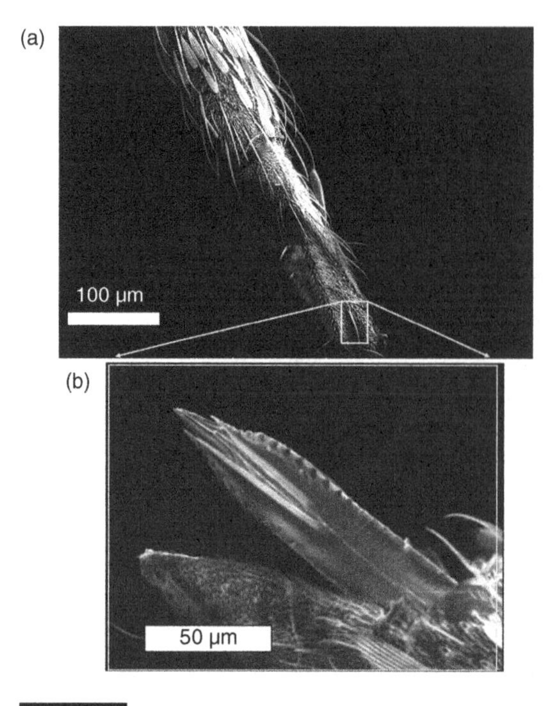

Figure 8.50.

SEM micrographs of a mosquito proboscis. (a) Proboscis covered with hairy sheath; (b) serrated stylet designed to section tissue for dual needle penetration.

this concept and developed a mosquito-inspired hypodermic needle, which is discussed in Section 12.1.17.

The stinger of the common bee (*Apis mellifera*) is yet another example of the efficiency of serrations, as shown in Fig. 8.51. It is equipped with reverse-facing barbs, which are used to propel the needle deep into the tissue of its prey. These barbs are approximately 10–20 μm long and run along the shaft of the stinger. When the insect has used the stinger, it stays embedded in the skin.

8.8 Other natural fibers

Natural vegetable fibers are also known as lignocellulosic fibers. Fig. 8.52 shows four natural fibers: sisal, bamboo, coconut fiber (known as coir), and piassava. They can have excellent tensile properties and have been used by humans since prehistory (for items such as baskets, ropes, fabric for clothes, floors, and roofing materials). More recently, they are being incorporated into components with high-tech applications such as automobiles and other transportation systems. In comparison with fiber-glass-reinforced composites, they present some definite advantages such as lower density, better thermal and sound insulation, and reduced skin irritation. In the life cycle, they also present an

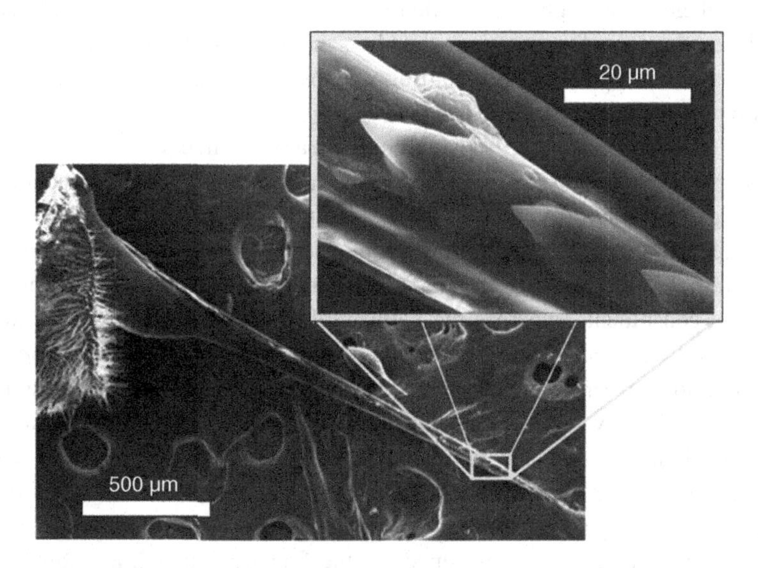

Figure 8.51.

Bee (*Apis mellifera*) stinger; note directional serrations that are responsible for the stinger staying embedded in epidermis, in contrast with mosquito proboscis, which is removed after blood extraction.

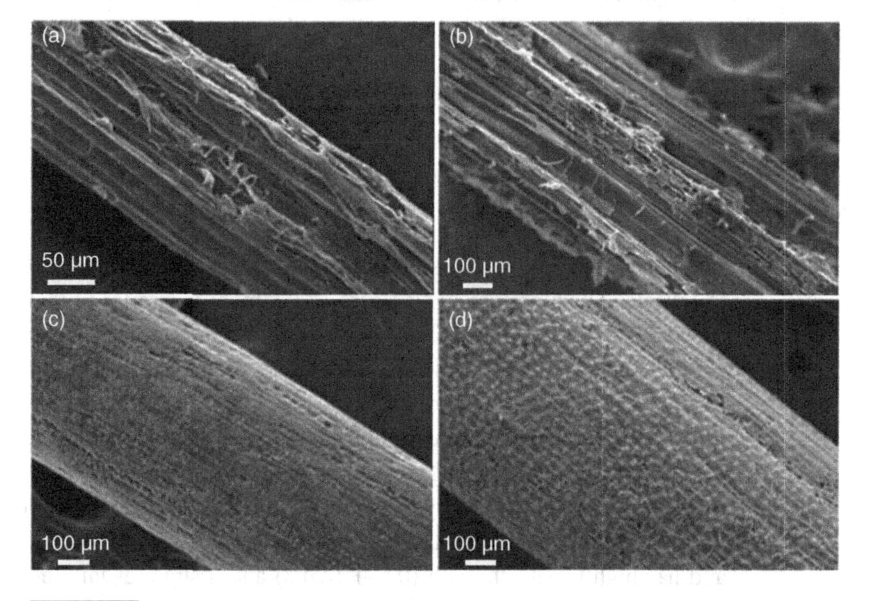

Figure 8.52.

Examples of natural lignocellulosic fibers from Brazil: (a) sisal, (b) bamboo, (c) coir (coconut fiber), and (d) piassava. (Figure courtesy S. N. Monteiro.)

advantage in the degradation. The fibers absorb water and can be degraded by bacteria. The cutting tools for processing the fibers undergo less erosion than with fiber glass. The construction industry is also starting to consider them for insulation and in reinforcing components in cement.

Bledzki and Gassan (1999) point out that a single fiber of a plant is a complex natural composite consisting of several cells. This can be seen in the cross-sectional photographs of Fig. 8.53, where several cells can be distinguished in some of the fibers. These cells are composed of cellulose microfibrils connected by lignin and hemicelluloses. As seen in Section 3.5, the lignin is an amorphous binder with both aliphatic and aromatic polymeric constituents, whereas the hemicellulose comprises polysaccharides that remain associated with the cellulose even after lignin is removed (Bledzki and Gassan, 1999). The ratio between cellulose and lignin/hemicelluloses as well as the spiral angle of the microfibrils vary from one natural fiber to another. The different arrangements of the lignocellulosic fibers and the diameter of the fibers determine their tensile strength, which can attain high values. Figure 8.54 shows that the diameter of the fiber can influence its strength dramatically. The explanation is given in Fig. 8.54(b). Each fiber is made of sub-fibers, as evidenced in the fractured curauá, as shown in Fig. 8.54(b). These sub-fibers have smaller flaws than the larger fibers and therefore a higher strength. Monteiro *et al.* (2011a, b) established tensile strength values above 1 GPa for three different fibers: curauá, sisal, and ramie at the smallest possible diameters. The strength of these lignocellulosic fibers is still considerably lower than the tensile strength of glass fibers, 2–3.5 GPa, but natural fibers present other advantages. Figure 8.55 demonstrates a potential application of natural fiber for automobile manufacturing (as components in Mercedes-Benz vehicles).

Summary

- The function of a ligament is to connect bone to another bone. Tendon connects bone to muscle. The hierarchical structure of ligament is similar to that of tendon and is based on collagen fibers. The mechanical behavior of tendon and ligament are similar. Collagen fibrils uncrimp (straighten), stretch elastically, slide, and eventually fracture, creating a J-curve response.
- Silk is composed of two proteins: fibroin (tough strands) and sericin (gummy glue). The mechanical properties (strength and maximum elongation) can vary widely, depending on the application intended by the animal. Spiders produce two types of silk. The first is the dragline, used in the radial components of the web. This is the structural component, and has high tensile strength (0.6–1.1 GPa) and a strain at failure of about 6%. It is produced in the major ampullate and is therefore often called MA silk. The tangential components, called spiral, are intended to capture prey, and are "soft" and "sticky." The strain at failure can exceed 16 (or 1600%), according to some sources.
- The stress–strain response of many tendons and ligaments, composed primarily of collagen fibers, has the characteristic J shape.

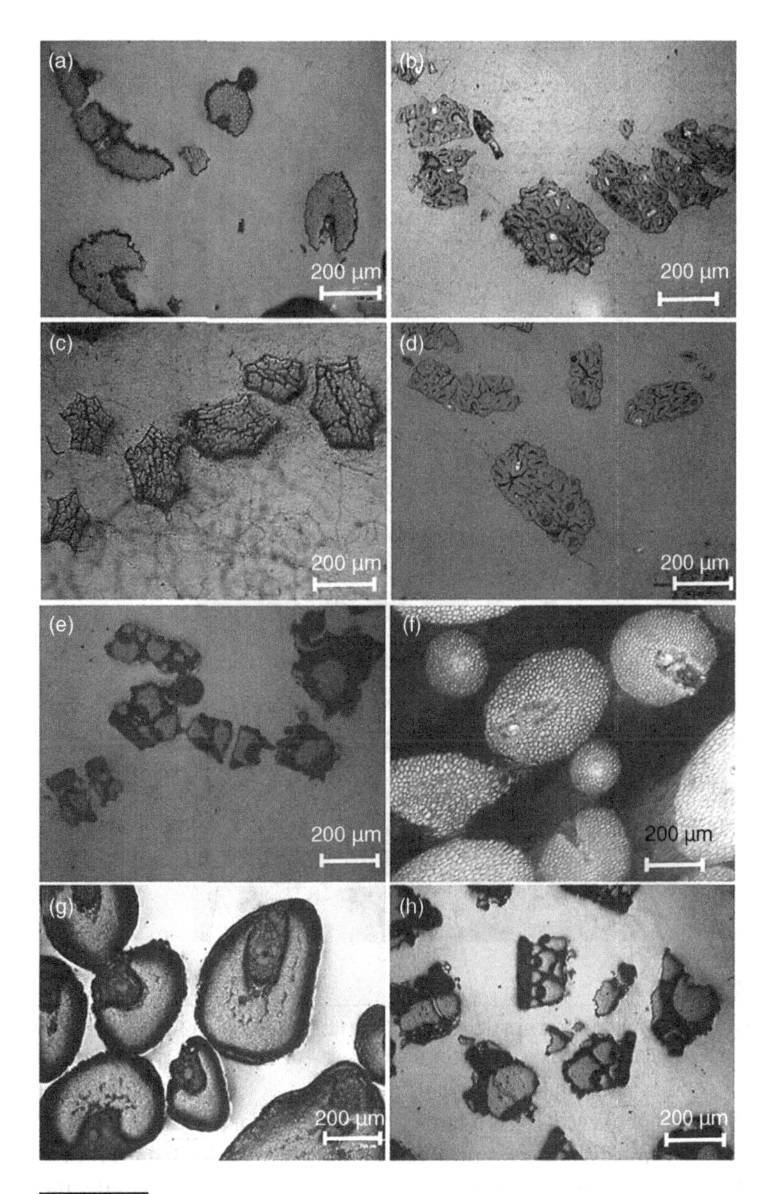

Figure 8.53.

Cross-sections of lignocellulose fibers: (a) sisal, (b) ramie, (c) curauá, (d) jute, (e) bamboo, (f) coir, (g) piassava, and (h) buriti. (Figure courtesy S. N. Monteiro.)

- Silk is composed of β-sheet (10–15 vol.%) nanocrystals embedded in a disordered matrix. It also has the J-shape stress–strain curve. Similar to collagen, the low-stress region corresponds to uncoiling and straightening of the protein strands. This region is followed by entropic unfolding of the amorphous strands and then stiffening from the β-sheets. Despite the high strength, the major molecular interactions in the

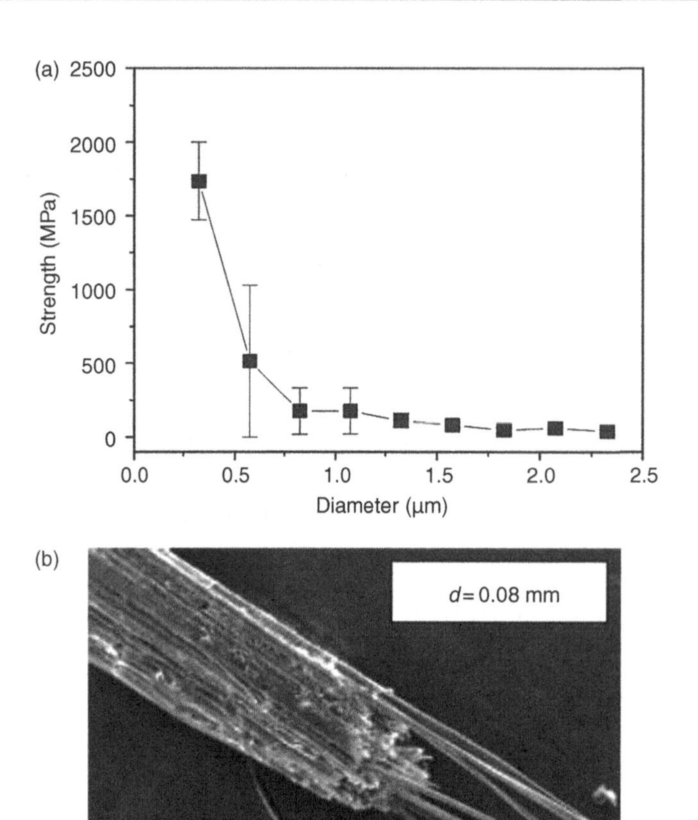

Figure 8.54.

(a) Effect of fiber diameter (piassava) on its strength. (Taken from Monteiro *et al.* (2011a); figure courtesy S. N. Monteiro.) (b) Fractured fiber (curauá) showing that it is composed of fibrils with ~10 μm diameter. (Figure courtesy Felipe Perisse Duarte Lopes.)

β-sheets are weak hydrogen bonds. Molecular dynamics calculations reveal the detailed mechanisms of the deformation inside the β-sheets. For a stack having a height $L \leq 3$ nm the shear stresses are more significant than the flexure stresses, and the hydrogen bonds contribute to the high strength obtained (1.5 GPa). However, if the stack of β-sheets is too high, it undergoes bending, with tensile separation between adjacent sheets.

- Arthropods are the largest phylum of animals including the trilobites, chelicerates (spiders, mites, and scorpions), mariapods (millipedes and centipedes), hexapods (insects), and crustaceans (crabs, shrimps, lobsters, and others). All arthropods are covered by a rigid exoskeleton, which is periodically shed as the animal grows. The arthropod exoskeleton is multifunctional: it not only supports the body and resists mechanical loads, but also provides environmental protection and resistance to

Figure 8.55.

Natural fiber components in Mercedes-Benz vehicle. (Used with kind permission of Mercedes-Benz.)

desiccation. It is a laminated biological composite composed of chitin fibers embedded in a protein matrix. In the crustacean species, the exoskeleton shows a high degree of mineralization, typically calcium carbonate in calcite form with small amounts of amorphous $CaCO_3$ deposited within the chitin–protein matrix. However, the mineral phase is often absent in the exoskeletons of insects.

- Arthropod exoskeletons have a well-defined hierarchical organization. At the molecular level is the polysaccharide chitin. Several chitin molecules arrange in an anti-parallel fashion forming α-chitin crystals. The next structure level consists of 18–25 such molecules, wrapped by proteins, forming nanofibrils of about 2–5 nm diameter and length about 300 nm. These nanofibrils further assemble into bundles of fibers of about 50–300 nm diameter. The fibers then arrange parallel to each other and form horizontal planes. These planes are stacked in a helicoid fashion, creating a twisted plywood or Bouligand structure. The Bouligand (helical stacking) arrangement provides structural strength that is in-plane isotropic in spite of the anisotropic nature of the individual fiber bundles. In crustaceans, the minerals are mostly in the form of crystalline $CaCO_3$, deposited within the chitin–protein matrix. In the vertical direction, the crab exoskeleton shows well-developed, high-density pore canals (spacing between canals is ~2 μm) containing tubules penetrating through the exoskeleton.
- Keratin consists of dead cells that are produced by biological systems (keratinocytes). The most common examples are hairs, nails, hooves, and horns in mammals, and beaks, talons, and feathers in birds. Whereas mammals have primarily alpha keratin, birds and reptiles have beta keratin.
- The hierarchical structure of horn from a desert bighorn sheep consists, at the lowest level, of two polypeptide chains (types I and II), which belong to the family of related proteins, and form two-strand coiled-coil molecules approximately 45 nm in length and 1 nm in diameter. These molecules are helically wound and assemble into microfibrils (called intermediate filaments, IFs) forming "superhelical" ropes of 7 nm diameter. The α-helices are mainly parallel to the long axis of the ropes.

These IFs are embedded in a viscoelastic protein matrix. These lamellae are flat sheets that are held together by other proteinaceous substances. Long tubules extend the length of the horn interspersed between the lamellae. The resulting structure is a three-dimensional laminated composite that consists of fibrous keratin, and has a porosity gradient across the thickness of the horn.

- Fish scales constitute flexible dermal armor and offer protection against predators at a low fractional increase in weight (~5%). They are classified into placoid (sharks and rays), ganoid (bony armor of alligator gar and bichir), elasmoid (the most common flexible scales), and cosmoid. In *Polypterus senegalus* (bichir), the scales consist of a thin layer of hard ganoine, akin to dental enamel, supported by dentin, which in turn rests on isopedine and a thicker bone basal plate. The surface layer is hardest, and the foundation provides toughness.
- Similar in design but different in structure is the elasmoid scale of the Arapaimas, the largest Amazon basin fish. The surface consists of a corrugated highly mineralized layer riding on a flexible foundation that is made of lamellae of parallel collagen mineralized fibrils. Each layer is characterized by a specific collagen orientation so that the scale has in-plane mechanical properties that are fairly isotropic. Arapaimas scales offer protection against piranha attacks and therefore they share the same habitat of riverine lakes. Other elasmoid scales have the same structure; however, the thickness of the lamellae scales with fish dimensions.
- Important parameters in the scales are the degree of imbrication (or overlap), the aspect ratio, and the scale length as a fraction of fish length.
- The squid is a mollusc belonging to the class of cephalopods. The necessity of a squid's beak to tear flesh from its prey in the absence of a biomineral poses a significant challenge. The beak somewhat resembles a parrot's beak, with two sharp parts. The beak is embedded into the softer tissues of the squid head. The beak is much darker, and the coloration gradually lightens as one moves away from its tip. This change in color (the dark part is called "tanned") is directly related to the presence of proteins in the structure. Whereas the lighter part is primarily a network of chitin in water, the tanned region contains hydrophobic post-translated amino acids that have been identified as "DOPA." When the entire beak and surrounding region is dried, the Young modulus is the same: 5–10 GPa. However, in the hydrated condition, the Young modulus of the tanned region corresponding to the beak tip is ~5 GPa, and it drops to 0.05 GPa in the farthest regions. This indicates that there is a gradient in mechanical stiffness provided by the gradual change in hydration of the chitin network.
- There are many natural lignocellulosic fibers that have outstanding mechanical properties and are being investigated for use in composites: sisal, bamboo, coconut fiber (known as coir), and piassava. In comparison with fiber-glass-reinforced composites, they present some definite advantages, such as lower density, better thermal and sound insulation, and reduced skin irritation. In the life cycle, they also present an advantage in the degradation. The fibers absorb water and can be degraded by

bacteria. In the case of the piassava, the strength increases up to 2 GPa when the fiber diameter is decreased.

Exercises

Exercise 8.1 An artery with dimensions inner diameter = 12 mm, outer diameter = 20 mm is subjected to increasing pressures. The tensile strength of the collagen–elastin matrix is equal to 10 MPa at a strain of 0.17 and 50 MPa at a strain of 0.25; assume an exponential Fung stress–strain relationship:

$$\sigma = Ae^{\alpha\varepsilon}.$$

The maximum stress is 80 MPa.

(a) Determine the parameters A and α.
(b) At what pressure will it burst? Assume that the volume remains constant during expansion ($\upsilon = 0.5$).

Exercise 8.2 Keratin from the beak of a toucan follows the Ramberg–Osgood law. The linear part of the stress–strain curve is shown in Fig. 8.36(b); the tests in Fig. 8.36(b) were conducted at 10^{-3} s^{-1}. Determine what the slopes should be at 10^{-1} and 10^{-5} s^{-1}. The Ramberg–Osgood exponent for beak keratin is equal to $d = 0.05$.

Exercise 8.3 Calculate the weight of a breast implant filled with silicone (density 1.3 g/cm^3) if the diameter is 12 cm and the height is 6 cm. Assume an oblate spheroid. What weight advantage would be obtained using a saline solution (density 1.1 g/cm^3)?

Exercise 8.4 The saline solution in breast augmentation surgery can be introduced in liquid form after the empty implant has been inserted. What happens to the solution when it is heated to room temperature?

Exercise 8.5 Early breast implants used a polyvinyl alcohol formaldehyde polymer sponge. What negative effects were caused by these implants?

Exercise 8.6 Do breast implants burst when a diver descends beyond a critical depth? Why? What effects would long-distance running have on the implants?

Exercise 8.7 What is the Bouligand structure? List at least three natural materials that have this structure and explain with sketches.

Exercise 8.8 Spiders can produce two types of silk: major ampullate (MA) silk and viscid silk. Compare MA and viscid silk based on their function, microstructure, and mechanical properties.

Exercise 8.9 What is BioSteel®? Explain how it is made and give at least two potential applications.

Table E8.10.

Pressure (psi)	Diameter (in)
20	0.0815
40	0.0845
60	0.0885
80	0.0910
100	0.0930
120	0.0950
140	0.0970
160	0.0990
180	0.1005
200	0.1037
220	0.1060
240	0.1085
260	0.1110
270	0.1125
277	burst

Exercise 8.10 The data in Table E8.10 were obtained by measuring the pressure applied to a lumen and its increase in radial diameter. The lumen is inserted into the arteries and inflated at the region of constriction. From these results, plot the radial stress vs. strain experienced by the lumen. The wall thickness is 0.014 in. What is the failure stress of the polymer (low-density polyethylene)?

9 Biological elastomers

Introduction

"Soft" biological materials have distinct applications in tissues and organisms. They provide tensile resistance and enable stretching under controlled conditions. In Chapter 3 we saw that they exhibit the characteristic J-curve response, with a stress–strain (or stress–stretch ratio) in which the slope increases with strain ($d^2\sigma/d\varepsilon^2 \geq 0$). This response is very similar to synthetic elastomers, and this is why we call them "biological elastomers." First we introduce the different constitutive equations that represent this response (Section 9.1). These equations have different origins and various purposes. Some of them are based on the fundamental physical phenomena whereas others are just convenient fitting. The Flory–Treloar, Mooney–Rivlin, Arruda–Boyce, and Ogden equations are for rubbers (synthetic elastomers), whereas the Fung and worm-like chain models were specifically designed for J-curve biological materials (principally collagen and elastin). Then we describe important biological materials that exhibit this response: skin and blood vessels. This is supplemented by two interesting materials, in which a phase transition creates an inflection point in the curve ($d^2\sigma/d\varepsilon^2 = 0$): whelk eggs and wool.

9.1 Constitutive equations for soft biopolymers

9.1.1 Worm-like chain model

We start with a single-protein molecule being pulled in tension. One way to visualize it is as a worm-like rod. A force is required to extend it, decreasing the curvature. The final state is when the worm becomes a straight line. Figure 9.1(a) shows a protein molecule being extended in an atomic force microscope (AFM). The curve has a jagged appearance due to the unfolding of the "bundles," but each of the teeth in the force–displacement curve corresponds to the well-known J-curve behavior.

For a single chain, the worm-like chain (WLC) model, based on the entropic effects of deformation, has the following form:

$$f = \frac{k_B T}{L_p}\left(\frac{1}{4\left(1 - \frac{z}{L}\right)^2} - \frac{1}{4} + \frac{z}{L}\right), \tag{9.1}$$

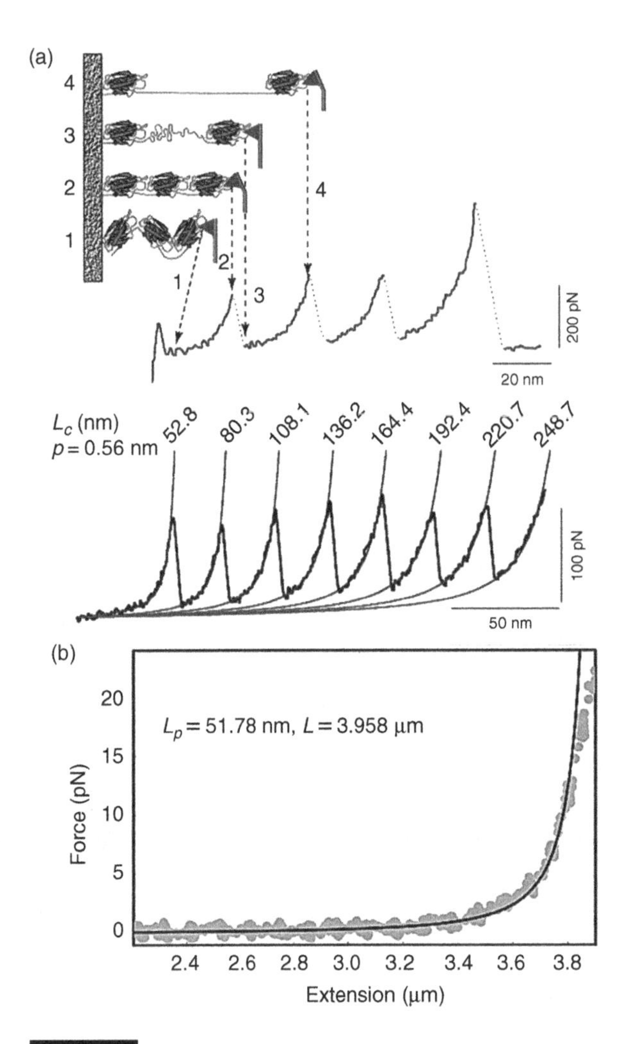

Figure 9.1.

(a) Domain deformation and unfolding of a multidomain protein under stretching with AFM. L_c and p are its contour length and persistence length, respectively. (From Fisher *et al.* (1999), with permission from Elsevier.) (b) WLC modeling of a DNA chain having a length L = 3.958 μm when completely stretched (contour length of molecule); L_p is the characteristic (persistence) length of the folds. (Taken from http://biocurious.com/2006/07/04/wormlike-chains, with kind permission of Philip Johnson.)

where k_B is the Boltzmann constant, T is the absolute temperature, L_p is the persistence (characteristic) length, and L is the contour (total) length of the fibers. This equation is especially useful in the prediction of DNA unfolding. Such a curve is shown in Fig. 9.1(b) and compared with experimental results. It can be seen that f, the force, is equal to zero for $z = 0$; as z approaches $L, f \rightarrow \infty$.

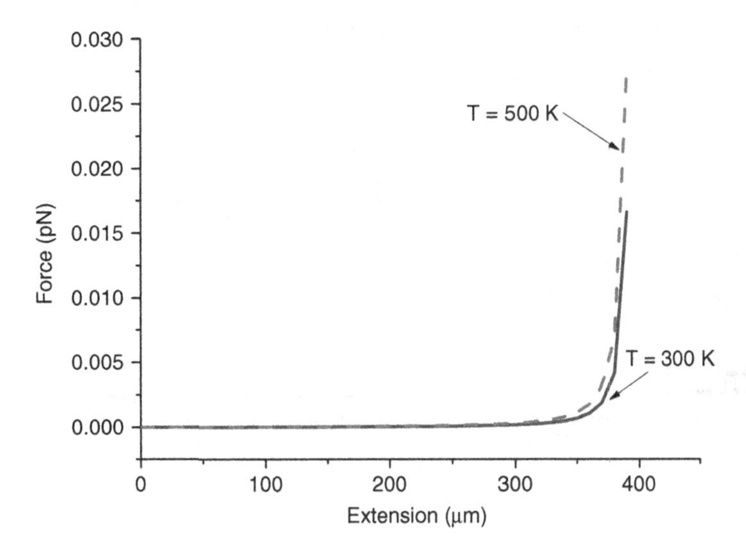

Figure 9.2.

Force (f) and extension (z) for a protein at 300 K and 500 K using the WLC model.

Example 9.1 Using the WLC model, plot the force f vs. extension z for a protein at 300 K and 500 K. Given: stretched length = 400 μm, persistence length = 100 μm.

Solution According to the WLC model,

$$f = \frac{k_B T}{L_p}\left(\frac{1}{4\left(1-\frac{z}{L}\right)^2} - \frac{1}{4} + \frac{z}{L}\right),$$

where $k_B = 1.3806 \times 10^{-23}$ J/K $= 1.3806 \times 10^{-5}$ pN μm/K, $L_p = 100$ μm, and $L = 400$ μm.

At 300 K:

$$f = 4.1418 \times 10^{-5}\left[\frac{1}{4\left(1-\frac{z}{400}\right)^2} - \frac{1}{4} + \frac{z}{400}\right].$$

At 500 K:

$$f = 6.9030 \times 10^{-5}\left[\frac{1}{4\left(1-\frac{z}{400}\right)^2} - \frac{1}{4} + \frac{z}{400}\right].$$

The plot is shown in Fig. 9.2.

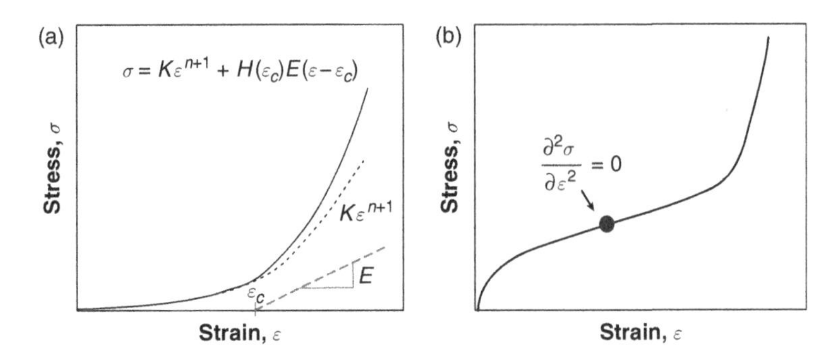

Figure 9.3.

Two types of tensile response exhibited by biological materials: (a) J curve ($d\sigma/d\varepsilon > 0$); (b) curve with inflection point $d^2\sigma/d^2\varepsilon = 0$.

9.1.2 Power equation

A simple formulation is given first; in a J curve, the slope of the stress–strain (σ–ε) curve increases monotonically with strain. Thus, one considers two regimes:

(a) unfurling and straightening of polymer chains:

$$\frac{d\sigma}{d\varepsilon} \propto \varepsilon^n \ (n > 1); \tag{9.2}$$

(b) stretching of the polymer chain backbones:

$$\frac{d\sigma}{d\varepsilon} \propto E, \tag{9.3}$$

where E is the elastic modulus of the chains. The combined equation is

$$\sigma = k_1\varepsilon^{n+1} + H(\varepsilon_c)E(\varepsilon - \varepsilon_c). \tag{9.4}$$

Here, k_1 is a parameter and H is the Heaviside function, which activates the second term at $\varepsilon = \varepsilon_c$, where ε_c is a characteristic strain at which collagen fibers are fully extended. Subsequent strain gradually becomes dominated by chain stretching. The computational results by Gautieri *et al.* (2010) on collagen fibrils predict a value for $n = 1$. This corresponds to a quadratic relationship between stress and strain ($\sigma \propto \varepsilon^2$) and has the characteristic J shape. However, experimental results on skin have a better fit with higher values of n. Figure 9.3(a) shows schematically a curve that follows Eqn. (9.4). Elastomers, on the other hand, exhibit a response that shows an inflection point ($d^2\sigma/d\varepsilon^2 = 0$). This is also the case for biological materials that exhibit a phase change.

9.1.3 Flory–Treloar equations

The initial portion of the tension curve for rubber, where the slope (stiffness) decreases, has been modeled by considering entropic terms (Flory, 1964; Treloar, 1975). The basic assumptions are that the extension is solely due to configurational changes of the randomly oriented chains (no chemical bond stretching, no change in enthalpy) and that there is a Gaussian distribution of the randomly oriented protein chains. As opposed to J-curve materials, the configurational entropy is expressed by the Boltzmann equation $S = k_B \ln\Omega$ (where S = entropy, k_B = Boltzmann's constant, Ω = number of ways to arrange a chain) and defining the force, $F = (d\Delta G/dx)$, where ΔG is the change in the Gibbs free energy and x is the displacement. Considering a cubic solid (ℓ_0^3) that is stretched uniaxially in tension maintaining a square cross-section, the true stress can be derived as (Treloar, 1975):

$$\sigma = nkT\left(\lambda - \frac{1}{\lambda^2}\right), \tag{9.5}$$

where n is the number of chain segments and $\lambda = l_1/l_0$ is the stretch ratio in the loading direction ($\varepsilon = \lambda - 1$). This is known as the Treloar equation and also assumes an isotropic material with no change in pressure, volume, or temperature (Treloar, 1975). This expression has been shown to predict fairly well the stress–strain behavior of rubbers for $\lambda < \sim 4$ ($\varepsilon < \sim 3$) (Treloar, 1975).

9.1.4 Mooney–Rivlin equation

Mooney arrived at an expression assuming an isotropic, isometric solid using a strain-energy function (Mooney, 1940) (which was later modified by Rivlin and Saunders (1951) to give the Mooney–Rivlin equation for the true stress):

$$\sigma = \left(2C_1 + \frac{2C_2}{\lambda}\right)\left(\lambda^2 - \frac{1}{\lambda}\right), \tag{9.6}$$

where C_1 and C_2 are materials constants. The model is valid for strains up to ~2 (e.g. auto tires, where C_1 and $C_2 \sim 0.2$ MPa).

9.1.5 Ogden equation

A phenomenological model that is widely used is the Ogden equation, developed initially for elastomeric materials such as rubber. It is based on the following one-term strain-energy density per unit volume:

$$\Phi = \frac{2\mu}{\alpha^2}(\lambda_1^\alpha + \lambda_2^\alpha + \lambda_3^\alpha - 3), \tag{9.7}$$

where λ_1, λ_2, and λ_3 are the principal stretch ratios, α is a strain-hardening exponent, and μ can be interpreted as a shear modulus. The physical interpretation of this equation is simple: the deformation energy is a power function of the stretch ratios. The change in strain-energy density under a uniaxial stress ($\sigma_1 \neq 0$ and $\sigma_2 = \sigma_3 = 0$) is defined as

$$d\Phi = \sigma_1 d\lambda_1 \quad \text{or} \quad \sigma_1 = \frac{d\Phi}{d\lambda_1}. \tag{9.8}$$

Assuming a constant volume (a reasonable approximation for biological materials at ambient pressure), the deformed volume V_1 is equal to the initial volume V_0. Letting $V_0 = V_1 = 1$, we have

$$V_0 = V_1 = (1 + \varepsilon_1)(1 + \varepsilon_2)(1 + \varepsilon_3) = \lambda_1\lambda_2\lambda_3 = 1. \tag{9.9}$$

Thus, the strain-energy density is, for uniaxial compression or extension with k being the ratio between the two stretch ratios orthogonal to the longitudinal one ($\lambda_2 / \lambda_3 = k$):

$$\Phi = \frac{2\mu}{\alpha^2}(\lambda_1^\alpha + k^{\alpha/2}\lambda_1^{-\alpha/2} + k^{-\alpha/2}\lambda_1^{-\alpha/2} - 3). \tag{9.10}$$

Substituting Eqn. (9.10) into (9.8) yields

$$\sigma = \frac{2\mu}{\alpha}\left(\lambda_1^{\alpha-1} - \frac{1}{2}(k^{\alpha/2} + k^{-\alpha/2})\lambda^{-1-\frac{\alpha}{2}}\right). \tag{9.11}$$

For the case of isotropy, we assume that $k = 1$ and have, for uniaxial tension or compression,

$$\sigma = \frac{2\mu}{\alpha}\left(\lambda_1^{\alpha-1} - \lambda_1^{-1-\frac{\alpha}{2}}\right). \tag{9.12}$$

When $\alpha = 2$, the equation reduces to the form of the Treloar equation (Eqn. (9.5)). The Ogden equation provides a good representation of the J-curve behavior (Ogden, 1972). This model fits well to any incompressible rubbery solid for strains up to ~7.

Another model is the Arruda–Boyce model (Arruda and Boyce, 1993), which has been shown to work well with silicone and neoprene for strains up to ~3.

9.1.6 Fung equation

The nonlinear elastic response was modeled by Fung (1967) and applied successfully to blood vessels and other soft biological tissues. This is the first model developed specifically for biological materials. The increase in slope is due to the extension of the collagen and elastin fibers. If they are stretched beyond a critical strain (or stretch ratio), failure takes place.

The stress–strain response of collagen can be represented mathematically, and indeed Fung (1990) proposed the following expression for the toe region:

$$\sigma = C(e^{\alpha\lambda} - e^{\alpha}) = C(e^{\alpha\varepsilon} - 1), \tag{9.13}$$

where C and α are parameters and λ is the stretch ratio ($= \varepsilon + 1$). This equation was also used by Sacks (2003).

It is instructive to plot the slope $d\sigma / d\varepsilon = E$ as a function of stress. This is done in Fig. 9.4 for the aorta of a dog (circumferential strip). The slope first increases by a relationship that can be described by a power function. Then, it reaches a linear range, in which the increase is more gradual. In Fig. 9.4, the following form of the Fung equation (Fung, 1967) was used:

$$\sigma = (\sigma^* + \beta)e^{\alpha(\varepsilon - \varepsilon^*)} - \beta, \tag{9.14}$$

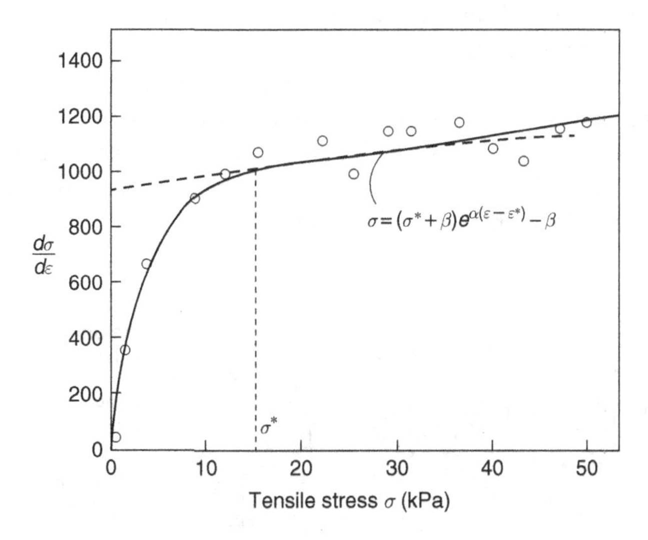

Figure 9.4.

Representation of mechanical response of a dog aorta (circumferential strip) in terms of tangent modulus (slope of stress–strain curve) vs. tensile stress; the slope and intercept provide parameters for the Fung equation. (Reproduced based on Fung (1993).)

where α and β are parameters defined in Fig. 9.4: α is the slope of the linear portion and β is related to the intercept; σ^* and ε^*correspond to the onset of the linear portion. If one compares Eqns. (9.13) and (9.14), one sees that Eqn. (9.13) is a particular case of Eqn. (9.14). There is a requirement of

$$(\sigma^* + \beta)e^{-\varepsilon^*} = \beta = C. \tag{9.15}$$

9.1.7 Molecular dynamics calculations

These are described in Chapter 3. This methodology is rapidly evolving, and the modeling of individual molecules has led to entire collagen fibrils. The results shown in Fig. 3.13 represent the state-of-the art. However, the modeling of entire fibers will follow as the grain coarsening and other techniques evolve.

9.2 Skin

The skin has multifunctionalities, such as temperature regulation, barrier between organism and environments, even camouflage from predators (Montagna and Parakkal, 1974). Basically, the layered structure of the skin is composed of the epidermis and dermis layers. The epidermis is a protective layer of the skin and consists of several layers, as shown in Fig. 9.5 (Fraser and Macrae, 1980). The outermost layer of the mammalian skin is *stratum corneum* made from thin soft keratin. The *stratum corneum* plays a role as a barrier between environment and organism (Montagna and Parakkal, 1974; Marks and Plewig, 1983). Dermis is a connective tissue between the epidermis and the organism. The primary proteins of the dermis are elastin and collagen. Elastin has outstanding elastic properties, and Fig. 9.6 shows the stress–strain curve. Collagen is the main source of the mechanical properties of skins and elastin is not significant. Elastin is responsible for the small deformation of the skin and helps it to recover to the original position (Oxlund, Manschot, and Viidik, 1988).

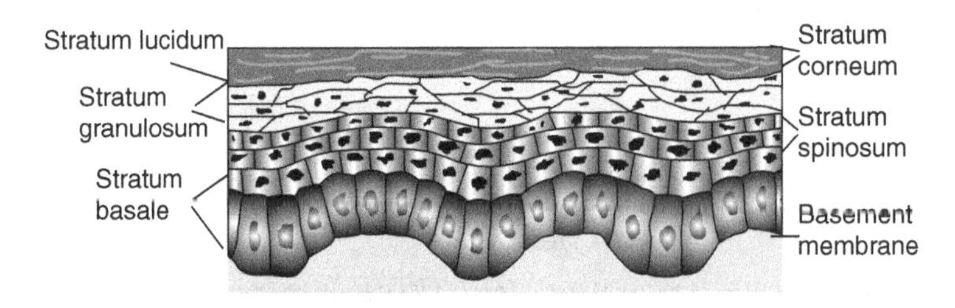

Figure 9.5.

Structure of mammalian skin. (From McKittrick *et al.* (2012), with kind permission from Springer Science and Business Media.)

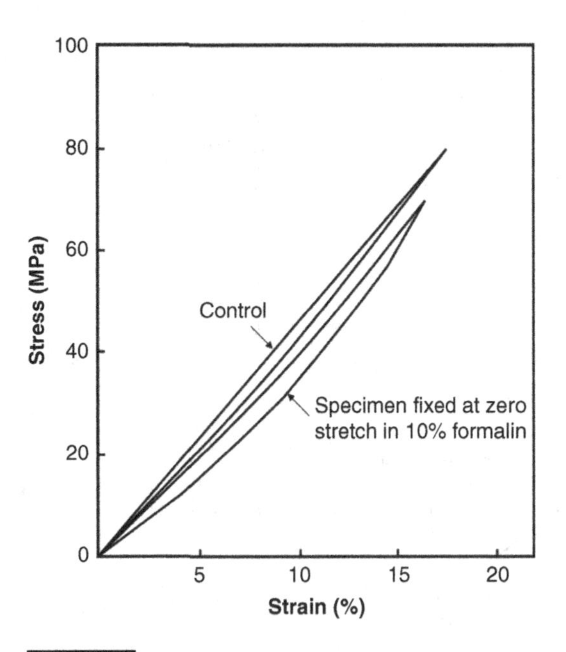

Figure 9.6.

Stress–strain curves of elastin. (From Fung (1993), with kind permission from Springer Science+Business Media B.V.)

Wu *et al.* (2006) studied mechanical properties of *stratum corneum* as a function of temperature and under different environmental conditions (humidity), as shown in Figs. 9.7(a) and (b). The strength decreases with an increase of humidity and temperature. They also studied the role of lipids in the *stratum corneum* and found that delipidized samples provide higher mechanical strength, as shown in Figs. 9.7(c) and (d). The maximum delamination energy of untreated *stratum corneum* is 8 J/m^2 and that of delipidized *stratum corneum* is 13 J/m^2 at low temperature. The mechanical properties of skin vary with age (Escoffier *et al.*, 1989).

Mechanical tests on the *stratum corneum* show the Young modulus to range from 0.01 to 9 GPa. It is highly dependent on the relative humidity and temperature (Park and Baddiel, 1972; Papir, Hsu, and Wildnauer, 1975). Failure strains up to 140% were found in 100% RH at room temperature for rat skin (Papir *et al.*, 1975). "Soft" keratin is formed by loosely packed bundles of IFs embedded in the amorphous matrix (Fraser *et al.*, 1986; Coulombe *et al.*, 2000; Coulombe and Omary, 2002), in contrast to "hard" keratin, which is formed by ordered arrays of IFs embedded in an amorphous alpha-keratin matrix. The importance of the *stratum corneum* on the properties of the skin is enormous, and the efforts to which cosmetic companies go to produce its enrichment and moisture maintenance are extreme. Levi *et al.* (2009, 2011) developed, using materials science and engineering analysis, an original approach to the calculation and measurements on the *stratum corneum* stresses (which reach levels of 4 MPa) and how these generate damage (cracks and chapping) in the *stratum corneum*. Figure 9.8 shows this in the most eloquent manner.

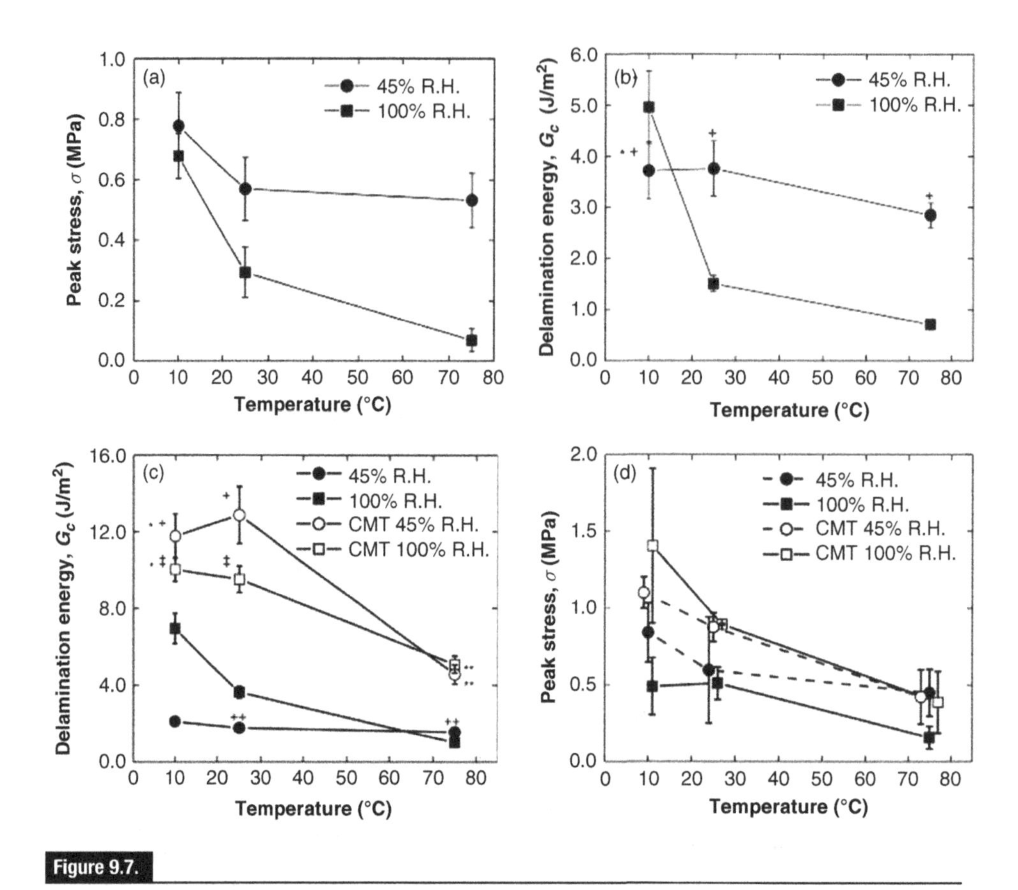

Figure 9.7.

Mechanical properties of *stratum corneum*: (a), (b) peak stress and delamination energy of *stratum corneum* with variation of temperature and humidity; (c), (d) delamination energy and peak stress of delipidized and untreated *stratum corneum* with variation of temperature and humidity. (Reprinted from Wu, van Osdol, and Dauskardt (2006), with permission from Elsevier.)

Figure 9.9 shows the ratio of the recovered deformation, U_R, and initial deformation of skin, U_E, as a function of age, which represents the elastic recovery of skin after deformation. The elasticity of skin continuously decreases with age (Escoffier *et al.*, 1989). The skin strength is also strain-rate sensitive (Vogel, 1972). Figure 9.10 shows the Young modulus plotted as a function of strain rate. As the strain rate increases, the elastic modulus also increases. This is typical of viscoelastic materials.

Both structure and mechanical property differences exist over the body and in all mammalian animals (Montagna and Parakkal, 1974). For example, the rhinoceros has an amazingly thick and tough skin for protection. Figures 9.11(a) and (b) show collagenous rhinoceros skin of flank and belly; the thicknesses are 25 mm and 15 mm, respectively (Shadwick, Russell, and Lauff, 1992). The tensile strength of the dorsolateral (flank) skin is 30.5 MPa and the compressive strength is 170 MPa. This is much higher than for other mammalian skin. The average toughness of dorsolateral skin of rhinoceros is 77 kJ/m², which is higher than the maximum toughness of rat skin, 30 kJ/m² (Purslow, 1983).

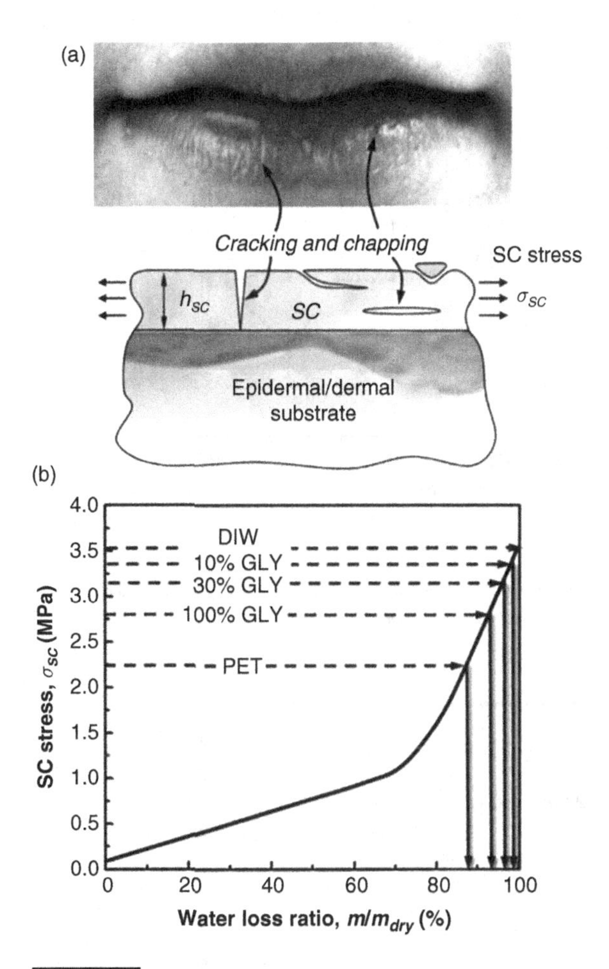

Figure 9.8.

(a) Typical dry skin and effect on *stratum corneum*: cracking and chapping. (b) *Stratum cornum* (SC) stresses developed as a function of water loss ratio. DIW: distilled water; GLY: glycerine; PET: petrolatum. (Reprinted from Levi *et al.* (2011), with permission from Elsevier.)

Figure 9.12 shows stress–strain curves of three collagenous materials; the schematic in Fig. 9.12 shows the orientation of the collagen fibers. The highly aligned collagen fibers in the tendon provide the highest strength and stiffness (~80 MPa). Cat skin, on the other hand, exhibits a large elastic strain of (~1), due to the loose arrangement of fibers. Its amazing properties are highly appreciated by Brazilians. The famous cuica, a unique Carnaval drum that makes a noise that resembles a sad cry, is made from cat skin. From January until Ash Wednesday, ladies in Brazil keep their cats inside! Rhinoceros skin has the same strength as cat skin, but much less elasticity because the collagen molecules are extended.

The skin is an orthotropic material, i.e. the mechanical response has defined and different trajectories for the perpendicular orientations determined by a coordinate

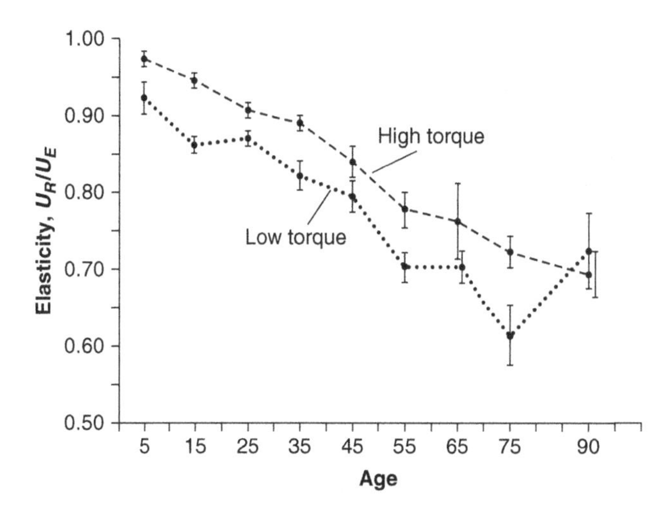

Figure 9.9.

Skin elasticity recovery/extensibility ratio (U_R/U_E) as a function of age for high and low torque. (Reprinted by permission from Macmillan Publishers Ltd.: *Journal of Investigative Dermatology* (Escoffier *et al.*, 1989), copyright 1989.)

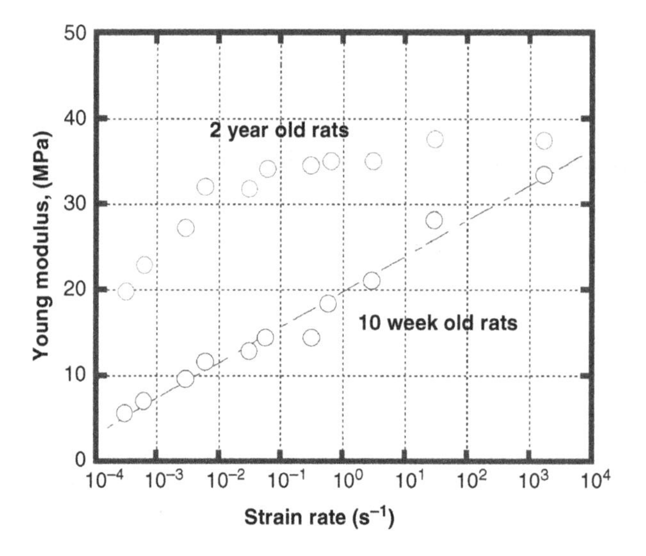

Figure 9.10.

Young modulus of rat skin plotted as a function of strain rate. (Adapted from Vogel (1972), with permission from Elsevier.)

system in which two axes are parallel to the skin surface and the third axis is perpendicular to it. Langer lines cover the entire skin and mark the directions along which the mechanical stiffness is highest. Figure 9.13 shows the Langer lines on the human body. The origin of the Langer lines can be traced to a nineteenth century physician, Baron Dupuytren, who was faced with a suicide attempt by a man using a pick. The puncture

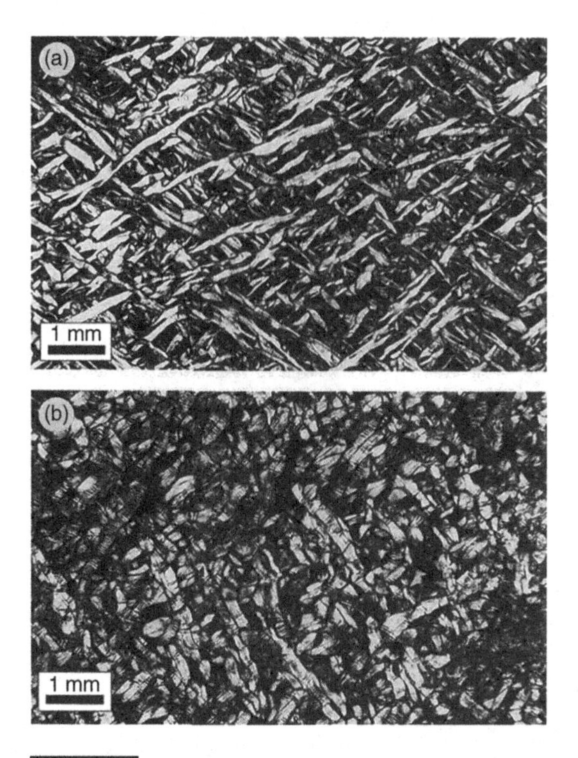

Figure 9.11.

Skin of the rhinoceros: (a) flank; (b) belly. (Used with kind permission of Professor R. E. Shadwick.)

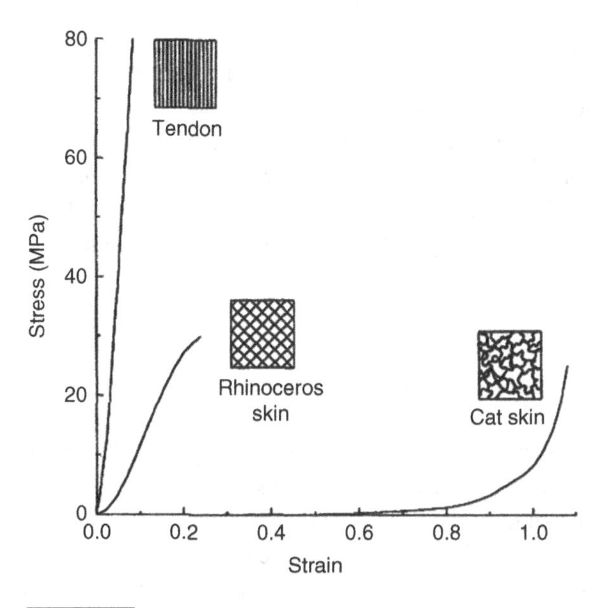

Figure 9.12.

Stress–strain curves of collagenous materials. (Used with kind permission of Professor R. E. Shadwick.)

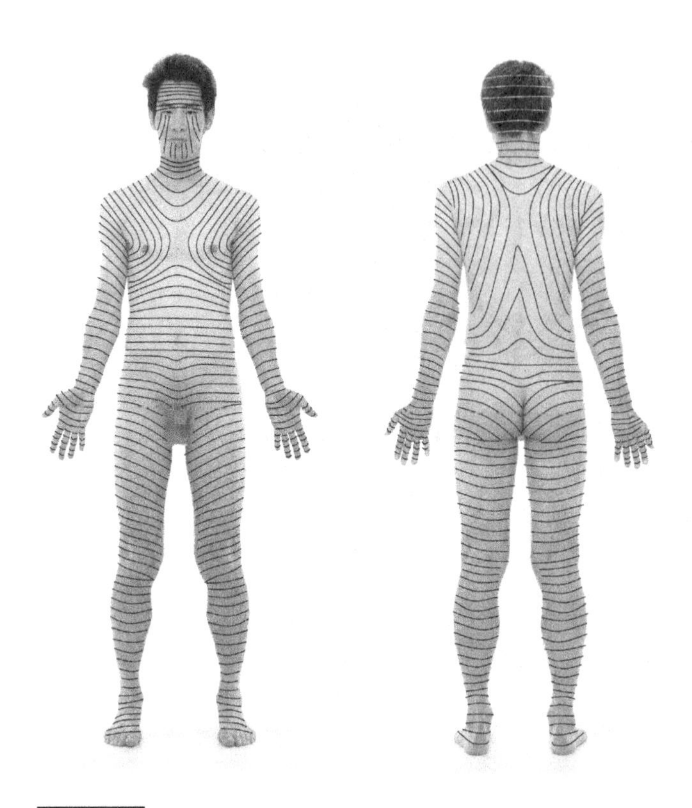

Figure 9.13.

Langer lines indicating the direction of alignment of the collagen fibrils.

wounds were not circular but elongated, leading the Baron to suspect foul play. Upon further examination, he confirmed that a sharp cylindrical pick could produce elongated punctures. Langer repeated the procedure, using an ice pick on cadavers, a most macabre procedure. He confirmed that the perforations were elongated and not circular. (Note to students: Don't try this experiment at home.) Plastic surgeons make the incisions along Langer lines to minimize scarring and help healing. The dermis can be envisaged as a wavy configuration of collagen fibers. Upon stretching, these fibers straighten. Figure 9.14 shows the collagen in chicken skin. The curved fibers are apparent and marked by the line. Upon closer examination, the fibers are revealed to consist of bundles of fibrils, each with a diameter of ~100 nm. The 67 nm pattern is seen very clearly at the highest magnification.

Figure 9.15 shows the tensile stress–strain curves for the skin of pig belly along two orientations: parallel and perpendicular to the spine. The Langer lines for pigs should be analogous to those of humans since they are close cousins in voracity and beauty. The stress–strain curves are of the J type. There is a considerable difference in the total elongation, increasing from 0.2 parallel to the spine to 0.8 perpendicular to it. The mechanical properties of mammalian skin are dominated by the dermis, which

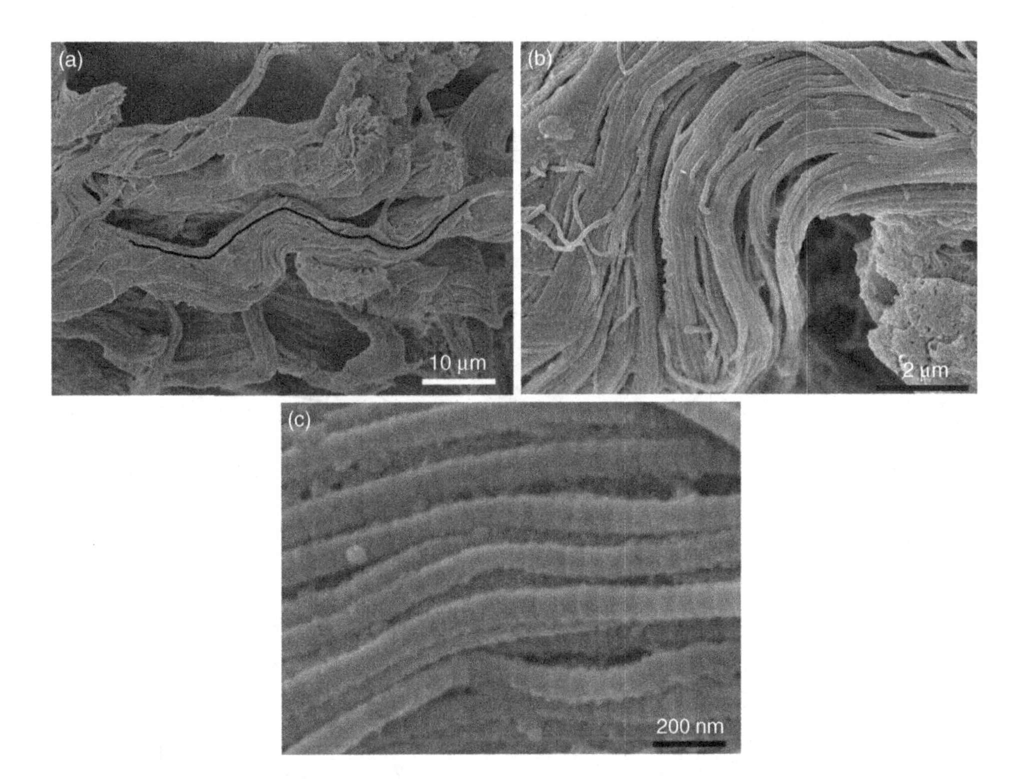

Figure 9.14.

Collagen fibers and fibrils in chicken skin: (a) curvy structure of fibers; (b) fibril assemblages in fibers; (c) characteristic 67 nm bands in fibrils. (Figure courtesy W. Yang.)

is 90–95% of the total. The dermis is composed primarily of collagen (60–80%), elastin, and a hydrated gel-like substance.

The Ogden equation, derived in Section 9.1.5, was applied to skin by Shergold *et al.* (2006). In the case of human and pig skin, Shergold *et al.* (2006) obtained a best fit with $\alpha = 9$ and $\alpha = 12$, respectively.

Figure 9.16 shows the same degree of anisotropy for rabbit skin. These curves are continued through the unloading stage, and it is clear that there is some viscoelastic behavior, expressed through the area enclosed by the loading and unloading paths, which do not match exactly.

As will be seen in the following, this is dependent on orientation, strain rate, and temperature. The effect of temperature can also be incorporated into the Ogden equation, leading to

$$\frac{2\mu}{\alpha} = \frac{2\mu_0}{\alpha_0}\left(\frac{T}{T_0}\right)^n \ln\left(\frac{\dot{\varepsilon}}{\dot{\varepsilon}_0}\right), \tag{9.16}$$

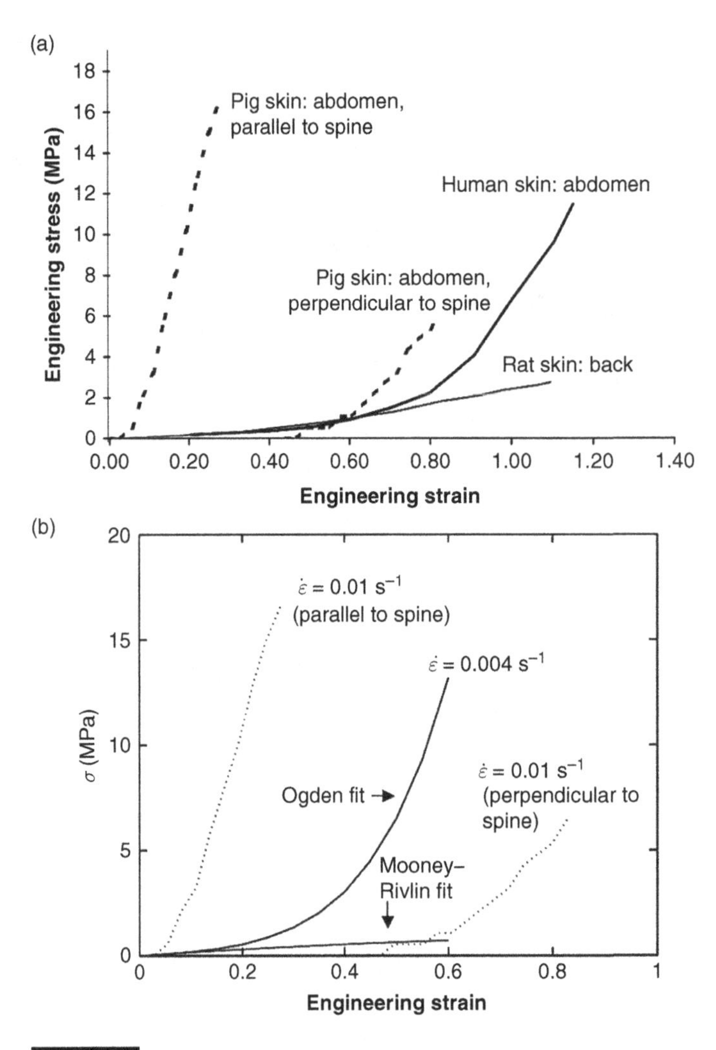

Figure 9.15.

(a) Stress–strain curves for pig belly skin parallel and perpendicular to the Langer lines and for human and rat skin. (Reprinted from Shergold *et al.* (2006), with permission from Elsevier.) (b) Comparison with Ogden and Mooney–Rivlin equations.

where n is a temperature softening parameter, $\dot{\varepsilon}$ is the strain rate, and the strain-rate sensitivity is expressed by the well-accepted logarithmic behavior; α_0, μ_0, T_0, and $\dot{\varepsilon}_0$ are reference values.

Another effect needs to be incorporated into the Ogden equation: the strain-rate sensitivity. Indeed, Fig. 9.17 shows the effect of the strain rate on the compressive strength of pig skin; both experimental and computational results are shown. Different values of the shear modulus μ were used to obtain a satisfactory fit. It is possible to incorporate the effect of strain rate into the Ogden equation.

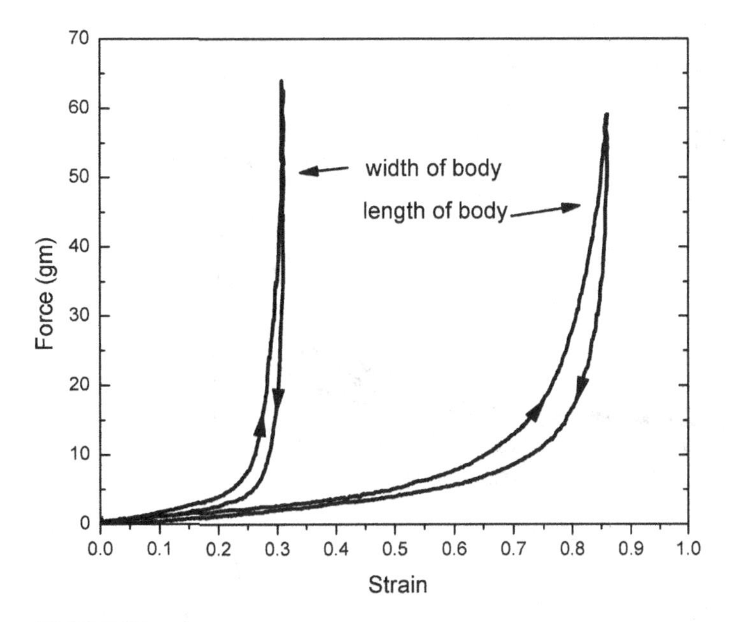

Figure 9.16.

Force–stretch ratio test in tension for rabbit skin along two directions: along width and along length of body.

Combining Eqns. (9.12) and (9.16),

$$\sigma = \frac{2\mu_0}{\alpha_0}\left(\frac{T}{T_0}\right)^n \ln\left(\frac{\dot{\varepsilon}}{\dot{\varepsilon}_0}\right)\left(\lambda_1^{\alpha-1} - \lambda_1^{-1-\frac{\alpha}{2}}\right). \tag{9.17}$$

Zhou *et al.* (2010) showed that there is a significant variation in skin tensile behavior with temperature and strain rate. They also used the Ogden model to represent the effects of strain rate and temperature on the measured constitutive response through two parameters (α and μ). Figure 9.17 shows the effects of strain rate and temperature on the tensile stress–strain (or stretch) curves. There is significant effect of strain rate in the range of 0.0025–0.1 s^{-1}. Similarly, the maximum strain increases with temperature. They also obtained a best fit with $\alpha \sim 9$. The "effective" shear modulus, μ, depends on strain rate. Shergold *et al.* (2006) quote values varying between 0.4 and 7.7 MPa for strain rates ranging from 0.004 s^{-1} to 4000 s^{-1}. In contrast, for an elastomer tested by Shergold *et al.* (2006) the value of α was much lower, ~3. This is much closer to the entropic equation for rubber, for which $\alpha = 2$. Thus, skin "stiffens up" much faster than rubber. More sophisticated constitutive models of skin can be developed, incorporating the mechanical response of the individual collagen and elastin fibers and their dependence on orientation.

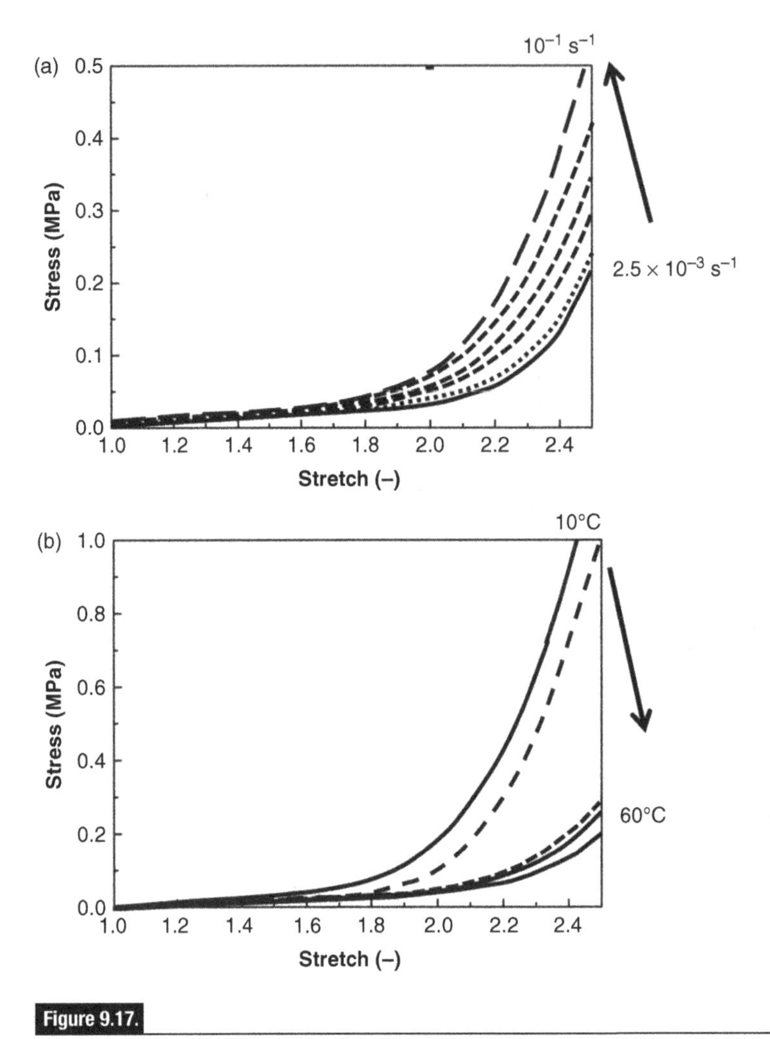

Figure 9.17.

Effect of (a) strain rate (0.0025–0.1 s^{-1}) at 45 °C and (b) temperature (10–60 °C) at 0.01 s^{-1} on the tensile mechanical response of pig belly skin. (Reproduced based on Zhou *et al.* (2010).)

The structure of *stratum compactum* in the skin of a toad (*Bufo marinus*) is shown in Fig. 9.18. The *stratus compactum* layer is found to have a thickness between 0.2 and 0.7 mm. The angles between the collagen fibers vary between 50° and 80°, and therefore it seems that they form a Bouligand pattern. This is a cross-ply structure, which provides the desired stiffness along different directions. There are significant differences in the stress–strain response of different frogs, and Schwinger, Zanger, and Greven (2001) attribute them to the environment. Whereas aquatic or semi-aquatic frogs (*Rans*

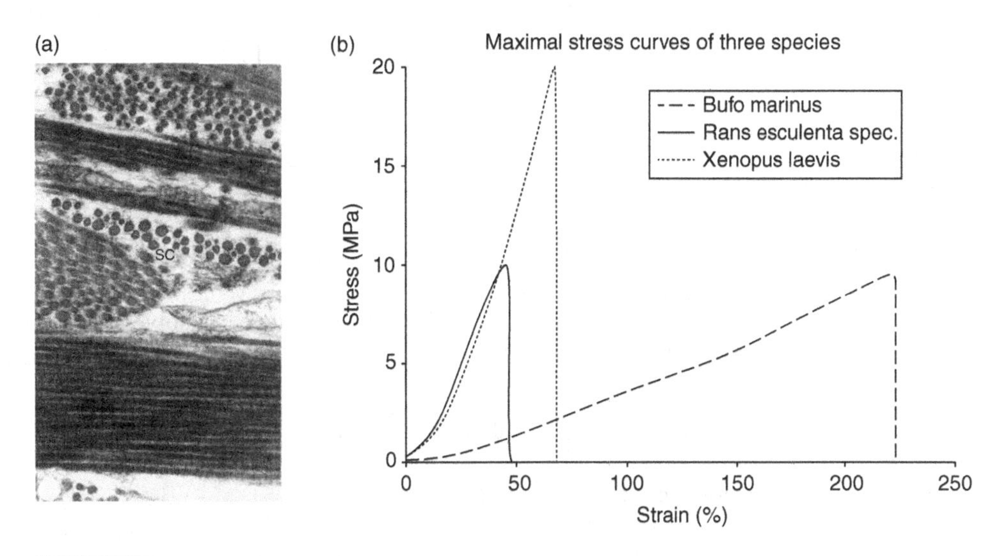

Figure 9.18.

(a) Arrangement of collagen fibers in *stratum compactum* (SC) of *Bufo marinus*, a terrestrial toad. (b) Comparison of tensile responses of *B. marinus* (terrestrial), *Rans esculenta* (semi-aquatic), and *Xenopus laevis* (fully aquatic). (Reprinted from Schwinger *et al.* (2001), with permission from Elsevier.)

esculenta and *Xenopus laevis*) have maximum strains of ~50%, the skin of *Bufo marinus* is much more compliant, with strains of up to 200%.

Example 9.2 Using the Maxwell model, determine the viscosity (η) of tendon stretched to a specified strain of 0.04 at room temperature, using the stress relaxation plot shown in Fig. 9.19. Given: $E = 0.1$ GPa.

Solution The Maxwell model is described by the following equation (see Chapter 2, Section 2.7):

$$\frac{d\varepsilon_{total}}{dt} = \frac{\sigma}{\eta} + \frac{1}{E}\frac{d\sigma}{dt}.$$

Then we set $d\varepsilon/dt = 0$:

$$\frac{\sigma}{\eta} + \frac{1}{E}\frac{d\sigma}{dt} = \frac{d\varepsilon_{total}}{dt} = 0 ,$$

$$\frac{d\sigma}{\sigma} = -\frac{E}{\eta}dt ,$$

$$\int_{\sigma_0}^{\sigma_f} \frac{d\sigma}{\sigma} = -\frac{E}{\eta}\int dt\,,$$

$$\ln\frac{\sigma}{\sigma_0} = -\frac{E}{\eta}t\,,$$

$$\sigma = \sigma_0 e^{-\frac{E}{\eta}t}\,.$$

By choosing two points on the plot (Fig. 9.19), we can solve for the ratio of E/η. At $t = 25$ s, $\sigma/\sigma_0 = 0.73$; at $t = 150$ s, $\sigma/\sigma_0 = 0.49$. We have two equations:

$$0.73 = e^{-25\frac{E}{\eta}} \quad \text{and} \quad 0.49 = e^{-150\frac{E}{\eta}}.$$

Combining the two equations, we obtain

$$\frac{0.73}{0.49} = e^{125\frac{E}{\eta}},$$

$$\frac{E}{\eta} = 3.2 \times 10^{-3}.$$

Because $E = 0.1$ GPa,

$$\eta = \frac{E}{3.2 \times 10^{-3}} = 3.125 \times 10^{10}\,\text{Pa s}.$$

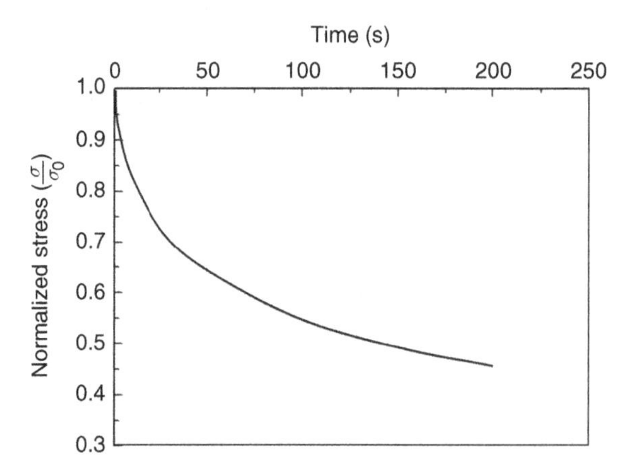

Figure 9.19.

Stress relaxation plot (normalized stress vs. time) of stretched tendon.

9.3 Muscle

The maximum force that a muscle fiber can generate depends on the velocity at which it is activated. Figure 9.20 shows the stress that can be generated as a function of strain rate for "slow-twitch" and "fast-twitch" muscles. We use slow-twitch muscles for long-range events (e.g. distance running) and fast-twitch muscles for explosive activities, such as sprinting or throwing a punch. Both muscles show a decreasing ability to generate stress as the strain rate is increased. However, the fast-twitch muscles show a lower decay.

The plot shown in Fig. 9.20 is only schematic and represents the rat *soleus* (slow-twitch) and extensor *digitorum longus* (fast-twitch). The equation that describes the response in Fig. 9.20 is called the Hill equation (Hill, 1938). It has the following form:

$$(\sigma + a)(\dot{\varepsilon} + b) = (\sigma_0 + a)b, \tag{9.18}$$

where σ_0 is the stress at zero velocity (equal to 200 MPa). The range of σ_0 is usually between 100 and 300 MPa. In Eqn. (9.18) a and b are parameters and $\dot{\varepsilon}$ is the strain rate (obtained from the velocity).

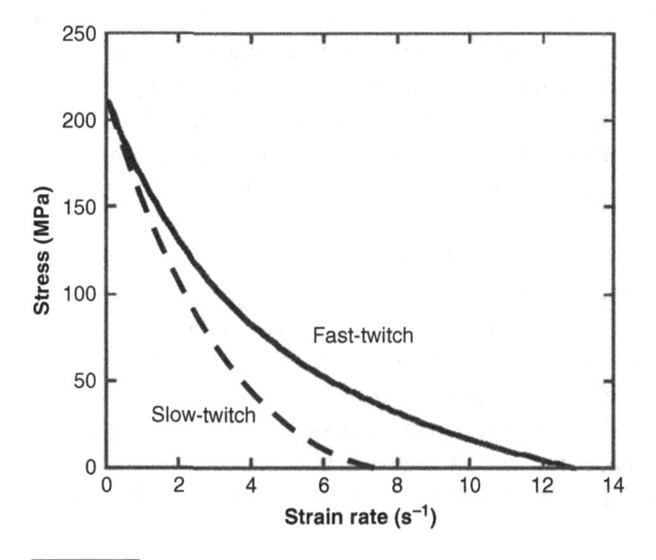

Figure 9.20.

Stress–strain rate for slow-twitch and fast-twitch muscles.

Box 9.1 Polymeric biomaterials

Various types of polymers are widely used in biomedical applications and represent the largest class of biomaterials. Polymers can be classified into two groups: natural and synthetic. Natural polymers can be obtained from plants, such as cellulose, starch, and natural rubber, and animal materials, including collagen, chitin/chitosan, elastin, keratin, abductin, resilin, silks, glycosaminoglycan (GAG), heparin, and hyaluronic acid. Synthetic polymers commonly used as biomaterials are polycarbonate (PC), polyethylene (PE), polyetheretherketone (PEEK), polyethylene glycol (PEG), polyethylene terephthalate (PET), polyhydroxyethyl methacrylate (PHEMA), copoly-lactic-glycolic acid (PLGA), polymethylmethacrylate (PMMA), polypropylene (PP), polytetrafluoroethylene (PTFE), polyvinylchloride (PVC), silicon rubber (SR), or poly-dimethyl siloxane (PDMS).

Polymers have a wide variety of compositions and physical and chemical properties, and can be fabricated into different forms (films, fabrics, and gels) with complex shapes and structures through various processes. It is of great importance for a materials or biomedical engineer to select polymeric materials with suitable properties that match those required for specific biomedical applications. The characteristics and properties of polymers depend on tacticity (arrangement of substituents around the extended polymer chain), molecular weight, crystallinity, cross-linking, and glass transition temperature (T_g). For example, increasing molecular weight and crystallinity usually results in enhancing mechanical properties. The mechanical properties of typical polymeric biomaterials are summarized in Table B9.1. One of the limitations of polymers is their relatively weak mechanical properties compared with those of ceramics and metals. Synthesis of polymer-based composites (incorporating ceramics and metals) can overcome many shortcomings of polymers. They range in bioreactivity from inert to bioactive and bioresorbable. The latter are finding applications in scaffolds that are absorbed by the body. Selected polymeric biomaterials are introduced as follows:

PE — High-density PE (HDPE) is used as tubing material for drains and catheters. Ultrahigh-molecular-weight PE (UHMWPE) has enhanced toughness and wear resistance and is applied in the cups of artificial hips and other orthopedic implants (see Box 3.1). The molecular weight varies from 30 000 to 50 000 g/mol for LDPE, to 200 000 to 500 000 g/mol for HDPE, and finally to 4 to 6 million g/mol for UHDPE. The changes in density, in spite of the name, are minimal: 0.91–0.925 g/cm^3 for LDPE to 0.941–0.980 g/cm^3 for HDPE. However, the changes in mechanical properties (strength, ductility, wear rate) are extraordinary (as can be seen in Table B9.1).

PEG — PEG is a water soluble, low-melting temperature polymer commonly used in drug delivery. When attached to various protein medications, PEG allows a slowed dissolution rate of the carried protein into the blood, ensuring longer dosing intervals and longer-acting medicinal effects and reduced toxicity.

PMMA — PMMA is a hydrophobic, transparent, thermoplastic polymer, and is often used as a lightweight or shatter-resistant alternative to glass. It has excellent light transmittance and can be used for replacement intraocular lenses (IOLs) and hard contact lenses. It is

Table B9.1. Mechanical properties of typical polymeric biomaterials[a]

Material	Young modulus (GPa)	Tensile strength (MPa)	Strain to fracture (%)
Polyethylene (low density)	0.096–0.26	6.9–9.6	400–700
Polyethylene (medium density)	0.24–0.62	8.2–24.1	50–600
Polyethylene (high density)	0.59–1.11	21–37	15–100
Nylon 6/6	–	62–82	60–300
Polyurethane	0.0018–0.009	28–40	600–720
Polytetrafluoroethylene	0.3–0.7	15–40	250–550
Polyacetal	1.4–2.8	44–90	40–250
Polymethylmethacrylate	1.8–3.3	38–80	2.6–5
Polyethylene terephthalate	2.2–3.5	42–80	50–300
Polyetheretherketone	3.6–13	70–208	1.3–50
Polysulfone	2.4–2.9	50–100	25–80
Dexon	4.85	560	22.6
Vicryl	5.96	580	18.4
Mersilene	4.68	370	8
Nurolon	1.85	335	18.2
Ethilon	1.76	550	33
Prolene	5.16	450	42

[a] Compiled from Black and Hastings (1998) and Meyers and Chawla (2009).

biocompatible, with good toughness and stability, and is one of the major ingredients in bone cement for orthopedic implants. In dental implants, dentures are often made of PMMA, and it is also used in the production of ocular prostheses. A large majority of white dental filling materials have PMMA as their main organic component.

PHEMA Poly-HEMA is a hydrophilic polymer that forms a hydrogel in water. Its common medical use is in soft contact lenses.

PP PP is a thermoplastic, isotactic crystalline polymer with high rigidity and good tensile strength. It is commonly used as synthetic, nonabsorbable suture. PP has been used in hernia and pelvic organ prolapse repair operations.

PTFE PTFE is a very hydrophobic, high-melting-temperature synthetic polymer, better known as Teflon (discovered by DuPont Co.). Its excellent lubricity is used to make catheters.

Box 9.1 (cont.)

The microporous form of PTFE, or e-PTFE, commonly recognized as Gore-Tex, is used in vascular grafts.

PVC — PVC is mainly used as a tubing material (e.g. in blood transfusion, feeding, and dialysis) and for blood storage bags. Pure PVC is hard and brittle; it can be made softer and more flexible by the addition of plasticizers, the most widely used being phthalates.

SR (PDMS) — SR, or PDMS, is the most widely used silicon-based organic polymer, and is known for its unusual rheological properties. It is optically clear, chemically inert, and nontoxic, and has very low T_g. It has been widely applied in the biomedical field. For example, it is used in a variety of prostheses, such as heart valves, joints, and breast implants, as well as in soft-tissue reconstructions (e.g. ear, nose, and face). Other applications include catheter and drainage tubing, membrane oxygenators, cosmetics, and biomedical MEMS and microfluidic devices. The major limitation of SR is its relatively poor mechanical strength, and this is often modified by adding reinforcing silica filler.

Bioresorbable polymers

The principal bioresorbable polymers are polylactic acid (PLA), polyglycolic acid (PGA), and copoly-lactic-glycolic acid (PLGA), a copolymer of lactic acid and glycolic acid. In the body, these polymers decompose into the monomers lactic acid and glycolic acid. These two monomers are, under normal physiological conditions, by-products of various metabolic pathways. Thus, the body recognizes these acids and can eliminate them without generating toxicity.

9.4 Blood vessels

The vascular system provides the transport of nutrients, oxygen, and other chemical signals to the various parts of the body. It is divided into two subsystems: pulmonary and circulatory. We will not go into any details of the pathology of these two subsystems. Rather, we will concentrate on their mechanical properties. Arteries (which carry blood from the heart to the various parts of the body) and veins (that collect blood back to the heart) exhibit some significant differences in structure. Arteries are exposed to higher pressures and fluctuations associated with the diastolic and systolic portions of the cardiac cycle. Figure 9.21(a) shows a network of veins and arteries in rabbit dermis. This network resembles a map of rivers. In the arteries, blood flows to increasingly narrow passages (arteries to arterioles to capillaries); in veins, the opposite is the case. Thus, the regions where the arterioles connect to the venules are small and cannot be seen with the naked eye. This is a critical linkage in the circulatory system that could not be understood for many centuries. It was the seminal discovery of William Harvey that for the first time enabled the full understanding of the mechanics of the circulatory system. It

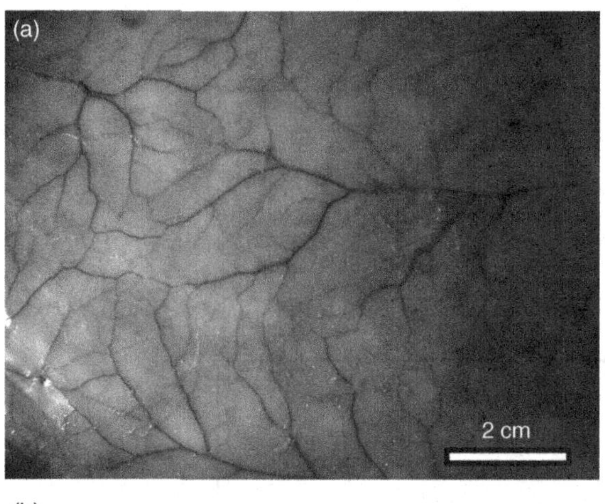

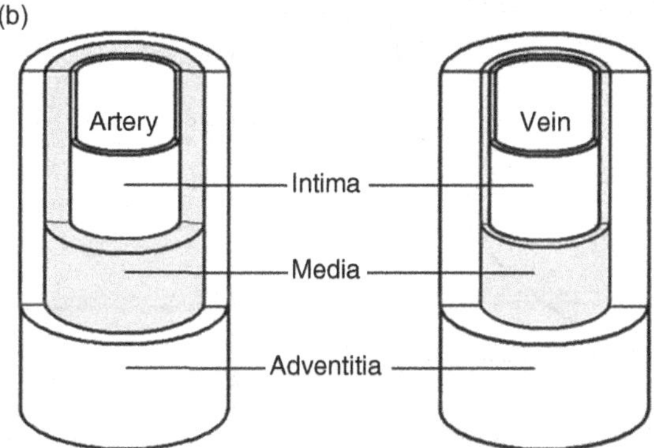

Figure 9.21.

(a) Network of arteries and veins in rabbit dermis. (b) Cross-section of an artery and a vein composed of the endothelium, tunica intima, tunica media, and tunica adventitia.

should be noted that prior to that (in the tenth century) Avicenna mentions that the Arabic scholar Ibn al-Nafis had already identified the system. Figure 9.21(b) shows the longitudinal and normal sections of an artery. The structure is layered, with three distinct regions: the tunica intima (innermost), the tunica media (middle), and the tunica adventitia (outermost).

The formation of an aneurysm (a tensile instability forming a local bulge) and bursting (longitudinal splitting) of blood vessels are highly undesirable but all too frequent events in humans. Figure 9.22(a) shows the formation of an aneurysm in an artery. If an artery in the brain bursts, we have a hemorrhagic stroke. In the arteries close to the heart, we have what is commonly known as a burst. There are two unique aspects of the mechanical response of arteries and veins that are instrumental in minimizing the chance of the aforementioned problems: nonlinear elasticity and residual stresses.

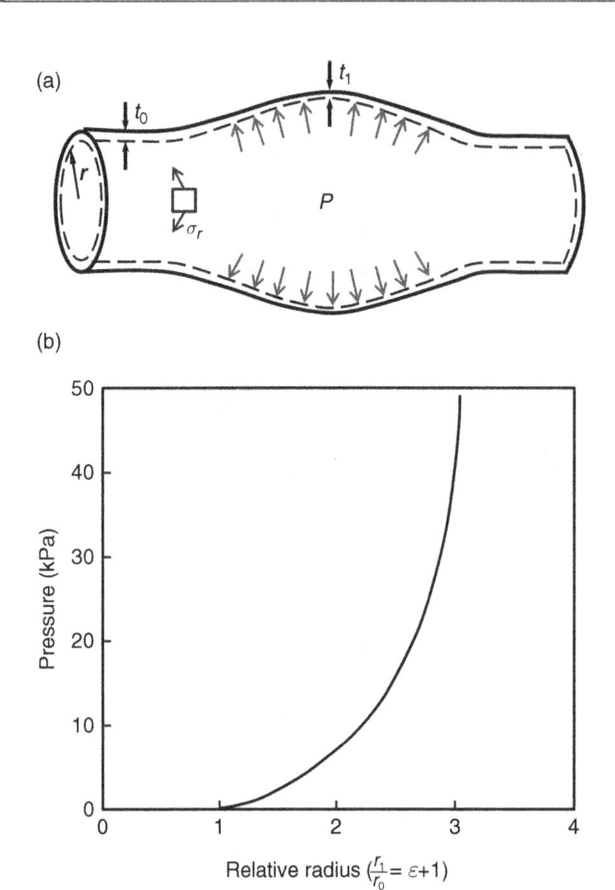

Figure 9.22.

(a) Aneurysm in artery. (b) Pressure vs. relative radius in artery; the J curve is due to the action of collagen. (Pressure conversion: 10 kPa = 75 mm Hg.) (Reproduced based on Ennos (2012).)

The pressure in a vessel, P, is related to the tensile radial stresses, σ_r, by a well-known expression, which can be easily derived by cutting the vessel longitudinally and replacing the missing part by forces. If the radius of the vessel is r and the thickness is t,

$$\sigma_r = \frac{Pr}{t}. \tag{9.19}$$

As r increases, t decreases. Assuming a constant volume:

$$r_0 t_0 = rt. \tag{9.20}$$

Thus,

$$\sigma_r = \frac{Pr^2}{r_0 t_0}. \tag{9.21}$$

The strain is given by

$$\varepsilon = \frac{2\pi r_1 - 2\pi r_0}{2\pi r_0} = \frac{r_1 - r_0}{r_0}. \tag{9.22}$$

Inserting Eqn. (9.22) into Eqn. (9.21), we obtain

$$\sigma_r = \frac{Pr_0(1+\varepsilon)^2}{t_0}. \tag{9.23}$$

The response of collagen, with a J curve, can be represented by a simple quadratic equation:

$$\sigma = k\varepsilon^2. \tag{9.24}$$

This is obtained from Eqn. (9.4) but setting $n = 1$ and ignoring the last term.The strain to which the artery expands is obtained by setting Eqn. (9.23) equal to Eqn. (9.24).

Thus, the formation of an aneurysm in an artery is often impeded by the J response of the artery walls, where collagen plays an important role. Such as in skin and other organs, the collagen is initially wavy and responds to strain by straightening. After it is stretched, its resistance increases significantly. Application of Eqns. (9.19) and (9.20) to the J-shaped stress–strain curve leads to the response shown in Fig. 9.22(b). If the artery wall is weakened in a certain region, or if the pressure is too high, it can develop an aneurism. It is possible to insert, surgically, a reinforcement around the artery. The "burst" is the result of a crack forming longitudinally when the radial stress exceeds the failure stress of the artery.

9.4.1 Nonlinear elasticity

The three layers comprising blood vessels have different functions and composition. Table 9.1 summarizes the similarities and differences between arteries and veins, including main vessels such as the aorta. The composition of arteries is made up primarily of elastic fibers (elastin), collagen, and smooth muscle. Compared to veins, arteries contain much more elastic material. Thicker arteries, such as the aorta, contain less smooth muscle than both smaller arteries and veins. These differences account for the ability of arteries to resist large pressure fluctuations during the cardiac cycle.

The mechanical response of blood vessels is shown in Fig. 9.23. This is the longitudinal stress–strain response of human *vena cava*. The response is nonlinear elastic. We

Table 9.1. Dimensions and composition of blood vessels

Vessel		Dimensions (mm)	Composition
Artery	aorta	diameter: 25 thickness: 2	Elastic fibers Endothelium Collagen Smooth muscle
	medium-sized artery	diameter: 4 thickness: 1	Elastic fibers Endothelium Collagen Smooth muscle
Vein		diameter: 20 thickness: 1	Elastic fibers Endothelium Collagen Smooth muscle

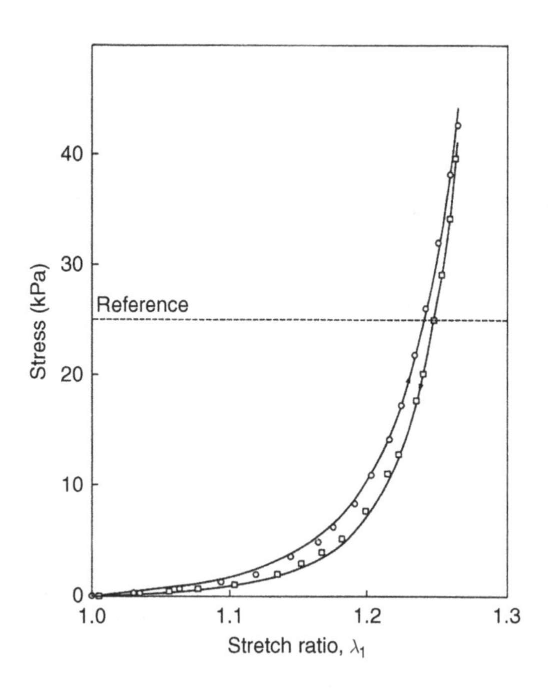

Figure 9.23.

Stress–strain response of human *vena cava*: circles, loading; squares, unloading. (Adapted from Fung (1993, p. 329), with kind permission from Springer Science+Business Media B.V.)

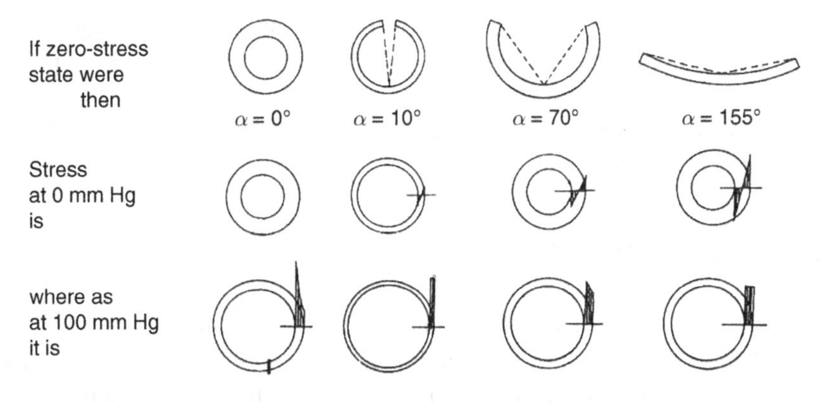

Figure 9.24.

Residual stresses in arteries; the artery is sliced longitudinally and the angle α is measured. (From Fung (1990, p. 389), with kind permission from Springer Science+Business Media B.V.)

know that it is elastic because on unloading the artery returns to its original dimension. However, there is a slight hysteresis on loading and unloading, due to viscoelastic processes. The slope approaches the elastic modulus of the fibers as the strain approaches 0.3. This increase in slope is due to the extension of the bonds in the collagen and elastin fibers. If they are stretched beyond this point, failure takes place. Instead of the strain, ε, the stretch ratio ($\lambda = \varepsilon + 1$) is used in the plot.

It is instructive to plot the slope, $d\sigma/d\varepsilon = E$, as a function of stress. This was done in Fig. 9.4 for the aorta of a dog (circumferential strip). The slope first increases by a relationship that can be described by a power function. Then it reaches a linear range, in which the increase is more gradual. This nonlinear elastic behavior is a characteristic feature of many soft tissues in the human body. It serves as an important function: as the pressure in the blood vessels is increased, the vessels become stiffer.

9.4.2 Residual stresses

Biological materials such as arteries contain residual stresses. In the case of a segment of artery that is not under internal blood pressure, the walls of the artery are under zero strain and therefore have residual stress. Fung (1990) showed that if one makes an axial cut in the wall of an artery, the artery will spontaneously open. This geometry is known as the zero-stress state. The angle by which the artery springs open is defined as the opening angle. As this opening angle increases, the stress distribution in the wall becomes more uniform. This makes sense since, under normal blood pressure, arteries inflate, causing higher strain on the inner wall of the artery (compared to the outer wall). In arteries, stress is an exponential or power function of strain, so the observed increase in strain at the inner wall will be accompanied by an increase in stress at the inner wall.

Four different arteries, with different zero-stress angles, are shown in Fig. 9.24: $\alpha = 0°$, 10°, 70°, and 155°. For the same arteries, the wall stresses at two values of the applied internal pressure are shown. For zero pressure, there is a detrimental effect on the stress distribution. However, this is not the critical condition. For 100 mm (Hg column) internal pressure (in the range of pressure of blood inside our body), the artery with the highest value of α has the lowest stress in the wall. Thus, the residual stress reduces the maximum stress in the artery walls.

9.5 Mussel byssus

The byssal threads of marine mussels act as the only anchor lines which attach the animals to reefs, rocks, and other fixtures. These thin hair-like fibers undergo repeated shock from pounding waves and changing currents and must survive without failure to prevent the animal from being swept out to sea. Thus, the toughness the byssal thread material is of the utmost importance. On the macro-scale the threads are composed of both a stiff tether region (the distal thread) which attaches to rock, and an elastic region (the proximal thread) which acts as a shock absorber for the animal (Smeathers and Vincent, 1979). The microstructure of byssal threads is roughly similar to that of tendons. The distal region consists of many bundles of collagen fibrils separated by fibrion-like domains and histidine-rich blocks, including possible β-sheet regions (Rudall, 1955; Gathercole and Keller, 1975; Qin, Coyne, and Waite, 1997). An illustration of the byssal thread micro- and macrostructures is presented in Fig. 9.25 (Rudall, 1955). Although these threads are similar to tendons in tensile strength, the distal region is remarkably more extensible and much tougher (Smeathers and Vincent, 1979).

The two regions of the thread serve different functions and thus have very different mechanical properties. The elastic modulus of the proximal thread is approximately an order of magnitude lower than that of the distal region (Smeathers and Vincent, 1979). However, failure almost exclusively occurs in the proximal region (Bell and Gosline, 1996). This implies that the distal region is considerably tougher (Bell and Gosline, 1996).

The mismatch in elastic modulus of the two regions presents an interesting materials science problem. Failure often occurs at the interface of joined materials with high modulus mismatches (Rabin *et al.*, 1995; Vaccaro and Waite, 2001). This does not

Box 9.2 Vascular implants

Cardiovascular disease is the curse of the modern industrialized society, and three factors contribute to it: sedentarism, smoking, and fatty diet. Genetics also plays a role, and the result is that cardiovascular disease is the leading cause of death in advanced societies. Atherosclerosis increases the thickness of blood vessels by plaque deposition, while at the same time embrittling them. Low-density cholesterol (LDL) plays a key role in the formation of arterial plaque. The opposite effect, the thinning of the vessels through excessive expansion (triggered, often, by excessive blood pressure), is called "aneurysm." As the blood vessel expands radially, its thickness has to decrease, assuming a constant volume. This is seen in Section 9.4. And, as the wall thickness decreases, the stress on it increases, leading to additional expansion.

Corrective action may be taken by different approaches; in the 1970s, aneurysms were corrected by wrapping the expanded site with cellophane.

The current technology employed to overcome cardiovascular disease uses a variety of approaches: grafts (autogenous, from either a cadaver or the same person, or synthetic), the opening of the vessel using a balloon, and the insertion of stents to ensure the opening and support of the vessel wall.

Various materials are used in stents, which comprise a metal mesh. Some have only a metal mesh, whereas others are a combination of metallic mesh and fabric. Many current generation metallic stents are coated with chemicals (drug-eluting stents). These drugs prevent blood clotting and the deposition of cells on the stent. Figure B9.1 shows a lumen with the stent on the outside. Expansion of the lumen, when it is in the correct position, opens the artery. The stent is placed at the end of a catheter that is guided to the correct place. The stent is subsequently or concomitantly placed and expanded. This specific stent is coated with Paclitaxel.

An alloy that has extraordinary properties is NITINOL, which is a Ni–Ti alloy. It exhibits two unique responses, which have found application in many modern devices: the shape memory and superelastic effects. The latter is used in stents. This enables approximately up to 6% elastic deformation through the movement of martensitic interfaces. Thus, the stent can be squeezed until it completely flattens, as shown in Fig. B9.2(a). Conventional alloys have elastic stresses that are one order of magnitude smaller. The stent shown in Fig. B9.2 is made by etching a tube, so no welds are required. Figure B9.2(b) demonstrates that the superelastic hoop force exerted by the stent is approximately constant. The stent is manufactured in the expanded condition and has a diameter larger than the vessel. This corresponds to point a in Fig. B9.2(b). It is then crimped so that the diameter is reduced (point b). The stent is released when in place and expands until it reaches the inner wall of the vessel (point c). At this

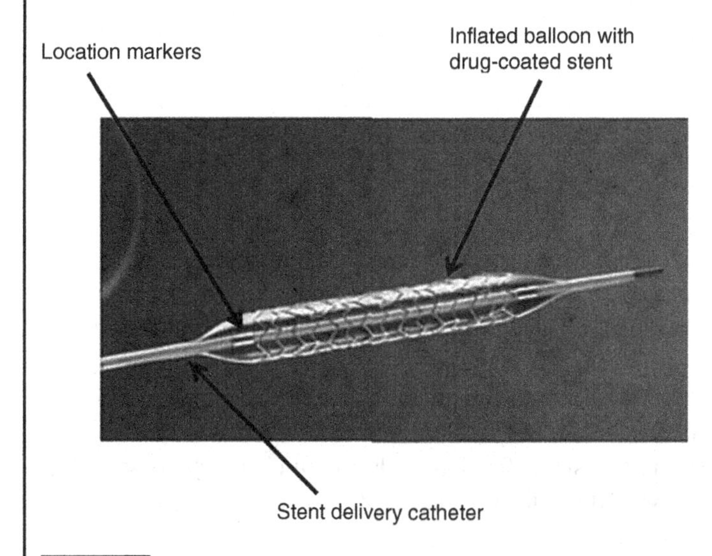

Figure B9.1.

Lumen at the extremity of a catheter with a stent. (From http://commons.wikimedia.org/wiki/File:Taxus_stent_FDA.jpg. Image by Food and Drug Administration (Public domain), via Wikimedia Commons.)

Box 9.2 (cont.)

(a)

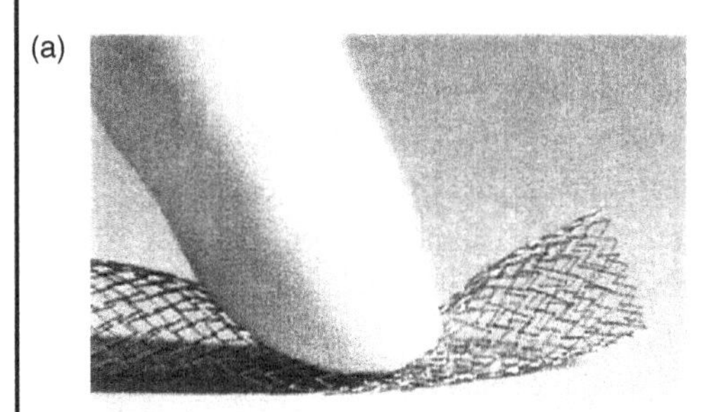

(b)

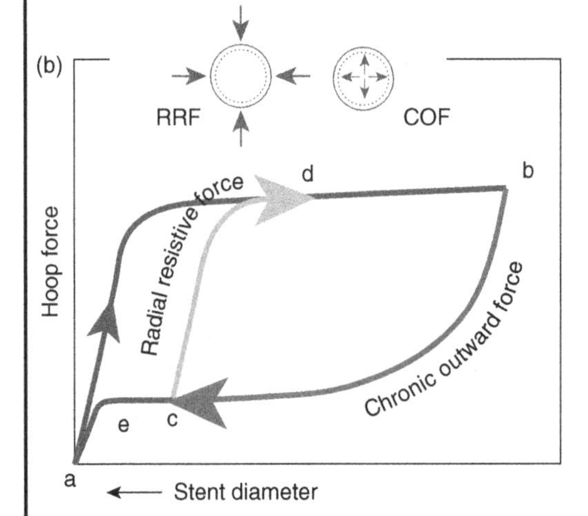

Figure B9.2.

(a) Demonstration of how the Cordis SMART stent can be deformed elastically. (b) Superelastic effect obtained through loading and unloading path a–d–b–c–e–a. The stent is placed, and a hoop force keeps the artery open. (From Stoeckel *et al.* (2004). With kind permission from Springer Science and Business Media.)

point, it has not returned to its original, stress-free diameter, and exerts a force that is known as "chronic outward force." It can be seen that this force is low. If the stent is solicited by an external force, it reacts with a force that is considerably larger (the radial resistive force). The great dimensional stability to the stent and low outward force make this system very favorable.

Experimental results have shown that NITINOL stents in femoral artery occlusion produced better results than angioplasty. In the case of extended (100 mm) occlusions, angioplasticity was only marginally successful after one year, with restenosis (defined as occlusion of over 50%) observed in 67% of patients. The insertion of long NITINOL stents increased the success considerably, and restenosis dropped to 27% (Schillinger *et al.*, 2006).

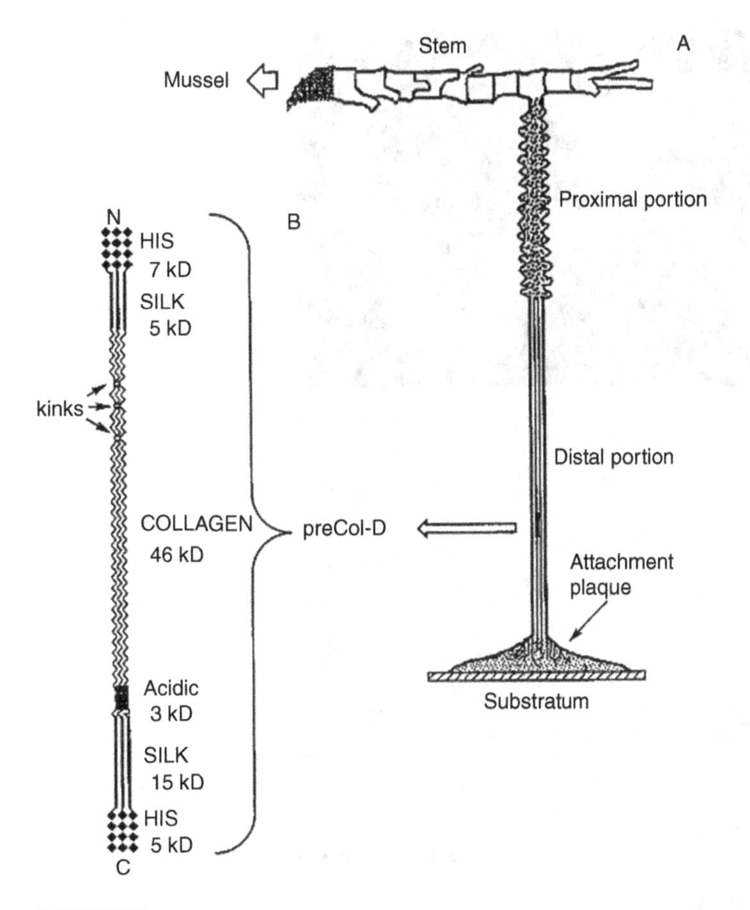

Figure 9.25.

Schematic representation of mussel byssal threads. (Reprinted from Qin *et al.* (1997), with kind permission via Copyright Clearance Center.)

occur in the case of byssal threads, and is likely the result of modulus management in which the distal region softens and the proximal portion stiffens close to the interface.

9.6 Whelk eggs

There is a large marine snail, *Busyon canaliculum*, which has peculiar eggs, forming a helical strand of about 1 m. This has been nicknamed "mermaid's necklace." These strands contain up to 160 capsules (see Fig. 9.26(a)) and are repeatedly hit by seashore waves. Thus, the capsules containing the whelk embryos have to have unique mechanical properties to resist the onslaught of coastal waves. They can undergo large reversible deformations akin to the synthetic elastomers (rubber). The mechanical response is analogous to the

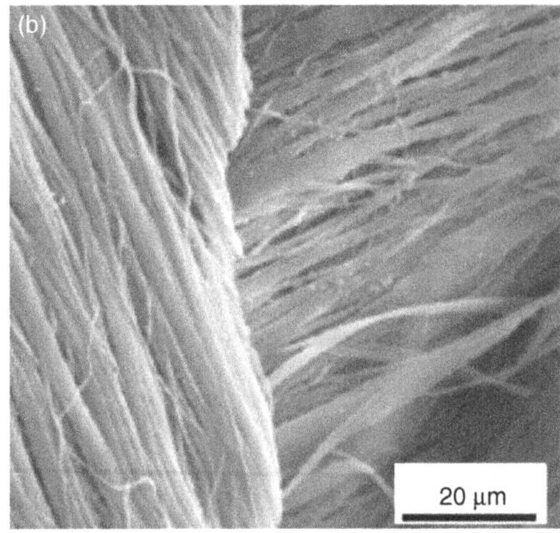

Figure 9.26.

(a) "Mermaid's necklace" of interconnected capsules forming a helical pattern around strand. (b) Cross-plywood structure of fibers from whelk egg capsules, each with 0.2–0.5 μm diameter. (Reprinted by permission from Macmillan Publishers Ltd.: *Nature Materials* (Miserez *et al.*, 2009b), copyright 2009.)

superelastic effect of the Ni–Ti (NITINOL) and other alloys. In NITINOL (see Box 9.2), the martensitic transformation induced by stress is reversible. Upon unloading, the alloy returns to its original structure. Reversible strains of up to 0.06 can be obtained in these shape-memory alloys. In contrast, in conventional metals, the reversible (elastic) strains rarely exceed 0.01. This type of response is shown in Fig. B9.2(b).

The same response is obtained when specimens from the capsule of whelk eggs are subjected to tension. The results of stress–strain tests in tension at temperatures ranging from −1 to 80°C are shown in Fig. 9.27(a), showing typical inflection point behavior (point at which $d\sigma/d\varepsilon = 0$, as discussed in Section 9.1). The stress–strain curves have three stages:

(a) an initial region with an elastic modulus of ~100 MPa;
(b) a plateau covering strains up to 0.7;
(c) a stiffening beyond 0.7.

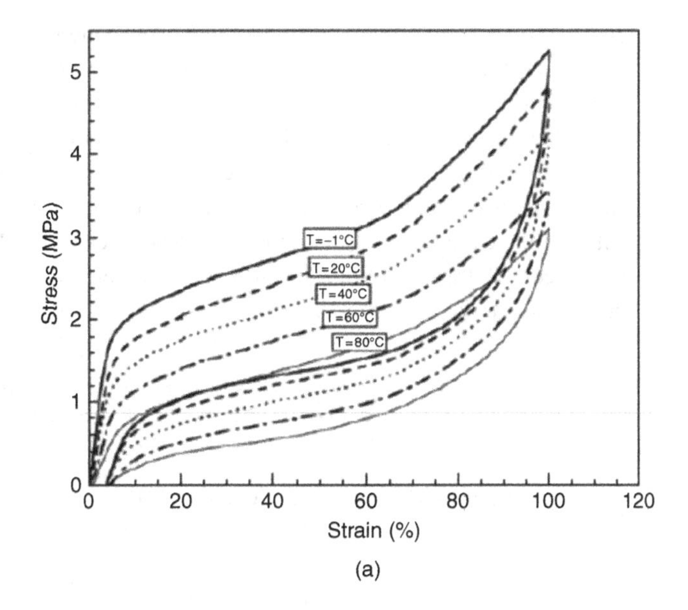

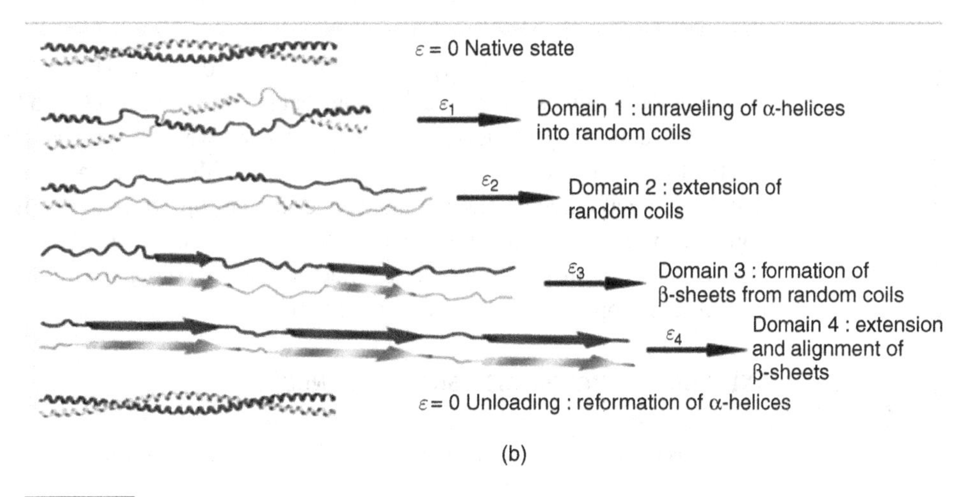

Figure 9.27.

(a) Engineering tensile stress–strain curves of whelk egg capsules in loading and unloading, up to a strain of ~1; note the superelastic effect and hysteresis; (b) schematic representation of presumed reversible structural transformation. (Reprinted by permission from Macmillan Publishers Ltd.: *Nature Materials* (Miserez *et al.*, 2009b), copyright 2009.)

Upon unloading, the three stages are again observed, albeit with a hysteresis. The region from $\varepsilon = 0.05$ to $\varepsilon = 0.7$ corresponds to the structural transition of the bioelastomer, which changes from an α-helix to a β-sheet configuration, in the same manner as wool upon being tested in tension. The structural change of the biopolymer is shown in Fig. 9.27(b). Initially, the fibers comprise α-helices (presented in Fig. 3.15). These helices open up (unravel) under the effect of stress, being bonded by van der Waals and hydrogen bonds. The stretched chains subsequently organize into β-sheets. Upon unloading, the opposite effect happens. Miserez *et al.* (2009b) analyzed this phenomenon thermodynamically and compared it with rubber elasticity. As the polymer chains in rubber are extended, they align themselves with the tensile axis, decreasing the disorder (entropy) of the system. The Treloar equation (Eqn. (9.5)) shows, based on entropic considerations, that the stiffness increases with increasing temperature at a fixed value of the extension ratio (strain). The experimental results shown in Fig. 9.27(a) do not agree with this behavior. The analysis by Miserez *et al.* (2009b) uses two terms: one due to the internal energy increase (σ_U) and one due to the entropy decrease (σ_S):

$$\sigma = \sigma_U + \sigma_S = \left(\frac{\partial U}{\partial l}\right)_{V,T} - T\left(\frac{\partial S}{\partial l}\right)_{V,T}. \tag{9.25}$$

The application of Maxwell's relationship yields

$$\sigma = \left(\frac{\partial U}{\partial l}\right)_{V,T} + T\left(\frac{\partial S}{\partial T}\right)_{V,l}. \tag{9.26}$$

The strains in the stress vs. strain plots are in the range 10–100%. The plots of Figure 9.27(a) at various temperatures unequivocally show that the stress decreases with temperature at all strains tested, contrary to rubbery elastomer behavior, which is governed by entropy.

The decrease in the strength at a constant strain with increasing temperature suggests that the internal energy term dominates over the entropy decrease in Eqn. (9.26). This argument is in agreement with Flory's theoretical analysis of elastic proteins (Flory, 1956), showing that strains can be due to internal-energy-dominant mechanisms. However, entropic effects can be and probably are present.

9.7 Extreme keratin: hagfish slime and wool

Keratin is not usually associated with a high elastic deformation. However, there are two fascinating exceptions that deserve description. The mechanism by which this large recoverable deformation is obtained is through the alpha- to beta-phase transformation, which can be reversible and returns the specimen to close to its original length, after extension. This mechanism is akin to the one responsible for the extraordinary elasticity of whelk eggs, presented in Section 9.6.

Hagfish produce a mucus-like, viscous substance from their body when startled (see Fig. 9.28(a)). This slime is composed of mucins and seawater, held together by long protein threads (Fudge *et al.*, 2005). The slime reacts with water and clogs the gills of the predator fish, an effective and unique defense mechanism (Lim, 2006). The slime is produced at an astonishing speed, and one hagfish can produce enough slime to "clog" a 20 liter bucket of water in minutes. The slime contains threads that have an alpha-keratin-like IF structure (Downing *et al.*, 1984; Fudge *et al.*, 2003). The thread bundles are

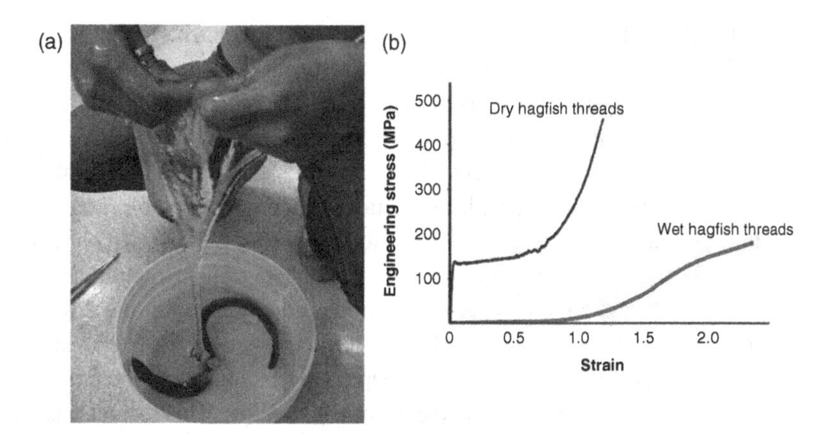

Figure 9.28.

Hagfish slime being poured into a bucket. (Used from http://www.people.fas.harvard.edu/~lim/research.htm, with kind permission from Jeanette Lim.) (b) Stress–strain curves for wet wool fibers and wet and dry hagfish threads. (Adapted from Fudge *et al.* (2003), with permission from Elsevier.)

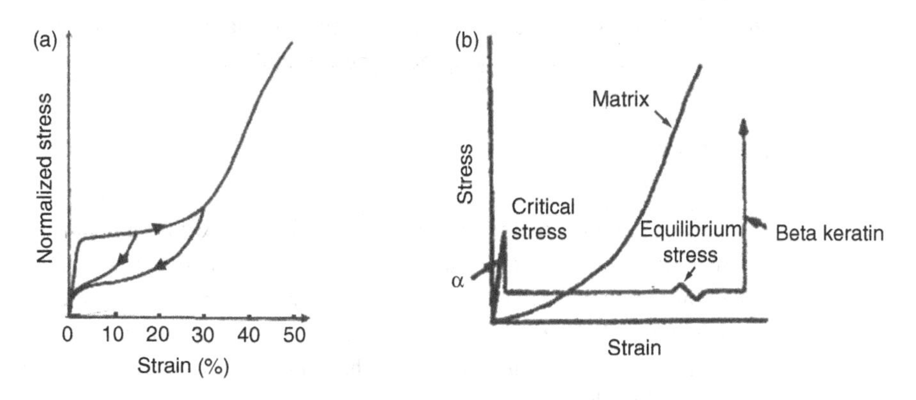

Figure 9.29.

(a) Tensile stress–strain response of wet wool showing evidence of alpha- to beta-keratin transformation with plateau. (b) Schematic representation of strain accompanied by the alpha–beta transformation in keratin and the J-curve response for matrix (M); the combination of the two curves produces the response shown in (a). (Reprinted from Hearle 2000), with permission from Elsevier.)

aligned, 1–3 μm in diameter, and several centimeters long (Downing *et al.*, 1981; Fernholm, 1981). Because the threads are not encased in a matrix, useful studies have been performed to evaluate the bulk mechanical properties of pure keratin IFs (Fudge *et al.*, 2003; Fudge and Gosline, 2004). Studies of these bundles are akin to studies on tendons, which are aligned nonmineralized collagen fibrils. Figure 9.28(b) shows a comparison between tensile stress–strain curves for wet and dry hagfish threads. The initial slope of wool fibers is orders of magnitude higher than that for the hagfish threads; however, the maximum failure strain is four times lower. The initial Young modulus of the hagfish slime is quite low, i.e. 6 MPa; this is attributed to significant direct hydration of the IFs, which are normally shielded by the matrix in hard alpha keratins. The mechanical response of dry slime is significantly different from that when it is wet. The Young modulus of the dry slime is 7.7 GPa, much higher than that of the wet slime and more similar to other keratin materials. This is indeed a characteristic of most biological materials, an extreme dependence on the degree of hydration.

Wet wool undergoes considerable strain, which is recoverable, as shown in Fig. 9.29(a). There is a first step with a high slope, followed by a plateau, which then gradually gives rise, at ~30% strain, to additional hardening. The transformation strain associated with the alpha–beta change in structure produces an extension in the direction of the keratin fibers. Figure 9.29(b) shows the keratin curve, with a clear plateau, and the curve for the matrix, having the characteristic J shape. Upon superposition of these two curves one obtains the real experimental response of Fig. 9.29(a). This response is clearly connected with the transition, which has a strain of ~30%. Beyond this strain, the β-sheet keratin is loaded elastically and follows a J-curve behavior. The different stages upon loading and unloading are seen in the sequence of Fig. 9.30. Alpha keratin is represented by springs and beta keratin is represented by a zig-zag structure. This corresponds to the structures shown in Fig. 3.15. Upon unloading, if conditions are right, the beta keratin reverts to alpha keratin and the specimen eventually shrinks to its original dimensions. Thus, a strain of ~35% can be achieved by this transformation. If the keratin does not revert, heating the specimen will bring back the original alpha structure.

Summary

- "Soft" biological materials have distinct applications in tissues and organisms. They provide tensile resistance and enable stretching under controlled conditions. They commonly exhibit the characteristic J-curve response, with a stress–strain (or stress–stretch ratio) in which the slope increases with strain ($d^2\sigma/d\varepsilon^2 \geq 0$).
- A number of constitutive equations have been used to represent the J-curve behavior. The worm-like chain model yields, for force f,

$$f = \frac{k_B T}{L_p}\left(\frac{1}{4\left(1-\frac{z}{L}\right)^2} - \frac{1}{4} + \frac{z}{L}\right),$$

  where k_B is the Boltzmann constant, T is the absolute temperature, L_p is the persistence length, L is the contour (total) length of the fibers, and z is the extension (displacement). It can be seen that f, the force, is equal to zero for $z = 0$; as z approaches $L, f \rightarrow \infty$.
- A two-term equation describes the principal phenomena in a J curve of (a) unfurling of chains and (b) stretching:

$$\sigma = k_1\varepsilon^{n+1} + H(\varepsilon_c)E(\varepsilon - \varepsilon_c),$$

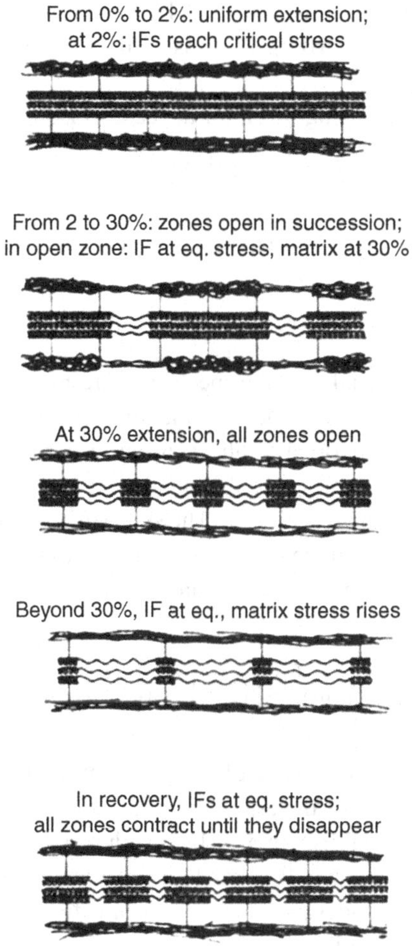

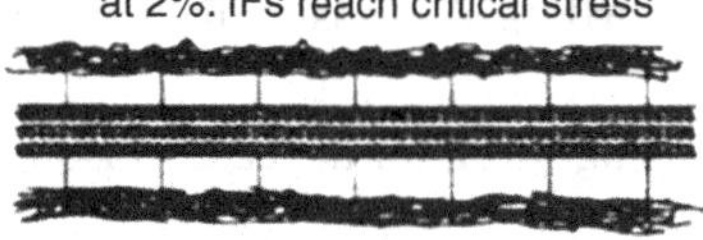

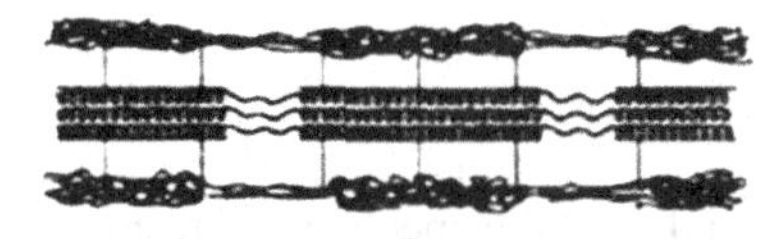

At 30% extension, all zones open

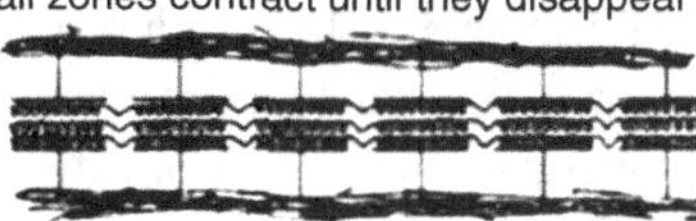

Figure 9.30.

Evolution of structure in tensile extension and recovery according to Chapman–Hearle theory; spring-like features represent α-helices, whereas zig-zag features represent β-sheets. (Reprinted from Hearle (2000), with permission from Elsevier.)

where E is the elastic modulus of the chains, k_1 is a parameter, and H is the Heaviside function, which activates the second term at $\varepsilon = \varepsilon_c$, where ε_c is a characteristic strain at which collagen fibers are fully extended. Subsequent strain gradually becomes dominated by chain stretching.

- The Mooney–Rivlin equation, developed initially for rubber, has the following form:

$$\sigma = \left(2C_1 + \frac{2C_2}{\lambda}\right)\left(\lambda^2 - \frac{1}{\lambda}\right),$$

where C_1 and C_2 are materials constants. The model is valid for strains up to ~2 (e.g. auto tires, where C_1 and $C_2 \sim 0.2$ MPa). A measurement of the looseness of the network is given by $2C_1/C_2$.

- The Ogden equation has the following form, for the case of isotropy:

$$\sigma = 2\mu\left(\lambda_1^{\alpha-1} - \lambda_1^{-1-\frac{\alpha}{2}}\right).$$

- The Fung equation, developed specifically for bioelastomers, has the following form:

$$\sigma = (\sigma^* + \beta)e^{\alpha(\varepsilon-\varepsilon^*)} - \beta.$$

- The skin is an orthotropic material, i.e. the mechanical response has defined and different trajectories for the perpendicular orientations defined by a coordinate system in which two axes are parallel to the skin surface and the third axis is perpendicular to it. The Langer lines cover the entire skin and mark the directions along which the mechanical stiffness is highest. The skin can be visualized as a wavy pattern of collagen fibers that straighten out when subjected to tension. By incorporating the effects of temperature and strain rate, one generalizes the Ogden equation as follows:

$$\sigma = \frac{2\mu_0}{\alpha_0}\left(\frac{T}{T_0}\right)^n \ln\left(\frac{\dot{\varepsilon}}{\dot{\varepsilon}_0}\right)\left(\lambda_1^{\alpha-1} - \lambda_1^{-1-\frac{\alpha}{2}}\right).$$

- Arteries contain residual stress. In the case of a segment of artery that is not under internal blood pressure, the walls of the artery are under strain, and therefore have residual stress. One way to establish this is by making a longitudinal cut and determining the angle that the walls will form. These residual stresses decrease the maximum tension that the wall undergoes during pressurization.
- Muscle fibers follow the Hill equation:

$$(\sigma + a)(\dot{\varepsilon} + b) = (\sigma_0 + a)b.$$

- There are biological materials that have responses more complex than the simple J response. Some of them have an inflection point: $d^2\sigma/d\varepsilon^2 = 0$.
- The whelk eggs exhibit a stress–strain behavior that is characterized by an inflection point $d^2\sigma / d^2\varepsilon = 0$. The stress–strain curves have three stages: an initial region with an elastic modulus of ~100 MPa; a plateau covering strains up to 0.7; a stiffening beyond 0.7. The region from $\varepsilon = 0.05$ to $\varepsilon = 0.7$ corresponds to the structural transition of the bioelastomer, which changes from α-helix to β-sheet configuration.
- Another biological material that undergoes such a phase transition is keratin. Keratin is not usually associated with a high elastic deformation. However, there are two fascinating exceptions that deserve description. The mechanism by which this large recoverable deformation is obtained is through the alpha- to beta-phase

transformation, which can be reversible and returns the specimen to close to its original length, after extension. This occurs in hagfish and wool. Wet wool undergoes considerable strain, which is recoverable. There is a first step with a high slope, followed by a plateau, which then gradually gives rise, at ~30% strain, to additional hardening. The transformation strain associated with the alpha–beta change in structure produces an extension in the direction of the keratin fibers. This response is clearly connected with the transition, which has a strain of ~30%. Beyond this strain, the β-sheet keratin is loaded elastically and follows a J-curve behavior.

Exercises

Exercise 9.1 What is the distribution of stresses in an artery that has internal stresses such that

(a) $\alpha = 180°$;
(b) $\alpha = 150°$?

At what internal pressure will the stress outside and inside the wall become the same? Assume (i) that the stress from the pressure decays linearly to zero at the external surface, and (ii) a linear elastic behavior with $E = 400$ MPa. Given: ID = 15 mm; OD = 22 mm.

Exercise 9.2 Read the following paper: M. J. Buehler, Nanomechanics of collagen fibrils under varying cross-link densities: atomistic and continuum studies, *J. Mech. Behav. Biomed. Mater.* **1**, 2008, 59–67.

(a) Referring to Fig. 3 of this paper, apply the Ogden equation and obtain parameters.
(b) Compare strength of one collagen fibril with the strength of skin. Discuss the difference. How do the Ogden parameters compare?

Exercise 9.3 An artery with the dimensions given below is subjected to increasing pressures. At what pressure will it burst if the maximum stress in the artery wall is 30 MPa? Given: ID = 15 mm; OD = 22 mm.

Exercise 9.4 Medical doctors have known for centuries that bullet wounds produce skin perforation that is not circular, but elongated. The same phenomenon is observed in skin perforations for the insertion of cylindrical percutaneous devices. Why?

Exercise 9.5 Derive Eqn. (9.19).

Exercise 9.6 Skin from the belly of a rabbit is governed approximately by the following equation:

$$\sigma = k\varepsilon^3 .$$

In the longitudinal direction, $k = 20$ MPa; in the transverse direction, $k = 10$ MPa.

(a) If a stress of 17 MPa is applied to the skin biaxially, what are the strains in the two directions?
(b) Sketch the orientation of the Langer lines in the belly.
(c) Interpret the results in terms of the waviness of the collagen molecules.

Exercise 9.7

(a) Using the Hill equation, plot the stress vs. strain curves for a muscle at strain rates of 10^{-2} s^{-1} and 1 s^{-1}. Given: $\sigma_0 = 200$ MPa; $a = 10$ MPa; $b = 1$ s^{-1}.
(b) Is this a fast- or slow-twitch muscle?

Exercise 9.8 Using the WLC model, plot the force f vs. extension z for a protein at 200 and 320 K. Given: stretched length = 300 μm, persistence length = 7 μm.

Exercise 9.9 Using the Mooney–Rivlin equation, calculate and plot the stress vs. stretch ratio curves up to $\lambda = 3$. Given: $C_1 = 0.2$ MPa; $C_2 = 0.2$ and 2 MPa. Which one has the loosest structure?

Exercise 9.10 An artery with dimensions OD = 20 mm and ID = 17 mm is subjected to pressures ranging from 80 mm Hg to 130 mm Hg.

(a) Determine the diameter of the artery of the systolic (highest) and diastolic (lowest) points. The material in the artery follows the relationship $\sigma = k\varepsilon^2$, with $k = 25$ MPa.
(b) Plot the pressure–radius curve due to the internal pressure from the stress–strain response of the material.

10 Biological foams (cellular solids)

Introduction

Many biological structures require light weight. The prime examples are birds, whose feathers, beaks, and bones are designed to maximize performance while minimizing weight. Other examples are plants, where the same requirements operate. Figure 10.1 illustrates two examples of structures designed to have significant flexure resistance.

Plant stalks are composed of cellulose and lignin arranged in cells aligned with the axis of growth. The giant bird-of-paradise stem is lightweight and has a structure consisting of cells which have, on their longitudinal section, a rectangular (but interconnected in a staggered manner) shape (Meyers *et al.*, 2013) (Fig. 10.1(a)). The cross-sectional shape is closer to elliptical. Thus, the cells have a cylindrical shape. If we look at the cell walls at a greater magnification, we observe that they are also cellular. This produces two levels of cellular structure, in a manner akin to the feather rachis in *Falco sparverius*, presented in Chapter 2 (see Fig. 2.1). The solution often found in nature is to use tailored structures to address the stress demands. The structure is designed to resist flexure stresses without buckling.

A second example are porcupine quills (Fig. 10.1(b)), which provide protection against predation. Their sharp points are a clear indication of their function: to keep a potential predator away. One of us (MAM) recalls hunting dogs returning from forays in the forest with quills embedded in their noses. Their extraction was a painful process. Porcupine quills are keratinous. The bending resistance (stiffness) is provided by a sandwich structure in which the core is cellular and the shell is solid, with a high flexural strength/weight ratio. The external shell (cortex) surrounds a cellular core that provides stability to the walls under compression.

First, we present the fundamental equations of mechanics that quantitatively predict the response of beams, and extend this treatment to cellular materials (foams), using largely the framework developed by Gibson and Ashby (1997). Then, we will illustrate these principles with a number of examples: bird beaks, bones, and feathers; wood; and cuttlefish bone.

10.1 Lightweight structures for bending and torsion resistance

Resistance to flexural and torsional tractions with a prescribed deflection is a major attribute of many biological structures. The student is referred to Section 4.5 for more

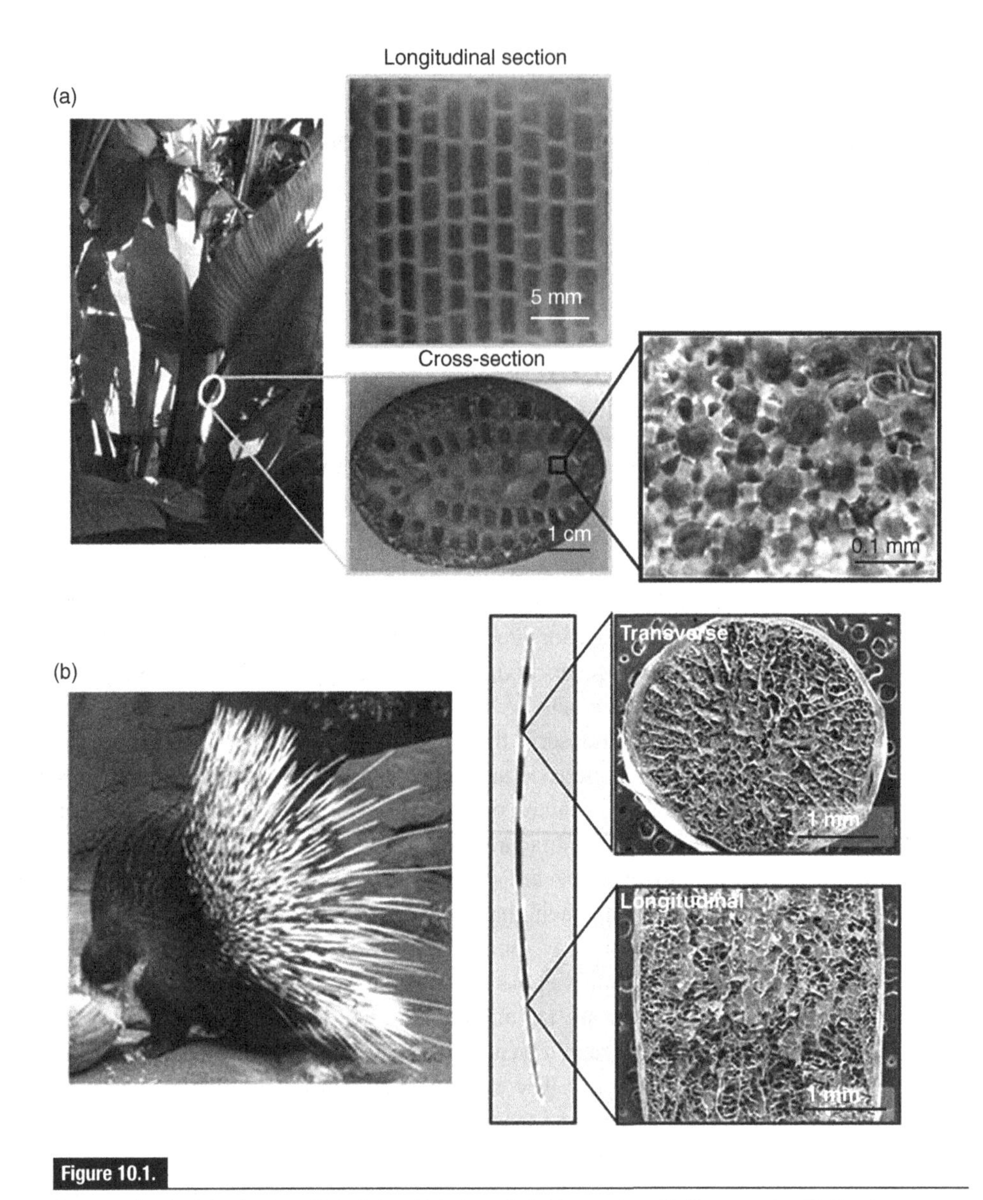

Figure 10.1.

Two examples of cellular structure: (a) plant stem (bird-of-paradise); (b) porcupine quill. (Figure courtesy W. Yang.)

detailed derivations. The fundamental mechanics of elastic (recoverable) deflection, as it relates to the geometrical characteristics of beams and plates, is given by three equations: the first relates the bending moment, M, to the curvature of the beam, d^2y/dx^2 (y is the deflection):

$$\frac{d^2y}{dx^2} = \frac{M}{EI}, \tag{10.1}$$

where E is the Young modulus and I is the moment of inertia, which depends on the geometry of the cross-section.

The second equation predicts the stress. The fundamental mechanics equation connecting the stresses to the radial distance, y, measured from the centroid, is as follows:

$$\sigma_y = \frac{My}{I}. \tag{10.2}$$

Thus, the stresses increase linearly with y, and the central core does not contribute significantly to the flexure resistance. This explains why low-density (and low-strength) cellular materials are used in the center of structures. Additional weight savings are achieved if the core is hollow.

This is of utmost importance: the curvature of a solid beam, and therefore its deflection, is inversely proportional to the fourth power of the radius R, because, in a cylindrical beam,

$$I = \pi R^4/4. \tag{10.3}$$

For a tubular beam with external radius R, internal radius R_0, and wall thickness t, we have

$$I = \frac{\pi}{4}(R^4 - R_0^4) \approx \pi R^3 t. \tag{10.4}$$

The diligent student can show that I increases significantly over that for a solid cylinder, for a fixed weight per unit length.

The third equation, commonly referred to as Euler's buckling equation, calculates the compressive load at which global buckling of a column takes place:

$$P_{cr} = \frac{\pi^2 EI}{(kL)^2}, \tag{10.5}$$

where k is a constant dependent on the column end conditions (i.e. pinned, fixed, free), and L is the length of the column. Resistance to buckling can also be accomplished by increasing I; however, the required I decreases with the second power of L for a certain load P. Both Eqns. (10.1) and (10.5) predict the principal design guideline for a lightweight/stiff structure: place the mass as far away from the neutral axis (which passes through the centroid of the cross-section). The same reasoning can be extended to torsion. This is readily accomplished by having a hollow tube with radius R and thickness t.

The local buckling (crimping) tendency, however, increases with decreasing thickness t of the tube and increasing R:

$$\sigma_{cr} = \frac{E}{\sqrt{3(1-\nu^2)}}\left(\frac{t}{R}\right), \tag{10.6}$$

where ν is Poisson's ratio. A compromise must be reached between I and t. Nature has addressed this problem with ingenious solutions: by creating a thin solid shell and filling the core with lightweight foam (Gibson *et al.*, 2010) or by adding internal reinforcing struts or disks. These stratagems provide resistance to local buckling (crimping) with a minimum weight penalty. Primary examples of these design principles are some skeletal and antler bones, which have a cellular core (cancellous bone) and a solid exterior (cortical bone). Bamboo has a hollow tube with periodic disks at prescribed separations. The wing bones of soaring birds use this strategy with internal struts and ribs. This subject has been treated in detail by Gibson and Ashby (1997) and Gibson *et al.* (2010).

10.2 Basic equations for foams

Many naturally occurring materials are not fully dense, i.e. they possess internal cavities. This type of design is intentional, because it reduces the density. Examples are cork, bone, wood, sponge, and plant stalks; examples of these structures are shown in Fig. 10.2 (Gibson and Ashby, 1988, 1997).

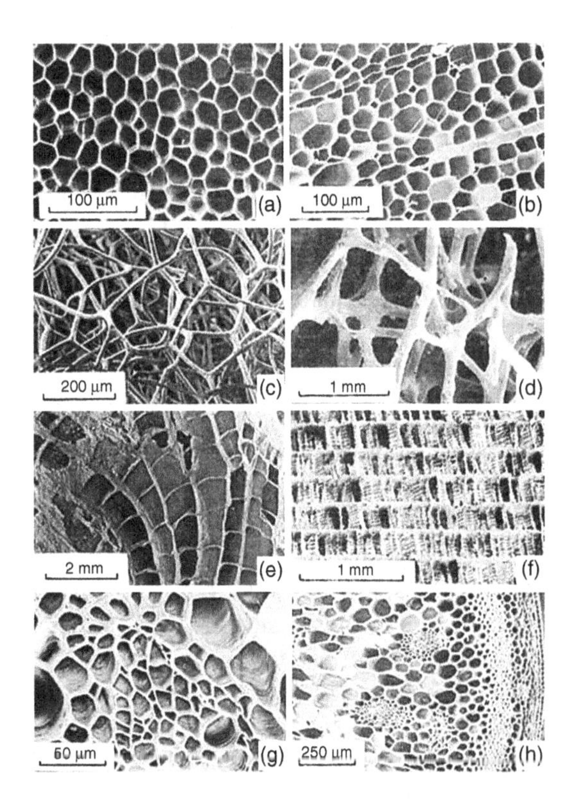

Figure 10.2.

Examples of cellular materials: (a) cork; (b) balsa; (c) sponge; (d) cancellous bone; (e) coral; (f) cuttlefish bone; (g) iris leaf; (h) stalk of plant. (From Gibson and Ashby (1997), p. 22. Copyright © 1997 Lorna J. Gibson and Michael F. Ashby. Reprinted with the permission of Cambridge University Press.)

(a)

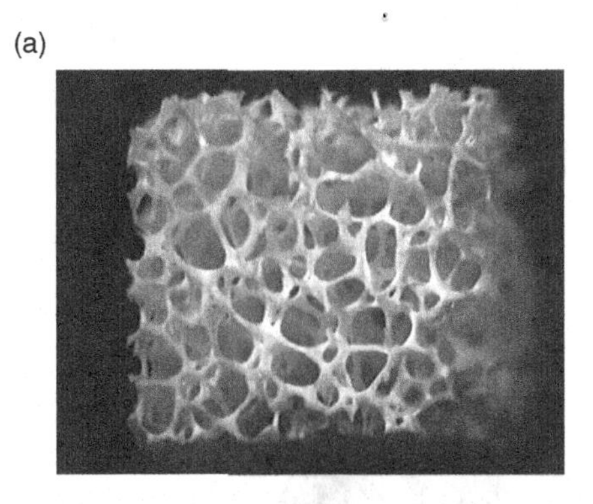

(b)

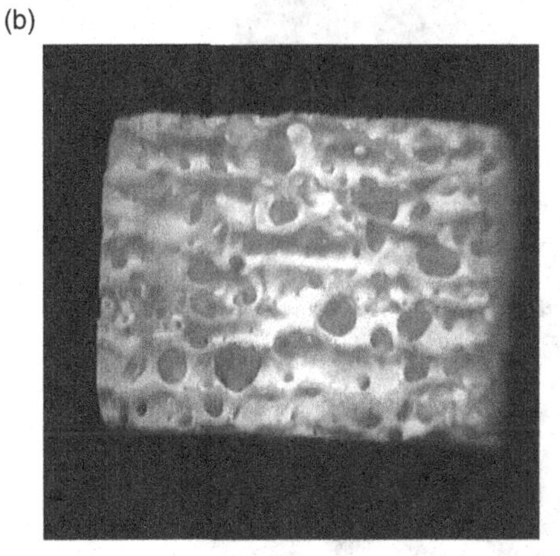

Figure 10.3.

Deproteinized cancellous antler bone: (a) cross-section; (b) longitudinal section.

Modern synthetic materials have also adopted this form, and we have metallic, ceramic, and polymeric foams. Some are common and in everyday usage, such as Styrofoam. Others are quite esoteric, such as the Space Shuttle tiles, which have a density of 0.141 g/cm^3 and a maximum working temperature of 1260 °C.

An example of a biological cellular material is cancellous bone. Bone is designed to have a variable density. Regions subjected to higher stress are denser. The outside surface is made of high-density material and is called compact bone. The inside of bone tends to have a lower density and is termed cancellous bone. We have seen in Fig. 7.1(a) the longitudinal section of a femur. The same occurs in antlers. Figure 10.3(a) shows the cross-section of an elk antler, and Fig. 10.3(b) shows the longitudinal section. The surface region is compact, whereas the center is cellular. This is also justified by the resistance to flexure: the inside is not subjected to such high stresses as the outside ($\sigma \propto r$, where r is radial distance). Thus, the strength requirement increases linearly from the neutral axis to the surface. This corresponds to the porosity distribution.

There are numerous other examples of cellular materials; they are used either by themselves or in sandwich arrangements. Sandwich structures range from common

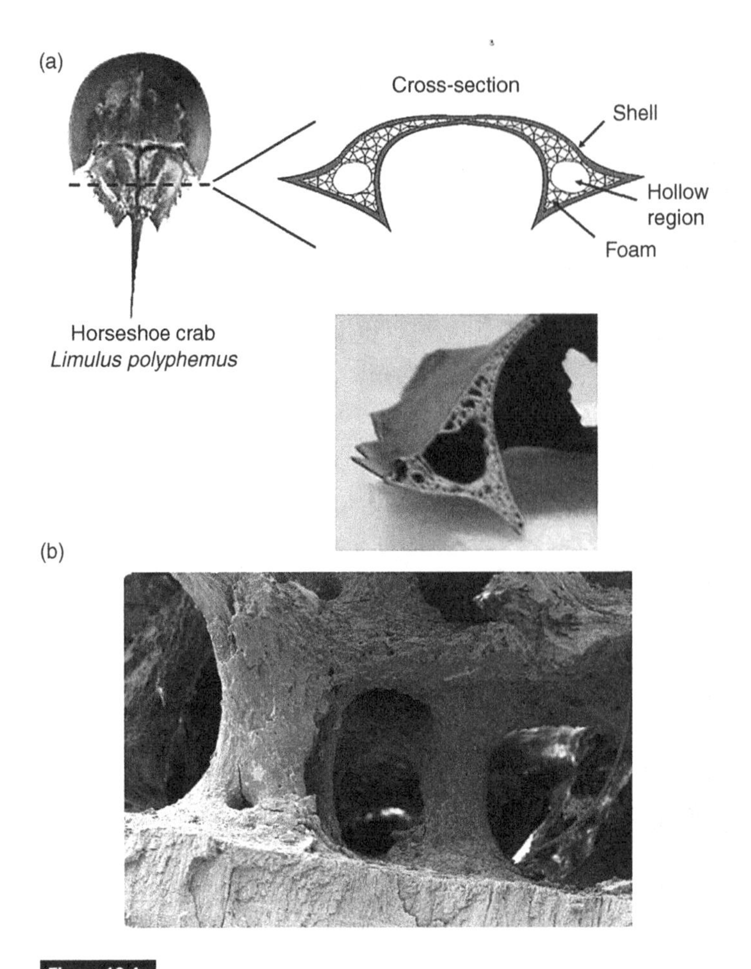

Figure 10.4.

(a) Cross-section image of horseshoe crab exoskeleton showing the foam structure. (b) SEM micrograph showing detail of the foam structure in horseshoe crab exoskeleton. (Reprinted from Chen *et al.* (2008b), with permission from Elsevier.)

cardboard used in packaging to important uses in the aircraft industry. The basic idea is to have a dense skin and a lightweight interior. Figure 10.4 shows a cross-section of a horseshoe crab; we can see that a cellular network provides the rigidity. Two other examples, wood and beak interior, are described in Sections 10.3 and 10.5, respectively.

The compressive stress–strain curves of cellular materials have three characteristic regions: (a) an elastic region, (b) a collapse plateau, and (c) a densification region. These are shown in Fig. 10.5 (Gibson and Ashby, 1997). The higher the initial density, expressed in Fig. 10.5 by ρ^*/ρ_s, the smaller the collapse plateau region. It also occurs at a higher stress.

We will develop expressions that predict this behavior. These are the Gibson–Ashby equations. They are developed here for an open-cell geometry that represents well cellular materials with a low relative density. Section 10.5.2 presents the equivalent

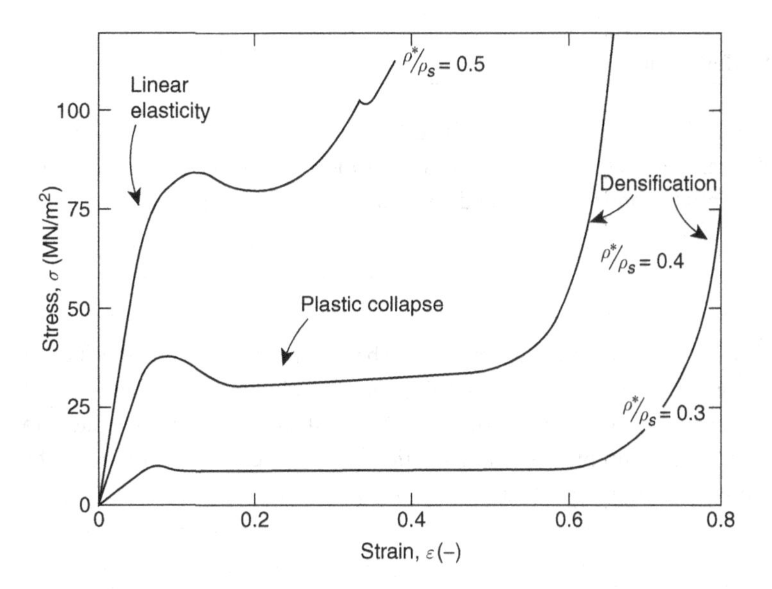

Figure 10.5.

Stress–strain curves for cancellous bone at three different relative densities: 0.3, 0.4, and 0.5. (Reprinted from Hayes and Carter (1976), with permission from John Wiley & Sons, Inc.)

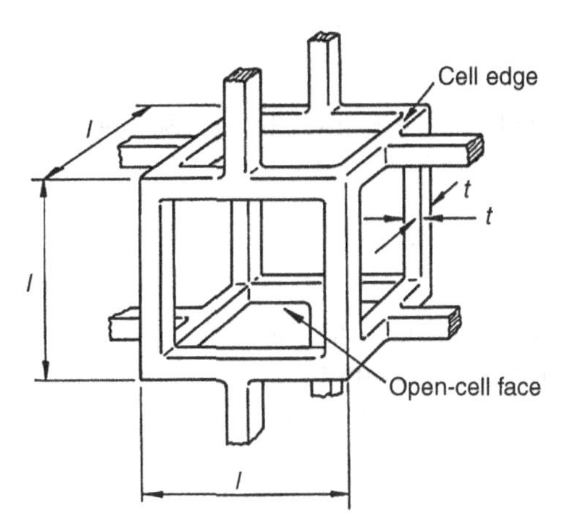

Figure 10.6.

Open-cell structure for cellular materials with low relative density. This is the structure upon which the Gibson–Ashby equations are based. (From Gibson and Ashby (1997), Copyright © 1997 Lorna J. Gibson and Michael F. Ashby. Reprinted with the permission of Cambridge University Press.)

equations for a closed-cell geometry. Figure 10.6 represents this open-cell structure. It consists of straight beams with a square cross-section (Gibson and Ashby, 1997). The model is very simple but captures the essential physics. There are two characteristic dimensions: the cell size, l, and the beam thickness, t.

10.2.1 Elastic region

Three elastic constants are defined for an isotropic foam: E^*, G^*, and υ^*. The density of the cellular material is ρ^*, and that of the solid material is ρ_s. From Fig. 10.6 we can obtain an expression for the density in terms of l and t:

$$\frac{\rho^*}{\rho_s} = C_1\left(\frac{t}{l}\right)^2; \tag{10.7}$$

C_1 is a proportionality constant that can be obtained by the student. When the cell is subjected to compressive loading, it will deflect as shown in Fig. 10.7 (Gibson and Ashby, 1997).

The vertical columns push on the horizontal beams and cause them to bend. A force F on each column produces a deflection δ in the beam. The moment of inertia of a beam with a rectangular section (sides b and h) is

$$I = \frac{bh^3}{12}. \tag{10.8}$$

For the square cross-section with side t:

$$I = \frac{t^4}{12}. \tag{10.9}$$

Beam theory states that the deflection, δ, is given by (see Section 4.5)

$$\delta = C_2\frac{Fl^3}{E_sI}, \tag{10.10}$$

where C_2 is a constant. The stress acting on the cell is related to the force F (each force F is shared by two neighboring cells) as follows:

$$\sigma = \frac{F}{l^2}.$$

The strain, ε, is related to the deflection by

$$\varepsilon = \frac{2\delta}{l}.$$

Thus, the Young modulus is given by

$$E^* = \frac{E_sI}{2C_2l^4} = \frac{E_st^4}{24C_2l^4}.$$

This can be expressed as a function of density (Eqn. (10.7)):

$$\frac{E^*}{E_s} = \frac{C_1}{24C_2}\left(\frac{\rho^*}{\rho_s}\right)^2.$$

Experimental measurements indicate that $C_1/24C_2$ should be approximately equal to unity. Thus,

$$\frac{E^*}{E_s} \sim \left(\frac{\rho^*}{\rho_s}\right)^2. \tag{10.11}$$

Similarly, an expression for the shear modulus can be obtained:

$$\frac{G^*}{E_s} = \frac{3}{8}\left(\frac{\rho^*}{\rho_s}\right)^2. \tag{10.12}$$

10.2.2 Plastic plateau

At a certain level of deformation, elastic behavior gives way to plastic deformation. The Gibson–Ashby equations are based on the formation of plastic hinges at the regions where the beams terminate. Four of these plastic hinges are circled in Fig. 10.7.

For the elastic case, the stresses increase linearly from the neutral axis. For the case of a plastic hinge, $\sigma = \sigma_s$, the stresses acting on the cross-section are uniform and tensile above

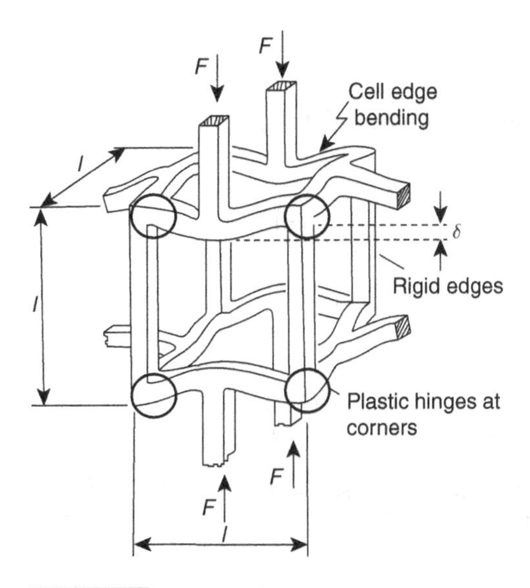

Figure 10.7.

Open-cell configuration under compressive loading. Note the deflection δ observed. (From Gibson and Ashby (1997) Copyright © 1997 Lorna J. Gibson and Michael F. Ashby. Reprinted with the permission of Cambridge University Press.)

the neutral axis and uniform and compressive below the neutral axis. Figure 10.7 shows the configuration.

The plastic moment, M_p, about the neutral axis is given by

$$M_p = F\frac{t}{2}. \tag{10.13}$$

The yield stress is related to F as follows:

$$\sigma_y = \frac{2F}{t^2}. \tag{10.14}$$

Thus, substituting Eqn. (10.14) into Eqn. (10.13), we obtain

$$M_p = \frac{1}{4}\sigma_y t^3. \tag{10.15}$$

Taking the beam with length $l/2$ and considering the force $F/2$ applied to each of the two hinges, we get

$$M_p = \frac{F}{2}\frac{l}{2} = \frac{1}{4}Fl. \tag{10.16}$$

The global stress acting on the foam is the force F divided by the area upon which it acts, l^2:

$$\sigma_{pl}^* = \frac{F}{l^2}. \tag{10.17}$$

Equating Eqns. (10.15) and (10.16) and applying Eqn. (10.17):

$$\frac{\sigma_{pl}^*}{\sigma_{ys}} = \left(\frac{t}{l}\right)^3. \tag{10.18}$$

Substituting Eqn. (10.7) into Eqn. (10.18):

$$\frac{\sigma^*_{pl}}{\sigma_{ys}} = C_1^{-3/2}\left(\frac{\rho^*}{\rho_s}\right)^{3/2}. \tag{10.19}$$

There are other expressions for the closed-walled cell, for cells that do not undergo plastic deformation, and for other cases.

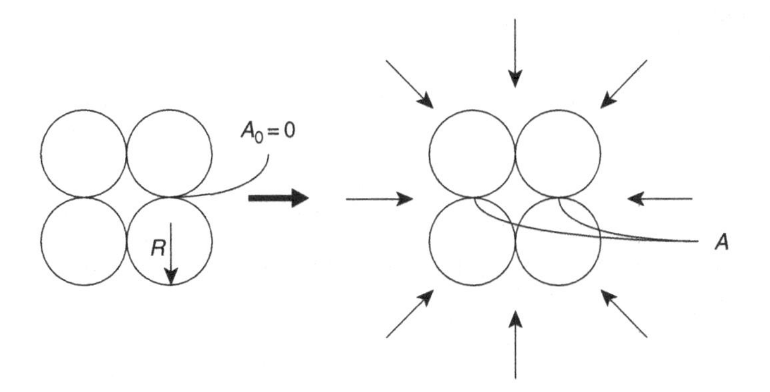

Figure 10.8.

Particle flattening (Fischmeister–Arzt) densification mechanism.

10.2.3 Densification

Densification starts when the plastic plateau comes to an end. This region is characterized by a complex deformation pattern. The stresses required for densification rise rapidly as the open spaces between the collapsed cell structure close up. The analytical treatment for the collapse of pores and voids will not be presented here. There are theories that address this problem. One of the best known, the Carroll–Holt–Torre theory (Torre, 1948; Carroll and Holt, 1972), assumes a spherical hole inside a solid sphere. By applying an external pressure it is possible to collapse the internal hole. The smaller the hole, the higher the stress. The Helle *et al.* model (Helle, Easterling, and Ashby, 1985) addresses the same problem. A third formulation is the Fischmeister–Arzt theory (Fischmeister and Arzt, 1982). Figure 10.8 shows the model envisaged in this formulation. Fischmeister and Arzt assumed that each particle (assumed to be initially spherical) has Z neighbors, and that the number of neighbors increases with density. Again, the relative density $D = \rho^*/\rho_s$. For the solid, this coordination number is taken as 12. This is equal to the coordination number for atoms in the FCC and HCP structures. Thus,

$$Z = 12D. \tag{10.20}$$

At D_0 (initial value of D) = 0.64, $Z = 7.7$; for $D = 1, Z = 12$. The average area of contact between neighbors, A, is shown in Fig. 10.8; it increases from $A_0 = 0$ to one-twelfth of the particle surface area, $4\pi R^2$, *because, as the contact points are flattened, their surface area increases. The maximum area is equal to the total sphere surface divided by the number of neighbors. A simple expression for* A *is as follows:*

$$A = \frac{\pi(D - D_0)}{3(1 - D_0)} R^2. \tag{10.21}$$

When $D = D_0$, $A = 0$; when $D = 1$, $A = 4\pi R^2/12 = (\pi/3)R^2$, where R is the radius of particles (Fig. 10.8(a)).

The force F applied to each contact region is related to the external pressure:

$$F = \frac{4\pi R^2}{ZD} P. \tag{10.22}$$

This force, divided by the contact area A, yields the average pressure on the particle at the contact region:

$$P_p = \frac{F}{A} = \frac{4\pi R^2}{AZD} P. \tag{10.23}$$

It is known (see, e.g., Meyers and Chawla (2009) that the stress required to make an indentation on a surface is ~$3\sigma_y$, where σ_y is the yield stress of the material. Thus,

$$3\sigma_y = \frac{4\pi R^2}{AZD} P, \tag{10.24}$$

and, substituting Eqn. (10.23) into Eqn. (10.20) yields

$$P = Z\sigma_y \frac{D(D - D_0)}{4(1 - D_0)}. \tag{10.25}$$

Example 10.1 A piece of equiaxed open-cell aluminum foam is designed and synthesized for energy absorbent application.

(a) Calculate its apparent density if the required elastic modulus of the foam is 8 GPa and you are given that $E_s = 72$ GPa, $\rho_s = 2.7$ g/cm^3, and $C_1 \sim 1$.

(b) If the cell wall thickness is 20 μm, estimate the average cell size of the aluminum foam by applying $C \sim 1$.

Solution (a) The relative elastic modulus for the foam depends on its relative density:

$$\frac{E^*}{E_s} = C_1 \left(\frac{\rho^*}{\rho_s}\right)^2 \cong \left(\frac{\rho^*}{\rho_s}\right)^2;$$

$$\rho^* = \rho_s \sqrt{\frac{E^*}{E_s}} = 2.7\sqrt{\frac{8}{72}} = 0.9 \text{ g/cm}^3.$$

(b) The relative density for the foam is

$$\frac{\rho^*}{\rho_s} = \left(\frac{t}{l}\right)^2;$$

$$\frac{t}{l} = \sqrt{\frac{\rho^*}{\rho_s}} = \sqrt{\frac{0.9}{27}} \cong 0.58;$$

$$l = \frac{t}{0.58} \cong 34.5\ \mu\text{m}.$$

Box 10.1 Cellular biomaterials and osteogenesis

Porous engineering scaffolds play an important role in regenerating tissues such as bone, cartilage, and skin. Scaffolds for bone regeneration (osteogenesis) have to satisfy critical criteria, including adequate mechanical properties, biocompatibility, biodegradability, proper porosity, and osteoconductivity. Pores are essential for bone-tissue regeneration by allowing migration, proliferation, and formation of osteoblasts, cells, and vascular systems. The minimum pore size required for osteogenesis is $\sim 100\ \mu m$, yet pore sizes $>300\ \mu m$ are recommended. However, the mechanical properties significantly decrease with porosity and pore size, so porous biomaterials are typically used as non-load-bearing implants. The interconnectivity of pores is also critical in osteogenesis. Cellular materials and scaffolds can be synthesized by various techniques, such as sintering, molding, electrospinning, freeze-drying, three-dimensional protocol printing, demineralization/deproteinization, etc. Biomedical scaffolds can be classified into metals, ceramics, polymers, and composites, with either natural or synthetic sources.

Stainless steel, titanium, and titanium alloys (Ti-6Al-4V) are common metal implants for osteogenesis. Titanium scaffolds with 86% porosity and $250\ \mu m$ pores have been fabricated and used. Another application of metallic implants is to create a porous surface by plasma-spraying, shot-blasting, or acid-etching. The lack of tissue adherence and biodegradability has limited the use of pure metal scaffolds. Ceramic scaffolds for osteogenesis are made mainly of hydroxyapatite, as well as tricalcium phosphate (TCP), Bioglass®, and glass ceramics. Natural sources include hydroxyapatite converted from coral (ProOsteon®) or animal bone (Endobon®) and deproteinized bovine trabecular bone (Bio-Oss®). Ceramic scaffolds are generally good in osteointegration yet limited due to the brittleness and slow degradation rates. Polymeric scaffolds have the advantage of biocompatibility and biodegradability. Natural polymers, such as collagen, GAG, hyaluronic acid, silk fibroin, and chitosan, as well as synthetic polymers (e.g. poly-lactide, poly-lactide-co-glycolide), are commonly used to synthesize scaffolds. The major limitation of polymer-based scaffolds is their weak mechanical properties.

Various approaches and vigorous investigations have been carried out on the synthesis of composite scaffolds. For example, coating hydroxyapatite scaffolds with a HAP/polymer composite improves the mechanical properties. Polymer-based scaffolds (collagen, collagen–GAG, chitosan) can be strengthened by ceramic coating or surface mineralization, as shown in Fig. B10.1. Coating porous titanium implants with calcium phosphate results in better bone ingrowth. Other novel advances include the addition of growth factors, such as bone morphogenetic proteins (BMPs),

Box 10.1 (cont.)

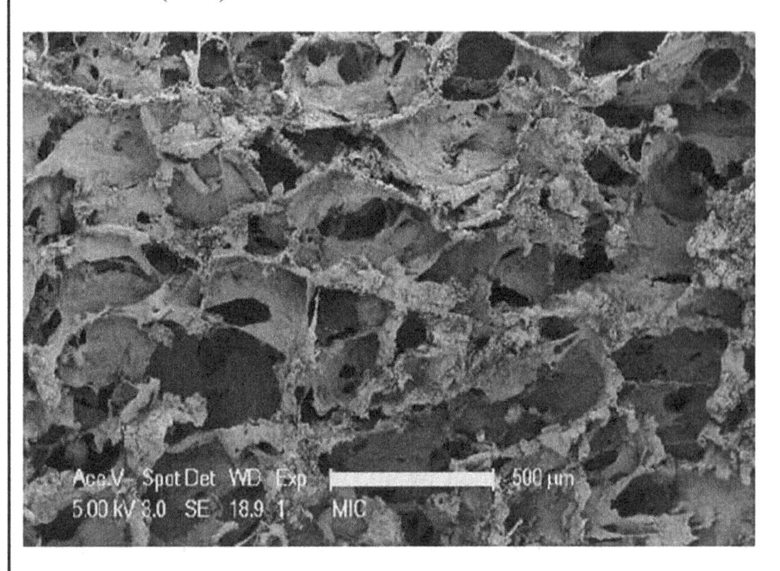

Figure B10.1.

SEM micrograph showing a calcium phosphate mineralized collagen–GAG scaffold synthesized by freeze-drying. (From Harley *et al.* (2010), with kind permission from John Wiley & Sons, Inc.)

insulin-like growth factors (IGFs), and transforming growth factors (TGFs). Collaborative studies at MIT (Professors Lorna Gibson and Ioannis Yannas) and Cambridge University (Professor William Bonfield) have led to the development of a bilayer, hybrid osteochondral scaffold for the simultaneous generation of superficial cartilage as well as underlying bone. This is described further in Chapter 12.

10.3 Wood

Wood is one of the most ancient structural materials in the world and has played an important role in civilization. It is still widely used in building, the construction of furniture, ships, and musical instruments, the manufacture of paper, and so on. Wood has a high specific stiffness (stiffness per unit weight) and specific strength that is comparable with steel (Wegst and Ashby, 2004). Its outstanding mechanical properties are mainly due to the hierarchical structure and optimized reinforcement orientation of cellulose fibrils.

Wood is a cellular composite with four levels of hierarchical structure: molecular, fibrillar, cellular, and macroscopic (Jeronimidis, 1980). Figure 10.9 shows the hierarchical structure of wood (Tirrell, 1994). The main structural constituents of wood are cellulose, a high-molecular-weight polysaccharide (see Chapter 3), which contributes to

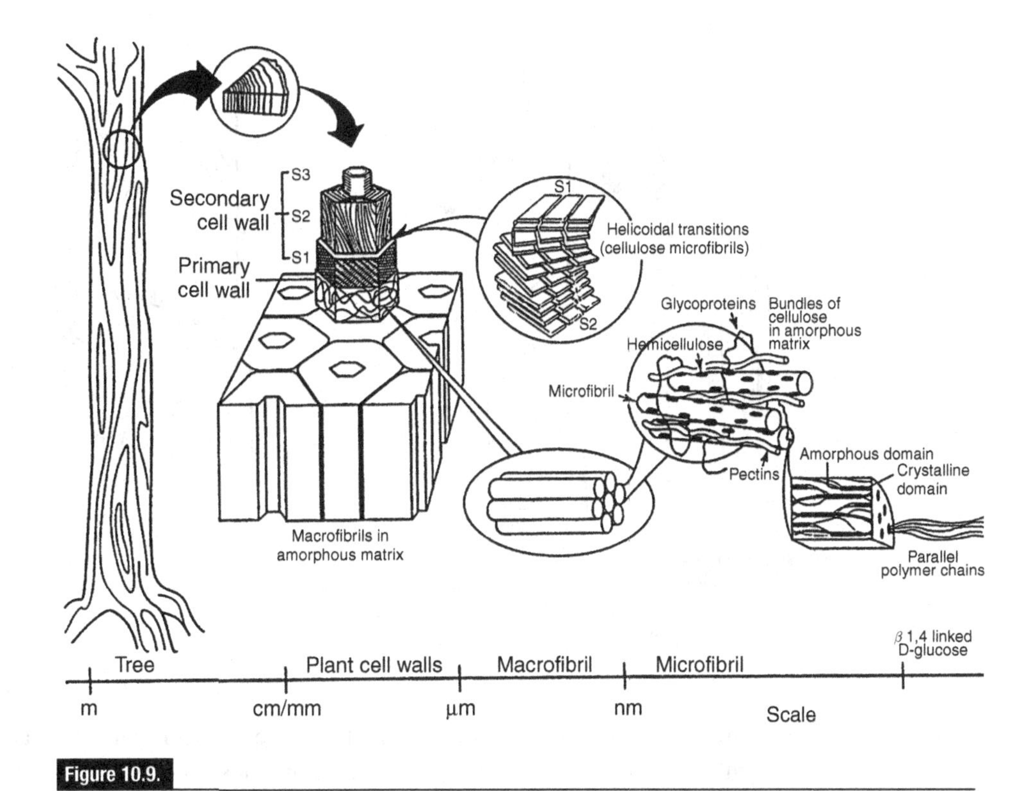

Figure 10.9.

Hierarchical structure of cellulose in wood. (Figure courtesy D. Kaplan.)

the stiffness and strength, and lignin. The cellulose is organized into microfibrils of about 10–20 nm diameter. The microfibrils consist of both crystalline and amorphous regions. Bundles of cellulose microfibrils form macrofibrils, which are embedded in an amorphous matrix of lignin, hemicellulose, and other compounds.

The most characteristic structural level is the cellular structure, or the wood tracheid. Mark (1967) and Preston (1974) carried out comprehensive studies on the structure of wood tracheids. Figure 10.10(a) shows a simplified general model of a wood tracheid (Young and Rowell, 1986). Cotton fibers, which consist mainly of cellulose, have a similar multilayer structure (Fig. 10.10(b)). From the outer surface to the center, the fiber has a cuticle, a primary wall, a secondary wall, a lumen wall, and a lumen. The cuticle is a waxy layer less than 0.25 μm thick. The primary wall is the original cell wall and consists of cellular fibrils in a randomly oriented network. The primary wall can restrict swelling of fibers. The secondary wall is composed of an outer layer (S_1), middle layer (S_2), and inner layer (S_3). The three layers of the secondary wall are built up by lamellae formed by almost parallel microfibrils stacking in a helicoidal pattern. The S_1 layer is roughly 0.2–0.3 μm and has an angle of 20–35° with respect to the fiber axis (parallel to the trunk). The S_3 layer, also known as lumen wall, is not always detected. It is 0.1 μm thin and has

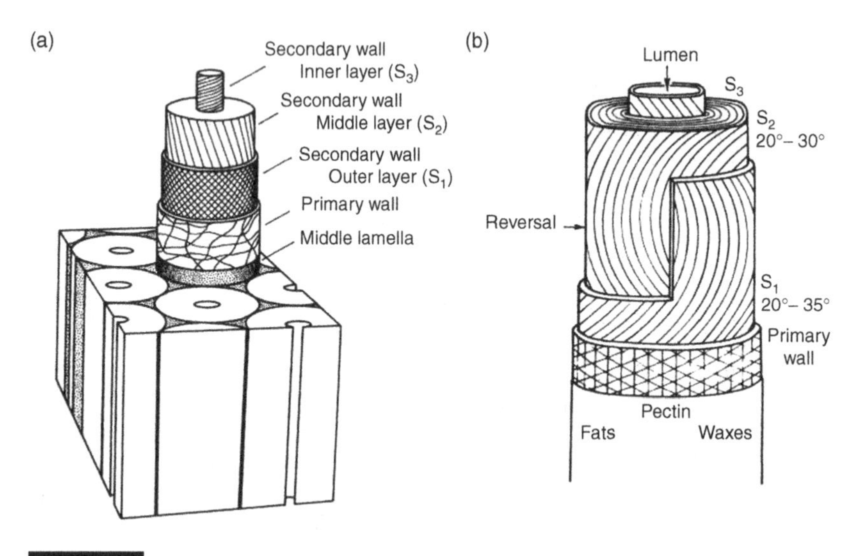

Figure 10.10.

Structure of (a) wood tracheid and (b) cotton fibers showing microfibril orientation and relative size of different layers of the cell wall. (Figure courtesy J. O. Warwicker, the Shirley Institute, Manchester, UK.)

an angle of 60–90° with the growth direction. The S_2 layer constitutes about 95% by weight of the wood tracheid and can be 4 to 5 μm thick. The microfibrils in the S_2 layer have an angle of 20–30° (Nevell and Zeronian, 1985). The relative cross-sectional area (roughly 80% of the total cell wall area) and the low microfibrillar angle make the S_2 layer the major load-bearing component (Jeronimidis, 1980). The macroscopic structure of wood can be seen with the naked eye. Growth rings can be observed in most species.

Some woods have a very low density, balsa wood being the extreme example (0.04 g/cm^3 –0.34 g/cm^3). Others are highly dense, such as lignum vitae, or ironwood, with 1.37 g/cm^3.

Palms and bamboos are known to give the maximum bending resistance per unit weight (Wegst and Ashby, 2004) while having great flexibility. This is due to the graded structure, with the core being either hollow or filled with a lower-density wood and the periphery being a denser and stronger wood. Trees, in contrast, tend to be more homogeneous. Figure 10.11(a) shows the cross-section of bamboo, which is hollow. This is called culm. The parenchyma and fiber bundles become progressively denser from the inner to the outer surface (Fig. 10.11(b)). The same phenomenon is characteristic of the palm. The periphery consists of denser wood whereas the center is lighter. This is an efficient way to design the tree, which has to resist high winds by undergoing significant bending (Fig. 10.11(c)). The attentive student will recall Eqn. (10.2).

The mechanical properties of wood are highly anisotropic due to the preferred orientation of cellulose fibrils (parallel to the trunk). Gibson and Ashby (1997) describe the mechanical properties of wood. Figure 10.12 shows three orthogonal planes of symmetry of wood: the radial, the tangential, and the axial directions. The stiffness

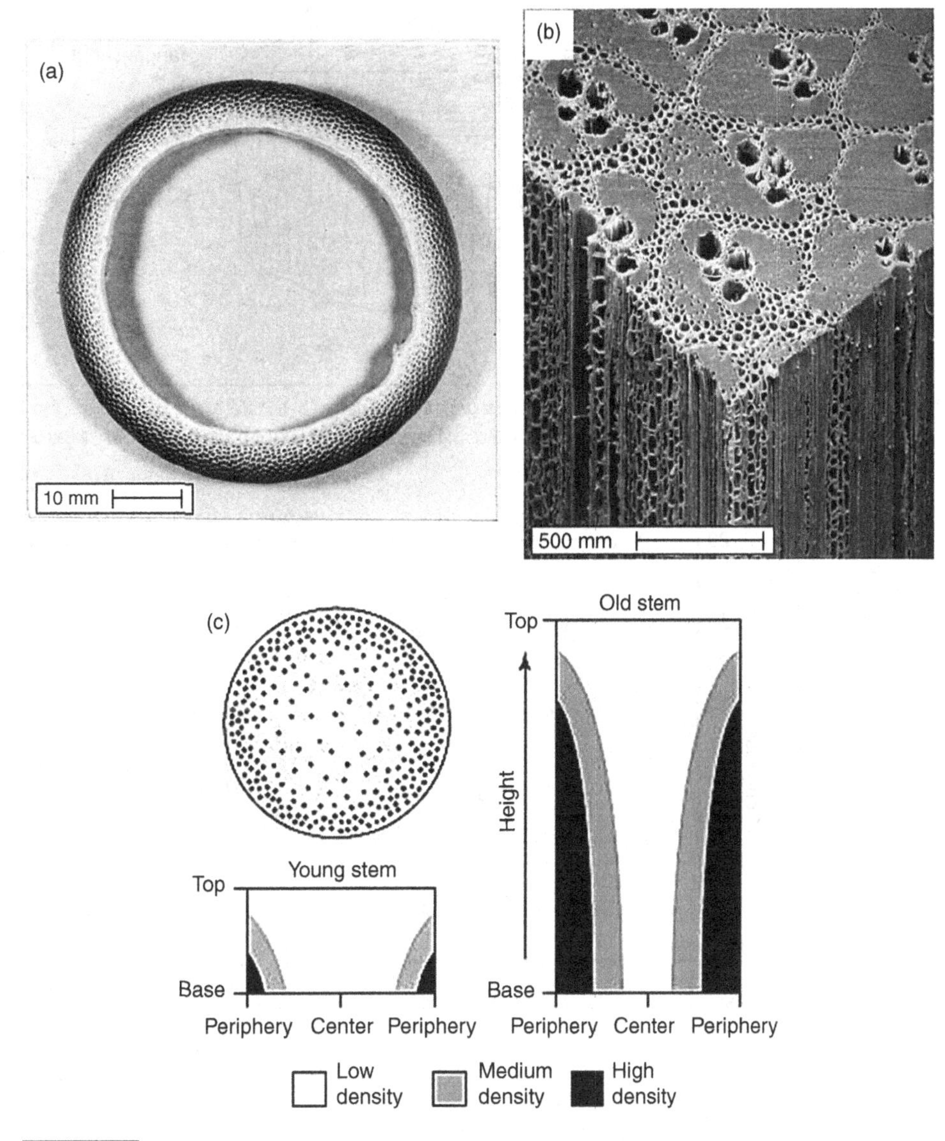

Figure 10.11.

(a) Transverse cross-section of a bamboo culm. (b) Wall of the bamboo culm, showing the radial distribution of vascular bundles (optical); note the density gradient from the inner wall to the outer wall. (c) Schematic cross-section of palm tree with greater density of fibers and higher density on the outside. (Reprinted from Wegst (2011), with permission from Elsevier.)

and strength are greater by a factor of 2–20 in the axial direction than they are in the radial and tangential directions, depending on the species.

The general compressive stress–strain curves for wood (balsa) in three directions are shown in Fig. 10.13(a). These curves have the same three stages discussed in Section 10.1 and are similar to the one for cancellous bone (Fig. 10.5). At small strains

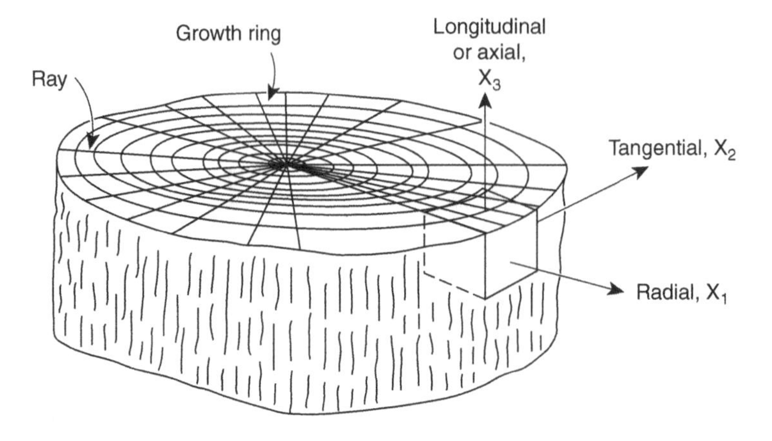

Figure 10.12.

Cross-section through the trunk of a tree showing the radial, tangential, and axial (longitudinal) directions. (From Gibson and Ashby (1997, p. 390) Copyright © 1997 Lorna J. Gibson and Michael F. Ashby. Reprinted with the permission of Cambridge University Press.)

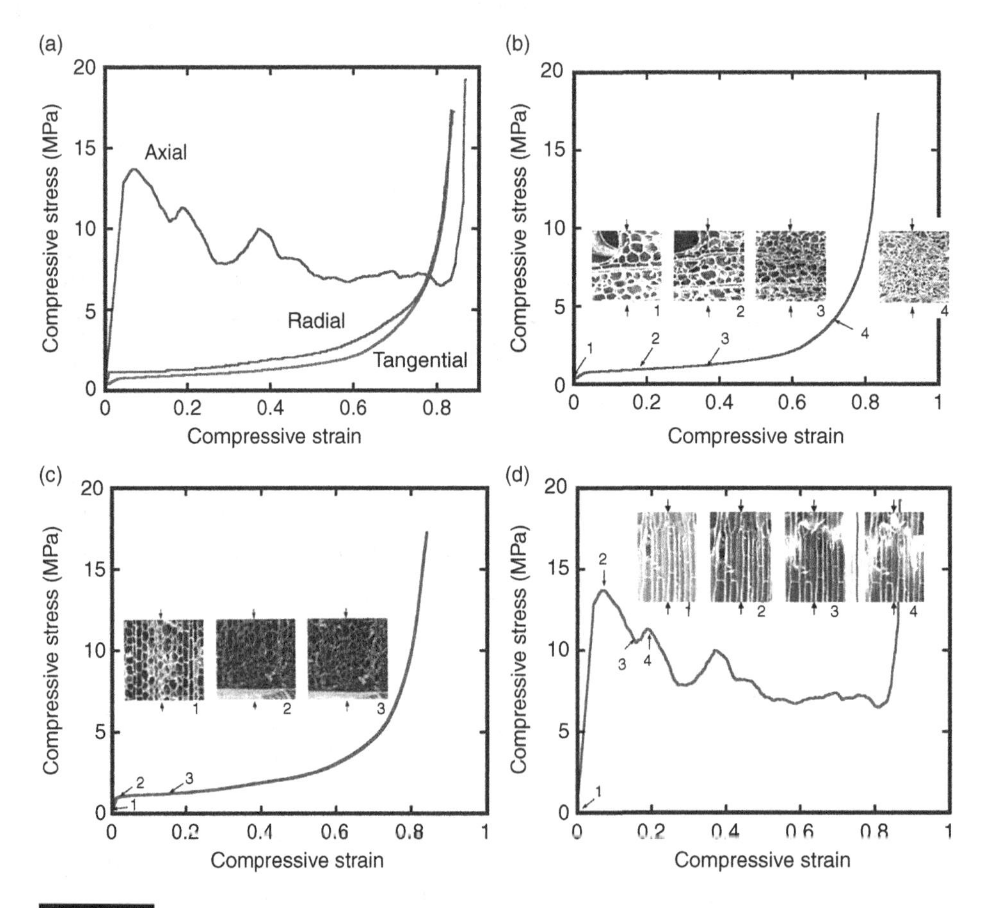

Figure 10.13.

(a) Compressive stress–strain curves for balsa tree. (b) Tangential compression, with scanning electron micrographs showing the deformation of the cells. (c) Radial and (d) axial compression with scanning electron micrographs showing the deformation of the cells. (Adapted from Easterling *et al.* (1982), with kind permission from Professor Michael Ashby, Cambridge University.)

(less than 0.02), the behavior is elastic in all three directions. The Young modulus in the axial direction is much larger than that in the tangential and radial directions. Beyond the elastic regime, the loading curves in three directions show extensive stress plateaus. The yield stress in the axial direction is much higher than that in the tangential and radial directions and is followed by a sharply serrated plateau. The plateaus in the tangential and radial directions are relatively flat and smooth. Compression in the tangential direction causes uniform bending and uniform collapse by plastic yielding of the cell walls (Fig. 10.13(b)) (Easterling *et al.*, 1982). Compression in the radial direction causes uniform bending and uniform plastic collapse (Fig. 10.13(c)). A schematic drawing of the evolution of deformation under compressive loading in the radial direction is shown in Fig. 10.14. This is the mechanism modeled in Section 10.2 (Fig. 10.6). The deformation in the tangential direction is the same.

The mechanism of compressive deformation in the axial direction is quite different (Fig. 10.13(d)). In woods of low density, the cell walls collapse by end-cap fracture (Fig. 10.15(b)). The stress drops until the next layer of end cap is intercepted; then, the stress increases until the second end cap breaks. The repeated deformation process gives the sharply serrated plateau. In woods of high density, the cell walls collapse by local buckling of the cell walls (Fig. 10.15(c)). As the density of the wood increases, both the Young modulus and strength increase.

A most remarkable property of wood is the fracture toughness; it is also highly anisotropic and its highest value is ten times greater than that of a fibrous composite with the same volume fraction of fibers and matrix. Jeronimidis (1976) proposed that the creation of new surface area by fiber pullout is responsible for fracture toughness. The cracks due to shearing will open and propagate longitudinally, which allows each cell wall to be pulled apart without being broken through. Simulation of the S_2 cell walls (see Fig. 10.10(a)) (Gordon and Jeronimidis, 1980) has shown that when the helical angle of the fibrils is 15–20°, there is an optimum compromise between energy absorption and reduction in axial stiffness. Figure 10.16 shows the crack propagation in wood along directions (a) parallel and (b) perpendicular to the grain. Whereas it is easy for a crack

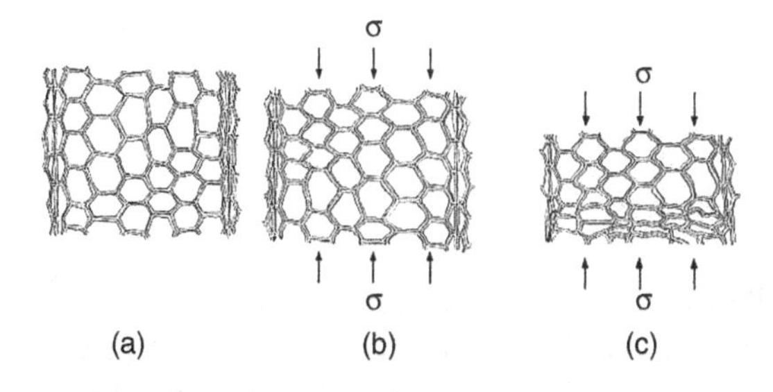

Figure 10.14.

Schematic drawing of the deformation of wood cells under loading in the radial direction. (a) No applied load. (b) Cell-wall bending in the linear elastic range. (c) Nonuniform cell collapse by plastic yielding. (From Gibson and Ashby (1997, p. 401) Copyright © 1997 Lorna J. Gibson and Michael F. Ashby. Reprinted with the permission of Cambridge University Press.)

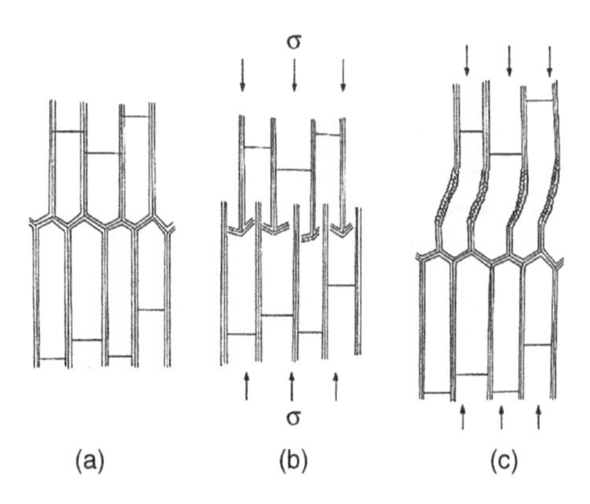

Figure 10.15.

Schematic of the deformation of wood cells under loading in the axial direction. (a) No applied load. (b) Cell collapse by end-cap fracture. (c) Cell collapse by local buckling of the cell walls. (From Gibson and Ashby (1997, p. 403) Copyright © 1997 Lorna J. Gibson and Michael F. Ashby. Reprinted with the permission of Cambridge University Press.)

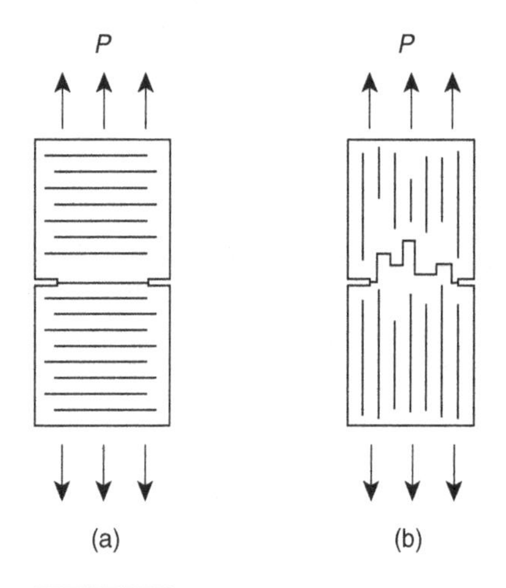

Figure 10.16.

Crack propagation in wood: (a) initial crack parallel to the cell; (b) initial crack perpendicular to the grain. (From Gibson and Ashby (1997, p. 424) Copyright © 1997 Lorna J. Gibson and Michael F. Ashby. Reprinted with the permission of Cambridge University Press.)

to propagate in a parallel direction, perpendicular propagation to the grain is very difficult. This fracture mode is analogous to that of nacre when the shell is loaded along the direction of its surface; the individual platelets are pulled out and this provides the high toughness. The sap channels can stop cracks from propagating, as shown in Fig. 10.17. When the crack propagates towards a sap channel, it can either enter it or run around its wall and then stop.

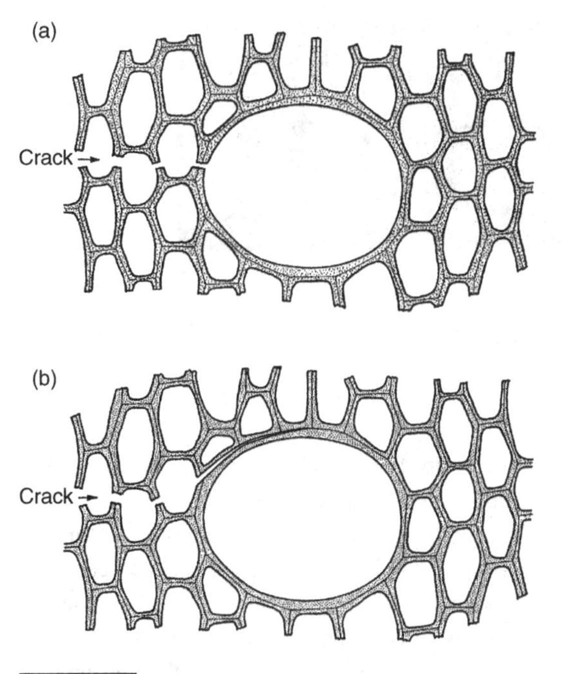

Figure 10.17.

The crack is arrested by a sap channel. Schematics of (a) crack breaking into a sap channel, (b) crack splitting the wall of a sap channel. (Figure courtesy Professor Michael Ashby, Cambridge University.)

10.4 Bird bones

Birds and other flying vertebrates (bats) have lightweight skeletons, which, in part, makes flight possible. Birds range in mass from the 2 g humming bird to the 100 kg ostrich, with over half of all birds weighing less than 40 g. It has been reported that, for bald eagles, the skeleton amounts to only 6% of the body mass; the feathers represent twice that (12%) (Brodkorb, 1955). For flight, other adaptations have evolved, such as a smaller number of bones than terrestrial vertebrates and the development of a lightweight beak instead of a jaw with teeth. As early as Galileo's time, bird skeletons were reported to possess hollow bones. Figure 10.18 illustrates the differences between bird and mammalian limb bones. Both cross- and longitudinal sections are shown for bird and dog bones. The bird bones are characterized by a much thinner sheath of cortical bone, relative to the diameter. Additionally, for the bird, thin rods of bone are seen to extend across the medullary cavity. The internal structure of the bones varies among species. The bones of birds capable of flight are more hollow (i.e. not marrow filled) than flightless birds (e.g. ostrich, penguin). Diving birds and humming birds have few hollow bones. Diving birds need to have a higher density skeleton to propel themselves through water. For humming birds, the weight savings involved with hollow bones would be minimal.

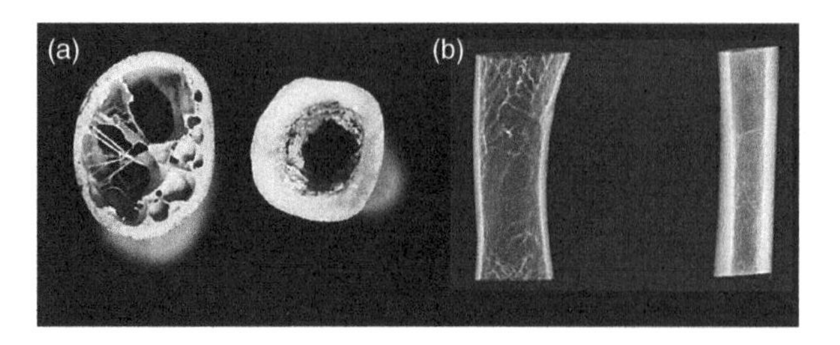

Figure 10.18.

Bird (left) compared with dog (right) bone: (a) photographic and (b) radiographic images. Bird bones are characterized by a much thinner cortical sheath. (Figure courtesy J. Kiang.)

The hollow bones have struts for reinforcement, similar to the wings on a biplane. Figure 10.19(a) shows a cut-away of a hollow bone from the femur of a California condor. A pattern of struts can be seen in the cutout, connecting the surrounding cortical bone. In Fig. 10.19(b), the cross-section of a bird bone is drawn schematically. As with mammalian bone, there is a periosteal and endiosteal sheath surrounding the cortical bone and a medullary core that is filled with less dense bone. The hollow bones are also classified as pneumatic bone, where air is forced into the bone, thereby increasing buoyancy. In flying birds, bones need to be sufficiently strong and stiff to withstand forces during take-off and landing, which necessitates some reinforcement in the bone interior. Wing bones have to resist both bending and torsion loads. Two reinforcing structures, struts and ridges, are shown in Figs. 10.19(c) and (d), respectively. The struts are thought to reduce ovalization, which can be simply visualized by taking a straw from a drink and subjecting it to bending (see Fig. 10.19(e)). The cross-section, initially circular, will become oval, with a decrease in the moment of inertia (see Section 4.5 and Table 4.1). The struts can inhibit this localization by not allowing the sides to expand. Another important reinforcing structure, the ridge, is shown in Fig. 10.19(d). It increases the resistance to torsion. This is explained in the schematic representation of Fig. 10.19(f). These ridges form a 45° pattern on the inside of the hollow tube.

One focus on the study of the mechanical properties of bird bones has been on pneumatized (hollow) vs. marrow-filled bones. Currey and Alexander (1985) found that the ratio of the internal to external diameter was lower for pneumatized bone. The mean bending strength and flexural modulus were significantly higher for marrow-filled than pneumatic bones (Cubo and Casinos, 2000). However, torsional resistance during flight has been proposed to be more significant than bending resistance (Swartz, Bennett, and Carrier, 1992). Torsional resistance is maximized in thin-walled hollow bones at the expense of bending resistance.

Despite bird skeletons being described as "lightweight," the bone itself is found to be overall more dense than that of mammals of similar size. Dumont (2010) measured the bone density from the cranium, humerus, and femur of song birds, bats, and rats, and found that, on average, the bird bone density was higher than the

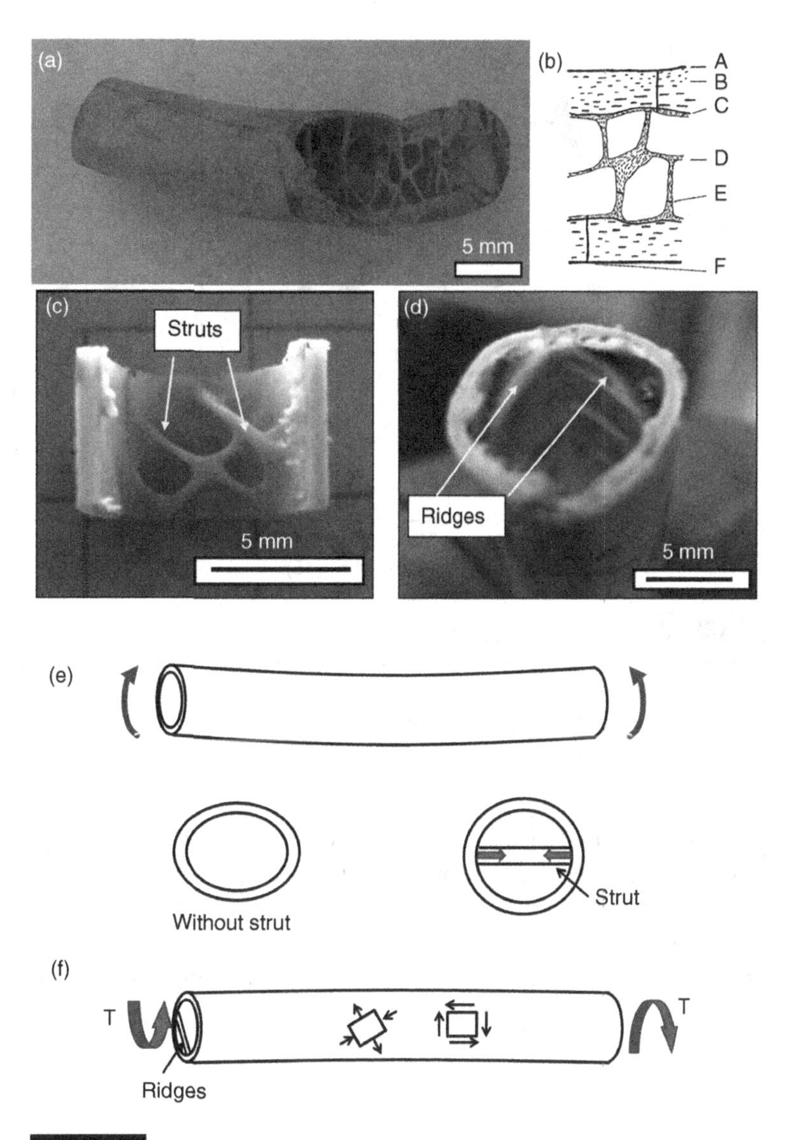

Figure 10.19.

(a) Interior of a condor femur bone, showing the struts in the hollow interior. (b) Cross-sectional schematic of a bird bone. A – periosteal surface; B – lamellar cortical bone; C – endiosteal surface; D – trabecular layer; E – pores/pneumatic or blood vessel openings. (From Davis (1998), with kind permission from John Wiley & Sons, Ltd.) (c) Reinforcing net of struts (truss) and (d) reinforcing ridges inside the Turkey vulture ulna. (e) Schematic showing how struts in tension can inhibit ovalization due to bending. (f) Schematic showing how ridges in a helicoidal pattern can inhibit tensile failure due to torsion.

other two (Fig. 10.20). It was proposed that additional strength and stiffness (provided by a higher mineral content) is required for bird bones for flight purposes. Thus, bone density and also bone shape form the basis for the stiffness and strength.

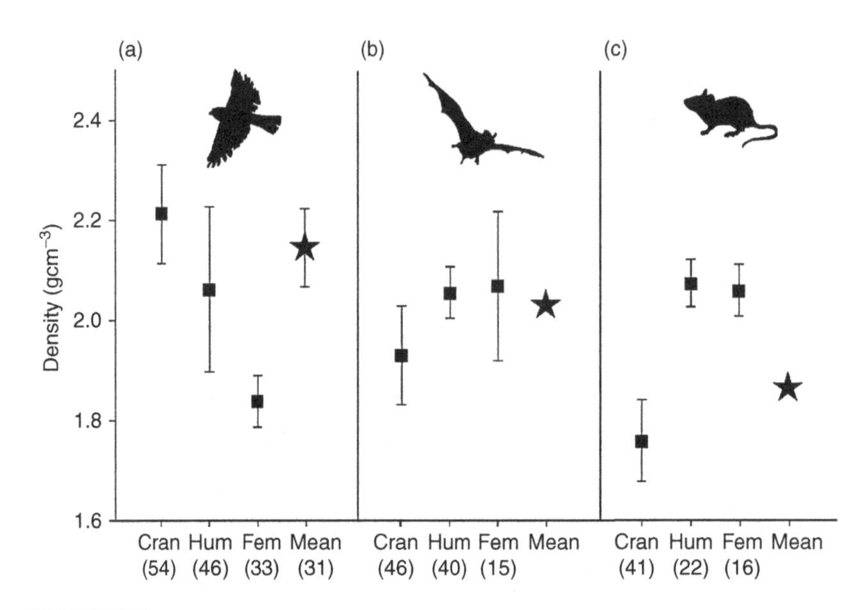

Figure 10.20.

Means and 95% confidence intervals for cranial (cran), humeral (hum), and femoral (fem) density in (a) birds, (b) bats, and (c) rodents. Sample sizes in parentheses. Stars for bats and rodents indicate weighted means of bone density for each group. The mean of bone density for birds is based on 31 specimens for which density was available for all three skeletal elements, and is accompanied by 95% confidence intervals. (From Dumont (2010), with kind permission of Professor B. Dumont.)

The skulls of birds can take on several configurations. Figures 10.21(a) and (b) show cross-sections of the skull of a pigeon and long-eared owl (*Asio otus*), respectively. In the pigeon, the trabecular bone is randomly oriented and is thickest at the base of the head. For the larger owl, a layered sandwich structure is observed with an oriented cell structure. The high density of the cranium of birds, relative to bats and rats, is associated with feeding habits (Dumont, 2010).

10.5 Bird beaks

10.5.1 Toucan and hornbill beaks

Bird beaks usually fall into two categories: short/thick and long/thin. The toucan is an exception. It has a long beak that is also thick, a necessity for food gathering in tall trees. The Toco toucan (*Ramphastos toco*) has the largest beak among the species. The large beaks help in picking up fruits at the tips of branches in the canopy and extracting prey (little baby birds!) from holes in trees. They also assist in combat and bill fencing. This is accomplished by an ingenious solution, enabling a low density and high stiffness: a composite structure consisting of an external solid keratin shell and a cellular core. Figure 10.22 shows the toucan and hornbill beaks in a schematic fashion. The toucan

(a)

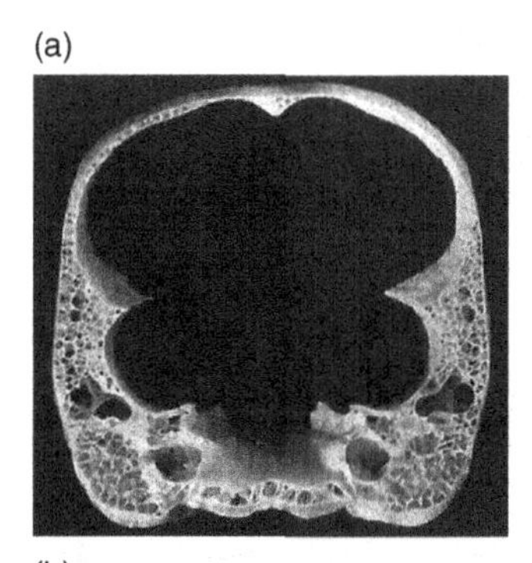

(b)

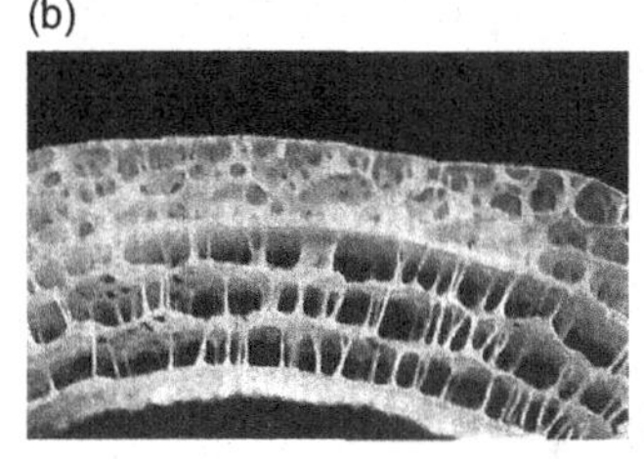

Figure 10.21.

Bird skull cross-sections. (a) Random foam core in a house pigeon; (b) oriented and multiple cellular structure of a long-eared owl (*Asio otus*). (Reproduced based on Buhler (1972).)

beak has a density of approximately 0.1 g/cm^3, which enables the bird to fly while maintaining a center of mass at the line of the wings. Indeed, the beak comprises 1/3 the length of the bird, yet makes up only about 1/20 of its mass. The hornbill beak, consisting of 1/4 of the total length, has a density of approximately 0.3 g/cm^3. A distinctive feature of the hornbill is its casque formed from cornified keratin layer. The mesostructure and microstructure of toucan and hornbill beaks reveal a material that is reminiscent of sandwich structures of functionally graded materials, with components made of foam covered by a hard surface layer. Therefore, this biological material serves as a useful source for research and as an inspiration for structural design in engineering. This is accomplished by an ingenious solution, enabling a low density and sufficient rigidity. The outside shell (integument) of beak consists of β-keratin, which is typically found in avian and reptilian species. The inside is filled with a cellular bone. This internal foam has a closed-cell structure constructed from bony struts with thin membranes. We summarize here the principal results obtained by our group (Seki *et al.*, 2005, 2006, 2010, 2012; Fecchio *et al.*, 2010).

Figures 10.23(a) and (b) show scanning electron micrographs of toucan and hornbill beaks. The structure is similar to that of cancellous bone, and the foams consist of asymmetric rod-like struts. Most of the cells in the toucan and the hornbill are sealed off by thin membranes. Thus, it can be considered a closed-cell foam. The cell sizes vary depending on location, and the closed-cell network comprises struts with a connectivity of normally three or four.

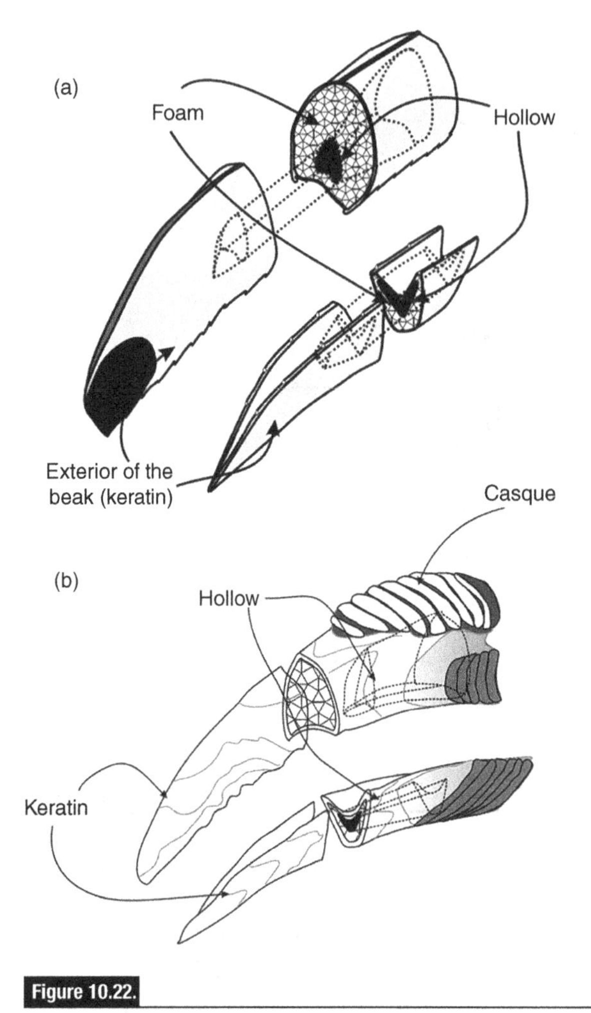

Figure 10.22.

(a) Toucan beak; (b) hornbill beak. (Figure courtesy Y. Seki.)

Energy disperse X-ray (EDX) analysis shows that toucan and hornbill keratins (Figs. 10.24(a) and (c)) contain principally carbon and oxygen, which are the main components of the protein. A relatively low content of sulfur in the chemical component of keratin seems to point to a low concentration of cystine, a sulfur-containing amino acid. The beak keratin also contains minerals, as indicated by the presence of calcium, potassium, sodium, and chlorine. The presence of calcium indicates a degree of mineralization that provides the hardness of the keratin. However, the content of calcium is low and is less than 1% in the keratin of the beak. Figures 10.24(b) and (d) are in stark contrast with Figs. 10.24(a) and (c). The trabeculae of the foams contain more minerals than the shell of the beak. Distinctively, the trabeculae contain a large amount of calcium, giving rise to the increased hardness. The EDX results can be compared with

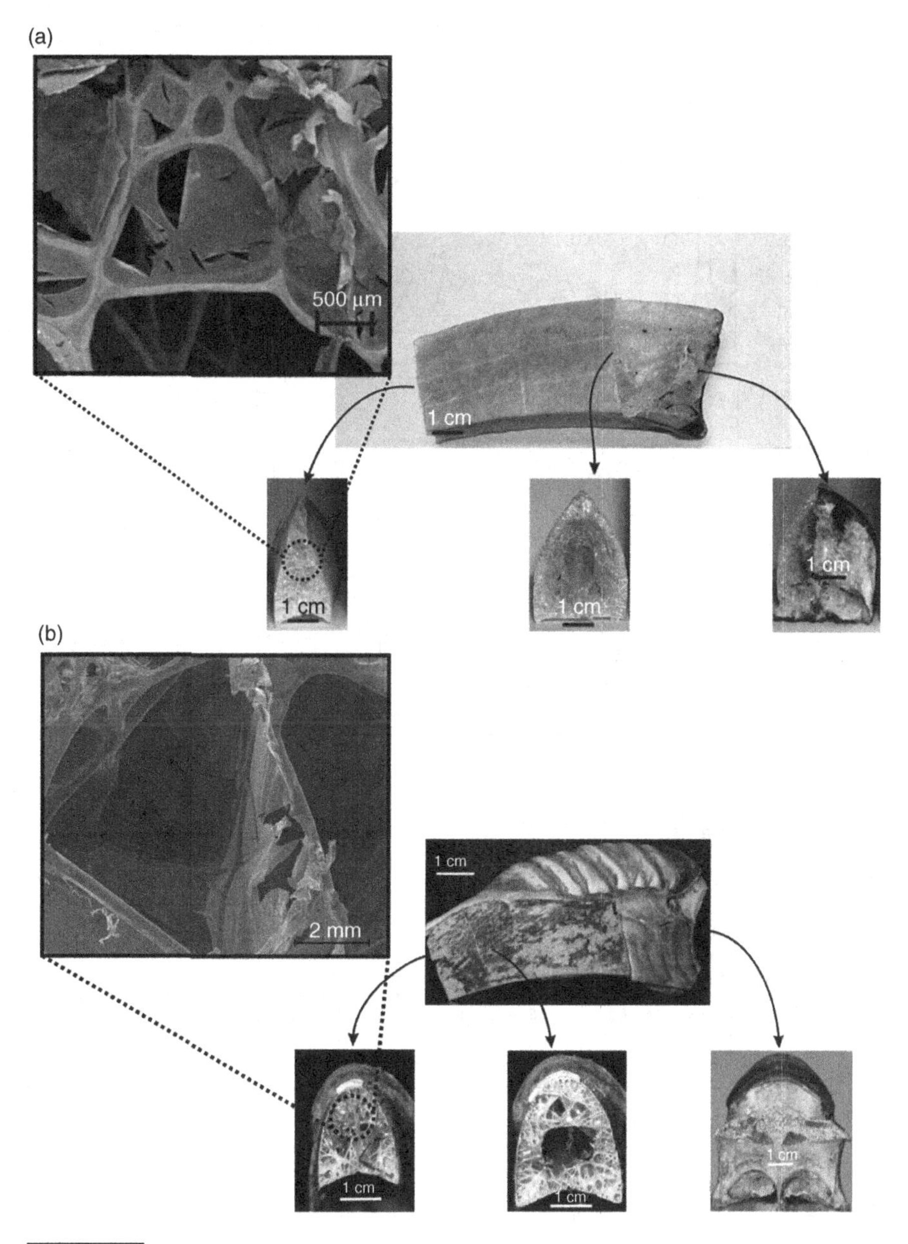

Figure 10.23.

Images of internal foam structure with three cross-sections and scanning electron micrographs of foams. (a) Toucan foam; (b) hornbill foam. (Figure courtesy Y. Seki.)

chromatographic compositional studies from several bird beaks by Frenkel and Gillespie (1976). It is confirmed here that the keratins of the Toco toucan and hornbill beaks appear to be similar to other bird species, with a low sulfur and mineral content. Pautard (1963) reported that 0.28% of the pigeon beak comprises calcium.

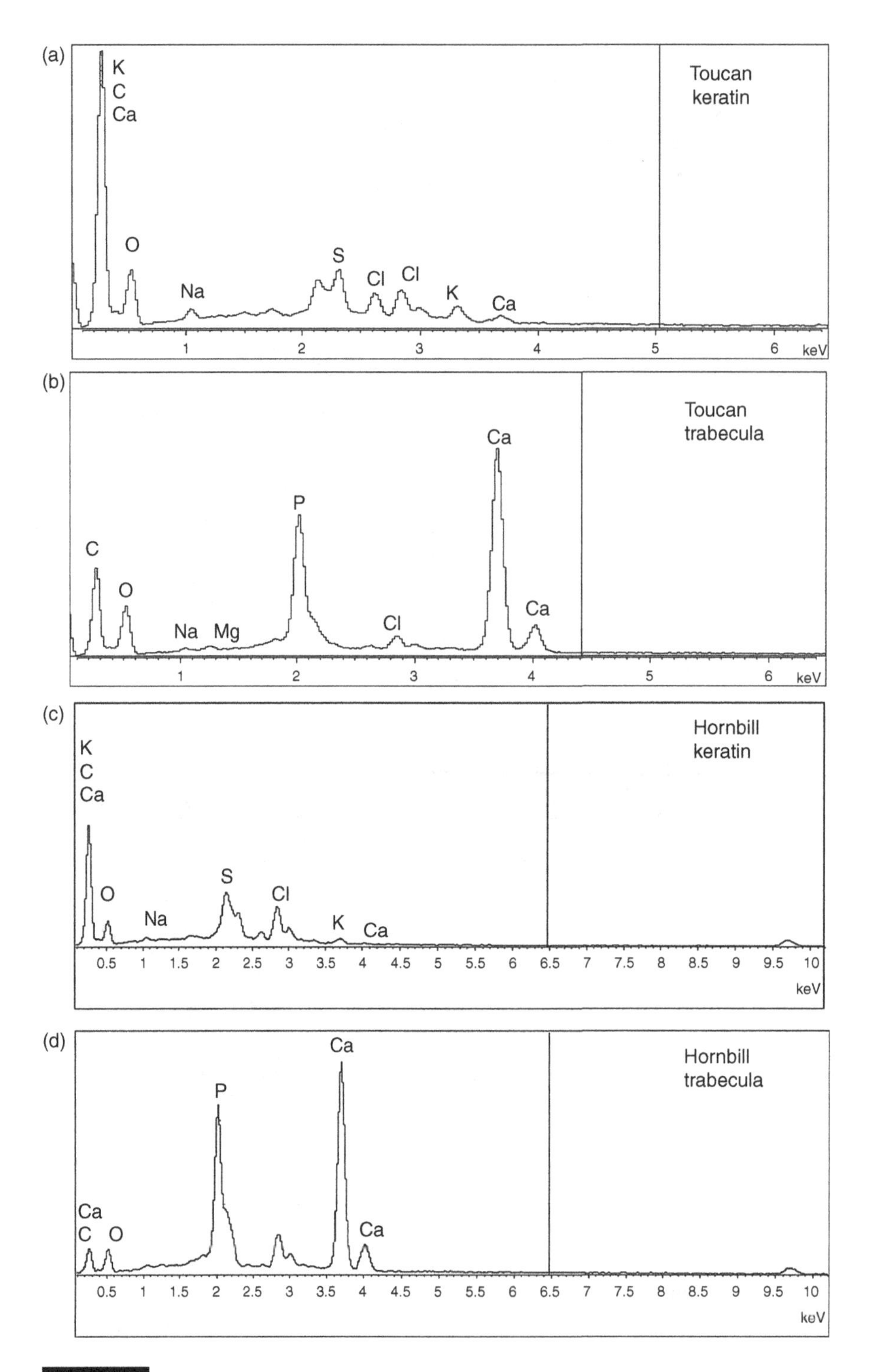

Figure 10.24.

Energy disperse X-ray results: (a) toucan keratin; (b) toucan trabecula; (c) hornbill keratin; (d) hornbill trabecula. (Figure courtesy Y. Seki.)

10.5.2 Modeling of interior foam (Gibson–Ashby constitutive equations)

The most significant feature of a cellular solid is the relative density, ρ^*/ρ_s (the density of the cellular material, ρ^*, divided by the density of the solid material, ρ_s). Gibson and Ashby (1997) provide an analytical treatment for the mechanical behavior of a broad range of cellular materials presented in Section 10.1.

The toucan beak foam can be considered as a closed-cell system. Deformation of the closed cells is more complicated than that of open cells. When open-cell foams are deformed, cell-wall bending occurs. The deformation of closed cells involves not only rotation of cell walls, but also stretching of the membranes and internal gas pressure.

The simplest closed-cell cubic model is introduced here to describe the deformation of the foam. Figure 10.25 shows undeformed (Fig. 10.25(a)) and deformed (Fig. 10.25(b)) cubic closed cells, as envisaged by Gibson and Ashby (1997). The linear elastic region is limited to small strains. The foams made from material possessing a plastic yield stress are subjected to plastic collapse when the load is beyond the linear elastic regime. When plastic

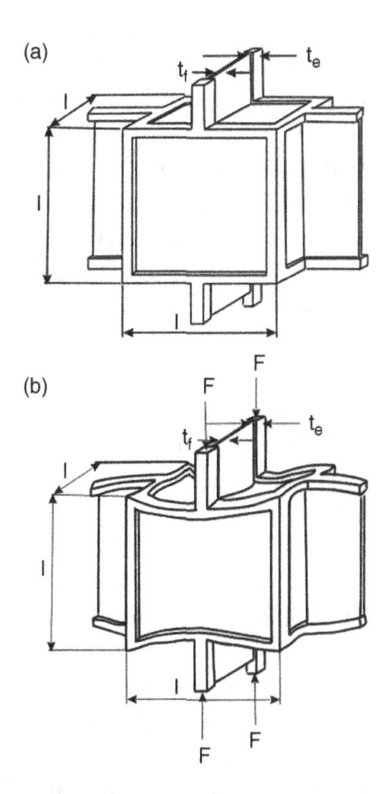

Figure 10.25.

(a) Gibson–Ashby model for closed-cell foam; (b) deformation of closed-cell foam. (Reprinted from Seki *et al.* (2006), with permission from Elsevier.) Note: t_e is the membrane thickness and t_f is the bean thickness (equivalent to the beam in the open-cell structure shown in Fig. 10.6).

collapse occurs, there is a long horizontal plateau in the stress–strain curve. Equation (10.26) represents the response of the closed-cell foam shown schematically in Fig. 10.25:

$$\frac{\sigma^*_{pl}}{\sigma_y} = C_5\left(\phi\frac{\rho^*}{\rho_s}\right)^{3/2} + (1-\phi)\frac{\rho^*}{\rho_s} + \frac{p_0 - p_{at}}{\sigma_y}, \tag{10.26}$$

where σ^*_{pl} is the plastic collapse stress of foam, σ_y is the yield stress of the solid portion, C_5 is a parameter, ϕ is the ratio of volume of face to volume of edge, p_0 is the initial fluid pressure, and p_{at} is the atmospheric pressure.

For the open-cell geometry, the parameter ϕ in Eqn. (10.26) is equal to 1. Additionally, the pressure is unchanged, i.e. $p_0 - p_{at} = 0$. Thus, Eqn. (10.26) is reduced to

$$\frac{\sigma^*_{pl}}{\sigma_y} = C_5\left(\frac{\rho^*}{\rho_s}\right)^{3/2}. \tag{10.27}$$

This is the open-cell equation from Gibson and Ashby (1997) (Eqn. (10.19)). The parameter C_5 is equivalent to the parameter $C_1^{-3/2}$ in Eqn. (10.19); it has an experimentally obtained value of 0.3 for plastic collapse and 0.2 for brittle crushing (where σ^*_{pl}/σ_y in Eqns. (10.19) and (10.27) are replaced by the normalized crushing stress $\sigma^*_{cr}/\sigma_{fs}$).

The cell shapes of toucan and hornbill foams are highly complex and consist of a combination of triangular, quadrilateral, and even higher numbers of polygonal edges. There are circular cells at the nodes, which are shared by several cells. To avoid the complexity of characterizing the cell geometry of toucan and hornbill foams, we measured the relative density. The relative density of the toucan foam is approximately 0.09 and that of the hornbill foam is 0.1. The yield stresses of the toucan and hornbill trabeculae, σ_y, are estimated from microindentation values ($H \approx 3\sigma_y$), which seem to be more accurate than the nanoindentation values due to the size effect. This gives values of $\sigma_s = 91$ MPa for toucan and $\sigma_s = 128$ MPa for hornbill.

Figure 10.26(a) shows the predictions from Eqns. (10.26) and (10.27) as well as experimental results for a number of materials (Matonis, 1964; Traeger, 1967; Thornton and Magee, 1975a,b). These equations bracket the experimental results quite well. A more detailed plot of the compressive strength for the toucan foam as a function of relative density is shown in Fig. 10.26(b). Although the relative densities of toucan and hornbill foams show a small difference, the relative yield strength of the hornbill trabeculae has more than a four-fold difference due to the high degree of mineralization. It should be noted that the membranes are not expected to contribute significantly to the mechanical response of the foam since many of them contain tears due to desiccation. However, one would not expect this to be the case for the live animal. Gibson and Ashby (1997) give values of $C_5 = 0.3$ and $C_5 = 0.2$ for plastic buckling and brittle crushing, respectively. The response of the toucan foam is intermediate between the two.

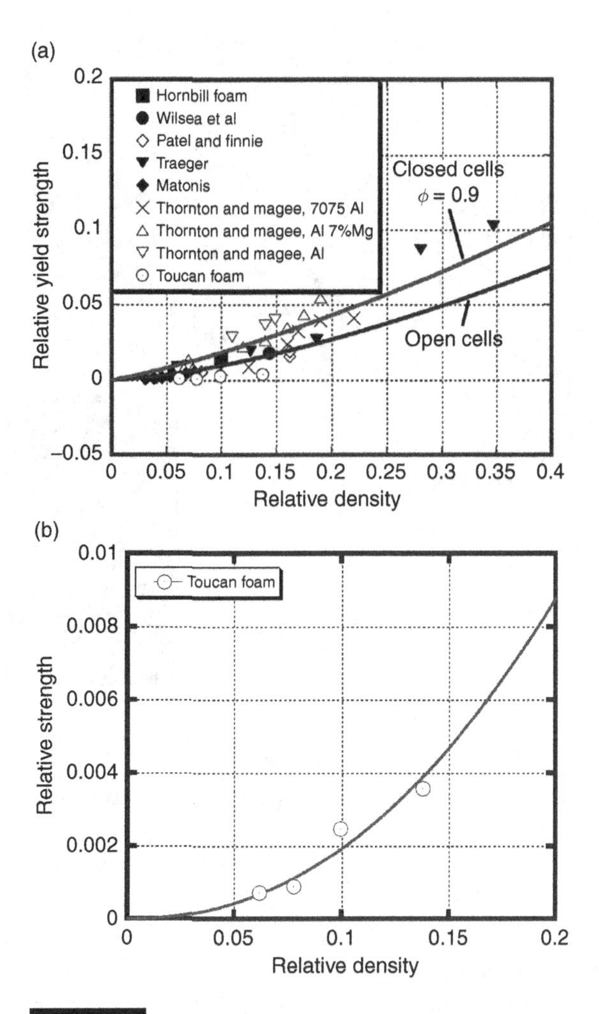

Figure 10.26.

(a) Experimental results (hollow circles) and Gibson–Ashby prediction for open- and closed-cell foams (continuous lines); (b) detailed plot. (Figure courtesy Y. Seki.)

Figure 10.27(a) shows the fracture pattern in the foam. It comprises a mixture of plastic deformation, and partial and total fracture of the trabeculae. The trabeculae have a fibrous structure similar to wood and can fracture partially when they are subjected to bending (Fig. 10.27(b)). In other locations, the trabeculae undergo total fracture; an example is shown in Fig. 10.27(c). The "green twig" appearance of the trabecula is evident in Fig. 10.27(b). Hence, the cellular material does not crumble when compressed to its maximum strain. Rather, it collapses in a semi-plastic manner.

Figure 10.27.

(a) Fracture of bone struts in cellular structure of toucan beak: (b) "green" fracture; (c) complete fracture. (Figure courtesy Y. Seki.)

The beak structure was found by Seki *et al.* (2005) to have a bending strength (Brazier moment) that is considerably higher than if all the mass were concentrated in the shell as a solid hollow cylinder by applying the analysis developed by Karam and Gibson (1995). Seki *et al.* (2005, 2006) showed that the internal cellular core serves to increase the buckling resistance of the beak and demonstrated a synergism between the two components that provides the stability in bending configuration. Thus, there is clearly an advantage in having an internal foam to support the shell. The nature of the internal foam structure is revealed in considerable detail by micro-computerized tomography.

The outer surface of the beak is covered by a thin horny sheath called the rhamphotheca, which consists of polygonal keratin tiles, as shown in Fig. 10.28(b). The wavy keratin tile boundaries (traced by black lines for greater clarity) are shown in the longitudinally sectioned beak keratin in Fig. 10.28(c). The intermediate filaments (IFs) are distributed in the keratin matrix, indicated by arrows in Fig. 10.28(c). The cross-sectional mosaic image of the lower beak was captured by confocal microscope in

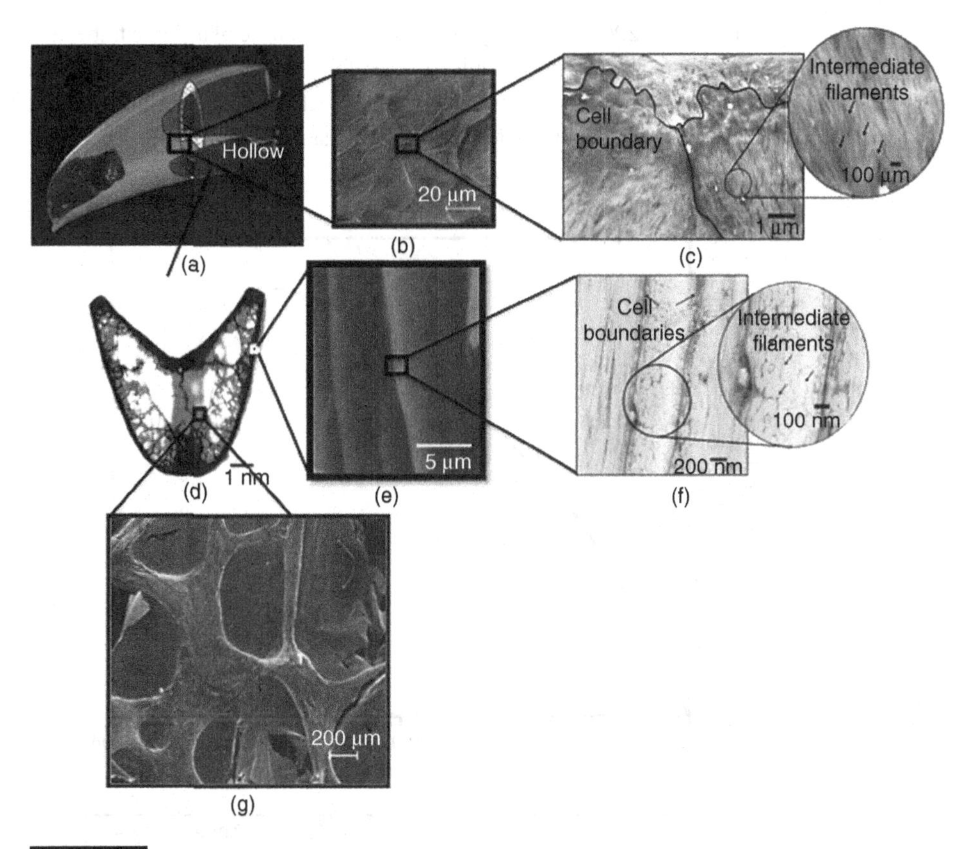

Figure 10.28.

Overall view of toucan beak showing details of outer keratin layer and inner cellular bone structure; the cells are sealed by membranes. (Reprinted from Seki *et al.* (2012), with permission from Elsevier.)

Fig. 10.28(d). The transverse section of the beak keratin exhibits a layered structure (Fig. 10.28(e)), and the tiles are connected by organic glue, as shown in Fig. 10.28(f). The IFs, embedded fibers in the keratin matrix, are either aligned along the cell boundaries or are in the spongy structure (see Fig. 10.28(f)). The toucan beak trabeculae are composed of cylindrical or elliptical rods ~200 μm in diameter. The porosity of the trabecular bone in beak is enclosed by thin membranes; the typical pore size is 1 mm (Fig. 10.28(g)). This is a different structure from human trabecular bone, which has no membranes. The average edge connectivity of pores is approximately 3. The trabecular bone is enclosed by a thin cortical shell.

We detail in the following an approach to understanding the structure-mechanical property relationships of the toucan beak that has greater generality, since it can be applied to other biological materials. This approach uses computerized tomography, in which the image is converted to a finite element mesh. Finite element calculations are performed using the structural element of the beak and compared with experiments.

Figure 10.29(a) shows a typical stress–strain curve of a single trabecula of the toucan beak in tension. The Young modulus ranges from 0.4 to 7 GPa. The average Young modulus, obtained from 29 measurements, is 3.0 ± (S.D. 2.2) GPa. Values reported in the literature have varied from 1 GPa to 19 GPa in tension (Ryan, 1989; Lucchinetti, Thomann, and

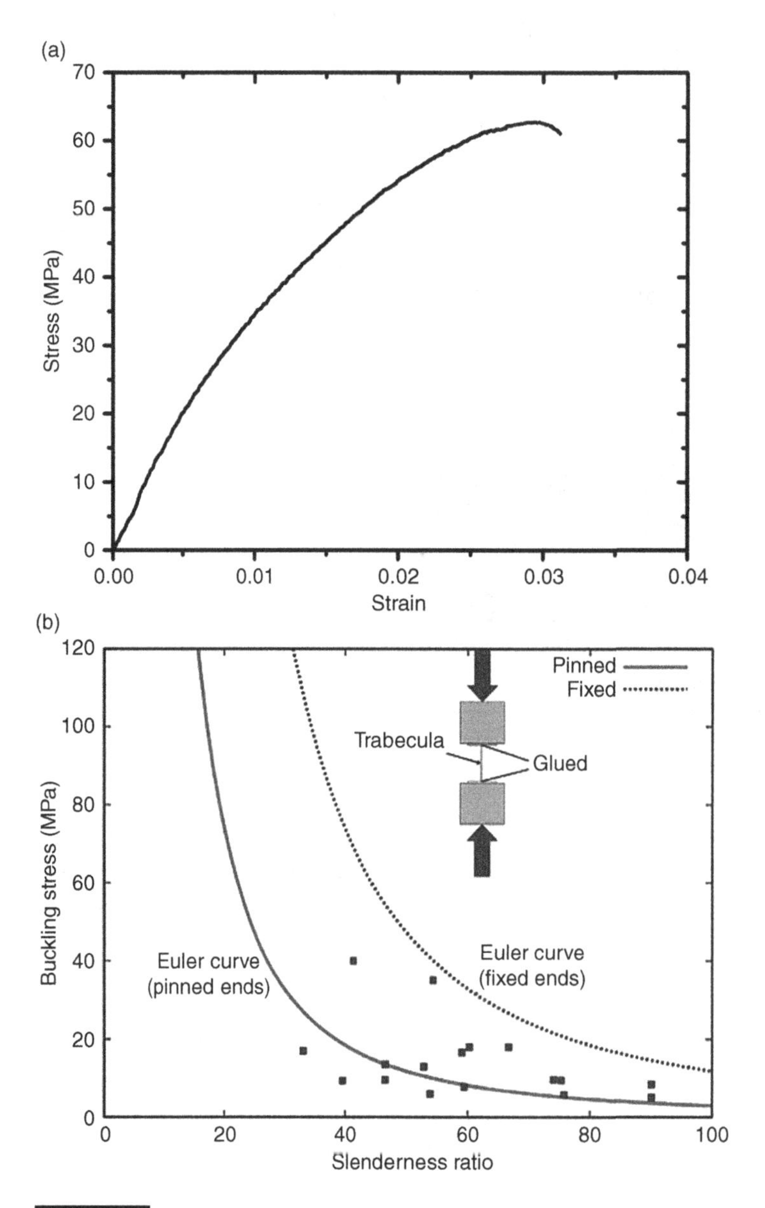

Figure 10.29.

(a) Typical stress–strain curve of toucan beak trabecula in tension. (b) Calculated Euler curves of toucan beak trabeculae with fixed ends (dotted curve) and pinned ends (continuous curve) and experimental compressive test results (squares). (Reprinted from Seki *et al.* (2012), with permission from Elsevier.)

Danuser, 2000; Bini *et al.*, 2002; Hernandez *et al.*, 2005). Lucchinetti *et al.* (2000) discussed the limitations and error sources associated with mechanical testing at the micrometer scale.

In most cases, the trabecular bone fails after buckling and bending of the struts. Therefore, it is important to evaluate the buckling of a single trabecula in compression to obtain material parameters for the beak foam. The elastic buckling load of a beam is given by the Euler equation (which is shown in its simplified ($n = 1$) form in Eqn. (10.5)):

$$P = \frac{n^2 \pi^2 EI}{L^2} \; (n = 1, 2, 3, \ldots), \tag{10.28}$$

where n is defined by the supports, E is the Young modulus, I is the moment of inertia, and L is the length of the beak. This equation is derived in Chapter 4. If both ends are pinned and fixed, $n = 1$ and $n = 2$, respectively.

The corresponding critical stresses σ_{cr} are, for pinned supports and fixed ends (Section 4.5), respectively,

$$\sigma_{cr} = \frac{\pi^2 E}{(L/r)^2} \quad \text{and} \quad \sigma_{cr} = \frac{4\pi^2 E}{(L/r)^2}, \tag{10.29}$$

where A is the cross-sectional area and $r = \sqrt{I/A}$ is the radius of gyration.

The equations for critical stresses are valid for ideal, relatively long columns. In this study, the column slenderness ratio (L/r) is less than 100, which leads to inelastic buckling. Therefore, the tangent modulus theory for inelastic buckling was applied (Seki, 2011).

Figure 10.29(b) shows that the two Euler curves with predicted values for pinned and fixed ends track experimental results. A Young modulus of 3 GPa (from tensile testing) was used to draw the Euler curves. It is simple to envisage that the supports can undergo damage during assembly of the test setup. If the ends are cracked, a fixed support essentially becomes pinned. The Euler curve for pinned ends tracks the lower buckling stresses and that for fixed ends represents the upper bound for the buckling stresses. The Euler equation suggests that the estimated Young modulus matches the assumed 3 GPa value used for elastic buckling.

The experimentally obtained material parameters of the trabeculae and cortical bone were used for the finite element method (FEM) calculations. Yield stresses of $\sigma_y = 20$ MPa for trabeculae and $\sigma_y = 10$ MPa for the cortical shell were used. The Young moduli were taken as 3 GPa for trabeculae and 0.3 GPa for the cortical shell. The trabecular and cortical bone were deemed to have failed at strains of 0.03 and 0.1, respectively.

In order to create an FE model, 150 images scanned at 93 μm resolution by micro-tomography (μCT) were used. With this resolution, we could capture the main structure of the trabeculae, which contributes to the stiffness of the foam. The CT images were converted to the tridimensional structure of the trabeculae and the cortical shell by a marching cube algorithm, shown in Fig. 10.30(a). The core part (trabeculae) and shell part (cortical shell) of the beak foam were created individually. The tridimensional model of the core and cortical shell was created. A piecewise linear plasticity model (material model 24) in LS-DYNA was used for the material model.

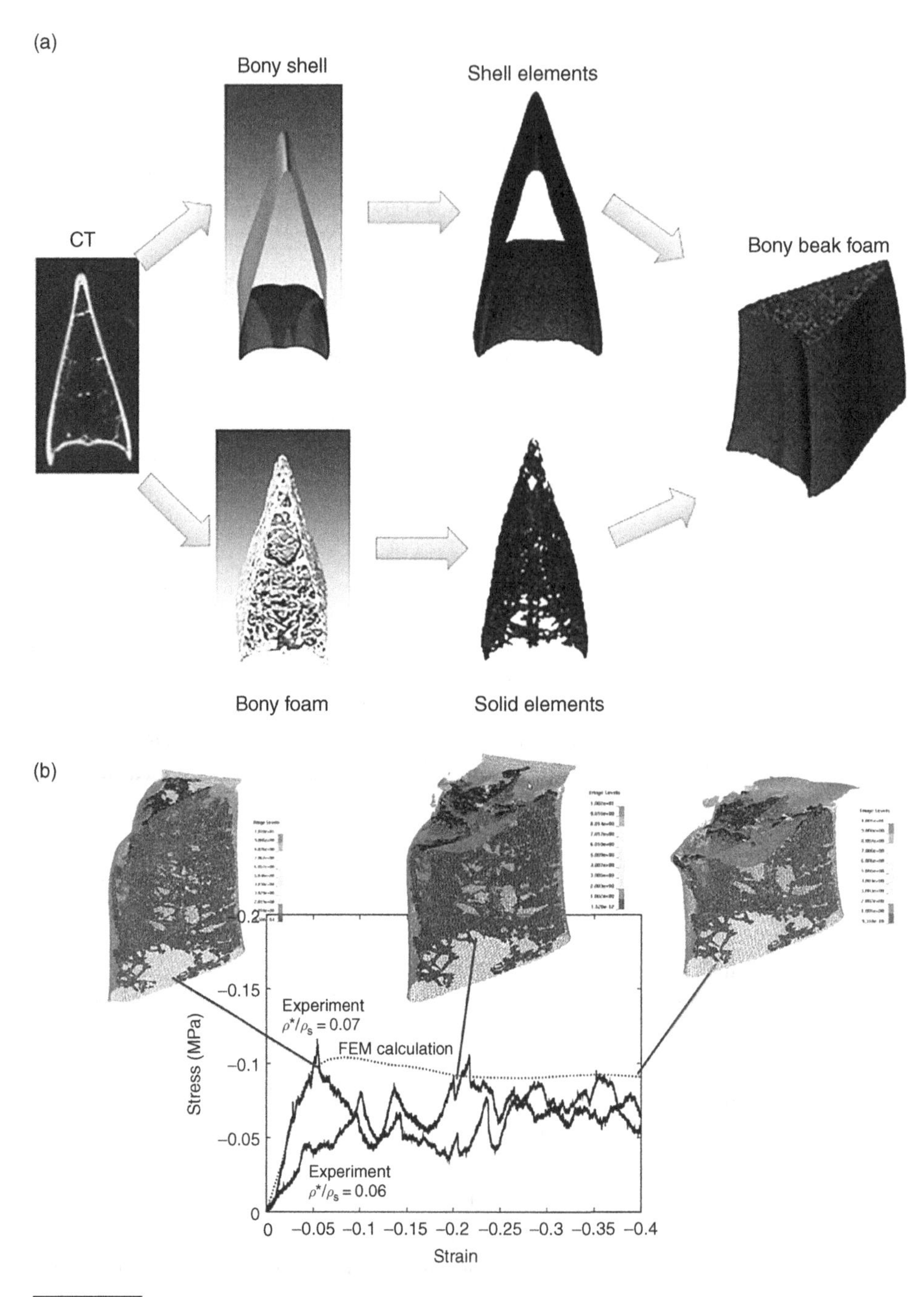

Figure 10.30.

(a) Conversion from CT images to FE model of foam core: 150 CT images were used to create tridimensional models of bony foam and shell by DDV. The models were converted to FE models by ANSYS ICEM CFD. Both models were assembled to make the beak foam core. (b) Stress–strain curves of beak foam under compression: comparison between experiments and FE calculation. Foams in experiments have relative densities (ρ^* / ρ_s density of foam core/density of foam material) of 0.06 and 0.07, respectively. (Reprinted from Seki *et al.* (2012), with permission from Elsevier.)

Figure 10.30(b) shows two experimental stress–strain curves of the compressive behavior of the beak foam and a computationally predicted curve with three FE deformation models at different strains. The two experimental stress–strain curves of foam represent the different relative densities, 0.06 (with a lower modulus) and 0.07 (with higher modulus). The shell folding or buckling is visible at the culmen (the upper ridge of a bird beak). At this stage, this deformation is at the end of the linear elastic region. At a strain of −0.20, the trabeculae are compressed and shell breakage is observed in simulation. The breakage of the shell becomes severe and the trabeculae undergo failure at a strain of −0.40. The calculation shows the average stress at the plateau after the linear elastic region; there is ~30% difference in the plateau stress and a good agreement in stiffness, in comparison with the experimental stress–strain curve for 0.07 initial relative density.

The mechanical behavior of bird beaks is governed by both the ductile keratin integument and the semi-brittle bony foam. Most of the mechanical loading on the beak is carried by the exterior keratin, whereas the foam increases the energy absorption and stabilizes the deformation of the beak to prevent catastrophic failure (Seki, 2005). In the case of the toucan, the beak is mainly used for apprehension of food so that it is designed to resist bending moments. Indeed, the beak, having a hollow core, exhibits a high bending resistance (Seki, 2005).

The stiffness of the beak keratin is mechanically isotropic in the transverse and longitudinal directions (Seki, 2005). The IFs are assumed to be homogeneously distributed in the beak keratin. The IF of beak keratin mechanically plays the same role as the fibers for fiber-reinforced composite materials. The branched structure of the IFs acts as an anchor in the keratin matrix and gives greater strength to the keratin composites. Indeed, the comparison between plain and branched fibers in composites theoretically proves that the branched fiber increases the strength of composites.

The study conducted by Seki *et al.* (2012) demonstrates that the merging of characterization (μCT and TEM), micromechanical testing, and finite element computation are converging to predict, in a realistic manner, the mechanical response of structures. This approach is being spearheaded by other groups, and is currently applied to bones, teeth, and other biological materials.

Example 10.2 The collapse strength of equiaxed closed-cell plastic foams is given by

$$\frac{\sigma^*_{pl}}{\sigma_y} \cong 0.3\left(\phi\frac{\rho^*}{\rho_s}\right)^{3/2} + 0.4(1-\phi)\frac{\rho^*}{\rho_s} + \frac{p_0 - p_{atm}}{\sigma_y},$$

where ϕ is the ratio of volume of face to volume of edge, p_0 is the initial fluid pressure, and p_{atm} is the atmospheric pressure. Answer the following questions.

Table 10.1.

ϕ	0	0.1	0.2	0.3	0.4	0.5	0.6	0.7	0.8	0.9	1
σ^*_{pl}	16.4	15.0	13.9	12.9	11.9	11.1	10.3	9.6	9.0	8.5	7.6

(a) Given $\phi = 0.3$, $\rho^*/\rho_s = 0.4$, $p_0 = 5p_{atm} = 5$ atm and $\sigma_y = 100$ MPa, calculate the collapse strength of the foam.
(b) Calculate the collapse strength of the foam if $\phi = 0.7$.
(c) Calculate the collapse strength of the foam if $\phi = 1$ and there is no pressure difference, which is the case for open-cell foams.
(d) Plot σ^*_{pl} as a function of ϕ and discuss how the face/edge volume ratios affect the collapse strength of closed-cell foams.

Solution (a) The relative strength for the foam is

$$\frac{\sigma^*_{pl}}{\sigma_y} \cong 0.3(0.3 \times 0.4)^{3/2} + 0.4(1 - 0.3)0.4 + \frac{(5 - 1) \times 1.013 \times 10^5 \text{Pa}}{100 \times 10^6 \text{Pa}}$$
$$= 0.0125 + 0.112 + 0.004 \cong 0.13;$$
$$\sigma^*_{pl} = 13 \text{ MPa}.$$

(b) The relative strength for the foam is

$$\frac{\sigma^*_{pl}}{\sigma_y} \cong 0.3(0.7 \times 0.4)^{3/2} + 0.4(1 - 0.7)0.4 + \frac{(5 - 1) \times 1.013 \times 10^5 \text{Pa}}{100 \times 10^6 \text{Pa}}$$
$$= 0.0444 + 0.048 + 0.004 \cong 0.0964;$$
$$\sigma^*_{pl} = 9.6 \text{ MPa}.$$

(c) The relative strength for the foam is

$$\frac{\sigma^*_{pl}}{\sigma_{ys}} \cong 0.3(1 \times 0.4)^{3/2} + 0 \cong 0.076;$$
$$\sigma^*_{pl} = 7.6 \text{ MPa}.$$

(d) The collapse strength decreases with increasing ϕ (closed-cell to open-cell foam). The results are given in Table 10.1.

10.6 Feather

Feathers are very light and sometimes stiff epidermal structures that distinguish the class of Aves. Feathers are lightweight, complex, multifunctional integumentary appendages that are entirely composed of beta keratin in the form of β-pleated sheets. Feathers have a variety of functions that include flight, camouflage, courtship, thermal insulation, and water resistance. Feathers form from follicles in the epidermis that are periodically replaced by molting. The basic structure of the bird feather is composed of the main shaft (which has two parts, the rachis and the calamus), and side branches called barbs or "rami," which are themselves composed of barbules. Figure 10.31(a) shows a schematic matched with scanning electron micrographs of a feather belonging to the Blue-and-gold Macaw (*Ara ararauna).* Closed-cell foam can be observed within both the rachis and barb structures (Bodde *et al.*, 2011). The tertiary structures extending from the barb are called barbules or "radii." They are tied together by "hooklets." Barbs, barbules, and hooklets extending from the rachis comprise the vane or vexillum of feather. The rachis consists of a hollow cylinder called a "cortex," and a supporting foam core called a "medulla." The rachis is completely filled with medullary foam at the distal end and has a hollow at the root section or "calamus," shown in Fig. 10.31(b) (Bodde *et al.*, 2011). X-ray diffraction studies report that the crystallinity and the diffraction pattern of feather keratin are very similar to that of other avian keratins, such as claw and beak (Fraser and Macrae, 1980). However, there are differences in composition between feather and other avian keratins (Brush and Wyld, 1982; Brush, 1986). Brush and Wyld (1982) measured the molecular weight of avian keratins and found 10 500 g/mol for feather, which is lower than claw and beak keratins, which range from 13 000 g/mol to 14 500 g/mol.

The bending behavior of foam-reinforced feather rachis is of great importance. The feather must have stiffness and flexibility in order to withstand some degree of bending caused by flight activity. The bending behavior of the rachis (feather shaft) has been studied by Purslow and Vincent (1978). They demonstrated that the bending behavior of the rachis depends on the size and geometry of the cortex. The calamus of the feather has a circular or square cross-section, which becomes elliptical toward the distal end. Bonser and Purslow (1995) found an average Young modulus of 2.5 GPa in tension. The Young modulus of volant bird feathers varies with the position in the rachis, while that of flightless birds is almost the same at any position along the length of the rachis (Cameron, Wess, and Bonser, 2003). Corning and co-workers investigated the uniaxial oscillatory strain pattern of feather during flight, and the peak compressive strain was found to be just over twice the peak tensile strain (Corning and Biewener, 1998).

Mechanical properties of feather keratin demonstrate humidity sensitivity. The stiffness of feather rachis decreases with an increased moisture content, which is similar to behavior of other keratinous tissue (Taylor *et al.*, 2004). Figure 10.32 shows stress–strain curves of feather rachis and claw keratin for three different humidity levels. The feather rachis is stiffer than claw keratin due to molecular orientation differences. The

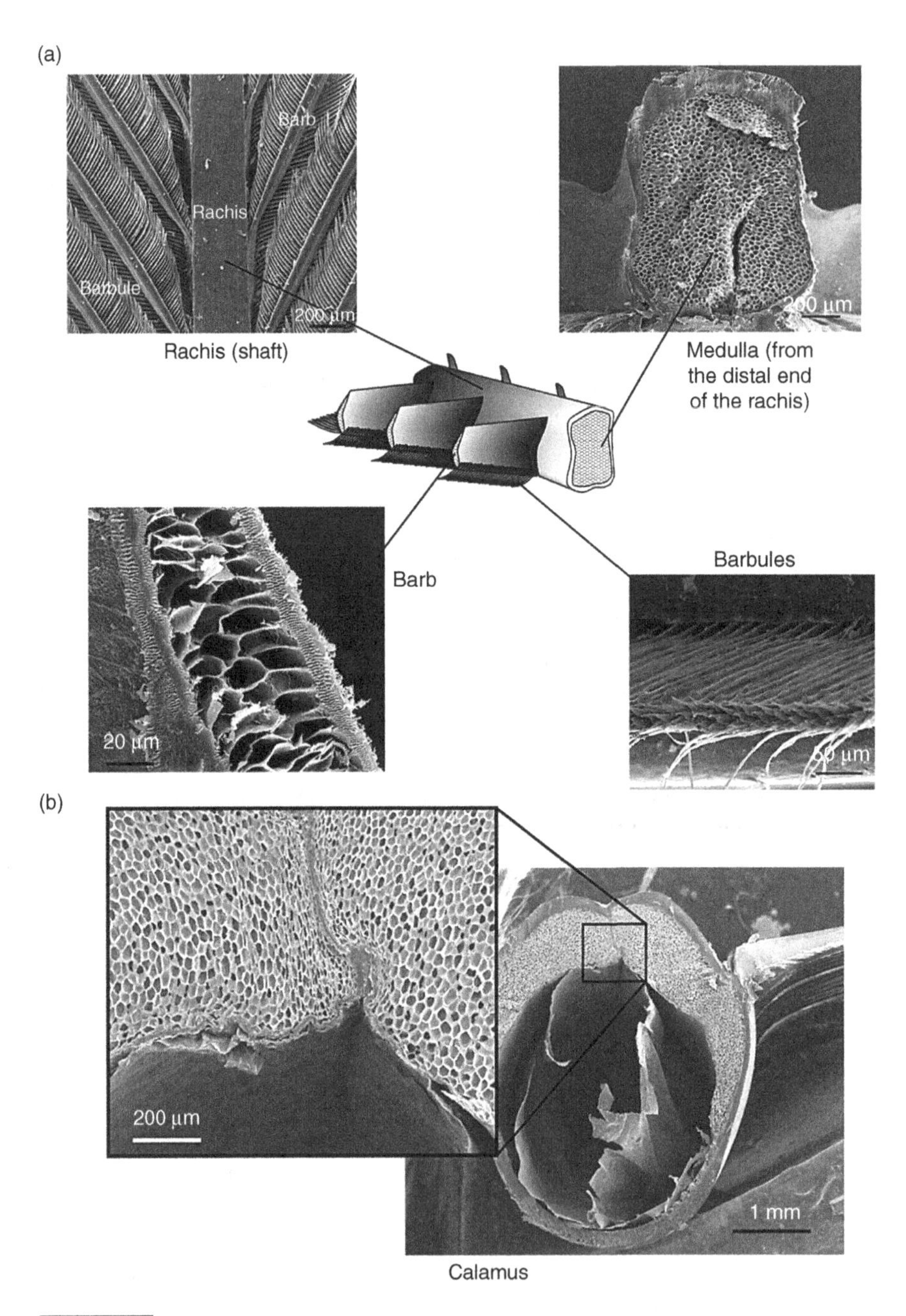

Figure 10.31.

(a) Schematics of feather structure with scanning electron micrographs of macaw feather. (b) Scanning electron micrographs of a hollow section of macaw rachis. (Figure courtesy S.G. Bodde.)

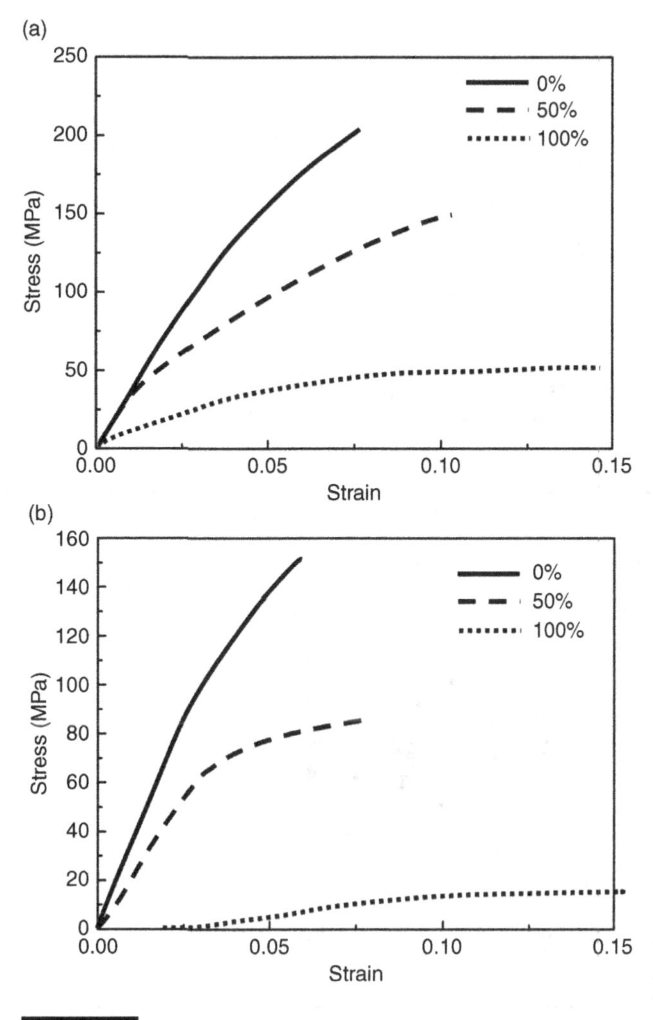

Figure 10.32.

(a) Stress–strain curve of feather at three different humidity conditions. (b) Stress–strain curves of claw keratin at three different humidity conditions. (From Taylor *et al.* (2004), with kind permission of Springer Science+Business Media B.V.)

compressive behavior of the medullary foam was studied and the Young modulus was found to increase with relative density (Bonser, 2001). Figure 10.33 shows the Young modulus revised from Bonser's results (Bonser, 2001) plotted as a function of relative density.

The relationship between melanin and the mechanical properties of avian keratin has also drawn attention, after Bonser and Witter (1993) described the increased hardness of the starling (*Sturnis vulgaris*) bill associated with seasonal melanization. Butler and Johnson (2004) studied melanized and nonmelanized barbs of feather. They considered not only color, but also the location along the rachis of the feather. Figure 10.34 shows the breaking stress as a function of positional distance of the barb along the length of the

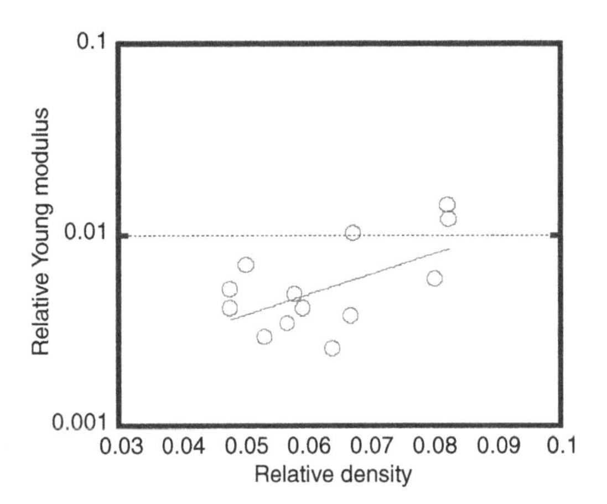

Figure 10.33.

Relative Young modulus vs. relative density of medulla. (Reproduced based on Bonser (2001).)

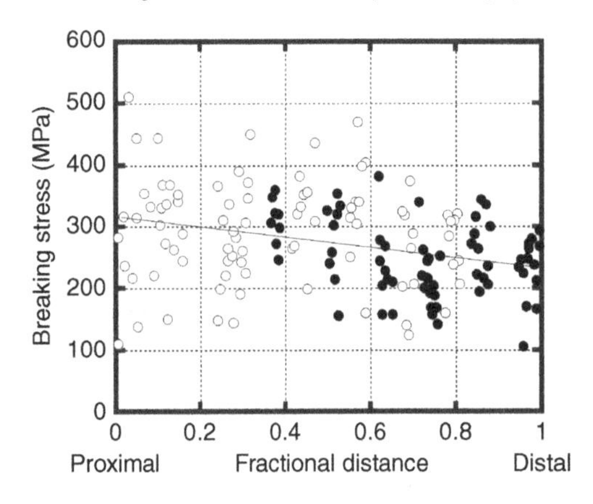

Figure 10.34.

Breaking stress as a function of fractional distance. (Reproduced based on Butler and Johnson (2004).)

rachis. Black dots represent melanic, and white dots represent nonmelanic regions. Variation in barb strength was observed along the rachis from the proximal end, enduring higher stresses during flight, to the distal end. The mechanical effects of melanin were found to be insignificant relative to the geometric variation in feather barb and position along the rachis.

The hierarchical structure of a feather of a toucan (*Ramphastos toco*) is shown in Fig. 10.35. The rachis is the long shaft that extends along the length of the feather and supports the barbs, which form the herringbone pattern along the rachis. It has a design that satisfies resistance to rupture during flexure without a proportionate increase in weight (Crenshaw, 1980). The barbs have smaller barbules extending from them. The cortex of the rachis is constructed of fibers, ~6 μm in diameter, which are aligned along the long axis and consist of 0.3–0.5 μm diameter bundles of barbule cells that are

Figure 10.35.

Hierarchical structure of a feather (Toco toucan). SEM of the surface microstructure of the cortex and (a) the cross-section of a distal section of rachis. This sandwich construction of the rachis is also repeated in the barbs. The dorsal (b) and ventral (c) cortical rachis keratin is smooth at the micron scale, while the lateral cortical rachis keratin (d) is fibrous and textured with ridges separated by 10–20 μm. Enclosed by a thin-walled shell or cortex is (e) a medullary core constructed of cells ranging from 20 to 30 μm in diameter. (Reprinted from Bodde *et al.* (2011), with permission from Elsevier.)

connected end to end. A cellular material is in the center of the rachis; thus, a feather has a sandwich construction that gives a maximum flexural strength to weight ratio. The SEM micrographs of dorsal (Fig. 10.35(b)) and ventral (Fig. 10.35(c)) surfaces reveal relatively smooth topography compared to that of lateral surfaces (Fig. 10.35(d)), which exhibit intersecting ridges with a spacing of 10 to 20 μm, with considerable overlap in diameter of cells of the medullary core (Fig. 10.35(e)) ranging from 20 to 30 μm. The smoothness, at micro-scale, of the dorsal and ventral surfaces is remarkable, as it is expected that surface roughness would serve to decrease drag (i.e. if the laminar-to-turbulent transition occurs before the point of boundary-layer separation), as proposed regarding micro-scale features observed on the surface of barbs (Chernova, 2005).

A more detailed schematic drawing of the fiber orientation is shown in Figs. 10.36(a) and (b). In Fig. 10.36(a), a cut-away shows the longitudinal orientation of the fibers along the cortex, and in Fig. 10.36(b) the thin layer of circumferentially wound fibers is illustrated. Lingham-Soliar, Bonser, and Wesley-Smith (2009) propose that the circumferentially wound fibers could control the hoop and longitudinal stresses thus preventing ovalization of the rachis when it is subjected to bending. The fibers in the common

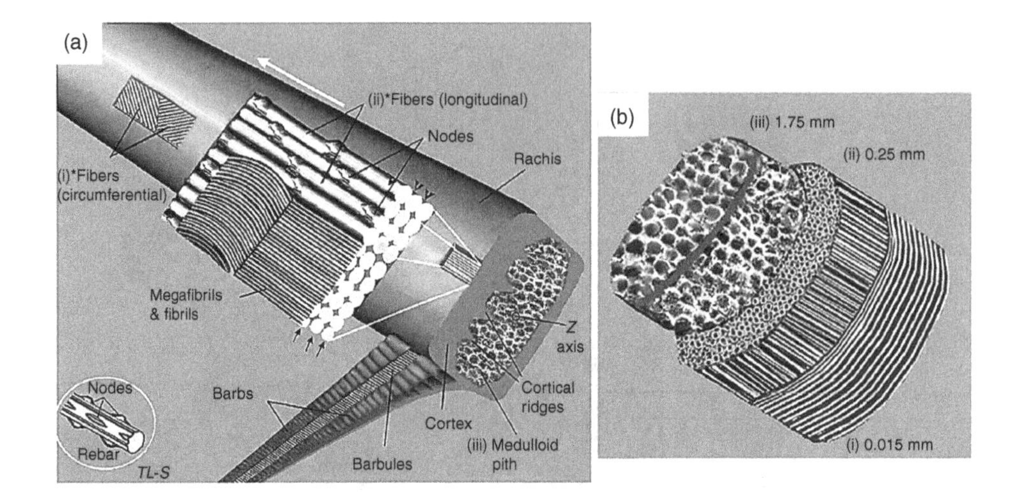

Figure 10.36.

(a) Schematic representation of fiber orientation. The majority of the fibers (~6 μm diameter) are oriented along the main axis of the shaft (rachis). (b) Illustration of a thin sheath of circumferentially oriented fibers on the surface of the rachis. (Figure courtesy Professor Solly Lingham-Soliar, University of KwaZulu-Natal, South Africa, with kind permission.)

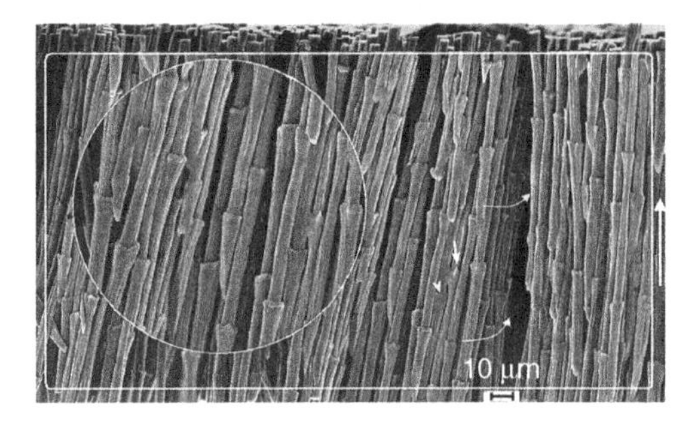

Figure 10.37.

SEM image of the aligned fibers (syncitial barbules) from the cortex of a chicken (*Gallus gallus*) rachis after fungal biodegradation. The hooked ends to the barbules or rings are denoted by the upward-pointing arrow. The densely packed fibers (curved arrows) show a staggered arrangement. (Figure courtesy of Professor Solly Lingham-Soliar, University of KwaZulu-Natal, South Africa, with kind permission.)

chicken (*Gallus gallus*) show protuberances similar to a bamboo pattern (Fig. 10.37). These protuberances may minimize sliding between fibers thereby increasing stiffness.

Figure 2.1 shows a schematic of feather foam and several SEM micrographs at increasing magnifications for a falcon (*Falco sparverius*). The internal foam has a fascinating structure. It consists of cells with approximate diameter of 10 μm. If we image the cell walls at a higher magnification, we recognize that they are not solid but are, in turn, composed of fibers with a diameter of ~200 nm. Thus, one has a two-level cellular structure that minimizes, as required, weight. Close observation of the cell walls reveals that they are made of a foam in a second level of porosity that further decreases density, as shown below. By applying the foam equation from Gibson and Ashby (1997) and assuming geometrical self-similarity, one has

$$\frac{\rho_f}{\rho_s} = C\left(\frac{t}{l}\right)^4, \tag{10.30}$$

where ρ_f is the density of the foam, ρ_s is the density of the solid, C is a constant, t is the thickness of the cell struts, and l is the length of the struts (either level I or II). This fourth-order dependency demonstrates that the decrease in density accomplished by hierarchical foam of levels I and II is dramatic, as illustrated by the cortex foams of the feather rachis and bird-of-paradise. Synthetic foams using this hierarchy of materials are only at the conceptual stage at present but can lead to significant weight reduction.

Typical stress–strain curves for the tail feathers of a Toco toucan are plotted in Fig. 10.38. As shown, there is considerable variability in mechanical properties. Most dorsal and ventral samples failed at strains higher than 5% (and some even in excess of 10%), and the majority of lateral samples failed at less than 5% strain. The average tensile modulus was 2.6 GPa and the average strength ranged from 33 to 141 MPa (Bodde, 2011).

Samples of dorsal and ventral cortex were found to be significantly stiffer and stronger than those of the lateral edges. Samples from the dorsal surface sustain both the highest stress and have the highest stiffness. The discrepancy in mechanical properties on the

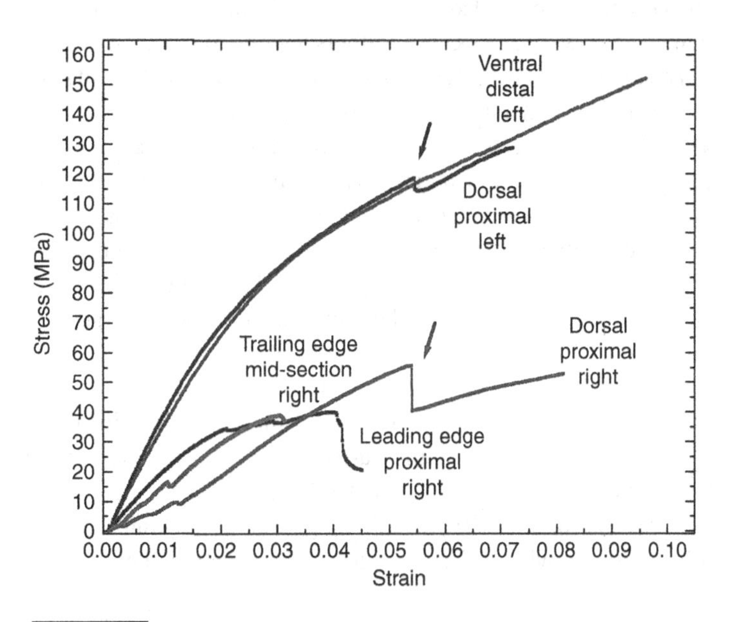

Figure 10.38.

Typical tensile stress–strain curves for Toco toucan cortical rachis keratin. The rachis cortex sample from dorsal and ventral surfaces exhibits linear elasticity over a larger strain range than lateral samples. Variation in stiffness and maximum tensile strength is high. Arrows denote localized failures. (Reprinted from Bodde *et al.* (2011), with permission from Elsevier.)

surfaces of the cortex may be related to the role of melanin, a pigment macromolecule that provides the black coloration in plumage in the form of rod-like granules (Filshie and Rogers, 1962). The dorsal surface appears to be uniformly and densely melanized; the ventral surface, based on visual inspection, is less melanized, appearing brown in color in some places, and the lateral surfaces are nonmelanized. Bonser (1995) reported a 39% increase in the microhardness of melanized dorsal cortex from the wing feather of a ptarmigan compared to nonmelanized ventral surfaces.

Distal samples of cortical rachis were weakest and least stiff compared to proximal and middle specimens, but not significantly so. The fracture strength and stiffness of proximal and middle specimens were approximately equivalent. This result is in contrast to the proximal to distal increase in stiffness reported by Bonser and Purslow (1995) for the case of the wing feathers of the swan. It was suggested that it was correlated to the increased proportion of the axially oriented fibers in the cortex. An increasing stiffness gradient from the proximal to the distal end may serve to compensate for the effect of decreased cross-sectional area along the length of the feather, if flexural stiffness, EI (I is the second moment of inertia), is to be conserved or compensated. However, this increase in strength and stiffness may be confounded by a temporal or aging effect of keratin, as the more distal cortex is more mature (Prum, 1999). The cortex sampled from distal regions of peacock tail coverts is significantly less crystalline than in other regions along the length of the feather (Pabisch *et al.*, 2010). Therefore, the cortex of the rachis, in addition to being a fiber-reinforced bi-laminate composite, may also be described as functionally graded.

The male peafowl (peacock) tail feathers are one of the most beautiful and colorful in nature. The fanned out feathers are used to attract peahens. Several structural features are required for these feathers: the fanned out feathers must support their own weight, but must be lightweight because peacocks can fly; a minimal amount of material must be used for energy conservation; and finally they must have sufficient strength and stiffness so that buckling during natural bending does not occur (Burgess, King, and Hyde, 2006). There are ~300 feathers on the tail and they range in length from a few centimeters up to 1.5 m. The rachis has ~300 barbs covered with 10^6 barbules. The barbules contain ~20 indentations along the length, and these produce the color (Burgess *et al.*, 2006). Weiss and Kirchner (2010) examined the strength and stiffness, and found the longitudinal stiffness of the cortex to be 3.3 GPa while the transverse value was 1 GPa. The medullary foam had a stiffness of 10 MPa. The foam played no significant role in the tensile properties, but provides 96% of the transverse compressive stiffness. It is unclear what the role of the circumferentially wound fibers are in the compressive stiffness. The fracture strengths of the cortex, both longitudinal and transverse, were found to be 120 MPa, similar to what is found for the Toco toucan feathers.

Example 10.3 Inspired by the hierarchical cellular structure in feathers, a materials scientist tried to design an ultra-lightweight foam in which the density is reduced to 1/1000 of its solid material. Assume self-similar geometry and that the beam thickness to length ratio t/l is kept at 0.5; the constant C is approximately 1. Estimate how many levels of hierarchy are required for the synthetic foam.

Solution Feather foam has a two-level cellular structure, and the relative density is given by

$$\frac{\rho_f}{\rho_s} = C\left(\frac{t}{l}\right)^4.$$

For a cellular structure with N levels of hierarchy, a general equation is as follows:

$$\frac{\rho_f}{\rho_s} = C\left(\frac{t}{l}\right)^{2N} \cong \left(\frac{t}{l}\right)^{2N}.$$

The density of the foam is reduced to 1/1000 of that of solid material, i.e.

$$\begin{aligned} 0.001 &= (0.5)^{2N}, \\ \log(0.001) &= 2N\log(0.5), \\ -3 &= 2N(-0.3), \end{aligned}$$

so $2N = 10$, $N = 5$. Five levels of hierarchy are required.

10.7 Cuttlefish bone

Cuttlefish are marine animals of the order *Sepiida* and the class *Cephalopoda* (which also includes squid, octopuses, and nautiluses). Cuttlefish possess an internal structure called the cuttlebone, which has a cellular structure and is mainly made of $CaCO_3$ in aragonite form with a small amount of chitin and other organic components. Photographs of cuttlefish *Sepia pharaonis* and the cuttlebone are shown in Fig. 10.39. The cuttlebone is a rigid buoyancy tank which enables the cuttlefish to maintain a fixed position in water with minimal effort. It must be strong enough to withstand hydrostatic pressure and lightweight so as not to sacrifice buoyancy (Birchall and Thomas, 1983; Sherrard, 2000). Thus, the demands of buoyancy and mechanical strength conflict: if the cuttlebone is too dense (or the chambered cells are too small), buoyancy decreases; if the cuttlebone has a very porous structure, its mechanical performance is too weak, increasing the risk of fractures at the animal's normal habitat depths.

Figure 10.40(a) illustrates the ventral view of the cuttlebone. The dorsal side of the cuttlebone is covered by a dense layer of exoskeleton. Beneath the exoskeleton is the endoskeleton, which consists of numerous chambers that serve as a buoyancy controlling device. Chambers are emptied through the siphuncular zone, maintaining buoyancy. It should be noted that the growth of cuttlebone starts from the tail region and continues

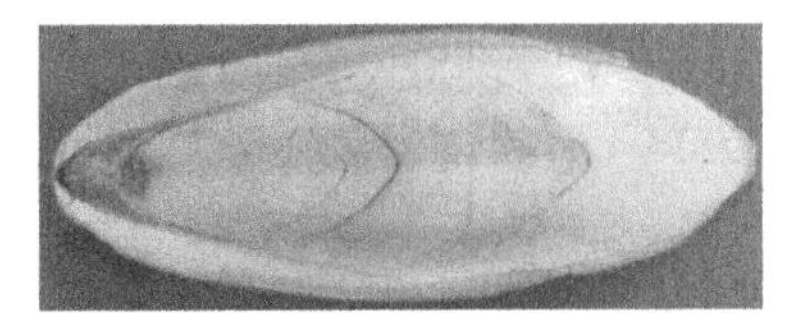

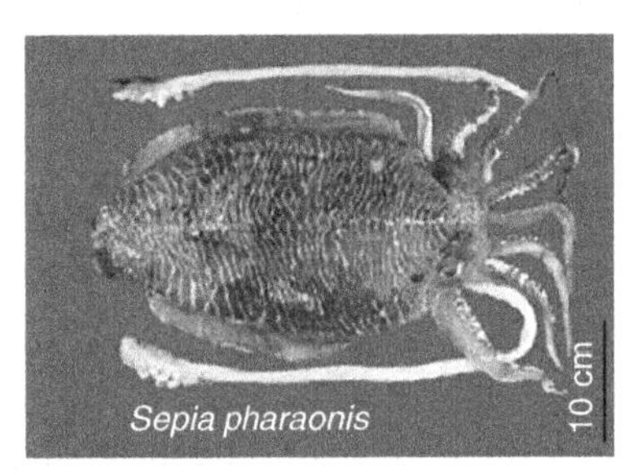

Figure 10.39.

Cuttlebone and cuttlefish *Sepia pharaonis*.

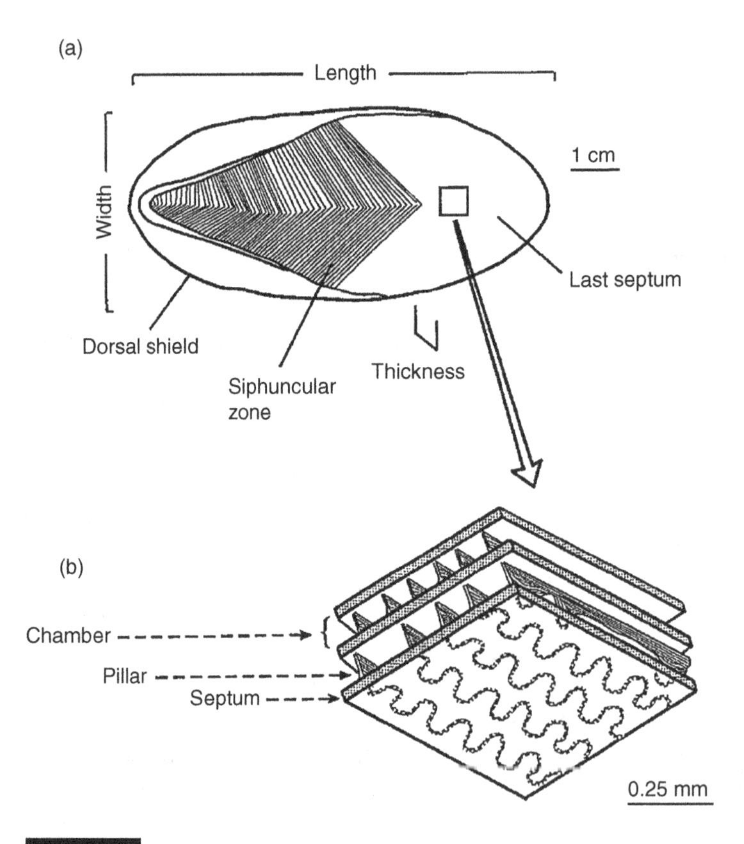

Figure 10.40.

(a) Schematic showing ventral view of the cuttlebone (anterior is to the right). Chambers are emptied through the siphuncular zone, maintaining buoyancy. (b) Microstructural features of cuttlebone, showing the chambers and the wavy pillars between the septa. (From Sherrard (2000). Reprinted with permission from the Marine Biological Laboratory, Woods Hole, MA.)

toward the head region. Figure 10.40(b) indicates the main microstructural components of cuttlebone. The chambers are constructed by "ceilings" called septums and "walls" named pillars. The pillars have a wavy shape and interconnect adjacent septums. The dimensions of the chambers, pillars, and septums vary with species and their habitat. Cuttlefish species that live at greater depths have thicker septums and less space between pillars than those that live at shallower depths (Sherrard, 2000). Figure 10.41(a) is a cross-sectional stereoscopic image showing the cellular structure of the cuttlebone. Figure 10.41(b) is a stereoscopic image, viewing from the top, revealing the wavy morphology and the arrangement of the pillars. The microstructure of cuttlefish *Sepia* is shown in the SEM micrograph in Fig. 10.42. Detailed structural features of the septum, pillar, and chambers can be observed. There are thin sheets made mainly of chitin connecting adjacent pillars laterally.

The representative stress–strain curve of cuttlebone under compressive loading in the dorsal–ventral direction is shown in Fig. 10.43. The compressive mechanical behavior of cuttlebone can be modeled as a typical cellular solid, with linear elastic, plastic plateau, and densification regions. The compressive strength of cuttlebone is on the order of 1 MPa. The fracture surface of cuttlebone examined under SEM is shown in Fig. 10.44(a).

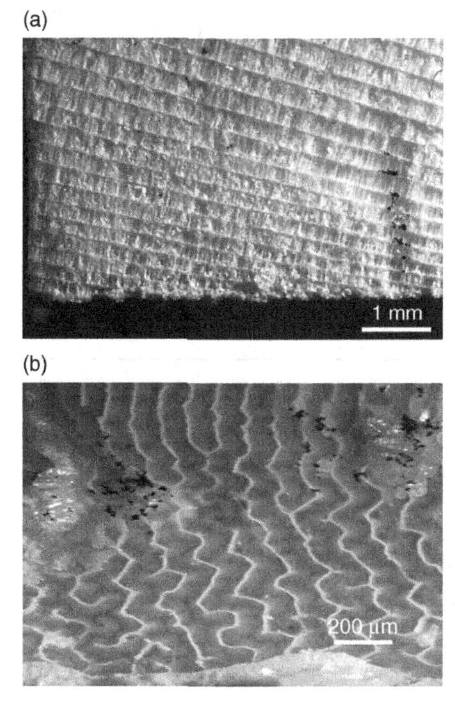

Figure 10.41.

(a) Stereoscopic image showing the cellular structure of cuttlebone (cross-sectional view). (b) Stereoscopic image showing the wavy pillars (top view).

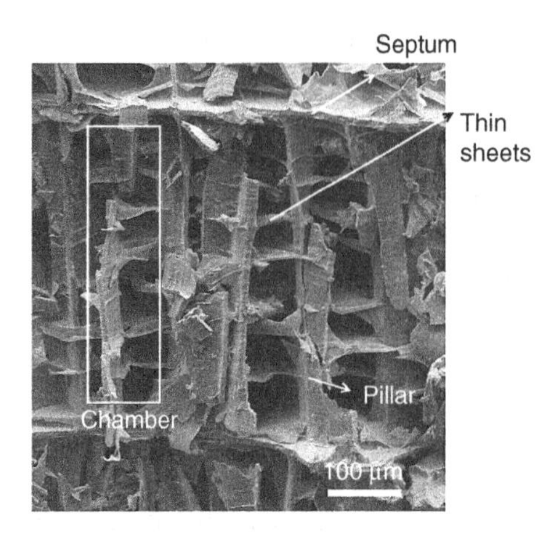

Figure 10.42.

SEM micrograph of cuttlebone revealing detailed microstructure of chamber, septum, pillar, and thin organic sheets between adjacent pillars.

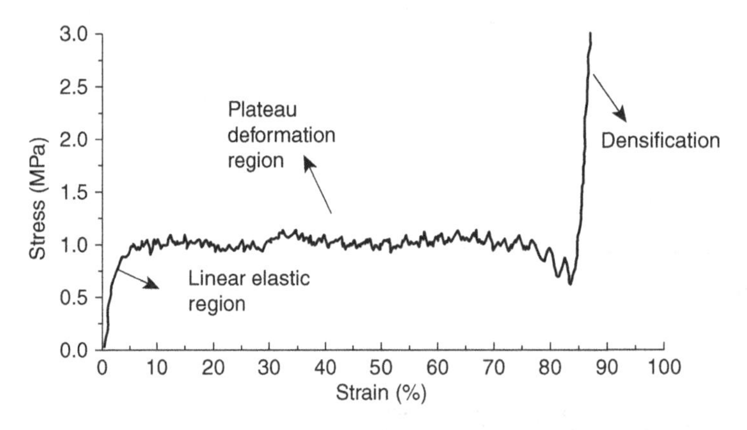

Figure 10.43.

Representative stress–strain curve of cuttlebone under compression showing the typical mechanical behavior of cellular solids (linear elastic region, plastic plateau region, and densification).

After 10% compressive strain, fracture occurs in the pillar regions and separation of thin sheets can be observed. Figure 10.44(b) is an SEM micrograph at higher magnification, showing the fracture of the pillars in detail.

Summary

- The Euler equation, which represents the resistance of a structure to compressive loading without buckling, is given by

(a)

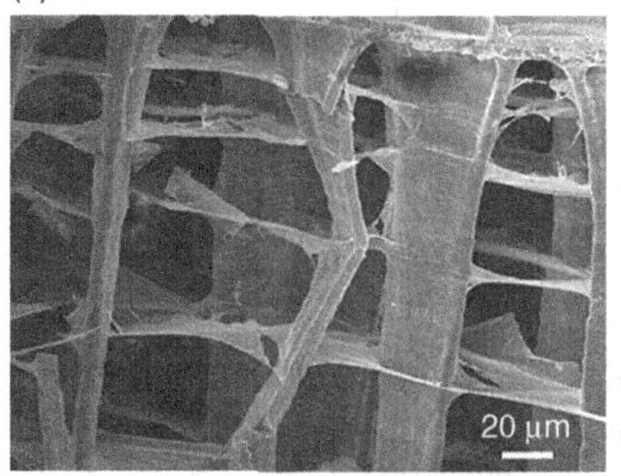

(b)

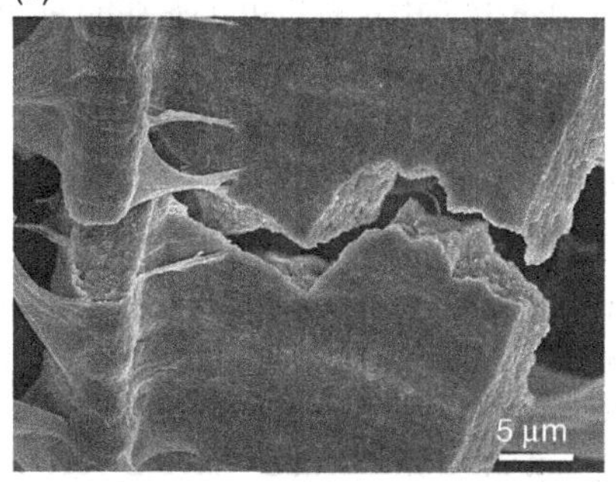

Figure 10.44.

SEM micrographs showing fracture morphology after 5% compressive deformation. The fracture of pillars can be observed.

$$P_{cr} = \frac{\pi^2 EI}{(kL)^2}.$$

- Densification of cellular materials (foams) takes place in three stages: (a) elastic loading up to the yield/fracture stress; (b) plastic plateau (for ductile) or fracture plateau (for brittle) foams; (c) densification.
- The density of an open-cell foam material, ρ^*, is related to that of the solid material, ρ_s, in terms of the cell dimensions l and t:

$$\frac{\rho^*}{\rho_s} = C_1 \left(\frac{t}{l}\right)^2.$$

- The elastic modulus is given by

$$\frac{E^*}{E_s} = \frac{C_1}{24C_2} \left(\frac{\rho^*}{\rho_s}\right)^2,$$

where $C_1/24C_2$ should be approximately equal to unity.
- The plastic plateau for an open cell has the following expression:

$$\frac{\sigma^*_{pl}}{\sigma_y} = C_1^{-3/2} \left(\frac{\rho^*}{\rho_s} \right)^{3/2}.$$

- The plastic response for a closed-cell foam is given by

$$\frac{\sigma^*_{pl}}{\sigma_y} = C_5 \left(\phi \frac{\rho^*}{\rho_s} \right)^{3/2} + (1 - \phi) \frac{\rho^*}{\rho_s} + \frac{p_0 - p_{at}}{\sigma_y},$$

  where σ^*_{pl} is the plastic collapse stress of foam, σ_y is the yield stress of the solid portion, C_5 is a parameter, ϕ is the ratio of the volume of face to volume of edge, p_0 is the initial fluid pressure, and p_{at} is the atmospheric pressure.
- The third stage in compression, densification, is well described by the Fischmeister–Arzt equation:

$$p = Z\sigma_y \frac{D(D - D_0)}{4(1 - D_0)},$$

  where D_0 and D are the initial and final relative densities; $D = \rho^*/\rho_s$.
- A most remarkable property of wood is the fracture toughness: it is highly anisotropic and its highest value is ten times greater than that of a fibrous composite with the same volume fraction of fibers and matrix. Some woods have a very low density, balsa wood being the extreme example (0.04–0.34 g/cm^3). Others are highly dense, such as lignum vitae, or ironwood, with 1.37 g/cm^3.
- Many bird bones are pneumatic (filled with air) and have to be light to enable flight. In order to provide stability and resist torsional loading, bird bones have internal reinforcements in the form of struts and ridges.
- Bird beaks are subjected to the same restrictions of weight minimization and bending strength maximization. This is accomplished by an ingenious solution, enabling a low density and high stiffness: a composite structure consisting of an external solid keratin shell and a cellular core. The mechanical behavior of bird beaks is governed by both the ductile keratin integument and the semi-brittle bony foam. The core is a closed cell made of bone. The toucan beak has a density of approximately 0.1 g/cm^3, which enables the bird to fly while maintaining a center of mass at the line of the wings.
- Bird feathers are lightweight, complex, multifunctional integumentary appendages that are entirely composed of beta keratin in the form of β-pleated sheets. They have a variety of functions that include flight, camouflage, courtship, thermal insulation, and water resistance. The basic structure of bird feather is composed of the main shaft, or "rachis" and side branches called barbs or "rami." These, in turn, separate into barbules. They are tied together by "hooklets." Barbs, barbules, and hooklets extending from the rachis comprise the vane or vexillum of feather. The rachis consists of a hollow cylinder called the "cortex" and a supporting foam core called "medulla." Mechanical properties of feather keratin demonstrate humidity sensitivity.

The stiffness of feather rachis decreases with an increased moisture content, which is similar to behavior of other keratinous tissue.

- Close observation of the cell walls in bird feathers reveals that they are themselves made of a foam in a second level of porosity that further decreases density. By applying the foam equation from Gibson and Ashby and assuming geometrical self-similarity,

$$\frac{\rho_f}{\rho_s} = C\left(\frac{t}{l}\right)^4.$$

where ρ_f is the density of the foam, ρ_s is the density of the solid, C is a constant, t is the thickness of the cell struts, and l is the length of the struts (either level I or II). This fourth-order dependency demonstrates that the decrease in density accomplished by hierarchical foam of levels I and II is dramatic, as illustrated by the cortex foams of the feather rachis and bird-of-paradise.

- Cuttlefish are marine animals of the class *Cephalopoda* (which also includes squid, octopuses, and nautiluses). Cuttlefish possess an internal structure called the cuttlebone, which has a cellular structure and is mainly made of $CaCO_3$ in aragonite form with a small amount of chitin and other organic components. The cuttlebone is a rigid buoyancy tank which enables the cuttlefish to maintain a fixed position in water with minimal effort.

Exercises

Exercise 10.1 Both Ti and Ta are being investigated for the replacement of compact bone. Given: E

(a) Determine the porosity of the foam needed to match that of the bone.
(b) What happens to stiffness as bone grows into metal foam?
(c) How would you change the design to eliminate this problem?

Exercise 10.2 An arterial graft is made of porous silicon. After three months the pores become filled. The inside diameter is 5 mm and the wall thickness is 1 mm. The initial porosity is equal to 0.2 (assume cylindrical pores).

(a) What is the function of the porosity?
(b) What should be the maximum size of pores for impeding blood from seeping out?
(c) If the pressure experienced by the patient is as high as 180 mmHg, determine the stress in the graft.
(d) The silicon rubber has a tensile strength of 25 MPa. Considering the weakening by the pores, what is the factor of safety of this prosthesis?

Exercise 10.3 The combined effects of porosity and strain rate on bone strength have been expressed by the following equation:

$$\sigma = K\rho^* \, \dot{\varepsilon}^m,$$

where ρ^* is the relative density. If the fracture stress of compact bone at $10^{-3}\,s^{-1}$ is equal to 120 MPa, what is the fracture strength of cancellous bone with 50 vol.% porosity at $10\,s^{-1}$?

Exercise 10.4 A titanium metal sponge is used to replace the inter-disc space in the spinal column. Calculate its density if we want to match the elastic modulus of bone, E = 20 GPa. Calculate the corresponding strength. Given: $E_{Ti} = 105$ GPa; $\sigma_y = 450$ MPa; $C_1 = 0.3$.

Exercise 10.5 Determine the Young modulus and strength of cancellous bone having a density equal to 40% of that of compact bone. Assume that compact bone strength is equal to 120 MPa and E = 20/GPa.

Exercise 10.6 Cellular solids and scaffolds have been widely applied in the biomedical field. Compare the advantages and disadvantages of metallic, ceramic, and polymeric foams, and explain the novel design strategies of composite scaffolds.

Exercise 10.7 Calculate the elastic modulus and compressive strength of the cellular structure of Fig. 10.1(a) in two directions: longitudinal and transverse. Assume E = 5 GPa and compressive strength of 20 MPa. Make necessary assumptions.

Exercise 10.8 The collapse strength of equiaxed closed-cell plastic foams is given by

$$\frac{\sigma_{pl}^*}{\sigma_y} \cong 0.3\left(\phi\frac{\rho^*}{\rho_s}\right)^{3/2} + 0.4(1-\phi)\frac{\rho^*}{\rho_s} + \frac{p_0 - p_{atm}}{\sigma_y},$$

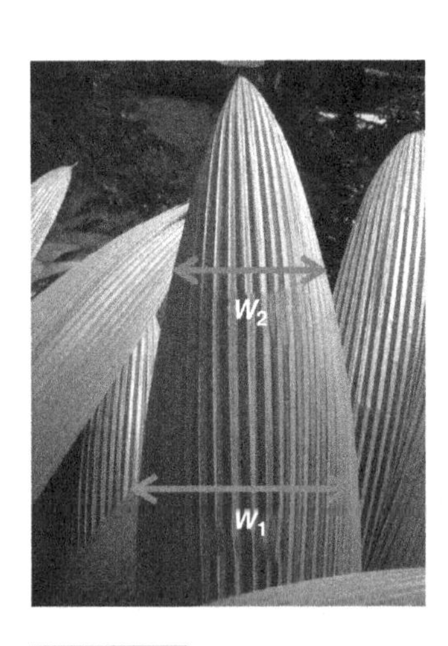

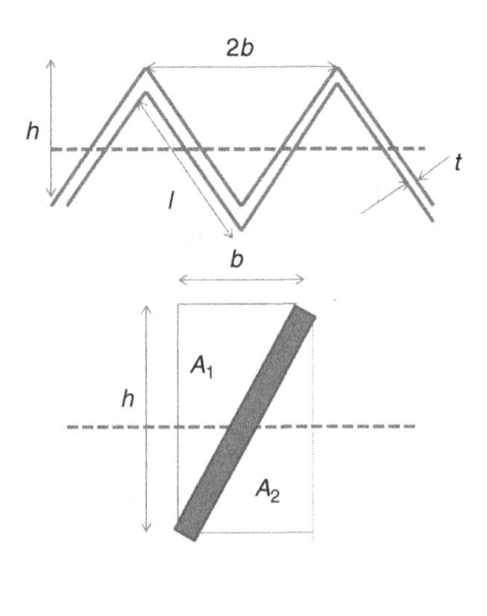

Figure E10.10.

Leaf with parallel veins forming a zig-zag pattern that increases stiffness.

where ϕ is the ratio of the volume of face to volume of edge, p_0 is the initial fluid pressure, and p_{atm} is the atmospheric pressure. Answer the following questions.

(a) Given $\phi = 0.2$, $\rho^*/\rho_s = 0.5$, $p_0 = p_{atm} = 2$ atm and $\sigma_y = 150$ MPa, calculate the collapse strength of the foam.
(b) Calculate the collapse strength of the foam if $\phi = 0.4$.
(c) Calculate the collapse strength of the foam if $\phi = 1$ and there is no pressure difference, which is the case for open-cell foams.
(d) Plot σ^*_{pl} as a function of ϕ and discuss how the face/edge volume ratios affect the collapse strength of closed-cell foams.

Exercise 10.9 Calculate the Young modulus and compressive strength of the cancellous bone of Fig. 10.3; the cell size is approximately 1 mm. Estimate the thickness of the trabeculae. Assume $E = 20$ GPa for HAP and a strength of 40 MPa. Make necessary assumptions.

Exercise 10.10 Cellular cores are not the only means to increase stiffness. Leaves have structural elements that provide them with controlled stiffness. These stiffening elements are called veins. When one principal vein runs the length of the leaf, the leaf is called pinnate. When the veins radiate out like the fingers in the hand, they are called palmate leaves. However, nature is ingenious and yet another strategy is shown in Fig. E10.10: parallel veins create a zig-zag structure that, in itself, provides structural rigidity at a very low weight penalty. This design shifts the weight of the leaf as far out of the neutral axis as possible. The curious student should take a sheet of paper and fold it into this pattern, checking the stiffness.

(a) Calculate the moment of inertia as a function of the angle between two adjacent segments (a zig and a zag). Given: $t = 1$ mm; $l = 15$ mm.
(b) Propose a bioinspired design where the zig-zag concept of stiffening can be applied.

11 Functional biological materials

Introduction

Materials can be divided into two classes: structural and functional. Structural materials are the ones where mechanical properties are of foremost importance: strength, hardness, toughness, ductility, and density are some of their most important characteristics. Functional materials, on the other hand, have other primary attributes: optical, magnetic, electrical, superconducting, and energy storage are some of the more important properties for which they are developed.

In biological materials, these boundaries are not so clear because many are multifunctional. Nevertheless, some have as primary function the sustainment of the structure, while others have other functions as principal attributes.

In this chapter, we present the principles of attachment with emphasis on gecko feet, which are, not coincidentally, the subject of the cover of this book. This is a fascinating subject, and several research groups are feverishly working on it with the goal of creating synthetic reversible attachment devices. Next we present the superhydrophobic effect through which the surface of the lotus plant remains clean. Water simply rolls off the surface, taking with it all particles. Another interfacial property that is important is the shark skin; the scales have a morphology and configuration that generate microvorticity in the water, decreasing drag. Biological materials and systems also have outstanding optical properties: chameleons change color through a complex interplay of chromatophores, for example. Some butterflies have wings with features that act as photonic arrays. The deep blue of the *Morpho* butterfly is the result of the interaction of the light with nano-scale arrays. The iridescent feathers of some birds are the result of the interference of light with arrays on the barbules, which have a periodicity close to the wavelength of light. The iridescent throat patch of humming birds is an eloquent example. We review these and some other aspects in this chapter. The presentation is by no means exhaustive, but illustrates the nature of functional biological materials.

11.1 Adhesion and attachment

A fascinating and challenging functionality of biological systems is the ability to attach to all kinds of surfaces under varying conditions. Barnacles (Berglin and Gatenholme, 2003) and mussels (Waite, 1987; Bell and Gosline, 1996) attach in a permanent manner through intricate chemical processes, and a number of animals have evolved reusable

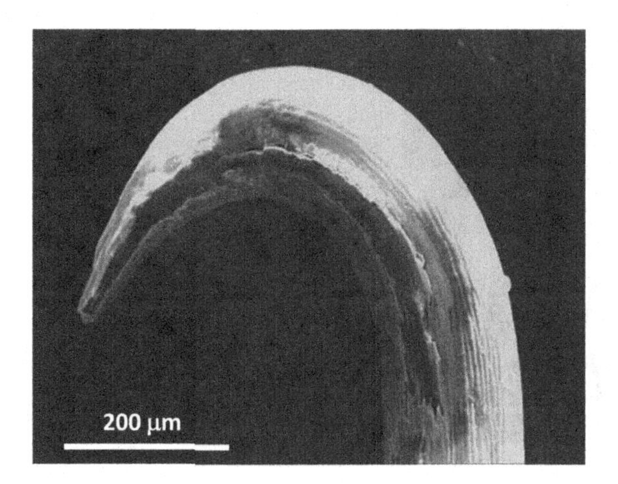

Figure 11.1.

Burr hook, which inspired George de Mestral to create VELCRO®.

attachment devices: insects, lizards, tree frogs, and molluscs, for example. Barnes (2007) classifies attachment mechanisms found in animals into the following categories.

- Interlocking. The mechanism by which felines climb trees. This is a strictly mechanical process and can be accomplished by penetrating the surface with sharp claws. A very simple mechanical attachment device are the hooks on burrs which attach themselves to hikers. This was the inspiration for the best known biomimicking application VELCRO®. Figure 11.1 shows one of these natural hooks. Each burr has thousands of these hooks that ensure attachment to wool loops in clothing and fur in animals. Thus, the seeds are carried away from their origin through attachment to a temporary host.
- Friction. This consists of the micro-scale interlocking of surfaces due to their roughness and the intermolecular forces between materials, and requires an angle below 90°.
- Bonding. This involves the formation of bonds between the animal and the surface and has three possible mechanisms acting separately or together: wet adhesion (capillarity), dry adhesion (van der Waals), and suction (through reduced internal pressure).

Gecko feet present a fascinating problem of adhesion (Autumn *et al.*, 2000). This attachment system was first recognized by Ruibal and Ernst (1965). Indeed, there are several biological systems in which the attachment to surfaces uses similar principles: flies, beetles, and spiders. It is possible that the tree frog might be included in this category. Again, novel experimental techniques coupled with analysis are revealing these mechanisms. Fly and gecko feet are made of a myriad of thin rods, called setae, terminated by spatulae, with submicron diameters. These are shown for the fly *Calliphora vicina* in Fig. 11.2. Figure 11.3(a) shows a cross-section of the gecko foot with setae marked (st). Each seta has, at its tip, a number of spatulae, marked (sp) in Fig. 11.3(b). Arzt and co-workers (Arzt *et al.*, 2003; Arzt, 2006) and Spolenak *et al.* (2005a, b) calculated the stress required to pull off a contact. Here, we will illustrate these

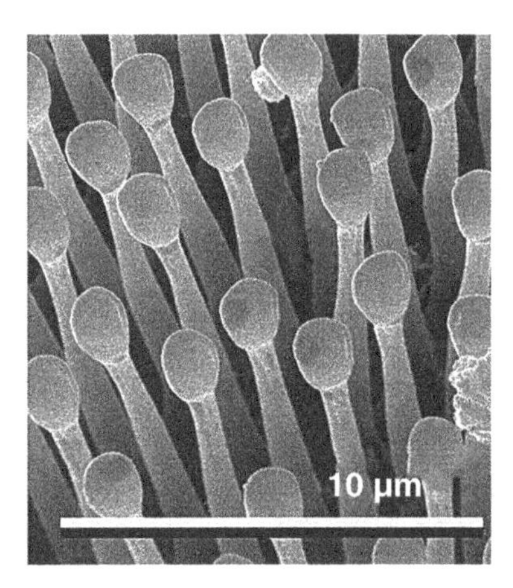

Figure 11.2.

Setae and distal spatulae in the fly *Calliphora vicina*. Van der Waals forces at the spatula–surface interface generate attachment forces that in the gecko can be as high as 20 N. (Reprinted from Spolenak, Gorb, and Arzt (2005b), with permission from Elsevier.)

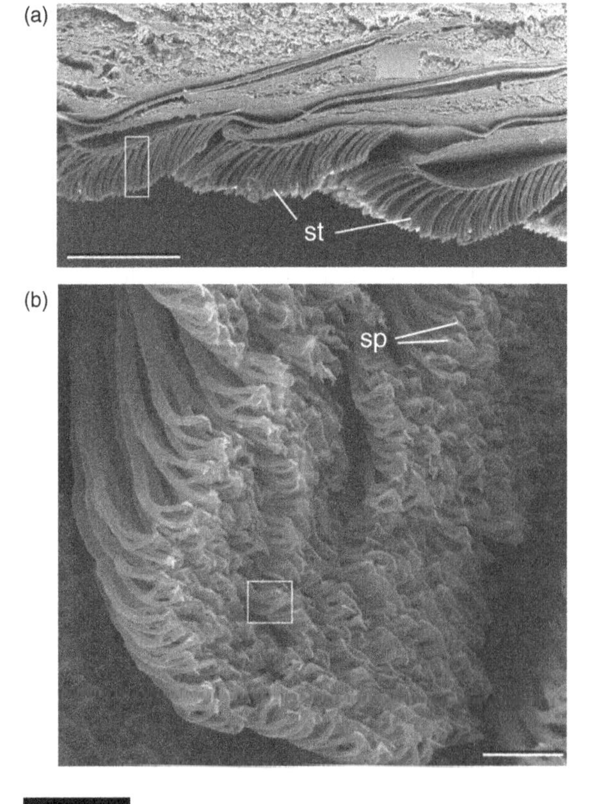

Figure 11.3.

SEM micrographs showing the detail of (a) setae (st) and (b) spatulae (sp). (Figure courtesy E. Artz and G. Huber.)

attachment mechanisms through five examples: the gecko, the beetle, and the ladybug, which primarily use van der Waals forces through setae having diameters that decrease with increasing body mass, and the tree frog (*Scynax perereca*) and the abalone foot, which use a combination of mechanisms.

11.2 Gecko feet

The gecko foot is one of the most fascinating examples of how animals can apparently defy gravity by using a unique structure in the foot pad. It has been intensively studied by several groups (Autumn *et al.*, 2002; Arzt *et al.*, 2003; Huber *et al.*, 2005; Spolenak *et al.*, 2005a,b). Figure 11.4 shows the hierarchical structure of the gecko foot. At the macrometer scale, the foot consists of "V"-shaped patterns. At higher magnification, the setae arrange into bundles that are approximately square with 10 μm sides arranged in a regular pattern with channels between them. At the distal ends, the setae terminate into a large number of nano-sized fibrils, called spatulae, with diameters of ~100 nm.

Arzt *et al.* (2003) and Huber *et al.* (2005) calculated the stress required to pull off a contact and explained the scale effect. This calculation is based on the van der Waals forces combined with Hertzian contact stresses. For simplicity, spatulae are assumed to have semispherical extremities as a first approximation. The contact radius, a, for a spherical cap of radius R in contact with a flat surface and subjected to a compressive force F is, according to Hertzian elasticity (see, e.g., Ugural and Fenster, 1981)

$$a = \left[\frac{3(1-v^2)RF}{4E}\right]^{1/3}, \tag{11.1}$$

where E is the Young modulus of the spatula. Johnson, Kendall, and Roberts (1971) inserted the attractive force between two bodies, and Arzt (2006) provides the following expression:

$$F = \frac{4}{3}E\frac{a^3}{R} - (4\pi E\gamma a^3)^{1/2}, \tag{11.2}$$

where the first term (repulsion) is obtained from Eqn. (11.1) by assuming that

$$(1-v^2)\sim 1.$$

The second term represents the van der Waals attraction, where γ is the work of adhesion of the two surfaces (~50 mJ/m^2) that produces the van der Waals attraction. The critical radius of contact, a_c, can be obtained by taking

$$\frac{\partial F}{\partial a} = 0. \tag{11.3}$$

We then find a_c and substitute it into Eq. (11.2):

$$F_c = -\frac{3}{4}\pi R\gamma.$$

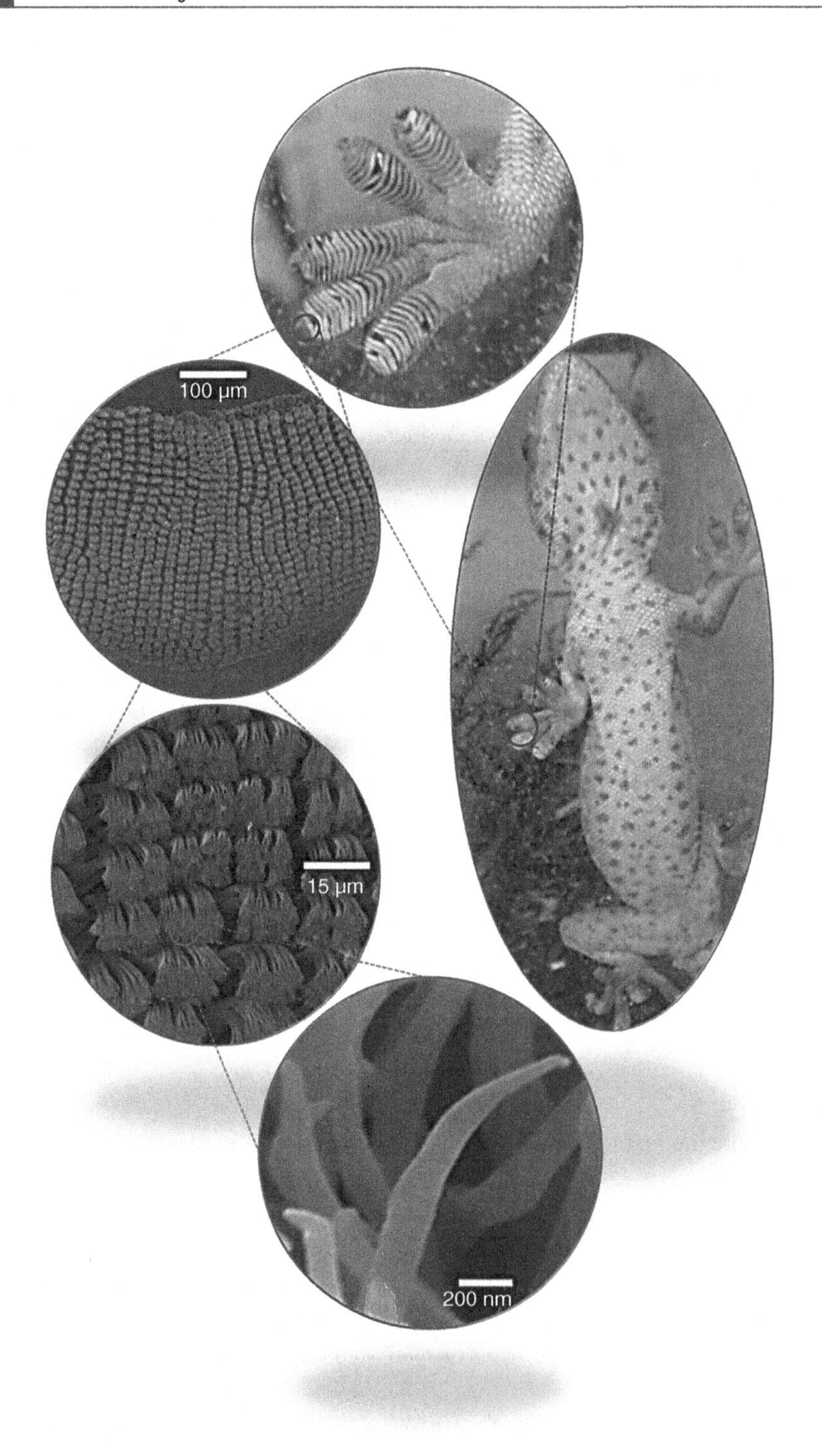

Figure 11.4.

Hierarchical structure of gecko foot shown at increasing magnifications, starting with setae in rectangular arrays terminating in spatulae with diameters of approximately 100–200 nm.

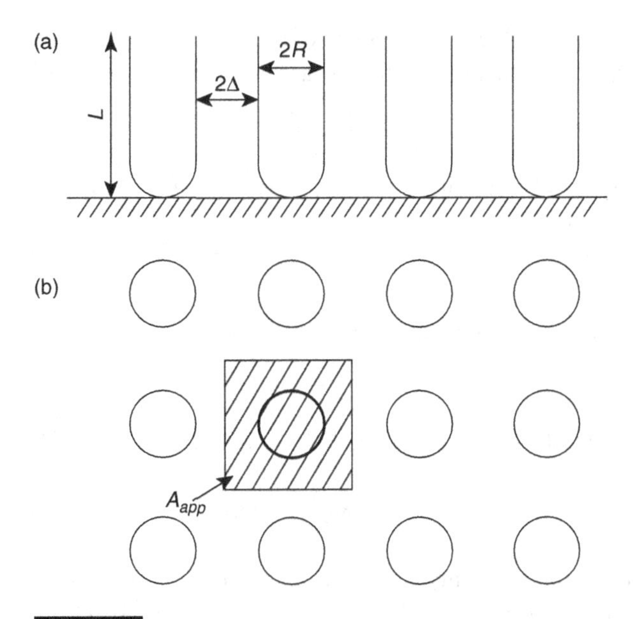

Figure 11.5.

Idealized arrangement of attachment system with spherical tip shape (radius R and spacing 2Λ). (Reprinted from Spolenak *et al.* (2005b), with permission from Elsevier.)

Johnson *et al.* (1971) obtained a slightly different expression, which is commonly used (note that we defined attraction as negative):

$$F_c = -\frac{3}{2}\pi R\gamma. \tag{11.4}$$

This is the famous Johnson–Kendall–Roberts (JKR) equation. This result is, surprisingly, independent of E. More complex analyses incorporate the elastic constants. In the case of attachments, only a fraction of the surface, f_s, is covered by spatulae. The adhesion stress, σ_{adh}, can be computed from F_c considering the area of contact of each spatula to be equal to πR^2. This is shown in Fig. 11.5. The stress required to pull off a spatula is the force F divided by the area πR^2. This can be corrected for the apparent area:

$$\sigma_{adh} = \frac{3}{2}\frac{f_s\gamma}{R}, \tag{11.5}$$

where f_s, the fraction of the area covered by setae, is given by

$$f_s = \frac{\pi R^2}{A_{app}}. \tag{11.6}$$

It can be seen that the pull-off stress in Eqn. (11.5) is inversely proportional to R. Thus, the larger the mass of the biological system, the smaller R has to be.

Arzt and colleagues (Arzt *et al.*, 2003; Arzt, 2006) arrived at a quantitative relationship. Considering a spherical body, the surface S varies with mass m as follows:

$$S \propto m^{2/3}. \tag{11.7}$$

Assuming geometrical similarity, the area covered by setae, A_s, is a constant fraction of the total body surface:

$$A_s = kS. \tag{11.8}$$

The force required for attachment is related to the mass:

$$F_A = mg. \tag{11.9}$$

This force is the product of σ_{adh} (Eqn. (11.5)) and A_s (Eqn. (11.8)). Thus,

$$\frac{3}{2}\frac{f_s\gamma}{R}km^{2/3} \propto mg. \tag{11.10}$$

This results in the proportionality between R, the spatula radius, and $m^{-1/3}$, the mass. This is obeyed to a first approximation. However, there are other aspects that come into play, as will be seen in Chapter 12.

It was shown (Eq. (11.5)) that the pull-off stress is inversely proportional to R. Thus, the larger the mass of the biological system, the smaller R has to be. This is confirmed by the experimental plot of Fig. 11.6. The number density of attachments, proportional to R^{-2}, increases with the mass. For geckos, which have a mass of approximately 100 g, it is equal

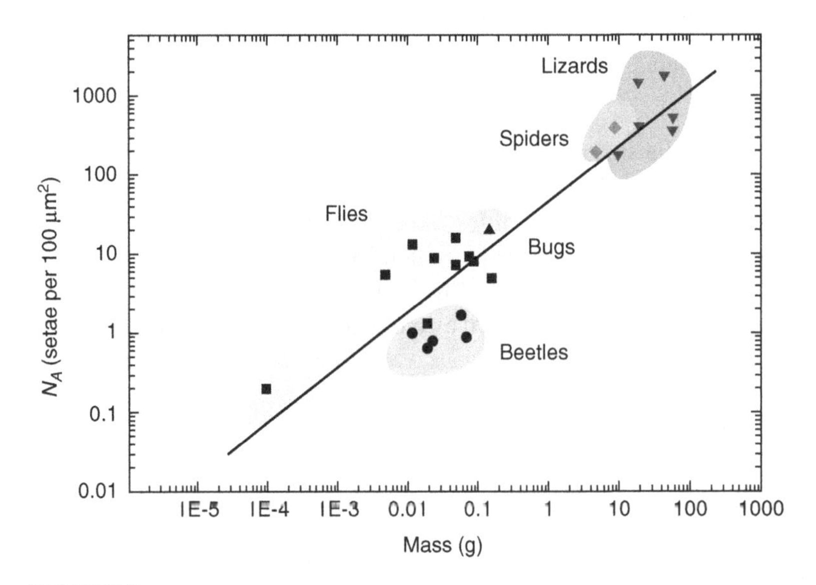

Figure 11.6.

Plot of areal density of setae (number per 100 μm^2 as a function of animal mass). (Reprinted from Arzt *et al.* (2003), with permission; copyright (2003) National Academy of Sciences, USA.)

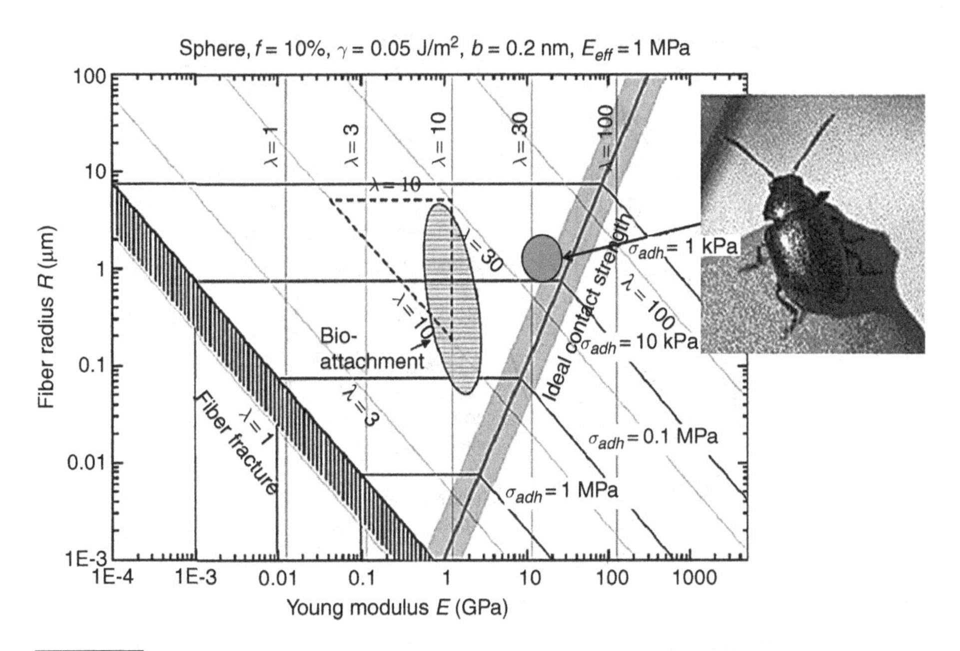

Figure 11.7.

Partial adhesion map for a spherical tip shape; thin lines are contours of equal apparent contact strength; oval section represents the regime of bioattachments. (Reprinted from Spolenak *et al.* (2005b), with permission from Elsevier.)

to 1000 setae/100 μm^2, or 10 setae/μm^2. This is in full agreement with Fig. 11.4, which shows spatulae having an approximate diameter of 0.2 μm.

Spolenak *et al.* (2005b) developed a design map that incorporates both the tensile strength of setae and the ideal contact strength. This plot is shown in Fig. 11.7: it represents the fiber radius along the ordinate plotted against the Young modulus on the abscissa. Two major lines define an inverted cone in which the system should be. The line with negative slope represents the failure of setae by tension, and is obtained from the application of the theoretical strength ($\sigma_{th} = E/10$) to Eqn. (11.5). This results in (assuming $f_s = 1$):

$$R \geq \frac{3\gamma}{2\sigma_{th}} = \frac{15\gamma}{E}. \tag{11.11}$$

The right-hand line represents the ideal contact strength and is given by

$$R \geq kE^2, \tag{11.12}$$

where k is a parameter incorporating several dimensions. Indeed, the biological systems fall within the V region of the plot, showing that the calculations bracket the requirements well.

The biomimicking of this attachment principle is being implemented in synthetic systems. Whereas the paws of a gecko can generate adhesion forces of tens of newtons, it is hoped that much greater forces will be achieved in synthetic systems, and then Spiderman is in the realm of reality (Section 12.1.7).

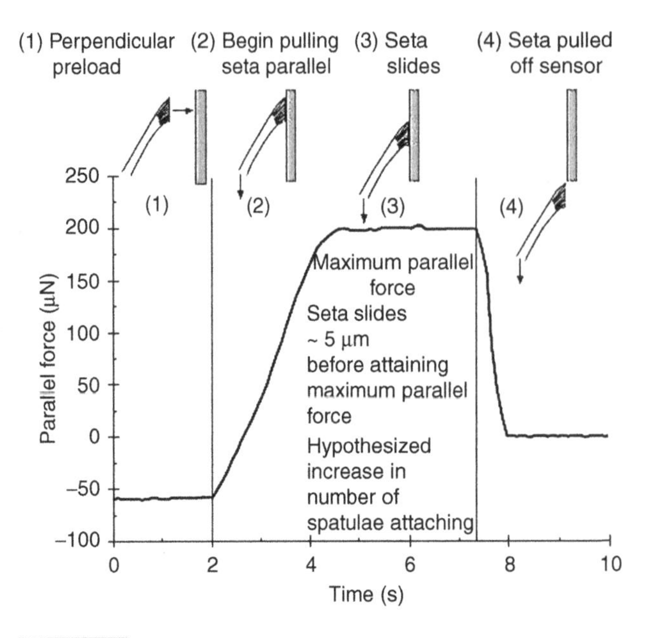

Figure 11.8.

Force of single seta pulled parallel to surface (preload of 15 μN). (Reprinted by permission from Macmillan Publishers Ltd.: *Nature* (Autumn *et al.*, 2000), copyright 2000.)

Figure 11.8 shows one procedure to measure the pull-off force (Autumn *et al.*, 2000). The force on a single seta is applied parallel to the surface, after a preload normal to the surface of 15 μN is applied. At a force of 200 μN the seta slips off. However, van der Waals forces are not the complete story, and capillarity plays a role. The adhesion force exerted by a single gecko spatula (much smaller than a seta!) was measured by Huber *et al.* (2005) after modifying the substrates. The seta of a gecko was glued to an AFM cantilever. Although work by Autumn *et al.* (2000) indicated that the pull-off force did not increase with humidity, this is clearly evident in the work by Huber *et al.* (2005). The results are expressed analytically as follows:

$$F = F_{drag}\left(1 + 1.22\,Hg\sqrt{\frac{A_W}{A_S}}\right), \tag{11.13}$$

where H is the humidity, g is a geometrical parameter (~1.2), and A_W (3.7×10^{-20} J) and A_S (6.5×10^{-20} J) are the Hamaker constants for water and the substrate, respectively.

Example 11.1 A gecko-inspired surface containing semispherical pillars with radius ~500 nm is synthesized. Assume that all the adhesion comes from the van der Waals interaction. Apply the Johnson–Kendall–Roberts equation and estimate the total adhesive force and adhesive stress. Given: adhesion energy per area, $\gamma = 15$ mJ/m^2; pillar density, $\rho = 10^6$ pillars/mm^2; size of the adhesive surface = 1 cm^2.

Solution From the JKR equation,

$$F_c = \frac{3}{2}\pi R\gamma = 1.5\pi \times 500 \times 10^{-9} \times 15 \times 10^{-3} \cong 35\,\text{nN}.$$

The total adhesive force is

$$F_{total} = F_c \times N = F_c \times \rho \times A \cong 35 \times 10^{-9} \times 10^6 \times 10^2 = 3.5\,\text{N}.$$

The total adhesive stress is then

$$\sigma_{total} = \frac{F_{total}}{A} = \frac{3.5\,\text{N}}{1 \times 10^{-4}\text{m}^2} = 35\,\text{kPa}.$$

11.3 Beetles

The relationship developed by Arzt (Arzt *et al.*, 2003; Arzt, 2006) predicts a spatular diameter that decreases with the body mass to the −1/3 power. This is obeyed in an approximate fashion, but there are significant differences between species. Often in a species the spatula diameter is constant in spite of significant changes in the size of the animal. Two examples are provided in Figs. 11.9 and 11.10, for the common beetle and the ladybug, respectively. The rain beetle (*Pleocoma puncticollis*) feet have hooks at the ends; the spatulae have a diameter ~11 μm. The ladybug (*Coccinellidae)* is much smaller, with a mass of ~0.2 g and a density of 0.05–0.2 setae/μm^2, the spatulae have a diameter of ~3 μm. These two examples go against the Arzt equation. Indeed, attachment is a complex biological process involving a number of mechanisms, and different species exhibit different plateaus of response. Figure 11.11 shows spatula radii as a function of body mass for several species (Peatte and Full, 2007). The Arzt equation is superimposed, and its generality and elegance are indeed striking. Qian and Gao (2006) discuss the complexities of attachment: the effect of contributions beyond van der Waals forces, different geometrical relationships between species, differences in spatulae construction and geometry, contributions of capillarity to adhesion, and other factors.

11.4 Tree frog toe pad

The toe pad of the Brazilian tree frog (*Scynax perereca*) provides a splendid example of functional adhesion in nature. This animal, which lives in the moist environment of the subtropical rain forest, is able to jump from surface to surface and attach itself effectively through a variety of electro/mechanical/chemical actions employed by the materials at

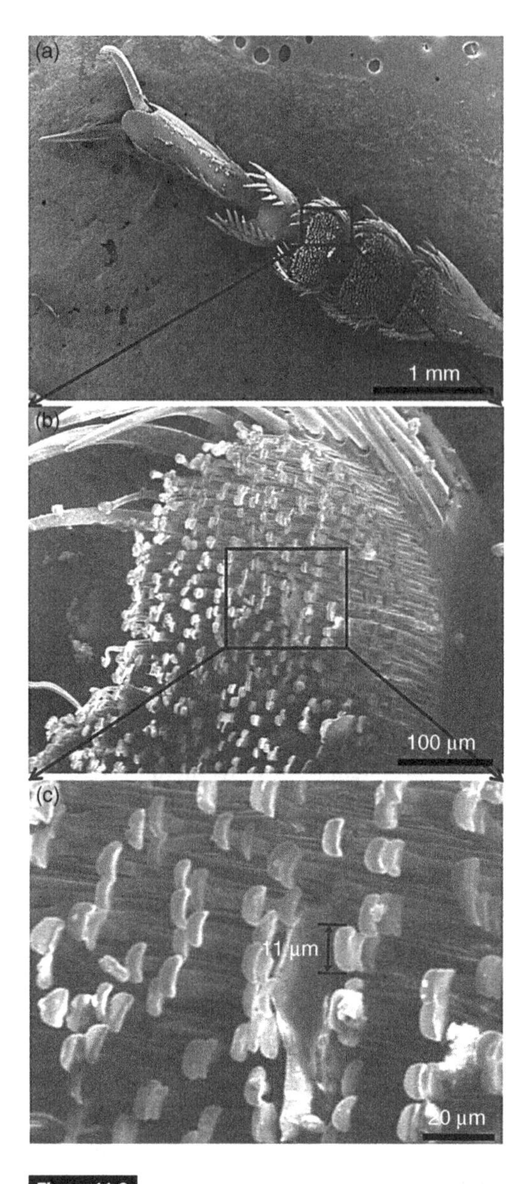

Figure 11.9.

(a) Rain beetle (*Pleocoma puncticollis*) foot with hooks at end. (b), (c) Views at increasing magnification of setae terminating in spatulae with diameter ~11 μm.

the surface of its toe. Its movements are much more dynamic than those of the gecko. We observed that the toe pad of the Brazilian tree frog is also composed of aligned nano-scale fibrils. The fibrils are sectioned into highly ordered hexagonal bundles (Meyers *et al.*, 2011). These bundles, described first by Ernst (1973a, b), Welsh, Storch, and Fuchs (1974) and studied later by many others (Green, 1979; McAllister and Channing, 1983; Green

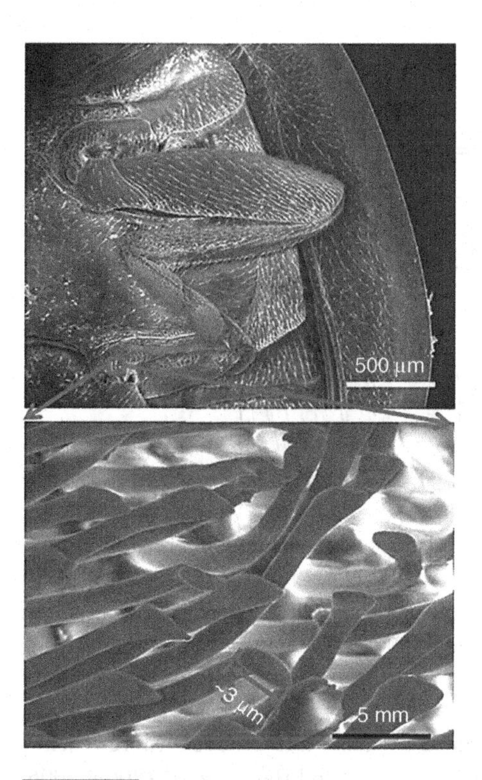

Figure 11.10.

Ladybug attachment device. Top: retracted leg; bottom: setae terminating in spatulae. Diameter ~3 µm.

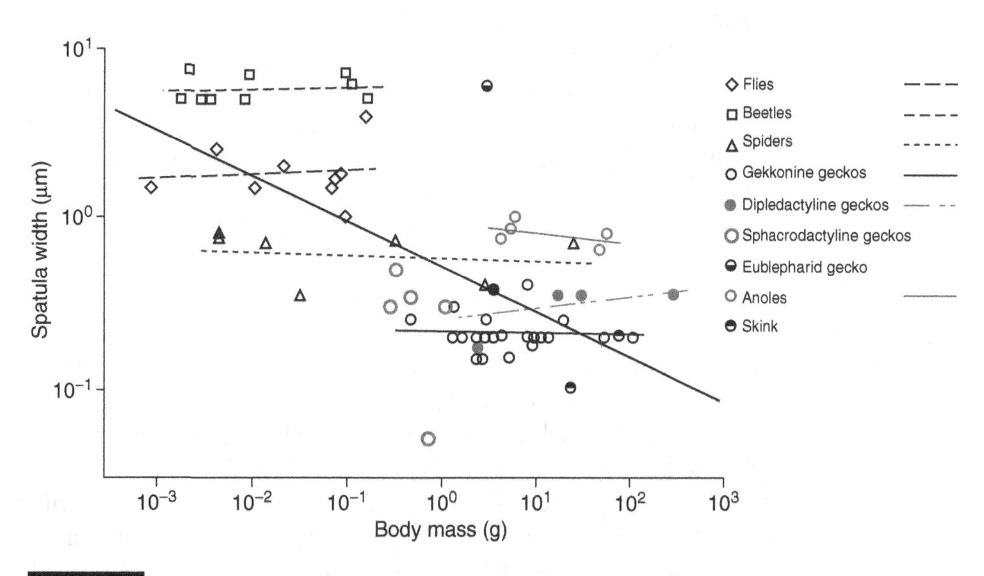

Figure 11.11.

Spatula width as a function of body mass for several species. The straight line with slope equal to −1/3 is the Arzt equation. It tracks the overall dependence well; however, within each species there is a much lower dependency on mass, as shown by the different lines. (Adapted from Peattie and Full (2007), with permission; copyright (2007) National Academy of Sciences, USA.)

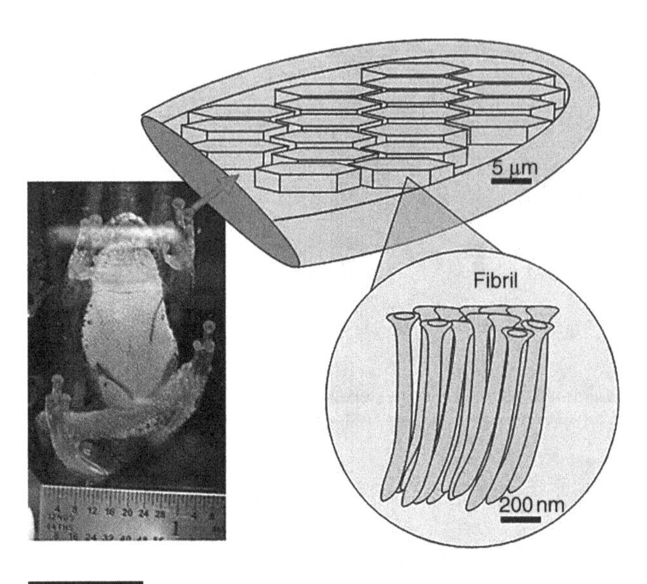

Figure 11.12.

Structural hierarchy found in the toe pad of a Brazilian tree frog (*Scynax perereca*).

and Simon, 1986) are separated by canal-like grooves. More recent studies have been carried out by Hanna and Barnes (1991) and Barnes *et al.* (2005, 2006).

The structure of the toe pad is depicted in the schematic diagram presented in Fig. 11.12. The diagram shows a tree frog attached to a glass substrate, with an illustration of the hexagonal subdivisions approximately 10 μm in diameter. These subdivisions comprise closely packed fibers approximately 100 nm in diameter. Each fiber terminates in cups approximately 200 nm in diameter (Meyers *et al.*, 2011). Each one of these cups is aligned beside its neighbor, forming a flat surface. Figure 11.13 provides a scanning electron micrograph of a single toe at low magnification with an expanded view of the surface of the toe pad showing hexagonal subdivisions. A well-defined circular pad roughly 2.5 mm in diameter can be seen in the center of the toe.

Barnes (2007) reported that the mucous glands excrete a viscous fluid which can be transported through the canals that exist between the hexagonal subsections. It had been suggested that the fluid plays an essential role in adhesion, indicating the dominance of the wet adhesion mechanism. It is proposed here that the contribution of molecular forces through van der Waals interactions between the nanofibril ends and a surface may have a place in the discussion of tree frog toe pad adhesion.

Figure 11.14 provides top and cross-sectional views of an individual bundle. The tightly packed well-aligned fibers comprising these bundles are shown in greater detail in Fig. 11.14(c). The terminating cups on neighboring fibers are aligned closely against each other to create a smooth and consistent surface.

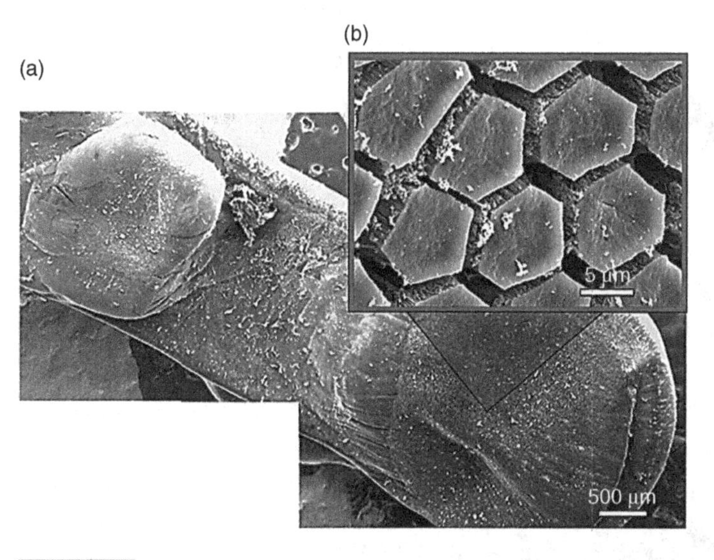

Figure 11.13.

Scanning electron micrograph of the toe pad of a Brazilian tree frog. (a) Low magnification view of single toe; (b) hexagonal subsections found on the contact surface of the toe pad.

11.5 Abalone foot: underwater adhesion

Our group has extended the study of the structure of abalone to the attachment forces required to separate it from a surface (Lin *et al.*, 2009). The detachment stresses were measured on live and healthy abalone and found to be on the order of 115 kPa. The pedal foot of red abalone (*Haliotis rufescens*) is shown in Fig. 11.15. The dark pedal folds, spaced approximately 0.5 mm apart, are the source of locomotion waves used in transportation (Trueman and Hodgson, 1990; Donovan and Carefoot, 1997). These locomotion waves have an analog in materials science in the dislocation. The fold in the pedal propagates along the foot, generating, after its passage, a displacement equal to the length of the surface at the fold. Figure 11.15(b) shows an abalone supporting its own weight via a single contact point (a human finger). Figure 11.16 shows a large-magnification SEM image of the cross-section of the soft tissue. Folds can be seen in greater detail as a mechanism in which the surface area of the foot can expand and contract, allowing an increase or decrease in contact surface area and providing the mechanism for the propagation of waves on the ventral surface of the pedal muscle.

At higher magnification, Fig. 11.17(a) shows setae lining the outer surface of the tissue with a thickness of 1–2 μm. At their extremities, (Fig. 11.17(b)), the setae separate into nanoscale probes with hemispherical ends (Fig. 11.17(c)), averaging 150 nm in diameter and uniaxially aligned perpendicular to the plane of the foot tissue. It is proposed that, as in the case of the gecko, these nanofibrils create intimate contacts at the molecular level to form van der Waals interactions, which can be accumulated into a formidable macro-scale effect.

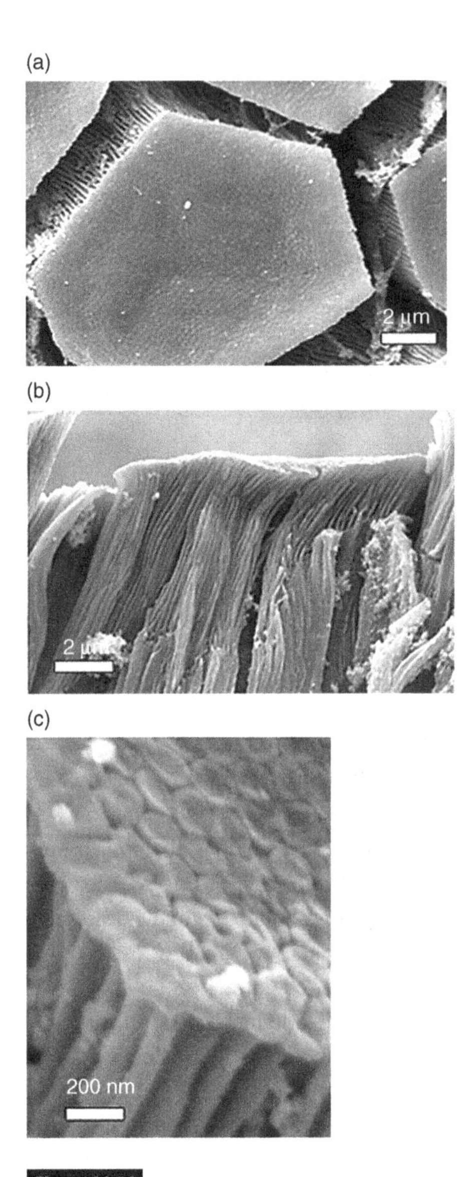

Figure 11.14.

Scanning electron micrographs of hexagonal subdivisions in toe pad of Brazilian tree frog (*Scynax perereca*): (a) top-down view; (b) side view; (c) high magnification showing individual fibers ~100 nm in diameter. (Reprinted from Meyers *et al.* (2011), with permission from Elsevier.)

The influence of a meniscus fluid between a fiber and a substrate is increasingly significant with decreased liquid-surface contact angle, i.e. a hydrophilic substrate would have more capillary interactions than a hydrophobic one. This is clearly seen in Fig. 11.18. In the case of the hydrophobic material (a carbon-coated substrate) the average pull-off force was determined to be 294 nN, remaining constant under varying

(a)

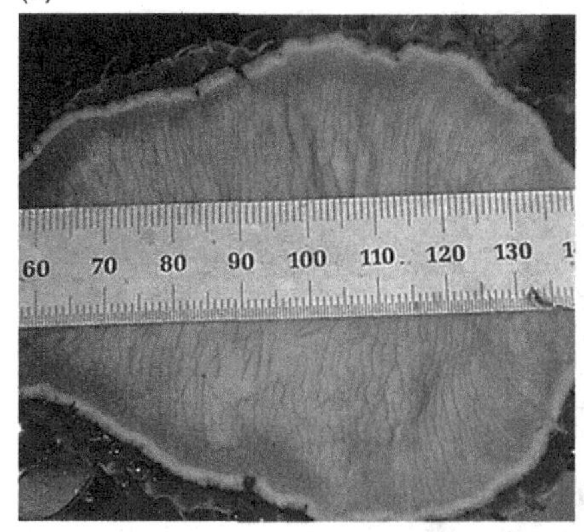

(b)

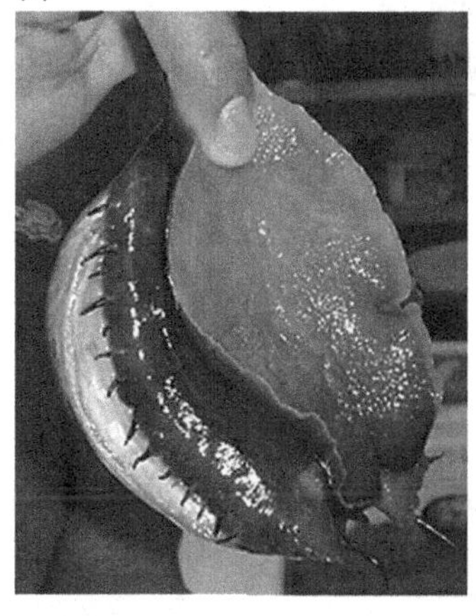

Figure 11.15.

The pedal foot of abalone: (a) optical image of bottom surface of foot; (b) abalone supporting its own hanging weight through a single contact point. (Reprinted from Lin *et al.* (2009), with permission from Elsevier.)

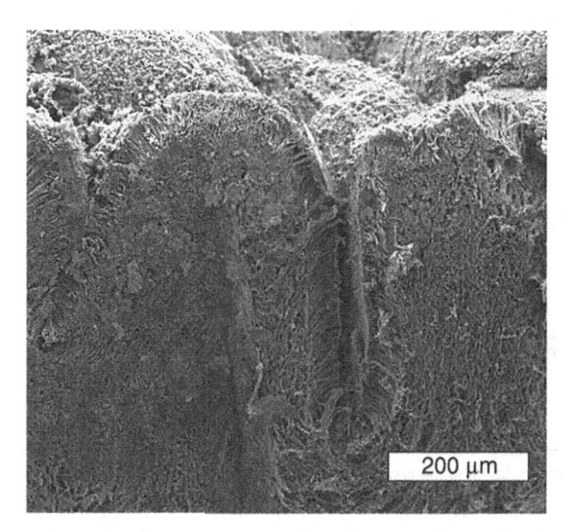

Figure 11.16.

Scanning electron microscopy of foot tissue of abalone cross-section; the top of the image represents the contact surface of the foot. (Reprinted from Lin *et al.* (2009), with permission from Elsevier.)

(a)

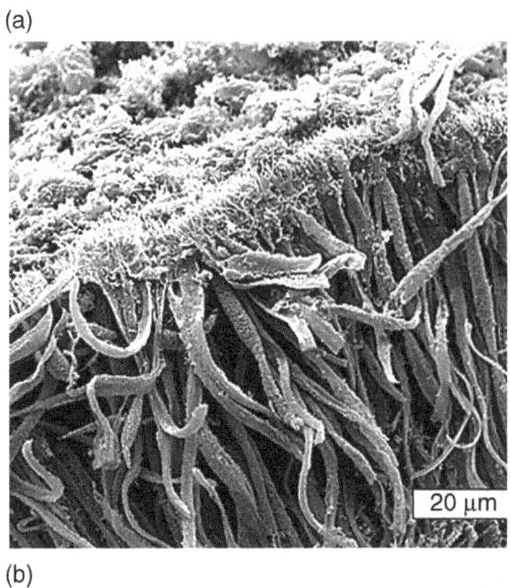

(b)

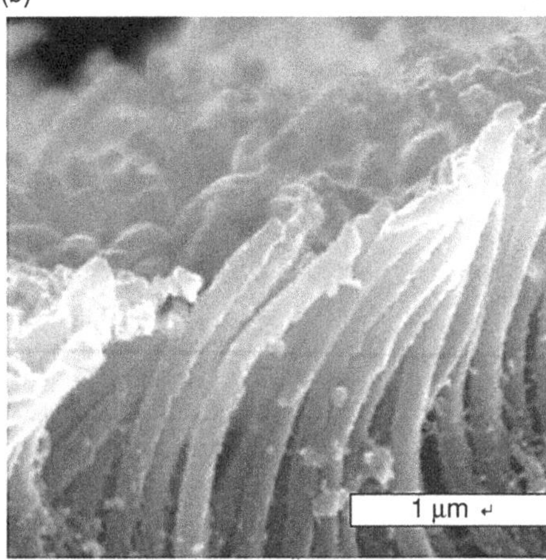

(c)

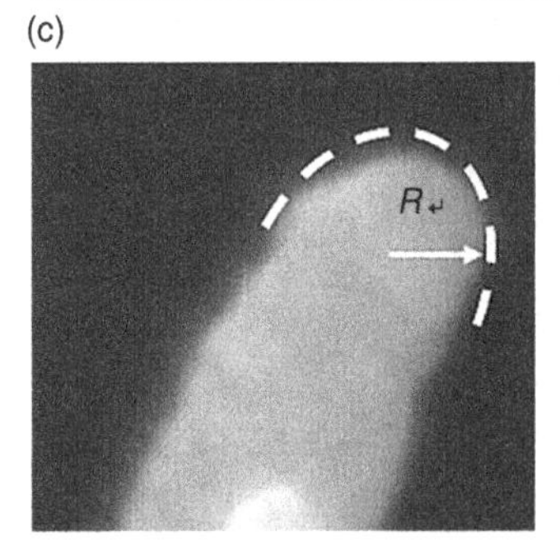

Figure 11.17.

SEM characterization of abalone foot tissue: (a) setae lining the outer surface of the foot; (b) nanofibers uniaxially aligned on setae; (c) single nanofiber with hemispherical tip. (Reprinted from Lin *et al.* (2009), with permission from Elsevier.)

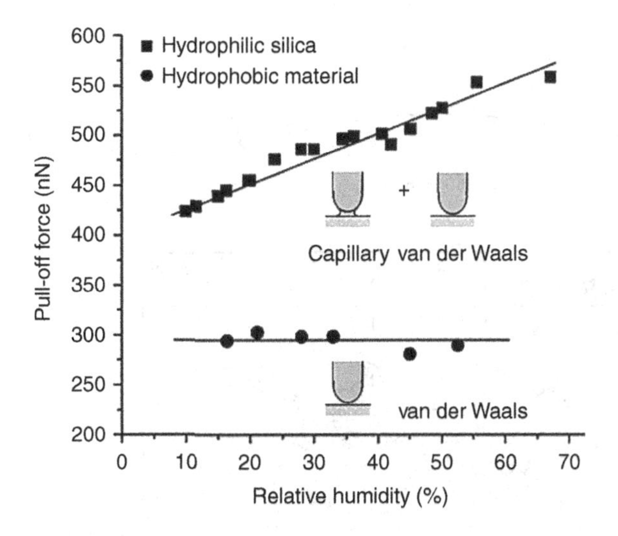

Figure 11.18.

Pull-off force as a function of relative humidity of a single seta on a hydrophobic (carbon-coated) and a hydrophilic (SiO_2) substrate. (Reprinted from Lin *et al.* (2009), with permission from Elsevier.)

humidity. If one assumes that 60 nanofibrils on a single seta are in contact with the surface, this would correspond to an adhesion force of approximately 5 nN per nanofibril. This estimate is in exact agreement with the theoretical results of 5 nN calculated using the Johnson–Kendall–Roberts equation. When the seta was tested on a hydrophilic substrate (SiO_2) at a relative humidity of 10%, the pull-off force was observed to be 424 nN before detachment. This represents an increased force of 130 nN relative to the test on the hydrophobic substrate, which can be partially explained by the variation in surface energies for the two substrates (20 mJ/m^2 and 55.5 mJ/m^2 for the carbon-coated disk and silicon oxide, respectively). However, raising the relative humidity to 67% resulted in an additional increase in pull-off force to 558 nN. Similar to predictions by Autumn *et al.* (2002) and work by Huber *et al.* (2005) for the gecko foot, this shows evidence of capillary interactions. The characterization of the abalone foot pedal and the mechanical tests suggest that the three mechanisms proposed by Barnes (2007) act cooperatively (and perhaps synergistically). Suction can generate attachment forces, as explained schematically in Fig. 11.19(a). It can be shown that the detachment force F_d is equal to

$$F_d = PA, \tag{11.14}$$

where P is the pressure and A is the projected area of the abalone foot on the plane of the surface of attachment. Assuming that the effect of the water column is negligible, i.e. $P = P_{atm}$, we obtain the mean attachment stress:

$$\sigma_d = P_{atm} = 101\,\text{kPa}. \tag{11.15}$$

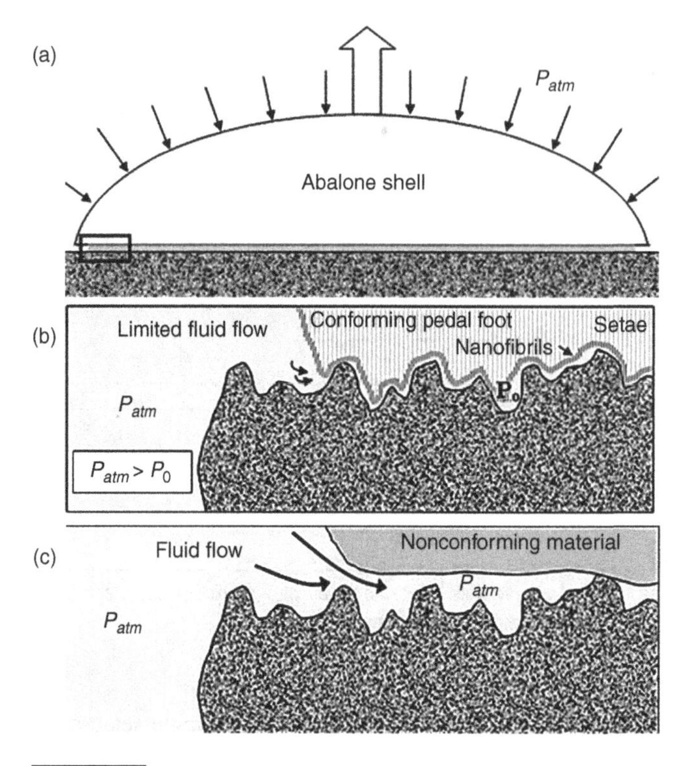

Figure 11.19.

Schematic representation of how suction might generate attachment forces. (Reprinted from Lin *et al.* (2009), with permission from Elsevier.)

Figures 11.19(b) and (c) show schematically how the three mechanisms can operate cooperatively to create the attachment stress on the same order of magnitude as the theoretical suction stress. The setae and nanofibrils maintain intimate contact with any irregular surface, closing possible channels and impeding water penetration. The pressure at the interface, P_0, is equal to P_{atm} when no external detachment force is applied. As F_d increases, P_0 decreases. Once it becomes zero, detachment occurs. Figure 11.19(c) shows the situation for a nonconforming material: a continuous fluid path to the interface region ensures pressure equilibration around the animal and effectively eliminates suction. It is proposed that capillarity and van der Waals forces can maintain the intimate contact between the ventral side of the foot pedal and the attachment surface; in this manner, the suction force can reach and even exceed $P_{atm}A$.

Box 11.1 Spinal plates, cages, and intervertebral disc implants

The spine is an extremely delicate part of the body, and the stresses and strains to which it is subjected are complex. It is said that evolutionary bipedalism has strained the spine beyond its capability. All we have to do is to observe the ease with which our four-legged friends climb and descend hills, hardly ever complaining of back pain! One set of statistics states that 26 million Americans of working age suffer from chronic back pain.

Therefore, back pain is an insidious problem, especially in our sedentary lifestyle. Different surgical procedures have been developed to correct and mitigate problems. Scoliosis, or the lateral curvature of the spine, is a congenital disease that can cause chronic pain, deformity, and other problems.

Spinal implants can be roughly divided into two groups: fixation devices and intervertebral disc implants. Although there are many designs of spinal implants, plates and cages are the most common. A plate is shown in Fig. B11.1. It is screwed to two vertebrae, although a screw could also be added to immobilize the central vertebra.

Spinal cages are introduced into the intervertebral space after the disc is removed. These discs can be damaged or simply wear out. The lumbar (lower back) portion of the spine experiences greater weight and therefore the problem is more prevalent there. This is the infamous "lower back pain," the curse of old professors. The cage is a hollow cylinder the size of the tip of a finger. It is threaded in the outside and has a hollow inside and lateral holes; thus, the name "cage." Two cages are used (one on each side) to re-establish the space between two adjacent vertebrae once the damaged disc is removed. The ends are screwed onto the vertebrae. This ensures space between vertebrae and provides room for the nerves. A cage manufactured with carbon-fiber-reinforced PEEK is commercially available.

A more advanced concept is the replacement of a disc by an intervertebral implant. The concept is to use a flexible material, such as ultra high-molecular-weight polyethylene (UHMWPE) sandwiched between two metal (Co-Cr or Ti) plates that are anchored in the adjacent vertebrae. The advantage of this device over a cage or simple fusing of the vertebrae is the mobility afforded by the polymeric component.

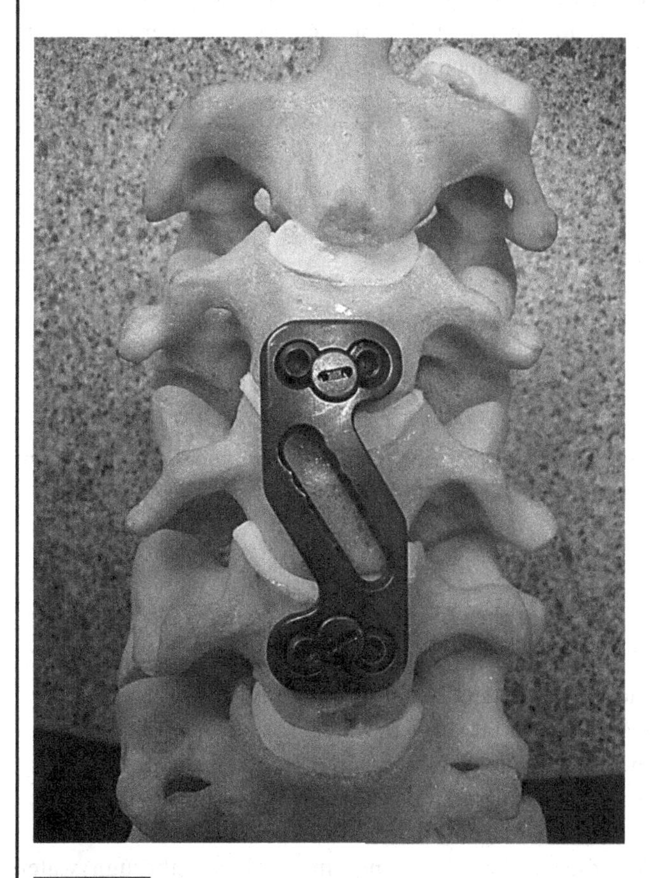

Figure B11.1.

Titanium plate and screws used in spinal fixation; an insertion of a screw in the center of the plate can be added to immobilize it.

11.6 Surfaces and surface properties

11.6.1 Multifunctional surface structures of plants

The surfaces of the plants serve as protective outer coverage and provide multifunctional interfaces to various environments. The surface structures, properties, and functionalities have been reviewed by Koch, Bhushan, and Barthlott (2009). Figure 11.20(a) shows the basic components and the layered stratification of the plant epidermis. The outermost layer is a thin extracellular membrane, called the cuticle. This thin membrane is the multifunctional protective interface between the plant tissues and the environment and is the most important part of the epidermis cell. Beneath the cuticle are a pectine layer, the cell wall, and the plasma membrane. The plant cuticle is a natural composite made mainly of cutin and hydrophobic waxes. The cuticular waxes can be classified into intracuticular and epicuticular waxes. Intracuticular waxes are integrated to the cutin network and epicuticular waxes are exposed on the surface. Whereas the intracuticular waxes function as a transport barrier, preventing water loss and leaching of molecules from living cells, the epicuticular waxes are of great importance as an interface layer. Figure 11.20(b) summarizes the multifunctional properties of the plant cuticle. Depending on the environment and habitat, the plant cuticle can exhibit the following functionalities:

(A) transport barrier;
(B) surface wettability;
(C) anti-adhesion, self-cleaning;
(D) signaling, sensing;
(E) optical properties: protection against harmful radiation;
(F) mechanical properties;
(G) reduction of surface temperature by increasing turbulent air flow.

In the following sections, we introduce plants in wetland and aquatic environments as well as those in hot and arid environments.

11.6.1.1 Lotus leaf

The lotus leaf was studied by Barthlott and co-workers (Barthlott and Ehler, 1977; Barthlott, 1990; Barthlott and Neinhuis, 1997) and recently reviewed by Koch *et al.* (2009). Figure 11.21(a) shows a lotus leaf floating in water. The water does not wet the top of the leaf, but rather concentrates into small areas by virtue of the hydrophobicity of the surface. Drops of water just roll off (as shown on the right-hand side of Fig. 11.21(a)). There are two main reasons that aquatic plants develop a water-repellent surface. First, CO_2 diffuses 10 000 times more slowly through water

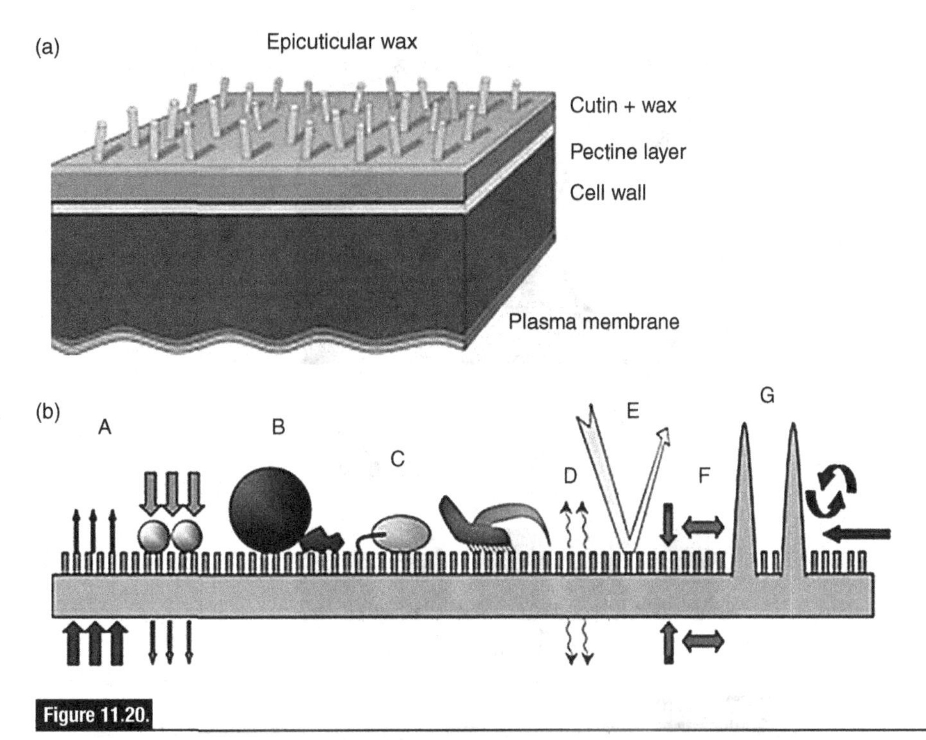

Figure 11.20.

(a) Multilayer structure of plant epidermis cells. The outermost layer is epicuticular wax, a composite made of cutin and wax. Beneath the epicuticular wax are the pectine layer, the cell wall, and the plasma membrane. (b) Most prominent functions of the plant surfaces: (A) transport barrier, (B) surface wettability, (C) anti-adhesive, self-cleaning properties, (D) signaling, (E) optical properties, (F) mechanical properties, (G) reduction of surface temperature by increasing turbulent air flow over the boundary air layer. (Reprinted from Koch *et al.* (2009), with permission from Elsevier.)

than air. A thin layer of water on leaves can significantly reduce the uptake of CO_2, which is necessary for photosynthesis. Second, the water-repellent ability limits the growth of micro organisms (bacteria and fungi) on the surface of the plant.

The surface of the lotus leaf has small pillars of a few micrometers height and spaced ~20 μm apart (Fig. 11.21(b)). These are covered by smaller-scale protrusions, with dimensions ~0.2–1 μm. These are covered with wax. The net result is that the angle of contact between the water droplets and the surface is dramatically increased. Hydrophilic surfaces have contact angles below 90°, whereas hydrophobic surfaces have angles above 90°. For the lotus, this angle can be as high as 160°, and it is therefore called superhydrophobic. This can be expressed by the equilibrium equation for interface tensions, $\gamma_{SG} = \gamma_{SL} + \gamma_{LG} \cos\theta$, where γ_{SG}, γ_{SL}, and γ_{LG} are the solid–gas, solid–liquid, and liquid–gas interfacial energies. This hydrophobicity has an important effect: water droplets, almost perfect spheres, can roll over the surface, picking up dust particles which decrease their surface energy by being absorbed into the water droplets. Thus, the surface cleans itself. In

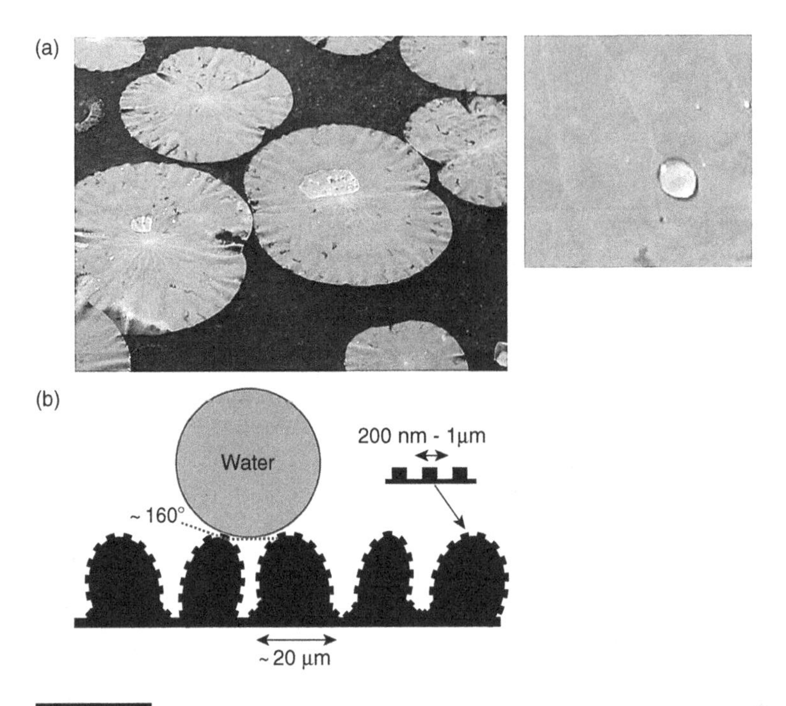

Figure 11.21.

The lotus leaf effect: (a) lotus leaf on water, which does not get wet; (b) surface of lotus leaf showing two scales of protrusions, which lead to an angle greater than 90°. (Reprinted from Meyers *et al.* (2011), with permission from Elsevier.)

1999, a commercial and very successful product was launched: the facade paint Lotusan has been applied on more than 500 000 buildings worldwide so far. Other applications are self-cleaning glass installed in the sensors of traffic control units on German highways and coatings applied to microwave antennas, which help to keep them dust free and decrease the buildup of ice and snow.

11.6.1.2 Plants in hot and arid environments

Plants in hot and arid environments (deserts, for example) have to adapt to extreme conditions, such as low water availability, high temperature, intense solar radiation, and sand abrasion. These extreme environmental conditions require specific physiological and/or morphological adaptations of the plants. One of the most classic and well-known examples is the cactus, which has needle-like spines that help prevent water loss by reducing air flow close to the cactus and providing some shade. Other important surviving strategies are surface modifications for efficient uptake and storage of water.

Some plant species are able to absorb water via the porous surface of their thorns. Figures 11.22(a) and (b) show the porous thorn surface of *Pelecyphora aselliformis*. Water storage can be achieved by developing water-storing cells and tissues. The water-storage vesicles of *Mesembryanthemum crystallinum* are shown in Figs. 11.22(c) and (d). These vesicles are large convex cells, several hundred micrometers in diameter, and can be seen with the naked eye.

There are various strategies utilized in plant leaves for the reduction of water loss. Water loss occurs through the cuticle, and most water evaporation occurs through the stomata, which have the main function of gas exchange. Desert plants show a reduction in the number of stomata per leaf area compared to those from habitats with more rainfall, to reduce the water loss. In succulent plants, stomata are often sunk into pits of the epidermis to reduce air turbulence and minimize evaporation during gas exchange. The sunken stomata of succulent cacti (*Rhipsalis*) are demonstrated in the SEM micrograph of Fig. 11.23(a). In aloe vera, the stomata are superimposed by epicuticular waxes, which significantly reduce the water loss, as shown in Fig. 11.23(b). *Colletia cruciata* has wax chimneys on the surface, forming a three-dimensional cave over the stomata (Fig. 11.23(c)). In order to reduce the uptake of radiation energy and prevent harmful solar radiation, some desert plants are covered by reflective structures, typically three-dimensional waxes or a dense layer of air-filled hairs. The

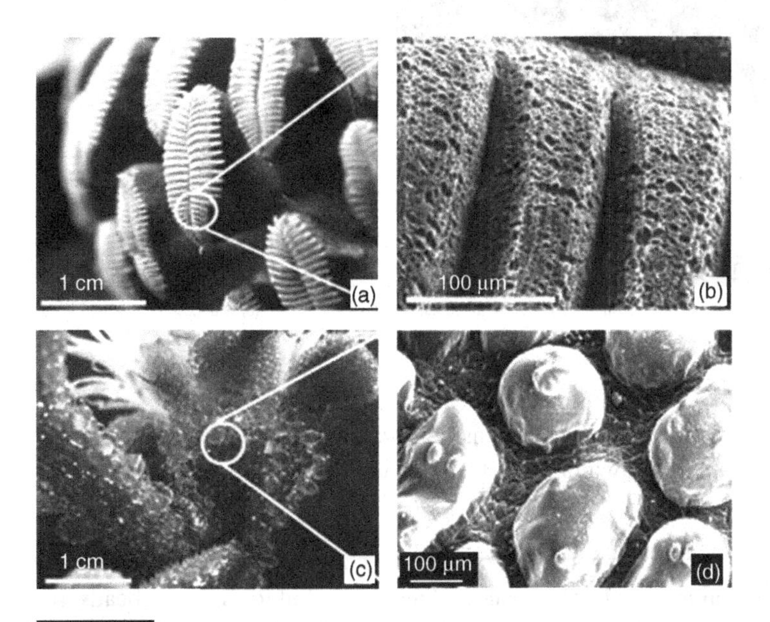

Figure 11.22.

(a), (b) Porous thorns for water absorption (*Pelecyphora aselliformis*). (c), (d) Water-storage vesicles (*Mesembryanthemum crystallinum*). (Reprinted from Koch *et al.* (2009), with permission from Elsevier.)

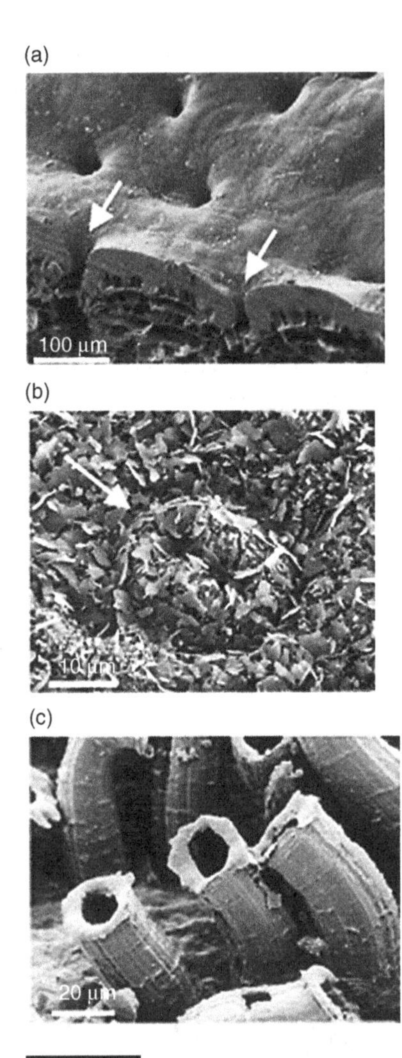

Figure 11.23.

Strategies to reduce water evaporation: (a) sunken stomata (*Rhipsalis*); (b) wax-covered stomata (aloe vera); (c) wax chimneys (*Colletia cruciata*). (Reprinted from Koch *et al.* (2009), with permission from Elsevier.)

microstructural three-dimensional waxes of *Kalanchoe pumila* and the hairy waxes of *Salvia* are shown in Figs. 11.24(a) and (b), respectively.

Example 11.2 If a car is waxed, the surface no longer wets but forms water beads. If the surface energy of wax is equal to 50 mJ/m^2 and the interfacial energy with water is 72 mJ/m^2, determine the angle that the droplets will form with the surface along the edge of the contact area.

(a)

(b)

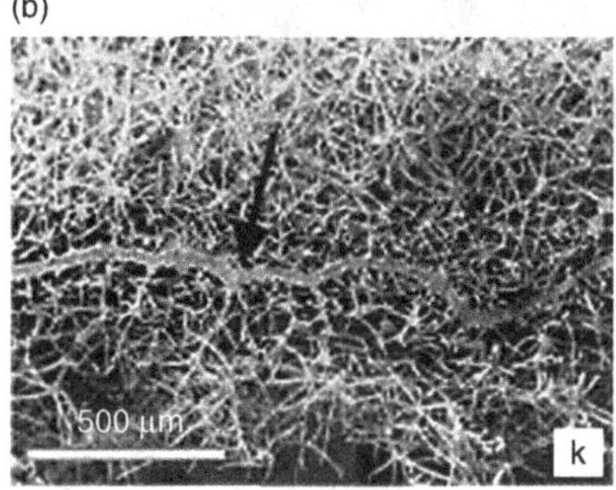

Figure 11.24.

Reflective surface structures prevent plants from harmful solar radiation: (a) three-dimensional waxes and wax crusts (*Kalanchoe pumila*); (b) dense layer of air-filled hairs (*Salvia*). (Reprinted from Koch *et al.* (2009), with permission from Elsevier.)

The surface tension of water is equal to 70 mJ/m^2. Explain how this angle can be increased to 160° in the lotus leaf.

Solution We have the following expression for the angle:

$$\cos\theta = \frac{\gamma_{SG} - \gamma_{SL}}{\gamma_{LG}} = \frac{50 - 72}{70}.$$

Thus: $\theta = 108°$.

The angle can be increased to 160° by introducing two levels of rugosity in the surface.

11.6.2 Shark skin

The skin of shark has some unique characteristics: its surface is covered by dermal "denticles" (placoid scales, shown in Section 8.5), which are optimized for minimizing resistance to flow, and they are singularly devoid of parasites and bacteria. Figure 11.25 shows the morphology of the dermal denticles, with micrometer-sized ridges. These

Figure 11.25.

Microscopic morphology of shark dermal denticles showing parallel ridges. (Used from http://australianmuseum.net.au/Placoid-scales/ Sue Lindsay © Australian Museum, with kind permission.)

ridges prevent the creation of eddies in the boundary layer, thereby reducing drag. The inspiration of shark skin has led to drag-reducing swimming suits, anti-bacterial films for biomedical application, and anti-fouling coatings on boats.

11.7 Optical properties

11.7.1 Structural colors

Structural color is a phenomenon of the wave properties of light. Biological systems are able to produce structural color using highly precise and sophisticated nanometer-scale architectures (Vukusic and Sambles, 2003). The coloration of birds has attracted many scientists and is related to sexual signaling, as is obvious in the case of nuptial plumage change, camouflage, and aggression. The mechanisms of coloration to be discussed are chemical or pigmentary and physical or structural. Among structural colors, scattering by photonic crystal arrays and interference in thin films will be addressed.

Pigment is greatly responsible for the coloration of birds (Bereiter-Hahn, Matoltsy, and Richards, 1986). Melanin is a pigment responsible for black, gray, or brown. Melanin is secreted by melanocytes embedded in the epidermis. It also serves to increase the hardness of bill keratin, as Bonser and Witter (1993) presented in the case of the seasonal change of bill color in the European starling (*Sturnis vulgaris*). Among other scenarios, Bonser and Witter proposed that the harder melanic bill keratin may resist increased abrasion and wear experienced during foraging in winter months. Carotenoids are diffuse organic pigments of which there are two classes: xanthophylls, or oxygen-containing, carotenoids, responsible for yellow coloration and carotenes; or oxygen-free carotenoids, producing red coloration. In avian coloration, however, there is no blue pigment. Blue and (generally) green colors are possible due either to structural color or to the interaction between carotenoid pigments and structurally produced color.

11.7.2 Photonic crystal arrays

Photonic structures are widely distributed in nature, such as in feathers, scales, or insect cuticles. In birds, a blue color is produced when light is scattered coherently from an array of melanin granules suspended in a matrix of keratin and air vacuoles in the feather parts, for example. Iridescence such as that observable in peacock (*Pavo muticus*) feather or humming bird throat patch is produced by interference. Figure 11.26(a) shows scanning electron micrographs of peacock barbules (Zi *et al.*, 2003). The barbules are composed of medullary core bound by a melanin-containing cortex encased by a thin keratinous cuticle. In this case, a periodic array of melanin rods is observable in the cortex; see Fig. 11.26(b). By varying the lattice constants in simulation of the two-dimensional photonic-crystal-like structure, Zi *et al.* were able to vary the wavelength of the scattered light.

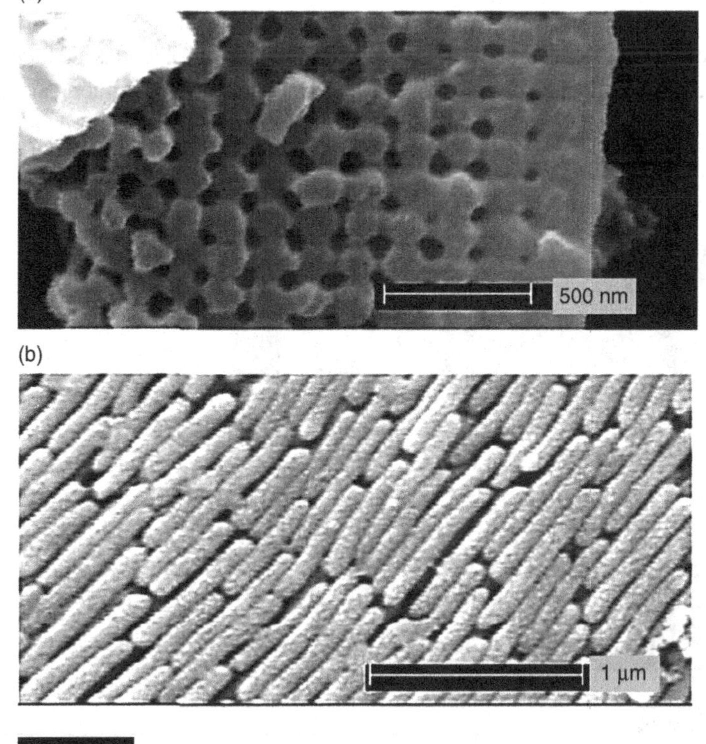

Figure 11.26.

(a) SEM of barbules of peacock; (b) SEM of melanin rod embedded in keratin layer. (From Zi *et al.* (2003),with permission; copyright (2003) National Academy of Sciences, USA.)

Structural coloration is not limited to the class of *Aves*; it is observed in fish and in insects such as butterflies (*Lepidoptera*), or beetles (*Celeoptera*). One of the most striking examples of color-producing structures is that present in the *Lepidoptera* (butterflies and moths) species, including the tropical *Morpho* family (Ghiradella, 1991; Nassau, 1998, 2001; Lawrence, Vukusic, and Sambles, 2002). The *Morpho* butterflies are among the largest in the world, with a wingspread of 7.5 to 20 cm. The males of *Morpho menelaus* have brilliant blue coloration on the dorsal side wings and a camouflage brown color on the ventral side of the wings. Such designs help them to elude potential predators, for example birds. When they fly, the top surfaces of the wings continuously change from metallic blue to dull brown as the angle of the light striking the wing changes. They seem to disappear and then reappear a distance away during flight. Coupled with the unpredictable pattern of flight, the ability to change color makes them difficult for predators to pursue. When the *Morpho* butterflies land, they close their wings completely, showing the camouflaged brown underside of wings that matches the surrounding environment and protects them from being found.

The blue color of the *Morpho* butterfly comes from the well-defined structure of the wings. This is another example of hierarchical structure from nanometer, to micrometer, to millimeter scale found in nature (Fig. 11.27; see Vukusic and Sambles (2003)). The wings of butterflies and moths consist of a translucent membrane covered by a layer of scales. Butterfly scales are composed of chitin, which is a white polysaccharide widely found in arthropod cuticles (Ghiradella, 1991). Each scale (Fig. 11.27(b)) is a flattened outgrowth of a single cell that fits into a socket on the wing and is about 100 μm long and 50 μm wide. The scales overlap like roof tiles and completely cover the membrane, appearing as dust

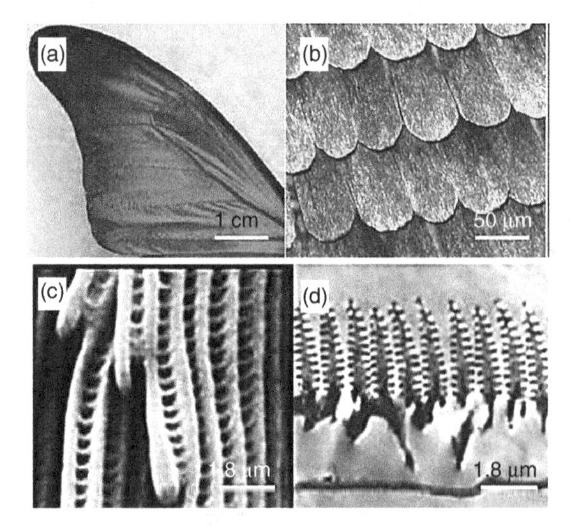

Figure 11.27.

Hierarchical structure of the wing of the butterfly *Morpho rhetenor*. (a) wing; (b) scales; (c) ridges (top view); (d) ridges (cross-section). (Reprinted by permission from Macmillan Publishers Ltd.: *Nature* (Vukusic and Sambles, 2003), copyright 2003.)

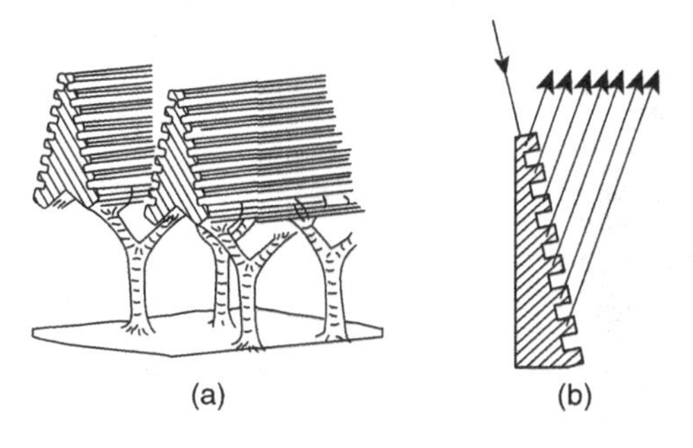

Figure 11.28.

(a) Structure of vanes on the surface of the scales of *M. rhetenor*. (b) Constructive interference in the ridges. (Reproduced based on Nassau (2001).)

to the naked eye. Each scale is covered by longitudinal ridges joined at intervals by cross-ribs, as shown in Fig. 11.28(a) (Nassau, 2001). Ridges and cross-ribs frame a series of windows that open into the interior of the scale, where the pigment granules are located. For more radiant color observed in other species, the ridges are much higher and have a very precise nanostructure. The cross-sectional view of the ridges reveals the discretely configured multiple slits, which are spaced 200 nm apart. Interference occurs when light waves striking the wing interact with light waves reflected by the wing. Though sunlight contains a full range of wavelengths, only high-energy waves survive scattering events. The periodicity of the nanostructure defines a reflected wavelength of blue light, ranging from 400 to 480 nm (Nassau, 1998, 2001). Figure 11.28(b) (Nassau, 2001) shows the detailed grating responsible for interference. There is frequently a background layer of a dark pigment such as melanin which absorbs low-energy waves, perhaps serving to intensify the reflected blue color.

11.7.3 Thin film interference

As another example of structural coloration, the feather of the Hadeda Ibis, *Bostrychia hagedash*, has been studied (Brink and van der Berg, 2004). At first glance, the plumage of this species appears dull brown in color. However, the color appears to "change" from blue to green to reddish. Interestingly, the Hadeda Ibis plumage also reflects ultraviolet and infrared, while it is known that only one animal, the common goldfish, can detect both. Figure 11.29(a) shows a scanning electron micrograph of an iridescent feather from the Hadeda Ibis, and Fig. 11.29(b) is a transmission electron micrograph of a cross-section of a single barbule. The barbules are rotated by 90° so that the flattened surface reflects light normal to the contour of the Hadeda Ibis's body. The thickness of a single barbule is 0.8 μm. Hollow melanosomes are visible in barbules via TEM. Unlike other bird species, in the Hadeda Ibis melanosomes only play a minor role, only to define the thick keratinous cortex in the production of structural colors. Interference at the

(a)

(b)

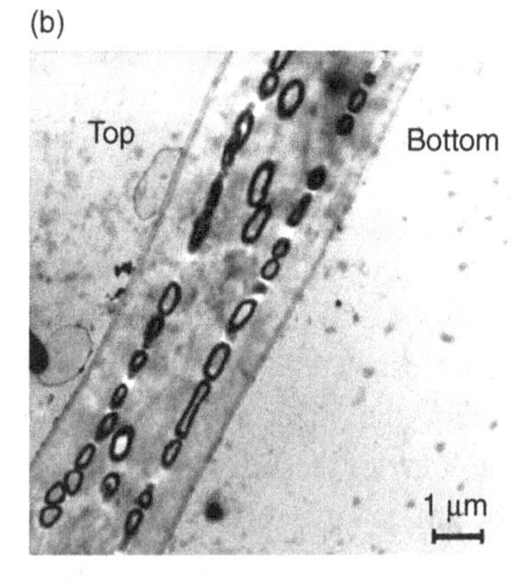

Figure 11.29.

(a) Scanning electron micrograph of Hadeda Ibis feather; (b) transmission electron micrograph of cross-section of a barbule. (From Brink and van der Berg (2004), with kind permission of D. J. Brink and N. G. van der Berg.)

boundaries of the unusually thick keratinous cortex is responsible for the iridescence of the feather (Brink and van der Berg, 2004).

Structural colors in biological systems are due to thin film interference involving a well-defined multilayer structure or else coherent scattering from photonic arrays. The layers may be composed of chitin, as in the wings of butterflies, and keratin, as in avian plumage coloration, as discussed. Other examples of structural coloration in nature include calcium carbonate films in mother of pearl, chitin films in the iridescent cuticle of beetles, and melanosome platelets in humming birds (Nassau, 1998).

11.7.4 Chameleon

We have seen earlier that alpha keratin is mammalian, whereas beta keratin is avian; the epidermis of the reptilians is unique, consisting of both alpha and beta keratin. Metachrosis (changing color) in lizards is the most famous and complex feature in the reptile family (Zoond and Eyre, 1934). The structure of chameleon skin consists of several layers. The thin outermost layer of oberhautchen (from the German, meaning "small surface skin") consists of cornified cells with spinules throughout its surface. This

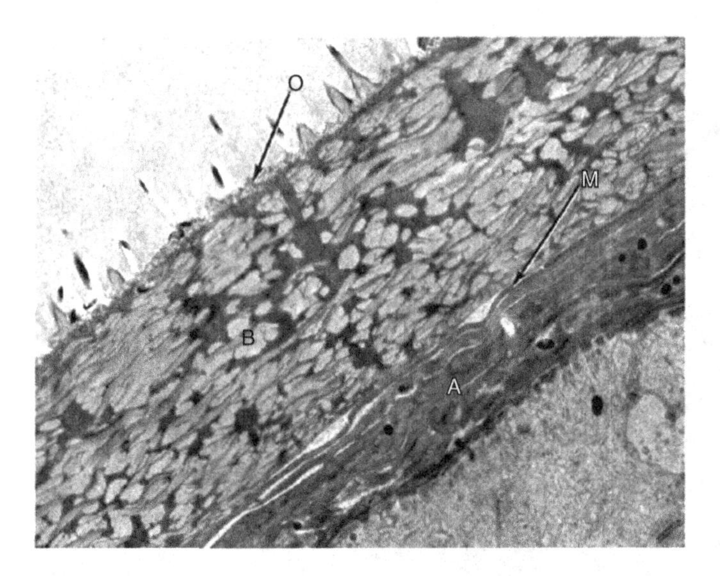

Figure 11.30.

Cross-section of epidermis of American chameleon. O: oberhautchen layer; B: beta layer; M: mesolayer, A: alpha layer. (From Alexander and Parakkal (1969), with kind permission of Springer Science+Business Media B.V.)

covers a thicker beta layer which decreases in thickness along its hinges (Alexander and Fahrenbach, 1969; Alexander and Parakkal, 1969; Alexander, 1970). Figure 11.30 shows the cross-section of the epidermis of the American chameleon (*Anolis carolinensis*) and the meso layer between the α-layer and β-layer (Alexander and Parakkal, 1969). Below the α-layer, the structure is similar to a mammalian epidermis. The chameleon has pigment-containing cells called chromatophores, embedded in the dermal layers. The chromatophores allow the chameleon to change its skin color. Two or three kinds of pigments have been detected in the body of the American chameleon (Carlton, 1903), while the African chameleon has five pigments. Figure 11.31 shows the vertically sectioned back scale of the American chameleon (Alexander and Fahrenbach, 1969). The hexagonal scale of the epidermis varies with position from 250 μm to 400 μm. There are two types of chromatophores containing yellow pigment (xanthophore) and red pigments (erythrophore). The xanthophore layer lies under the stratum germinativum layer, providing 10 μm of the dermal layer (Alexander and Fahrenbach, 1969). There is a 10 to 20 μm thick iridophore layer below the xanthophore layer. This contains inorganic crystalline pellets that reflect blue or white light. Erythrophores have been found in the basal zone of the iridophore layer of the American chameleon (Alexander and Fahrenbach, 1969). The crystalline structure of iridophore platelets yields blue-green light (Rohrlich and Rubin, 1975). The melanophores, which contain melanin, are the largest of the chromatophores. These produce black or brown colors. This layer is followed by the iridophore layer, which meets the xanthophore layer (Alexander and Fahrenbach, 1969). The collagenous basement lamella lies at the bottom of the dermal layer. The skin color is controlled by the expansion or contraction of the

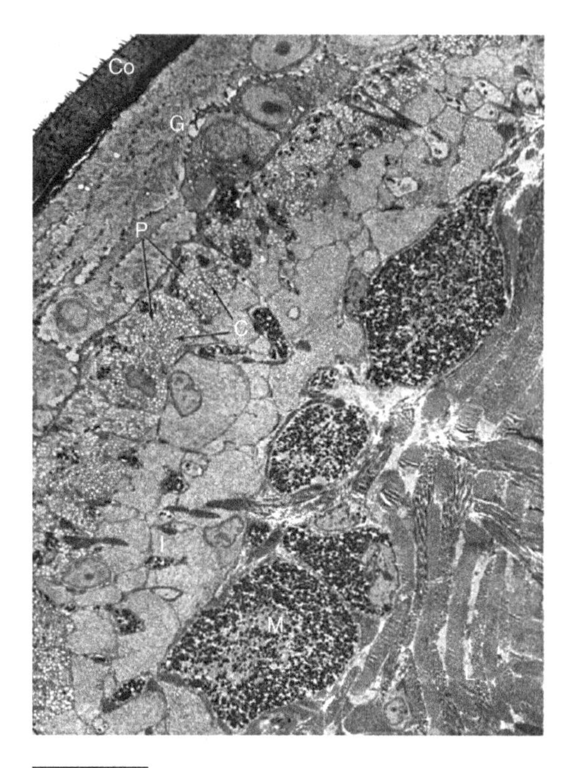

Figure 11.31.

Cross-section of integument of American chameleon. Co: stratum corneum; G: stratum germinativum; P: xanthophore layer; C: cartenoid containing cell; I: iridophore; M: melanophore layer; L: collagenous basement lamella. (Reprinted from Alexander and Fahrenbach (1969), copyright 1969. This material is reproduced with permission of John Wiley & Sons, Inc.)

chromatophores, producing a variety of colors from the different combinations of chromatophores. The light modulation of color is provided by the melanophores. Although the chameleon does not contain any green pigmentation, the yellow pigment and the reflecting blue light at the iridophore layer produce the green color of the skin.

11.7.5 Echinoderms

Both the brittle sea star and the sea urchin are echinoderms. They are composed of the calcite form of calcium carbonate. There are two unique aspects of these echinoderms that are inspiring researchers.

The brittle-star surface contains protuberances that act as lenses. Figure 11.32 shows the wavy surface. The brittle-star possesses an internal skeleton of calcite plates and a complex vascular system. It has five arms. The surface is parallel to the rhombohedral *c*-axis. For this orientation there is perfect refraction. These lenses, which have

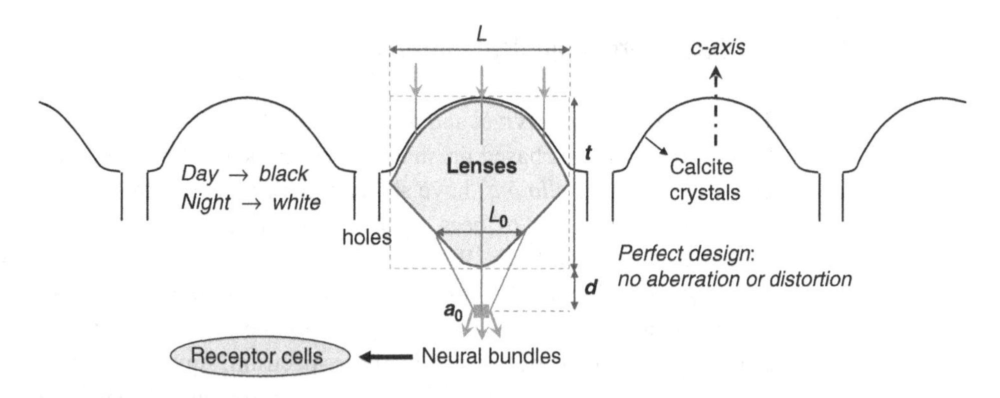

Figure 11.32.

Surface of brittle-star showing lenses and holes through which pigment is released.

Figure 11.33.

Sea urchin surface consisting of spines and microflorets (pedicullaria); note larger spines. (From Aizenberg (2010). Copyright © 2010 Materials Research Society. Reprinted with the permission of Cambridge University Press.)

dimensions of ~10–40 μm, are perfectly designed and there is no aberration. The parallel light rays penetrate the surface and converge to the focal point of the lenses, where neural bundles capture the energy. The valleys have holes, whose function is to release the pigment that coats the brittle-star during the day, making it black. At night, the brittle-star is white.

The sea urchin is covered with calcite spines that can pivot around their attachments, as shown in Fig. 11.33. Under these spines is a network of microflorets that is thought to protect the surface from parasites by continuously moving.

11.8 Cutting: sharp biological materials

There are important survival and predation mechanisms in a number of plants, insects, fishes, and mammals based on sharp edges and serrations. Some plants (e.g. pampas grass, *Cortaderia selloana*) have sharp edges covered with serrations. The mosquito proboscis and stingers of bees are examples in insects. Serrations are a prominent feature in many fish teeth. We discuss the teeth of the piranha and various sharks, focusing on the hierarchical aspects. The chiton radula is composed of magnetite, the hardest biomineral with Vickers hardness number, VHN = 9–12 GPa (Weaver *et al.*, 2010, p. 643). On the opposite side of the spectrum, squid do not have minerals but do have sharp beaks that use an ingenious tanning process, providing a graded structure with VHN up to 5 GPa. Rodents have teeth that are sharpened continuously, ensuring continuous efficacy. Insect stingers and ovipositors, which penetrate into the dermis and wood, respectively, are also presented.

We illustrate these unique aspects by focusing on one characteristic of biological materials: their ability to puncture, cut, and shred. The fact that serrations and needles are present in many species and in diverse configurations is direct evidence that they developed independently, by a mechanism that biology calls convergent evolution. This chapter contains extensive material from Meyers *et al.* (2008c).

11.8.1 Plants

Figure 11.34 shows a blade of pampas grass (*Cortaderia selloana*) with serrations along its outer edge. Each serration is in the shape of a thorn protruding upward along the side of the blade. The serrations extend approximately 50 μm from the body of the leaf and form sharp points with an apex angle of roughly 20°. This sharp cutting edge was evolutionarily designed as a defense mechanism against grazing animals. This feature is also prominent in other grasses, such as *Hypolitrium sharaderenium*.

Figure 11.34.

Pampas grass leaf; note serrations at edges. These serrations, with approximately 150 μm barbs, protect grasses against herbivores.

Another example can be found in cacti, the bodies of which are covered in thorns for protection.

11.8.2 Fish teeth

We discuss a few fish teeth with emphasis on their function. The "big teeth" of the Amazon dogfish have a piscivorous evolutionary design. They are used to puncture and hold prey and are thus designed in a hook-like fashion facing inward toward the mouth of the fish. This can be seen in Fig. 11.35.

Figure 11.36(a) shows the structural hierarchy of the cutting mechanisms found in the jaw of a piranha (*Serrasalmus manueli*). The jaw is designed with sharp triangular teeth aligned so that as the mouth of the fish closes the initial points of puncture of both the lower and upper jaws are superimposed. Each tooth exhibits microserrations along its cutting edge, seen in the detail of Fig. 11.36(a). These serrations, which are approximately 10–15 μm in wavelength, are used to create a highly efficient cutting effect which converts some of the dragging force into a normal force at localized points. As the jaw closes further, any tissue caught in the trough of the aligned teeth is trapped in a guillotine-like confinement of teeth. This is shown in Figs. 11.36(b) and (c). There is a superimposed compression stress P and a shear stress S which effectively cut through the skin and muscle fibers. The mechanics of cutting is rather complex and is presented by Atkins (2009).

The teeth of the great white shark (*Carcaradon carcharias*) evolved from the scales of its antecedents. The shark uses these extremely sharp teeth to perform a very specific

Figure 11.35.

Dogfish and teeth. A unique characteristic of this fish is that the teeth from the lower mandible penetrate through the skull. Their tips actually protrude out of the skull when the mouth is closed. They do not have serrations because this fish is primarily piscivorous.

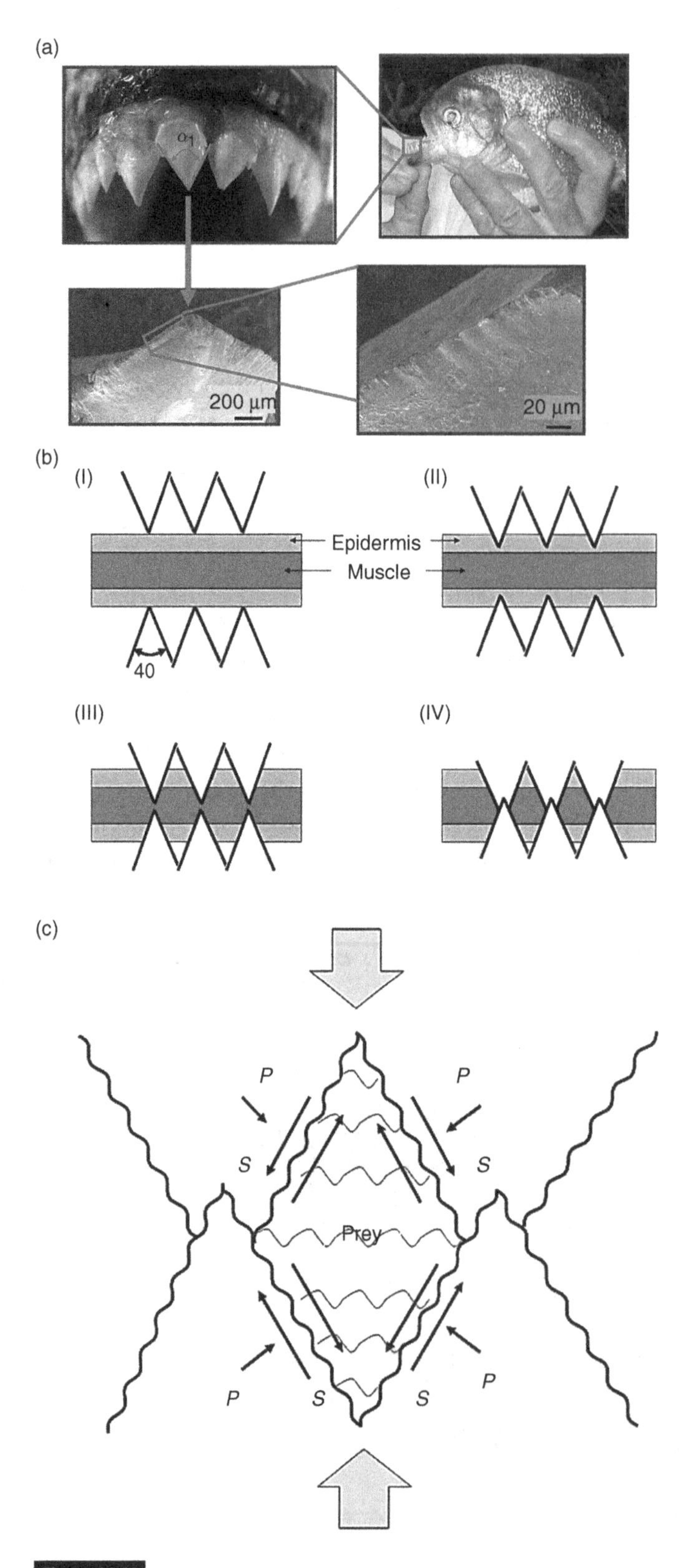

Figure 11.36.

Piranha, teeth, and biting action. (a) Progressively higher magnifications showing SEM micrographs of a single tooth; the higher magnification micrograph on the right shows small serrations on cutting edge. (b) Diagram of guillotine-like confinement of material during the biting action. I–IV: teeth closing and slicing through epidermis and muscle. (c) Applied force from jaws generating pressure *P* and shear stress *S*.

function: to cut through the skin and flesh of mammals. To avoid self-injury, the great white shark takes one efficient large bite out of its prey then retreats and waits for its victim to undergo shock or hemorrhage before final consumption. This bite takes only one second to complete (Diamond, 1986), and thus extremely sharp teeth are required. Each tooth is outfitted with a line of large serrations, with up to 300 μm between points. The serrations are perfectly aligned along the cutting edge of the tooth, each creating a mini tooth on the side of its parent tooth. Similar to the piranha tooth, the serrations on this edge maximize the efficiency of the drag force and convert it into points of normal force summed along the side of each serration. Figure 11.37 shows an optical image of the overall jaw of a great white shark, with multiple rows of teeth and scanning electron micrographs of the cutting edge of the tooth with large serrations and a higher magnification view of the serrations. It is interesting that these serrations are well represented by a cycloid equation, which is shown in Fig. 2.9(a) in comparison with the actual profile of the tooth. Equations (2.8) and (2.9) describe the trajectory. The cycloid is a curve generated by a point on the surface of a circle rolling on a plane. The radius of the circle that defines the cycloid is r, and x and y are the coordinates of the points on the cycloid.

In contrast with the great white shark, there are no serrations on the edge of the shortfin mako shark (*Isurus oxyrinchus*) tooth (Frazzetta, 1988). The teeth are slender and slightly curved in a hook-like fashion. The function of these teeth is primarily to puncture and capture prey, whereas in the great white shark the teeth are used more as cutting tools. It is clear in Fig. 11.38 that the angle of the apex of the tooth of a mako shark is much smaller than that of the great white shark. Again, this sharp angle, similar to that of the dogfish, is used to puncture the prey. Piscivorous fish do not have to penetrate the tough skin of mammals against which serrations are needed. The analogy with knives is applicable here. A butter knife is smooth whereas a steak knife is more effective if it has serrations.

Sharks that consume hard-shelled prey have teeth that have a unique morphology and microstructure. Figure 11.39(a) shows the jaw of a zebra shark (*Stegostoma fasciatum*), which consists of multiple rows and series of tricuspid-shaped teeth, forming an ideal crushing surface. Zebra sharks are able to consume the sturdiest molluscs and crustaceans efficiently. Optical micrographs show the microstructure of the tooth, which consists of an external layer of enameloid and internal dentin (Fig. 11.39(b)). It should be noted that the enameloid layer is much thicker on the grinding edge. As previously discussed in Section 6.3, the chiton radula also has tricuspid-shape teeth, which seem to be the optimized adaptation for crushing and grinding. Figure 11.40 depicts the hardness mapping of zebra shark tooth measured by nanoindentation in rehydrated condition. The results indicate that the tooth possesses graded mechanical properties. The outmost enameloid has the highest hardness (>5 GPa) and the hardness values gradually decrease toward the dentin–enameloid junctions (~2 GPa), much higher than the hardness of dentin (~0.3 GPa). The thick and hard enameloid layer covering the outmost surface of zebra shark teeth enables efficient grinding and crushing.

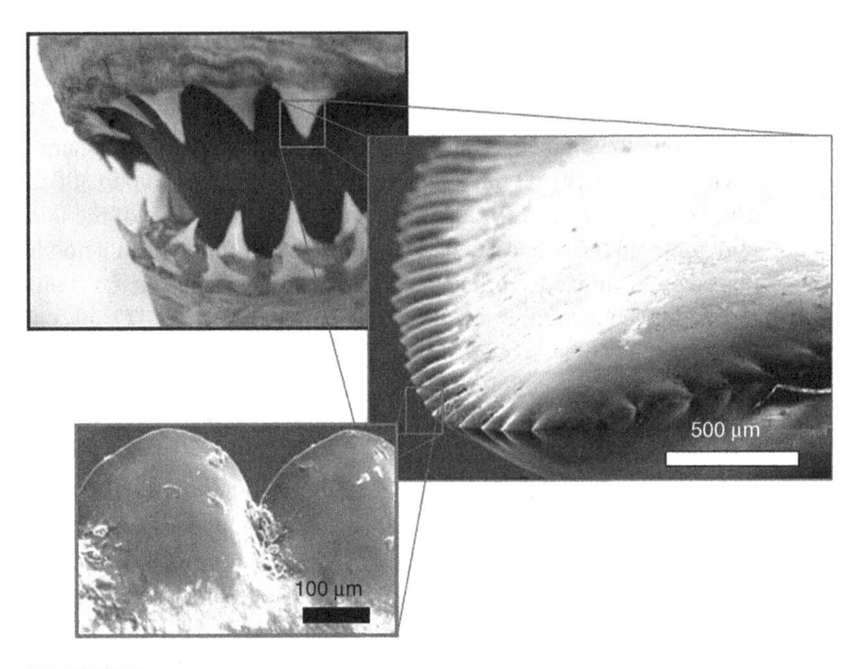

Figure 11.37.

Teeth of great white shark (*Carcaradon carcharias*). Shark mouth and detail of teeth at increasing magnifications, showing serrations.

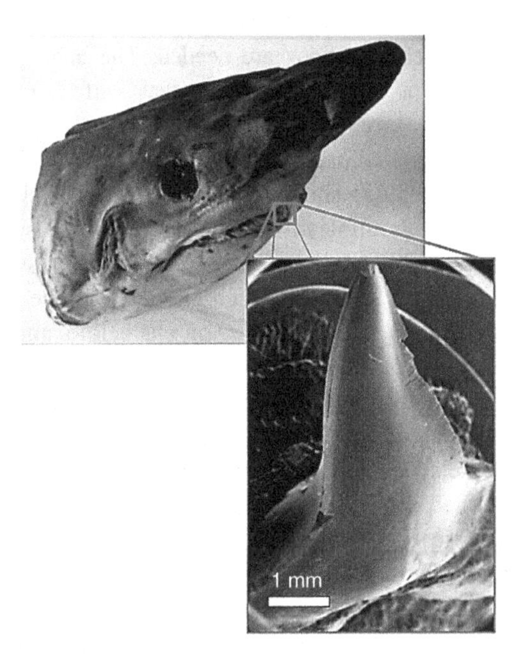

Figure 11.38.

Mako shark and tooth. The teeth are sharp and do not contain serrations, in contrast with great white shark teeth.

(a)

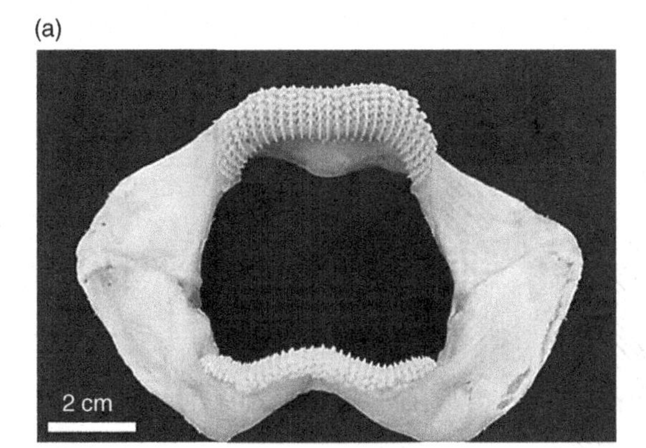

(b)

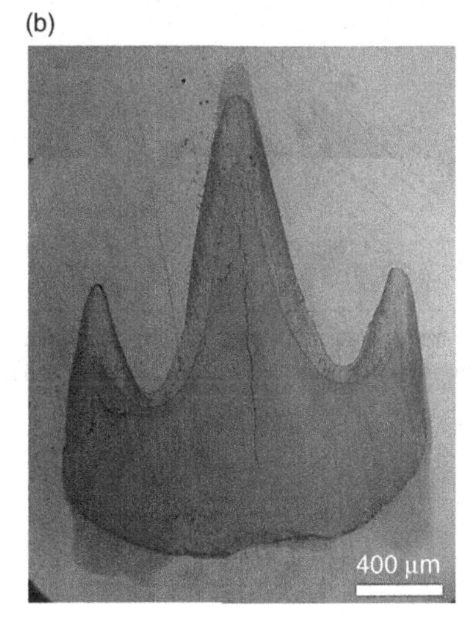

Figure 11.39.

(a) The jaw of a zebra shark, showing multiple rows and series of tricuspid-shaped teeth, forming a crushing surface. (b) Optical micrograph showing the cross-section of a zebra shark tooth, which consists of an enameloid layer enclosing an internal dentin region. The tip of the tooth has the thickest layer of enameloid.

11.8.3 Rodent incisors

The incisor teeth of animals such as the rabbit and rat (Figs. 11.41(a) and (b), respectively) have been evolutionarily designed to "self-sharpen" via a process which takes advantage of natural wear and discrepancies in wear rates depending on the hardness of certain materials. These teeth are designed so that a softer dentin backing is worn away at a faster rate than the hard enamel cutting edge. This action continuously exposes new sections of the enamel material, creating a self-sharpening effect. The teeth of rodents continuously grow. Hence, the selective wear is necessary to maintain their sharpness.

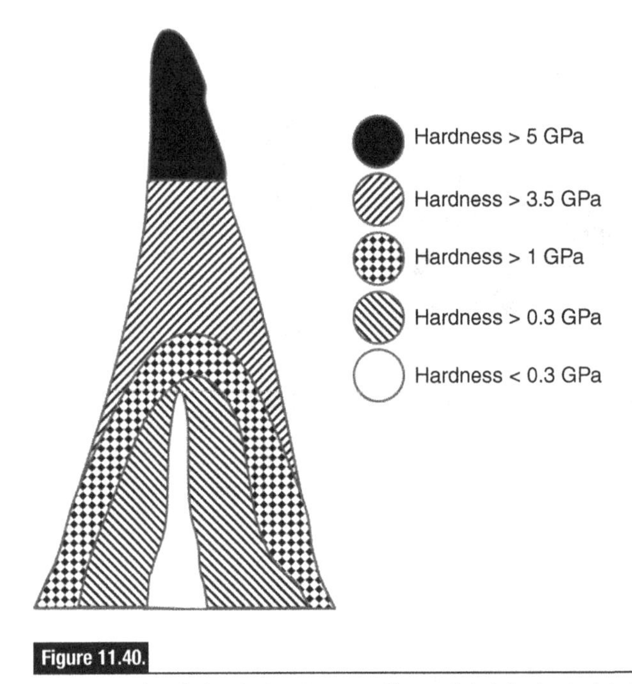

Figure 11.40.

Hardness mapping of zebra shark tooth measured by nanoindentation in rehydrated condition indicates graded mechanical properties of the tooth. The outmost enameloid has the highest hardness, and the hardness values gradually decrease toward the dentin–enameloid junctions.

11.8.4 Wood wasp ovipositor

The wood wasp uses its ovipositor (egg-laying organ) to drill holes in wood to deposit eggs. The ovipositor can be as long as 10 cm, as in the parasitic wasp (*Megarhyssa ichneumon*). The wasps can drill to depths of over 20 mm without causing much damage to the plant. Vincent and King (1995) examined the ovipositors of two wood wasps: the *Sirex nocitilo* (10 mm length) and the *Megarhyssa nortoni* (50 mm length), both with diameters of ~200 μm. The ovipositor has an unusual structure – there are two halves, called valves, that slide relative to each other. No rotary motion is involved, rather the wood is scraped out with "teeth" that surround each valve. Figure 11.42 illustrates the method of drilling: one valve penetrates the wood, creating a tensile force, while the other valve is placed in compression. Once the tensile side is stabilized in the wood, the compressive side can then be inserted and the motion is repeated. Because the tensile and compressive forces are balanced, there is no net force on the ovipositor, or the net force is below the critical load for Euler buckling (Vincent and King, 1995). Figure 11.43 shows an SEM micrograph of the ovipositor of the *S. nocitilo*. The valve separation is shown as the light colored line bisecting the ovipositor. The push and pull teeth are at the proximal and distal ends, respectively.

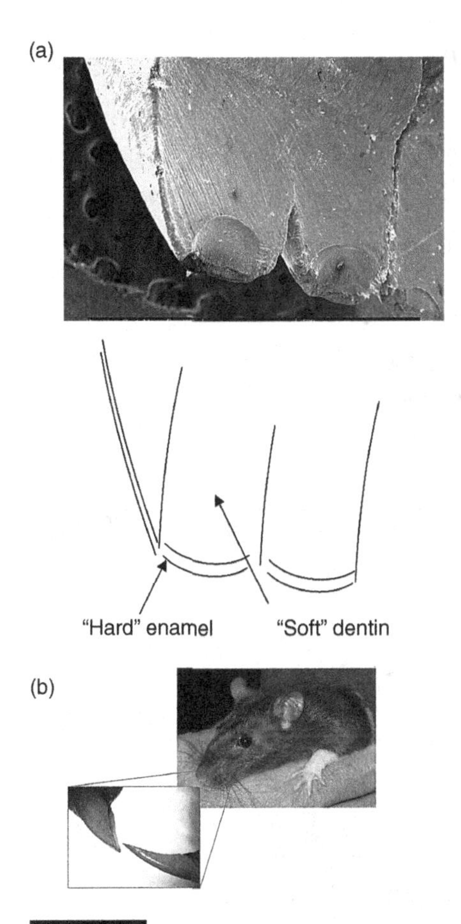

Figure 11.41.

(a) Rabbit tooth (incisor). The outside is hard enamel and the inside comprises much softer dentin. The rabbit keeps the teeth sharp by wearing down the dentin, thereby maintaining a sharp protruding enamel layer. (b) Rat incisors. The enamel is the outside layer and the softer dentin is the inside. The action of incisors rubbing against each other ensures that the teeth remain sharp.

Summary

- Materials science divides materials into two classes: structural and functional. Structural materials are the ones where mechanical properties are of foremost importance. Functional materials, on the other hand, have other primary attributes: optical, magnetic, electrical, superconducting, and energy storage are some of the more important properties for which they are developed. In biological materials, these boundaries are not so clear because many are multifunctional. Nevertheless, some have as primary function the sustainment of the structure, while others have other functions as principal attributes.

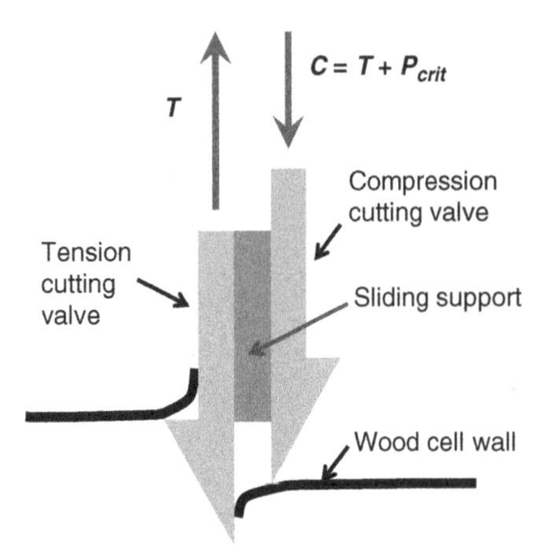

Figure 11.42.

Three-part ovipositor cutting into wood. The left valve pulls with a force T. The right valve is under a compressive force C, which is enough such that Euler buckling at P_{crit} does not occur.

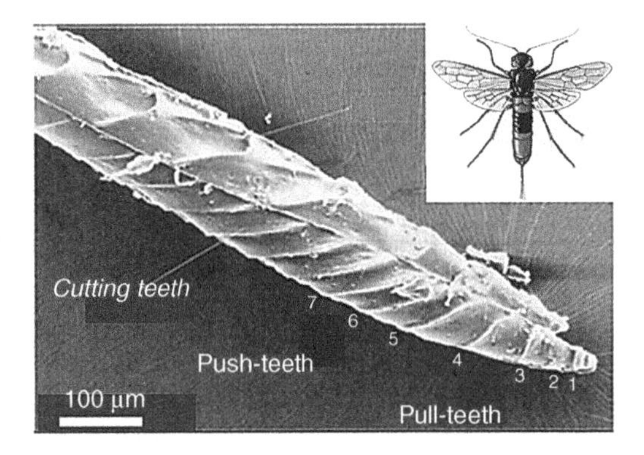

Figure 11.43.

Ovipositor of the wood wasp *S. nocitilo*. The push teeth are pointed down the shaft and the pull teeth point up the shaft, allowing for cutting in both directions. (Reproduced from Vincent and King (1995).)

- Adhesion and attachment: there are several means by which this is achieved in nature, for example interlocking, friction, bonding, suction, and gluing.
- Gecko feet present a fascinating problem of adhesion. The soles of the feet are composed of keratin setae. They are arranged into bundles, approximately square with 10 μm sides, in a regular pattern with channels between them. At the distal ends, the setae terminate into a large number of nano-sized fibrils, called spatulae, with diameters of ~100 nm.
- The Johnson–Kendall–Roberts (JKR) equation provides the magnitude of the van der Waals attractive forces between a spatula with radius R and a surface:

$$F_c = \frac{3}{2}\pi R\gamma .$$

From there, we obtain the pull-off stress, which varies with $1/R$:

$$\sigma_{app} = \frac{3}{2}\frac{f_s \gamma}{R}.$$

- If we consider that the surface of an animal's feet is a fixed fraction of the entire animal, we can arrive at an equation that predicts the spatula size required to generate attachment. There is a proportionality between R, the spatula radius, and $m^{-1/3}$, the mass.
- The abalone foot provides an underwater attachment that requires stresses on the order of 115 kPa to separate it. Setae line the outer surface of the tissue with a thickness of 1–2 μm. At their extremities, the setae separate into nano-scale fibrils with hemispherical ends, averaging 150 nm in diameter and uniaxially aligned perpendicular to the plane of the foot tissue.
- The characterization of the abalone foot pedal and the mechanical tests suggest that three mechanisms act cooperatively (and perhaps synergistically): van der Waals, capillarity, and suction. Suction can generate attachment forces that create stresses $\sigma_d = P_{atm} = 101$ kPa. This is on the same order as the forces required to separate the abalone from a host.
- The lotus leaf exhibits extreme hydrophobicity; water does not wet it, but concentrates into small areas by virtue of the surface of the leaf. Drops of water will just roll off the leaf (shown in the right photograph of Fig. 11.21(a)). There are two main reasons that aquatic plants develop a water-repellent surface. First, CO_2 diffuses 10 000 times more slowly through water than in air. A thin layer of water on leaves can significantly reduce the uptake of CO_2, which is necessary for photosynthesis. Second, the water-repellent ability limits the growth of micro organisms (bacteria and fungi) on the plant surface.
- The surface of the lotus leaf has small pillars a few micrometers in height and spaced ~20 μm apart. These are covered by smaller-scale protrusions, with dimensions ~0.2–1 μm. These are covered with wax. The net result is that the angle of contact between the water droplets and the surface is dramatically increased. Hydrophilic surfaces have contact angles below 90°, whereas hydrophobic surfaces have angles above 90°. For the lotus, this angle can be as high as 160°, and it is therefore called superhydrophobic. This hydrophobicity has an important effect: water droplets, almost perfect spheres, can roll over the surface. When they do this, they pick up dust particles which decrease their surface energy by being absorbed into the water droplets. Thus, the surface cleans itself.

Exercises

Exercise 11.1 Applying the equation for a cycloid, determine the parameters of the serrations in the tooth of the Komodo dragon, shown in Fig. 2.9(b).

Exercise 11.2 Using the Johnson–Kendall–Roberts equation, calculate the weight that four paws of a gecko can sustain. The spatula diameter is 100 nm and their spacing is 300 nm.

(a) Use measurements that you can extract from Fig. 11.4, knowing that the gecko pictured has a body length of 20 cm (not including the tail). Assume that the work of adhesion is: $\gamma = 15$ mJ/m^2.
(b) Continuing this reasoning, could a mature Nile crocodile climb the walls of the temples of the pharaohs to stalk tourists? Explain quantitatively.

Exercise 11.3 Calculate the attachment force of a gecko, knowing that the spatulae have semispherical caps with diameters of 100 nm. The fraction of the foot occupied by the spatulae is 30%. The contact area of each foot is 1 cm^2, the Hamaker constant is taken as 6×10^{-2} J, and $D_0 = 0.16$ nm. Use the paper by Lin *et al.* (2009) to provide the equation for interfacial energy.

Exercise 11.4 In hydrophobic surfaces, the angle θ is higher than 90°. Using the following equation, explain what values of the interfacial energy favor this effect:

$$\gamma_{SG} = \gamma_{SL} + \gamma_{LG} \cos\theta.$$

Exercise 11.5 Explain the principles of the cochlear implant that assists hearing. Can piezoelectric ceramics assist this function? Name and briefly describe two piezoelectric ceramics with wide technological applications. Note: obtain information from the web to answer this.

Exercise 11.6 What are the attachment mechanisms in nature? Give at least one example for each mechanism.

Exercise 11.7 What is the major adhesive mechanism of the gecko foot? Briefly describe the hierarchical structure of the gecko foot. Plot pull-off strength as a function of fiber radius.

Exercise 11.8 What are the major types of coloration in biological systems? Explain with examples.

Exercise 11.9 Explain the origin of the blue color in *Morpho* butterfly wings. Use illustrations.

Exercise 11.10 What is a photonic crystal? Give three examples of a photonic crystal structure in biological systems.

Exercise 11.11 Explain how a chameleon changes its color to adapt to its environment. What is the microstructural design in skin when a chameleon turns green?

Part III Bioinspired materials and biomimetics

The ultimate goal for a materials engineer is to learn from the lessons of nature and to apply this knowledge to new materials and design. This is not a new quest, and humans have sought inspiration from nature since prehistory. The early materials used by humans were primarily natural: stones, bones, wood, skins, bark. The accelerating pace of the civilizing process has been attributed to the introduction of new synthetic materials; thus, the bronze and iron ages followed the stone age. We have now entered the brave new world of the silicon age, which is bound to produce unimaginable change. Homo silicensis, connected 24/7 to computerized contraptions, can already be seen on campuses and elite coffee houses, sipping lattes.

The constant quest for new materials and designs is leading us to a systematic inquiry of nature in order to unravel its secrets. This is the field of biomimetics, and VELCRO® is the standard example of bioinspired design. It is inspired by the burrs of plants that contain small hooks that attach themselves to animal wool or our clothes.

Other examples of bioinspired designs abound, and they are divided into two classes (described in Chapter 12): structural/functional and medical. The first models of gliders mimicked bats and birds. The famous Leonardo da Vinci concept is clearly patterned after the wings of a bat. The monk Eilmer of Malmesbury is reported (~1100 AD) to have flown off the roof of his Abbey in Malmesbury, England, gliding about 200 m before crashing and breaking his legs. Apparently, feathers were used in the wings. In modern times (late nineteenth century), Otto von Lilienthal made over 2000 successful flights using a glider before crashing and fracturing his spine.

The frontier in biomimetics resides, however, at the molecular level, and not in structures that are inspired at the meso- or structural level by nature. Can we reproduce the natural processes of living organisms? Can we self-assemble structures in the manner of nature? The quest for creating a single cell might be beyond our capability, since each contains millions of different proteins.

However, humans are ingenious as well as mischievous, and we are developing methods to manipulate DNA. By extracting it from the host, modifying it, and reintroducing it, we are opening a cornucopia of possibilities. We are at the verge of a new age in which we will have the capability of dictating and directing our own evolution. Pandora's box will open itself and reveal her secrets.

Embedded in the chromosomes that reside in the nucleus of the cell is the genome which contains the DNA carrying the entire genetic information of an organism. The discipline of genomics uses recombinant DNA and DNA sequencing methods to identify and manipulate the genome. Thus, mapping the human genome establishes an individual's predisposition for disease. Proteomics goes beyond genomics: the genes give orders to the various parts of the organism that are carried out by proteins; proteomics consists of identifying these proteins and establishing their functions. This area is at the frontier of biological research.

The blending of natural and synthetic processes by forming hybrid constructs is another area of active research. Virus–DNA arrays are being used to build lithium-ion batteries; ATP motors are being combined with a metallic propeller that rotates and forms a nanomotor. This expands the field of mechatronics and nanotronics into the biological domain. In the field of medicine, there are many possibilities and applications of new devices that can be created, inspired by nature. These aspects are the realm of Chapter 13, Molecular-based biomimetics.

12 Bioinspired materials: traditional biomimetics

Introduction

It is a misunderstanding to think that the ultimate goal of biomimetics is to reproduce living organisms. There are essential differences, as we list in the following.

(a) Organisms are composed of cells (in the case of humans, 10^{13} of them). They direct a great deal of the activity. As seen in Chapter 4, each mammalian cell contains 10 000 different proteins, for a total of 500 million. There are numerous types of cells in each organism, and they have complex lives.

(b) Organisms are, for the most part, composed of a limited number of elements: C, Ca, H, O, P, N, and S. Synthetic materials have, on the other hand, a cornucopia of elements, made possible by the synthesis and processing techniques developed by humans.

What we try to do is to emulate the design and assembly principles used in natural materials. In this book we have seen many examples where superior properties are obtained through a hierarchical design and ingenious solutions. Bone, nacre, and dentin have toughnesses significantly superior to those of the mineral constituents, hydroxyapatite and calcium carbonate. Silk reaches strengths higher than 1 GPa using the weak hydrogen bond, through the existence of nano-scale β-sheet crystals with proper dimensions. Bioinspiration requires identification, understanding, and quantification of natural design principles and their replication in synthetic materials, taking into account the intrinsic properties (Studart, 2012). This approach is being pursued not only for structural materials, but also for functional materials and devices. The areas of sensing, optics, architecture, and robotics are exploring biological solutions.

Natural selection provides a tool with which nature has processed, improved, and refined biologically based elements over millions of years. As scientists we can learn from these evolutionary refinements and use them with the intention of creating novel improvements in technology. Materials science is beginning to explore the resources presented by nature and use them to create novel advances in technology (Zhou, 1996; Weiner, Addadi, and Wagner, 2000; Vincent and Mann, 2002; Sarikaya *et al.*, 2003). This interdisciplinary synergy between materials science, engineering, and biology is the field of *Biomimetics* (Srinivasan *et al.*, 1991; Sarikaya, 1994).

There have been two approaches taken in generating bioinspired materials and structures. It can be argued that probably all early attempts at flying were inspired by birds. The Greek legend of Icarus comes to mind. It was Leonardo who studied in detail the wings of bats to propose the design of the first airplane. This traditional approach has utilized synthetic materials to produce performance that mimics biological materials. The current frontier in bioinspired materials goes further and uses bioinspired processes at the molecular level to generate materials and structures. Such is the case of tissue engineering and other molecular-based approaches. Some of these efforts are presented in Chapter 13, in which we will describe a new methodology based on molecular-level design.

The production of inorganic materials in nature generally occurs at ambient temperature and pressure under isothermal and isobaric conditions. Yet the simple organisms through which these inorganic materials are formed create extremely precise and complex structures with advanced functionality (Carlton, 1903). Furthermore, the constituent materials provided in nature are often brittle and weak. Thus, the strength of hard tissue such as shell or bone is derived from the structure rather than through materials selection. As material selection reaches its limitations, engineers may look toward nature's examples of structural optimization for the next generation of technology.

When synthetic materials are fabricated, they are usually designed to optimize one property (e.g. strength or conductivity). Generally this is for a specific length scale (e.g. at the dislocation density for strength, or by considering grain boundary properties for conductivity). As a consequence, a material may have one superior property while others are inferior. What seems to be effortless in a biological system that produces complex, multifunctional materials can, indeed, be translated to processing of synthetic materials. Layered structures, such as abalone nacre, exhibit high toughness values despite low fracture toughness constituents. For example, it has been shown that for metal/metal, polymer/polymer, and glass/polymer laminates, thinner layers produce a tougher composite than thicker layers (Ma *et al.*, 1990a, b; Syn *et al.*, 1993: Mayer, 2005). This suggests that the scale and hierarchy are important considerations in materials design.

Concurrent synthesis and assembly of constituents in biological systems is a hallmark of biological fabrication. This should be considered in the design of biomimicked structures, and several factors should be deliberated (Tirrell, 1994):

- concurrent materials synthesis and structural assembly;
- processes to fabricate highly specific synthetic membranes and filters;
- the use of cells to synthesize and/or deposit materials;
- biosynthetic pathways to the cost-effective manufacturing of new classes of shaped, hybrid composites;
- biosynthetic concepts and materials for self-repair of critical components and devices.

It is clear that the design of materials and structures inspired by nature involves special challenges not encountered before. Traditional design has followed disciplinary lines, but bioinspired design will require multidisciplinary teams of engineers (design and materials) and scientists (biochemistry, biology, physiology, anatomy, and molecular biology) to develop materials with complex, hierarchical structures. In this chapter, we present biologically inspired materials and designs using traditional, existing technology. There is a cornucopia of bioinspired materials and designs, and we cannot cover all of them in this chapter. For instance, termite mounds have sophisticated temperature regulation systems such that drastic variations in outside temperature, common in Africa, have little effect on the inside. This concept is being adapted to modern air conditioning systems in buildings. A number of bioinspired bone substitutes have been or are being developed: ProOsteon® and HAPEX are two examples. We divide this chapter into two parts: structural and medical applications.

12.1 Structural and functional applications

12.1.1 VELCRO®

One of the most successfully commercialized biomimetic materials literally stuck itself to its inventor's pants. In the summer of 1948, the electrical engineer George de Mestral was walking with his dog when he realized his pants had become covered in a plethora of seed-bearing burrs. As he sat in frustration, tediously pulling burr after burr from the fabric of his clothes, curiosity grew in George's mind.

When he investigated the structure of the burr under his microscope, Mestral realized the secret of the seeds' ability to attach to wool: tiny little hooks which covered the seed, as shown in Fig. 11.1 (Armstrong, 1979). From this velours and crochet, or "VELCRO®" was born, giving rise to a multi-billion dollar industry. However, it was not an immediate success story, and it took George several years to develop the process to produce the tiny hooks and loops. It can be seen from Fig. 12.1(b) that the hooks are made from loops that

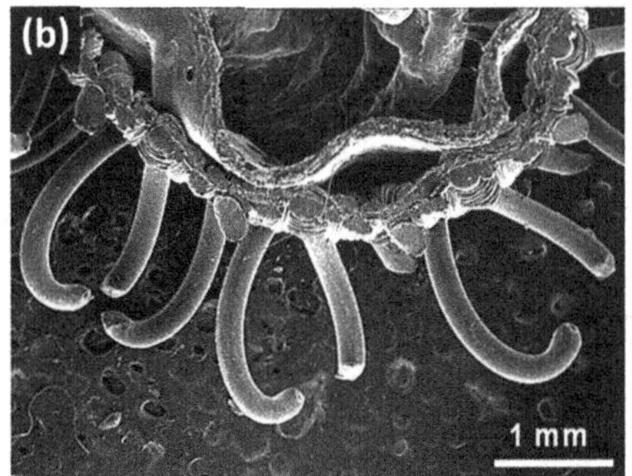

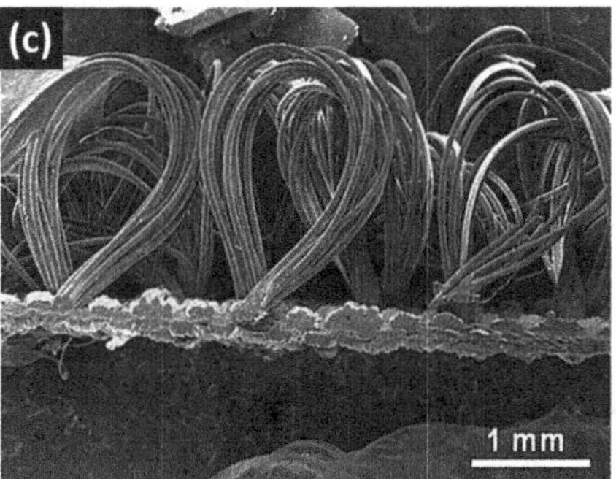

Figure 12.1.

(a) Burrs of a seed similar to the one that gave birth to VELCRO®; (b) VELCRO® hooks and (c) loops. (Figure courtesy W. Yang.)

are cut. The material for the hooks is stiffer than for the loops. This can be immediately gathered by rubbing the hand over VELCRO®: the loop part is smooth, whereas the hook side scratches the hand.

Example 12.1 (a) Calculate the force required to separate each hook from its loop in VELCRO®. Given: $E = 10\,\text{GPa}$ (for the hook material). Obtain dimensions from Fig. 12.2.

(b) What force would be necessary to open a 5 cm wide strip of VELCRO® assuming that we are pulling the loops perpendicular to the layer containing hooks (see Fig. 12.3)?

Solution (a) In order to solve this problem, we have to make some simplifying assumptions. Emulating the physicist, we can say… let us assume a cubic cow (i.e. we reduce a complex animal to a simpler shape in order to render the problem tractable mathematically). We can estimate the following parameters from Fig. 12.2:

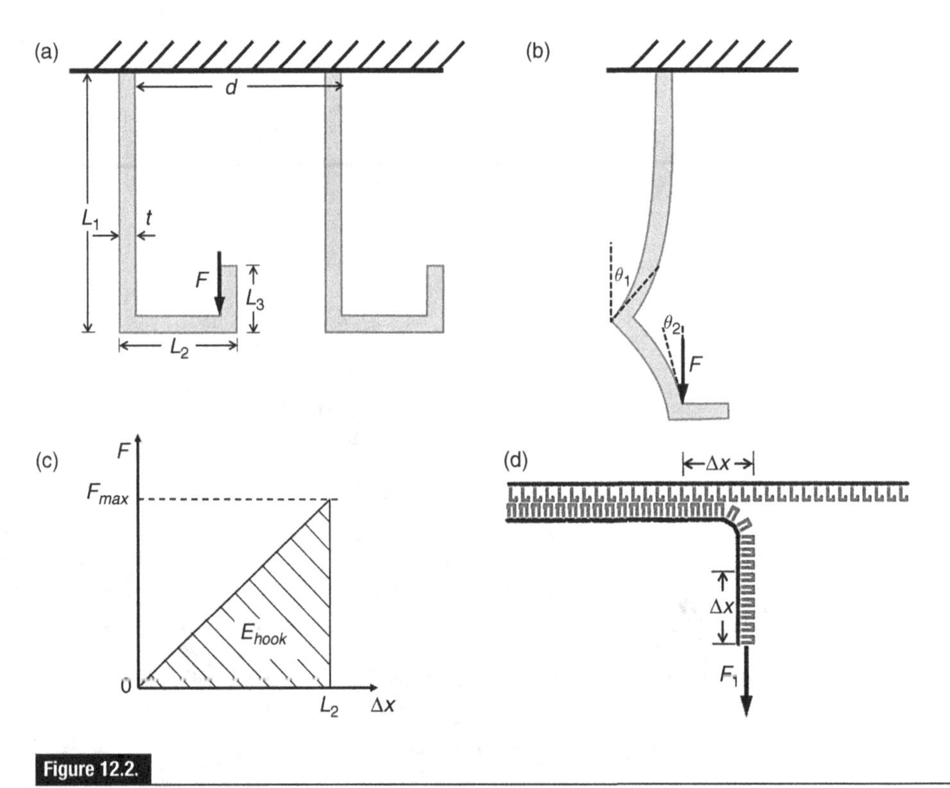

Figure 12.2.

Idealized shape of hook in (a) the original and (b) loaded condition just prior to slipping off the loop. (c) Elastic energy stored in hook. (d) Relationship between applied force on strip and displacement.

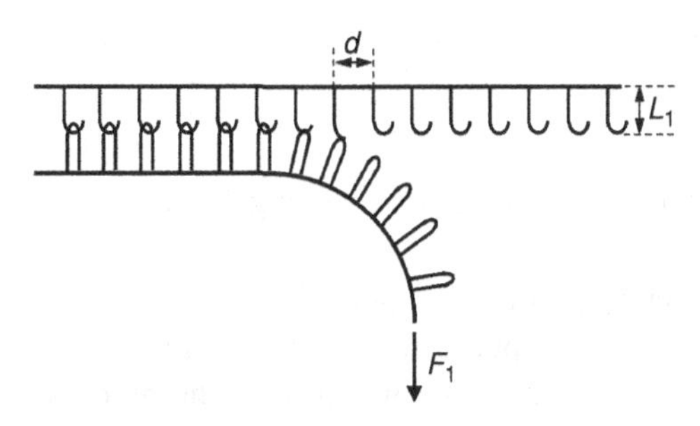

Figure 12.3.

VELCRO® being opened with force F_1.

$L_1 = 3$ mm,
$L_2 = 1$ mm,
$t = 0.25$ mm,
$d = 2$ mm.

We start by assuming that the hooks have an idealized shape, as shown in Fig. 12.2. We consider that the loop will slip off the hook once the segment of length L_3 becomes horizontal. The beams of length L_1 and L_2 are subjected to bending moments due to the force F.

These moments produce deflections which reach their maximum at the extremities of the beams. The slip-off condition is given by

$$\theta_1 + \theta_2 = \frac{\pi}{2}.$$

The moment for beam L_1 is

$$M_1 = \left(L_2 - \frac{3t}{2}\right)F.$$

The moment for beam L_2 is

$$M_2 = FL_2.$$

The beam deflections v (see equations in Chapter 4) are related to the angle θ:

$$\theta = \frac{dv}{dx} = \frac{M}{EI}x.$$

The two beams have slightly different loading configurations. The total angle is

$$\theta_1 + \theta_2 = \frac{M_1}{EI}L_1 + \frac{M_2}{EI}L_2 = \frac{\pi}{2}.$$

The moment of inertia is (see Table 4.1) given by

$$I = \frac{\pi c^4}{4}.$$

Substituting the values for the moments and solving for F yields $F = 0.83$ N, less than 1 N.

(b) We first estimate the energy required to pull a loop until it slides off, assuming that the maximum deflection is $\Delta = L_2$ at pull-off. We assume a linear elastic material with a straight force–deflection curve. The energy involved in the process is the area of the triangle which is given by

$$E_{hook} = \frac{1}{2} F L_2 = 0.83 \times 10^{-3} \text{ N m}.$$

In order to calculate the force required to "peel off" a strip with a width $w = 5$ cm, we use the first law of thermodynamics, which equates the internal energy increase in the system to the sum of the work done on the system and the heat added to the system:

$$dE = \delta W + \delta q.$$

We neglect the heat term and therefore we obtain

$$dE = \delta W.$$

The work done on the system to unzip a strip of length Δx by an external force F_1 applied normal to the VELCRO® is $\Delta W = F_1 \Delta x$. The energy change involved in separating all the hooks from their respective loops in an unzipping of Δx is given by

$$\Delta E = E_{hook} \frac{w \cdot \Delta x}{d^2},$$

where d is the spacing between hooks and $1/d^2$ is the number of hooks per unit area. Equating the work to the change in internal energy and canceling the Δx term, we obtain

$$F_1 = E_{hook} \frac{w}{d^2} = 0.83 \times 10^{-3} \frac{5 \times 10^{-2}}{4 \times 10^{-6}} = 10.4 \text{ N}.$$

12.1.2 Aerospace materials

The effect of gravity applies to both technology and nature, as can be seen in the design characteristics of biological avionic materials such as bird beaks and bones. There is an apparent optimization of weight to strength through sandwich structures consisting of solid shells filled with compliant cellular cores (Seki *et al.*, 2005, 2006). The core significantly increases the buckling capabilities of the entire system while maintaining the light weight

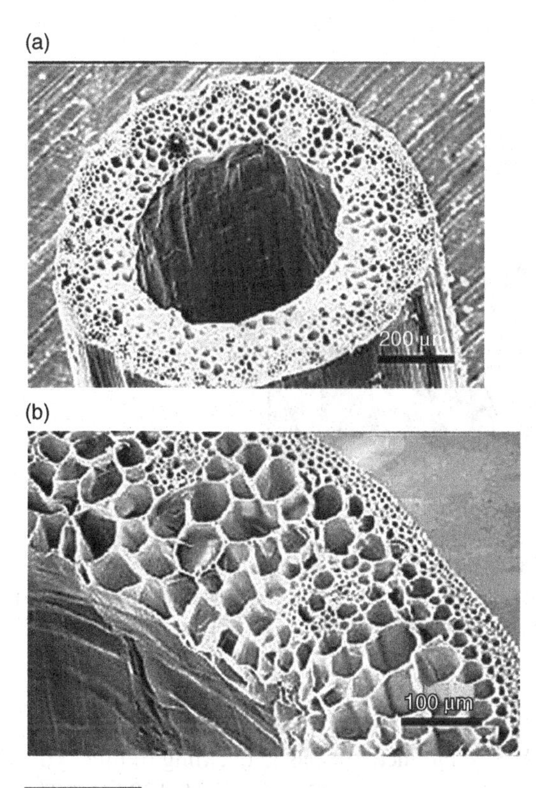

Figure 12.4.

Cross-section of grass stem (*Elytrigia repens*) showing a shell structure with a foam-like core. (Reprinted from Dawson and Gibson (2007), with permission from Elsevier.)

required for flight. This synergism between a hard shell and a compliant core is also exhibited in other biological structures which require resistance against axial buckling. Plant stems are an example of this, often consisting of thin-walled cylindrical structures filled with a cellular core (Karam and Gibson, 1994; Dawson and Gibson, 2007). Their large aspect ratio creates an interesting engineering problem, which has been posed both in nature and technology. Figure 12.4 provides an example of this structure in nature.

Birds have structures that are optimized for weight; Fig. 12.5(a) shows hollow bone from a red-tailed hawk. The bone is strengthened by V-shaped struts. This is also described in Chapter 10 (Fig. 10.19). Other bone structures are also shown in Section 10.4. This bone has a unique structure composed of two layers connected by a tridimensional array of inclined struts. This structure provides the stiffness required by minimizing the weight, since most of the mass is displaced away from the neutral plane. Interestingly, researchers are developing structures that are very similar. The example shown in Fig. 12.5(b) was developed for an experimental multifunctional material. This is a clear example of modern research finding solutions that have been applied by biological systems for millions of years.

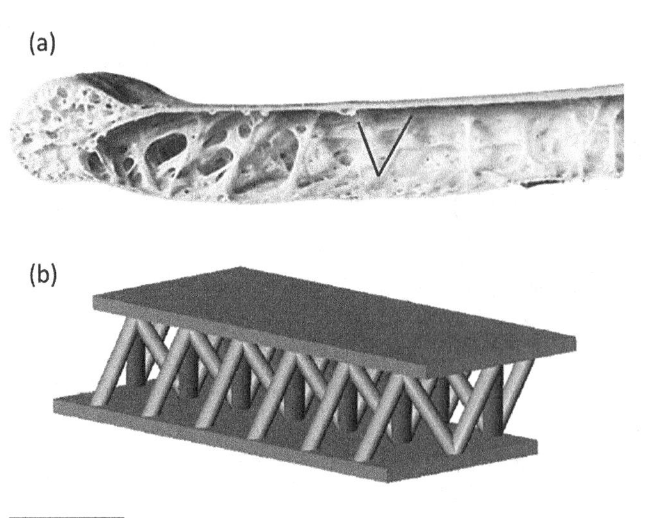

Figure 12.5.

(a) Hollow bone from a red-tailed hawk. The structure is stiffened by V-shaped internal struts in a three-dimensional configuration. (From http://platospond.com/WatsonsBlog/?p=22 with kind permission of Fred Andrews.) (b) CAD image of the truss core structure. The multifunctional cellular metals have a similar structure to the metacarpal bone of birds. (Reprinted from Evans *et al.* (2001b), with permission from Elsevier.)

Karam and Gibson have examined the elastic buckling of thin cylindrical shells filled with compliant elastic cores (Karam and Gibson, 1994, 1995a,b). Their results confirm the benefit of the hard shell – elastic core synergy. Again, nature's design stood up to the test, showing significant mechanical improvement with limited weight gain. Furthermore, the problem of axial buckling in structures with high aspect ratios has to be addressed as such materials and structures are commonly found in both nature and technology.

12.1.3 Building designs

Other examples of buckling resistance can be found throughout the natural world; even the deepest parts of the ocean can accommodate organisms with interesting biomimetic potential. The skeleton of a sea sponge, for example, exhibits amazing hierarchical levels of complexity, each providing the essential components of structural design necessary for the conversion of the otherwise brittle constituent material (silica) into a sophisticated masterpiece of architectural evolution. This structure has been studied to expose some of the engineering lessons which have stood the test of time over millions of years (Aizenberg *et al.*, 2005; Mayer, 2005). Figure 12.6 shows the skeleton of the sponge *Euplectella* in its entirety (Aizenberg *et al.*, 2005). It is also presented in Section 6.1. The cylindrical cage extends 20–25 cm in length and 2–4 cm in diameter. The frame of the cage consists of long vertical struts running the entire length of the structure. Horizontal struts form a regular square lattice with the vertical struts. This structure is reinforced by external ridges that extend from the surface of the cylinder and spiral the cage at an angle of 45° to the cross-sectional plane. A brittle material loaded in torsion will fail along the surface where tensile stresses are the maximum. This occurs 45° to the cross-sectional plane, resulting in helical fractures (Meyers and Chawla, 2009). The helical reinforcement of the cage is no accident of nature; the spiraling struts offer important strengthening mechanisms to combat the destructive forces applied through the ocean's currents. Many similar reinforcements can be seen in advanced structural engineering masterpieces, including the Swiss Re Tower (or the "Gherkin") in London, the Hotel De Las Artes in Barcelona, and the Eiffel Tower in Paris (see Fig. 12.6). The Burj Khalifa in Dubai, the highest man-made structure in the world, has a cross-section of three leaves inspired by a local desert flower.

The architect Andres Harris has designed several structures based on bird skulls, which have double-layered shells. These open atria allow for an uninterrupted flow of air across the top, while maintaining a lightweight materials design (Fig. 12.7). A striking

Figure 12.6.

Skeleton of the deep sea sponge *Euplectella*. (Taken from Bell Laboratories.)

Figure 12.7.

Bioinspired architectural shells based on bird skulls. (Courtesy Dr. Wen Yang.)

Figure 12.8.

Feather-inspired architecture of the Zayed National Museum, Abu Dhabi, that is saving energy by drawing cooling air currents through the building.

building in Abu Dhabi is the Zayed National Museum shown in Fig. 12.8. This energy-efficient feather-inspired design uses "thermal chimneys" that draw cooling air currents through the building and force the hot air out to keep the building ventilated.

12.1.4 Fiber optics and microlenses

The brittle-star is a marine animal that has great sensitivity to light. It was discovered by Aizenberg and Hendler (2004) that this organism, which does not have specialized eyes, has a set of lenses that channel light. These lenses focus the light 4–7 μm below the array into the neural bundles that exist in the animal. Figure 11.32 shows these lenses. They have many features that are superior to synthetic microlenses. They minimize spherical aberration, have a crystallographic orientation that eliminates birefringence, and contain an organic–inorganic composite that prevents the brittle calcium carbonate from fracturing easily. This brittle-star lens design was used at Bell Laboratories to produce synthetic lenses that mimic these properties. These synthetically produced microlens arrays have pores less than 10 μm and a controlled crystallographic orientation. They are shown in Fig. 12.9.

Aizenberg (2010) mimicked this structure by combining microlenses and a fluidic network. She used three-beam interference lithography to create synthetic polymeric (noncalcitic) lenses surrounded by a porous network. Two such structures are shown in Fig. 12.9.

Structural reinforcement of the deep sea sponge is not limited to its macro-level design described earlier. Each strut of the cage is composed of bundles of silica spicules. These spicules, which were discussed in detail in Section 6.1, consist almost entirely of silica and exhibit remarkable flexibility and toughness. Unlike their commercial counterparts, they can be bent and tied into knots without fracture. This is attributed to the

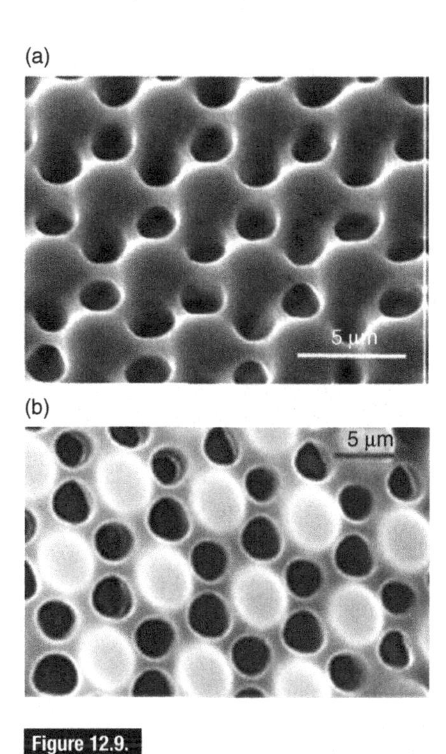

Figure 12.9.

Brittle-star-inspired synthetic structure. (From Aizenberg (2010) copyright © 2010 Materials Research Society. Reprinted with permission of Cambridge University Press.)

concentric lamellar structure. Cylindrical layers, 0.2–1.5 μm thick, are separated by extremely thin organic inter-layers (see Fig. 6.6) (Aizenberg *et al.*, 2005). This style of interplay between hard and soft materials can be seen throughout the natural world. In almost all cases, the laminate design of brittle inorganic layers separated by thin elastic regions of organic material leads to a dramatic increase in toughness of otherwise weak materials. These organic regions are usually a very small percentage of the entire composite. The optimization of composition and microstructure in this natural material may inspire some novel improvements for its man-made equivalent.

The similarities between these two materials is not limited only to composition. The variation of the reflective index of the various sections of the spicule results in wave-guiding properties similar to those found in modern fiber optics (Sundar *et al.*, 2003). Described by Aizenberg *et al.* (2005), the cross-section in Fig. 6.6(a) shows three distinct regions: a core of pure silica surrounding an organic filament; a central cylinder of high organic content; and the laminate layers with decreasing organic content. Interferometric refractive index profiling reveals a high reflective index at the core surrounded by a low reflective index in the central cylinder. The "core–cladding" profile in Fig. 12.10(a) shows how light can be confined within the core and guided through the spicule. It is not clear whether the organism uses the optical properties of

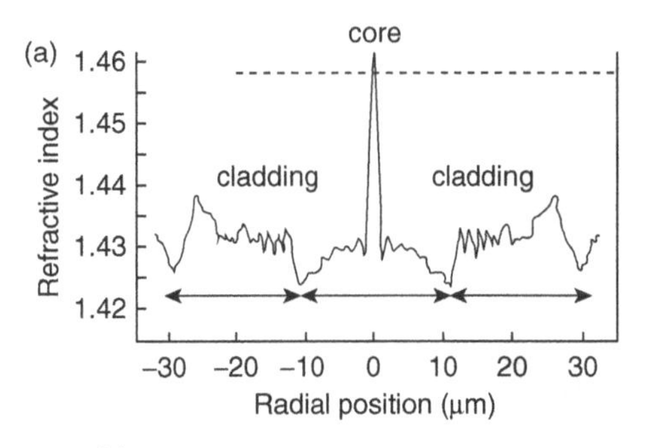

(b)

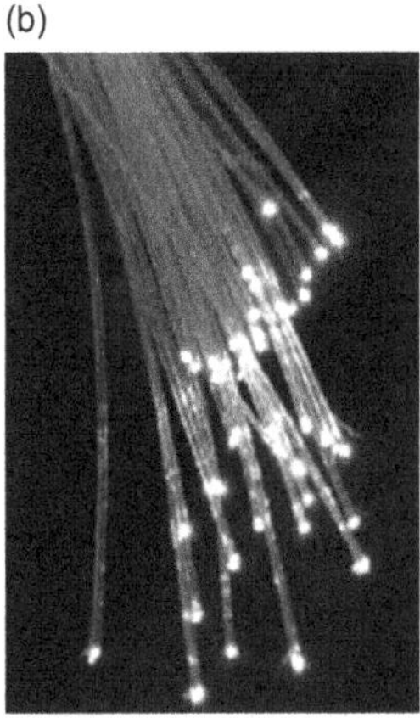

Figure 12.10.

Sponge spicule with (a) "core–cladding" profile showing wave-guiding properties. (Reprinted by permission of Macmillan Publishers Ltd.: *Nature* (Sundar *et al.*, 2003), copyright (2003).) (b) Commercial fiber optics. (From http://commons.wikimedia.org/wiki/File:Fibreoptic.jpg. Image by BigRiz. Licensed under CC-BY-SA-3.0, via Wikimedia Commons.)

the spicule; however, in comparison to man-made counterparts, the technical advantages are impressive.

12.1.5 Manufacturing

The incredible control of nano-scaled material design in nature is unparalleled in current technology (Mann *et al.*, 1993; Belcher *et al.*, 1996; Mann, 2001). As mentioned previously, the sophisticated structures and materials which are created in nature are done so under ambient temperatures and pressures. The man-made counterparts of these materials are typically produced under extreme conditions, resulting in a myriad of engineering obstacles, which often drive up the cost of production. Biomimetically inspired manufacturing techniques may result in the technical evolution of the way in which we think of manufacturing. Morse and co-workers (Kisailus *et al.*, 2006) developed a manufacturing method for semiconductor thin films using inspiration from spicule formation. They found that by putting enzymes, similar to those of marine sponges, onto gold surfaces, they could create templates on which semiconductor films could grow. Using biomimetic catalysts they were able to grow films at room temperature (Schwenzer, Gomm, and Morse, 2006). Additional information will be given in Chapter 13.

12.1.6 Water collection

Even the most extreme climates in the world can host animal life. The desert beetle, for example, survives in one of the driest environments known to man. Extreme daytime temperatures, howling winds, and almost nonexistent rainfall leave desert areas almost entirely uninhabited. Yet every morning the tenebrionid beetle *Stenocara sp.* walks through a transitory fog, collecting essential drinking water on its back (Hamilton and Selly, 1976). The bumpy back of the beetle is designed to pull moisture from the air along alternating regions of hydrophobic and hydrophilic surfaces (Parker and Lawrence, 2001). Macro-scaled bumps of approximately 0.5 mm diameter pepper the back of the beetle, as shown in Fig. 12.11(a). At the peak of each bump a hydrophilic region protrudes into the surrounding atmosphere (Fig. 12.11(b)) The troughs between bumps

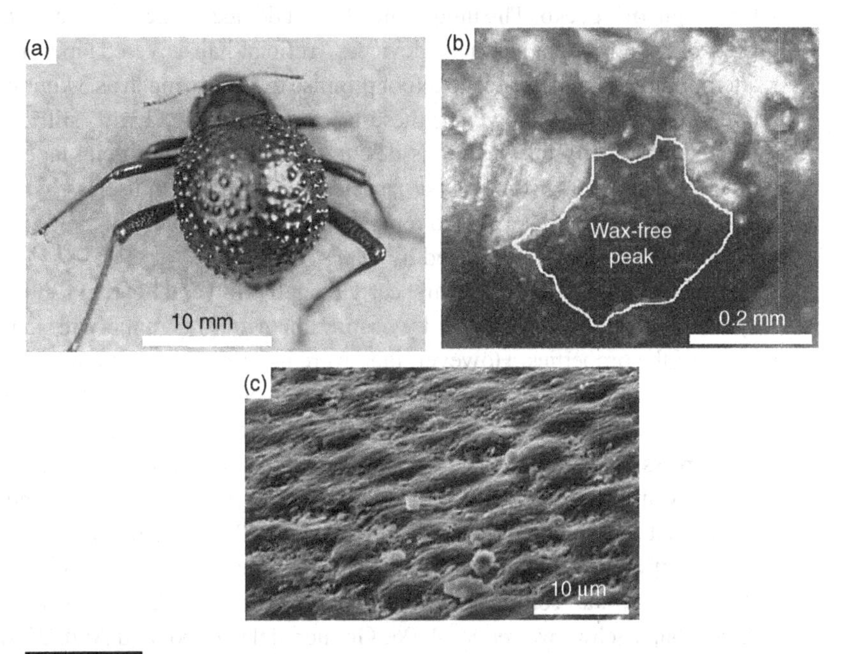

Figure 12.11.

(a) Adult female tenebrionid beetle, dorsal view; peaks and troughs are evident on the surface. (b) "Bump" showing hydrophilic wax-free region. (c) Scanning electron micrograph of the textured surface of the depressed areas. (Reprinted by permission from Macmillan Publishers Ltd.: *Nature* (Parker and Lawrence, 2001), copyright (2001).)

and the sloping sides which descend from each peak are covered in wax. This area is equipped with a microstructure of tightly packed flattened hemispheres 10 μm in diameter (Fig. 12.11(c)). As a result, the surface of the back, with the exception of the peak of each bump, is superhydrophobic. The hydrophilic peaks act as nucleation points for droplets to form from the surrounding moisture of the fog. As droplets grow to a critical size they slide off their unstable position at the peak of each bump and into the hydrophobic troughs where they roll down towards the mouth of the beetle.

As a biomimetic material, this type of functionality could be easily reproduced for commercial use (Parker and Lawrence, 2001). Sheets of a similar structure could provide the essential drinking water to the inhabitants of the extreme corners of the world.

12.1.7 Gecko feet

As seen in Section 11.1, tiny nano-scaled hairs known as spatulae line the bottom of a gecko's foot; they are used to create a bond with any surface they touch (Vincent and Mann, 2002; Arzt *et al.*, 2003). A van der Waals attraction between individual atoms on the tip of each hair and the atoms on a surface is created. When millions of these hairs are all creating an attraction, the combined force is great enough to keep the gecko tightly secured to a wall. In fact the combined force theoretically could hold much more than just the weight of a gecko. The unique attachment devices of geckos (Section 11.2) are being used as inspiration for synthetic devices. Artificial-hair-covered tapes made by engineers attempting to mimic the gecko's foot promise to hold as much as 3 kg/cm^2 of surface area (Geim *et al.*, 2003). However, these tapes, seen in Fig. 12.12, still lack the ability to recreate absolutely the ingenious design of nature. The artificial tapes tend to become quickly laden with water or dust particles, rendering them useless, while the individual spatulae of the gecko are able to remain clean and reusable. This self-cleaning ability has yet to be successfully mimicked by man.

This biomimicry work was first carried out by Sitti and Fearing (2003), who produced synthetic polyethylene bristles by casting them from ceramic with nanoholes that had some of the properties. However, they were marred with problems.

The design principle of the gecko foot has given rise to extensive research in the hopes of creating structures that have reusable and effective attachment properties. Arzt and co-workers (Arzt *et al.*, 2003; Huber *et al.*, 2005; del Campo *et al.*, 2007; Greiner, Arzt, and del Campo, 2009) have made significant progress in mimicking the structure of the gecko foot by producing polydimethylsiloxane (PDMS) surfaces, patterned with fibrils having different termination shapes – flat and mushroom – as shown in Fig. 12.14. The experimental data (continuous lines) apply to synthetic pillar arrays manufactured by the Arzt group (Schwenzer *et al.*, 2006; Greiner, del Campo, and Arzt, 2007). These arrays are shown in the photomicrographs at the corners of the plot. Arrays with different diameters were produced, as shown in the photographs. The mushroom shape has a pull-off force approximately 20 times that for surfaces patterned with fibrils having simple

(a)

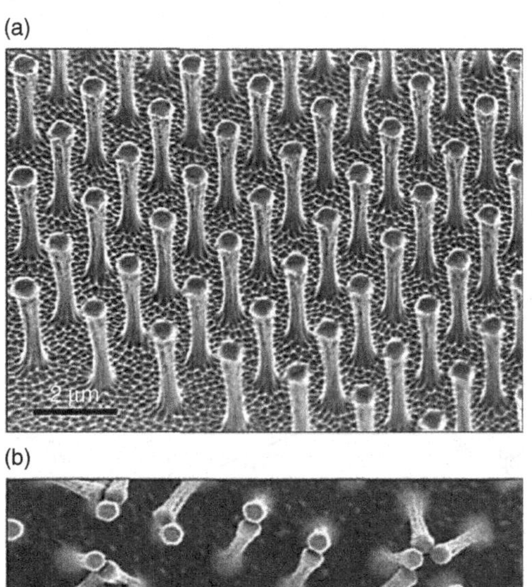

(b)

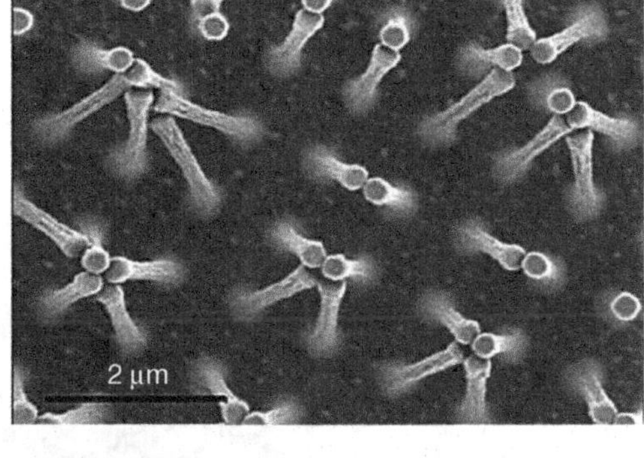

Figure 12.12.

Biomimetic tape using gecko feet as inspiration. (Reprinted by permission from Macmillan Publishers Ltd.: *Nature Materials* (Geim *et al.*, 2003), copyright (2003).)

semispherical cap geometry. For mushroom-shaped fibrils, exceptionally high adhesion strength values, approaching and exceeding that of the natural attachment of the gecko, were found. The sizes of the pillars were also changed, and it was found that the pull-off stress increased with the decrease in radius according to $R^{-1/2}$, as shown in Fig. 12.13. The JKR equation (Johnson *et al.*, 1971) (Section 11.2, Eqn. (11.4)) predicts a stress proportional to R^{-1}. The dependence on the radius is a function of the shape of the spatula.

Mahdavi *et al.* (2008) have developed a gecko-inspired adhesive for biomedical application. They demonstrated that gecko-inspired arrays of sub micrometer pillars of PGSA (polyglycerol sebacate acrylate) (Fig. 12.15) provided attachment under water. This tape is proposed as a biocompatible and biodegradable polymer adhesive to tissue. The adhesion was tested *in vitro* to porcine intestine and *in vivo* to rat abdominal muscle. Carbon nanotube arrays were also proposed to form fibrillar arrays and have self-cleaning properties due to an extreme hydrophobicity (Sethi *et al.*, 2008).

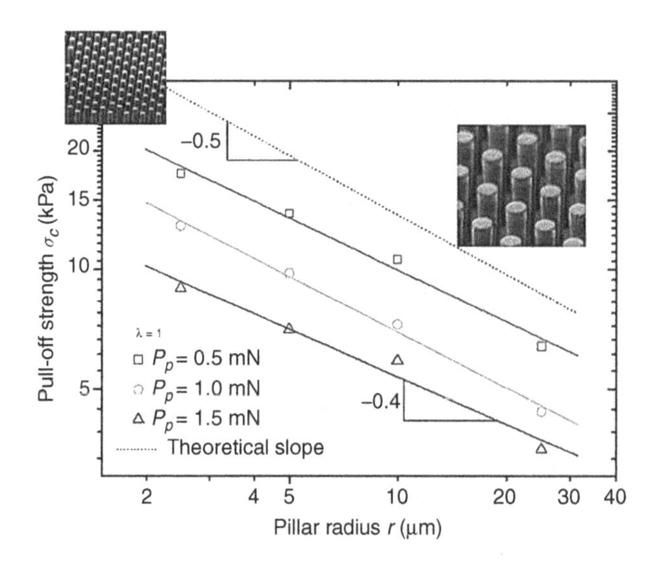

Figure 12.13.

Pull-off strength as a function of pillar radius. (Figure courtesy E. Arzt.)

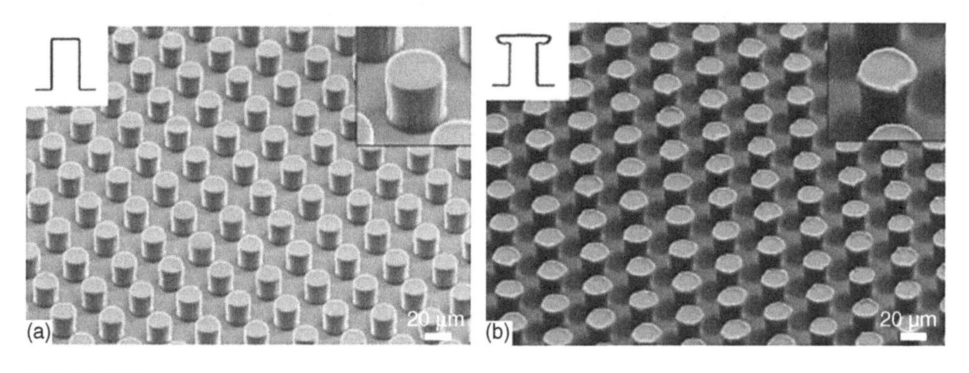

Figure 12.14.

SEM micrographs showing micro-pillar arrays with cylindrical (a) and mushroom-like (b) geometries. Pull-off strength increases with decreasing pillar radius. (Figure courtesy E. Arzt.)

12.1.8 Nacre-inspired structures

Given the widespread interest in the structure and mechanical properties of abalone shell, there have been many attempts to fabricate nacre-like structures. Various *potential* methods have been outlined to fabricate bioinspired materials based on the structure of nacre (Heuer *et al.*, 1992; Mann *et al.*, 1993; Calvert, 1994; Heywood and Mann, 1994;

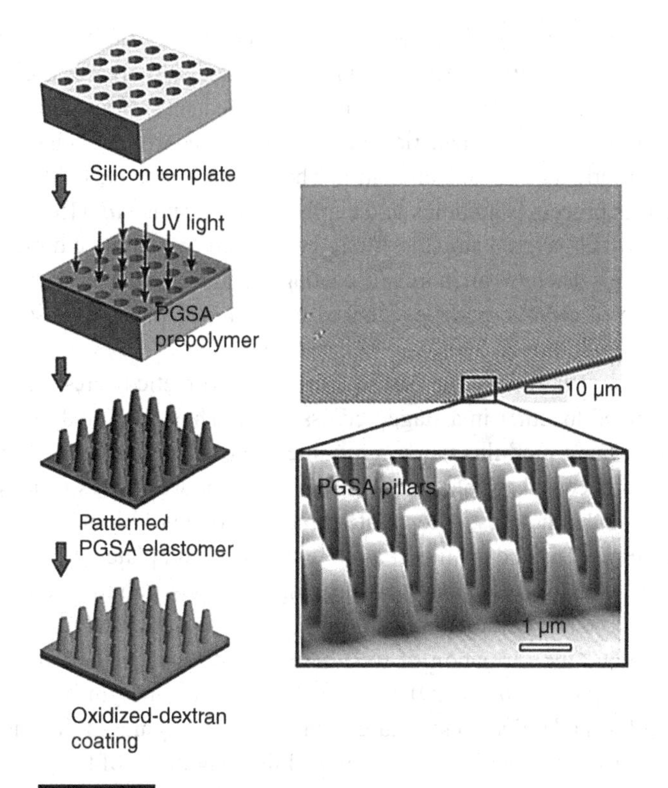

Figure 12.15.

The fabrication procedure of biodegradable gecko-inspired adhesives. SEM micrographs showing PGSA pillar arrays with tip pillar diameter ~500 nm and height ~1 μm. (From Mahdavi *et al.* (2008), with permission; copyright (2008) National Academy of Sciences, USA.)

Arias and Fernández, 2003; Mayer, 2006), and numerous research projects report on the results from successful techniques. Unfortunately, the terms "nacre-like" or "mimic nacre" have been so generously applied that nearly all papers that claim a bioinspired nacre design are simply reports on the fabrication of a one-layer $CaCO_3$/organic structure. These formations are not unexpected considering that this has been the topic of research in the older ceramics literature on ceramic/organic materials. Other papers report on layered nanocomposites of SiO_2/organic, where a three-dimensional structure is obtained. Because of the particularities of SiO_2 gels, and the easy chemical linkages to other molecules and substances, it is not surprising that uniform, layered structures are obtained. Some papers actually report a nacre-like arrangement from nacre itself (demineralization of shell, then synthetic remineralization of the protein).

Heuer *et al.* (1992) proposed to cast successive layers of polymer plus mineral precursor and initiate mineral precipitation through the use of an external field. Other suggestions involve sputtering or physical vapor deposition of ceramic films, but the incorporation of the inorganic phase was not discussed. Arias and Fernández (2003) proposed an elaborate method to fabricate the structure of a mollusc shell. First, an inorganic is chosen so that the nucleation of calcium carbonate occurs on well-defined sites. Nucleation sites are formed by proteins (aspartate or glutamate rich) and a gel is formed from proteins, which will serve as the calcium carbonate nucleation matrix. The growth is stopped by the deposition of an inert layer. This process is complex and cumbersome. Mann *et al.* (1993) have outlined a procedure in which layered structures may grow from two to three dimensions with the inclusion of lipids that have an inorganic coating to self-assemble on the inorganic surface. Laser sintering of ceramic particles stacked between polymers has also been suggested (Heuer *et al.*, 1992; Mayer, 2005).

Jackson *et al.* (1989) were the first to fabricate a synthetic nacre-like structure. Glass slides were glued together in a staggered assembly. They used analytical techniques to calculate the Young modulus using a modified rule of mixtures and found a good fit. Tensile failure occurred by plate pullout. Meldrum and Ludwigs (2007) have formed three-dimensional structures. A template-directed growth of biominerals resulted in sea sponge structures. To produce these, a sea urchin skeleton plate was first infused with a polymer. The skeleton was dissolved, leaving a porous polymer template. Solutions containing Ca ions were filled into the mold, which nucleated and formed $CaCO_3$. They also demonstrated the formation of well-defined $CaCO_3$ crystals deposited on various substrates. Almqvist *et al.* (1999) fabricated several talc/polymer mixtures that were processed to form a laminated structure with 10 wt.% polymer in the end product. Talc has a platy morphology, and it was speculated that alignment of the plates could mimic nacre. Several techniques were used to treat the mixtures: centrifuging, spinning cylinder, spinning plate, shearing plate, and dip coating by a glass slide. A maximum orientation factor of ~50% was found. These composites were weak, and their fabrication was not pursued.

A laminated structure of Al–B_4C was produced by Sarikaya and Aksay (1992). They demonstrated significant increases in toughness. In Fig. 12.16, an increase in the fracture toughness of B_4C ($K_{Ic} \sim 3$ MPa m$^{1/2}$) to about 16 MPa m$^{1/2}$ was shown for the Al–B_4C laminate. Although improvements were achieved in the mechanical properties of synthetic laminate composites (Yasrebi *et al.*, 1990; Khanuja, 1991; Chen *et al.*, 2004) based on biological architecture, these have not been as extraordinary as the one nacre provides in comparison with monolithic $CaCO_3$. This may be due to limited laminate thickness in synthetic composites and the still not yet clearly identified composition and structure, especially of the complex nanolaminated structure in the organic layer. Thus, the potential benefits of complex architectures have not been fully explored yet. Nevertheless, research into bioinspired (specifically, abalone) armor continues, in spite of the severe problems that occurred because of the reaction of Al with B_4C forming Al_4C_3, a very brittle material.

Almqvist *et al.* (1999) tried to reproduce the structure of the abalone by using tiles with the addition of a synthetic polymer for the bonding. The tiles have the composition $Mg_3(Si_4O_{10})(OH)_2$. These particles have dimensions on the order of the $CaCO_3$ tiles in abalone: a thickness of a few hundred nanometers and diameter of 3–10 μm.

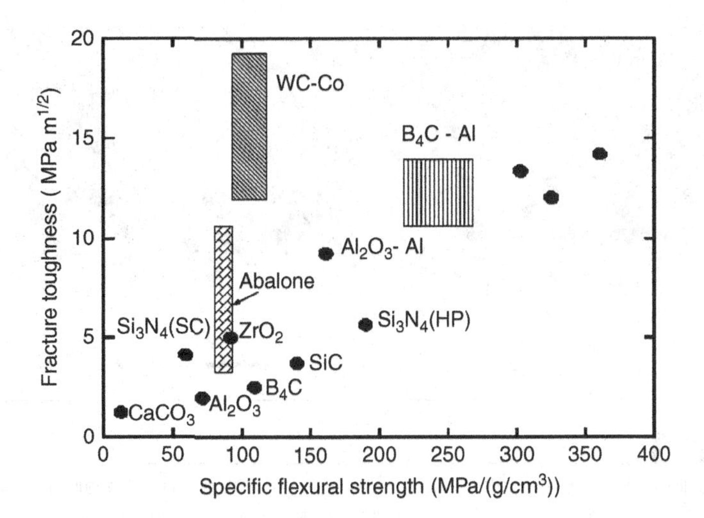

Figure 12.16.

Fracture toughness versus fracture strength of the nacre section of an abalone shell compared to some high-technology ceramic material. (From Sarikaya *et al.* (1990). Copyright © 1989 Materials Research Society. Reprinted with the permission of Cambridge University Press.)

There is considerable industrial application of tile polymer composites, with the polymer being more than 60 wt.%. During extrusion, the particles orient themselves along the polymer flow direction. Almqvist *et al.* (1999) attempted to reduce the polymer content to less than 10% by aligning the tile particles using several techniques including sedimentation, centrifugation, shearing, and spinning. They encountered significant problems in wetting the tile particles with the polymer. They were not able to reproduce the high degree of order existing in nacre, although significant alignment of the tablets was achieved. They concluded that the poor flexure strength achieved (in comparison with nacre) could be increased by using a polymer with greater similarity to the viscoelastic organic component in nacre.

There have been many other attempts at making microlaminates. Tang *et al.* (2003) used an electrical potential to deposit sequentially clay (montmorillonite) platelets and polymer chains, and in this manner produced thin layers of a glue with the tiles arranged much more regularly than the ones produced by Almqvist *et al.* (1999). However, the sequential deposition on a silicon wafer substrate created layers that were only a few micrometers thick, since the individual montmorillonite bricks had thicknesses of only 0.9 nm. This is seen in Fig. 12.17.

There have also been macro-scale attempts at biomimicry of abalone nacre. Tang *et al.* (2003) prepared a laminate that most closely resembles nacre. Through a sequential deposition method, they prepared a laminate of layers of oriented clay platelets (0.9 nm thick) and alternate layers of absorbed poly(diallyldimethylammonium) chloride polycation (MW = 200 000). Building 200 layers resulted in a film 5 μm thick, which could be tested in a modified load frame. Figure 12.18 shows an image of the film along with SEM

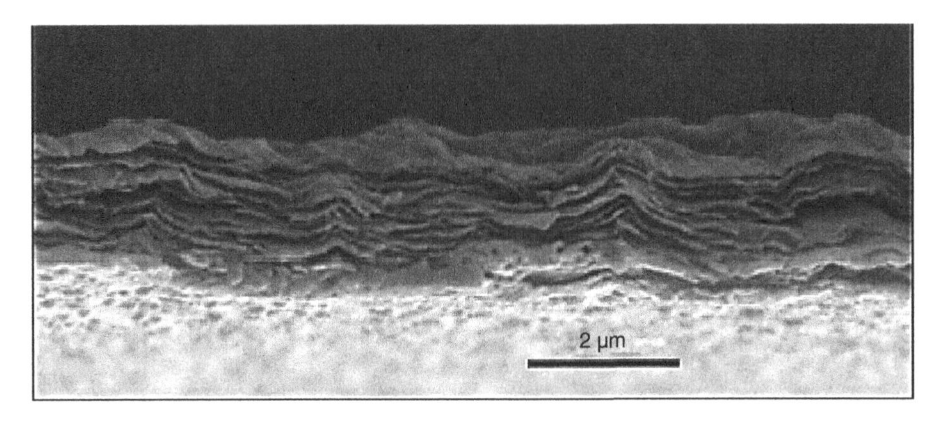

Figure 12.17.

Scanning electron microscopy of an edge of a $(P/C)_{100}$ film. $(P/C)_{100}$= 100 layers of poly (diallyldimethylammonium) chloride/clay laminate. (Reprinted by permission from Macmillan Publishers Ltd.: *Nature Materials* (Tang *et al.*, 2003), copyright (2003).)

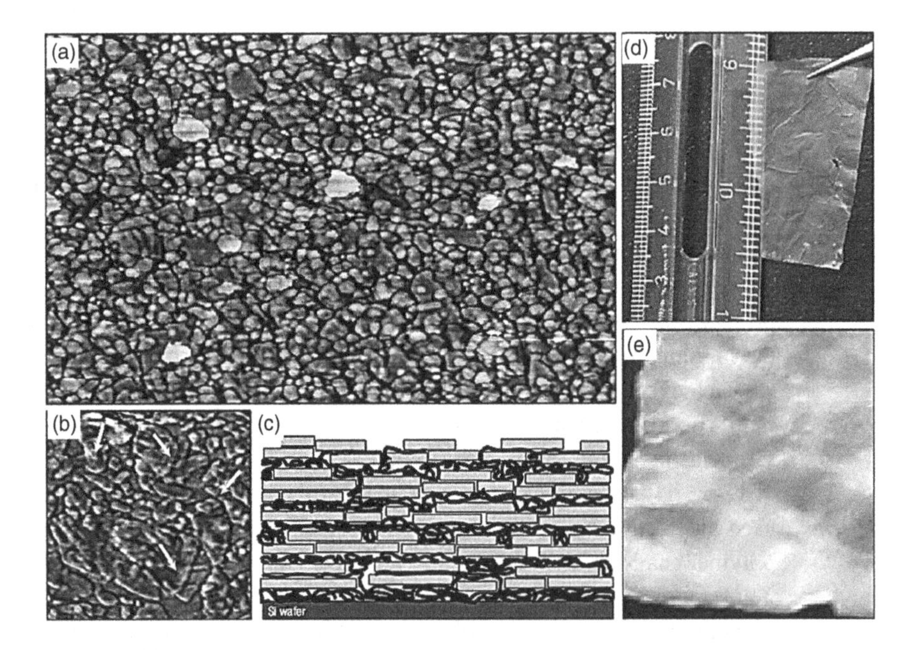

Figure 12.18.

Microscopic and macroscopic description of $(P/C)_n$ multilayers. (a) Phase-contrast AFM image of a $(P/C)_1$ film on Si substrate. (b) Enlarged portion of the film showing overlapping clay platelets (marked by arrows). (c) The $(P/C)_n$ film structure. The thickness of each clay platelet is 0.9 nm. (d) Free-standing $(P/C)_{50}$ film after delamination. (e) Close up of the film in (d) under side illumination. (Reprinted by permission from Macmillan Publishers Ltd.: *Nature Materials* (Tang *et al.*, 2003), copyright (2003).)

micrographs of the cross-section. The volume fraction of polymer was stated to be high, but no values were given. If a 0.9 nm clay layer thickness is assumed and there are 200 layers, this results in a total clay layer thickness of 180 nm, which is 36 vol.% clay. It appears that this material is more of a platelet-reinforced polymer than something that is nacre-like. Another approach with aligned clay particles in a polyimide was reported by Chen *et al.* (2008c). Aligned layers of the mineral separated by the polymer were fabricated by a centrifugal deposition process, which resulted in a composite with an organic fraction of 20 wt.% polymer, higher than for nacre (5 wt.%). Figure 12.19 shows an SEM micrograph of a cross-sectional area along with illustrations outlining the synthesis process. Films of thicknesses of up to 200 μm could be fabricated. The resulting mechanical properties (from nanoindentation) were comparable to the results of Tang *et al.* (2003).

None of these designs incorporates the most important features that contribute to the toughness of nacre: the nanostructural features at the organic/inorganic interface, namely the mineral bridges and surface asperities on the ceramic tile. LAST® armor tiles, built by Foster-Miller (foster-miller.com), consist of SiC or B_4C hexagonal tiles covered in a

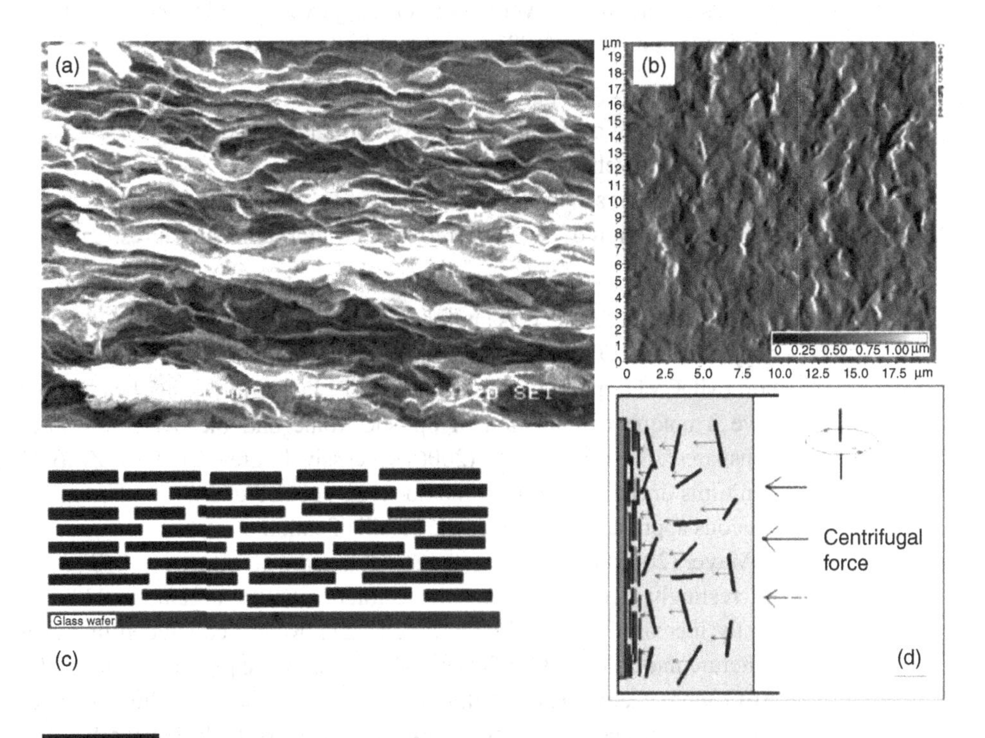

Figure 12.19.

Nacre-bioinspired layered clay/polyimide composites fabricated from centrifugal force. (a) Cross-section of the clay platelets in the composite. (b) Organic platelets inset in polymer. (c) Schematic representation of the ordered nanostructure. (d) Automatic alignment of clay platelets under the centrifugal force. (From Chen *et al.* (2008c). With kind permission of Springer Science and Business Media.)

Figure 12.20.

LAST® armor built by Foster-Miller (foster-miller.com). (Reprinted from Mayer (2006), with permission from Elsevier.)

Kevlar®/thermoset laminate held together with a VELCRO®-type adhesive. This structure provides energy absorption and toughening through many of the same mechanisms as its natural counterpart. The armor, shown in Fig. 12.20, has been implemented onto various ground and air vehicles, including over 1000 Humvees for the US Marines.

Manne and Aksay (1997) considered other approaches to synthesize microlaminates:

(a) Langmuir–Blodgett (LB) deposition – the particles are floated on a water surface after being coated with a hydrophobic layer;
(b) covalent self-assembly;
(c) alternating sequential absorption;
(d) intercalation of organics into layered inorganic structures.

As seen in Section 6.1, marine sponges and other organisms produce silica via aqueous biomediated methods. This approach inspired Schwenzer *et al.* (2006) to grow ZnO thin films. These ZnO films are conventionally grown by metal oxide chemical vapor deposition (MOCVD), pulsed laser deposition (PLD), or molecular beam epitaxy (MBE); they have a unique combination of optoelectronic and piezoelectronic properties, and are transparent. Schwenzer *et al.* (2006) successively grew ZnO and $Zn_5(OH)_8(NO_3)_3 \cdot 2H_2O$ thin films (including indium tin oxide (ITO) coated glass) on different substrates using an aqueous approach that mimics the biosilicification of marine sponges.

Mayer (2005) fabricated a staggered, layered structure of Al_2O_3 and a polymer glue. Interestingly, decreasing the volume fraction of the organic to 11% from 18% significantly increased the toughness (Fig. 12.21). Mayer concluded that even in this simple structure multiple modes of energy absorption were present. Zhao *et al.* (2003) prepared Al_2O_3/epoxy laminates with thin plates (0.6 mm) that were dipped into epoxy vinyl ester resin then stacked to produce a laminate five layers thick. The work of fracture was found to be six times greater for the laminates than for pure Al_2O_3, as observed by Mayer (2005), and the shear strength improved with additions of filler to the epoxy resin.

By far the most successful method has been adopted by Tomsia and co-workers (Deville *et al.*, 2006a,b; Munch *et al.*, 2008). A very innovative approach was

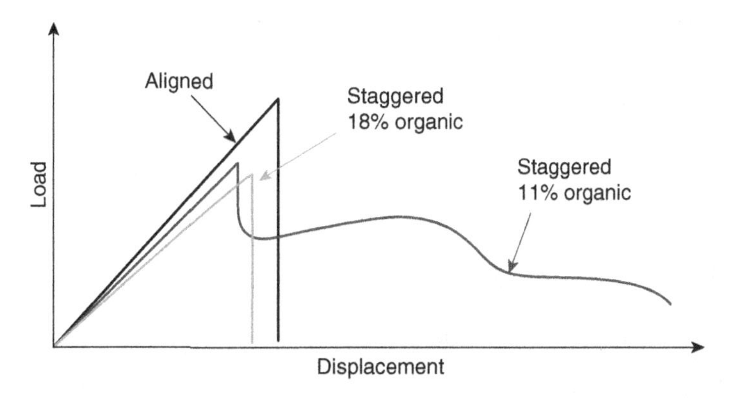

Figure 12.21.

Three-point-bend tests on an aligned, stacked laminate of Al_2O_3 and polymer glue, and staggered laminates with different amounts of the glue. (Figure courtesy G. Mayer.)

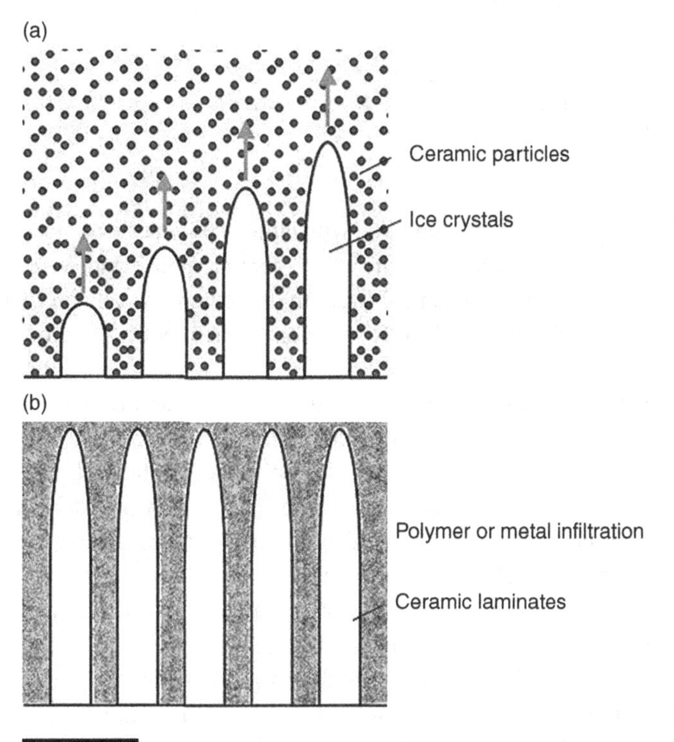

Figure 12.22.

Lawrence Berkley Laboratory synthesis of laminates through the growth of ice crystals (ice templating).

demonstrated by this group. They used the directional growth of ice crystals (dendrites) from a surface as the basis of their synthesis technique. As the ceramic slurry is frozen, the ice crystals expel the ceramic particles. Figure 12.22 shows the ice crystals forming columns or lamellae. The layers are as thin as 1 μm. The porous ceramic scaffolds formed

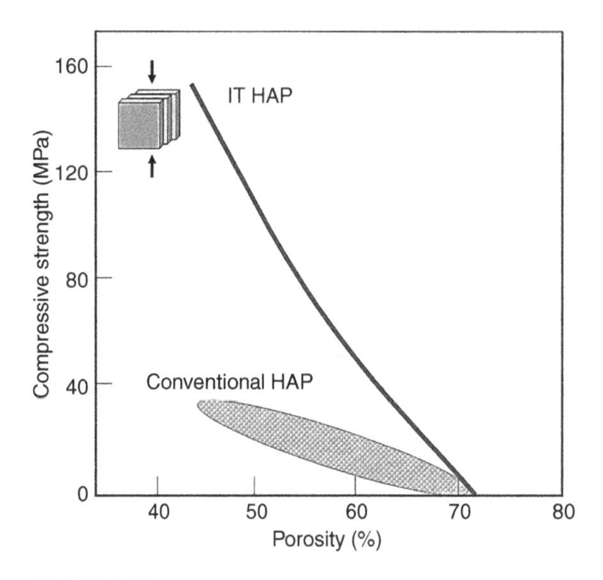

Figure 12.23.

Comparison of conventional porous hydroxyapatite and material produced by ice template (IT) method. (Reproduced based on Deville *et al.* (2006a, b).)

by sublimation of the ice are then filled with either a polymeric or metallic phase. Deville *et al.* (2006a, b) used hydroxyapatite as the ceramic phase and were able to achieve a controlled porosity. The mechanical strength of the porous skeleton (without infiltration of metal or ceramic) was significantly superior to that of porous hydroxyapatite produced by other techniques (Fig. 12.23).

A laminated structure resulted, including "mineral bridges" that formed from dendrites that grew perpendicular to the main growth direction. After the water was sublimated, the scaffold was filled with epoxy or metals. Increased strength and fracture toughness were obtained. Figure 12.24 shows a schematic of the ice-templating process, a comparison of the microstructure of nacre with the artificial layered composite (PMMA and alumina) and stress–strain curves that compare nacre with the layered composite. The sintered laminate is shown in Fig. 12.25(a) with high-magnification micrographs of the bridge area in Figs. 12.25(b) and (c). Filling the empty spaces with PMMA resulted in a final product of a solid ceramic-based laminate composite with a high yield strength and fracture toughness. For an 80% Al_2O_3–20% PMMA laminate, a tensile strength of 200 MPa and fracture toughness of 30 MPa $m^{1/2}$ are obtained (Fig. 12.26(a)). These values represent specific properties comparable to those of aluminum alloys. It should be noted that this is accomplished with the majority being alumina. The resistance to crack propagation (R-curve behavior) is shown in Fig. 12.26(b). The increase in toughness with crack propagation is obvious. The K_J vs. crack extension curve rises to K_{Ic} = 30 MPa $m^{1/2}$. This is in

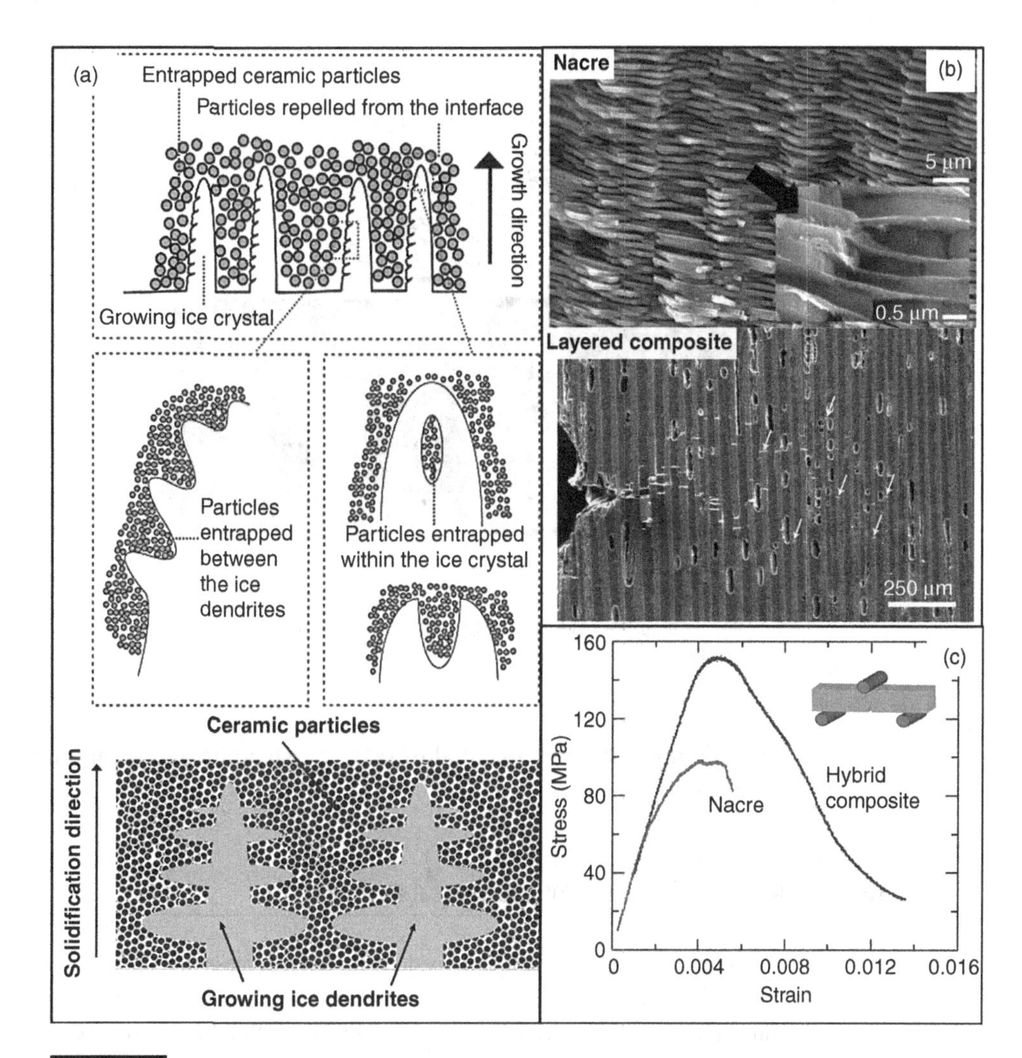

Figure 12.24.

Ice-templating procedure and results. (a) Directional solidification of an aqueous ceramic (Al_2O_3) slurry results in ice dendrites that expel the particles. After solidification, the ice is removed and filled with a polymer (PMMA). (b) SEM micrographs comparing nacre with the synthetic layered structure. (c) Stress–strain curves for nacre and the layered composite material. (Adapted from Deville (2006a) and Munch *et al.* (2008); with kind permission from Dr. Deville and Professor Ritchie.)

stark contrast with monolithic alumina, which has a flat curve with $K_{IC} = 1$ MPa m$^{1/2}$. One unique aspect of the architecture of this bioinspired material is that when it has the abalone brick-and-mortar structure with bridges between the tiles, the toughness is increased over the purely lamellar structure (~15 MPa m$^{1/2}$). The more conventional simple laminate structure is definitely less tough than the new material. This

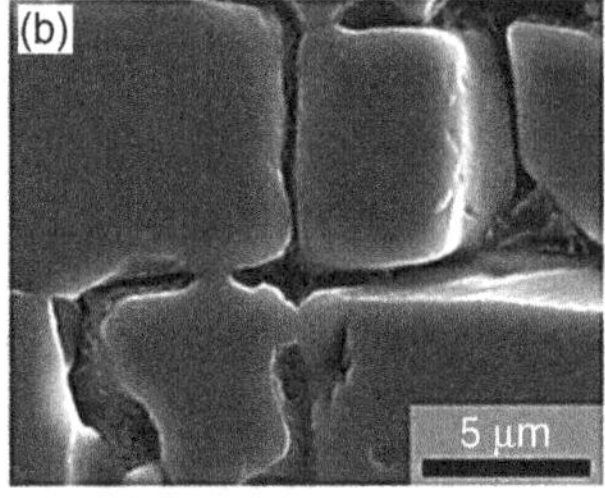

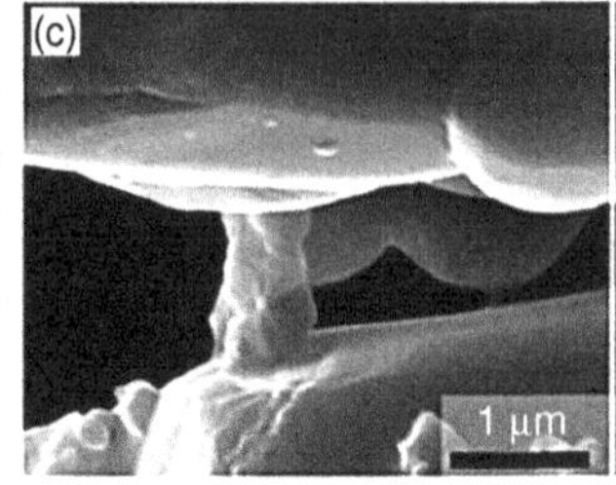

Figure 12.25.

Microstructure of abalone nacre-inspired Al_2O_3/PMMA composites in (a) lamellar and (b) brick-and-mortar forms produced by freeze-casting techniques. (c) Bridge between two tile layers. (Reprinted from Launey *et al.* (2009), with permission from Elsevier.)

example illustrates the potential of (a) understanding and (b) mimicking biological materials.

12.1.9 Marine adhesives: mussel byssal attachment

When developing an anchoring system for marine vessels, why not look at the systems which have truly stood the test of time? The tides of the oceans have provided a tumultuous environment for marine life since the beginning of life on this earth. Organisms such as the common mussel have been forced to develop remarkable methods of attaching themselves onto structures in order to avoid being swept away. The mussel byssus threads seen in Fig. 12.27 are 5 times tougher and 16 times more extensible than the human tendon (Holl *et al.*, 1993; Waite *et al.*, 2004); furthermore, the natural glue which attaches each thread to a surface is stronger than any man-made marine adhesive (Smeathers and Vincent, 1979). The creation of commercial materials with these properties could be possible after first understanding the mechanisms with which nature produces them.

Mussels attach themselves to rocks through byssal threads. These threads have special mechanical properties. They have two regions: the proximal (stiffer) and the distal (adjacent to the attachment site). Figure 12.27(a) shows a California mussel (*Mytilus californianus*) attached to Plexiglas®. The elongated pads (marked by arrows) constitute

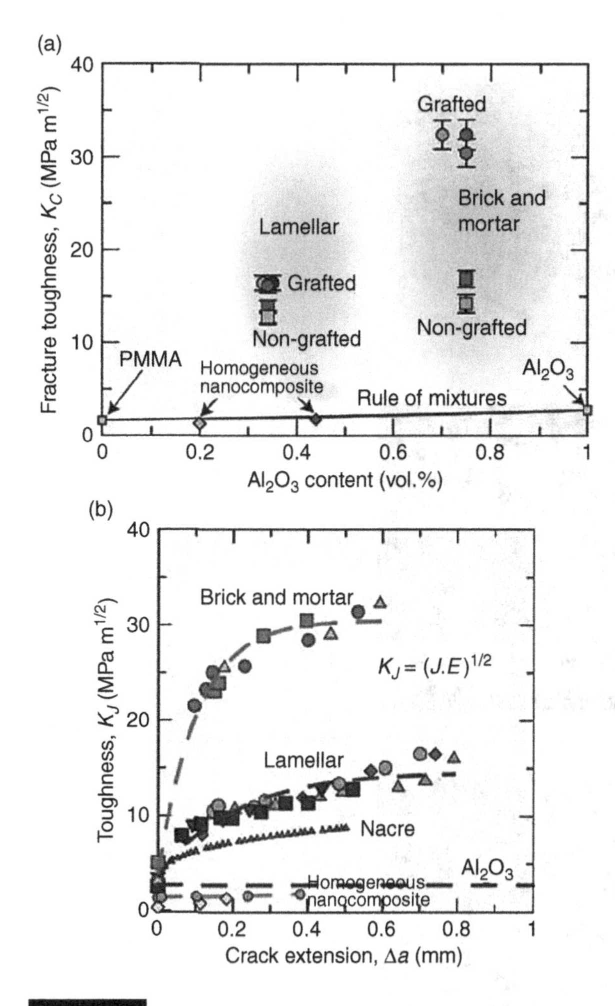

Figure 12.26.

(a) Fracture toughness for lamellar and brick-and-mortar composites compared with that for Al_2O_3 and PMMA. Both types of composites are significantly tougher than their constituents. (b) The composites show significant rising R-curve behavior due to micro- and nano-scale toughening mechanisms. (Adapted from Munch *et al.* (2008); with kind permission from Professor Ritchie.)

the attachment plaques. The attachment of the byssal threads to rock constitutes a fascinating and still only partially understood phenomenon.

Water is an impediment for most glues, which use a combination of mechanical interlocking with noncovalent interactions: salt bridges, hydrogen bonding, and van der Waals bonding. The attachment of the mussel byssus to rocks occurs underwater. A freeze-fractured section of the attachment plaque is shown in Fig. 12.27(b). The core is porous, and this porous region acts as a shock absorber, distributing the stresses evenly over the interface. Waite *et al.* (2005) identified the principal factors in the

(a)

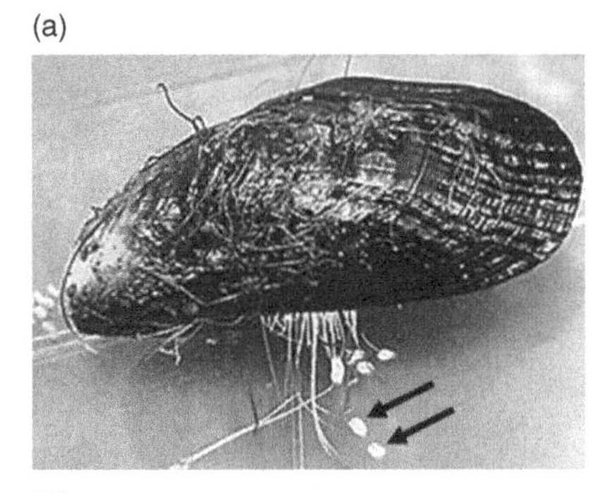

(b)

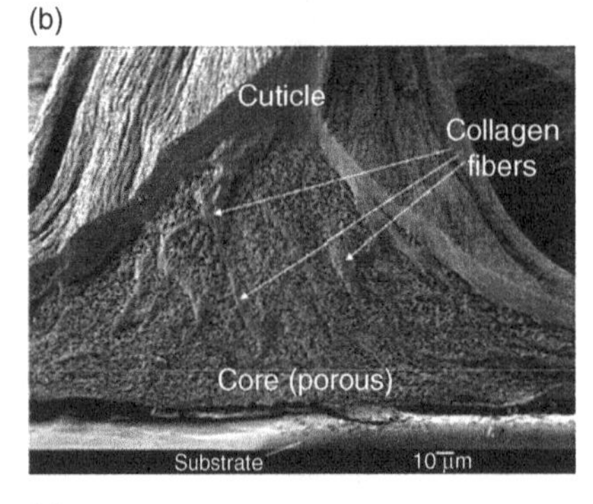

(c)

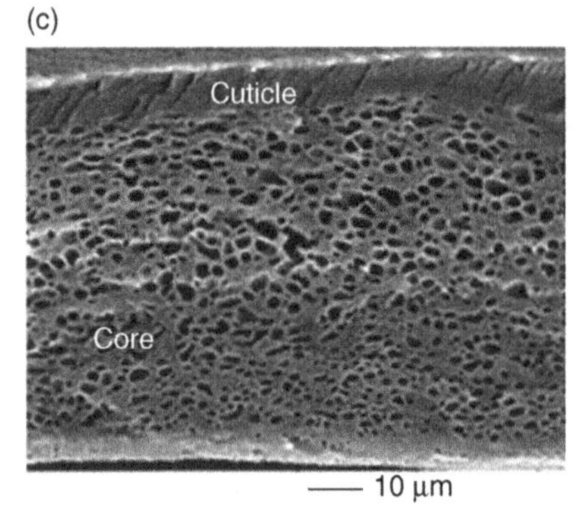

Figure 12.27.

(a) California mussel (*Mytilus californianus*) attached to Plexiglas®. (b) Attachment plaques. (c) Cross-section of attachment plaque showing the porous core. (Adapted from Waite *et al.* (2005), reprinted by permission of the publisher (Taylor & Francis Ltd.).)

attachment of the mussel. The formation of this plaque proceeds by the squirting of adhesive from a gland near the tip of the mussel foot. This squirt is in the form of granules with diameters 1–2 μm that coalesce at the surface, forming the adhesive and creating the foaming plaque. It is also believed that a plunger-like action squeezes the water away from the interface. This still does not explain the high strength of the bond. The answer seems to lie in the presence (30 mol.%) of 3,4-dihydroxyphenylalanine (DOPA) in the plaque. Guvendiren *et al.* (2009) have conducted AFM (atomic force microscopy) experiments using DOPA in the cantilever tip and found that the attachment energy to TiO_2 is very high: 130 kJ/mole. Such a high energy considerably exceeds noncovalent interactions and suggests that some type of charge transfer mechanism occurs at the surface.

There has been considerable effort made to mimic the mussel byssal thread adhesion by Allied Signal, but with only limited success. Collaborative launched a product containing DOPA secreted from *Mytilus edulis* under the trade name BD Cell-Tak®. This product is used to attach cells or tissue sections to many types of surfaces: glass, metals, and polymers.

Lee *et al.* (2007a) reported a method to form multifunctional polymer coatings through the simple dip-coating of objects in an aqueous solution of dopamine. Inspired by the composition of adhesive proteins in mussels, they used dopamine self-polymerization to form thin, surface-adherent polydopamine films onto a wide range of inorganic and organic materials, including noble metals, oxides, polymers, semiconductors, and ceramics. They further synthesized a reversible wet/dry adhesive inspired by mussels and geckos named "Geckel" by coating DOPA-like polymer onto the nanofabricated polymer pillar arrays (Lee, Lee, and Messersmith, 2007b), as shown in Fig. 12.28. Underwater adhesive strength increased nearly 15-fold when coated with mussel-mimetic polymer. The system maintains its adhesive performance after over 1000 contact cycles in both dry and wet environments. This hybrid adhesive, which combines the design elements of both gecko nanostructure and mussel adhesives, may lead to applications in biomedicine, construction, robotics, and marine engineering.

12.1.10 Sonar-enabled cane inspired by bats

It is well established that bats use a sonar-like system in the dark for navigation and to prevent collisions. A similar approach has been used in the design of a cane for visually impaired people (the UltraCane). It is a regular cane supplemented by an echolocation device. The person receives the information into vibrating buttons in the handle and can establish the location and distance of obstacles.

12.1.11 Butterfly wings

The beautiful, iridescent wings of butterflies have fascinated both nature lovers and, more recently, materials scientists. The genus *Morpho* has over 80 species of butterfly that are found mainly in Latin America. Additional detail is presented in Chapter 11. The blue color arises from an intricate, periodic nanostructure on the surface of the wing, not from pigmentation. Figure 12.29(a) shows a TEM cross-sectional micrograph of *Morpho*

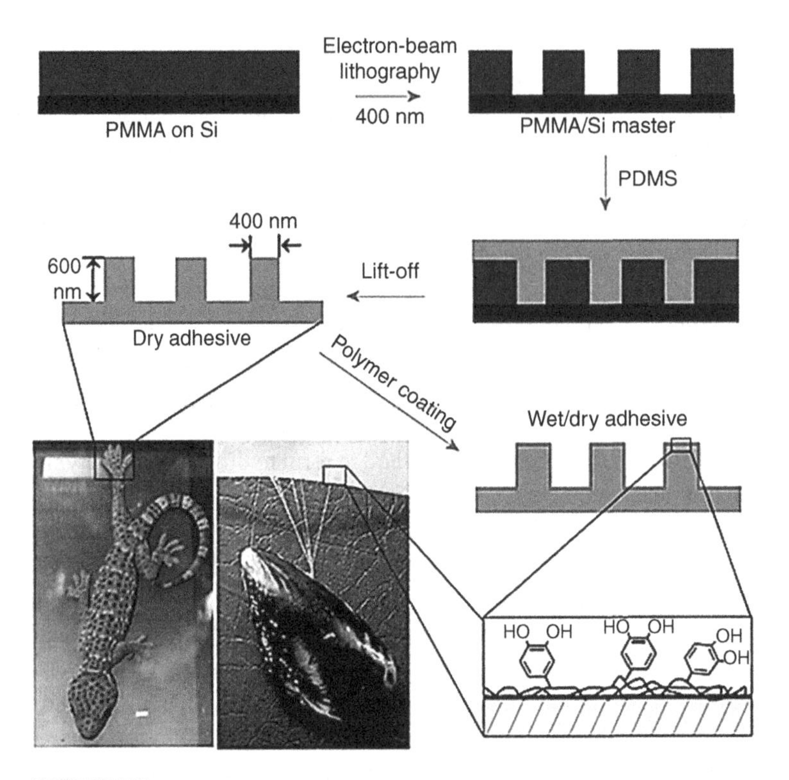

Figure 12.28.

Design and fabrication of wet/dry hybrid nanoadhesive. Electron-beam lithography was used to create an array of holes in a PMMA thin film supported on Si (PMMA/Si master). PDMS casting onto the master is followed by curing, and lift-off resulted in gecko-foot-mimetic nanopillar arrays. Finally, a mussel adhesive protein-mimetic polymer is coated onto the fabricated nanopillars. The topmost organic layer contains catechols, a key component of wet adhesive proteins found in mussels. (Reprinted by permission from Macmillan Publishers Ltd.: *Nature* (Lee *et al.*, 2007b), copyright (2007).)

butterfly wing (Potyrailo *et al.*, 2007). The dorsal surface has longitudinal ridges (R) that have cross-ribs that create air pockets. The ridges are ~0.1 μm thick and spaced apart by 0.7 μm, forming a lamellar structure. The ridges combined with the cross-ribs have a structure similar to that of a tree – the cross-rib length increases from the top to the base. Figure 12.29(b) shows an SEM micrograph of a fracture surface of the wing that points out the various micro- and nanostructural features.

The surface morphology of the wings has been investigated for chemical sensing properties and for use as templates for photonic crystal growth. Due to the large surface area, General Electric is developing sensors based on the *Morpho* butterfly wing. Figure 12.30(a) shows a schematic diagram of the wing scales. The wings were exposed to various chemicals with different concentrations and the differential reflectance spectra were obtained. In Fig. 12.30(b), the spectral response to water, and methanol and ethanol

(a)

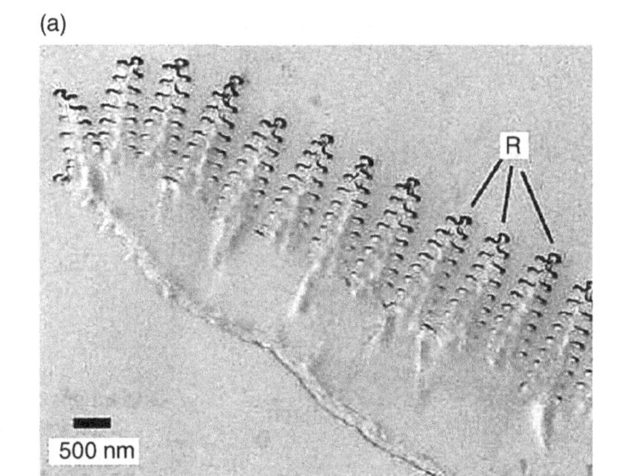

(b)

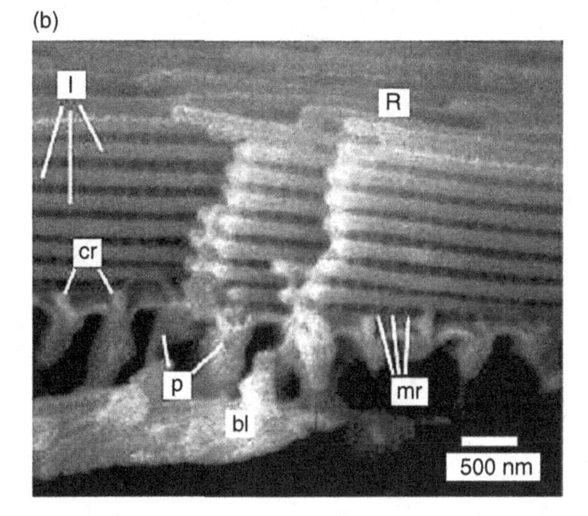

Figure 12.29.

Morpho butterfly wing structure: (a) TEM showing ridges (R) and associated lamellae; (b) longitudinal SEM view of fractured surface showing the ridges (R), lamellae (l), microribs (mr), pillars (p), and cross-ribs (cr). (Reprinted by permission from Macmillan Publishers Ltd.: *Nature Photonics* (Potyrailo *et al.*, 2007), copyright (2007).)

vapors, was obtained, indicating that this structural morphology is very sensitive to chemical species.

Several groups have attempted to mimic the structure. Figure 12.31 shows the structure produced by focused-ion-beam etching combined with chemical vapor deposition (Watanabe 2005a, b), and Fig. 12.32 shows a ZnO network structure produced by dip-coating *Ideopsis similis* butterfly wings in an aqueous precursor solution (Zhang *et al.*, 2006). Other methods include alternate TiO_2 and SiO_2 (different indices of refraction) thin films from electron-beam deposition (Saito *et al.*, 2006). Al_2O_3 nanostructures were fabricated by coating wings through low-temperature atomic layer

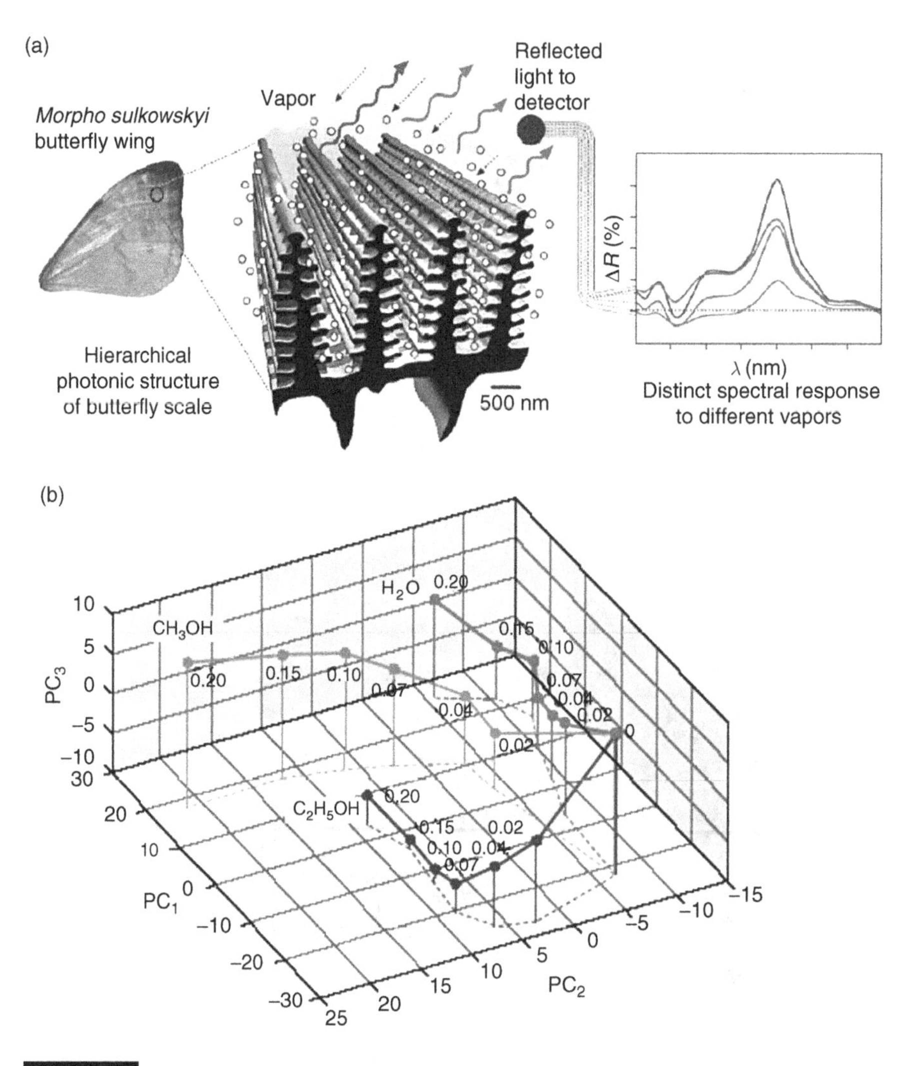

Figure 12.30.

(a) Highly selective vapor response based on hierarchical photonic structure of the wings of the *Morpho sulkowskyi* butterfly. Differential reflectance spectra were used to measure the response to different vapors and concentration of vapors. (b) Spectral response of water, and methanol and ethanol vapors, using principal components (PCs) analysis of reflectivity spectra. Numbers represent partial pressures. (Reprinted by permission from Macmillan Publishers Ltd.: *Nature Photonics* (Potyrailo *et al.*, 2007), copyright (2007).)

deposition (ALD). Changing the thickness resulted in changing the reflection spectrum and thus the color observed (Huang, Wang, and Wang, 2006). Al_2O_3/TiO_2 multilayers on polymer films were also produced by ALD that had monodispersed micron-sized pores formed from a templated assembly (Crne *et al.*, 2011). This resulted in reflections of blue and yellow light that when mixed produced the green color observed on the

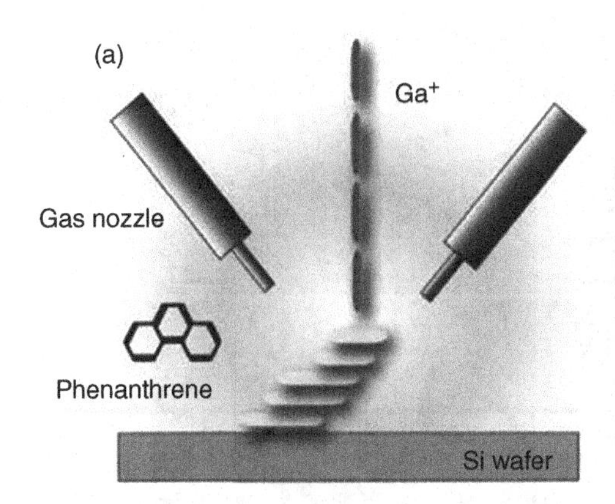

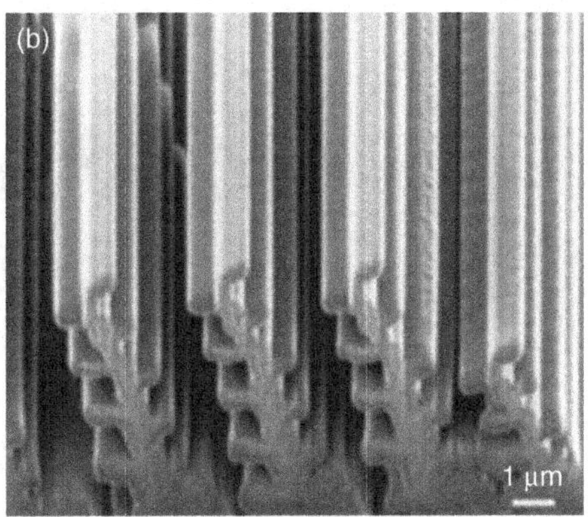

Figure 12.31.

(a) FIB-CVD process used to fabricate biomimetic wing scales. (b) Inclined SEM micrograph of the nanostructure. (Reprinted with permission from Watanabe *et al.* (2005b). Copyright (2005), American Vacuum Society.)

wing of the *Papilio palinurus* butterfly. The color mixing effect was also found by Kolle *et al.* (2010) who combined layer deposition methods (colloidal self-assembly, sputtering, and ALD) that created photonic structures mimicking the color mixing effect found on the wings of the *Papilio blumei*. Kustandi *et al.* (2009) used nano-imprint lithography and shear patterning to stamp out tiled, multilayered nanostructures on a polymer film. The nanostructural features are shown in Figs. 12.33(a)–(c) at different magnifications, demonstrating that this more cost-effective approach can replicate the wing nanostructure. The periodic ridge structure creates a diffraction grating, and the multilayers create interference and scattering. This results in tunable reflectance properties, as shown in Figs. 12.33(d)–(f).

The reflective properties of wings have been used for Qualcomm's Mirasol® display, a microelectromechanical systems (MEMS) based technology that is using the reflective properties of materials, similar to butterfly wings. The display is a low-power device that is readable in sunlight.

12.1.12 Origami structures

In spring, the leaves of the chestnut tree come out of the cocoon and unfold in a wonderful manner. Sitting on a bench in the Jardin du Luxembourg in Paris, one can observe these mysterious ways of nature as the cocoons open and leaves unfurl. Similarly, the wings of beetles unfold as their carapace is lifted. These wings fold back when the beetle returns to its ambulatory configuration. This process of folding and unfolding of leaves is known as Ha-ori in Japanese and has inspired researchers. Biruta

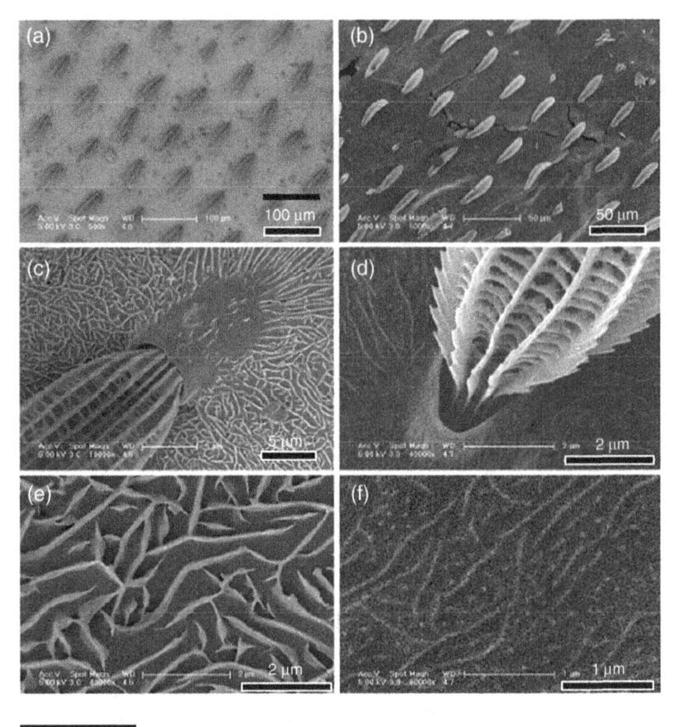

Figure 12.32.

SEM images of *Ideopsis similis* butterfly wings (left) and ZnO replica (right) at increasing magnification. (Taken from Zhang *et al.* (2006), with kind permission of Dr. Zhang.)

Kriesling was inspired by the chestnut trees in Paris and teamed up with Koryo Miura. This has led to the development of deployable solar panels in space vehicles based on Miura's origami pattern. The collapse of cans enabled by a pattern of pseudo-cylindrical concave polyhedrons is well known in Japan.

12.1.13 Self-healing composites

Self-healing is a property of bone, and new research is directed toward developing composite materials that can repair cracks after they have formed. The basic concept is to embed two different liquid precursors that will polymerize when in contact. If a crack propagates through both agents, they will leak out into the crack, polymerize, and thereby hold the crack faces together. This concept has been used with glass tubes filled with the healing agent and small pockets of catalyst and with microcapsules housing the healing agent with the pockets of catalyst (White *et al.*, 2001). Figure 12.34(a) shows the biological mechanism of self-healing in bone. Osteoclasts remove bone,

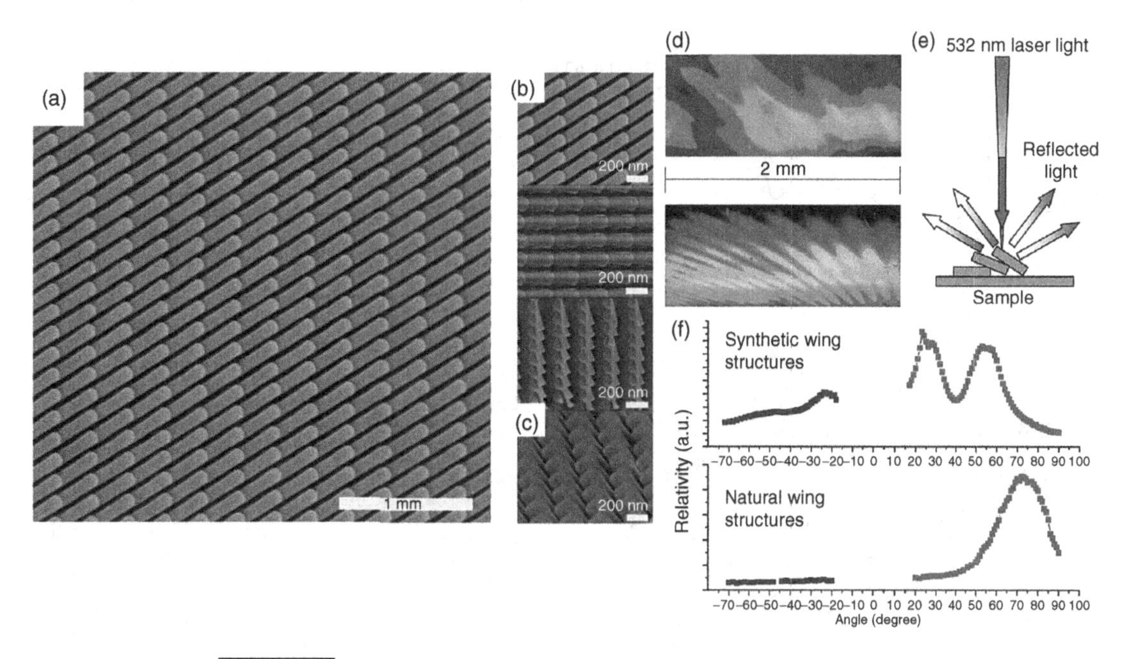

Figure 12.33.

Synthetic butterfly wing nanostructures produced by nanoimprint lithography and shear patterning. (a) Stacked polycarbonate nanostructures oriented by a horizontal force during the demolding process. (b) Higher-magnification SEM images of the overlapping nanostructures (top view). (c) Cross-sectional view of tiled multilayered structures comprising 50 nm thick air gaps. Tunable iridescent effects of synthetic butterfly wing nanostructures. (d) Strong single and rainbow-type color patterns when the incident collimated white light beam is turned 360° around the vertical axis. (e) Measurement system used to obtain a reflectivity spectrum of the natural and synthetic butterfly wings. (f) Reflection intensity distribution versus angle along the ridge direction. (Reprinted from Kustandi *et al.* (2009); with kind permission from John Wiley & Sons, Inc.)

forming resorption cavities; at the same time, osteoblasts deposit new bone, which is most effective under an applied load. This continuous remodeling of bone allows for self-healing of breaks and cracks in bone. Figure 12.34(b) shows the microcapsule concept illustrating how the healing agent leaks out into the crack but polymerizes to a hard polymer when the catalyst is impinged upon by the crack front. Figure 12.34(c) demonstrates evidence of self-healing. The epoxy resin with the microcapsules and catalyst has a higher strength than the resin alone. After a crack has been introduced and allowed to propagate, the self-healed composite does not immediately break in half, rather it retains significant structural integrity to have a strength over that of the pure epoxy resin.

A composite with microvascular channels that can self-heal has been developed by Toohey *et al.* (2007). This assembly behaves like skin, as surface cracks that develop can repair, in a process similar to what has been developed for bone-inspired self-healing composites. Figure 12.35(a) shows a schematic diagram of the

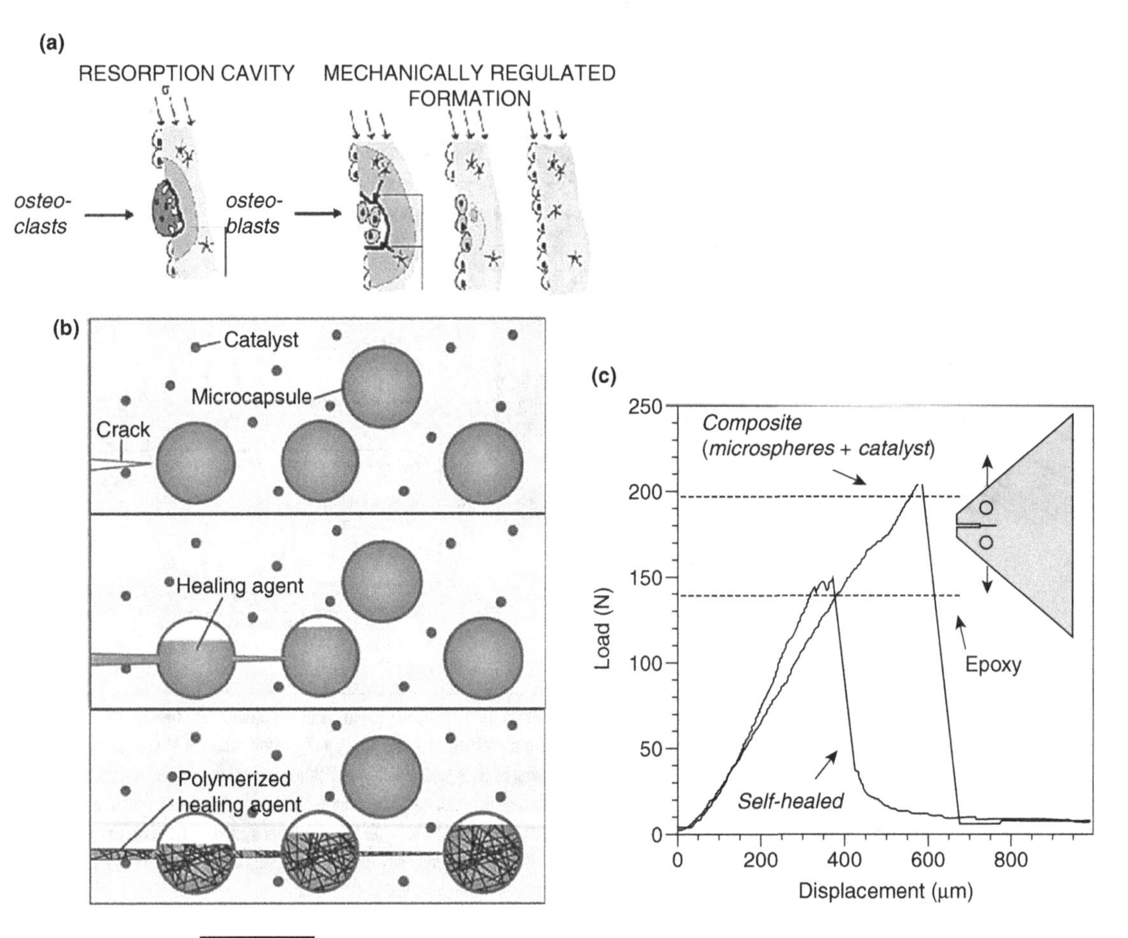

Figure 12.34.

(a) Biological mechanism of self-healing in bone. Osteoclasts break down bone, creating resorption cavities, and osteoblasts deposit new bone. Stress is needed for efficient operation. (b) Design of a self-healing composite. Microcapsules of a healing agent are embedded in an epoxy matrix along with a catalyst. The crack ruptures the microcapsules and releases the healing agent. Polymerization occurs when the catalyst is contacted, thereby bonding the crack faces. (c) Mechanical integrity of the self-healed composite. ((b) and (c) taken from White *et al.* (2001). Reprinted by permission from Macmillan Publishers Ltd.: *Nature* (White *et al.*, 2001), copyright (2001).)

dermis, illustrating the vascular network at the surface. Figure 12.35(b) shows the skin-mimicked structure consisting of a hard epoxy layer that is embedded with the catalyst on top of an array of channels that contain the healing agent. Cracks of sufficient length that form on the surface layer and impinge on the channel array are healed (Figs. 12.35(c) and (d)) through the polymerization of the healing agent with the catalyst. The stress–strain curves are shown in Fig. 12.35(e), demonstrating that the initial strength and stiffness of the original composite can be retained, even after crack development.

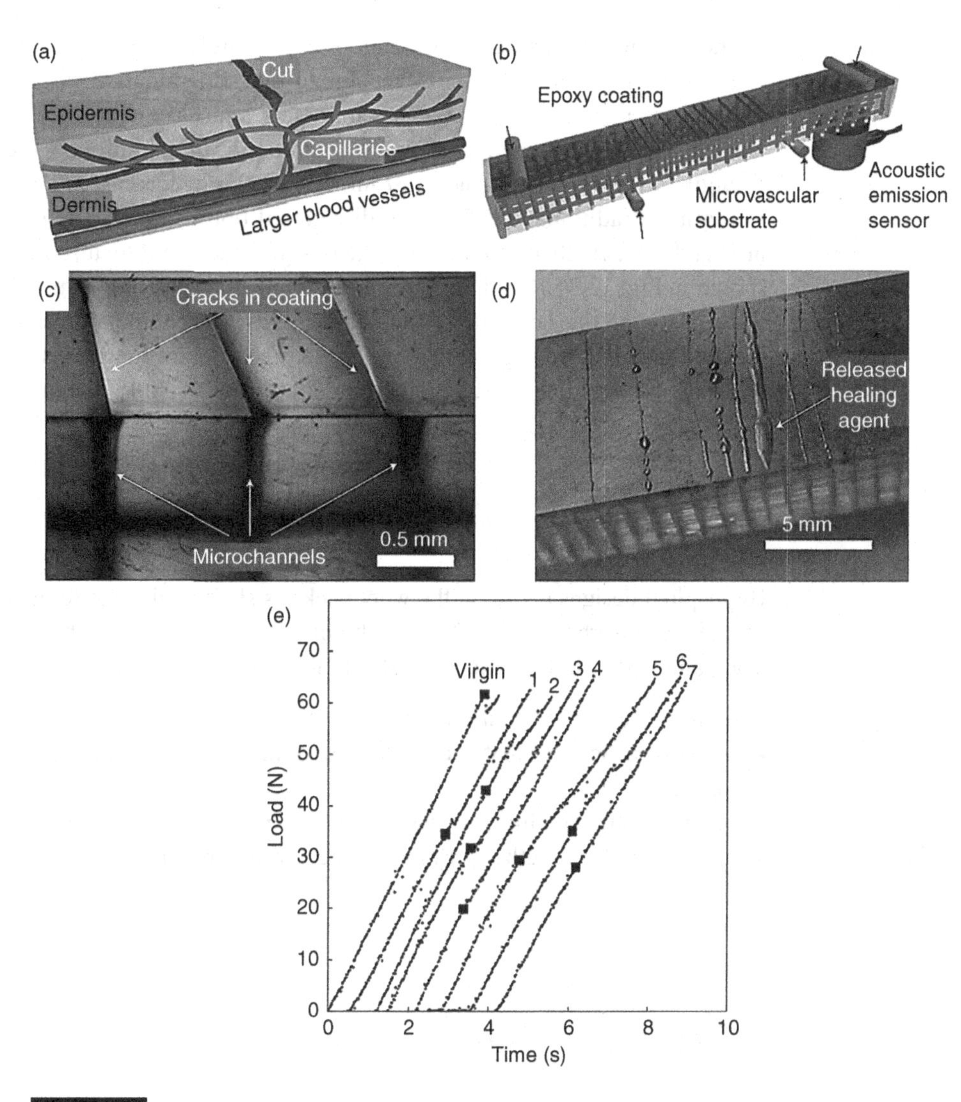

Figure 12.35.

Self-healing skin-like composites. (a) Skin dermal layer, illustrating the vascularity. (b) Bioinspired skin-like composite consisting of an epoxy layer with the catalyst embedded in it on top of an array of microchannels that contain the healing agent. (c), (d) When cracks develop and impinge into the microchannel array, the healing agent and catalyst combine and polymerize to close the crack. (e) Stress–strain curves (offset for clarity) of the virgin epoxy and the self-healed composites that demonstrate that the strength and stiffness can be retained in the self-healed material. (Reprinted by permission from Macmillan Publishers Ltd.: *Nature Materials* (Toohey *et al.*, 2007), copyright (2007).)

12.1.14 Sheep-horn-inspired composites

The sheep horn has keratin filaments that are not only embedded parallel to the growth direction but also extend from one layer to the next (Section 8.4). These cross-ply fibers aid in decreasing delamination by strongly holding together the layers. Composite materials companies recognize that delamination is the most common failure mode for layered composite materials and have fabricated composites that are cross-stitched together. However, although this improves the delamination strength, the presence of holes from the stitching decreases the overall fracture strength. A traditional 0°/90°/0° unidirectional fibrous composite structure is shown in Fig. 12.36. A novel composite that is similar to the structure of animal horns is shown in Fig. 12.37 (Veedu *et al.*, 2006). A "forest" of carbon nanotubes is grown on the surface of the laminate, which then holds the composite plies together. This resulted in enhanced mechanical properties: the fracture toughness was increased by ~350%, the flexural modulus increased by ~100%, and the flexural toughness increased by ~525% over the base composite.

12.1.15 Shock absorbers based on woodpecker's head

Bioinspired designs based on the woodpecker's skull could lead to the development of new shock absorbers for airplane black boxes and helmets and other safety equipment. Yoon and Park (2011) fabricated a shock-absorbing system that consisted of:

- a high-strength external layer, which acts like the beak;
- a viscoelastic layer that evenly distributes mechanical vibration, which acts like the hyoid;
- a porous structure that suppresses the transmission of vibrations;
- another high-strength layer that has the porous structure of the skull bone.

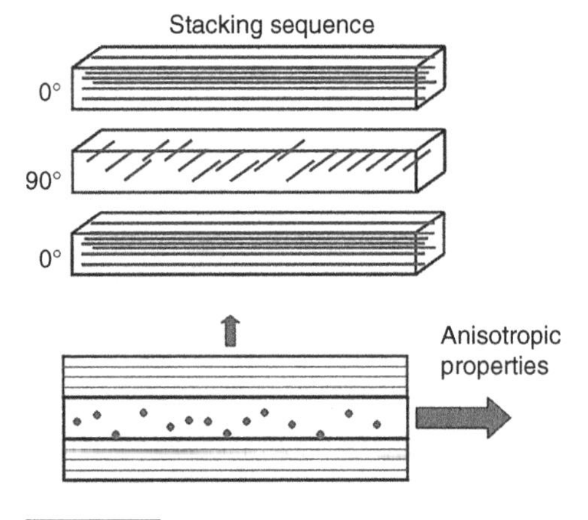

Figure 12.36.

Traditional stacking sequence of 0°/90°/0° unidirectional fibrous composites. The properties are highly anisotropic, with strength and stiffness optimized in one direction.

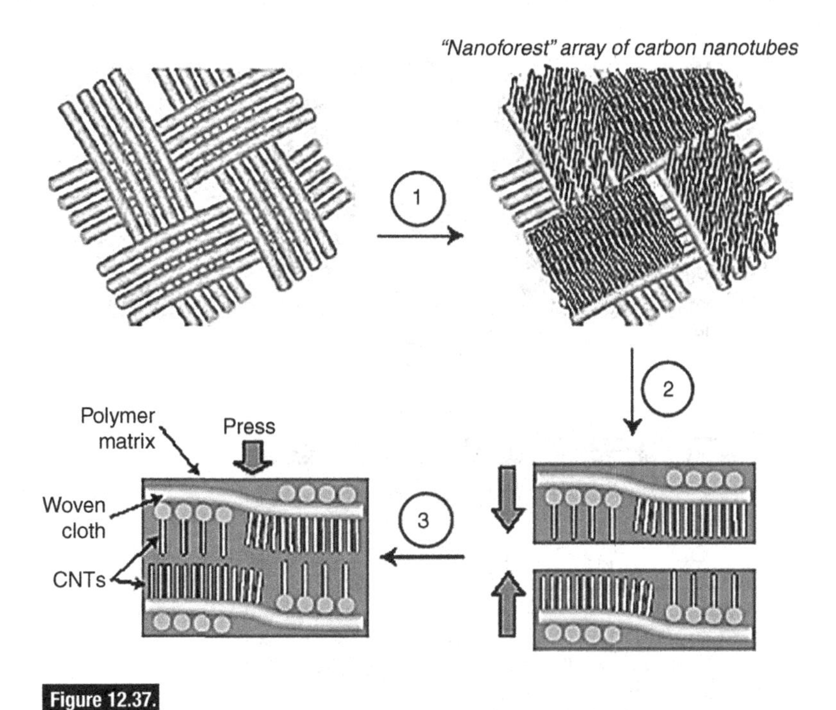

Figure 12.37.

Using a carbon nanotube forest, the laminates in this composite are stitched together to form a stronger, stiffer composite. (Reprinted by permission from Macmillan Publishers Ltd.: *Nature Materials* (Veedu *et al.*, 2006), copyright (2006).)

In Fig. 12.38, the "beak" is fabricated from steel, the "hyoid" from an elastomer, the "spongy bone" from glass microspheres, the "skull bone" and "cerebral fluid" from aluminum, and the "brain" from micromachined devices. The system was fired from a gas gun and survived up to 60 000 g, indicating that a superior shock-absorbing device was fabricated.

12.1.16 Natural graded and sandwich structures (osteoderms)

Many plant stems comprise cylinders with graded densities. These structures use an ingenious method of increasing the flexural strength/weight ratio. The design principle is similar to that for beams, but with an added complexity. In beams, the moment of inertia is increased by moving matter as far away from the neutral axis as possible. In biological structures, the density is varied using cellular materials that are graded, the density increasing with distance from the neutral axis. One extreme of lightness is the absence of matter, and many structures have a hollow core, such as bamboo. In some cases, there is a clear outside shell that sandwiches an inner cellular core, such as the toucan beak, the human skull, porcupine quills, and hedgehog spines.

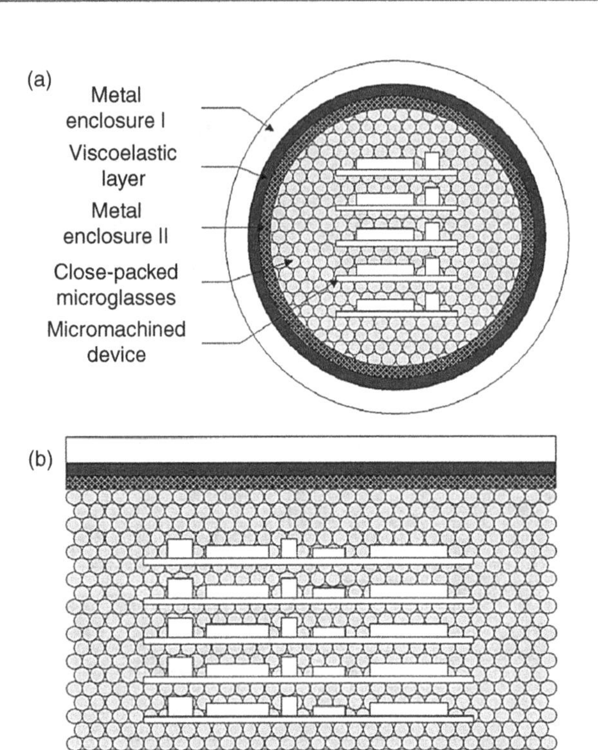

Figure 12.38.

Bioinspired shock absorption system based on the head of the woodpecker. A metal sheath encloses a viscoelastic layer and another layer. The cylinder is filled with microglass spheres. (a) Cross-section; (b) longitudinal section. (Reprinted from Yoon and Park (2011); with kind permission from IOP Publishing Ltd.)

The introduction of sandwich and graded structures has been a major development in structural design, and their use in numerous applications can be attributed to bioinspiration. The English longbow, for instance, uses both the inside (dark wood) and the outside of the yew tree, which confer different properties: strength and flexibility, respectively. The Mongolian bow, on the other hand, is also a composite design, but using completely different materials: dried tendon (sinew) on the outside, bamboo in the middle, and horn on the inside. Collagen-based glue (*cola* is the Greek name for glue; thus, collagen) was used to keep the bow together. The WW2 Mosquito airplane used wood in a sandwich arrangement: lightweight balsa wood was sandwiched between denser wood veneer faces. The ubiquitous corrugated cardboard is another example of a sandwich structure.

We present in the following three bioinspired concepts. Milwich *et al.* (2006) copied the structure of the horsetail and giant reed stem using textile weaving techniques: a fiber pultrusion process using a braiding machine. Milwich *et al.* used these stems as models, suggesting that the methodology could be applied to a wide range of cylindrical designs. Figure 12.39 shows the cross-sections of a horsetail stem, a bioinspired polyurethane foam, and a bioinspired woven composite. The hollow external rim of the hollow cylinder contains holes. The horsetail stem can be reproduced at different scales.

The lotus root, characterized by a geometrical pattern of holes, has been mimicked by Utsunomiya *et al.* (2008). This was done by co-extruding copper and aluminum and by leaching out the aluminum; in this way, a cylindrical copper beam was created with a hole pattern mimicking the lotus root.

An ingenious biomimicking method applied to rattan is described by Zampieri *et al.* (2005) and Zollfrank *et al.* (2005). It consists of pyrolizing wood in an inert atmosphere at around 800 °C. This preserved the cellular structure with a characteristic size of ~0.5 mm. Silicon gas was subsequently infiltrated into the structure, generating, through reaction, SiC. A similar process can be used to produce other carbides, such as TiC and ZrC. This process has also been applied to other woods (Griel, Lifka, and Kaindl, 1998). The porous structure can also be infiltrated with a metal, such as aluminum, creating a fully dense composite.

12.1.17 Cutting edges

Figure 12.40 shows schematic representations of some successful and possible biomimetic approaches to devices inspired from the sharp objects described in Section 11.8. Figure 12.40(a) represents a hypodermic needle, inspired by the proboscis of a mosquito, which was developed by Oka *et al.* (2002). The serrations in the needle help it to slice through the epidermis in a painless fashion. This mimics the action of the mosquito, whose proboscis enters the virtually unsuspecting host. This hypodermic needle has dimensions comparable with the mosquito proboscis, but, more importantly, uses the serrated edges (one on each side) to slice through the tissue. The syringe manufactured by Oka *et al.* (2002) has a built-in reservoir and is equipped with jagged edges that mimic the mosquito stylet. It is made from SiO_2 using a silicon micromachining technology.

Figure 12.40(b) is a schematic of a cutting tool designed to self-sharpen using the same mechanism as rodent teeth. The outside is a high-hardness ceramic composite, whereas the inside is a high-strength alloy. This equipment, inspired by rat and rabbit incisors, was successfully manufactured in Germany (Berling and Rechberger, 2007). It uses a hard titanium nitride ceramic reinforced with nanoparticles as the hard "enamel" portion of the cutting blade. The soft "dentin" part of the knife is made by a tungsten-carbide–cobalt alloy. The titanium nitride layer is twice as hard as the alloy. In Fig. 12.40(b) are shown the inner regions of the three blades of the shredder, which rub against the materials to be cut and wear out, keeping the outer layer (the hard titanium nitride) exposed and sharp.

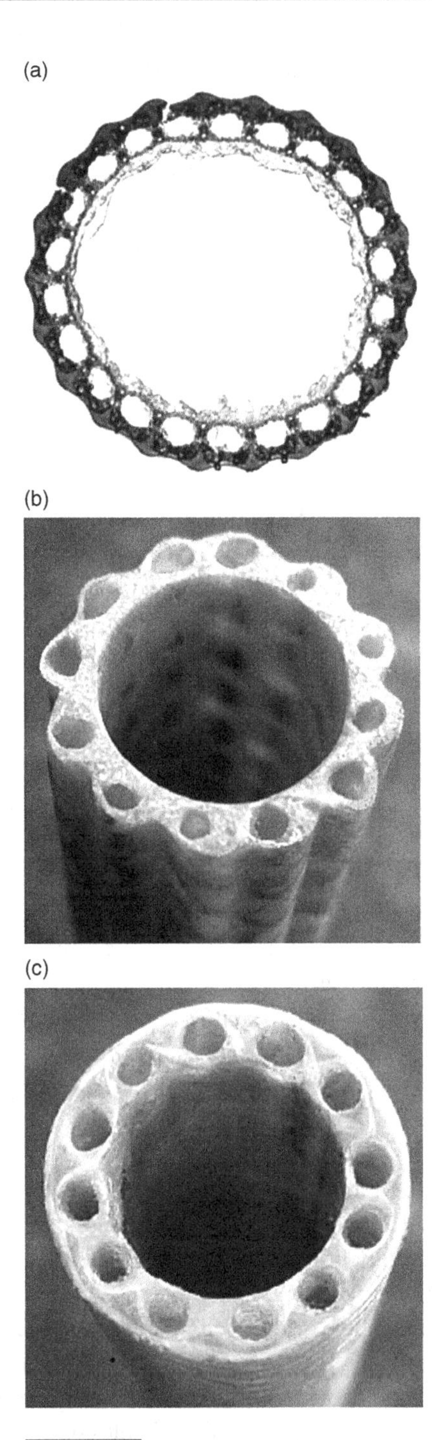

Figure 12.39.

(a) Cross-section of horsetail stem. (Courtesy Plant Biomechanics Group, University of Freiburg, with kind permission.) Bioinspired woven composite using (b) polymeric (polyurethane) foam and (c) a double-braided composite construction. (Courtesy Dr. Markus Milwich, with kind permission.)

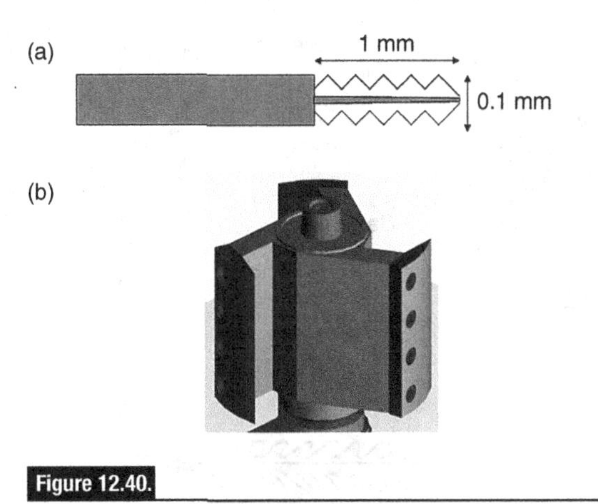

Figure 12.40.

Possible biomimetic devices: (a) syringe inspired by mosquito proboscis; (b) cutting blade inspired by rabbit incisors. (Adapted from Meyers *et al.* (2008c).With kind permission from Springer Science and Business Media.)

12.1.18 Ovipositor drill

The mechanism of wood wasp ovipositor drilling (Section 11.8.4) has been applied to several designs. One is for a planetary probe and the other is for a surgical cranial drill. In both designs, no forces (impact or normal) are applied, thereby minimizing damage to the surrounding material. Planetary probes (e.g. Mars exploration), must operate in a vacuum, so excessive pressures that arise in a conventional rotary drill cannot be employed. The ovipositor biomimetic drill is ideal, since holes can be drilled with a minimal amount of damage.

Figure 12.41(a) shows a diagram of a planetary probe along with details of the drill bit (Gao *et al.*, 2006). Figure 12.41(b) shows a diagram of the probe tip for a surgical drill. The successful development of a flexible, steerable probe for neurosurgery is shown in Fig. 12.41(c). Since there is no net force in the ovipositor-inspired assembly, there are no stability problems, and there is no theoretical limit on its length.

12.1.19 Birds

The 500 Series Shinkansen bullet train has a maximum speed of ~200 mph. It takes inspiration from birds in two ways. The train's nose (Fig. 12.42) is based on the beak of the kingfisher (see inset), which has a shape that allows it to dive into water with the minimum amount of resistance. In the case of the bullet train, shock waves are emitted when the train exits a tunnel, and the sonic boom is eliminated by this efficient design. The second way in which birds are mimicked is in a component that connects the train to the electrical power supply. Serrations similar to the ones in an owl's feathers (which allow the owl to fly stealthily as it moves in on its prey) were introduced to reduce noise and vibration.

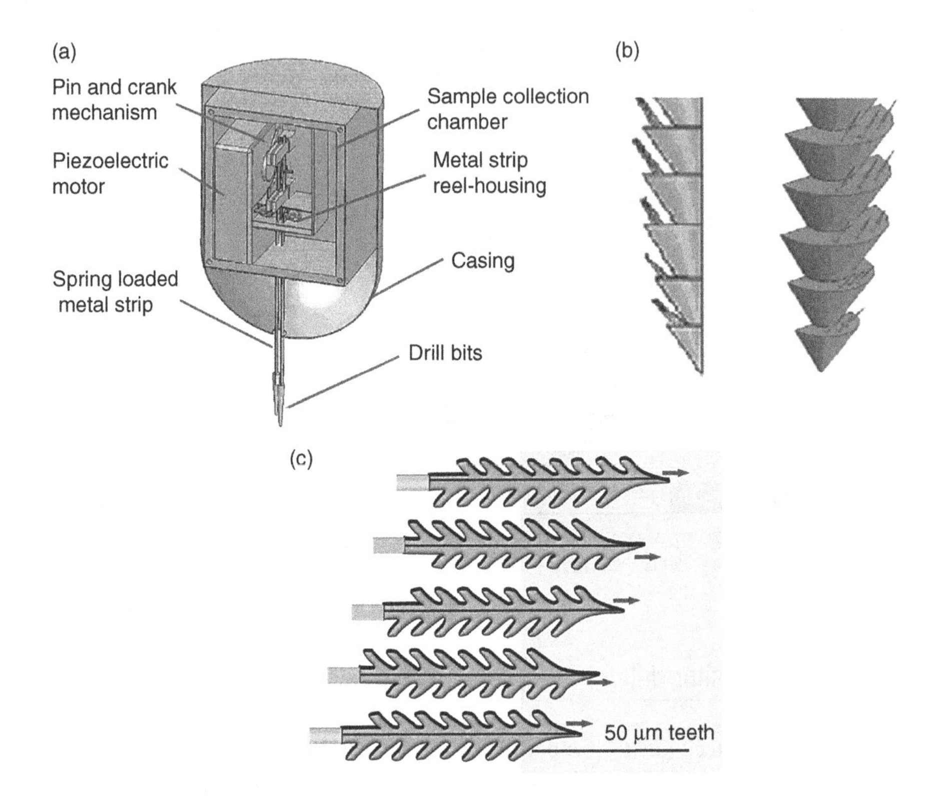

Figure 12.41.

(a) Bioinspired drill; (b) close up of the drill bit configuration. (From Gao *et al.* (2006). With kind permission from Professor Gao.) (c) Mechanism of movement of the two-valve drill from a cranial drill. (Courtesy Imperial College, London, Research in Mechatronics in Medicine.)

Figure 12.42.

Shinkansen train nose and kingfisher's beak. (From http://commons.wikimedia.org/wiki/File:500_Series_Shinkansen_JR_Shin-Osaka_Sta01bs5s2400.jpg by 663highland. Licensed under CC-BY-SA-3.0, via Wikimedia Commons. From http://commons.wikimedia.org/wiki/File:Alcedo_azurea_-_Julatten.jpg by J. J. Harrison. Licensed under CC-BY-3.0, via Wikimedia Commons.)

12.1.20 Fish

Bartol *et al.* (2003, 2005) studied the fluid dynamics of the bony carapace of boxfish using digital particle image velocimetry (DPIV), pressure distribution, and force balance measurements. The results showed that the boxfish carapace effectively generates self-correcting forces for pitching and yawing motions that contribute to the hydrodynamic stability of the fish during swimming. Mercedes-Benz introduced a new car design at the Shanghai and New York auto shows in 2011 based on the body design of the boxfish. Figure 12.43 shows the boxfish and a morphological shape, which has provided the design model for the car. It should be noted that this shape does not optimize the aerodynamics, but is a compromise between internal space and aerodynamic stability.

Shark skin, shown in Fig. 11.25, has served as inspiration for the famous swimsuits that have caused havoc during the Olympic Games in 2012. The swimwear company Speedo claims that its Fastskin FSII swimsuit can reduce friction in water by up to 4%. Additionally, the diamond-shaped bumps play an important role in keeping colonies of microbes from forming on the material.

Figure 12.43.

Bioinspired Mercedes-Benz, a class car based on the shape of the boxfish. (a) Boxfish *Ostracion cubicus.* (From http://commons.wikimedia.org/wiki/File:Side_view_of_juvenile_yellow_boxfish_Ostracion_cubicus_(5821904497).jpg by Paul Asman and Jill Lenoble. Licensed under CC-BY-2.0, via Wikimedia Commons.) (b) Mercedes-Benz bionic car. (From http://commons.wikimedia.org/wiki/File:Bioniccar_11.jpg by NatiSythen. Licensed under CC-BY-SA-3.0, via Wikimedia Commons.)

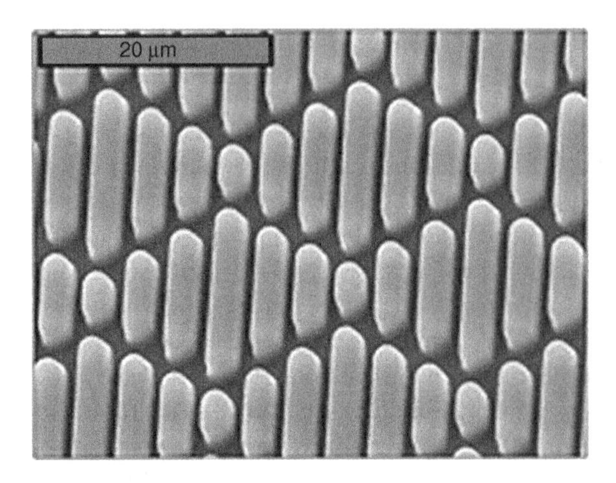

Figure 12.44.

Sharklet® Technologies surface for use in hospitals. (Figure courtesy The Biomimicry Institute.)

The US Navy was interested in finding anti-fouling paints for submarine hulls to reduce algae coating the surfaces. This resulted in the development of Sharklet®, a film that replicates shark skin ridges. Figure 12.44 shows a synthetic surface inspired by shark skin. It has proven effectivity in reducing algae settlement by 85% and preventing the formation of *E. coli* and *Staphylococcus aureus* colonies in a hospital. Bacteria attach singly or in small groups to establish large colonies. The Sharklet® surface keeps these films from forming because the patterned surface produces an energy barrier for colony formation.

12.1.21 Structures from diatoms

There are on the order of 10 000 different species of diatoms. Diatoms are single-cell microalgae having rigid silica-based walls, called frustules. Thus, there are approximately 10 000 shapes of frustules. Each has a unique structure, with ridges, pores, and features on the order of ~100 nm. They are commonly used as filters for water. These cells reproduce several times per day and can be considered as biofactories of three-dimensional structures with nano-scale features. Sandhage *et al.* (2002) developed a method to react the diatoms with metals to displace the Si, thereby creating ceramic structures where the shape and features of the frustules are retained. The following two reactions are used:

$$2\mathrm{Mg(g)} + \mathrm{SiO_2(s)} \rightarrow 2\mathrm{MgO(s)} + [\mathrm{Si}],$$

$$\frac{4}{3}\mathrm{FeF_3(g)} + \mathrm{SiO_2(s)} \rightarrow \frac{2}{3}\mathrm{Fe_2O_3(s)} + \mathrm{SiF_4(g)}.$$

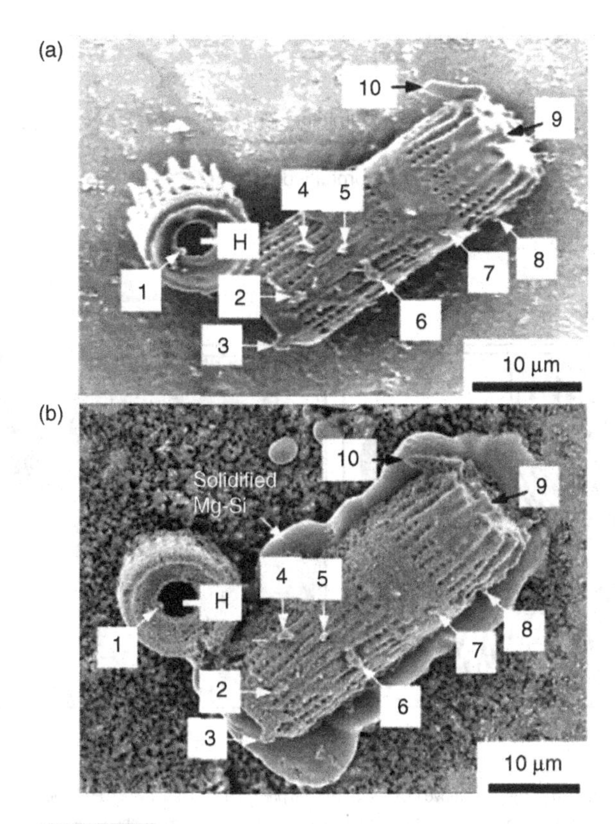

Figure 12.45.

SEM images of the same diatom frustules (a) before and (b) after reaction with Mg(g) for 4 h at 900 °C. Ten features that are retained after reaction are marked, as are a larger hole (H) and finer pores and ridges on the surfaces of the frustules. (Reprinted from Sandhage *et al.* (2002); with kind permission of John Wiley & Sons, Inc.)

Figures 12.45(a) and (b) show *Aulacoseira* frustules before and after reaction with Mg(g) for 4 h at 900 °C. Several well-preserved surface features are identified, and it can be seen that they are retained after the reaction. There are other possible reactions with Ca, Li, Nb, Sr, Ta, Ti, and Zr. It is envisaged that, in the future, this approach could be used to produce genetically engineered devices such as sensors, optical gratings, catalysts, and actuators.

12.1.22 Structures based on echinoderms

One can envisage bioinspired mechanisms that move spines, and Aizenberg (2010) has created such a device, shown in Fig. 12.46(a). The spines are made of silicon by a photo-etch method and have a diameter of ~200 nm. These nano-scale spines have one extremity embedded in a hydrogel. This hydrogel swells in the presence of water and

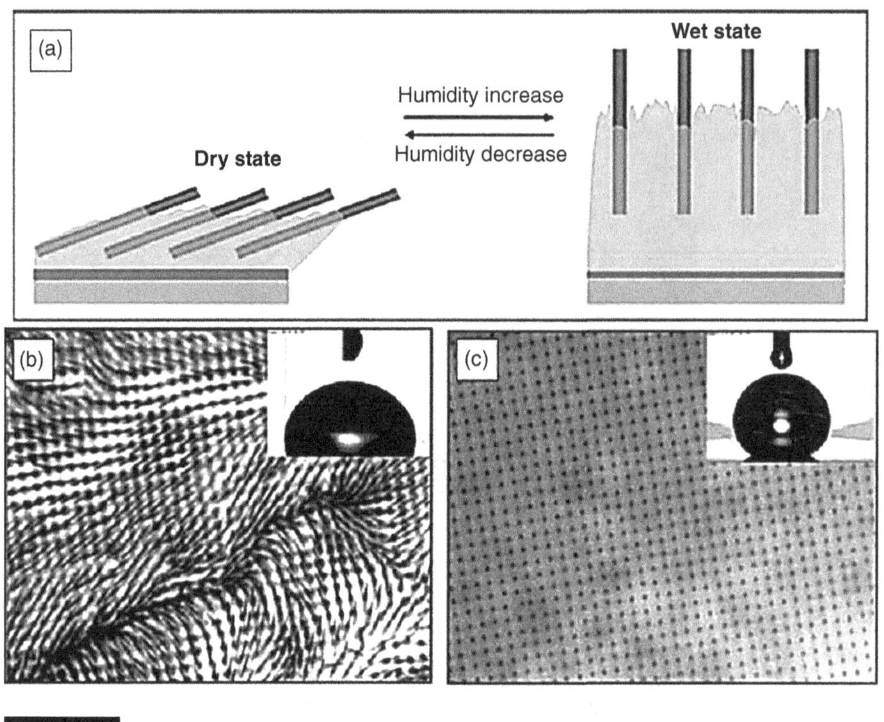

Figure 12.46.

(a) Schematic showing how nanospines can move in a controlled and predictable manner when embedded in a hydrogel that swells in the presence of water. (b) Hydrophilic surface created when hydrogel is dry. (c) Hydrophobic surface created when surface is wet. (From Aizenberg (2010). Copyright © 2010 Materials Research Society. Reprinted with the permission of Cambridge University Press.)

this causes the spines to stand up. When the hydrogel shrinks, the spines "lay down." The actuation time is very small, around 60 ms. The surface with the spines upright is hydrophobic, i.e. water droplets do not "stick" to it (Fig. 12.46(c)). When the spines are down, the surface is hydrophilic (Fig. 12.46(b)). Aizenberg (2010) suggested that this and similar structures can be used for a variety of actuators.

12.1.23 Whale-fin-inspired turbine blades

The frontal part of whale fins is not smooth but has a characteristic irregular pattern. This puzzled researchers until they discovered that this waviness decreased drag. This concept is being applied to water turbines in order to increase their efficiency. Figure 12.47(a) is a drawing representing the front part of a whale fin; the equivalent bioinspired turbine blade is shown in Fig. 12.47(b).

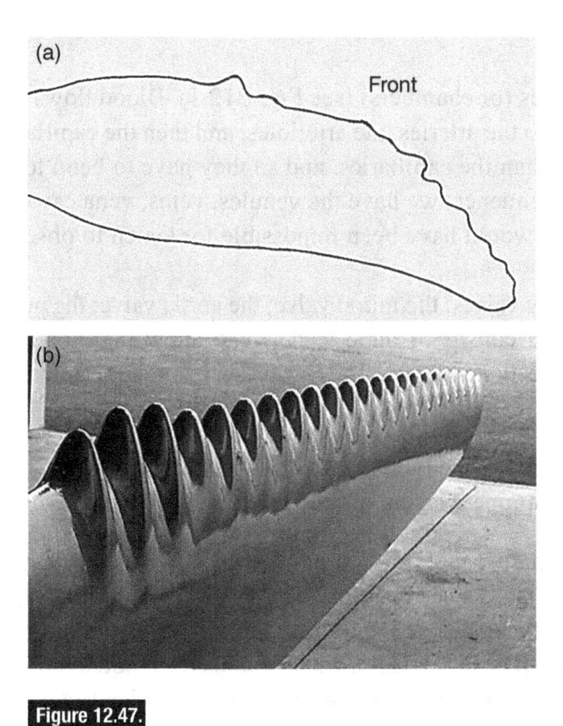

Figure 12.47.

Whale fin (a) inspiring front portion of turbine blade (b). (Figure courtesy Windpower Engineering and Development.)

12.2 Medical applications

Box 12.1 Artificial heart valves and heart-assist devices

The momentous discovery by William Harvey in 1628 that the heart is the pump through which blood flows is one of the foundations of modern medical science. Prior to that, medicine was strongly influenced by Galen, a medical researcher and philosopher who was born in Pergamum 129 BC and became prominent in Rome, writing incessantly throughout his life. Galen believed that the circulatory system consisted of two separate one-way systems of distribution, instead of a single system of circulation. He correctly identified veins and arteries by experimenting with dead and live animals. He practiced both the dissection and vivisection of pigs and monkeys, since these experiments were prohibited on humans. According to his findings, venous blood was generated in the liver, from where it was distributed and consumed by all organs of the body; arterial blood originated in the heart, from where it was distributed and also consumed by the organs. The blood was regenerated in either the liver or the heart, completing the cycle. The absence of good microscopes prevented Galen from observing the connection, by means of arterioles and capillaries, between the arterial and venous systems.

Box 12.1 (cont.)

The human heart has four ventricles (or chambers) (see Fig. B12.1). Blood flows out of the heart through the aorta, which conducts it to the arteries, the arterioles, and then the capillaries. Red blood cells have a larger diameter (8 μm) than the capillaries, and so they have to bend to travel through them. On the other side of the flow sequence, we have the venules, veins, vena cava, and entry into the heart through the right atrium. It would have been impossible for Galen to observe this crucial transition.

The four ventricles are connected by valves: the mitral valve; the aortic valve; the pulmonary valve; and the tricuspid valve. These valves consist of three leaflets and allow the blood to flow in one direction only, closing if the blood direction reverses. These valves are therefore called tricuspid. They can develop two problems:

(a) stenosis, in which the valve diameter, and therefore the blood flow, diminishes;
(b) regurgitation, which is due to improper closing of the leaflets and allows back-flow of blood.

The replacement of heart valves started in 1960 and has grown into a very important endeavor. Millions of heart valves have been replaced since that time. The survival rate for this type of surgery is approximately 97%. The valve most often replaced is the aortic valve, followed by the mitral valve. The two valves in the right ventricle (the left of Fig. B12.1), which pumps the blood to the lungs, are of no great concern. Several heart valve concepts have evoloved. They are divided into mechanical and biological valves (Fig. B12.2).

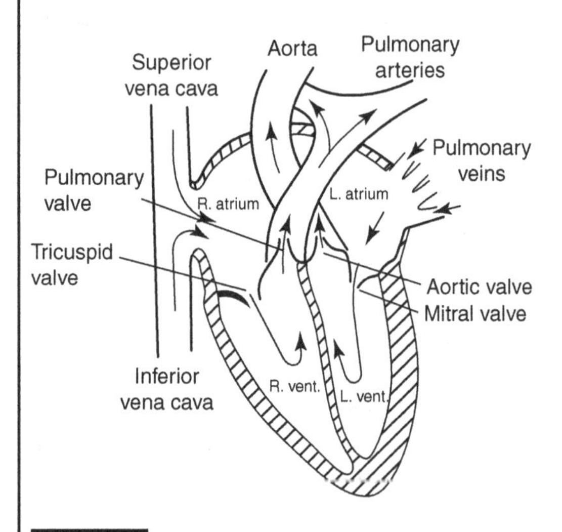

Figure B12.1.

Structure of heart with four ventricles and four valves: mitral valve; aortic valve; pulmonary valve; tricuspid valve. (From Park and Lakes (2007), with kind permission from Springer Science+Business Media B.V.)

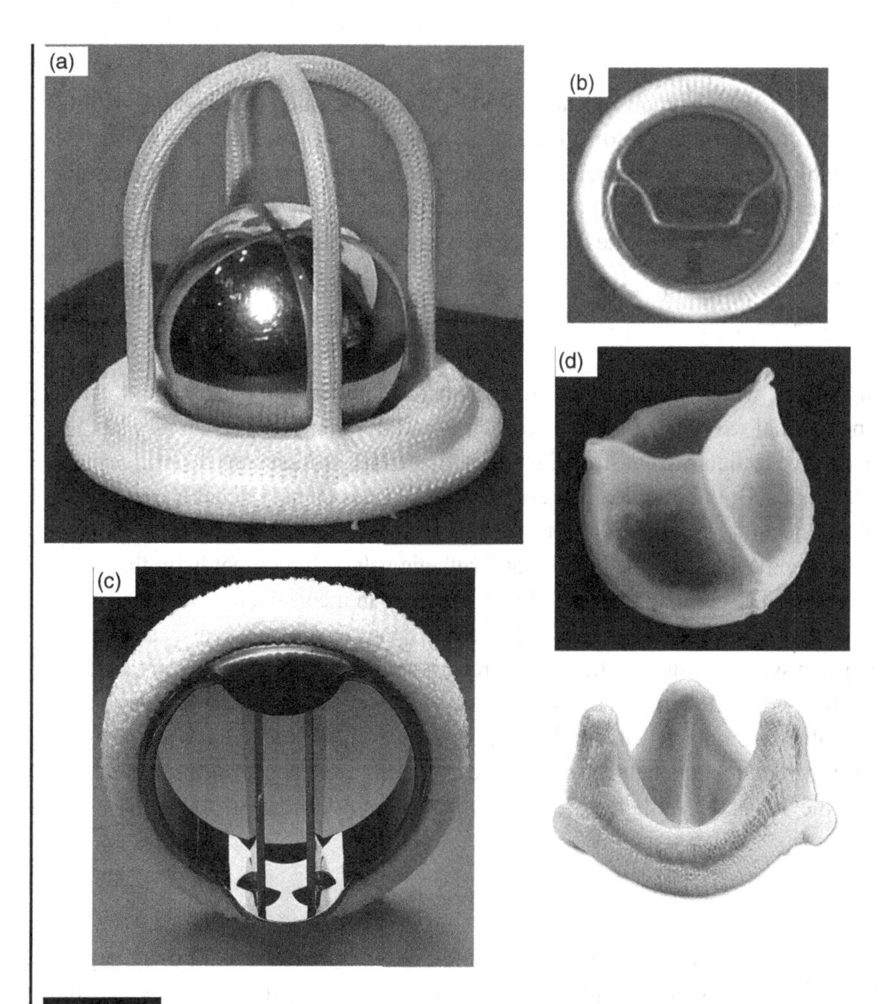

Figure B12.2.

Replacement valves. (a) Ball-in-cage (Starr–Edwards). (From http://commons.wikimedia.org/wiki/File:Starr-Edwards-Mitral-Valve.jpg by Dr. Mirko Junge. Licensed under CC-BY-3.0, via Wikimedia Commons.) (b) Medtronic-Hall tilting disc. (From http://en.wikipedia.org/wiki/File:Heartvalve.jpg#filelinks. The image has become public domain by the copyright holder.) (c) St. Jude bi-leaflet. (From Grunkemeier *et al.* (2002), with kind permission from Springer Science+Business Media B.V.) (d) Biological valves. Top: a replaceable model of cardiac biological valve prosthesis. (From http://en.wikipedia.org/wiki/File: BiologicalValves.JPG by Robertolyra. Licensed under the Creative Commons Attribution-Share Alike license version 3.0.) Bottom: aortic porcine bioprosthesis. (Figure courtesy Edwards Lifesciences LLC, Irvine, CA, whose trademarks include Edwards, Carpentier-Edwards, and Carpentier-Edwards S.A.V.)

Mechanical valves

Three systems dominate this market. The first is the ball-in-cage, or Starr–Edwards valve (Fig. B12.2(a)), in which the flow of blood moves the ball away from the socket. When the pressure gradient inverts, the ball returns to the socket, closing the valve. The same mechanism operates in the tilting disc. (Bjork–Shiley and Medtronic-Hall) and bi-leaflet (St. Jude) valves. The Bjork–Shiley was removed from the market in 1986, but by then an estimated 86 000 people worldwide had received the valve. Fatigue failure of this valve caused great problems for the FDA and the manufacturer because 600 failures resulted in approximately

Box 12.1 (cont.)

400 deaths. The bi-leaflet is the most common mechanical valve used currently. The use of pyrolitic graphite, which has exceptional compatibility to blood, is a great advantage. Pyrolytic carbon is resistant to wear, strong, durable, is highly resistant to blood clotting, and causes little damage to blood cells. Nevertheless, this brittle material is subject to rapid crack propagation once a crack is formed. Thus, an accidental nick by a surgeon's scalpel can prove to be fatal. Pyrolitic graphite has poor fatigue properties and has to be treated with great care.

Biological valves

There are two sources: pig valves, which are the closest to humans, and pericardial tissue valves, taken from bovine hosts. The pericardial valve is manufactured, whereas the porcine valve already has the final shape. The pericardial valve leaflets are cut from the pericardial tissue, which consists primarily of collagen layers making 60° angles. Both biological materials have to be completely denatured to kill all the cells that would create a rejection response from the host.

Biological valves are more compatible with the blood than synthetic valves and do not harm the blood cells or elicit a thrombic response. The compatibility with blood is called, in medical terms, hemodynamic performance. Thus, anti-coagulant medication is not required. However, they calcify more easily and have a shorter life span (5–7 years) than synthetic valves. They are not recommended for younger patients. Ball-in-cage valves were the first to be used, but the impact of the ball on the cage causes considerable injury to the red blood cells. Tilting disc valves were much better in this respect. However, they were removed from the market in 1986 after a number of fatigue failures caused deaths. This was due to improper design modifications introduced in the 1980s. Nevertheless, as already stated, an estimated 86 000 persons had them implanted.

Heart-assist devices

These devices assist the heart function and are implanted in the body with an external power supply (battery). After the great problems (primarily rejection) encountered with heart transplants and the limitations of the Jarvik heart machine, heart-assist devices are a success story, with approximately 25 000 patients in the USA. Figure B12.3 shows a left ventricle heart-assist device. Two pump mechanisms have been used: an alternating pump mimicking the heart function, and a turbine. The latter seems to produce better results. A big problem in all devices that involve blood flow is damage to the red blood cells. In a healthy individual, red blood cells have an average life of 28 days. However, trauma in an external device can damage red blood cells and reduce their life expectancy. The rotary pump reduces the life of red blood cells to 24 days, not a dramatic loss. Both right and left ventricular assist devices are fabricated. These devices decrease the load in the heart and actually help it to regenerate.

The following story illustrates how successful the device is. A patient was shopping in a high-risk area, keeping his battery in a well-conceived shoulder pack. A purse snatcher ran by and grabbed his backpack. The patient immediately chased him at full speed, recovered the backpack, and reconnected the battery. He survived, demonstrating that the heart-assist device had improved the performance in his heart and enabled him to carry out "the run of his life."

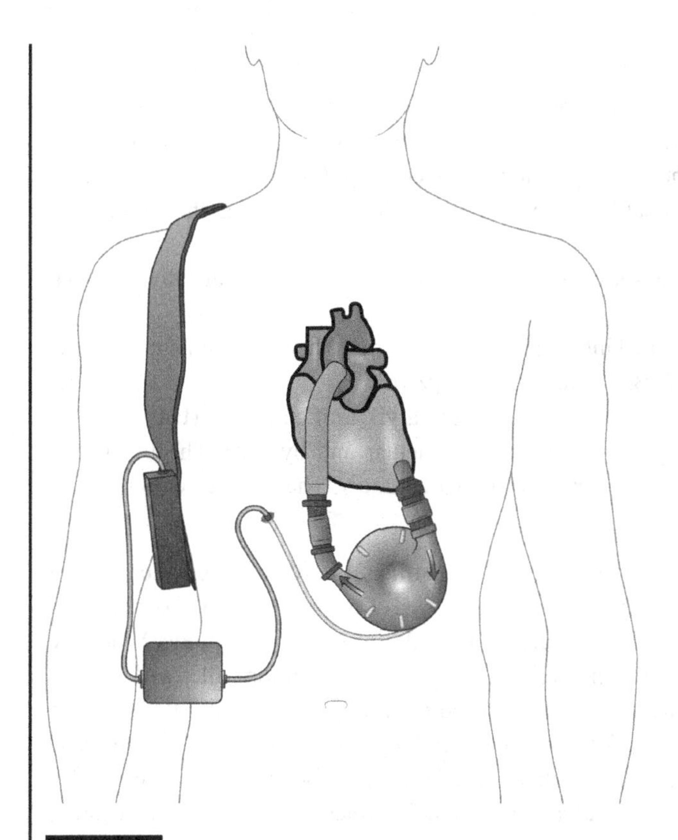

Figure B12.3.

Ventricular assist device and battery pack. (From http://commons.wikimedia.org/wiki/File:Ventricular_assist_device.png by Madhero88. Licensed under the CC-BY-SA-3.0, via Wikimedia Commons.)

Box 12.2 Other bioinspired devices

Beyond materials, devices are more complex, consisting of a number of components and performing mechanical, electrical, or chemical functions and assisting or replacing (in some cases) organ function. These devices mimic the functions of organs using synthetic technology and, in some cases, biological systems. The materials and functionalities are achieved through existing or novel engineering. There is significant effort toward manufacturing artificial organs worldwide.

The most important devices used or under investigation are as follows.

- Kidney: artificial dialysis machines, which use synthetic semi-permeable membranes to perform the filtering. Blood is extracted from an artery and reintroduced through a vein.
- Liver: liver dialysis machine. This is still an open field, and there is no widespread use yet. The objective is similar to the kidney dialysis machine; it is used to filter albumin and other toxins

Box 12.2 (cont.)

(ammonia, bile acids, bilirubin, phenols, copper, and iron) out of the blood. One system developed in Germany is the molecular adsorbents recirculating system (MARS). There are approximately 28 000 deaths from liver failure in the USA per annum since only 6000 patients can receive transplants. An artificial liver would enable these individuals to survive until they receive a transplant.

- Urinary incontinence: an artificial urethral sphincter that is an inflatable jacket activated by a manual pump inserted in the scrotum.
- Hydrocephalitis: a shunt valve that is drained into the body. One of the authors (MAM) witnessed its development by one of his professors, J. B. Newkirk, in the 1970s.
- Heart: there is a range of devices, ranging from the ill-fated Jarvik artificial heart (University of Utah, 1982) to heart-assist devices (e.g. de Bakey left ventricular and bypass). The heart-lung machine replaces both functions temporarily during open-heart surgery. The pacemaker is the least complex and invasive of these devices.

As an illustration of biomedical devices, we describe the artificial cardiac pacemaker, which provides the electrical pulse that triggers the contraction of the heart. The sinoatrial node of the heart generates the electrical pulse that passes through a bundle of His to the atrioventricular node. When the generation of the pulse or its conduction is deficient, patients suffer from a condition called arrhythmia. The discovery of the electrical nature of the heart can be traced to Kolliker and Müller in 1856. We know of the classic Volterra experiments in which the motion of frog legs was activated by an electrical pulse. Kolliker and Müller went further and connected a frog's leg to a beating frog's heart. The activation of the leg muscle by the beating of the heart demonstrated the electrical nature of the beating of the heart.

Many public buildings in the USA are equipped with defibrillators that are able to provide a hefty electrical pulse to a person that has suffered heart failure. We know that 60 mA of AC current (60 Hz) or 300–500 mA of DC can cause fibrillation. The current provided by a pacemaker is on the order of 7.5–14 mA, with a pulse duration of only 1.5 to 2 ms (Bilitch, Lau, and Cosby, 1967), and is thus insufficient, by a wide margin, to produce fibrillation.

The first artificial pacemakers were introduced in 1960: currently there are over half a million people in the USA with these implantable devices. The pacemaker is implanted subcutaneously over the clavicle and has two or more insulated wires, with the entrance to the casing being made of a polymer, such as polypropylene (see Fig. B12.4(a)). The casing is most often titanium and the power source is a lithium-iodine battery that has to be replaced every five years. Needless to say, the case has to be hermetically sealed. The wires enter the heart through arteries, and at their ends they contain the electrodes that are implanted into the heart muscle. Different shapes for the electrode tips have been developed: a ball-point tip (diameter ~1 mm), a screw-in helical shape, and a porous tip are three common ones. One of the problems with the tip is that scar formation (collagen) decreases the electrical conductivity. Another problem is fatigue of the electrodes. This contributes to the need to replace the electrodes periodically (approximate average life span of ten years). Figure B12.4(b) shows a screw-in electrode. Modern adaptive-rate pacemakers have sensing devices that adjust the pulse to the physiological needs of the patient.

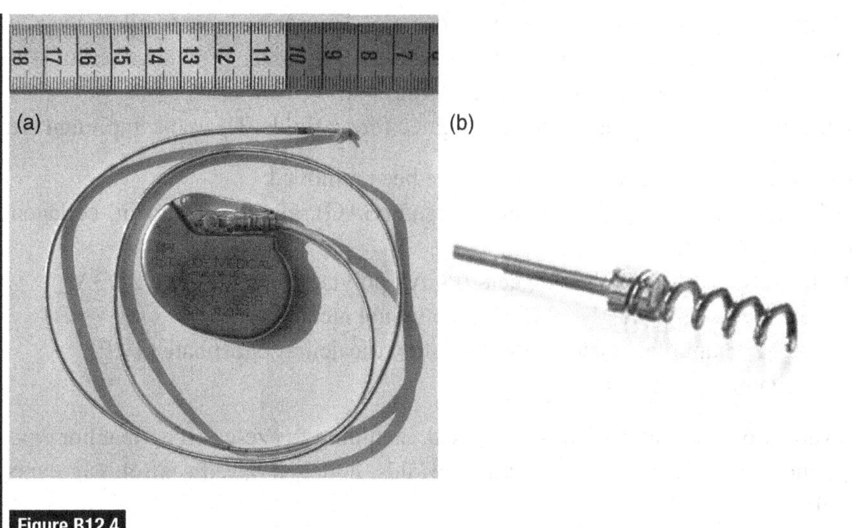

Figure B12.4.

(a) Implantable cardiac pacemaker (St. Jude). (From http://commons.wikimedia.org/wiki/File:St_Jude_Medical_pacemaker_with_ruler.jpg by Steven Fruitsmaak. Licensed under the Creative Commons Attribution 3.0 Unported license, via Wikipedia Commons.) (b) One of the available electrode designs.

12.2.1 Bioglass®

Bioglass® was developed by Hench (1999, 2006) during the Vietnam War as a result of the need to replace parts of bone destroyed in combat. This special glass was designed by adding CaO and P_2O_5 to SiO_2 to simulate better the composition of hydroxyapatite. Wide ranges of bioglasses are currently available, and their compositions depend on the function. Some of them bond to soft tissue and bone, some to bone, and some (with more than 65 mol.% SiO_2) do not bond at all and become encapsulated by fibrous tissue. One of the most important bioglasses, Bioglass® 45S5, has 45 mol.% SiO_2 and a CaO/P_2O_5 ratio of 5/1. Its Young modulus is 30–35 GPa, close to that of cortical bone (~20 GPa). The flexural strength is quite low at 40–60 MPa, and therefore it is not a good structural material. More information is provided in Box 5.1.

12.2.2 Tissue engineering scaffolds

One of the most active research areas in biomedical sciences is tissue scaffold engineering. Biological structures can be envisaged as comprising cells and an extracellular matrix (ECM). These scaffolds are, for the most part, porous. They provide the environment that mimics the body's extracellular matrix and enable cells to attach themselves, multiply, differentiate, and exercise their functions. One of the important functions is to generate the extracellular matrix. The porosity must be sufficiently high to enable the cells to migrate. The scaffold must also be biodegradable, i.e. it must dissolve in the body when its function is gone. This process must occur at a prescribed rate to allow the new ECM generated by the cells to take on the mechanical functions. Typical values of pore sizes are: skin = 20–125 μm, bone = 100–500 μm.

A broad range of materials has been investigated for scaffolds. The most important are:

- denatured native ECM, where all cells have been removed;
- biopolymers – collagen, glycosaminoglycans (GAG), alginate, chitosin, collagen-glycosaminoglycans;
- synthetic polymers – polyglycolic acid (PGA), polylactic acid (PLA), PLGA;
- hydrogels – polyethylene glycol (PEG), polyvinyl alcohol (PVA);
- ceramics – calcium phosphate, hydroxyapatite, tricalcium phosphate (TCP);
- metals – titanium, tantalum.

A broad variety of fabrication methods are used, including freeze-casting, fiber bonding, foaming, and rapid prototyping. For skin scaffolds, collagen-based scaffolds are very successful.

A recent development that has considerable promise, and is indeed already commercialized (by Orthomimetics), is an osteochondral scaffold that has two sides and regenerates the bone–cartilage interface. It consists of a tape with a thickness of ~5 mm. Some of the aspects are described by Pek *et al.* (2004). This scaffold was developed as a collaborative work involving MIT (L. Gibson), Cambridge University (W. Bonfield (Bonfield and Datta, 1974)), and Addenbrooke's Hospital, Cambridge (N. Rushton). One side enables bone growth (type I collagen and mineral) while the other cartilage growth (type II collagen + glycosaminoglycans – GAG). The chondrocytes penetrate the cartilage side, whereas osteoblasts and osteoclasts penetrate the other side.

12.2.3 Bioinspired scaffolds

The traveler that goes to Asia is surprised to find that modern buildings are erected using bamboo scaffolds. In the USA and Europe, steel pipes with connections using fasteners are the standard. Bamboo scaffolds are lighter than steel ones, and the connections are made by rapidly tying strong nylon straps. This is an example of a bioinspired design. The same approach is being implemented at the cellular level. A number of approaches combining biological and synthetic materials is yielding promising results. The rattan structure, described in Section 12.1.16, is strong and has a porosity compatible with cellular dimensions. It can be used to produce carbon scaffolds after burning in the absence of oxygen; Ca, P, and O are infiltrated into the structure, creating a porous hydroxyapatite This scaffold has been colonized by osteoblasts and is biocompatible. It could find applications in bone implants.

Another interesting development is the use of electrospinning of a chitosan–synthetic polymer mixture. In electrospinning a high electrical field is applied to a spinnerette (or syringe), from which a tiny jet of liquid escapes and impacts a collector, accumulating there and forming a dense mesh of thin fibers. Neural cells grow and multiply in the network. This material can be used to connect severed nerves.

Corneal scarring is an eye condition that can lead to blindness. The limbal cells on the surface of the cornea carry out the repair of the cornea. If they are eliminated by scarring, the cornea will deteriorate. Limbal stem cells can be extracted from a healthy eye, grown in human amniotic fluid, and then grafted to the scarred cornea. Another approach has been successfully implemented: by electrospinning PLGA, a scaffold is constructed which can be implanted into the cornea. The PLGA will create the environment for the limbal stem cells to grow and multiply.

A new scaffold named STAR® is being developed; it has a porosity such that macrophages, the large white blood cells (~20 μm), can be inserted into the holes. It is made using glass microspheres and filling the space with silicone, then dissolving away the glass. The pores are on the order of 25–45 μm.

12.2.4 Vesicles for drug delivery

Cell boundaries are made of a lipid membrane forming a bilayer. One side of this lipid membrane is hydrophobic and the other is hydrophilic. The lipid bilayer is formed by the connection between the hydrophobic sides of the lipid membranes, as shown in Fig. 12.48. In the march toward recreating life by synthetic means, the formation of vesicles, "empty" cells, has already been accomplished. They are called liposomes.

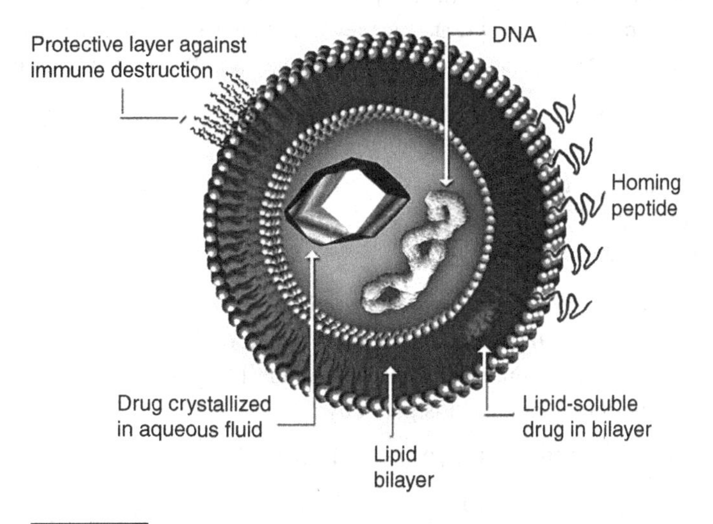

Figure 12.48.

Biologically inspired vesicle (liposome) consisting of lipid bilayer for drug or DNA delivery; peptides are designed to home in on targeted sites where the drug/DNA is released. (From http://en.wikipedia.org/wiki/File:Liposome.jpg#filelinks by Kosi Gramatikoff, who grants the image to be used.)

Novel uses for these vesicles are being found, and an important one is to deliver drugs to targeted places in the body.

12.2.5 The blue blood of the horseshoe crab

The horseshoe crab (*Limulus polyphemus*) is a living fossil. Unlike common crabs, its shell is not mineralized. Its blood is not red, but blue, by virtue of copper ions, and not iron ions, being in the cells. It lives in shallow waters infested with microbes. Not having an immune system like that of mammals, i.e. one that produces antibodies which "attack" invading bacteria that can cause disease (pathogens), it uses another system. This was discovered by Dr. Federick Bang, when he injected marine bacteria into the blood stream of the horseshoe crab, and found that the blood clotted extensively. The clotting source in the blood was identified as amebocytes. This component was isolated by a lysate process, and it is widely used currently for testing injectable drugs: if they contain bacteria or their toxins, the clotting reaction takes place. The product is called LAL (limulus amebocyte lysate). If an injectable drug contains bacteria, this can reproduce in the blood, leading to sepsis and, finally, death. Thus, this product is of great importance. The blue blood is extracted after the horseshoe crab is fished in shallow waters of the Cape Cod area; the animals are subsequently released without harm and their system replenishes the blood supply.

Example 12.2 Enthusiastic cyclists, the authors like to bike up and down Torrey Pines Hill in California, which has an elevation of 150 m. They do this three times as part of their triathlon routine. How many grams of carbohydrates do they have to ingest to compensate for the energy lost in the exercise? Assume an efficiency of conversion in the human body of 50%. Given:

fat: 1 g = 9 kcal;
protein: 1 g = 4 kcal;
carbohydrate: 1 g = 4 kcal;
alcohol: 1 g = 7 kcal;
1 cal = 4.18 J.

The potential energy is $E = mgh$, where m is the mass of the cyclist and the bicycle, g is the acceleration due to gravity, and h is the height of the hill.

Taking the mass of the rider to be 70 kg plus 15 kg for that of the bicycle, we have $m = 85$ kg. So the energy used is given by

$$E = 90 \times 9.8 \times 150 = 1.3 \times 10^5 \text{ J} = 31 \text{ kcal}.$$

For three ascents, $E = 93$ kcal.

Assuming 50% efficiency, the authors should ingest 186 kcal each. This is equivalent to 44 g of bread (about one slice) each. Alternatively, they could have consumed a meager ration of vodka (~50 g assuming that it is 100 proof, which would correspond to 50% alcohol).

Exercises

Exercise 12.1 Leonardo's magical sketches of flying machines are clearly inspired by birds and bats. However, as the size (and therefore mass) of a bird increases, so do the forces on its wings. Explain, using simple geometry, how the size of wings should increase with the mass of a bird. (This is known as allometric scaling.)

Exercise 12.2 Leonardo's airplane had a wing span of approximately 7 m; the wings had a width of approximately 2 m. The rule of thumb for birds and other low-velocity flying machines is 5 kg/m^2. Would Leonardo's plane glide?

Exercise 12.3 On 4 August 2012, Oscar Pistorius became the first amputee (double, below-the-knee) runner to compete at an Olympic Games. In the 400 m race, he took second place in the first heat of five runners, finishing with a time of 45.44 s. He used an Ossur Cheetah prosthesis, which has the characteristic J shape. It is manufactured with carbon-fiber-reinforced composite blades. The blades return about 92% of the energy stored into them elastically, whereas for biological legs this number is 93–95%. Did he have an advantage over athletes with "conventional" legs? Explain.

Exercise 12.4 Explain the shape-preserving chemical conversion of diatoms and give at least three potential applications.

Exercise 12.5 List at least two methods used to synthesize abalone nacre-like composites and briefly explain the procedures.

Exercise 12.6 What is bacteriophage? Describe how it invades bacteria and reproduces.

Exercise 12.7 Calculate the pull-off forces for 1 cm^2 of PDMS tape produced by the Arzt group if it is made with pillars with 2 and 20 μm. Use the theoretical curve in Fig. 12.13.

Exercise 12.8

(a) Calculate the force required to separate a hook from a loop in VELCRO®. Use Fig. 12.2 and the following parameters: $E = 8$ GPa, $t = 0.2$ mm, $L_1 = 4$ mm, $L_2 = 2$ mm, $L_3 = 1$ mm, $d = 3$ mm.

(b) What is the force required to pull of 1 cm^2 of this tape, assuming that all loops are equally loaded? Loop spacing $d = 3$ mm.

Exercise 12.9 Pyrolitic graphite is a wonderful material in terms of biocompatibility and resistance to blood clotting. Thus, it is an excellent candidate for heart valves and is used in bi-leaflet and tilting disc models. During an operation, the surgeon, a careless chap, nicks the leaf with his ultrasharp surgical scalpel during insertion. The scratch has a depth of 0.3 mm. The maximum stress acting on the valve is 1 MPa. The thickness of the blade is 1 mm. Use the values in Fig. E12.9. Given: K_{Ic} (pyrolitic graphite) = 1.35 MPa m$^{1/2}$; K_{Ic} (titanium) = 30 MPa m$^{1/2}$; Y = 1.12.

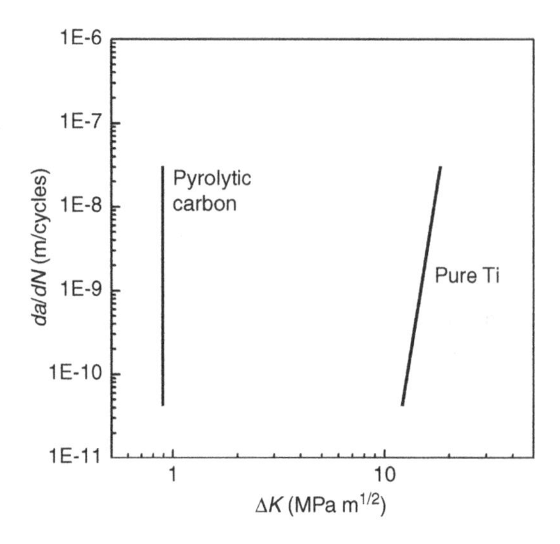

Figure E12.9.

Crack growth rate per cycle as a function of maximum stress intensity factor for titanium and pyrolytic graphite. (Reproduced based on Ritchie (1999).)

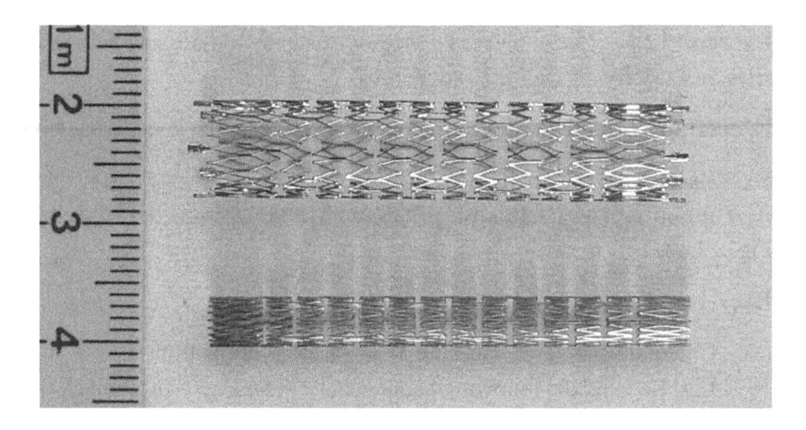

Figure E12.10.

Stent shown in normal and expanded states. (From http://commons.wikimedia.org/wiki/File:Stent3_fcm.jpg by Frank C. Müller. Licensed under CC-BY-SA-2.5, via Wikimedia Commons.)

(a) Calculate the number of cycles and time required for the crack to grow catastrophically in pyrolytic graphite. Assume 60 beats/min for the heart.
(b) What is the time required for the crack to grow catastrophically for titanium?
(c) Comment on the different slopes for pyrolytic carbon and titanium. The data are extracted from Ritchie (1999).

Hint: The Paris–Erdogan law has the following form:

$$\frac{da}{dN} = C\Delta K^m$$

This equation should be integrated from the intial crack size of $a = 3$ mm to the final size at which failure occurs catastrophically $(K_{Ic} = Y\sigma\sqrt{\pi a})$. Solution of the Paris–Erdogan equation is found in Meyers and Chawla (2009), p. 739, Example 14.3.

Exercise 12.10 Calculate the strain that is undergone by a stent structure when it is expanded from 2 to 5 mm. Assume the wire diameter is 0.1 mm. Make other necessary assumptions from Fig. E12.10.

13 Molecular-based biomimetics

Introduction

The quest to mimic nature is now reaching a new chapter. Whereas in the past, structural and functional characteristics served as inspiration for designs and materials, with attention given to mesoscopic and, perhaps, microscopic aspects, the arsenal of new experimental techniques and computational methods is descending to the nanometer scale. This is the scale at which atoms assemble into molecules, and molecules form molecular arrangements such as DNA, RNA, plasmids, and proteins.

Prominent characterization techniques available to modern researchers are (e.g. Gronau *et al.*, 2012):

- transmission and scanning electron microscopy at higher and higher resolutions;
- micro and nano computerized tomography;
- atomic force microscopy;
- fluorescence spectroscopy;
- Fourier transform infrared spectroscopy (FTIR);
- dynamic light scattering;
- X-ray diffraction.
- small-angle X-ray scattering (SAXS);
- nanomechanical testing (nanoindentation);

The principal computational methods are:

- density functional theory (DFT);
- quantum-mechanical and *ab initio* methods;
- coupled cluster method (CCSD);
- reactive and nonreactive molecular dynamics;
- coarse graining;
- finite element method;
- computational fluid mechanics.

Thus, there is intense research at the "bottom" of the spatial scale. The discoveries by Hershey and Chase (1952) that DNA had a role in heredity and by Watson and Crick (1953) of the double helix structure of DNA opened the door to the brave new world of genetic engineering, a term coined previously (in 1950) by Jack Williamson in his novel *Dragon's Island*. The complexities and challenges in this rapidly evolving field are immense, and we can only briefly touch on some aspects. We describe how self-assembly, a biological process, is being used to produce DNA scaffolds. Self-assembled monolayers (SAMs) have been made, creating highly hydrophilic surfaces. Genetic engineering has had great success in agriculture, drug manufacture, the synthesis of biopolymers, and medicine. It has evolved into a formidable field.

Genetically engineered peptides for inorganics (GEPIs) are produced via phage display, and can generate surfaces with different functionalities. Lithium-ion batteries have been produced using viruses that can help to assemble nanowires. They can also recognize carbon nanotubes and attach themselves to them.

Microelectromechanical systems (MEMS) and nanoelectromechanical systems (NEMS) can be designed using bacteria to create nano-sized rotary devices.

The biopolymers silk, elastin, resilin, and collagen have been synthetically mimicked.

13.1 Self-assembly structures

Aragonite and hydroxyapatite growth are, as seen in Chapter 5, complex processes including proteins and self-assembly. The current frontier in bioinspired materials goes beyond copying the structural elements. It starts at the nanometer level, with genetically engineered proteins; it also seeks to form the structures by self-assembly of its components. This field of research is still in its infancy but has enormous potential.

The process of self-assembly is the essential component of the bottom-up approach through which biological systems are formed. This process can be one, two, or three dimensional. Figure 13.1(a) illustrates how a two-dimensional network of branched DNA molecules, in the form of a cross with "sticky" ends (left-hand side) can self-assemble to form a reticulated network (right-hand side) (Seeman and Belcher, 2002). This is just an illustration, and many more complex arrays are possible, depending on the functionality of the bonds.

Whitesides and co-workers have pioneered self-assembled structures and used a number of approaches. The early experiments on self-assembled monolayers (SAMs) by Nuzzo and Allara (1983) inspired Bain and Whitesides (Bain and Whitesides, 1988; Whitesides, 2002) to use molecules with sulfur terminations. The sulfur atoms bond to the gold substrate, and the molecules therefore form a monolayer. This is shown in Fig. 13.1(b). The end of the molecule opposite to the sulfur can have a functional termination that changes the properties of the gold surface. For instance, if the end is a methyl group, the surface is hydrophobic. If the end is a carboxylic acid group, the surface is highly hydrophilic.

The Whitesides group (Bain and Whitesides, 1988; Bowden *et al.*, 1997; Whitesides, 2002) has also shown that 5 mm diameter balls (octahedrons with the corners cut off) with solder connections can be self-assembled. After self-assembly, the solder comprising the junctions can be made to fuse them together. This is accomplished by placing the balls in a warm solution and tumbling them. If the solder dots, placed at specific facets, encounter one another, they fuse. If they encounter a surface with a light-emitting diode (LED), they do not fuse. After the assembly is complete, a current is passed through and the LEDs are lit. This approach illustrates the potential of using self-assembly to create

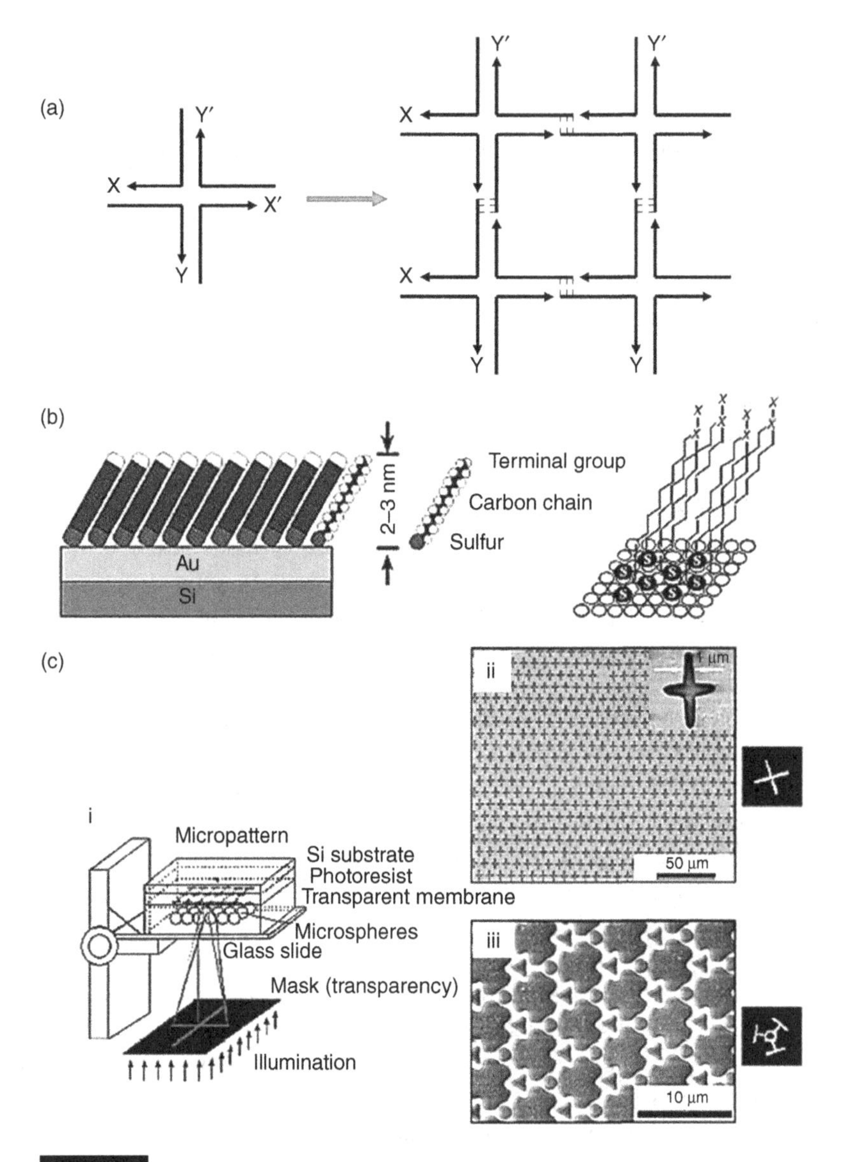

Figure 13.1.

(a) Formation of a two-dimensional lattice from a junction with sticky ends; X and Y are the sticky ends and X′ and Y′ are their complements. Four of the monomers assemble the structure on the right. (Reprinted from Seeman and Belcher (2002), with permission; copyright (2003) National Academy of Sciences, USA.) (b) Self-assembled monolayers (SAMs) of alkanethiols can be formed on gold evaporated onto a solid flat substrate such as silicon or glass. The sulfur groups interact covalently with the gold, the poly(methylene) chains pack tightly to form the monolayer, and the head groups are exposed. (c) Near-field photolithography with self-assembled microlenses. (From Whitesides (2002); with kind permission from Professor Whitesides.)

electrical networks and represents a possible prototype for the future. A photolithographic technique where the lenses are self-assembled is shown in Fig. 13.1(c).

13.2 Phage-enabled assembly

The bacteriophage (nicknamed "phage") is a microbe that has nano-scale dimensions. Phages, like viruses, do not have reproductive organs. The T4 phage injects DNA into *Escherischia coli* bacteria. Figure 13.2(a) shows the schematic representation of a phage that has an icosahedral head (capsid), a cylindrical tail sheath, and six legs. The scale of a T4 phage is approximately 70 nm in diameter and 200 nm in length. It has unique properties, which we briefly discuss. The T4 phage, resembling a Mars Lander, connects to the membrane of a bacterium through the six legs. At that point, the tail sheath contracts and penetrates the wall (Fig. 13.2(b)). It enters the *E. coli* bacterium through an ingenious process of contraction of the tail sheath, which forces the tail core to penetrate the *E. coli* wall prior to injecting the DNA (Meyers, 2008b). This contraction is accomplished by a martensitic-like transformation in the tail sheath, as shown in Fig. 13.2(c) and as identified by Olson and Hartman (1982). The contraction of the tail sheath propels a needle (tail core) through the bacterium wall. This enables the release of the DNA into the interior, where it uses the genetic machinery of the bacterium to reproduce itself.

Phages are very simple: they consist of geometrical assemblies of DNA and a few proteins. There is no fatty tissue, no blood, no reproductive system; these components, once manufactured inside the bacterium, self-assemble. As the bacterium dies, the phages spew out and continue the cycle (Forbes, 2007). Phages can reassemble in a test tube after being broken up in a blender. If a T4 macrophage is broken up into its constituent parts (as shown in Fig. 13.3(b)), it can reassemble to form the complete structure as shown in Fig. 13.3(a). This ability of phages to self-assemble in a quasi-mechanical way without any DNA templating or using proteins is inspiring scientists.

The process of mimicking nature is being pursued through the use of viruses. Viruses do not have their own reproductive systems and are therefore the most primitive form of life. They inject their DNA into cells or bacteria and use their reproductive machinery to take charge of carrying out viral replication. Viruses can enter cells in a number of ways.

Common types of phages used are M13, T4, T7, and λ. Smith (1985) demonstrated that peptides could be displayed on a filamentous phage by fusing them onto the gene of the filamentous phage. In phage display technology, a variety of peptides, small antibodies, or proteins can be obtained from a large library, and we select from them the variants which attach to a phage.

The technique has many variants, but in principle we start with a large library of proteins or DNA. Some of them will bind to the phages. The ones that do not bind are removed by washing. Those that remain are eluted. The repeated cycling is called "panning" from the mining method of gold panning. Thus, the eluted phages can be injected into bacteria, which reproduce them.

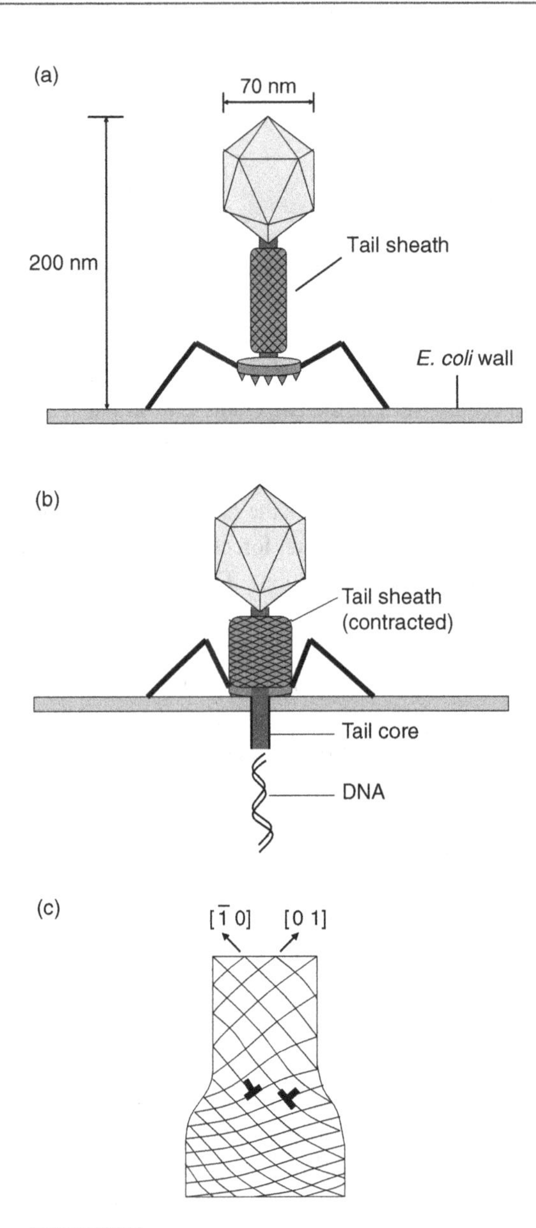

Figure 13.2.

(a) T4 phage attaches to *E. coli* cell wall. (b) Insertion of the genetic material by a T4 phage. (c) Cylindrical crystal structure of virus tail sheath during contraction by martensitic transformation. Interface is described by coherency dislocations which spiral up helical close-packed crystal rows.

Sarikaya *et al.* (2003) and Tamerler and Sarikaya (2007, 2008) used phage display methodology to obtain genetically engineered polypeptides for inorganics (see Section 13.3). They were able to select specific peptides that adhered to desired surfaces. In this way, they developed a procedure to assemble nanostructures from the bottom up.

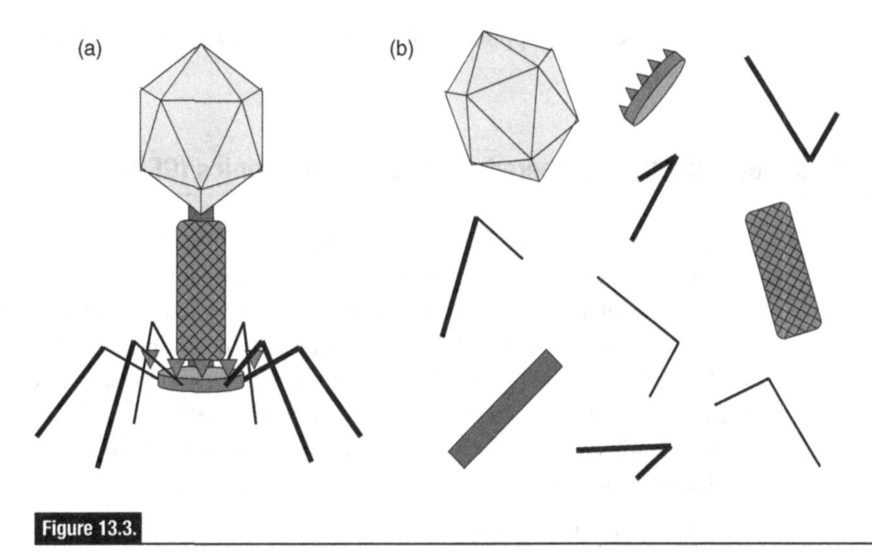

Figure 13.3.

(a) The structure of the T4 bacteriophage (b) The phage broken up into its constituent parts can reassemble spontaneously.

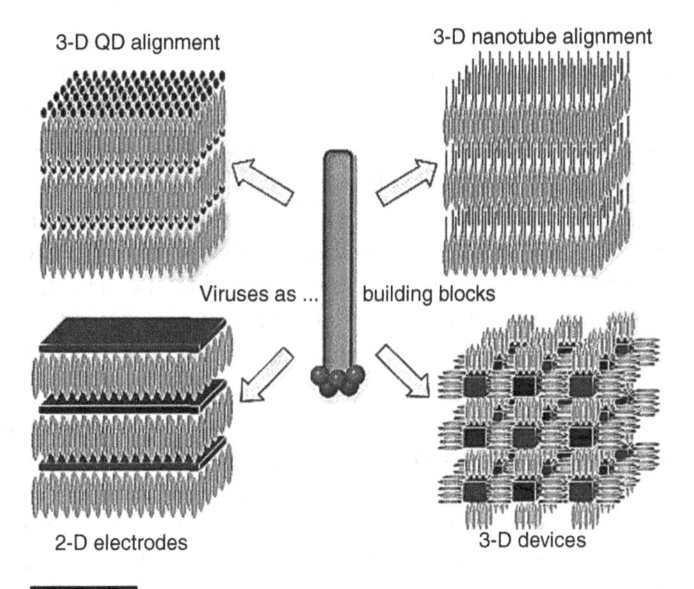

Figure 13.4.

Self-assembled structures using one-dimensional viruses with zero-dimensional quantum dots (QD) (top left), one-dimensional nanowires/nanotubes (top right), two-dimensional plate-shaped devices (bottom left), and three-dimensional components (bottom right). (Reprinted from Flynn *et al.* (2003), with permission from Elsevier.)

Viruses can also be used directly as building blocks for nanostructures assembled from the bottom, as demonstrated by Flynn *et al.* (2003). The ability of phages to bind to nanoparticles (quantum dots), single-wall carbon nanotubes, and tridimensional components was used to create the structures illustrated in Figure 13.4. These arrays have a number of potential applications, such as nanowires, semiconductors, and magnetic materials.

13.3 Genetically engineered peptides for inorganics (GEPIs)

Sarikaya and co-workers (Sarikaya *et al.*, 2003; Tamerler and Sarikaya, 2007) used phages to find the proteins required for specific surface interactions. Of special interest to materials scientists are protein interactions with inorganic substrates, such as silicon or gold. They used a technique from the molecular biology field called phage display (PD) to select proteins that bind to selected inorganic surfaces. This was based on earlier work by Brown (1997), who found that some proteins selectively bind to specific inorganic surfaces. Sarikaya and co-workers engineered new polypeptides that selectively bound to different inorganic substrates:

noble metals (Pt, Au, Ag, Pd);
metal oxides (ZnO, Al_2O_3, SiO_2);
semiconductors (GaN, Cu_2O, TiO_2, ITO, sulfides, selenides);
other functional materials.

This method, that they call GEPI (genetically engineered polypeptides for inorganics), is represented in Fig. 13.5. The first step consists of obtaining random sequences of amino acids by breaking up DNA. These are then incorporated into phages. These coated phages are shown in step 2, Fig. 13.5. These phages are then exposed to the selected surface (step 4) and washed away (step 5). Several washes are required. The polypeptides that bind to the selected surface cannot be removed by the washes, and are eluted (step 6). These polypeptides are then replicated by having the phage inject the DNA into a bacterium. An attractive methodology to PD is cell-surface display, in which the polypeptides are incorporated into cell surfaces. These genetically engineered polypeptides can potentially be used for generating a variety of nano-scale arrays on the surfaces of inorganics. These arrays have a range of applications connected with the anchoring, coupling, branching, displaying, and assembling of functional molecules, nanoparticles, and structures. The basic methodology is illustrated in Fig. 13.6. Two different GEPIs are used: GEPI-A and GEPI-B. They are assembled on a patterned substrate. Each GEPI is connected to an inorganic, which in turn, attaches to functionalized monomers. These functionalized monomers can serve a variety of functions, such as conductors, photonic devices, or others. Thus, the GEPIs can hypothetically serve as nano-scale platforms for nanoarrays.

Belcher's research group (Lee *et al.*, 2002) took the process one step further by using the phages directly to bind to ZnS particles. The phages held the ZnS particles in position. Nano-scale ZnS particles can, by this stratagem, produce regular arrays. The group's intended application for this approach is to create quantum dot devices.

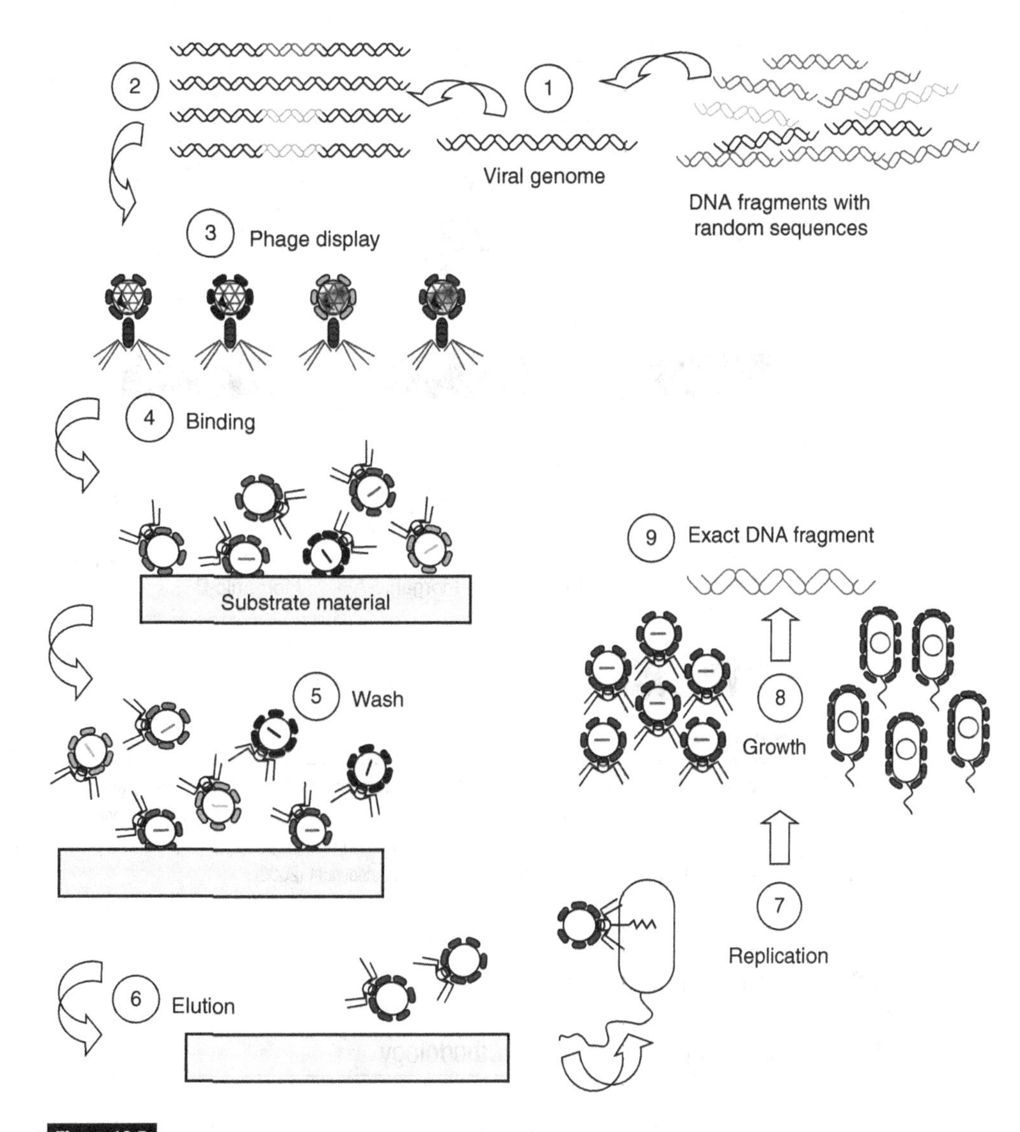

Figure 13.5.

Method for obtaining genetically engineered polypeptides for inorganics (GEPIs) by phage display: (1) DNA is broken up into random sequences; (2) a random polypeptide sequence is incorporated within a protein residing on the surface of the organism; (3) sequences are incorporated into phages; (4), (5) phages are put into contact with inorganic surface and washed sequentially; (6) phages that stick to surface are eluted and have their DNA injected into bacterium for expression replication (7); (8) cells are allowed to grow; (9) exact DNA fragments (polypeptides) that stick to surface are thus obtained; these are the famous GEPIs. (Reprinted by permission from Macmillan Publishers Ltd.: *Nature Materials* (Sarikaya *et al.*, 2003), copyright (2003).)

In summary, the range of biomimetic materials has barely been explored. The development of novel technological advances could greatly benefit from the lessons taught by nature. Millions of years of trial and error have resulted in designs which are incomparably refined.

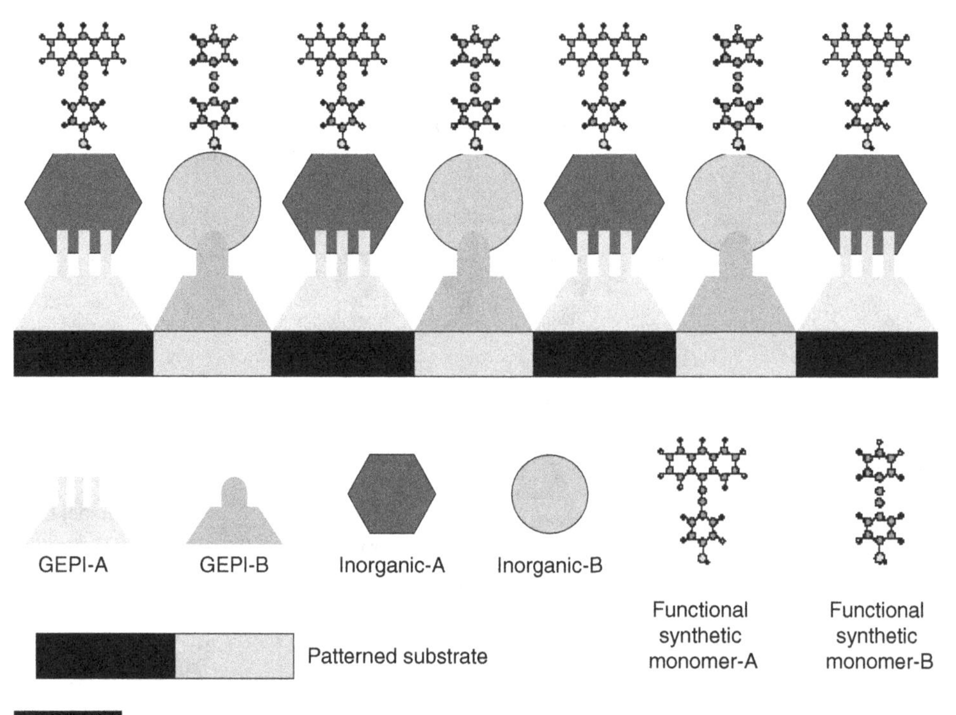

Figure 13.6.

The potential of using GEPI as "molecular erector" sets. Two different GEPI proteins (A and B) are assembled on ordered molecular or nano-scale substrates. Inorganic particles A and B are immobilized selectively on GEPI-A and GEPI-B, respectively. Synthetic molecules are assembled using functionalized side-groups on the nanoparticles. (Reprinted by permission from Macmillan Publishers Ltd.: *Nature Materials* (Sarikaya *et al.*, 2003), copyright (2003).)

13.4 Genetic engineering

13.4.1 General principles and methodology

In contrast with some contemporary researchers that use hyperbolae and bombastic claims to describe their accomplishments, Charles Darwin spent the first chapter of *On the Origin of Species* arguing that artificial selection was widely practiced and well recognized by breeders. He affirmed that evolution by selection was a well-known practice and downplayed his contributions. Indeed, genetic engineering continues this process, applied by humans for thousands of years to plants and animals, and elevates it to another level, enabled by nanotechnology. In simple words, we remove genetic material from the host, modify it, and reinsert it. This genetic manipulation was used in the past by breeders through selection. Thus, some dogs have short tails because clever breeders bred tailless dogs that appeared through spontaneous mutation. Genetic manipulation of seeds has been practiced for centuries: arid places were chosen to create hardy seeds that then exhibited superior qualities in more fertile soils. The great grandfather of one of the authors (MAM) sent seeds from Luxembourg to Turkey, where they were planted in arid soil. The surviving plants, having resisted the harsh conditions, had seeds that were hardier than the ones produced in Luxembourg, and were reimported with excellent results. This shows that genetic selection of seeds resulting from harsher environments can produce favorable genetic characteristics.

Genetic engineering was inaugurated when Jackson, Symons, and Berg (1972) combined DNA from the SV40 and lambda viruses. A transgenic mouse was created shortly thereafter by Jaenisch and Mintz (1974). Genetech was created in 1976, and genetically engineered human insulin was introduced a few years later. The commercialization of genetically modified organisms (GMOs) in agriculture is widespread, and numerous characteristics are conferred to plants, a prominent one being pest and pesticide resistance.

There are many variants to the processes used, called recombinant expression or recombinant DNA techniques. Generally, it involves the following sequence of events.

(1) Isolation of the gene that will be modified by extracting DNA from the host.
(2) Breaking of DNA into fragments; use of gel electrophoresis to separate different lengths.
(3) Cloning of the gene of interest by polymerase chain reaction (PCR); the genes can also be synthetically produced if the sequence is known.
(4) Insertion of the gene into a plasmid, which is a double-stranded DNA. The process of bonding is called "ligation." The plasmid is called the "vector."
(5) The vector is inserted into bacteria, such as *E. coli*; this process is called "transformation." About 1% of the bacteria accept the vector. This number can be increased by electric or heat shock, which renders the membrane more permeable to the plasmid. Bacteria are unicellular and reproduce clonally. Thus, the new bacteria contain the vector. In plants, the vector plasmid is usually introduced into *Agrobacterium*.
(6) The bacteria that contain the vector are cultured in a proper medium, leading to their "overexpression" (multiplication).
(7) Cells are "lysed" in order to reduce the proteins.
(8) The target protein is purified, separating it from the others.
(9) The modified genetic material is inserted into the host genome.

13.4.2 Applications

The applications fall into three groups.

Agriculture Twenty-five countries use genetically modified agricultural products. The principal ones are the USA, Brazil, Argentina, and Canada. Crops have been genetically modified to resist pesticides so that they can be sprayed. All other plants (and animals) perish. There is also a tomato, Flavr Savr, that has a thicker skin and therefore a longer shelf life. Genetically modified tobacco is pest resistant. A virus has been developed that can be sprayed onto crops to prevent freezing (*Pseudomonas syringae*). The concern that environmentalists have is that ingestion by humans can generate higher levels of cancer, since our bodies did not evolve defense mechanisms.

Drug manufacture: Since 1978, the following drugs have been manufactured by genetic engineering:

- human growth hormone;
- urofollitropin (follicle-stimulating hormone used in infertility treatment);
- human albumin;
- antibodies;
- vaccines.

Medicine/research: Genetically modified animals have ben produced for research. Mice with altered DNA are commonly used to study cancer (oncomouse), obesity, diabetes, substance abuse, and other diseases. Genetically modified pigs are being developed to have organs that are not rejected by humans, thus paving the way for pig-to-human transplants. In gene therapy, defective genes are replaced by functioning ones. This poses a broad ethical question, because humans could be altered to enhance their intelligence, athletic performance, and other characteristics. Some of the methods used to introduce genes can be passed to progeny. This leads to the revival of the much maligned field of eugenics.

An example of the effect that genetic modification can have is the work by Snead *et al.* (2006). They demonstrated that in dental enamel the self-assembly of proteins called amelogenin into nanospheres provides an environment from which hydroxyapatite crystals grow in the woven pattern shown in Chapter 7 (see Fig. 7.28(a)). By changing the amelogenin, transgenic mice were produced that had teeth with enamel that exhibited a different structure and was less hard. This was accomplished by genetic engineering, which consisted of altering or deleting amino acids in the two highly conserved domains of mouse amelogenin.

Resilin is a unique elastomer that demonstrates extremely long fatigue life and high resilience. It is found exclusively in the wings of insects, and thus it is impractical to try to harvest enough material to employ in an engineering situation. Researchers have isolated the genome for resilin and have inserted it into *E. coli* to proliferate. Figure 13.7 shows the microstructural and mechanical response of synthetic resilin. The high elasticity (tied into a knot) and diameter of the synthetic resilin are similar to those of the natural elastomer. The resilience is higher than that of two common rubbers: chlorobutyl and polybutadiene.

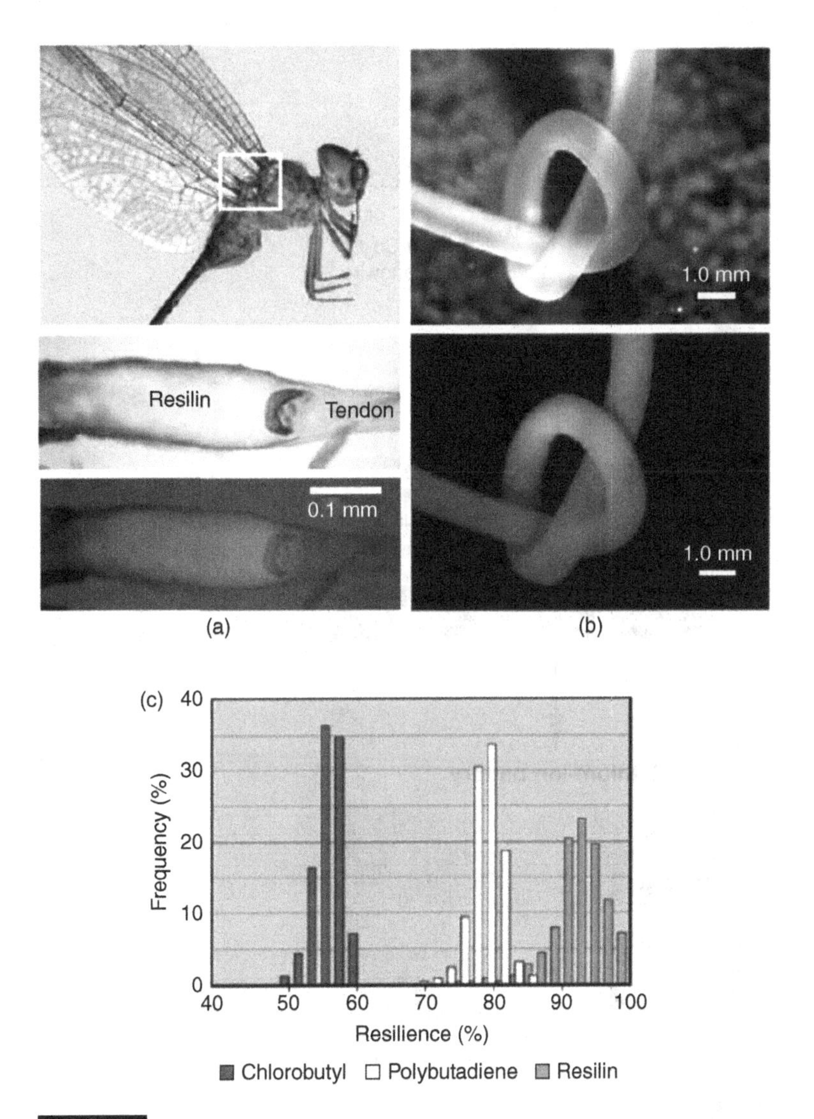

Figure 13.7.

Genomic DNA from the fruit fly, *Drosphilia melanogaster* was used as a template for PCR and inserted into *E. coli* for growth. (a) The resilin tendon extracted from the wing under ambient and ultraviolet light. (b) Synthetic resilin under ambient and ultraviolet light. (c) Resilience of two common rubbers (chlorobutyl rubber and polybutadiene rubber) and artificial resilin. (Reprinted by permission from Macmillan Publishers Ltd.: *Nature* (Elvin *et al.*, 2005), copyright (2005).)

13.5 Virus-assisted synthetic materials

Lithium-ion batteries are an important component in electric cars as well as in numerous applications in consumer electronics and aerospace. In a lithium-ion battery, lithium ions flow between the positively charged anode, usually graphite, and the negatively charged cathode, usually cobalt oxide or lithium iron phosphate. It was demonstrated by Belcher and

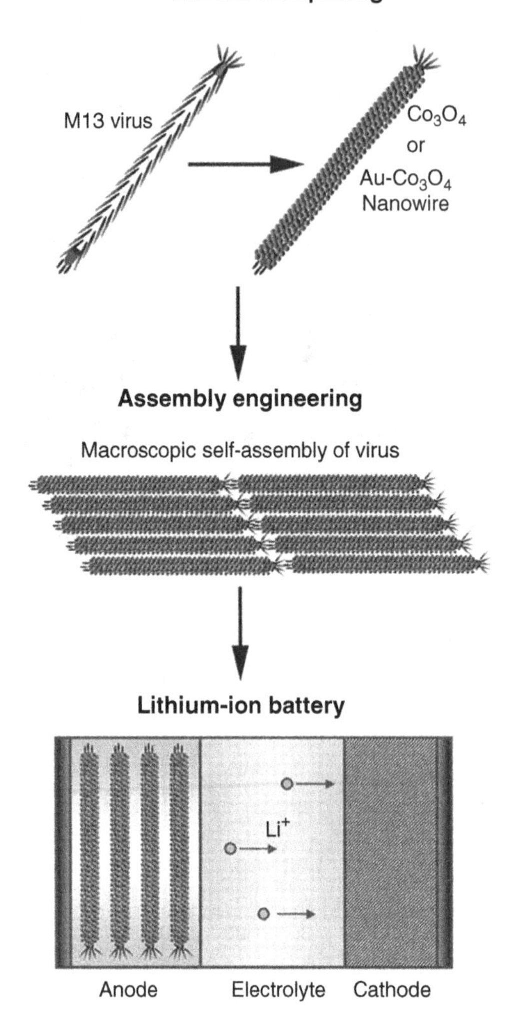

Figure 13.8.

Fabrication of lithium-ion battery using M13 virus onto which Co_3O_4 or Au-Co_3O_4 nano-scale particles are attached, forming a nanowire (top); these nanowires undergo self-assembly, enabled by the M13 virus (middle); the nanowire assembly is then used in the fabrication of a flexible lithium-ion battery. (Taken from Nam *et al.* (2006); with kind permission of Professor A. M. Belcher, MIT.)

co-workers that viruses can be engineered to build both the negatively (Nam *et al.*, 2006) and positively (Lee *et al.*, 2009) charged ends of a lithium-ion battery. The new virus-produced batteries have the same energy capacity and power performance as state-of-the-art rechargeable batteries being considered to power plug-in hybrid cars, and they could also be used to power a range of personal electronic devices. The new batteries can potentially be manufactured with a cheap and environmentally benign process, inspired by nature.

Figure 13.8 shows how M13 virus can assemble nanowires that are incorporated into the anode materials.

Figure 13.9.

M13 virus "grabbing" a single-wall carbon nanotube. (Taken from http://news.sciencemag.org/sciencenow/2009/04/03–01.html; with kind permission from Professor A. M. Belcher, MIT.)

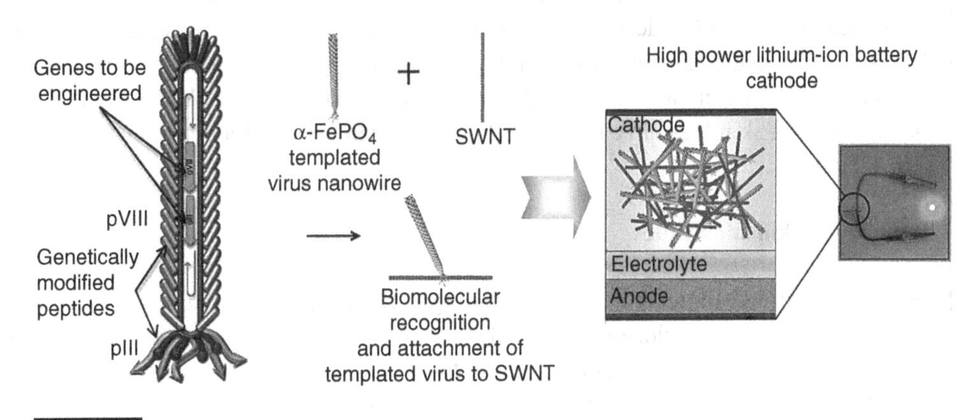

Figure 13.10.

M13 virus, with peptides which attach to nanocrystalline α-$FePO_4$, and the gene III protein (pIII), which is engineered to have a binding affinity for the single-wall nanotubes (SWNTs); this leads to a lithium-ion battery cathode that has a nanostructure and therefore enables greater current density. (Taken from Lee *et al.* (2009); with kind permission from Professor A. M. Belcher, MIT.)

Further development was reached when it was demonstrated that viruses recognize and bind specifically to carbon nanotubes. Figure 13.9 shows dramatically how an M13 virus can grab a single-wall carbon nanotube. In this case, each iron phosphate nanowire can be electrically "wired" to conducting carbon nanotube networks. Electrons can travel along the carbon nanotube networks, percolating throughout the electrodes to the iron phosphate and transferring energy in a very short time. By incorporating carbon nanotubes, these arrays can be added to the cathode. This increases the cathode's conductivity without adding too much weight to the battery. In laboratory tests, batteries with the new cathode material could be charged and discharged at least 100 times without losing any capacitance. That is fewer charge cycles than in currently available lithium-ion batteries, but the Belcher group expects them to be able to go much longer. Figure 13.10 shows this design and assembly (Lee *et al.*, 2009). The prototype is packaged as a typical coin cell battery, but the technology allows for the assembly of very lightweight, flexible, and conformable batteries that can take the shape of their container.

Box 13.1 Regenerative/synthetic skin

The skin is the largest organ in the body. It has a multilayered structure composed of an outer epidermis and an inner dermis, which can be subdivided into papillary dermis and reticular dermis. The skin provides essential functions, such as thermoregulation, desiccation barrier, mechanical and radiation (UV) protection, microbial defense, wound repair, sensing, and cosmetic appearance. Severe damage to large areas of skin leads to dehydration and infections that can result in death. Burn injuries result in loss of skin structure and functions to four degrees: first degree (epidermal layer), second degree (epidermal plus superficial dermal layers), third degree (epidermal plus entire dermal layers), and occasionally fourth degree (extending to the subcutaneous tissues). The epidermis is capable of regenerating, but the dermis cannot; so, it must be replaced. Several types of surgical management of skin loss have been developed and can be classified into temporary and permanent skin substitutes.

Temporary skin substitutes

Temporary skin substitutes are used as a dressing to clean superficial wounds, control pain, and provide temporary physiological closure. The ideal skin substitute is the patient's own skin, harvested from an uninjured area (known as an autograft). A split-thickness skin graft consists of epidermis plus a thin layer of dermis. However, an autograft is not always practical and has limitations. Allografts from cadavers remain the mainstay of temporary skin substitutes. Other approaches include xenografts (graft from a different species, typically pigs), synthetic membrane, and hydrocolloid dressings. Biobrane® (Mylan Laboratories, Inc.) is a bilayer synthetic membrane constructed of an outer silicon rubber film and an inner Nylon mesh cross-linked with type-I porcine collagen. The silicon rubber film acts as a barrier to water loss and microbial invasion, analogous to the epidermis. The complex three-dimensional Nylon mesh and collagen layer imitate the elastin–collagen network in the dermis. Biobrane® is flexible, has adequate mechanical properties, and is widely used as a temporary closure material as it is not biodegradable. Hydrocolloid dressings combine a synthetic membrane or foam with a hydrocolloid. These dressings not only prevent water loss and bacterial infection, but also absorb exudate produced by wounds.

Permanent skin substitutes

Keratinocyte is the principal cell of the epidermis and can be cultured and grown in large numbers from a small piece of skin. The biopsy of a patient's unburned skin can be sent to a company for keratinocyte culture (for example, Epicel®, Genzyme BioSurgery Inc.). Cultured epidermal sheets have been widely used for the treatment of large-area burns and damage.

Yannas, Burke, and co-workers (Burke *et al.*, 1981; Yannas *et al.*, 1982; Schultz, Tompkins, and Burke, 2000) developed a bilayer, synthetic skin substitute (Integra®, Integra LifeSciences Inc.) made of an outer silicon rubber membrane and an inner collagen–GAG extracellular matrix. The outer silicon rubber membrane functions as the epidermis, preventing evaporation and infection. The collagen–GAG matrix allows the growth of fibroblasts and endothelial cells and the formation of the neodermis. Once the neodermis has partially vascularized, the silicon rubber layer can be removed and replaced by a thin layer of epidermal skin autograft. Consequently, the donor site heals rapidly without significant scar tissue. The procedure of Integra® dermal regeneration is shown schematically in Fig. B13.1. However, Integra® has limitations: (1) it has no intrinsic immunologic defenses; (2) the silicone rubber is not biodegradable and must be removed; (3) it is difficult to apply on certain locations (e.g. axilla, back, groin).

A revolutionary approach in allografts is the development of acellular (decellulized) cadaver skin, called AlloDerm® (LifeCell, Inc.). Split-thickness cadaver skin is treated in solutions to remove dermal

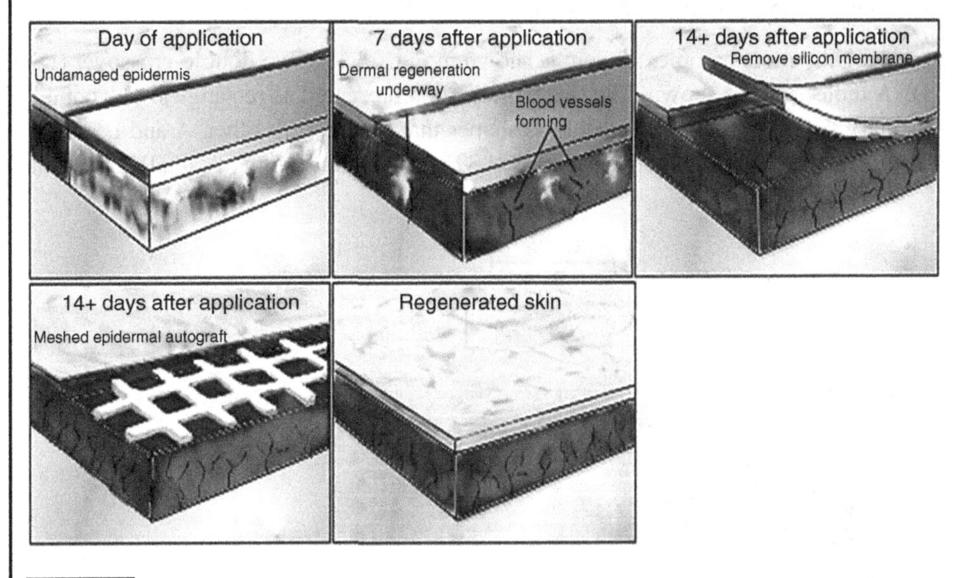

Figure B13.1.

Integra® dermal regeneration procedure.

Box 13.1 (cont.)

cells completely. The resulting acellular skin consists merely of collagen and elastin fibers. AlloDerm® is well tolerated by and compatible with patients, as expected. Acellular dermis has been used in oral and reconstructive surgery to replace small skin defects.

13.6 Bioinspiration from the molecular level: the bottom-up approach

Biological structures have molecular recognition. This property enables them to self-assemble, which leads to the templating of atomic and molecular assemblages. This is the principle of the bottom-up approach, in contrast with many synthetic processes that use a top-down approach. The simplest example is how DNA strands will pair and form a double helix. Some of these concepts were discussed in Meyers *et al.* (2008b), and we expand on recent developments here.

As illustrations, two examples from Seeman and Belcher (2002) are described. Synthetic molecules with "sticky ends" can be designed to build motifs in one, two, or three dimensions. A junction of four molecule "sticky ends" was shown in Fig. 13.1(a). These "sticky ends" in turn attach to each other, forming the regular pattern shown on the right-hand side of Fig. 13.1(a). These arrays can serve as:

- scaffolding to crystallize biological macromolecules;
- organization of compounds in nanoelectronics;
- quantum dots (bioinspired nanocomponent chips).

A slightly more complex system is shown in Fig. 13.11. Two double-crossover (DX) DNA molecules are shown. The two helices are represented as rectangles. The complementary sticky ends are represented as shapes that fit into each other. A and B* are not identical. The repeat unit in the assembly is 32 nm, twice the width of each DX molecule.

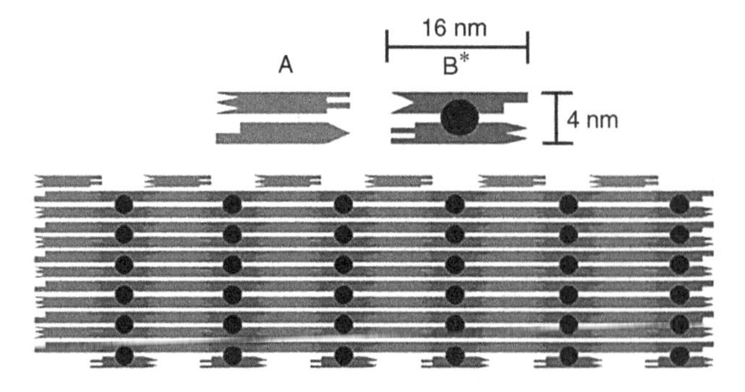

Figure 13.11.

More complex patterns of DNA. (Reprinted from Seeman and Belcher (2002), with permission; copyright (2003) National Academy of Sciences, USA.)

They build a two-dimensional array that has been demonstrated to construct calculators. It was envisaged (Seeman and Belcher, 2002) that these DNA systems, created by self-assembly, would be used in combination with inorganic nanoparticles such as carbon nanotubes, to generate different functionalities.

Box 13.2 Tissue engineering

This is a new approach in biomaterials: instead of using synthetic materials, a bioinspired process is used to synthesize parts of tissues using cells harvested from humans and inserted into an appropriate scaffold that mimics the organ to be reproduced. The cells are nourished and multiply, and tissue is created as the scaffold dissolves away. This conceptually simple approach presents immense challenges because it is at the confluence of medicine, nanomaterials engineering, and cellular biology. Figure B13.2 shows, in schematic fashion, the most commonly used methodology. Cells are harvested from an individual, then cultured and inserted into a scaffold. They form a tissue, which is then reintroduced into the individual.

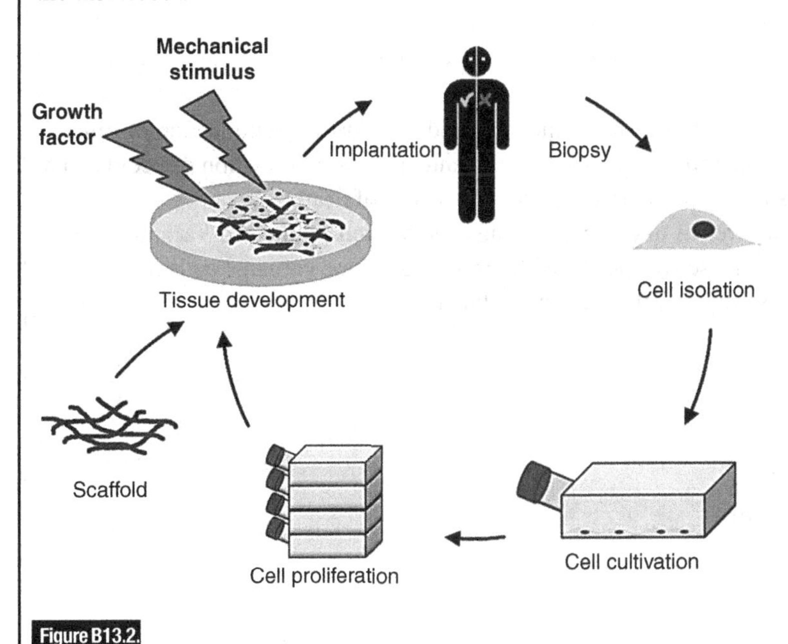

Figure B13.2.

Basic principles of tissue engineering. (From http://commons.wikimedia.org/wiki/File:Tissue_engineering_english.jpg. Image by HIA. Licensed under CC-BY-3.0, via Wikimedia Commons.)

There are strict requirements for the scaffolding, and the size and shape of the pores are important. It is difficult to provide nutrition to the cells if the scaffold is too thick. Thus, two-dimensional scaffolds were an early success. Skin is a primary example, as are vascular grafts. If the scaffold is thicker, vascularization might be needed to bring nutrients to the cells.

Box 13.2 (cont.)

The most important synthetic scaffold materials are:

- polyglycolic acid (PGA) – it loses its strength 2–4 weeks after implantation;
- polylactic acid (PLA);
- PLA-PGA copolymers – they are manufactured in foam form.

The most important biological scaffolds are:

- collagen;
- chitin and chitosan.

Biological scaffolds have to be denatured by removing all the cells. In this manner, rejection is avoided. There are special procedures to accomplish this.

Skin can be generated by harvesting cells from foreskins removed from infants during circumcision. Newborn cells do not activate the immune system. These cells are inserted into appropriate scaffolds and nourished; they can be used in cases of severe burns. The German laboratory Fraunhofer has developed a machine to produce large amounts of skin. This skin can replace animal skin used in the cosmetics industry. The skin cells can also be harvested from the person that will receive the implant.

Collagen is resorbed in the body. An early example is catgut suture, which is actually taken from sheep intestine.

Current research in tissue (or regenerative) medicine is geared towards the manufacture of organs, such as the liver. There is a great need for livers in the world, and only a small fraction of people with diseased livers can receive transplants. Organ failure is a major cause of death.

One of the most spectacular pictures of tissue engineering was the well-publicized Vacanti or "ear" mouse shown in Fig. B13.3. This mouse was produced by inserting bovine cartilage cells into an ear-shaped scaffold and surgically inserting it onto the back of the mouse.

Figure B13.3

A pensive Vacanti mouse. (Figure courtesy Professor Meyers.)

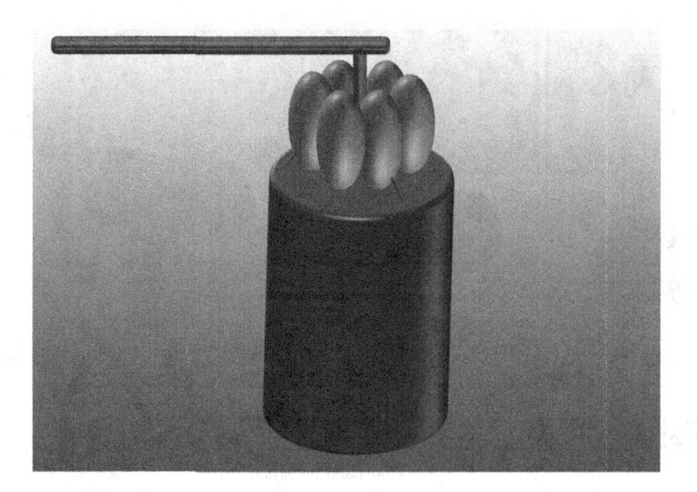

Figure 13.12.

Hybrid bionanomechanical rotary motor made of an ATP synthase and a nickel nanopropeller.

13.7 MEMS and NEMS

Microelectromechanical systems (MEMS) and nanoelectromechanical systems (NEMS) are two fascinating and promising research fields. MEMS, a field that emerged in the mid 1980s, utilizes a top-down semiconductor process to manufacture fine, micro-sized products. NEMS, on the other hand, adopts a bottom-up approach using nano-scale building blocks self-assembled into nano-sized devices. The bottom-up approach has attractive advantages over the top-down approach considering how biomolecular motors and machines (e.g. cells) are made in nature (Ozin *et al.*, 2005).

One approach to making artificial bionanomachines is to integrate biomolecules with synthetic nanostructures. Inspired by the flagella rotary motor of bacteria (*E. coli*) and sperms, a nano-sized rotary device has been assembled from an ATP synthase and a nickel propeller (Soong *et al.*, 2000). ATP synthase, or ATPhase, is an assembly of proteins embedded in the cell lipid bilayer responsible for the synthesis of ATP. When an ATP molecule is synthesized, the ATP synthase rotates 120°, making this protein a tiny and efficient nanopropeller. Figure 13.12 shows a schematic of the hybrid bionanomechanical device.

The second example is an electrically driven linear nanomotor assembled from an indium nanocrystal RAM between two carbon nanotube level arms, as shown in Fig. 13.13 (Regan *et al.*, 2005). Another indium nanocrystal is attached to one of the lever arms and serves as a reservoir of indium atoms. The voltage bias applied to the lever controls the direction of electromigration of indium atoms between the reservoir and RAM, making the RAM grow or shrink, and thereby prying the lever arms apart or bringing

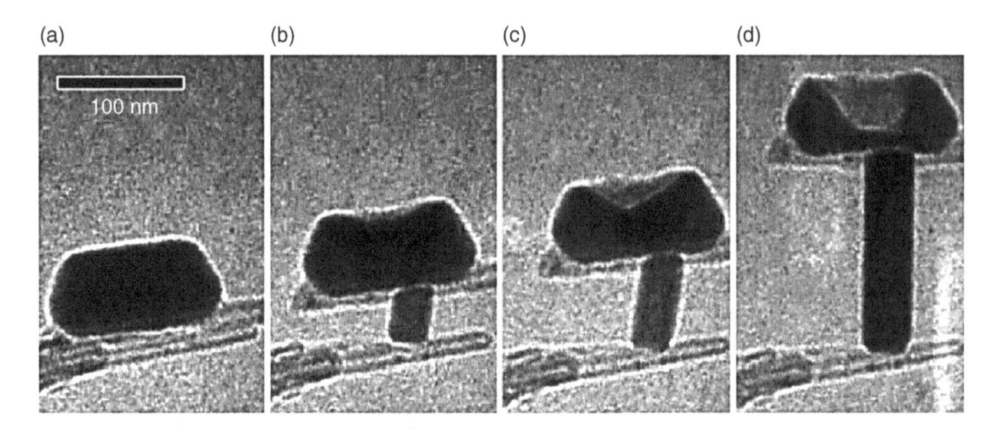

Figure 13.13.

Electrically driven linear nanomotor made from an indium nanocrystal and two carbon nanotube levers. (Reprinted with permission from Regan *et al.* (2005). Copyright (2005) American Chemical Society.)

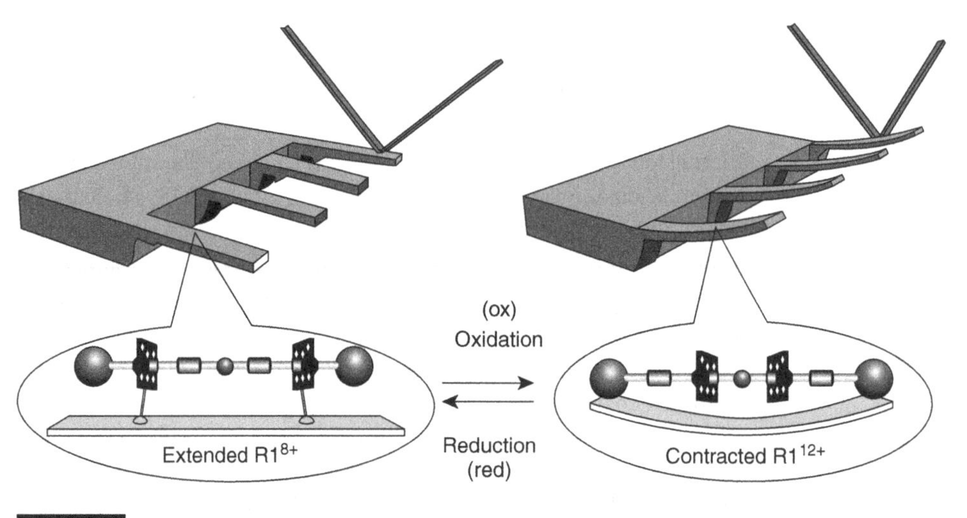

Figure 13.14.

Nanochemomechanical devices based on a redox-active rotaxane tethered to a microcantilever, causing reversible beam flexing in chemical oxidation–reduction cycles. (Reprinted with permission from Huang *et al.* (2004). Copyright (2004), AIP Publishing LLC.)

them together. The nanomachines can function as electrically activated transistors, or as light-emitting diodes, lasers, and other devices.

Nanochemomechanical-induced bending of an array of microcantilevers has been developed by a redox-powered linear molecular rotaxane motor, as shown in Fig. 13.14 (Huang *et al.*, 2004). Cycles of chemical oxidation and reduction of the tethered rotaxane molecule induce a contractile strain in the surface of the microcantilever, causing it to flex. This is analogous to how muscle filaments work.

13.8 Bioinspired synthesis and processing of biopolymers

Considerable biomimetic efforts have been devoted to reproducing the principal biopolymers that comprise the extracellular matrix (ECM): collagen, elastin, and silk. In their synthesis, a number of techniques have been used, ranging from conventional polymer chemistry methods to more advanced approaches using recombinant DNA technology, in which genetic templates are used via bacterial expression. This technology has already yielded products that have gained FDA approval.

The advantage of recombinant technology is that synthesis is accomplished at ambient temperature and in an aqueous environment. The disadvantage is the time taken for the assembly of genetic constructs and the inability to process some amino acids.

Silk-like and silk–elastin-like copolymers have been synthesized by the methods developed by Kaplan's group (Krishnaji *et al.*, 2011; Rabotyagova, Cebe, and Kaplan, 2011). The processing methods for these biopolymers are as follows.

- Electrospinning: a large electric field is applied between a polymer solution and a collection plate, and a jet is expelled from the solution. Fibers with diameters from a few nanometers to micrometers can be produced by electrospinning. This technique is effective in producing scaffolds.
- Wet spinning: this uses an extrusion nozzle to produce the fibers, which can be used for three-dimensional constructs.
- Microfluidic processing; the polymer solution is forced through a microfluidic channel (Bevelander and Nakahara, 1969).
- Biopolymer films: a solution is cast into a mold and allowed to dry. This produces a two-dimensional film.
- Self-assembly: this has been used to produce hydrogel, which is a tridimensional network of polymers. Hydrogels swell with the addition of water but are not dissolved. The self-assembled particles can be used in drug delivery. Hydrogels are used in contact lenses, artificial organs, and controlled drug delivery.

Box 13.3 Electronic medical implants

"Intelligent implants" containing computer chips are being developed and present a bright potential. We briefly describe some concepts that are finding their way in applications. The pacemaker is an example of such a device, and modern versions have added features enabled by microchips.

Pharmacy on a chip The idea for a programmable, wirelessly controlled microchip implanted in a patient's body to deliver drugs is becoming a reality. Implantable devices can be controlled externally through radio signals and can administer daily doses of, for instance, an osteoporosis drug normally given by injection.

PillCam Noninvasive imaging of the gastrointestinal (GI) tract is now possible with a pill containing sensors, which collect information as the pill passes through the GI tract. Wireless capsule endoscopy

Box 13.3 (cont.)

(WCE) has rapidly become the preferred method of small intestine diagnosis; by 2009, over one million PillCam capsules had been used clinically.

Proteus smart pill This pill contains physiological sensors combined into a medication pill. Gastric acids in the stomach activate an energy source, allowing the sensor to send signals to a skin patch electrode, which wirelessly transmits information to a remote monitor, such as a smart phone.

Continuous glucose monitoring device This is composed of a sensor, embedded in the body, a transmitter, and a telemonitor. The glucose level is continuously monitored.

Future generations of electronic medical implants will have components that are dissolved in the body (edible electronics) and other capabilities. Special power supplies (batteries) that are biocompatible need to be developed in tandem.

Summary

- Molecular-based biomimetics requires knowledge of the structure of biological materials at the nano-scale. Nanotechnology plays a vital role because it provides the experimental and computational tools that enable the understanding and manipulation of the structure at the molecular level.
- The process of self-assembly is the essential component of the bottom-up approach through which biological systems are formed. This process can be one, two, or three dimensional.
- A two-dimensional network of branched DNA molecules, in the form of a cross with "sticky" ends, can self-assemble to form a reticulated network. Possible applications are: scaffolding to crystallize biological macromolecules; organization of compounds in nanoelectronics; bioinspired nanocomponent chips (quantum dots).
- The bacteriophage (nicknamed "phage") is a microbe that has nano-scale dimensions. Phages, like viruses, do not have reproductive organs. They have been commonly used in biochemistry and pharmaceutical fields since the momentous discovery of a new technique (in 1985) for the study of protein–protein, protein–peptide, and protein–DNA interactions. Phages infect bacteria but are harmless to humans.
- Viruses can also be used directly as building blocks for nanostructures assembled from the bottom up. The ability of phages to bind to nanoparticles (quantum dots), single-wall carbon nanotubes, and tridimensional components has been used to create structures.
- Genetic engineering is a new frontier in research and manufacturing that involves many facets and numerous steps, which vary according to the intended application. Nevertheless, there are some common traits: the removal of genes from the host; manipulationg by gene cloning using PCR or de novo synthesis; insertion of gene into a plasmid vector; introduction of vector into bacteria (transformation); over-expression of vector containing bacteria; lysing of cells and release of proteins; purification of protein and reinsertion of genes into host.

- The applications of genetic engineering have led to GMOs in agriculture (pest- and pesticide-resistant crops; the Flavr Savr tomato; *P. syringae* for protection of crops against frost), drugs (human insulin, follicle-stimulating hormones, human albumin, vaccines).
- Viruses can be engineered to build both the negatively and positively charged ends of a lithium-ion battery. They have the same capacity and power as conventional batteries and can potentially be manufactured with a cheap and environmentally benign process, inspired by nature.
- The biopolymers collagen, elastin, and silk are being synthesized by a number of techniques, ranging from conventional polymer chemistry methods to more advanced approaches using recombinant DNA technology. They are processed by using techniques such as electrospinning, wet spinning, microfluidics, casting, and self-assembly.

Exercises

Exercise 13.1 Compare the performance of the virus-derived batteries made by Belcher's group with conventional lithium-ion batteries by researching the literature.

Exercise 13.2 By researching the literature, describe the steps that were required to produce the first transgenic mouse.

Exercise 13.3 Using sketches and explanations, describe electrospinning.

Exercise 13.4 C. T. Lim, at NUS in Singapore, used microfluidics to separate healthy red blood cells from those contaminated with malaria. This procedure was later adapted to isolate selectively cancer cells in the blood stream. By researching the literature, find the sources and write a description of the technique with sketches.

Exercise 13.5 Describe the sequence of procedures required to produce a genetically modified plant.

Exercise 13.6 Find the original paper by Crick and Watson, copy the figure, and comment on the significance of the results and the length of the contribution.

Exercise 13.7 Give three examples of how a virus can be used to synthesize inorganic materials and list the potential applications.

Exercise 13.8 What are GEPIs? Explain with illustrations/sketches. Give at least three potential applications of GEPIs.

References

Aaron, B. B. and Gosline, J. M. (1981) Elastin as a random-network elastomer: a mechanical and optical analysis of single elastin fibers. *Biopolymers* **20**: 1247–1260.

Achrai, B. and Wagner, H. D. (2013) Micro-structure and mechanical properties of the turtle carapace as a biological composite shield. *Acta. Biomater.* **9**: 5890–5902.

Addadi, L. and Weiner, S. (1985) Interaction between acidic proteins and crystals: stereochemical requirements in biomineralization. *Proc. Natl. Acad. Sci. USA* **82**: 4110–4114.

Addadi, L., Moradian, J., Shay, E., Maroudas, N. G., and Weiner, S. (1987) A chemical model for the cooperation of sulfates and carboxylates in calcite crystal nucleation: relevance to biomineralization. *Proc. Natl. Acad. Sci. USA* **84**: 2732–2736.

Addadi, L., Raz, S., and Weiner, S. (2003) Taking advantage of disorder: amorphous calcium carbonate and its roles in biomineralization. *Adv. Mater.* **15**: 959–970.

Addadi, L., Joester, D., Nudelman, F., and Weiner, S. (2006) Mollusk shell formation: a source of new concepts for understanding biomineralization processes. *Chem. Eur. J.* **12**: 980–987.

Adharapurapu, R. R., Jiang, F., and Vecchio, K. S. (2006) Dynamic fracture of bovine bone. *Mater. Sci. Eng. C* **26**: 1325–1332.

Aizenberg, J. (2010) New nanofabrication strategies: inspired by biomineralization. *MRS Bull.* **35**: 323–330.

Aizenberg, J. and Hendler, G. (2004) Designing efficient microlens arrays: lessons from nature. *J. Mater. Chem.* **14**: 2066–2072.

Aizenberg, J., Weaver, J. C., Thanawala, M. S., Sundar, V. C., Morse, D. E., and Fratzl, P. (2005) Skeleton of Euplectella sp.: structural hierarchy from the nanoscale to the macroscale. *Science* **309**: 275–278.

Aladin, D. M., Cheung, K. M., Ngan, A. H. *et al.* (2010) Nanostructure of collagen fibrils in human nucleus pulposus and its correlation with macroscale tissue mechanics. *J. Orthop. Res.* **28**: 497–502.

Alexander, N. J. (1970) Composition of α-and β-keratin in reptiles. *Cell. Tissue Res.* **110**: 153–165.

Alexander, N. J. and Fahrenbach, W. F. (1969) The dermal chromatophores of *Anolis carolinensis* (Reptilia, Iguanidae). *Am. J. Anat.* **126**: 41–55.

Alexander, N. J. and Parakkal, P. F. (1969) Formation of α- and β-type keratin in lizard epidermis during the molting cycle. *Cell. Tissue Res.* **101**: 72–87.

Almqvist, N., Thomson, N. H., Smith, B. L., Stucky, G. D., Morse, D. E., and Hansma, P. K. (1999) Methods for fabricating and characterizing a new generation of biomimetic materials. *Mater. Sci. Eng. C* **7**: 37–43.

Altman, G. H., Diaz, F., Jakuba, C. *et al.* (2003) Silk-based biomaterials. *Biomater.* **24**: 401–416.

Argon, A. S. (1972) *Fracture of Composites*. Treatise of Materials Science and Technology. New York: Academic Press, p. 1.

Arias, J. L. and Fernández, M. S. (2003) Biomimetic processes through the study of mineralized shells. *Mater. Charact.* **50**: 189–195.

Armbrust, E. V., Berges, J. A., Bowler, C. *et al.* (2004) The genome of the diatom *Thalassiosira pseudonana*: ecology, evolution, and metabolism. *Science* **306**: 79–86.

Armstrong, W. P. (1979) Nature's hitchhikers. *Environ. Southwest* **486**: 20–23.

Arruda, E. M. and Boyce, M. C. (1993) A three-dimensional model for the large stretch behavior of rubber elastic materials. *J. Mech. Phys. Solids* **41**: 389–412.

Arzt, E. (2006) Biological and artificial attachment devices: lessons for materials scientists from flies and geckos. *Mater. Sci. Eng. C* **26**: 1245–1250.

Arzt, E., Gorb, S., and Spolenek, R. (2003) From micro to nano contacts in biological attachment devices. *Proc. Natl. Acad. Sci. USA* **100**: 10603–10606.

Ashby, M. F. (1989) On the engineering properties of materials. *Acta Metal.* **37**: 1273–1293.

Ashby, M. F. (1992) *Materials Selection in Mechanical Design*. Oxford: Butterworth–Heinemann.

ASTM E399–09e2 Standard test method for linear elastic plane strain fracture toughness K_{Ic} of metallic materials.

Atkins, A. G. (2009) *The Science and Engineering of Cutting*. Oxford: Butterworth–Heinemann.

Autumn, K., Liang, Y. A., Hsieh, S. T. *et al.* (2000) Adhesive force of a single gecko foot-hair. *Nature* **405**: 681–684.

Autumn, K., Sitti, M., Liang, Y. A. *et al.* (2002) Evidence for van der Waals adhesion in gecko setae. *Proc. Natl. Acad. Sci. USA* **99**: 12252–12256.

Aveston, J., Cooper, G. A., and Kelly, A. (1971) Single and multiple fracture. In *Proc. Conf. Properties of Fiber Composites*. Guildford: IPC Science and Technology Press, pp. 15–26.

Baer, E., Hiltner, A., and Morgan, R. J. (1992) Biological and synthetic hierarchical composites. *Phys. Today* **45**: 60–67.

Baillie, C. and Fitford, R. (1996) The three-dimensional composite structure of cow hoof wall. *Biomimetics* **4**: 1–22.

Baillie, C., Southam, C., Buxtin, A., and Pavan, P. (2000) Structure and properties of bovine hoof horn. *Appl. Composite Lett.* **9**: 107–115.

Bain, C. D. and Whitesides, G. M. (1998) Molecular-level control over surface order in self-assembled monolayer films of thiols on gold. *Science* **240**: 62–63.

Ballarini, R., Kaycan, R., Ulm, F-J., Belytschko, T., and Heuer, A. H. (2005) Biological structures mitigate catastrophic fracture through various strategies. *Int. J. Fracture* **135**: 187–197.

Bao, G. and Suresh, S. (2003) Cell and molecular mechanics of biological materials. *Nature Mater.* **2**: 715–725.

Barnes, W. J. P. (2007) Functional morphology and design constraints of smooth adhesive pads. *MRS Bull.* **32**: 479–485.

Barnes, W. J. P., Perez-Goodwyn, P., and Gorb, S. N. (2005) Mechanical properties of the toe pads of the tree frog, Litoria caerulea. *Comp. Biochem. Physiol. A* **141**: S145.

Barnes, W. J. P., Oines, C., and Smith, J. M. (2006) Whole animal measurements of shear and adhesive forces in adult tree frogs: insights into underlying mechanisms of adhesion obtained from studying the effects of size and shape. *J. Comp. Physiol. A* **192**: 1179–1191.

Barthelat, F., Li, C. M., Comi, C., and Espinosa, H. D. (2006) Mechanical properties of nacre constituents and their impact on mechanical performance. *J. Mater. Res.* **21**: 1977–1986.

Barthlott, W. (1990) Scanning electron microscopy of the epidermal surface in plants. In Claugher, D., ed. *Application of the Scanning EM in Taxonomy and Functional Morphology*, Systematics Association Special Volume. Oxford: Clarendon Press, pp. 69–94.

Barthlott, W. and Ehler, N. (1977) Raster-Elektronenmikroskopie der Epidermis-Oberflächen von Spermatophyten. *Tropische und subtropische Pflanzenwelt (Akad. Wiss. Lit. Mainz)* **19**: 110.

Barthlott, W. and Neinhuis, C. (1997) The purity of sacred lotus or escape from contamination in biological surfaces. *Planta* **202**: 1–8.

Bartol, I. K., Gharib, M., Weihs, D., Webb, P. W., Hove, J. R., and Gordon, M. S. (2003) Hydrodynamic stability of swimming in ostraciid fishes: role of the carapace in the smooth trunkfish Lactophrys triqueter (Teleostei: Ostraciidae). *J. Exp. Biol.* **206**: 725–744.

Bartol, I. K., Gharib, M., Webb, P. W., Weihs, D., and Gordon, M. S. (2005) Body-induced vortical flows: a common mechanism for self-corrective trimming control in boxfishes. *J. Exp. Biol.* **208**: 327–344.

Bechtle, S., Ang, S. F., and Schneider, G. A. (2010) On the mechanical properties of hierarchically structured biological materials. *Biomater.* **31**: 6378–6385.

Behiri, J. C. and Bonfield, W. (1980) Crack velocity dependence of longitudinal fracture in bone. *J. Mater. Sci.* **15**: 1841–1849.

Belcher, A. M. (1996) Spatial and temporal resolution of interfaces: phase transitions and isolation of three families of proteins in calcium carbonate-based biocomposite materials. Unpublished Ph.D. Thesis, University of California, Santa Barbara.

Belcher, A. M. and Gooch, E. E. (1998) In Bauerlein, E., ed. *Biomineralization*. Weinheim: Wiley-VCH.

Belcher, A. M., Wui, X. H., Christensen, R. J., Hansma, P. K., Stucky, G. D., and Morse, D. E. (1996) Control of crystal phase switching and orientation by soluble mollusk-shell proteins. *Nature* **381**: 56–58.

Belcher, A. M., Hansma, P. K., Stucky, G. D., and Morse, D. E. (1997) First steps in harnessing the potential of biomineralization as a route to new high-performance composite materials. *Acta Mater.* **46**: 733–736.

Bell, E. C. and Gosline, J. M. (1996) Mechanical design of mussel byssus: material yield enhances attachment strength. *J. Exp. Biol.* **199**: 1005–1017.

Ben-Yosef, E., Levy, T. E., Higham, T., Najjar, M., and Tauxe, L. (2010) The beginning of Iron Age copper production in the southern Levant: new evidence from Khirbat al-Jariya, Faynan, Jordan. *Antiquity* **84**: 724–746.

Bereiter-Hahn, J., Matoltsy, A. G., and Richards, K. S., eds. (1986) *Biology of the Integument, Vol. 2: Vertebrates*. Berlin: Springer-Verlag.

Berglin, M. and Gatenholm, P. (2003) The barnacle adhesive plaque: morphological and chemical differences as a response to substrate properties. *Colloids Surf.* **28**: 107–117.

Berling, J. and Rechberger, M. (2007) Knives as sharp as rat's teeth. Fraunhofer Institute for Environmental, Safety and Energy Technology, Research News 1, Topic 3, http://www.archiv.fraunhofer.de/archiv/pi-en-2004-2008/EN/press/pi/2005/01/Mediendienst012005Thema3.html.

Bertram, J. E. A. and Gosline, J. M. (1986) Fracture toughness design in horse hoof keratin. *J. Exp. Biol.* **125**: 29–47.

Bertram, J. E. A. and Gosline, J. M. (1987) Functional design of horse hoof keratin: the modulation of mechanical properties through hydration effects. *J. Exp. Biol.* **130**: 121–136.

Best, S. M., Porter, A. E., Thian, E. S., and Huang, J. (2008) Bioceramics: past, present and for the future. *J. Euro. Ceramic Soc.* **28**: 1319–1327.

Bevelander, G. and Nakahara, H. (1969) An electron microscope study of formation of nacreous layer in shell of certain bivalve molluscs. *Calcif. Tiss. Res.* **3**: 84–87.

Bigliana, L. U., Pollock, R. G., Soslowsky, L. J., Flatow, E. L., Pawluk, R. J., and Mow, V. C. (1992) Tensile properties of the inferior glenohumeral ligament. *J. Orthop. Res.* **10**: 187–197.

Bilitch, M., Lau, F. Y. K., and Cosby, R. S. (1967) Recent advances in artificial pacemakers. *Calif. Med.* **107**: 164–170.

Bini, F., Marinozzi, A., Marinozzi, F., and Patanè, F. (2002) Microtensile measurements of single trabeculae stiffness in human femur. *J. Biomech.* **35**: 1515–1519.

Birchall, J. D. and Thomas, N. L. (1983) On the architecture and function of cuttlefish bone. *J. Mater. Sci.* **18**: 2081–2086.

Black, J. and Hastings, G. W. (1998) *Handbook of Biomaterials Properties*. London: Chapman and Hall.

Bledzki, A. K. and Gassan, J. (1999) Composites reinforced with cellulose based fibres. *Prog. Polymer Sci.* **24**: 221–274.

Boal, D. (2012) *Mechanics of the Cell*, 2nd edn. Cambridge: Cambridge University Press.

Bodde, S. G., Meyers, M. A., and McKittrick, J. (2011) Correlation of the mechanical and structural properties of cortical rachis keratin of rectrices of the Toco Toucan (*Ramphastos toco*). *J. Mech. Behav. Biomed. Mater.* **4**: 723–732.

Bonfield, W. and Datta, P. K. (1974) Young's modulus of compact bone. *J. Biomech.* **7**: 147–149.

Bonser, R. H. C. (1995) Melanin and the abrasion resistance of feathers. *The Condor* **95**: 590–591.

Bonser, R. H. C. (2001) The mechanical performance of medullary foam from feathers. *J. Mater. Sci. Lett.* **20**: 941–942.

Bonser, R. H. C. and Purslow, P. P. (1995) The Young's modulus of feather keratin. *J. Exp. Biol.* **198**: 1029–1033.

Bonser, R. H. C. and Witter, M. S. (1993) Indentation hardness of the bill keratin of the European starling. *The Condor* **95**: 736–738.

Boskey, A. (2003) Bone mineral crystal size. *Osteoporos. Int.* **14**: 16–21.

Bouligand, Y. (1970) Aspects ultrastructuraux de la calcification chez les Crabes. *7th Int. Cong. Electron Microscopy*, Grenoble, France, 31 Aug. 1970, pp. 105–106.

Bouligand, Y. (1972) Twisted fibrous arrangements in biological materials and cholesteric mesophases. *Tiss. Cell* **4**: 189–217.

Bowden, N., Tamerler, A., Carbech, J., and Whitesides, G. M. (1997) Self-assembly of meso-scale objects into ordered two-dimensional arrays. *Science* **276**: 233–235.

Brånemark, P. I. (1972a) Rehabilitation with intra-osseous anchorage of dental prosthesis. *Tandlakartidningen* **844**: 662–663.

Brånemark, P. I. (1972b) Rehabilitation with a denture anchored to the jawbone. *Lakartidningen* **69**: 4813–4814.

Brånemark, P. I. and Breine, U. (1964) Formation of bone marrow in isolated segment of rib periosteum in rabbit and dog. *Blut* **10**: 236–252.

Brånemark, P. I. and Eriksson, E. (1972) Method for studying qualitative and quantitative changes of blood flow in skeletal muscle. *Acta Physiol. Scand.* **84**: 284–288.

Brånemark, P. I., Breine, U., Johansson, B., and Roylance, P. J. (1964) Regeneration on bone marrow. *Acta Anat.* **59**: 1–46.

Bricteux-Grègoire, S., Florkin, M., and Grègoire, C. H. (1968) Prism conchiolin of modern or fossil molluscan shells: an example of protein paleization. *Comp. Biochem. Physiol.* **24**: 567–572.

Brink, D. J. and van der Berg, N. G. (2004) Structural colours from feathers of the bird Bostrychia hagedash. *J. Phys. D: Appl. Phys.* **37**: 813–818.

Brodkorb, P. (1955) Number of feathers and weights of various systems in a bald eagle. *Wilson Bull.* **67**: 142.

Brown, C. H. (1975) *Structural Materials in Animals*. London: Pitman.

Brown, S. (1997) Metal-recognition by repeating polypeptides. *Natl. Biotechnol.* **15**: 269–272.

Bruet, B. J. F., Qi, H. J., Boyce, M. C. *et al.* (2005) Nanoscale morphology and indentation of individual nacre tablets from the gastropod mollusc Trochus niloticus. *J. Mater. Res.* **20**: 2400–2419.

Bruet, B. J. F., Song, J., Boyce, M. C., and Ortiz, C. (2008) Materials design principles of ancient fish armor. *Nat. Mater.* **7**: 748–756.

Brush, A. H. (1986) Tissue specific protein heterogeneity in keratin structures. *Biochem. Syst. Ecol.* **14**: 547–551.

Brush, A. H. and Wyld, J. A. (1982) Molecular organization of avian epidermal structures. *Comp. Biochem. Physiol. B* **73**: 313–325.

Budiansky, B. (1983) *Micromech.* **16**: 3–12.

Buehler, M. J. (2008) Hierarchical nanomechanics of collagen fibrils: atomistic and molecular modeling. In Fratzl, P., ed. *Collagen: Structure and Mechanics*. New York: Springer.

Buehler, M. J. and Wong, S. Y. (2007) Entropic elasticity controls nanomechanics of single tropocollagen molecules. *Biophys. J.* **93**: 37–43.

Buhler, P. (1972) Sandwich structures in the skull capsules of various birds: the principles of light-weight structures in organisms. *Inf. Inst. Lightweight Struct. (Stuttgart)* **4**: 39–50.

Bulter, D. L., Grood, E. S., Noyes, F. R., Zernicke, R. F., and Brackett, K. (1984) Effects of structure and strain measurement technique on the material properties of young human tendons and fascia. *J. Biomach.* **17**: 579–596.

Burgess, S. C., King, A., and Hyde, R. (2006) An analysis of optimal structural feathers in the peacock tail feather. *Opt. Laser Technol.* **38**: 329–334.

Burke, J. F., Yannas, I. V., Quinby, W. C., Bondoc, C. C., and Jung, W. K. (1981) Successful use of a physiologically acceptable artificial skin in the treatment of extensive burn injury. *Ann. Surg.* **194**: 413–428.

Burr, D. B., Schaffler, M. B., and Frederickson, R. G. (1988) Composition of the cement line and its possible mechanical role as a local interface in human compact bone. *J. Biomech.* **21**: 939–945.

Butler, M. and Johnson, A. S. (2004) Are melanized feather barbs stronger? *J. Exp. Biol.* **207**: 285–293.

Byrom, D. (1991) *Biomaterials: Novel Materials from Biological Sources.* New York: Macmillan.

Calvert, P. (1994) Strategies for biomimetic mineralization. *Scripta Met.* **31**: 977–982.

Cameron, G. J., Wess, T. J., and Bonser, R. H. C. (2003) Young's modulus varies with differential orientation of keratin in feathers. *J. Struct. Biol.* **143**: 118–123.

Cao, J. (2002) Is the α-β transition of keratin a transition of α-helices to β-pleated sheets? II Synchroton investigation for stretched single specimens. *J. Molec. Struct.* **607**: 69–75.

Carlton, F. C. (1903) The color changes in the skin of the so-called Florida Chameleon. Anolis carolinensis Cuv. *Proc. Am. Acad. Arts Sci.* **39**: 259–276.

Carroll, M. and Holt, A. C. (1972) Suggested modification of the P-α model for porous materials. *J. Appl. Phys.* **43**: 759–761.

Cartwright, J. H. E. and Checa, A. G. (2007) The dynamics of nacre self-assembly. *J. R. Soc. Interface* **4**: 491–504.

Cha, J. N., Shimizu, K., Zhou, Y. *et al.* (1999) Silicatein filaments and subunits from a marine sponge direct the polymerization of silica and silicones in vitro. *Proc. Natl. Acad. Sci. USA* **96**: 361–365.

Cha, J. N., Stucky, G. D., Morse, D. E., and Deming, T. J. (2000) Biomimetic synthesis of ordered silica structures mediated by block copolypeptides. *Nature* **403**: 289–292.

Chazal, J., Tangguy, A., Bourges, M. *et al.* (1985) Biomechanical properties of spinal ligaments and a histological study of the supraspinal ligament in traction. *J. Biomech.* **18**: 167–176.

Checa, A. G., Cartwright, J. H., and Willinger, M. G. (2009) The key role of the surface membrane in why gastropod nacre grows in towers. *Proc. Natl. Acad. Sci USA* **106**: 38–43.

Chen, B., Peng, X., Wang, J. G., and Wu, X. (2004) Laminated microstructure of Bivalva shell and research of biomimetic ceramic/polymer composite. *Ceram. Intl.* **30**: 2011–2014.

Chen, I., Chen, P. -Y., Meyers, M. A., and McKittrick, J. (2011) Armadillo armor: mechanical testing and microstructural evaluation. *J. Mech. Behav. Biomed. Mater.* **4**: 713–722.

Chen, I. H., Yang, W., and Meyers, M. A. (2014) Alligator osteoderms: mechanical behavior and hierarchical structure, *MSEC* **35**: 441–448.

Chen, P. -Y., Lin, A. Y. M., McKittrick, J., and Meyers, M. A. (2008a) Structure and mechanical properties of crab exoskeleton. *Acta Biomater.* **4**: 587–596.

Chen, P. -Y., Lin, A. Y. M., Lin, Y. S. *et al.* (2008b) Structure and mechanical properties of selected biological materials. *J. Mech. Behav. Biomed. Mater.* **1**: 208–226.

Chen, P. -Y., Lin, A. Y. M., Lin, Y. S. *et al.* (2008c) Structural biological materials: overview of current research. *JOM* **60**: 23–32.

Chen, P. -Y., Stokes, A. G., and McKittrick, J. (2009) Comparison of the structure and mechanical properties of bovine femur bone and antler of the North American elk (*Cervus elaphus canadensis*). *Acta Biomater.* **5**: 693–706.

Chen, P. -Y., McKittrick, J., and Meyers, M. A. (2012) Biological materials: functional adaptations and bioinspired designs. *Prog. Mater. Sci.* **57**: 1492–1704.

Chernova, O. F. (2005) Polymorphism of the architectonics of definitive contour feathers. *Doklady Akademii Nauk* **404**: 280–285.

Clark, C. and Petrie, L. (2007) Fracture toughness of bovine claw horn cattle with and without vertical fissures. *Veterinary J.* **173**: 541–547.

Clutton-Brock, T. H. (1982) The function of antlers. *Behavior* **79**: 108–124.

Cohen, M., Kear, B. H., and Mehrabian, R., eds. (1980) *Rapid Solidification Processing: Principles and Technologies*. Baton Rouge, LA: Claitor's Publishing Division, p. 1.

Corning, W. R. and Biewener, A. A. (1998) In vivo strains in pigeon flight feather shafts: implications for structural design. *J. Exp. Biol.* **201**: 3057–3065.

Coulombe, P. A. and Omary, M. B. (2002) 'Hard' and 'soft' principles defining the structure, function and regulation of keratin intermediate filaments. *Curr. Opin. Cell. Biol.* **14**: 110–122.

Coulombe, P. A., Bousquet, O., Ma, L., Yamada, S., and Wirtz, D. (2000) The 'ins' and 'outs' of intermediate filament organization. *Trends Cell. Biol.* **10**: 420–428.

Crenshaw, D. G. (1980) Design and materials of feather shafts: very light, rigid structures. *Symp. Soc. Exp. Biol.* **34**: 485–486.

Cribb, B. W., Stewart, A., Huang, H. *et al.* (2008) Insect mandibles – comparative mechanical properties and links with metal incorporation. *Naturwissenschaft.* **95**: 17–23.

Cribb, B. W., Lin, C. -L., Rintoul, L., Rasch, R., Hasenpusch, J., and Huang, H. (2010) Hardness in arthropod exoskeletons in the absence of transition metals. *Acta Biomater.* **6**: 3152–3156.

Crne, M., Sharma, V., Blair, J., Park, J. O., Summers, C. J., and Srinivasarao, M. (2011) Biomimicry of optical microstructures of *Papilo palinurus*. *EPL* **93**: 14001(1–4).

Cubo, J. and Casinos, A. (2000) Incidence and mechanical significance of pneumatization in the long bones of birds. *Zool. J. Linnean Soc.* **130**: 499–510.

Cunniff, P. M., Fossey, S. A., Auerbach, M. A. *et al.* (1944) Mechanical and thermal properties of dragline silk from the spider *Nephila clavipes*. *Polymer Adv. Technol.* **5**: 401–410.

Currey, J. D. (1976) Further studies on mechanical properties of mollusk shell material. *J. Zool.* **180**: 445–453.

Currey, J. D. (1977) Mechanical properties of mother-of-pearl in tension. *Proc. R. Soc. Lond. B* **196**: 443–463.

Currey, J. D. (1979) Mechanical properties of bone tissues with greatly differing functions. *J. Biomech.* **12**: 313–319.

Currey, J. D. (1980) Mechanical properties of mollusc shell. In Vincent, J. F. V. and Currey, J. D., eds. *The Mechanical Properties of Biological Materials*, Symp. Soc. Exp. Biol. 34. Cambridge: Cambridge University Press, pp. 73–87.

Currey, J. D. (1984a) Effects of differences in mineralization on the mechanical properties of bone. *Phil. Trans. R. Soc. Lond. B* **304**: 509–518.

Currey, J. D. (1984b) *The Mechanical Adaptations of Bones*. Princeton, NJ: Princeton University Press.

Currey, J. D. (1989) Strain rate dependence of the mechanical properties of reindeer antler and the cumulative damage model of bone fracture. *J. Biomech.* **22**: 469–475.

Currey, J. D. (1999) The design of mineralized hard tissues for their mechanical functions. *J. Exp. Biol.* **202**: 3285–3294.

Currey, J. D. (2002) *Bones: Structure and Mechanics*. Princeton, NJ: Princeton University Press.

Currey, J. D. (2010) Mechanical properties and adaptations of some less familiar bony tissues. *J. Mech. Behav. Biomed. Mater.* **3**: 357–372.

Currey, J. D. and Alexander, R. M. (1985) The thickness of the walls of tubular bones. *J. Zool. A* **206**: 453–468.

Currey, J. D. and Brear, K. (1992) Fractal analysis of compact bone and antler fracture surfaces. *Biomimetics* **1**: 103–118.

Currey, J. D. and Kohn, A. J. (1976) Fracture in crossed-lamellar structure of conus shells. *J. Mater. Sci.* **11**: 1615–1623.

Currey, J. D. and Taylor, J. D. (1974) The mechanical behaviour of some molluscan hard tissues. *J. Zool. Lond.* **173**: 395–406.

Currey, J. D., Nash, A., and Bonfield, W. (1982) Calcified cuticle in the stomatopod smashing limb. *J. Mater. Sci.* **17**: 1939–1944.

Currey, J. D., Brear, K., and Zioupos, P. (1996) The effects of aging and changes in mineral content in degrading the toughness of human femora. *J. Biomech.* **29**: 257–260.

Currey, J. D., Zioupos, P., Davis, A., and Casinos, A. (2001) Mechanical properties of nacre and highly mineralized bone. *Proc. R. Soc. Lond. B* **268**: 107–111.

Dai, Z. and Yang, Z. (2010) Macro-/micro-structures of elytra, mechanical properties of the biomaterial and the coupling strength between elytra in beetles. *J. Bionic Eng.* **7**: 6–12.

Dao, M., Lim, C. T., and Suresh, S. (2003) Mechanics of the human red blood cell deformed by optical tweezers. *J. Mech. Phys. Solids* **51**: 2259–2280.

Dao, M., Li, J., and Suresh, S. (2006) Molecularly based analysis of deformation of spectrin network and human erythrocyte. *Mater. Sci. Eng. C* **26**: 1232–1244.

Davis, P. G. (1998) The bioerosion of bird bones. *Intl. J. Osteoarch.* **7**: 388–401.

Dawson, M. A. and Gibson, L. J. (2007) Optimization of cylindrical shells with compliant cores. *Intl. J. Solids Struct.* **44**: 1145–1160.

de Leeuw, N. H. and Parker, S. C. (1998) Surface structure and morphology of calcium carbonate polymorphs calcite, aragonite, and vaterite: an atomistic approach. *J. Phys. Chem. B* **102**: 2914–2922.

De Villiers, J. P. R. (1971) Crystal structures of aragonite, strontianite, and witherite. *Am. Mineral.* **56**: 758–767.

Del Campo, A., Greiner, C., Alvarez, I., and Arzt, E. (2007) Patterned surfaces with pillars with 3D tip geometry mimicking bioattachment devices. *Adv. Mater.* **19**: 1973–1977.

Denny, M. (1976) The physical properties of spider's silk and their role in the design of orb-webs. *J. Exp. Biol.* **65**: 483–506.

Deville, S., Saiz, E., Natta, R. K., and Tomsia, A. P. (2006a) Freezing as a path to build complex composites. *Science* **311**: 515–518.

Deville, S., Saiz, E., and Tomsia, A. P. (2006b) Freeze casting of hydroxyapatite scaffolds for bone tissue engineering. *Biomater.* **27**: 5480–5489.

Diamond, J. M. (1986) How great white sharks, saber-toothed cats and soldiers kill. *Nature* **322**: 773–774.

Dickinson, M. (2008) Mechanical properties of an arthropod exoskeleton – nanoindentation of the beetle *Scarites subterraneus*. Hysitron™ Nanoindetation Application Note.

Donovan, D. A. and Carefoot, T. H. (1997) Locomotion in the abalone *Haliotis kamtschatkana*: pedal morphology and cost of transport. *J. Exp. Biol.* **200**: 1145–1153.

Downing, S. W., Spitzer, R. H., Salo, W. L., Downing, J. S., Saidel, L. J., and Koch, E. A. (1981) Threads in the hagfish slime gland thread cells: organization, biochemical features, and length. *Science* **212**: 326–328.

Downing, S. W., Spitzer, R. H., Koch, E. A., and Salo, W. L. (1984) The hagfish slime gland thread cell. I. A unique cellular system for the study of intermediate filaments and intermediate filament-microtubule interactions. *J. Cell. Biol.* **98**: 653–669.

Druhala, M. and Feughelman, M. (1974) Dynamic mechanical loss in keratin at low temperatures. *Colloid Polymer Sci.* **252**: 381–391.

Dumont, E. R. (2010) Bone density and the lightweight skeletons of birds. *Proc. Roy. Soc. B* **277**: 2193–2198.

Easterling, K. E., Harrysson, R., Gibson, L. J., and Ashby, M. F. (1982) On the mechanics of balsa and other woods. *Proc. Roy. Soc. A* **383**: 31–41.

Ehrlich, H. and Worch, H. (2007) Collagen, a huge matrix in glass sponge flexible spicules of the meter-long *Hyalonema sieboldi*. In Baüerlein, E., Behrens, P., and Epple, M., eds. *Handbook of Biomineralization*. Weinheim: Wiley–VCH, chap. 3.

Ehrlich, H., Maldonado, M., Spindler, K. D. *et al.* (2007a) First evidence of chitin as a component of the skeletal fibers of marine sponges. Part I. Verongidae (demospongia: Porifera). *J. Exp. Zool. B Mol. Dev. Evol.* **308**: 347–356.

Ehrlich, H., Krautter, M., Hanke, T. *et al.* (2007b) First evidence of the presence of chitin in skeletons of marine sponges. Part II. Glass sponges (Hexactinellida: Porifera). *J. Exp. Zool. B Mol. Dev. Evol.* **308**: 473–483.

Elias, C. N. (2011) Factors affecting the success of dental implants. In Turkyilmaz, I., ed. *Implant Dentistry – A Rapidly Evolving Practice*. Rijeka: InTech.

Elias, C. N., Lima, J. H. C., Valiev, R., and Meyers, M. A. (2008) Biomedical applications of titanium and its alloys. *JOM* **60**: 46–49.

Elias, C. N., Meyers, M. A., Valiev, R. Z., and Monteiro, S. N. (2013) Ultrafine grained titanium for biomedical applications: an overview of performance. *J. Mater. Res. Technol.* **2**: 340–350.

Elices, M. (2000) *Structural Biological Materials: Design and Structure-Property Relationships*. Oxford: Pergamon.

Elices, M., Perez-Rigueiro, J., Plaza, G. R., and Guinea, G. V. (2005) *J. Metals* **57**: 60–66.

Elvin, C. M., Carr, A. G., Huson, M. G. *et al.* (2005) Synthesis and properties of crosslinked recombinant pro-resilin. *Nature* **437**: 999–1002.

Ennos, R. (2012) *Solid Biomechanics*. Princeton, NJ: Princeton University Press.

Erben, H. K. (1972) On the structure and growth of the nacreous tablets in gastropods. *Biomineral.* **7**: 14–27.

Ernst, V. V. (1973a) The digital pads of the tree frog Hyla cinerea. I. The epidermis. *Tissue Cell* **5**: 83–96.

Ernst, V. V. (1973b) The digital pads of the tree frog *Hyla cinerea*. II. The mucous glands. *Tissue Cell* **5**: 97–104.

Escoffier, C., de Rigal, J., Rochefort, A., Vasselet, R., Lévêquel, J. -L., and Agache, P. G. (1989) Age-related mechanical properties of human skin: an in vivo study. *J. Invest. Dermatol.* **93**: 353–357.

Evans, A. G. and Charles, E. A. (1976) Fracture toughness determination by indentation. *J. Am. Ceramic Soc.* **59**: 371–372.

Evans, A. G., Suo, Z., Wang, R. Z., Aksay, I. A., He, M. Y., and Hutchinson, J. W. (2001a) Model for the robust mechanical behavior of nacre. *J. Mater. Res.* **16**: 2475–2493.

Evans, A. G., Hutchinson, J. W., Fleck, N. A., Ashby, M. F., and Wadley, H. N. G. (2001b) The topological design of multifunctional cellular metals. *Prog. Mater. Sci.* **46**: 309–327.

Falini, G., Albeck, S., Weiner, S., and Addadi, L. (1996) Control of aragonite or calcite polymorphism by mollusk shell macromolecules. *Science* **271**: 67–69.

Fantner, G. E., Hassenkam, T., Kindt, J. H. *et al.* (2005) Sacrificial bonds and hidden length dissipate energy as mineralized fibrils separate during bone fracture. *Nature Mater.* **4**: 612–616.

Fecchio, R. S., Seki, Y., Bodde, S. G. *et al.* Mechanical behavior of prosthesis in toucan beak (Ramphastos toco). *Mater. Sci. Eng. C* **30**: 460–464.

Fernholm, B. (1981) Thread cells from the slime glands of hagfish (Myxinidae). *Acta Zool.* **62**: 137–145.

Feughelman, M. (1997) *Mechanical Properties and Structure of α-Keratin Fibres: Wool, Human Hair and Related Fibres*. Sydney: University of New South Wales Press.

Filshie, B. K. and Rogers, G. E. (1962) An electron microscope study of the fine structure of feather keratin. *J. Cell. Biol.* **13**: 1–12.

Fine, M. E. and Marcus, H. L. (1994) Materials science and engineering, an educational discipline. *Annu. Rev. Mater. Sci.* **24**: 1–17.

Fischmeister, H. and Arzt, E. (1982) Densification of powders by particle deformation. *Powder Metall.* **26**: 82–88.

Fisher, T. E., Oberhauser, A. F., Carrion-Vazquez, M., Marszalek, P. E., and Fernandez, J. M. (1999) The study of protein mechanics with the atomic force microscope. *Trends Biochem. Sci.* **24**: 379–384.

Fleck, N. A., Deng, L., and Budiansky, B. (1995) Prediction of kink width in compressed fiber composites. *J. Appl. Mech.* **62**: 329–337.

Flory, P. J. (1956) Theory of elastic mechanisms in fibrous proteins. *J. Am. Chem. Soc.* **78**: 5222–5235.

Flory, P. J. (1964) *Principles of Polymer Chemistry.* New York: Cornell University Press.

Flynn, C. E., Lee, S. -W., Peelle, B. R., and Belcher, A. M. (2003) Viruses as vehicles for growth, organization, and assembly of materials. *Acta Mater.* **51**: 5867–5880.

Forbes, P. (2007) *The Gecko's Foot.* London: Fourth Estate.

Franck, A., Cocquyt, G., Simoens, P., and De Belie, N. (2006) Biomechanical properties of bovine claw horn. *Biosys. Eng.* **93**: 459–467.

Franke, O., Göken, M., Meyers, M. A., Durst, K., and Hodge, A. M. (2011) Dynamic nano-indentation of articular porcine cartilage. *Mater. Sci. Eng. C* **31**: 789–795.

Franzblau, C. (1971) Elastin. In Florkin, M. and Stotz, E. H., eds. *Comparative Biochemistry.* Amsterdam: Elsevier, pp. 659–712.

Fraser, R. D. and MacRae, T. P. (1980) Molecular structure and mechanical properties of keratin. In Vincent, J. F. V. and Currey, J. D., eds. *The Mechanical Properties of Biological Materials*, Symposium of the Society of Experimental Biology. Cambridge: Cambridge University Press, pp. 211–246.

Fraser, R. D. B. and Parry, D. A. D. (1996) The molecular structure of reptilian keratin. *Int. J. Biol. Macro.* **19**: 207–211.

Fraser, R. D. B., MacRae, T. P., and Rogers, G. E. (1972) *Keratins: Their Composition, Structure, and Biosynthesis*. Springfield: Thomas.

Fraser, R. D., MacRae, T. P., Parry, D. A., and Suzuki, E. (1986) Intermediate filaments in alpha keratins. *Proc. Natl. Acad. Sci. USA* **83**: 1179–1183.

Fratzl, P., ed. (2008) *Collagen: Structure and Mechanics*. New York: Springer.

Fratzl, P. and Weinkamer, R. (2007) Nature's hierarchical materials. *Prog. Mater. Sci.* **52**: 1263–1334.

Fratzl, P., Groschner, M., Vogl, G., Plenk, H., Eschberger, J., and Fratzl-Zelman, N. (1992) Mineral crystals in calcified tissues: a comparative study by SAXS. *J. Bone Miner. Res.* **7**: 329–334.

Fratzl, P., Misof, K., Zizak, I., Rapp, G., Amenitsch, H., and Bernstorff, S. (1998) Fibrillar structure and mechanical properties of collagen. *J. Struct. Biol.* **122**: 119–122.

Fratzl, P., Gupta, H. S., Paschalis, E. P., and Roschger, P. (2004) Structure and mechanical quality of the collagen-mineral nano-composite in bone. *Mater. Chem.* **14**: 2115–2123.

Frazzetta, T. H. (1988) The mechanics of cutting and the form of shark teeth (Chondrichthyes, Elasmobranchii). *Zoomorphol.* **108**: 93–107.

Frenkel, M. J. and Gillespie, J. M. (1976) The proteins of the keratin component of bird's beaks. *Austral. J. Biol. Sci.* **29**: 467–479.

Fritz, M. and Morse, D. E. (1998) The formation of highly organized biogenic polymer/ceramic composite materials: the high-performance microaluminate of molluscan nacre. *Curr. Opin. Colloid Int. Sci.* **3**: 55–62.

Fritz, M., Belcher, A. M., Radmacher, M. *et al.* (1994) Flat pearls from biofabrication of organized composites on inorganic substrates. *Nature* **371**: 49–51.

Fu, G., Valiyaveettil, S., Wopenka, B., and Morse, D. E. (2005) $CaCO_3$ biomineralization: acidic 8-kDa proteins isolated from aragonitic abalone shell nacre can specifically modify calcite crystal morphology. *Biomacromol.* **6**: 1289–1298.

Fudge, D. S. and Gosline, J. M. (2004) Molecular design of the alpha-keratin composite: insights from a matrix-free model, hagfish slime threads. *Proc. Roy. Soc. Lond. B* **271**: 291–299.

Fudge, D. S., Gardner, K. H., Forsyth, V. T., Riekel, C., and Gosline, J. M. (2003) The mechanical properties of hydrated intermediate filaments: insights from hagfish slime threads. *Biophys. J.* **85**: 2015–2027.

Fudge, D. S., Levy, N., Chiu, S., and Gosline, J. M. (2005) Composition, morphology and mechanics of hagfish slime. *J. Exp. Biol.* **208**: 4613–4625.

Fung, Y. C. (1967) Elasticity of soft tissues in simple elongation. *Am. J. Physiol.* **213**: 1532–1544.

Fung, Y. C. (1990) *Biomechanics: Motion, Flow, Stress, and Growth*. New York: Springer-Verlag.

Fung, Y. C. (1993) *Biomechanics: Mechanical Properties of Living Tissues*, 2nd edn. New York: Springer.

Fung, Y. C. (1997) *Biomechanics: Circulation*, 2nd edn. New York: Springer-Verlag.

Gao, H. J. (2006) Application of fracture mechanics concepts to hierarchical biomechanics of bone and bone-like materials. *Int. J. Fracture* **138**: 101–137.

Gao, H. J. and Klein, P. A. (1998) Numerical simulation of crack growth in an isotropic solid with randomized internal cohesive bonds. *J. Mech. Phys. Solids* **46**: 187–218.

Gao, H. J., Ji, B. H., Jäger, I. L., Arzt, E., and Fratzl, P. (2003) Materials become insensitive to flaws at nanoscale: lessons from nature. *Proc. Natl. Acad. Sci. USA* **100**: 5597–5600.

Gao, Y., Ellery, A., Jaddou, M., and Vincent, J. (2006) Deployable wood wasp drill for planetary subsurface sampling. IEEEAC Paper #1591, Version 1, IEEE, pp. 1–8.

Garcia, A. (2005) Get a grip: integrins in cell-biomaterial interactions. *Biomater.* **26**: 7525–7529.

Garrido, M. A., Elices, M., Viney, C., and Perez-Riguerio, J. (2002) The variability and interdependence of spider drag line tensile properties. *Polymer* **43**: 4495–4502.

Gathercole, L. J. and Keller, A. (1975) Light microscopic waveforms in collagenous tissues and their structural implications. In Atkins, E. D. T., ed. *Structure of Fibrous Biopolymers*. London: Butterworth.

Gautieri, A., Ionita, M., Silvestri, D. *et al.* (2010) Computer-aided molecular modeling and experimental validation of water permeability properties in biosynthetic materials. *J. Comput. Theor. Nanos*. **7**: 1287–1293.

Gautieri, A., Vesentini, S., Redaelli, A., and Buehler, M. J. (2011) Hierarchical structure and nanomechanics of collagen microfibrils from the atomistic scale up. *Nano. Lett.* **11**: 757–766.

Geim, A. K., Dubonos, S. V., Grigorieva, I. V., Nvoselov, K. S., Zhukov, A. A., and Shapoval, S. Y. (2003) Microfabricated adhesive mimicking gecko foot-hair. *Nat. Mater.* **2**: 461–463.

Ghiradella, H. (1991) Light and colour on the wing: structural colours in butterflies and moths. *Appl. Opt*. **30**: 3492–3500.

Gibson, L. J. and Ashby, M. F. (1988) *Cellular Solids*, 1st edn. Oxford: Pergamon Press Ltd.

Gibson, L. J. and Ashby, M. F. (1997) *Cellular Solids: Structure and Properties*, 2nd edn. Cambridge: Cambridge University Press.

Gibson, L. J., Ashby, M. F., and Harley, B. A. (2010) *Cellular Materials in Nature and Medicine*. Cambridge: Cambridge University Press.

Giraud-Guille, M. M. (1984) Fine structure of the chitin-protein system in the crab cuticle. *Tissue Cell* **16**: 75–92.

Giraud-Guille, M. M. (1990) Chitin crystals in arthropod cuticles revealed by diffraction contrast transmission electron microscopy. *J. Structur. Biol*. **103**: 232–240.

Giraud-Guille, M. M. (1998) Plywood structures in nature. *Curr. Opin. Solid State Mater. Sci*. **3**: 221–228.

Giraud-Guille, M. M. and Bouligand, Y. (1995) Crystal growth in a chitin matrix: the study of calcite development in the crab cuticle. In Karnicki, Z. S., ed. *Chitin World*. Bremerhaven: Wirtschaftsverlag NW, pp. 136–144.

Goffredi, S. K., Warén, A., Orphan, V. J., Van Dover, C. L., and Vrijenhoek, R. C. (2004) Novel forms of structural integration between microbes and a hydrothermal vent gastropod from the Indian Ocean. *Appl. Environ. Microbiol.* **70**: 3082–3090.

Gordon, J. E. and Jeronimidis, G. (1980) Composites with high work of fracture. *Phil. Trans. Roy. Soc. Lond. A* **297**: 545–550.

Gosline, J. M., Denny, M. W., and DeMont, M. E. (1984) Spider silk as rubber. *Nature* **309**: 551–552.

Gosline, J. M., DeMont, M. E., and Denny, M. W. (1986) The structure and properties of spider silk. *Endeavour* **10**: 37–43.

Gosline, J. M., Guerette, P. A., Ortlepp, C. S., and Savage, K. N. (1999) The mechanical design of spider silks. *J. Exp. Biol.* **202**: 3295–3303.

Gray, W. R., Sandberg, L. B., and Forster, J. A. (1973) Molecular model for elastin structure and function. *Nature* **246**: 461–466.

Green, D. M. (1979) Tree frog toe pads: comparative surface morphology using scanning electron microscopy. *Can. J. Zool.* **57**: 2033–2046.

Green, D. M. and Simon, P. (1986) Digital microstructure in ecologically diverse sympatric microhylid frogs, genera Cophixalus and Sphenophryne (Amphibia: Anura) Papua New Guinea. *Austral. J. Zool.* **34**: 135–145.

Grègoire, C. (1957) Topography of the organic components in mother-of pearl. *J. Biophys. Biochem. Cytol.* **3**: 797–808.

Grègoire, C. (1961) Structure of the conchiolin cases of the prisms in *Mytilus edulis linne*. *J. Biophys. Biochem. Cytol.* **9**: 395–400.

Grègoire, C., Duchateau, G., and Florkin, M. (1954) La trame protidique des nacres. *Experientia* **10**: 37–40.

Greiner, C., del Campo, A., and Artz, E. (2007) Adhesion of bioinspired micropatterned surfaces: effects of pillar radius, aspect ratio, and preload. *Langmuir* **23**: 3495–3502.

Greiner, C., Arzt, E., and del Campo, A. (2009) Hierarchical gecko-like adhesives. *Adv. Mater.* **21**: 479–482.

Griel, P., Lifka, T., and Kaindl, A. (1998) Biomorphic cellular silicon carbide ceramics from wood: I. *J. Eur. Ceram. Soc.* **18**: 1961–1973.

Gronau, G., Krishnaji, S. T., Kinahan, M. E. *et al.* (2012) A review of combined experimental and computational procedures for assessing biopolymer structure-process-property relationships. *Biomater.* **33**: 8240–8255.

Grunkemeier, G., Rahimtoola, S., Starr, A., and Braunwald, E. (2002) *Atlas of Heart Diseases*, Vol. 11. New York: Springer, chap. 13.

Guvendiren, M., Brassa, D. A., Messersmith, P. B., and Shull, R. (2009) Adhesion of DOPA-functionalized model membranes to hard and soft surfaces. *J. Adhesion* **9**: 631–645.

Hall, S. J. (2003) *Basic Biomechanics*, 4th edn. Boston, MA: McGraw-Hill.

Hamilton, W. J. and Selly, M. K. (1976) Fog basking by the Namib Desert beetle, *Onymacris unguicularis*. *Nature* **262**: 284–285.

Hamm, C. E., Merkel, R., Springer, O. *et al.* (2003) Architecture and material properties of diatom shells provide effective mechanical protection. *Nature* **421**: 841–843.

Hanna, G. and Barnes, W. J. P. (1991) Adhesion and detachment of the toe pads of treefrogs. *J. Exp. Biol.* **155**: 103–125.

Hansma, H. G., Pietrasanta, L. I., Auerbach, I. D., Sorenson, C., Golan, R., and Holden, P. A. (2000) Probing biopolymers with the atomic force microscope: a review. *J. Biomater. Sci. Polymer Ed.* **11**: 675–683.

Hansma, P. K., Fantner, G. E., Kindt, J. H., *et al.* (2005) Sacrificial bonds in the interfibrillar matrix of bone. *J. Musculoskel. Neuron. Inter.* **5**: 313–315.

Harley, B. A., Lynn, A. K., Wissner-Gross, Z., Bonfield, W., Yannas, I. V., and Gibson, L. J. (2010) Design of a multiphase osteochondral scaffold II: fabrication of a mineralized collagen-GAG scaffold. *J. Biomed. Mater. Res.* **92**: 1066–1077.

Hassenkarm, T., Fantner, G. E., Cutroni, J. A., Weaver, J. C., Morse, D. E., and Hansma, P. K. (2004) High-resolution AFM imaging of intact and fracture trabecular bone. *Bone* **35**: 4–10.

Hayes, W. C. and Carter, D. R. (1976) Postyield behavior of subchondral trabecular bone. *J. Biomed. Mater. Res. Symp.* **7**: 537–544.

Hearle, J. W. S. (2000) A critical review of the structural mechanics of wool and hair fibres. *Int. J. Biol. Macromol.* **27**: 123–138.

Helle, A. S., Easterling, K. E., and Ashby, M. F. (1985) Hot isostatic pressing diagrams: new developments. *Acta Metall.* **33**: 2163–2174.

Hench, L. L. (1991) Bioceramics – from concept to clinic. *J. Am. Ceram. Soc.* **74**: 1487–1510.

Hench, L. L. (1999) Bioactive glasses and glass-ceramics. *Bioceram.* **293**: 37–63.

Hench, L. L. (2006) The story of Bioglass®. *J. Mater. Sci. Mater. Med.* **17**: 967–978.

Henshaw, J. (1971) Antlers – the unbrittle bones of contention. *Nature* **231**: 469.

Hepburn, H. R., Joffe, I., Green, N., and Nelson, K. J. (1975) Mechanical properties of a crab shell. *Comp. Biochem. Physiol.* **50**: 551–554.

Hernandez, C. J., Tang, S. Y., Baumbach, B. M. *et al.* (2005) Trabecular microfracture and the influence of pyridinium and non-enzymatic glycation-mediated collagen cross-links. *Bone* **37**: 825–832.

Herrick, W. C., Kingsbury, H. B., and Lou, D. Y. S. (1978) A study of the normal range of strain, strain rate and stiffness of tendon. *J. Biomed. Mater. Res.* **12**: 877–894.

Hershey, A. D. and Chase, M. (1952) Independent functions of viral protein and nucleic acid in growth of bacteriophage. *J. Gen. Physiol.* **36**: 39–56.

Heuer, A. H., Kink, D. J., Laraia, V. J. *et al.* (1992) Innovative materials processing strategies: a biomimetic approach. *Science* **255**: 1098–1105.

Heywood, B. and Mann, S. (1994) Template directed nucleation and growth of inorganic materials. *Adv. Mater.* **6**: 9–19.

Hieronymus, T. L., Witmer, L. M., and Ridgely, R. C. (2006) Structure of white rhinoceros (*Ceratotherium simum*) horn investigated by x-ray computed tomography and histology with implications for growth and external form. *J. Morphol.* **267**: 1172–1176.

Hight, T. K. and Brandeau, J. F. (1983) Mathematical modeling of the stress-strain-strain rate behavior of bone using the Ramberg-Osgood equation. *J. Biomech.* **16**: 445–450.

Hildebrand, M. (2003) Biological processing of nanostructured silica in diatoms. *Pro. Org. Coat.* **47**: 256–266.

Hildebrand, M. (2005) Prospects of manipulating diatom silica nanostructure. *J. Nanosci. Nanotech*. **5**: 146–157.

Hildebrand, M. (2008) Diatoms, biomineralization processes, and genomics. *Chem. Rev.* **108**: 4855–4874.

Hildebrand, M., York, E., Kelz, J. I. *et al*. (2006) Nanoscale control of silica morphology and three-dimensional structure during diatom cell wall formation. *J. Mater. Res*. **21**: 2689–2698.

Hill, A. V. (1938) The heat of shortening and the dynamic constants of muscle. *Proc. Roy. Soc. Lond. B* **126**: 136–195.

Hillerton, J. E. and Vincent, J. F. V. (1982) The specific location of zinc in insect mandibles. *J. Exp. Biol.* **101**: 333–366.

Hillerton, J. E., Reynolds, S. E., and Vincent, J. F. V. (1982) On the indentation hardness of insect cuticle. *J. Exp. Biol.* **96**: 45–52.

Holl, S. M., Hansen, D., Waite, J. H., and Shaefer, J. (1993) Solid-state NMR analysis of crosslinking in mussel protein glue. *Arch. Biochem. Biophys*. **302**: 255–258.

Homsy, C. (1970) Biocompatibility in selection of materials for implantation. I. *Biomed. Mater. Res*. **4**: 341–356.

Hörnschemeyer, T., Beutel, R. G., and Pasop, F. (2002) Head structures of *Priacma serrata* Leconte (Coleptera, Archostemata) inferred from x-ray tomography. *J. Morphol*. **252**: 298–314.

Hou, D. F., Zhou, G. S., and Zheng, M. (2004) Conch shell structure and its effect on mechanical behaviors. *Biomater.* **25**: 751–756.

Huang, J., Wang, X., and Wang, Z. L. (2006) Controlled replication of butterfly wings for achieving tunable photonic properties. *Nano. Lett.* **6**: 2325–2331.

Huang, T. J., Brough, B., Ho, C.-M. *et al*. (2004) A nanomechanical device based on linear molecular motors. *Appl. Phys. Lett.* **85**: 5391–5393.

Huber, G., Mantz, H., Spolenak, R. *et al*. (2005) *Proc. Natl. Acad. Sci. USA* **102**: 16293–16296.

Hughes, P. M. (1987) Insect cuticular growth layers seen under the scanning electron microscope: a new display method. *Tissue Cell* **19**: 705–712.

Ikeshoji, T. (1993) *The Interface Between Mosquitoes and Humans*. Tokyo: University of Tokyo Press (in Japanese).

Ikoma, T., Kobayashi, H., Tanaka, J., Wals, D., and Mann, S. (2003) Microstructure, mechanical, and biomimetic properties of fish scales from *Pagrus major*. *J. Structur. Biol.* **142**: 327–333.

Imbeni, V., Nalla, R. K., Bosi, C., Kinney, J. H., and Ritchie, R. O. (2003) In vitro fracture toughness of human dentin. *J. Biomed. Mater. Res. A* **66**: 1–9.

Imbeni, V., Kruzic, J. J., Marshall, G. W., Marshall, S. J., and Ritchie, R. O. (2005) The dentin–enamel junction and the fracture of human teeth. *Nature Mater.* **4**: 229–232.

Jackson, A. P., Vincent, J. F. V., and Turner, R. M. (1988) The mechanical design of nacre. *Proc. R. Soc. Lond. B* **234**: 415–440.

Jackson, A. P., Vincent, J. F. V., and Turner, R. M. (1989) A physical model of nacre. *Comp. Sci. Tech*. **36**: 225–266.

Jackson, D. A., Symons, R. H., and Berg, P. (1972) Biochemical method for inserting new genetic information into DNA of Simian Virus 40: circular SV40 DNA molecules containing lambda phage genes and the galactose operon of *Escherichia coli*. *Proc. Natl. Acad. Sci. USA* **69**: 2904–2909.

Jackson, S. A., Cartwright, A. G., and Lewis, D. (1978) The morphology of bone mineral crystals. *Calcif. Tissue Res.* **25**: 217–222.

Jaenisch, R. and Mintz, B. (1974) Simian virus 40 DNA sequences in DNA of healthy adult mice derived from preimplantation blastocysts injected with viral DNA. *Proc. Natl. Acad. Sci.* **71**: 1250–1254.

Jäger, I. and Fratzl, P. (2000) Mineralized collagen fibrils: a mechanical model with a staggered arrangement of mineral particles. *Biophys. J.* **79**: 1737–1746.

Jelf, P. M. and Fleck, N. A. (1992) Compression failure mechanisms in unidirectional composites. *J. Comp. Mater.* **26**: 2706–2726.

Jeronimidis, G. (1976) The work of fracture of wood in relation to its structure. In Baas, P., Bolton, A. J., and Catling, D. M., eds. *Wood Structure in Biological and Technological Research*. Leiden: The University Press, pp. 253–265.

Jeronimidis, G. (1980) Wood, one of nature's challenging composites. *Symp. Soc. Exp. Biol.* **34**: 169–182.

Ji, B. H. and Gao, H. J. (2004) Mechanical properties of nanostructure of biological materials. *J. Mech. Phys. Solid* **52**: 1963–1990.

Ji, B. H. and Gao, H. J. (2010) Mechanical properties of biological composites. *Ann. Rev. Mater. Res.* **40**: 77–100.

Ji, B. H., Gao, H. J., and Hsia, K. J. (2004) How do slender mineral crystals resist buckling in biological materials? *Phil. Mag. Lett.* **84**: 631–641.

Joffe, I., Hepburn, H. R., Nelson, K. J., and Green, N. (1975) Mechanical properties of a crustacean exoskeleton. *Comp. Biochem. Physiol. A* **50**: 545–549.

Johnson, K. L., Kendall, K., and Roberts, A. D. (1971) Surface energy and the contact of elastic solids. *Proc. Roy. Soc. Lond.* **324**: 301–313.

Johnson, W., Soden, P. D., and Trueman, E. R. (1972) A study in jet propulsion: an analysis of the motion of the squid, *Loligo vulgaris*. *Exp. Biol.* **56**: 155–165.

Kahler, G. A., Fisher, F. M., and Sass, R. L. (1976) The chemical composition and mechanical properties of the hinge ligament in bivalve mollusks. *Biol. Bull.* **151**: 161–181.

Kamat, S., Su, X., Ballarini, R., and Heuer, A. H. (2000) Structural basis for the fracture toughness of the shell of the conch Strombus gigas. *Nature* **405**: 1036–1040.

Kamat, S., Kessler, H., Ballarini, R., Nassirou, M., and Heuer, A. H. (2004) Fracture mechanisms of the *Strombus gigas* conch shell: II – micromechanics analyses of multiple cracking and large-scale crack bridging. *Acta Materialia* **52**: 2395–2406.

Kaplan, D. and McGrath, K. (1997) *Protein-Based Materials*. Boston, MA: Birkhäuser.

Kaplan, D. L., Lombardi, S. J., Muller, W. S., and Fossey, S. A. (1991) In Byrom, D., ed. *Biomaterials: Novel Materials from Biological Sources*. New York: Stockton Press.

Kaplan, D., Adams, W. W., Farmen, B., and Viney, C. (1994) Silk: biology, structure, properties, and genetics. *Am. Chem. Soc. Symp.* **544**: 2–16.

Karam, G. N. and Gibson, L. J. (1994) Biomimicking of animal quills and plant stems: natural cylindrical shells with foam cores. *Mater. Sci. Eng. C* **2**: 113–132.

Karam, G. N. and Gibson, L. J. (1995a) Elastic buckling of cylindrical shells with elastic cores I: Analysis. *Intl. J. Solids Struct.* **32**: 1259–1283.

Karam, G. N. and Gibson, L. J. (1995b) Elastic buckling of cylindrical shells with elastic cores II: Experiments. *Intl. J. Solids Struct.* **32**, 1285–1306.

Kasapi, M. A. and Gosline, J. M. (1996) Strain-rate-dependent mechanical properties of the equine hoof wall. *J. Exp. Biol.* **199**: 1133–1146.

Kasapi, M. A. and Gosline, J. M. (1997) Design complexity and fracture control in the equine hoof wall. *J. Exp. Biol.* **200**: 1639–1659.

Kasapi, M. A. and Gosline, J. M. (1998) Exploring the possible functions of equine hoof wall tubules. *Equine Vet. J.* **26**: 10–14.

Kasapi, M. A. and Gosline, J. M. (1999) Micromechanics of the equine hoof wall: optimizing crack control and material stiffness through modulation of the properties of keratin. *J. Exp. Biol.* **202**: 377–391.

Katz, J. L. (1971) Hard tissue as a composite material. 1. Bounds on elastic behavior. *J. Biomech.* **4**: 455–473.

Kelly, R. E. and Rice, R. V. (1967) Abductin: a rubber-like protein from the internal triangular hinge ligament of pectin. *Science* **155**: 208–210.

Keten, S., Xu, Z. P., Ihle, B., and Buehler, M. J. (2010) Nanoconfinement controls stiffness, strength and mechanical toughness of beta-sheet crystals in silk. *Nature Mater.* **9**: 359–367.

Khanuja, S. (1991) Processing of laminated B4C-polymer laminated composites. M.S. Thesis, University of Washington.

Kisailus, D., Truong, Q., Amemiya, Y., Weaver, J. C., and Morse, D. E. (2006) Self-assembled bifunctional surface mimics an enzymatic and templating protein for the synthesis of a metal oxide semiconductor. *Proc. Natl. Acad. Sci.* **103**: 5652–5657.

Kitchener, A. (1987a) Fracture toughness of horns and a reinterpretation of the horning behaviour of bovids. *J. Zool.* **213**: 621–639.

Kitchener, A. (1987b) Effect of water on the linear viscoelasticity of horn sheath keratin. *J. Mater. Sci. Lett.* **6**: 321–322.

Kitchener, A. (1988) An analysis of the forces of fighting of the blackbuck (Antilope cervicapra) and the bighorn sheep (*Ovis canadensis*) and the mechanical design of horns of bovids. *J. Zool.* **214**: 1–20.

Kitchener, A. C. (1991) The evolution and mechanical design of horns and antlers. In Rayner, J. M. V. and Wootton, R. J., eds. *Biomechanics and Evolution*. Cambridge: Cambridge University Press, pp. 229–253.

Kitchener, A. C. (2000) Fighting and the mechanical design of horns and antlers. In Domenici, P. and Blake, R. W., eds. *Biomechanics in Animal Behavior*. Oxford: BIOS Scientific Publishers, pp. 291–314.

Kitchener, A. and Vincent, J. F. V. (1987) Composite theory and the effect of water on the stiffness of horn keratin. *J. Mater. Sci.* **22**: 1385–1389.

Kobayashi, I. (1969) Internal microstructure of shell of bivalve mollusks. *Am. Zool.* **9**: 633–672.

Kobayashi, I. and Samata, T. (2006) Bivalve shell structure and organic matrix. *Mater. Sci. Eng. C* **26**: 692–698.

Koch, K., Bhushan, B., and Barthlott, W. (2009) Multifunctional surface structures of plants: an inspiration for biomimetics. *Prog. Mater. Sci.* **54**: 137–178.

Koester, K. J., Ager, J. W. III, and Ritchie, R. O. (2008) The true toughness of human cortical bone measured with realistic short cracks. *Nature Mater.* **7**: 672–676.

Kohr, E. (2001) *Chitin: Fulfilling a Biomaterials Promise*. Oxford: Elsevier Science.

Kokubo, T. (1991) Bioactive glass ceramics: properties and applications. *Biomater.* **12**: 155–163.

Kolle, M., Salgard-Cunha, P. M., Scherer, M. R. J. *et al.* (2010) Mimicking the colourful wing scale structure of the Papilio blumei butterfly. *Nature Nanotech.* **5**: 511–515.

Krajewska, B. (2004) Application of chitin- and chitosan-based materials for enzyme immobilizations: a review. *Enzyme Microbiol. Tech.* **35**: 126–139.

Krauss, S., Monsonego-Orman, E., Zelzer, E., Fratzl, P., and Shahar, R. (2009) Mechanical function of a complex three-dimensional suture joining the bony elements in the shell of the red-eared slider turtle. *Adv. Mater.* **21**: 407–412.

Krishnaji, S. T., Huang, W., Rabotyagova, O. *et al.* (2011) Thin film assembly of spider silk-like block copolymers. *Langmuir* **27**: 1000–1008.

Kruzic, J. J., Nalla, R. K., Kinney, J. H., and Ritchie, R. O. (2003) Crack blunting, crack bridging and resistance-curve fracture mechanics in dentin: effect of hydration. *Biomater.* **24**: 5209–5221.

Kuhn-Spearing, L. F., Kessler, H., Chateau, E., Ballarin, R., Heuer, A. H., and Spearing, S. M. (1996) Fracture mechanisms of the Strombus gigas conch shell: implications for the design of brittle laminates. *J. Mater. Sci.* **31**: 6583–6594.

Kulchin, Y. N., Bezverbny, A. V., Bukin, O. A. *et al.* (2009) Optical and nonlinear optical properties of sea glass sponge spicules. *Prog. Molec. Subcell. Biol.* **47**: 315–340.

Kulin, R. M., Chen, P.-Y., Jiang, F., McKittrick, J., and Vecchio, K. S. (2010) Dynamic fracture resilience of elk antler: biomimetic inspiration for improved crashworthiness. *JOM* **62**: 41–46.

Kulin, R. M., Chen, P.-Y., Jiang, F., and Vecchio, K. S. (2011) A study of the dynamic compressive behavior of elk antler. *Mater. Sci. Eng. C* **31**: 1030–1041.

Kustandi, T. S., Low, H. Y., Teng, J. H., Rodrizuez, I., and Yin, R. (2009) Mimicking domino-like photonic nanostructures on butterfly wings. *Small* **5**: 574–578.

Laraia, V. J. and Heuer, A. H. (1989) Novel composite microstructure and mechanical behavior of mollusk shell. *J. Am. Ceram. Soc.* **72**: 2177–2179.

Launey, M. E., Munch, E., Alsem, D. H. *et al.* (2009) Designing highly toughened hybrid composites through nature-inspired hierarchical complexity. *Acta Mater.* **57**: 2919–2932.

Launey, M. E., Buehler, M. J., and Ritchie, R. O. (2010a) On the mechanistic origins of toughness in bone. *Annu. Rev. Mater. Res.* **40**: 25–53.

Launey, M. E., Chen, P.-Y., McKittrick, J., and Ritchie, R. O. (2010b) Mechanistic aspects of the fracture toughness of elk antler bone. *Acta Biomater.* **6**: 1505–1514.

Lawrence, C., Vukusic, P., and Sambles, J. R. (2002) Grazing-incidence iridescence from a butterfly wing. *Appl. Opt.* **41**: 437–441.

Lee, G. Y. H. and Lim, C. T. (2007) Biomechanics approaches to studying human diseases. *Trends Biotechnol.* **25**: 111–118.

Lee, H. (2010) Biomaterials: intelligent glue. *Nature* **465**: 298–299.

Lee, H., Dellatore, S. M., Mille, W. M., and Messersmith, P. B. (2007a) Mussel-inspired surface chemistry for multifunctional coatings. *Science* **318**: 426–430.

Lee, H., Lee, B. P., and Messersmith, P. B. (2007b) A reversible wet/dry adhesive inspired by mussels and geckos. *Nature* **448**: 338–341.

Lee, S., Reyante, B., Tsukasa, T. *et al.* (2011) Impact testing of structural biological materials. *Mater. Sci. Eng. C* **31**: 730–739.

Lee, S. W., Mao, C. B., Flynn, C. E., and Belcher, A. M. (2002) Ordering of quantum dots using genetically engineered viruses. *Science* **296**: 892–895.

Lee, Y. J., Yi, H., Kang, K. *et al.* (2009) Fabricating genetically engineered high-power lithium-ion batteries using multiple virus genes. *Science* **324**: 1051–1055.

Levi, C., Barton, J. L., Guillemet, C., Le Bras, E., and Jehuede, P. J. (1989) A remarkably strong natural glassy rod: the anchoring spicule of the Monorhaphis sponge. *Mater. Sci. Lett.* **8**: 337–339.

Levi, K., Weber, R. J., Do, J. Q., and Dauskardt, R. H. (2009) Drying stresses and damage in human stratum corneum. *Inl. J. Cosmet. Sci.* **32**: 276–293.

Levi, K., Kwan, A., Rhines, A. S., Gorcea, M., Moore, D. J., and Dauskardt, R. H. (2011) Effect of glycerin on drying stresses in human stratum corneum. *J. Dermatol. Sci.* **61**: 129–131.

Levi-Kalisman, Y., Falini, G., Addadi, L., and Weiner, S. (2001) Structure of the nacreous organic matrix of a bivalve mollusk shell examined in the hydrated state using Cryo-TEM. *J. Struct. Biol.* **135**: 8–17.

Levy, T. E., Najjar, M., and Higham, T. (2010) Ancient texts and archaeology revisited radiocarbon and Biblical dating in the southern Levant. *Antiquity* **84**: 834–847.

Li, V. C., Stang, H., and Krenchel, H. (1993) Micromechanics of crack bridging in fibre-reinforced concrete. *Mater. Struct.* **26**: 486–494.

Liao, J., Yang, L., Grashow, J., and Sacks, M. (2005) Molecular orientation of collagen in intact planar connective tissues under biaxial stretch. *Acta Biomater.* **1**: 45–54.

Lichtenegger, H. C., Schöberl, T., Bartl, M. H., Waite, H., and Stucky, G. D. (2002) High abrasion resistance with sparse mineralization: copper biomineral in worm jaws. *Science* **298**: 389–392.

Lim, C. T. (2006) Single cell mechanics study of the human disease malaria. *J. Biomech. Sci. Eng.* **1**: 82–92.

Lim, C. T., Dao, M., Suresh, S., Sow, C. H., and Chew, K. T. (2004) Large deformation of living cells using laser traps. *Acta Mater.* **52**: 1837–1845.

Lim, C. T., Zhou, E. H., Li, A., Vedula, S. R. K., and Fu, H. X. (2006a) Experimental techniques for single cell and single molecule biomechanics. *Mater. Sci. Eng. C* **26**: 1278–1288.

Lim, C. T., Zhou, E. H., and Quek, S. T. (2006b) Mechanical models for living cells – a review. *J. Biomech.* **39**: 195–216.

Lin, A. and Meyers, M. A. (2005) Growth and structure in abalone shell. *Mater. Sci. Eng. A* **390**: 27–41.

Lin, A. Y. M. and Meyers, M. A. (2009) Interfacial shear strength in abalone nacre. *J. Mech. Behav. Biomed. Mater.* **2**: 607–612.

Lin, A. Y. M, Meyers, M. A., and Vecchio, K. S. (2006) Mechanical properties and structure of Strombus gigas, Tridacna gigas and Haliotis rufescens sea shells: a comparative study. *Mater. Sci. Eng. C* **26**: 1380–1389.

Lin, A. Y. M., Chen, P. -Y., and Meyers, M. A. (2008) The growth of nacre in the abalone shell. *Acta Biomater.* **4**: 131–138.

Lin, A. Y. M., Brunner, R., Chen, P. -Y., Talke, F. E., and Meyers, M. A. (2009) Underwater adhesion of abalone: the role of van der Waals and capillary forces. *Acta Mater.* **57**: 4178–4185.

Lin, K. L., Chen, L., and Chang, J. (2012) Fabrication of dense hydroxyapatite nanobioceramics with enhanced mechanical properties via two-step sintering process. *Int. J. Appl. Ceramic Tech.* **9**: 479–485.

Lin, Y. S., Wei, C. T., Olevsky, E. A., and Meyers, M. A. (2011) Mechanical properties and laminate structure of Arapaima gigas scale. *J. Mech. Behav. Biomed. Mater.* **4**: 1145–1156.

Lincoln, G. A. (1972) The role of antlers in the behaviour of red deer. *J. Exp. Zool.* **182**: 233–249.

Lincoln, G. A. (1992) Biology of antlers. *J. Zool. Lond.* **226**: 517–528.

Lingham-Soliar, T., Bonser, R. H. C., and Wesley-Smith, J. (2009) Selective biodegradation of keratin matrix in feather rachis reveals classic bioengineering. *Proc. Roy. Soc. B* **277**: 1161–1168.

Lopez, M. I., Chen, P. -Y., McKittrick, J., and Meyers, M. A. (2011) Growth of nacre in abalone: seasonal and feeding effects. *Mater. Sci. Eng. C* **31**: 238–245.

Lowenstam, H. A. (1962) Magnetite in denticle capping in recent chitons (polyplacophora). *Bull. Geol. Soc. Am.* **73**: 435.

Lowenstam, H. A. (1981) Minerals formed by organisms. *Science* **211**: 1126–1131.

Lowenstam, H. A. and Weiner, S. (1989) *On Biomineralization*. New York: Oxford University Press.

Lucas, G. L., Cooke, F. W., and Friis, E. A. (1999) *A Primer on Biomechanics*. New York: Springer.

Lucchinetti, E., Thomann, D., and Danuser, G. (2000) Review: micromechanical testing of bone trabeculae-potentials and limitations. *J. Mater. Sci.* **35**: 6057–6064.

Lynn, A. K., Nakamura, T., Patel, N. *et al.* (2005) Composition-controlled nanocomposites of apatite and collagen incorporating silicon as an osseopromotive agent. *J. Biomed. Mater. Res. A* **74**: 447–453.

Ma, M., Vijayan, K., Hiltner, A., Baer, E., and Im, J. (1990a) Thickness effects in microlayer composites of polycarbonate and poly-(styrene-acrylonitrile). *J. Mater. Sci.* **25**: 2039–2046.

Ma, M., Im, J., Hiltner, A., and Baer, E. (1990b) Fatigue crack propagation of polycarbonate and poly-(styrene-acrylonitrile). *J. Appl. Poly. Sci.* **40**: 669–684.

Magdans, U. and Gies, H. (2004) Single crystal structure analysis of sea urchin spine calcites: systematic investigations of the Ca/Mg distribution as a function of habitat of the sea urchin and the sample location in the spine. *Eur. J. Mineral.* **16**: 261–268.

Mahdavi, A., Ferreira, L., Sundback, C. *et al.* (2008) Biodegradable and biocompatible gecko inspired adhesive. *Proc. Natl. Acad. Sci. USA* **105**: 2307–2312.

Mahoney, E., Holt, A., Swain, M., and Kilpatrick, N. (2010) The hardness and modulus of elasticity of primary molar teeth: an ultra-micro-indentation study. *J. Dentistry* **28**: 589–594.

Malik, C. L., Gibeling, J. C., Martin, R. B., and Stover, S. M. (2003) Equine cortical bone exhibits rising R-curve fracture mechanics. *J. Biomech.* **36**: 191–198.

Mann, S. (1988) Molecular recognition in biomineralization. *Nature* **332**: 119–124.

Mann, S. (2001) *Biomineralization: Principles and Concepts in Bioinorganic Materials Chemistry*. New York: Oxford University Press.

Mann, S., Archibald, D. D., Didymus, J. M. *et al.* (1993) Crystallization at inorganic-organic interfaces: biominerals and biomimetic synthesis. *Science* **261**: 1286–1292.

Manne, S. and Aksay, I. A. (1997) Thin films and nanolaminates incorporating organic/inorganic interfaces. *Curr. Opin. Sol. State Mater. Sci.* **2**: 358–364.

Marin, F. and Luquet, G. (2005) Molluscan biomineralization: the proteinaceous shell constituents of *Pinna nobilis* L. *Mater. Sci. Eng. C* **25**: 105–111.

Mark, R. E. (1967) *Cell Wall Mechanics of Wood Tracheids*. New Haven: Yale University Press.

Marks, R. and Plewig, G. (1983) *Stratum Corneum*. New York: Springer-Verlag.

Marshall, C. and Gillespie, J. M. (1977) The keratin proteins of wool, horn and hoof from sheep. *Austr. J. Bio. Sci.* **30**: 389–400.

Martin, R. B. and Burr, D. B. (1982) A hypothetical mechanism for the stimulation of osteonal remodeling by fatigue damage. *J. Biomech.* **15**: 137–139.

Matonis, V. A. (1964) Elastic behavior of low density rigid foams in structural applications. *Soc. Plast. Eng. J.* **20**: 1024–1030.

Mayer, G. (2005) Rigid biological systems as models for synthetic composites. *Science* **310**: 1144–1147.

Mayer, G. (2006) New classes of tough composite materials – lessons from natural rigid biological systems. *Mater. Sci. Eng. C* **26**: 1261–1268.

Mayer, G. and Sarikaya, M. (2002) Rigid biological composite materials: structural examples for biomimetic design. *Exp. Mech.* **42**: 395–403.

McAllister, A. and Channing, L. (1983) Comparisons of toe pads of some Southern African climbing frogs. *S. Afr. J. Zool.* **18**: 110–114.

McBride, E. D. (1938) Absorbable metal in bone surgery: a further report on the use of magnesium alloys. *J. Am. Med. Assoc.* **111**: 2464–2467.

McElhaney, J. H. (1966) Dynamic response of bone and muscle tissue. *J. Appl. Physiol.* **21**: 1231–1236.

McKittrick, J., Chen, P.-Y., Tombolato, L. *et al.* (2010) Energy absorbent natural materials and bio-inspired design strategies: a review. *Mater. Sci. Eng. C* **30**: 331–342.

McKittrick, J., Chen, P.-Y., Bodde, S. G., Yang, W., Novitskaya, E. E., and Meyers, M. A. (2012) The structure, functions, and mechanical properties of keratin. *JOM* **64**: 449–468.

Meldrum, F. C. and Ludwigs, S. (2007) Template-directed control of crystal morphologies. *Macromol. Biosci.* **7**: 152–162.

Melnick, C. A., Chen, S., and Mecholsky, J. J. (1996) Hardness and toughness of exoskeleton material in the stone crab Menippe mercenaria. *J. Mater. Res.* **11**: 2903–2907.

Melvin, J. W. and Evans, F. G. (1973) Crack propagation in bone. *ASME Biomaterials Symp.* Detroit, MI 1973.

Menezes, G. C., Elias, C. N., Attias, M., and Silva-Filho, F. C. (2003) Osteoblast adhesion onto titanium dental implants. *Acta Microsc.* **12**: 13–19.

Menig, R., Meyers, M. H., Meyers, M. A., and Vecchio, K. S. (2000) Quasi-static and dynamic mechanical response of *Haliotis rufescens* (abalone) shells. *Acta Mater.* **48**: 2383–2398.

Menig, R., Meyers, M. H., Meyers, M. A., and Vecchio, K. S. (2001) Quasi-static and dynamic mechanical response of *Strombus gigas* (conch) shells. *Mater. Sci. Eng. A* **297**: 203–211.

Mercer, E. H. (1961) *Keratin and Keratinization: An Essay in Molecular Biology*. New York: Pergamon Press.

Meyers, M. A. and Chawla, K. C. (2009) *Mechanical Behavior of Materials*, 2nd edn. Cambridge: Cambridge University Press.

Meyers, M. A., Lin, A. Y. M., Seki, Y., Chen, P. -Y., Kad, B. K., and Bodde, S. (2006) Structural biological composites: an overview. *JOM* **58**: 35–41.

Meyers, M. A., Lin, A. Y. M., Chen, P. -Y., and Muyco, J. (2008a) Mechanical strength of abalone nacre: role of the soft organic layer. *J. Mech. Behav. Biomed. Mater.* **1**: 76–85.

Meyers, M. A., Chen, P. -Y., Lin, A. Y. M., and Seki, Y. (2008b) Biological materials: structure and mechanical properties. *Prog. Mater. Sci.* **53**: 1–206.

Meyers, M. A., Lin, A. Y. M., Lin, Y. S., Olevsky, E. A., and Georgalis, S. (2008c) The cutting edge: sharp biological materials. *JOM* **60**: 21–26.

Meyers, M. A., Lim, C. T., Li, A. *et al.* (2010) The role of organic layer in abalone nacre. *Mater. Sci. Eng. C* **29**: 2398–2410.

Meyers, M. A., Chen, P. -Y., Lopez, M. I., Seki, Y., and Lin, A. Y. M. (2011) Biological materials: a materials science approach. *J. Mech. Behav. Biomed. Mater.* **4**: 626–657.

Meyers, M. A., Lin, Y. S., Olevsky, E. A., and Chen, P. -Y. (2012) Battle in the Amazon: Arapaima versus Piranha. *Adv. Eng. Mater.* **14**: B1–B10.

Meyers, M. A., McKittrick, J., and Chen, P. -Y. (2013) Structural biological materials: critical mechanics-materials connections. *Science* **339**: 773–779.

Milwich, M., Speck, T., Speck, O., Stegmaier, T., and Planck, H. (2006) Biomimetics and technical textiles: solving engineering problems with the help of nature's wisdom. *Am. J. Botany* **93**: 1455–1465.

Miserez, A., Schneberk, T., Sun, C., Zok, F. W., and Waite, J. H. (2008) The transition from stiff to compliant materials in squid beaks. *Science* **319**: 1816–1819.

Miserez, A., Weaver, J. C., Pedersen, P. B. *et al.* (2009a) Microstructural and biochemical characterization of the nanoporous sucker rings from *Dosidicus gigas*. *Adv. Mater.* **21**: 401–406.

Miserez, A., Wasko, S. S., Carpenter, C. F., and Waite, J. H. (2009b) Non-entropic and reversible long-range deformation of an encapsulating bioelastomer. *Nature Mater.* **8**: 910–916.

Mitchison, T. J. and Cramer, L. P. (1996) Actin-based cell motility and cell locomotion. *Cell* **84**: 371–379.

Moir, B. G. (1990) Comparative-studies of fresh and aged *Tridacna gigas* shell – preliminary investigations of a reported technique for pretreatment of tool material. *J. Archaeol. Sci.* **17**: 329–345.

Montagna, W. and Parakkal, P. F. (1974) *The Structure and Function of Skin*, 3rd edn. New York: Academic Press.

Monteiro, S. N., Lopes, F. P. D., Barbosa, A. P., Bevitori, A. B., Da Silva, I. L. A., and Da Costa, L. L. (2011a) Natural lignocellulosic fibers as engineering materials – an overview. *Metall. Mater. Trans.* **42a**: 2963–2974.

Monteiro, S. N., Satyanarayana, K. G., Ferreira, A. S., Nascimento, D. O. C., and Lopes, F. P. D. (2011b) Selection of high strength natural fibers. *Revista Matéria* **15**: 488–505.

Mooney, M. (1940) A theory of large elastic deformation. *J. Appl. Phys.* **11**: 582–592.

Morais, L. S., Glaucio, G., Serra, G. C. *et al.* (2007) Titanium alloy mini-implants for orthodontic anchorage: immediate loading and metal ion release. *Acta Biomater.* **3**: 331–339.

Munch, E., Launey, M. E., Alsem, D. H., Saiz, E., Tomsia, A. P., and Ritchie, R. O. (2008) Tough, bio-inspired hybrid materials. *Science* **322**: 1516–1520.

Murr, L. E. and Ramirez, D. A. (2012) The microstructure of the cultured freshwater pearl. *JOM* **64**: 469–474.

Nachemson, A. and Evans, J. H. (1968) Some mechanical properties of the third human lumbar interlaminar ligament (ligamen tum flavum). *J. Biomech.* **1**: 211–220.

Nakahara, H. (1991) Nacre formation in bivalve and gastropod mollusks. In Suga, S. and Nakahara, H., eds. *Mechanisms and Phylogeny of Mineralization in Biological Systems*. New York: Springer, pp. 343–350.

Nakahara, H., Kakei, M., and Bevelander, G. (1982) Electron microscopic and amino acid studies on the outer and inner shell layers of *Haliotis rufescens*. *Venus Jpn. J. Malac.* **41**: 33–46.

Nalla, R. K., Kinney, J. H., and Ritchie, R. O. (2003a) Mechanistic fracture criteria for the failure of human cortical bone. *Nature Mater.* **2**: 164–168.

Nalla, R. K., Kinney, J. H., and Ritchie, R. O. (2003b) Effect of orientation on the in vitro fracture toughness of dentin: the role of toughening mechanisms. *Biomater.* **24**: 3955–3968.

Nalla, R. K., Kruzic, J. J., Kinney, J. H., and Ritchie, R. O. (2004) On the origin of the toughness of mineralized tissue: microcracking or crack bridging? *Bone* **34**: 790–798.

Nalla, R. K., Kruzic, J. J., Kinney, J. H., and Ritchie, R. O. (2005) Mechanistic aspects of fracture and R-curve behavior of human cortical bone. *Biomater.* **26**: 217–231.

Nalla, R. K., Kruzic, J. J., Kinney, J. H., Balooch, M., Ager, J. W., and Ritchie, R. O. (2006a) Role of microstructure in the aging-related deterioration of the toughness of human cortical bone. *Mater. Sci. Eng. C* **26**: 1251–1260.

Nalla, R. K., Kinney, J. H., Tomsia, A. P., and Ritchie, R. O. (2006b) Role of alcohol in the fracture resistance of teeth. *J. Dent. Res.* **85**: 1022–1026.

Nam, K. T., Kim, D. W., Yoo, P. J. *et al.* (2006) Virus-enabled synthesis and assembly of nanowires for lithium ion battery electrodes. *Science* **312**: 886–888.

Nassau, K. (1998) *Color for Science, Art, and Technology*. New York: Elsevier.

Nassau, K. (2001) *The Physics and Chemistry of Color*, 2nd edn. New York: Wiley.

Nelson, D. L. and Cox, M. M. (2005) *Lehninger Principles of Biochemistry*, 4th edn. New York: W.H. Freeman.

Nevell, T. P. and Zeronian, S. H. (1985) *Cellulose Chemistry and its Applications*. New York: Wiley.

Neville, A. C. (1975) *Biology of the Arthropod Cuticle*. New York: Springer-Verlag.

Neville, A. C. (1993) *Biology of Fibrous Composites*. Cambridge: Cambridge University Press.

Nicolis, G. and Prigogine, I. (1989) *Exploring Complexity*. New York: W. H. Freeman.

Nikolov, S., Petrov, M., Lymperakis, L. *et al.* (2010) Revealing the design principles of high-performance biological composites using ab initio and multiscale simulations: the example of lobster cuticle. *Adv. Mater.* **22**: 519–526.

Novitskaya, E., Chen, P. Y., Lee, S., *et al.* (2011) Anisotropy in the compressive mechanical properties of bovine cortical bone and the mineral and protein constituents. *Acta Biomater.* **7**: 3170–3177.

Nudelman, F., Gotliv, B. A., Addadi, L., and Weiner, S. (2006) Mollusk shell formation: mapping the distribution of organic matrix components underlying a single aragonitic tablet in nacre. *J. Struct. Biol.* **153**: 176–187.

Nuzzo, R. G. and Allara, D. L. (1983) Adsorption of bifunctional organic disulfides on gold surfaces. *J. Am. Chem. Soc.* **105**: 4481–4483.

Ogden, R. W. (1972) Large deformation isotropic elasticity – on the correlation of theory and experiment for incompressible rubberlike solids. *Proc. Roy. Soc. Lond. A* **326**: 565–584.

Oka, K., Aoyagi, S., Hashiguchi, G., Isono, Y., and Fujita, H. (2002) Fabrication of a micro needle for a trace blood test. *Proc. Sensor. Actuat. A* **97–98**: 478–485.

Olson, G. B. and Hartman, H. (1982) Martensite and life: displacive transformations as biological processes. *J. de Phys.* **43**: (C4) 855–865.

Olson, P. and Watabe, N. (1980) Studies on formation and resorption of fish scales. *Cell Tissue Res.* **211**: 303–316.

Onozato, H. and Watabe, N. (1979) Studies on fish scale formation and resorption. *Cell Tissue Res.* **201**: 409–422.

Orme, C. A., Noy, A., Wierzbicki, A. *et al.* (2001) Formation of chiral morphologies through selective binding of amino acids to calcite surface steps. *Nature* **411**: 775–779.

Oxlund, H., Manschot, J., and Viidik, A. (1988) The role of elastin in the mechanical properties of skin. *J. Biomech.* **3**: 213–218.

Ozin, G. A., Manners, I., Fournier-Bidoz, S., and Arsenault, A. (2005) Dream nanomachines. *Adv. Mater.* **17**: 3011–3018.

Pabisch, S., Puchegger, S., Kirchner, H. O. K., Weiss, I. M., and Peterlik, H. (2010) Keratin homogeneity in the tail feathers of *Pavo cristatus* and *Pavo cristatus mut.* Alba. *J. Struct. Biol.* **172**: 270–275.

Papir, Y. S., Hsu, K. H., and Wildnauer, R. H. (1975) The mechanical properties of stratum corneum: I. The effect of water and ambient temperature on the tensile properties of newborn rat stratum corneum. *Biochim. Biophys. Acta* **399**: 170–180.

Park, A. C. and Baddiel, C. B. (1972) Rheology of stratum corneum. Part I. A molecular interpretation of the stress-strain curve. *J. Soc. Cosmet. Chem.* **23**: 3–12.

Park, J. and Lakes, R. S. (2007) *Biomaterials: An introduction*, 3rd edn. New York: Springer.

Parker, A. R. and Lawrence, C. R. (2001) Water capture from desert fogs by a Namibian beetle. *Nature* **414**: 33–34.

Parry, D. A. D. and North, A. C. T. (1998) Hard α-keratin intermediate filament chains: substructure of the N- and C-terminal domains and the predicted structure and function of the C-terminal domains of type I and type II chains. *J. Struct. Biol.* **122**: 67–75.

Patek, S. N., Baio, J. E., Fisher, B. L., and Suarez, A. V. (2006) Multifunctionality and mechanical origins: ballistic jaw propulsion in trap-jaw ants. *Proc. Natl. Acad. Sci.* **103**: 12787–12792.

Pautard, F. G. E. (1963) Mineralization of keratin and its comparison with the enamel matrix. *Nature* **199**: 531–535.

Peattie, A. M. and Full, R. J. (2007) Phylogenetic analysis of the scaling of wet and dry biological fibrillar adhesives. *Proc. Natl. Acad. Sci. USA* **104**: 18595.

Pek, Y. S., Spector, M., Yanna, I. V., and Gibson, L. J. (2004) Degradation of a collagen-chondroitin-6 sulfate matrix by collagenase and chondroitinase. *Biomater.* **25**: 472–482.

Peña, E., Martinsh, P., Mascarenhas, T. *et al.* (2011) Mechanical characterization of the softening behavior of human vaginal tissue. *J. Mech. Behav. Biomed.* **4**: 275–283.

Perez-Rigueiro, J., Viney, C., Llorca, J., and Elices, M. (2000) Mechanical properties of silkworm silk in liquid media. *J. Appl. Polymer Sci.* **75**: 1270–1277.

Pins, G. D., Christiansen, D. L., Patel, R., and Silver, F. H. (1977) Self-assembly of collagen fibers. Influence of fibrillar alignment and decorin on mechanical properties. *Biophys. J.* **73**: 2164–2172.

Pollock, R. G., Soslowsky, L. J., Bigliani, L. U., Flatow, E. L., and Mow, V. C. (1990) The mechanical properties of the inferior glenohumeral ligament. *Trans. Orthop. Res. Soc.* **15**: 510.

Potyrailo, R. A., Ghiradella, H., Vertiatchikh, A., Dovidenko, K., Cournoyer, J. R., and Olson, E. (2007) Morpho butterfly wing scales demonstrate highly selective vapor response. *Nature Photon.* **1**: 123–128.

Poulsen, N., Sumper, M., and Kröger, N. (2003) Biosilica formation diatoms: characterization of native silaffin-2 and its role in silica morphogenesis. *Proc. Natl. Acad. Sci.* **100**: 12075–12080.

Presser, V., Schultheiβ, S., Berthold, C., and Nickel, K. G. (2009) Sea urchin spines as a model-system for permeable, light-weight ceramics with graceful failure behavior. Part I. Mechanical behavior of sea urchin spines under compression. *J. Bionic Eng.* **6**: 203–213.

Preston, R. D. (1974) *The Physical Biology of Plant Cell Walls*. London: Chapman and Hall.

Prigogine, I. (1962) *Non Equilibrium Statistical Mechanics*. New York: Wiley-Interscience.

Pruitt, L. A. and Chakravartula, A. M. (2011) *Mechanics of Biomaterials*. Cambridge: Cambridge University Press.

Prum, R. O. (1999) Development and evolutionary origin of feathers. *J. Exp. Zool.* **285**: 291–306.

Purslow, P. P. (1983) Measurement of the fracture toughness of extensible connective tissues. *J. Mater. Sci.* **18**: 3591–3598.

Purslow, P. P. and Vincent, J. F. V. (1978) Mechanical properties of primary feathers from the pigeon. *J. Exp. Biol.* **72**: 251–260.

Qian, J. and Gao, H. (2006) Scaling effects of wet adhesion in biological attachment systems. *Acta Biomater.* **2**: 51–58.

Qin, X. X., Coyne, K. J., and Waite, J. H. (1997) Tough tendons: mussel byssus has collagen with silk-like domains. *J. Biol. Chem.* **272**: 32623–32627.

Quicke, D. L. J., Wyeth, P., Fawke, J. D., Basibuyuk, H. H., and Vincent, J. F. V. (1998) Manganese and zinc in the ovipositors and mandibles of hymenopterous insects. *Zool. J. Linn. Soc.* **124**: 387–396.

Raabe, D., Al-Sawalmih, A., Romano, P. *et al.* (2005a) *Mater. Sci. Forum* **495–497**: 1665–1674.

Raabe, D., Romano, P., Sachs, C. *et al.* (2005b) Discovery of a honeycomb structure in the twisted plywood patterns of fibrous biological nanocomposite tissue. *J. Crystal Growth* **283**: 1–7.

Raabe, D., Sachs, C., and Romano, P. (2005c) The crustacean exoskeleton as an example of a structurally and mechanically graded biological nanocomposite material. *Acta Mater.* **53**: 4281–4292.

Raabe, D., Romano, P., Sachs, C. *et al.* (2006) Microstructure and crystallographic texture of the chitin–protein network in the biological composite material of the exoskeleton of the lobster Homarus americanus. *Mater. Sci. Eng. A* **421**: 143–153.

Rabin, B. H., Williamson, R. L., and Suresh, S. (1995) Fundamentals of residual stresses in joints between dissimilar material. *Mater. Res. Soc. Bull.* **20**: 37–39.

Rabotyagova, O. S., Cebe, P., and Kaplan, D. L. (2011) Protein-based block copolymers. *Biomacromol.* **12**: 269–289.

Ratner, B. D., Hoffman, A. S., Schoen, F. J., and Lemons, J. E. (2005) *Biomaterials Science: An Introduction to Materials in Medicine*. New York: Academic Press.

Regan, B. C., Aloni, S., Jensen, K., Ritchie, R. O., and Zettl, A. (2005) Nanocrystal-powered nanomotor. *Nano Lett.* **5**: 1730–1733.

Ren, D., Meyers, M. A., Zhou, B., and Feng, Q. (2013) Comparative study of carp otolith hardness: lapillus and asteriscus. *Mater. Sci. Eng. C* **33**: 1876–1881.

Rhee, H., Horstemeyer, M. F., Hwang, Y., Lim, H., El Kadiri, H., and Trim, W. (2009) A study on the structure and mechanical behavior of the *Terrapene carolina* carapace: a pathway to design bio-inspired synthetic composites. *Mater. Sci. Eng. C* **29**: 2333–2339.

Rhee, H., Horstemeyer, M. F., and Ramsay, A. (2011) A study on the structure and mechanical behavior of the *Dasypus novemcinctus* shell. *Mater. Sci. Eng. C* **31**: 363–369.

Rho, J. Y., Kuhn-Spearing, L., and Zioupos, P. (1998) Mechanical properties and the hierarchical structure of bone. *Med. Eng. Phys.* **20**: 92–103.

Rinaudo, M. (2006) Chitin and chitosan: properties and applications. *Prog. Polymer Sci.* **31**: 603–632.

Ritchie, R. O. (1988) Mechanisms of fatigue crack propagation in metals, ceramics and composites: role of crack tip shielding. *Mater. Sci. Eng. A* **103**: 15–28.

Ritchie, R. O. (1999) Mechanisms of fatigue-crack propagation in ductile and brittle solids. *Int. J. Fracture* **100**: 55–83.

Ritchie, R. O., Kinney, J. H., Kruzic, J. J., and Nalla, R. K. (2006) Cortical bone fracture. In Akay, M., ed. *Wiley Encyclopedia of Biomedical Engineering*. Hoboken, NJ: John Wiley & Sons Inc., pp. 1–18.

Rivlin, R. S. and Saunders, D. W. (1951) Large elastic deformations of isotropic materials. VII. Experiments on the deformation of rubber. *Phil. Trans. Roy. Soc. Lond. A* **243**: 251–288.

Rogers, G. J., Milthorpe, B. K., Muratore, A., and Schindhelma, K. (1990) Measurement of the mechanical properties of the ovine anterior cruciate ligament bone-ligament-bone complex: a basis for prosthetic evaluation. *Biomater.* **11**: 89–96.

Rohrlich, S. T. and Rubin, R. W. (1975) Biochemical characterization of crystals from the dermal iridophores of a chameleon Anolis carolinensis. *J. Cell. Biol.* **66**: 635–645.

Romano, P., Fabritius, H., and Raabe, D. (2007) The exoskeleton of the lobster *Homarus americanus* as an example of a smart anisotropic biological material. *Acta Biomater.* **3**: 301–309.

Rosewater, J. R. (1965) The family Tridacnidae in the Indo-Pacific. *Indo-Pacific Mollusca* **1**: 347–396.

Rudall, K. M. (1955) The distribution of collagen and chitin. *Symp. Soc. Exp. Biol.* **9**: 49–71.

Ruibal, R. and Ernst, V. (1965) The structure of the digital setae of lizards. *J. Morphol.* **117**: 271–293.

Runnegar, B. and Bengtson, S. (1992) Origin of hard parts: early skeletal fossils. In Briggs, D. E. G. and Crowther, P. R., eds. *Palaeobiology: A Synthesis*. Oxford: Wiley-Blackwell, pp. 24–29.

Ryan, S. D. and Williams, J. L. (1989) Tensile testing of rodlike trabeculae excised from bovine femoral bone. *J. Biomech.* **22**: 351–355.

Ryder, M. L. (1962) Structure of the rhinoceros horn. *Nature* **193**: 1199–1201.

Sachs, C., Fabritius, H., and Raabe, D. (2006a) Hardness and elastic properties of dehydrated cuticle from the lobster. *J. Mater. Res.* **21**: 1987–1995.

Sachs, C., Fabritius, H., and Raabe, D. (2006b) Experimental investigation of the elastic-plastic deformation of mineralized lobster cuticle by digital image correlation. *J. Structur. Biol.* **155**: 409–425.

Sacks, M. (2003) Incorporation of experimentally-derived fiber orientation into a structural constitutive model for planar collagenous tissues. *Trans. ASME* **125**: 280–287.

Sahni, V., Blackledge, T. A., and Dhinojwala, A. (2010) Viscoelastic solids explain spider web stickiness. *Nature Commun.* **1**: 19.

Saito, A., Miyamura, Y., Nakajima, M. *et al.* (2006) Reproduction of the Morpho blue by nanocasting lithography. *J. Vac. Sci. Tech. B* **24**: 3248–3251.

Sanchez, C., Arribart, H., and Giraud-Guille, M. M. (2005) Biomimetism and bioinspiration as tools for the design of innovative materials and systems. *Nature Mater.* **4**: 277–288.

Sandhage, K. H., Dickerson, M. B., Huseman, P. M. *et al.* (2002) Novel, bioclastic route to self-assembled, 3D, chemically tailored meso/nanostructures: shape-preserving reactive conversion of biosilica (diatom) microshells. *Adv. Mater.* **14**: 429–433.

Sarikaya, M. (1994) An introduction to biomimetics: a structural viewpoint. *Micros. Res. Tech.* **27**: 360–375.

Sarikaya, M. and Aksay, I. A. (1992) Nacre of abalone shell: a natural multifunctional nano-laminated ceramic-polymer composite material. In Case, S. T., ed. *Results and Problems in Cell Differentiation – Biopolymers*. Berlin: Springer-Verlag, pp. 1–26.

Sarikaya, M., Gunnison, K. E., Yasrebi, M., and Aksay, I. A. (1990) Mechanical property-microstructural relationships in abalone shell. In Rieke, P. C., Calvert, P. D., and Alper, M., eds. *Materials Synthesis Utilizing Biological Processes*, MRS Symp. Proc. Vol. 174. Pittsburgh, PA: Materials Research Society, pp. 109–116.

Sarikaya, M., Fong, H., Sunderland, N. *et al.* (2001) Biomimetic model of a sponge – spicular optical fiber-mechanical properties and structure. *J. Mater. Res.* **16**: 1420–1428.

Sarikaya, M., Tamerler, C., Jen, A. K. Y., Schulten, K., and Baneyx, F. (2003) Molecular biomimetics: nanotechnology through biology. *Nature Mater.* **2**: 577–585.

Sasaki, N. and Odajima, S. (1996) Elongation mechanism of collagen fibrils and force-strain relations of tendon at each level of structural hierarchy. *J. Biomech.* **29**: 1131–1136.

Sass, R. L. and Vidale, R. (1957) Interatomic distances and thermal anisotropy in sodium nitrate and calcite. *Acta Crystall.* **10**: 567–570.

Schäffer, T. E., Zanetti, C. I., Proksch, R. *et al.* (1997) Does abalone nacre form by heteroepitaxial nucleation or by growth through mineral bridges? *Chem. Mater.* **9**: 1731–1740.

Schillinger, M., Sabet, S., Loewe, C. *et al.* (2006) Balloon angioplasty versus implantation of Nitinol stents in the superficial femoral artery. *New Engl. J. Med.* **354**: 1879–1888.

Schneider, A. S., Heiland, B., Peter, N. J., Guth, C., Arzt, E., and Weiss, I. M. (2012) Hierarchical super-structure identified by polarized light microscopy, electron microscopy and nanoindentation: implications for the limits of biological control over the growth mode of abalone sea shells. *BMC Biophys*. **5**: 19.

Schofield, R. M. S., Nesson, M. H., and Richardson, K. A. (2002) Tooth hardness increases with zinc-content in mandibles of young adult leaf-cutter ants. *Naturwissenschaft*. **89**: 579–583.

Schultz, H. (2006) *Sea Urchins*. Hemdingen: Heinke & Peter Schutz Partner.

Schultz, J. T., Tompkins, R. G., and Burke, J. F. (2000) Artificial skin. *Annu. Rev. Med*. **51**: 231–244.

Schwenzer, B., Gomm, J. R., and Morse, D. E. (2006) Substrate-induced growth of nanostructured zinc oxide films at room temperature using concepts of biomimetic catalysis. *Langmuir* **22**: 9829–9831.

Schwinger, G., Zanger, K., and Greven, H. (2001) Structure and mechanical aspects of the skin of Bufo marinus (Anura, Amphibia). *Tissue Cell* **33**: 541–547.

Seeman, N. C. and Belcher, A. M. (2002) Emulating biology: building nanostructures from the bottom up. *Proc. Natl. Acad. Sci. USA* **99**: 6451–6455.

Seki, Y., Schneider, M. S., and Meyers, M. A. (2005) Structure and mechanical behavior of a toucan beak. *Acta Mater*. **53**: 5281–5296.

Seki, Y., Kad, B., Benson, D., and Meyers, M. A. (2006) The toucan beak: structure and mechanical response. *Mater. Sci. Eng. C* **26**: 1412–1420.

Seki, Y., Bodde, S. G., and Meyers, M. A. (2010) Toucan and hornbill beaks: comparative study. *Acta Biomater*. **6**: 331–343.

Seki, Y., Mackey, M., and Meyers, M. A. (2012) Structure and micro-computed tomography-based finite element modeling of toucan beak. *J. Mech. Behav. Biomed. Mater*. **9**: 1–8.

Selden, P. A. (1989) Orb-web weaving spiders in the early Cretaceous. *Nature* **340**: 711–712.

Serra, G., Morais, L. S., Elias, C. N. *et al.* (2010) Sequential bone healing of immediately loaded mini-implants: histomorphometric and fluorescence analysis. *Am. J. Orthodont. Dentofac. Orthop*. **137**: 80–90.

Sethi, S., Ge, L., Ajayan, P. M., and Dhinojwala, A. (2008) Gecko-inspired carbon nanotube based self cleaning adhesives. *Nano. Lett*. **8**: 822–825.

Sethman, I., Hinrichs, R., Wörheide, G., and Putnis, A. (2006) Nano-cluster composite structure of calcitic sponge spicules – a case study of basic characteristics of biominerals. *J. Inorg. Biochem*. **100**: 88–96.

Shadwick, R. E., Russell, A. P., and Lauff, R. F. (1992) The structure and mechanical design of rhinoceros dermal armour. *Phil. Trans. Roy. Soc. Lond. B* **337**: 419–428.

Shear, W. A., Palmer, J. M., Coddington, J. A., and Bonamo, P. M. (1989) A devonian spinneret: early evidence of spiders and silk use. *Science* **246**: 479–481.

Shen, X., Belcher, A. M., Hansma, P. K., Stucky, G. D., and Morse, D. E. (1997) Molecular cloning and characterization of Lustrin A, a matrix protein from shell and pearl nacre of Haliotis rufescens. *J. Biol. Chem*. **272**: 32472–32481.

Shen, Z. L., Dodge, M. R., Kahn, H. *et al.* (2008) Stress-strain experiments on individual collagen fibrils. *Biophys. J*. **95**: 3956–3963.

Shen, Z. L., Kahn, H., Ballarini, R., and Eppelli, S. J. (2011) Viscoelastic properties of isolated collagen fibrils. *Biophys. J.* **100**: 3008–3015.

Shepherd, S. A., Avalos-Borja, M., and Ortiz Quintanilla, M. (1995) Towards a chronology of *Haliotis fulgens*, with a review of abalone shell microstructure. *Mar. Freshwater Res.* **46**: 607–615.

Shergold, E. A., Norman, A., Fleck, N. A., and Radford, D. (2006) The uniaxial stress versus strain response of pig skin and silicone rubber at low and high strain rates. *Int. J. Impact Engng.* **32**: 1384–1402.

Sherrard, K. M. (2000) Cuttlebone morphology limits habitat depth in eleven species of Sepia (Cephalopoda: Sepiidae). *Biol. Bull.* **198**: 404–414.

Sikoryn, T. A. and Hukins, D. W. L. (1990) Mechanism of failure of the ligamentum flavum of the spine during in vitro tensile tests. *J. Orthop. Res.* **8**: 586–591.

Silyn-Roberts, H. and Sharp, R. M. (1988) Crystal growth and the role of the organic network in eggshell biomineralization. *Proc. R. Soc. Lond. B* **227**: 303–324.

Simkiss, K. and Wilbur, K. M. (1989) *Biomineralization: Cell Biology and Mineral Deposition*. San Diego: Academic Press.

Sitti, M. and Fearing, R. S. (2003) Synthetic gecko foot-hair micro/nano-structures as dry adhesives. *J. Adhes. Sci. Tech.* **17**: 1055–1073.

Skalak, R. J., Farrow, D. A., and Hoger, A. J. (1997) Kinematics of surface growth. *J. Math. Biol.* **35**: 869–907.

Skedros, J. G., Durand, P., and Bloebaum, R. D. (1995) Hypermineralized peripheral lamellae in primary osteons of deer antler: potential functional analogues of cement lines in mammalian secondary bone. *J. Bone Min. Res.* **10** (Suppl. 1): 441.

Smeathers, J. E. and Vincent, J. F. V. (1979) Mechanical properties of mussel byssus threads. *J. Mollusc. Stud.* **49**: 219–230.

Smith, G. P. (1985) Filamentous fusion phage: novel expression vectors that display cloned antigens on the virion surface. *Science* **228**: 1315–1317.

Snead, M. L., Zhu, D., Lei, Y. *et al.* (2006) Protein self-assembly creates a nanoscale device for biomineralization. *Mater. Sci. Eng. C* **26**: 1296–1300.

Song, F., Bai, X. H., and Bai, Y. I. (2002) Microstructure and characteristics in the organic matrix layers of nacre. *J. Mater. Res.* **17**: 1567–1570.

Song, F., Soh, A. K., and Bai, Y. L. (2003) Structural and mechanical properties of the organic matrix of nacre. *Biomater.* **24**: 3623–3631.

Song, J., Ortiz, C., and Boyce, M. C. (2011) Threat-protection mechanics of an armored fish. *J. Mech. Behav. Biomed. Mater.* **4**: 699–712.

Sonntag, R., Reinders, J., and Kretzer, J. P. (2012) What's next? Alternative materials for articulation in total joint replacement. *Acta Biomater.* **8**: 2434–2441.

Soong, R. K., Bachand, G. D., Neves, H. P., Olkhovets, A. G., Craighead, H. G., and Montemagno, C. D. (2000) Powering an inorganic nanodevice with a biomolecular motor. *Science* **290**: 1555–1558.

Spolenak, R., Gorb, S., Gao, H., and Arzt, E. (2005a) Effects of contact shape on the scaling of biological attachments. *Proc. Roy. Soc. A* **461**: 305–319.

Spolenak, R., Gorb, S., and Arzt, E. (2005b) Adhesion design maps for bio-inspired attachment systems. *Acta Biomater.* **1**: 5–13.

Srinivasan, A. V., Haritos, G. K., and Hedberg, F. L. (1991) Biomimetics: advancing man-made materials through guidance from nature. *Appl. Mech. Rev.* **44**: 463–482.

Stoeckel, D., Pelton, A., and Duering, T. (2004) Self-expanding Nitinol stents – material and design consideration. *Eur. Radiol.* **14**: 292–301.

Studart, A. R. (2012) Towards high-performance bioinspired composites. *Adv. Mater.* **24**: 5024–5044.

Su, X., Belcher, A. M., Zaremba, C. M., Morse, D. E., Stucky, G. D., and Heuer, A. H. (2002) Structural and microstructural characterization of the growth lines and prismatic micro-architecture in red abalone shell and the microstructures of abalone "flat pearls". *Chem. Mater.* **14**: 3106–3117.

Sun, C.-Y., and Chen, P.-Y. (2013) Structural design and mechanical behavior of alligator (*Alligator mississippiensis*) osteoderms. *Acta Biomater.* **9**: 9049–9064.

Sundar, V. C., Yablon, A. D., Grazul, J. L., Han, M., and Aizenberg, J. (2003) Fibre-optical features of a glass sponge. *Nature* **424**: 899–900.

Suresh, S. (2007) Biomechanics and biophysics of cancer cells. *Acta Biomater.* **3**: 413–438.

Suresh, S., Spatz, J., Mills, J. P. *et al.* (2005) Connections between single-cell biomechanics and human disease states: gastrointestinal cancer and malaria. *Acta Biomater.* **1**: 15–30.

Swartz, S. M., Bennett, M. E., and Carrier, D. R. (1992) Wing bone stresses in free flying bats and the evolution of skeletal design for flight. *Nature* **359**: 726–729.

Syn, C. K., Lesuer, D. R., Wolfenstine, J., and Sherby, O. D. (1993) Layer thickness effect on ductile tensile fracture of ultrahigh carbon steel-brass laminates. *Met. Trans. A* **24**: 1647–1653.

Tamerler, C. and Sarikaya, M. (2007) Molecular biomimetics: utilizing nature's molecular ways to practical engineering. *Acta Biomater.* **3**: 289–299.

Tamerler, C. and Sarikaya, M. (2008) Molecular biomimetics: genetic synthesis, assembly, and formation of materials using peptides. *MRS Bull.* **33**: 504–510.

Tang, Z., Kotov, N. A., Magonov, S., and Ozturk, B. (2003) Nanostructured artificial nacre. *Nature Mater.* **2**: 413–419.

Taylor, A. M., Bonser, R. H. C., and Farrent, J. W. (2004) The influence of hydration on the tensile and compressive properties of avian keratinous tissues. *J. Mater. Sci.* **39**: 939–942.

Taylor, J. D. and Layman, M. (1972) The mechanical properties of bivalve (Mollusca) shell structures. *Palaeontol.* **15**: 73–87.

Taylor, J. R. A. and Patek, S. N. (2010) Ritualized fighting and biological armor: the impact mechanics of the mantis shrimp's telson. *J. Exp. Biol.* **213**: 3496–3504.

Teilhard de Chardin, P. (1970) *Le Phénomène Humain*. Paris: Seuil.

Teng, H. H., Dove, P. M., Orme, C. A., and De Yoreo, J. J. (1998) Thermodynamics of calcite growth: baseline for understanding biomineral formation. *Science* **282**: 724–727.

Thompson, D. W. (1917) *On Growth and Form*. Cambridge: Cambridge University Press.

Thompson, D. W. (1968) *On Growth and Form*, 2nd edn., reprinted. Cambridge: Cambridge University Press.

Thompson, J. B., Kindt, J. H., Drake, B., Hansma, H. G., Morse, D. E., and Hansma, P. K. (2001) Bone indentation recovery time correlates with bone reforming time. *Nature* **414**: 773–775.

Thornton, P. H. and Magee, C. L. (1975a) The deformation of aluminum foams. *Met. Trans.* **6A**:1253–1263.

Thornton, P. H. and Magee, C. L. (1975b) Deformation characteristics of zinc foam. *Met. Trans.* **6A**: 1801–1807.

Tirrell, M., ed. (1994) *Hierarchical Structures in Biology as a Guide for New Materials*. Committee on Synthetic Hierarchical Structures, Commission on Engineering and Technical Systems, National Research Council. Washington D.C.: The National Academies Press, NMAB-464.

Tombolato, L., Novitskaya, E. E., Chen, P.-Y., Sheppard, F. A., and McKittrick, J. (2010) Microstructure, elastic and fracture properties of horn keratin. *Acta Biomater.* **6**: 319–330.

Tong, W., Glimcher, M. J., Katz, J. L., Kuhn, L., and Eppell, S. J. (2003) Size and shape of mineralites in young bovine bone measured by atomic force microscopy. *Calcif. Tiss. Int.* **72**: 592–598.

Toohey, K. S., Sottos, N. R., Lewis, J. A., Moore, J. S., and White, S. R. (2007) Self-healing materials with microvascular networks. *Nature Mater.* **6**: 581–585.

Torre, C. (1948) Theorie und Verhalten zusammengepresster Pulver. Berg.-u Huttenmann. *Monatsch. Montan. Hochschule Leoben* **93**: 62.

Torres, F. G., Troncoso, O. P., Nakamatsu, J., Grande, C. J., and Gomez, C. M. (2008) Characterization of the nanocomposite laminate structure occurring in fish scales from *Arapaima gigas*. *Mater. Sci. Eng. C* **28**: 1276–1283.

Traeger, R. K. (1967) Physical properties of rigid polyurethane foams. *J. Cell. Plast.* **3**: 405–418.

Traub, W., Arad, T., and Weiner, S. (1989) Three-dimensional ordered distribution of crystals in turkey tendon collagen fibers. *Proc. Natl. Acad. Sci. USA* **86**: 9822–9826.

Treloar, L. R. G. (1944) Stress-strain data for vulcanised rubber under various types of deformation. *Trans. Faraday Soc.* **40**: 59–70.

Treloar, L. R. G. (1975) *The Physics of Rubber Elasticity*, 3rd edn. Oxford: Oxford University Press.

Trim, W., Horstemeyer, M. F., Rhee, H. *et al.* (2011) The effects of water and microstructure on the mechanical properties of bighorn sheep (*Ovis canadensis*) horn. *Acta Biomater.* **7**: 1228–1240.

Trueman, E. R. and Hodgson, A. N. (1990) The fine structure and function of the foot of *Nassarius krausslanus*, a gastropod moving by ciliary locomotion. *J. Moll. Stud.* **56**: 221–228.

Ugural, A. C. and Fenster, S. K. (1981) *Advanced Strength and Applied Elasticity*, 2nd SI edn. New York: Elsevier.

Urry, D. W., Harris, R. D., Long, M. M., and Prasad, K. U. (1986) Polytetrapeptide of elastin: temperature-correlated elastomeric force and structure development. *Int. J. Peptide Protein Res.* **28**: 649–660.

Utsunomiya, H., Koh, H., Miyamoto, J., Skai, T. (2008) High strength porous copper by cold extrusion. *Adv. Eng. Mater.* **10**: 826–829.

Vaccaro, E. and Waite, J. H. (2001) Yield and post-yield behavior of mussel byssal thread: a self-healing biomolecular material. *Biomacromol.* **2**: 906–911.

Vashishth, D. (2004) Rising crack-growth-resistance behavior in cortical bone: implication for toughness measurements. *J. Biomech.* **37**: 943–946.

Vashishth, D., Behiri, J. C., and Bonfield, W. (1997) Crack growth resistance in cortical bone: concept of microcrack toughening. *J. Biomech.* **10**: 763–769.

Vashishth, D., Tanner, K. E., and Bonfield, W. (2000) Contribution, development and morphology of microcracking in cortical bone during crack propagation. *J. Biomech.* **33**: 1169–1174.

Vashishth, D., Tanner, K. E., and Bonfield, W. (2003) Experimental validation of a microcracking-based toughening mechanism for cortical bone. *J. Biomech.* **36**: 121–124.

Veedu, V. P., Cao, A., Li, X. *et al.* (2006) Multifunctional composites using reinforced laminate with carbon-nantube forests. *Nature Mater.* **5**: 457–462.

Verbrugge, J. (1934) Le matériel métallique résorbable en chirurgie asseuse. *La Presse Medicale* **23**: 460–465.

Vincent, J. F. V. (1990) *Structural Biomaterials*, rev. edn. Princeton, NJ: Princeton University Press.

Vincent, J. F. V. (1991) *Structural Biomaterials*. Princeton, NJ: Princeton University Press.

Vincent, J. F. V. (2002) Survival of the cheapest. *Mater. Today* **5**: 28–41.

Vincent, J. F. V. and Currey, J. D., eds. (1980) *The Mechanical Properties of Biological Materials*, Symposia of the Society for Experimental Biology, no. 34. Cambridge: Cambridge University Press.

Vincent, J. F. V. and King, M. J. (1995) The mechanism of drilling by wood wasp ovipositors. *Biomimetics* **3**: 187–201.

Vincent, J. F. V. and Mann, D. L. (2002) Systematic technology transfer from biology to engineering. *Phil. Trans. Roy. Soc. Lond. A* **360**: 159–173.

Vincent, J. F. V. and Wegst, U. G. K. (2004) Design and mechanical properties of insect cuticle. *Arthropod Struct. Develop.* **33**: 187–199.

Vogel, H. G. (1972) Influence of age, treatment with corticosteroids and strain rate on mechanical properties of rat skin. *Biochim. Biophys. Acta* **286**: 79–83.

Vollrath, F. (2000) Strength and structure of spiders' silks. *Rev. Mol. Biotechnol.* **74**: 67–83.

Vukusic, P. and Sambles, J. R. (2003) Photonic structures in biology. *Nature* **424**: 852–855.

Wada, K. (1958) The crystalline structure on the nacre of pearl oyster shell. *Bull. Jpn. Soc. Sci. Fish* **24**: 422–427.

Wada, K. (1959) On the arrangement of aragonite crystals in the inner layer of the nacre. *Bull. Jpn. Soc. Sci. Fish* **25**: 342–345.

Wagner, I. P., Hood, D. M., and Hogan, H. A. (2001) Comparison of bending modulus and yield strength between outer stratum medium and stratum medium zone alba in equine hooves. *Am. J. Vet. Res.* **62**: 745–751.

Wainwright, S. A., Biggs, W. D., Currey, J. D., and Gosline, J. M. (1976) *Mechanical Design in Organisms*. Princeton, NJ: Princeton University Press.

Waite, J. H. (1987) Nature's underwater adhesive specialist. *Intl. J. Adhes.* **7**: 9.

Waite, J. H., Lichtenegger, H. C., Stucky, G. D., and Hansma, P. (2004) Exploring the molecular and mechanical gradients in structural bioscaffolds. *Biochem.* **43**: 7653–7662.

Waite, J. H., Holten-Andersen, N., Jewhurst, S., and Sun, C. J. (2005) Mussel adhesion: finding tricks worth mimicking. *J. Adhesion* **81**: 297–317.

Wang, B., Gao, J., Wang, L., Zhu, S., and Guan, S. (2012) Biocorrosion of coated Mg-Zn-Ca alloy under constant compressive stress close to that of human tibia. *Mater. Lett.* **70**: 174–176.

Wang, H., Estrin, Y., and Zuberova, Z.(2008) Bio-corrosion of a magnesium alloy with different processing histories. *Mater. Lett.* **62**: 2476–2479.

Wang, R. Z., Suo, Z., Evans, A. G., Yao, N., and Aksay, I. A. (2001) Deformation mechanisms in nacre. *J. Mater. Res.* **16**: 2485–2493.

Warburton, F. L. (1948) Determination of the elastic properties of horn keratin. *J. Textile Inst.* **39**: 297–307.

Warén, A., Bengtson, S., Goffredi, S. K., and Van Dover, C. L. (2003) A hot-vent gastropod with iron sulfide dermal sclerites. *Science* **302**: 1007.

Watabe, N. and Wilbur, K. M. (1960) Influence of the organic matrix on crystal type in molluscs. *Nature* **188**: 334.

Watanabe, K., Hoshino, T., Kanada, K., Haruyama, Y., and Matsui, S. (2005a) Brilliant blue observation from a Morpho-butterfly-scale quasi-structure. *Jpn. J. Appl. Phys.* **44**: L48–L50.

Watanabe, K., Hoshino, T., Kanda, K., Haruyama, Y., Kaito, T., and Matsui, S. (2005b) Optical measurement and fabrication from a Morpho-butterfly-scale quasistructure by focused ion beam chemical vapor deposition. *J. Vac. Sci. Technol. B* **23**: 570–574.

Watchtel, E. and Weiner, S. (1994) Small-angle X-ray scattering study of dispersed crystals from bone and tendon. *J. Bone Miner. Res.* **9**: 1651–1655.

Watson, J. D. and Crick, F. H. (1953) Molecular structure of nucleic acids; a structure for deoxyribose nucleic acid. *Nature* **171**: 737–738.

Weaver, J. C., Wang, Q., Miserez, A. *et al.* (2010) Analysis of an ultra hard magnetic biomineral in chiton radular teeth. *Mater. Today* **13**: 42–52.

Weaver, J. C., Milliron, G. W., Miserez, A. *et al.* (2012) The stomatopod dactyl club: a formidable damage-tolerant biological hammer. *Science.* **336**: 1275–1280.

Weertman, J. and Weertman, J. R. (1970) Mechanical properties, strongly temperature dependent. In Cahn, R. W., ed., *Physical Metallurgy.* Amsterdam: North Holland.

Wegst, U. G. K. (2011) Bending efficiency through property gradients in bamboo, palm, and wood-based composites. *J. Mech. Behav. Biomed.* **4**: 744–755.

Wegst, U. G. K. and Ashby, M. F. (2004) The mechanical efficiency of natural materials. *Phil. Mag.* **84**: 2167–2181.

Weibull, W. (1951) A statistical distribution function of wide applicability. *J. Appl. Mech.* **18**: 293–297.

Weiner, S. (1980) X-ray-diffraction study of the insoluble organic matrix of mollusk shells. *FEBS Lett.* **111**: 311–316.

Weiner, S. (1984) Organization of organic matrix components in mineralized tissues. *Am. Zool.* **24**: 945–951.

Weiner, S. and Addadi, L. (2002) At the cutting edge. *Science* **298**: 375–376.

Weiner, S. and Hood, L. (1975) Soluble protein of the organic matrix of mollusk shells: a potential template for shell formation. *Science* **190**: 987–989.

Weiner, S. and Price, P. A. (1986) Disaggregation of bone into crystals. *Calcif. Tiss. Int.* **39**: 365–375.

Weiner, S. and Wagner, H. D. (1998) The material bone: structure-mechanical function relations. *Annu. Rev. Mater. Sci.* **28**. 271–298.

Weiner, S., Talmon, Y., and Traub, W. (1983) Electron diffraction of mollusc shell organic matrices and their relationship to the mineral phase. *Int. J. Bio. Macromol.* **5**: 325–328.

Weiner, S., Traub, W., and Parker, S. B. (1984) Macromolecules in mollusc shells and their functions in biomineralization. *Phil. Trans. R. Soc. Lond. B* **304**: 425–434.

Weiner, S., Traub, W., and Wagner, H. D. (1986) Lamellar bone: structure–function relations. *J. Structur. Biol.* **126**: 241–255.

Weiner, S., Addadi, L., and Wagner, H. D. (2000) Materials design in biology. *Mater. Sci. Eng. C* **11**: 1–8.

Weis-Fogh, T. (1961a) Thermodynamic properties of resilin. *J. Mol. Biol.* **3**: 520–531.

Weis-Fogh, T. (1961b) Molecular interpretation of the elasticity of resilin, a rubber-like protein. *J. Mol. Biol.* **3**: 648–667.

Weiss, I. M. and Kirchner, H. O. K. (2010) The peacock's train (Pavo cristatus and Pavo cristatus mut. alba) I. Structure, mechanics, and chemistry of the tail feather coverts. *J. Exp. Zool.* **313A**: 690–703.

Weiss, I. M. and Schönitzer, V. (2006) The distribution of chitin in larval shells of the bivalve mollusk Mytilus galloprovincialis. *J. Struct. Biol.* **153**: 264–277.

Weissbuch. I., Addadi, L., and Leiserowitz, L. (1991) Molecular recognition at crystal interfaces. *Science* **253**: 637–645.

Welsh, U., Storch, V., and Fuchs, W. (1974) The fine structure of the digital pads of rhacophorid tree frogs. *Cell Tiss. Res.* **148**: 407–416.

White, S. R., Sottos, N. R., Geubelle, P. H. *et al.* (2001) Autonomic healing of polymer composites. *Nature* **409**: 794–797.

Whitesides, G. M. (2002) Organic material science. *Mater. Res. Soc. Bull.* **27**: 56–65.

Wilt, F. W. (2005) Developmental biology meets materials science: morphogenesis of biomineralized structures. *Devel. Biol.* **280**: 15–25.

Wise, S. W. (1970) Microarchitecture and deposition of gastropod nacre. *Science* **167**: 1486–1488.

Witte, F., Kaese, V., Haferkamp, H. *et al.* (2005) In vivo corrosion of four magnesium alloys and the associated bone response. *Biomater.* **26**: 3557–3563.

Woesz, A., Weaver, J. C., Kazanci, M. *et al.* (2006) Micromechanical properties of biological silica in skeletons of deep-sea sponges. *J. Mater. Res.* **21**: 2068–2078.

Wren, T. A. L., Yerby, S. A., Beaupre, G. S., and Carter, D. R. (2001) Mechanical properties of human Achilles tendon. *Clin. Biomech.* **11**: 245–251.

Wright, T. M. and Hayes, W. C. (1977) Fracture mechanics parameters for compact bone – the effects of density and specimen thickness. *J. Biomech.* **10**: 419–430.

Wu, H., Thalladi, V. R., Whitesides, S., and Whitesides, G. M. (2002) Using hierarchical self-assembly to form three-dimensional lattices of spheres. *J. Am. Ceram. Soc.* **124**: 14495–14502.

Wu, K. S., van Osdol, W. W., and Dauskardt, R. H. (2006) Mechanical properties of human stratum corneum: effects of temperature, hydration, and chemical treatment. *Biomater.* **27**: 785–795.

Wu, T.-M., Fink, D. J., Arias, J. L., Rodriguez, J. P., Heuer, A. H., and Caplan, A. I. (1992) The molecular control of avian egg shell mineralization. In Slavkin, H. C. and Price, P., eds. *Chemistry and Biology of Mineralized Tissues*. NewYork: Elsevier, pp. 133–141.

Wulff, G. (1901) Zur frage der geschwindigkeit des wachstums und derauflösung der kristallflächen. *Z. Kristall.* **34**: 449–530.

Yang, Q. D., Cox, B. N., Nalla, R. K., and Ritchie, R. O. (2006a) Re-evaluating the toughness of human cortical bone. *Bone* **38**: 878–887.

Yang, Q. D., Cox, B. N., Nalla, R. K., and Ritchie, R. O. (2006b) Fracture length scales in human cortical bone: the necessity of nonlinear fracture models. *Biomater.* **27**: 2095–2113.

Yang, W., Kashani, N. M., Li, X. W., Zhang, G. P., and Meyers, M. A. (2011a) Structural characterization and mechanical behavior of a bivalve shell (*Saxidomus purpuratus*). *Mater. Sci. Eng. C* **31**: 724–729.

Yang, W., Zhang, G. P., Zhu, X. F., Li, X. W., and Meyers, M. A. (2011b) Structure and mechanical properties of *Saxidomus purpuratus* biological shells. *J. Mech. Behav. Biomed. Mater.* **4**: 1514–1530.

Yang, W., Chao, C., and McKittrick, J. (2013a) Axial compression of a hollow cylinder filled with foam: a study of porcupine quills. *Acta Biomater.* **9**: 5297–5305.

Yang, W., Chen, I. H., Gludovatz, B., Zimmermann, E. A., Ritchie, R. O., and Meyers, M. A. (2013b) Natural flexible dermal armor. *Adv. Mater.* **25**: 31–48.

Yang, W., Gludovatz, B., Zimmermann, E. A., Bale, H. A., Ritchie, R. O., and Meyers, M. A. (2013c) Structure and fracture resistance of alligator gar (*Atractosteus spatula*) armored fish scales. *Acta Biomater.* **9**: 5876–5889.

Yannas, I. V., Burke, J. F., Orgill, D. P., and Skrabut, E. M. (1982) Wound tissue can utilize a polymeric template to synthesize a functional extension of skin. *Science* **215**: 174–176.

Yao, H. and Gao, H. J. (2007) Multi-scale cohesive laws in hierarchical materials. *Int. J. Solids Struct.* **44**: 8177–8193.

Yao, H. and Gao, H. (2008) Multi-scale cohesive laws in hierarchical materials. *Int. J. Solids Struct.* **45**: 3627–3643.

Yao, H., Dao, M., Imholt, T. *et al.* (2010) Protection mechanisms of the iron-plated armor of a deep-sea hydrothermal vent gastropod. *Proc. Natl. Acad. Sci. USA* **107**: 987–992.

Yasrebi, M., Kim, G. H., Gunnison, K. E., Milius, D. L., Sarikaya, M., and Aksay, I. A. (1990) Biomimetic processing of ceramics and ceramic-metal composites. *Mater. Res. Soc.* **180**: 625–635.

Yeni, Y. N. and Norman, T. L. (2000) Calculation of porosity and osteonal cement line effects on the effective fracture toughness of cortical bone in longitudinal crack growth. *J. Biomed. Mater. Res.* **51**: 504–509.

Yeni, Y. N., Brown, C. U., Wang, Z., and Norman, T. L. (1997) The influence of bone morphology on fracture toughness of the human femur and tibia. *Bone* **21**: 453–459.

Yeni, Y. N., Brown, C. U., and Norman, T. L. (1998) Influence of bone composition and apparent density on fracture toughness of the human femur and tibia. *Bone* **22**: 79–84.

Yoon, S.-H. and Park, S. (2011) A mechanical analysis of woodpecker drumming and its application to shock-absorbing systems. *Bioinsp. Biomim.* **6**: 1–12.

Young, R. A. and Rowell, R. M. (1986) *Cellulose: Structure, Modification, and Hydrolysis.* New York: John Wiley and Sons.

Zampieri, A., Sieber, H., Selvam, T. *et al.* (2005) Biomorphic SiSiC/zeolite ceramic composites: from rattan palm to bioinspired structured monoliths for catalysis and sorption. *Adv. Mater.* **17**: 344–349.

Zaremba, C. M., Belcher, A. M., Fritz, M. *et al.* (1996) Critical transitions in the biofabrication of abalone shells and flat pearls. *Chem. Mater.* **8**: 679–690.

Zhang, W., Zhang, D., Fan, T. *et al.* (2006) Biomimetic zinc oxide replica with structural color using butterfly (*Ideopsis similis*) wings as templates. *Bioinsp. Biomim.* **1**: 89–95.

Zhao, S., Zhang, J., Zhao, S., Li, W., and Li, H. (2003) Effect of inorganic-organic interface adhesion of mechanical properties of Al_2O_3/polymer laminate composites. *Comp. Sci. Tech.* **63**: 1009–1014.

Zhou, B. L. (1996) Some progress in the biomimetic study of composite materials. *Mater. Chem. Phys.* **45**: 114–119.

Zhou, B., Xu, F., Chen, C. Q., and Lu, T. J. (2010) Strain rate sensitivity of skin tissue under thermomechanical loading. *Phil. Trans. Roy. Soc. A* **368**: 679–690.

Zhu, Q. and Asaro, R. J. (2008) Spectrin folding versus unfolding reactions and RBC membrane stiffness. *Biophys. J.* **94**: 2529–2545.

Zi, J., Yu, X., Li, Y. *et al.* (2003) Coloration strategies in peacock feathers. *Proc. Natl. Acad. Sci. USA* **100**: 12576–12578.

Zioupos, P. and Currey, J. D. (1998) Changes in the stiffness, strength, and toughness of human cortical bone with age. *Bone* **22**: 57–66.

Zioupos, P., Currey, J. D., and Sedman, A. J. (1994) An examination of the micromechanics of failure of bone and antler by acoustic emission tests and laser scanning confocal microscopy. *Med. Eng. Phys.* **16**: 203–212.

Zioupos, P., Wang, X. T., and Currey, J. D. (1996) Experimental and theoretical quantification of the development of damage in fatigue tests of bone and antler. *J. Biomech.* **29**: 989–1002.

Ziv, V. and Weiner, S. (1994) Bone crystal sizes: a comparison of transmission electron microscopic and X-ray diffraction width broadening techniques. *Connect. Tissue Res.* **30**: 165–175.

Zollfrank, C., Travitzky, N., Sieber, H., Selchert, T., and Greil, P. (2005) Biomorphous SiSiC/Al-Si ceramic composites manufactured by squeeze casting: microstructure and mechanical properties. *Adv. Eng. Mater.* **7**: 743–746.

Zoond, A. and Eyre, J. (1934) Studies in reptilian colour response. I. The bionomics and physiology of the pigmentary activity of the chameleon. *Phil. Trans. Roy. Soc. Lond. B* **223**: 27–55.

Zylberberg, L. and Nicolas, G. (1982) Ultrastructure of scales in a teleost (Carassius auratus L.) after use of rapid freeze-fixation and freeze-substitution. *Cell Tissue Res.* **223**: 349–367.

Zylberberg, L., Bereiter-Hahn, J., and Sire, J, Y. (1988) Cytoskeletal organization and collagen orientation in the fish scales. *Cell Tissue Res.* **253**: 597–607.

Zylberberg, L., Bonaventure, J., Cohen-Solal, L., Hartmann, D. J., and Bereiter-Hahn, J. (1992) Organization and characterization of fibrillar collagens in fish scales in situ and in vitro. *J. Cell. Sci.* **103**: 273–285.

Index